Chemistry In The Economy

An American Chemical Society Study
supported in part by the
National Science Foundation
Washington, D.C.

1973

Published October 1973

Price: $6.50

Order from: American Chemical Society
1155 Sixteenth St., N.W.
Washington, D.C. 20036

Library of Congress Card Catalog 73-86535

One of the earliest decisions of the then new Committee on Chemistry and Public Affairs was that the nation and its policymakers as well as its chemists and chemical engineers would benefit by a broad and thoughtful look at the place of chemistry in the U.S. national economy. As a consequence, a study project on "Chemistry in the Economy" was initiated. Progress on this project was slow for some years until the commitment of the present co-chairmen, Milton Harris and Max Tishler, was obtained. Under the leadership of these two, a major study effort involving the volunteer participation of a large number of ACS members and others was developed. Financial support was provided by the National Science Foundation and the American Chemical Society. The substantial and fact-filled volume which has resulted is a tribute to this financial support, to the volunteer efforts of a large group of chemists, chemical engineers, and others, and especially to the dedication and effort of the two co-chairmen of the study. The Committee on Chemistry and Public Affairs of the ACS welcomes the appearance of this volume and is confident that it will make important contributions to the better understanding of chemistry.

Franklin A. Long
Chairman
ACS Committee on
Chemistry and Public Affairs

ACS Committee on Chemistry and Public Affairs*

Dr. Franklin A. Long, Chairman
Henry R. Luce Professor of Science
and Society
Cornell University

Dr. Mary H. Aldridge
Professor of Chemistry
The American University

Dr. William O. Baker
President
Bell Telephone Laboratories

Dr. Robert W. Cairns
Chairman, Board of Directors
American Chemical Society

Dr. Melvin Calvin
Director
Laboratory of Chemical
Biodynamics
University of California at Berkeley

Dr. Thomas P. Carney
Executive Vice President
G. D. Searle & Company

Mr. Richard A. Carpenter
Executive Director
Commission on Natural Resources
National Academy of Sciences

Dr. Herbert E. Carter
Coordinator of Interdisciplinary
Programs
University of Arizona

Dr. Lloyd M. Cooke
Corporate Director of Urban Affairs
Union Carbide Corp.

Dr. James D. D'Ianni
Director of Research
Research Division
Goodyear Tire & Rubber Co.

Dr. George S. Hammond
Vice Chancellor of Natural Sciences
University of California at Santa
Cruz

Dr. Milton Harris
Suite 500
3300 Whitehaven Street, N.W.
Washington, D.C. 20007

Dr. Alan C. Nixon
Wells Fargo Building, Room 511
2140 Shattuck Avenue
Berkeley, Calif. 94704

Dr. Charles G. Overberger
Vice President for Research
University of Michigan

Dr. Charles C. Price
Benjamin Franklin Professor of
Chemistry
University of Pennsylvania

Dr. Max Tishler
University Professor of the Sciences
Wesleyan University

Committee consultant
Dr. George B. Kistiakowsky
Professor Emeritus
Harvard University

Committee staff director
Dr. Stephen T. Quigley
Director, Department of Chemistry
and Public Affairs
American Chemical Society

* Committee membership as of 1972, the year the study was completed. The study had its inception in 1966, however. Persons serving in the period 1966-71 and not listed in the foregoing were:

Dr. Arthur M. Bueche, 1967–68
Vice President
Research & Development Center
General Electric Co.

Dr. Theodore L. Cairns, 1966–70
Director, Central Research
  Department, Experimental Station
E. I. du Pont de Nemours & Co., Inc.

Dr. William E. Hanford, 1967–69
Vice President
Olin Corp.

Dr. William R. Kieffer, 1967–71
Chairman, Department of Chemistry
College of Wooster

Dr. Glenn T. Seaborg, 1966–70
University Professor of Chemistry
  and Associate Director
Lawrence Berkeley Laboratory
University of California

Dr. William J. Sparks (Retired),
  1966–67
Former ACS President
5129 Granada Boulevard
Coral Gables, Fla. 33146

Mr. Joseph Stewart (Deceased),
  1968–70

Dr. C. A. VanderWerf, 1966–69
Dean, College of Arts and Sciences
University of Florida

Dr. Robert W. Van Dolah, 1969–71
Research Director
Pittsburgh Mining and Safety
  Research Center
U. S. Bureau of Mines

Chemistry In The Economy

Original (1966–68) Task Force and Steering Committee

Dr. Robert L. Hershey, Chairman
(Deceased)

Mr. Theodore A. Burtis
Vice President for Marketing
Sun Oil Co.

Dr. Theodore L. Cairns
Director, Central Research
  Department, Experimental Station
E. I. du Pont de Nemours and Co.,
  Inc.

Dr. Harold A. Dewhurst
Director, Corporate Research and
  Development
Owens-Corning Fiberglas Technical
  Center

Dr. Vladimir Haensel
Vice President, Science and
  Technology
Universal Oil Products Co.

Dr. William E. Hanford
Vice President
Olin Corp.

Dr. Milton Harris
Suite 500
3300 Whitehaven Street, N.W.
Washington, D.C. 20007

Dr. William S. Johnson
Executive Head
Department of Chemistry
Stanford University

continued

v

Contents

Correction

To the considerable distress of all concerned, the late Dr. Robert L. Hershey is erroneously identified as "William" in the Foreword (page ix). Needless to say, we sincerely regret this error. Dr. Hershey served as the chairman of the original task force and steering committee during the study's formative years (1966-68). Formerly vice president of E. I. du Pont de Nemours & Co., Dr. Hershey provided invaluable guidance to the subsequent work, and his resignation from the project left a void that proved hard to fill.

Milton Harris
Max Tishler
Co-Chairmen

Foreword

This report deals with the manifold contributions of chemistry to the ways in which we work and live. It is the fruit of a scientific and technological inquiry on the one hand and an economics and social inquiry on the other. It is largely derived from contributions by experts in industry, education, and government. Many of these people are named in the lists of contributing and advisory committees printed at the publication's end. All rendered not only very considerable service but service in a period when their employers were struggling with the severe economic problems that have afflicted technically-based organizations in the past few years. To them and to those who assisted us informally in reviewing manuscripts and offering counsel we extend our sincerest thanks.

In any endeavor of this type, of course, there are a few persons to whom special debts are owed. The idea of the study developed in discussions by the American Chemical Society's Committee on Chemistry and Public Affairs (CCPA) under the chairmanship of Dr. Charles Price, and the task was begun under the sponsorship of that committee. We appreciate the guidance and assistance CCPA provided under Dr. Price and under his successor, Dr. Franklin A. Long. We are also grateful to a special subcommittee of CCPA under the chairmanship of Dr. William O. Baker, which monitored the study, and to Dr. William Hershey, who guided some of the preliminary steps.

Although much of the work was performed by what eventually totaled several hundred volunteers, the study also needed direct and continuing effort. We have benefited from such effort by three persons: Gordon Bixler, Kenneth M. Reese, and Dr. Irving Siegel, consulting economist.

Mr. Bixler served originally as project manager and made many of the early investigations that helped set up the objectives and procedures for Part I of this book. He collaborated with committees of experts representing various sectors of industry in the preparation of 13 chapters in that part. Mr. Reese collaborated with similar committees in the preparation of seven chapters in Part I. Dr. Siegel served as full-time professional economist for the study. He provided invaluable help in bringing the entire study into focus and in giving it cohesiveness. He wrote the chapters in Part III dealing with chemistry's place in the economy, and he worked with the Committees on Economics, Education, and Manpower in adapting material required for the report.

We also owe a special debt to Dr. Stephen T. Quigley, Director of the Society's Department of Chemistry and Public Affairs. As staff liaison with CCPA, Dr. Quigley was associated with the study from its inception. He was particularly helpful in coordinating views and in drafting the original proposal to the National Science Foundation. Ms. Sally Grozzo also merits special thanks for tirelessly typing manuscript after manuscript, for keeping track of accumulating files of material, and for otherwise coping with many of the daily details associated with the study.

Chairmen

Milton Harris
Washington, D.C.

Max Tishler
Middletown, Conn.

Goals and Scope

In this publication we present the results of a study conducted by the American Chemical Society with major financial support from the National Science Foundation.[1] The study was motivated in part by an earlier one, "Chemistry: Opportunities and Needs," prepared under the direction of the Committee on Science and Public Policy of the National Academy of Sciences–National Research Council. That study, published in 1965 and accorded an excellent reception, dealt only with "basic research in chemistry departments of U.S. universities and in those laboratories of industry, government, and private foundations that are staffed for the most part by scientists holding advanced degrees in chemistry." The study noted the explosive growth of chemistry in the service of man—in medicine, in industry, in agriculture, and in national defense. But perhaps most important, it made clear a need for a comprehensive inquiry into the contributions of chemistry to the economy and to the manner in which we work and live in the modern world.

This report presents the findings of such an inquiry. In it we have sought to meet the following objectives:

• To document past accomplishments of chemistry in many industrial sectors of the U.S. economy, including ones not normally considered "chemical."

• To describe statistically the nation's chemistry resources of trained manpower and to consider the demand for chemists and chemical engineers for the remainder of this decade.

• To analyze the extensive and ramified participation of chemistry in the nation's economy.

• To discern some of the plausible directions and the broader implications of future developments in chemistry.

• To suggest ways in which the chemical education system may more effectively contribute in meeting visible and still-to-emerge social and economic needs.

The entire report is intended to serve a wide range of informational needs and a multiplicity of audiences in industry, government, and education. It should also be especially helpful in equipping chemists and chemical engineers with a better insight into the relationships of their work to the requirements of the economy and of society.

CHEMISTRY'S PLACE IN INDUSTRY

To most people, the labels "chemistry" and "chemical technology" are largely associated with the production of (somewhat) simple chemicals—

---

1. The original study proposal was drafted under the aegis of the American Chemical Society's Committee on Chemistry and Public Affairs. It was submitted to the National Science Foundation April 1, 1969, with a request for partial support over a two-year period. NSF approved the proposal June 27, 1969, and later granted a one-year extension (to June 30, 1972) as the study progressed.

sulfuric acid, ammonia, ethylene glycol, and similar inorganic and organic "chemicals." A much smaller number of people when pressed might add to that list the products that are also "chemical" but that are usually thought of as having other primary functions, such as plastics and resins, synthetic rubber and fibers, drugs, soaps and detergents, cosmetics and toiletries, paints and varnishes, and fertilizers and pesticides.

Together there are some one dozen categories of chemicals and chemical products that comprise the output of what is more formally known as the "chemical and allied products industry." Yet there is considerable dependence on chemistry in many other manufacturing industries that are not normally thought of as "chemical." Since chemistry by definition deals with the molecular transformations that matter undergoes, any treatment of chemistry must also include the activities that occur in other industries classified by different names. Included from such a perspective would be industries dealing with ferrous and nonferrous metals; petroleum; food; coal and coke; pulp and paper; stone, clay, and glass; and electronic equipment.

Such a view of the pervasiveness of chemistry is not offered with the thought of improperly aggrandizing chemistry beyond its due. Other disciplines are equally and critically important to the proper functioning of the total U.S. economy. The point about chemistry's pervasiveness is made, however, because it may not be sufficiently appreciated. Yet the need to view chemistry's role in the widest perspective possible is perhaps the most basic thesis of this study. Any national policy affecting chemistry that does not rest on a broad perspective view of chemistry will at best be inadequate and at worst ineffective or even harmful.

It is with these views in mind, therefore, that in this report we do not use "chemistry" as a synonym for "the chemical industry." As the report demonstrates, chemistry is widely used throughout the economy. As noted earlier, chemical techniques are often used in establishments principally engaged in other kinds of processing and accordingly classified in nonchemical industries. Moreover, such nonchemical industries employ more than half of all chemists and chemical engineers working in industry. Both are strong reasons for not identifying chemistry solely with "the chemical industry."

When the report does refer to "the chemical industry" or to "the conventional chemical industry," it refers to "the chemical and allied products industry."[2] That industry makes chemicals and also such chemical products as plastics and resins, synthetic rubber and fibers, pharmaceuticals, soaps and detergents, cosmetics and toiletries, paints and varnishes, fertilizers, and pesticides. When any subsectors are omitted in particular discussions, the fact is appropriately noted. For example, if establishments engaged primarily in making drugs are omitted, this truncated version of chemicals and allied products is called "chemicals, except drugs."

When the term "chemical process industries" is used it signifies the enlargement of chemicals and allied products to include other manufacturing industries with a strong chemical base, such as petroleum refining, pulp and paper production, and the like. The added industries are generally specified in the report to avoid ambiguity.

We consider chemistry to embrace the practice of industrial chemistry and chemical engineering as well as the practice of chemical research and

---

2. Using the system devised by the Office of Statistical Standards, Bureau of the Budget, and published in "Standard Industrial Classification Manual—1967."

development and the academic preparation of professional manpower. Moreover, the practice of chemistry in industry is recognized to reach far beyond research and development and into production, inspection and testing, sales and purchasing, and many other areas.

The term "research" is often used as a short equivalent of "research and development." It also covers both "basic" and "applied" investigations. Where distinctions are important, however, more precise terminology is used. In any event, "research" does not include market, economic, and other social science research; nor does it include quality control and product testing normally associated with manufacturing, legal work incident to patent protection, and so forth.

Wherever possible, this report includes chemical engineers as well as chemists in the statistics for the professional manpower of chemistry. The scope of the data is usually reflected by the designations used, but the term "chemists" may sometimes be used for simplicity if no misunderstanding is likely to result. "Biochemists" are included with "chemists" in the employment figures, but they are commonly treated separately in educational and occupational source statistics.

CONTENTS OF THE REPORT

The study opens with a discussion of some of chemistry's possible future developments and their implications, and it then presents recommendations for action emerging from the study. This beginning is followed by the main body of the report, which is organized into three parts:

• The Accomplishments of Chemistry.

• The People of Chemistry.

• The Economics of Chemistry.

The first part records some of the accomplishments of chemists and chemical engineers that have provided avenues for significant economic and social change. In addition to demonstrating by example the widespread utility of chemical technology, it hopefully will serve as a reference document for persons seriously interested in chemistry. In addition, it should also provide a point of departure for other inquiries. Among them might be assessments of factors pertinent to the conversion of science into technology and production or to the transfer of technology among individuals, companies, industries, and nations. Moreover, a study of the discussions on prospective developments should disclose trends in research and development that have important implications for the future training and supply of professional chemical manpower.

Part II deals with the nation's human resources in chemistry. The opening chapter analyzes the concepts underlying the statistics on chemically trained manpower and estimates the effective supply of chemistry's qualified manpower. The three succeeding chapters deal with education, concentrating on the future prospective supply and demand of chemists and chemical engineers, the need and techniques of continuing education, and the adjustments that might be made to provide graduates with better training, especially for work in industrial organizations.

Part III of the report is devoted to a portrait of chemistry in action in the nation's economy. It considers the industrial and functional distributions of chemists and chemical engineers; trends in chemical research, patents, output, and productivity; input-output relations between the chemical industry and the rest of the economy; and employment projections to 1980.

Opportunities and Recommendations

Americans are passing through a period of clear-cut change in their attitudes toward their personal lives, their social and economic organizations, and their governments. Historians and foreign observers used to comment on the optimism, even the enthusiasm, with which people in the U.S. contemplated their future, and especially in the past half century or so their technological future. It now appears, however, that uncertainty and even apprehensiveness beset the national mood. Providing equal rights, curbing inflation, and reducing unemployment are topics of major concern. Disillusionment with the changes associated with present technology and fear of possible harmful effects from technology yet to be developed are widely expressed.

Yet controversy characterizes any viable society. While the controversies of today doubtless are often no less contentious than many of earlier times, they are nonetheless the ones that must be dealt with now. A study of chemistry's role in the U.S. might have been a useful addition at any time in history, but such a study becomes particularly important in a time when old practices are being closely questioned and new directions are being sought, especially when the practices and directions have a strong technological base.

Chemistry does not provide a large component to all of the challenges to the country's social and economic system, either as a cause or as a tool for change. Chemistry, however, is a component of a significant part, and an analysis of its role is germane to the country's future. Dismay over a conglomerate of ills lumped under the label "environment" is one case in point. This dismay ranges from the simple discomfort (to say nothing of the possibly more long-range harm) of driving behind a smelly bus to the more potentially serious consequences to health and life of air and water pollution generally, DDT in wildlife, and some heavy metals in water. All share a large chemical component, both as to cause and as to possible elimination.

Dissatisfaction and even fear of other products of technology supply additional cases in point. The most extreme adherents express high pessimism about food additives that allegedly damage the food supply; drugs that cause unwanted side effects; and apparel, appliances, and vehicles that directly or indirectly may disfigure or maim their users. These ills—real or imagined—also share a large chemical component, again both as to cause and as to possible elimination.

As if the foregoing apprehensions over technical aspects of American life were not enough, the economy began slowing as the 1960's ended. Jobs, even professional ones, began disappearing, and inflation began eroding real income. Government and industry both slowed the growth in spending for research and development, and they even reduced spending in some cases. In addition, education began receiving support much less readily. These stresses surely accentuate the popular concerns already felt about drugs, foods, pesticides, and the environment. Furthermore, they make remedial action to overcome these concerns more difficult.

Any reading of the national mood, of course, must be considered tentative. Otherwise, reaction to sharp calls for "action now" may stampede the unwary into ill-considered and possibly ultimately costly action. What now appears to be a rise of technological skepticism and even negativism may be seen by later historians as natural growth in ability to discriminate

1

among alternatives presented by an expanding technology. Moreover, no instant litmus test exists that will show the weighted average of meaningful opinion on which action should be based. Persons in policymaking positions, therefore, must always be mindful that opinions that are the most publicly and apparently widely held may in actuality suffer by being considerably distorted from what the facts warrant.

Conditions are doubtless not as bad as the most vocal would have everyone believe, but they clearly are not good enough to permit the country to rationalize taking no action. To the extent the present mood persists, the pattern and the degree of adoption of new technology, including that provided by chemistry, will surely be affected. The change in the odds or the entries that the mood might bring, however, is different from the cancellation of the race, as some of the extreme assertions about the future of technology suggest.

OPPORTUNITIES

Projections made for the 1970's, and indeed this decade's experience thus far, indicate a more discriminating and a more critical climate for science and technology than prevailed in the preceding three decades. This growing public skepticism plus spreading federal surveillance over choices traditionally considered reserved to private enterprise are making the industrial community less venturesome. It is showing more circumspection and placing greater emphasis on payoffs in evaluating research and capital investment proposals. Among other results, these attitudes and actions are slowing significantly the growth in employment of chemists and chemical engineers. Meantime, the system for graduate education can accommodate very many more qualified students than are likely to apply. This overhanging excess capacity may itself provide pressure for encouraging graduate enrollments regardless of job prospects.

While the outlook may appear dimmed for the growth and application of science in general and chemical science in particular, the bright features should not be ignored. In Part III of this study, for example, patent statistics amply demonstrate a profound shift toward chemical technology since World War I. This shift is not likely to be halted abruptly nor to be easily reversed. Furthermore, a remarkable gain in chemical output has been recorded during the past seven decades. Even more phenomenal is the chemical industry's gain with respect to productivity (output per man-hour) during the same period. The productivity achievement has in turn helped to keep the industry's unit labor costs and prices in check. Finally, the industry's favorable price experience has moderated the general deterioration in the nation's foreign trade balance.

Also impressive for a favorable outlook are the many hints given in Part I of the multiplying opportunities for advances in chemical technology. These new potentials in processes, materials, and products will not be translated automatically into reality, of course, but the momentum of a trillion-dollar economy weighs in their favor. Following are some of the highlights from Part I to show what only some of the prospects are.

Food. Chemistry has already contributed enormously to the volume, variety, and efficiency at all stages from planting and growing to harvesting, processing, and distributing. In addition to further gains in farm productivity and in crop and livestock yields, we should expect:

• Development of more formulated foods that increase consumer convenience, make diets more varied, and provide proper nutrition.

- Enrichment of all basic foods with nutrients as necessary, so that an adequate caloric intake would automatically provide adequate nutrition.
- Design of foods and diets for better control of diseases, such as the cardiac-related ones.
- Foods especially developed for the very young and the very old and for persons with dietary problems.
- Widening acceptance of meat analogs made of spun protein fibers impregnated with other ingredients.
- Production of some feeds and foods from nonliving raw materials, such as petroleum, leading to reduced dependence on the biosphere.
- Design of economical insecticides and herbicides that are highly specific as to target species and that degrade to ecologically acceptable residues within a reasonable time.

Clothing. Wardrobes are more varied and more easily cared for than ever, yet they account for a smaller share of the spending for personal consumption than in the past. Among future developments we may expect:

- Even greater variety in the assortment of fabrics appropriate for work, home, and leisure through such techniques as chemical treatment of fibers and textiles and proper blending of natural and man-made fibers.
- Greater protection by chemical treatment against flammability, soiling, wrinkling, and deterioration by environmental and biological attack.
- Easier maintenance of wardrobes and furnishings with quick, effective, and safe cleaning preparations.
- Increases in the variety of disposable apparel and furnishings made of paper for one-time use in hospitals, factories, restaurants, and homes.

Shelter. The kinds and versatility of structural materials and coatings will continue to increase. For example:

- Plastics will continue to be substituted for metal, wood, and other conventional materials that may be more costly for similar service.
- Improvements in steel, especially in production techniques but also in properties, will strengthen its competitive utility.
- New paints and other coatings will be formulated to protect structures and equipment better against the environment and against fire.
- Development of malleable, ductile glasses that yield rather than fracture under excessive load will open new applications as a material of construction for buildings and also for products and equipment.
- Plastic fibers will be used to produce paper with superior dielectric properties, heat resistance, and flex and tensile strength.

Health. The potential chemical contributions are enormous, and new opportunities will be afforded by:

- Better understanding of how small molecules interact with the large ones that make up the genetic and enzymatic systems, leading among other results to the design of potent drugs having greater specificity.
- Devising drug delivery systems in which products administered to patients find their ways, in the main, to the target tissues.
- Manipulation of the body's immune responses for defense against viral diseases, perhaps including cancers.

• Application of expanding knowledge about the newer hormones that will provide additional approaches to treating such ills as cardiovascular diseases, reproductive disorders, and degenerative processes.

• Developing a basic biochemical understanding of addiction and devising better therapeutic measures of managing or curing it.

• Intensified research on aging and the accompanying degenerative diseases, for which no cures now exist.

• Enlargement of the variety of materials usable in medical engineering and prosthesis and development of safe and specific immuno agents that will reduce the rejection rates for organ transplants.

• More sensitive and earlier detection of the presence or onset of major killing diseases.

Energy. The country has become suddenly and increasingly aware of energy shortages. Chemistry will be involved in helping meet the increasing demand by the following means among others:

• Coal gasification.

• Greater use of such "clean" fuels as propane and hydrogen.

• Methodical exploitation, when warranted by price competitiveness, of oil shale and tar sand reserves.

• Heavier reliance on nuclear power and the eventual design and operation of practical breeder reactors.

• More effective harnessing of solar and geothermal energy.

Transportation and Communication. Chemistry is also destined to expand the scope of economic choice for satisfying private and social objectives. For example, it will contribute to:

• Design of safer automobile tires, bodies, and windshields and of safer airplane interiors.

• Provision of safer containers for long-distance shipping of toxic, explosive, and other hazardous substances.

• Development of optical wave guides and laser beam communication to expand tremendously the number of communication channels available.

• More compact and more economic microcircuits and memory units for telephone, TV, radio, and other communication equipment, computers, and many other electronic applications.

• Efficient data storage and retrieval systems based on microfilm and Microcards.

• Improved recording media for sound and pictures and for their reproduction.

Ecology. In helping to protect and recondition man's physical environment, chemical science is bound to play a significant part. Included on the technical agenda are:

• Better control of factory stack gases through recycling and through fuel improvement.

• Greater attention to sewage treatment, the economic use of sludge, and the recovery of chemicals from effluents.

• More systematic recovery and reuse of scrap paper, wood, plastics, and metals.

• Possible replacement of phosphates in detergents with biodegradable alternatives.

• Protection of water supplies from contamination by agricultural and other runoffs.

• Quicker and more effective elimination of contaminations that do occur despite the best of control systems.

These technical opportunities facing chemists and chemical engineers have been presented as discrete entities, but it is important to realize they will become parts of an exceedingly complex system. Until very recent times, the effect of a new product or practice has not often been considered beyond that of its immediate application. Yet the ramifications of new products or practices can be extensive. Consider, for example, only a few of the ramifications stemming from concern over the environment. An energy shortage has been caused partly because of that concern, which has frequently inspired strong resistance to the construction of nuclear power facilities. Concern over environmental quality is also forcing a reduction and possibly the eventual elimination of lead antiknock compounds from gasoline. Reduction or removal of lead compounds, in turn, is causing refiners to modify their processes for meeting the octane requirements of modern engines, while engine manufacturers are redesigning engines to operate on low- or no-lead fuels. Meantime, the possible treatment of vehicle exhausts is affecting the markets for such precious metals as platinum and palladium, which the electronics industry depends on for circuits. Changes to refining will affect the nature and cost of feedstocks available to producers of industrial organic chemicals, while diversion of platinum and palladium to catalytic mufflers would force the electronics industry to develop suitable substitutes.

Some note is taken of such secondary, tertiary, and higher order effects in the main body of this study, especially in the chapters of Part I. The tracings are not exhaustive there, but obviously it is increasingly important that ramifications of technological change be given more attention in the future.

RECOMMENDATIONS

Three kinds of recommendations are offered as a result of this study. The first group (of three) is directed toward strengthening the information base for the appraisal and enhancement of chemistry's contribution to society. The second group (also of three) is concerned with improvement of the climate for continuing progress in the application of chemistry to the needs of society. The final group (of seven) relates to improving chemical education and especially to meeting the needs of industry for professionally trained manpower.

The first group of recommendations deals with manpower statistics, the process of translating science into technology and into production for domestic use or for trade, and the further illumination of productivity change.

Recommendation 1: Integrated Supply-Demand Statistics

The personal, institutional, and social stakes in avoiding excess production of professional manpower suggest a need for past, current, and projective statistics bearing on **supply and demand** and officially gathered and sanctioned. Authoritative and consistent quantitative information would be helpful to students, counselors, and recruiters as well as to planners, policymakers, and administrators serving government, industry, academia, and the professional societies. It is therefore recommended that:

> **The National Science Foundation, jointly with the Bureau of Labor Statistics, Bureau of the Census, Office of Education, National Academy of Sciences, the American Chemical Society, and other relevant organizations develop a comprehensive manpower accounting system relating to chemists and chemical engineers.**

A comprehensive manpower accounting system would permit persons dealing with manpower policy to assess more readily the situation with respect to **supply and demand,** to note developing trends in categories relating to **supply and demand,** and to be better prepared to devise policies to counteract unfavorable developments. The system should provide both historical data and frequently revised estimates for the future, and it should deal on an annual basis with such parameters relating to chemists and chemical engineers as:

• The numbers employed and unemployed.

• The numbers not employed in the profession but recallable to it with suitable pay incentives or in the event of a national emergency.

• Gains through graduation from formal education, on-the-job training, and return to the profession and losses from death, retirement, nonentry of new graduates into the profession, and transfers out of the profession.

Recommendation 2: Nonquantitative Studies of Science in Society

To enhance the contribution of chemistry to society, a better understanding is needed of the ways in which individual scientists, engineers, managers, organizations, and other pertinent entities interact with each other and with the larger external environments. The interactions of prime interest to this study are those occurring in the process of achieving or failing to achieve the translation of chemical science into technology and actual production. The mechanisms, aids, and impediments that help to explain the kinds and degrees of translation effected are typically slurred in quantitative studies. Such studies tend to feature the correlation of a small number of measurable key variables, such as research expenditures, patent awards, and output and productivity growth, but they tell nothing about the structure and details of the process of converting science into technology and sales. It is therefore recommended that:

> **Federal granting agencies, such as the National Science Foundation, fund a significant number and variety of anecdotal, literary, and descriptive inquiries to show the ingredients, relationships, decisions, and principles involved in the progression from chemical science to marketable processes, goods, and services.**

This report's Part I presents considerable evidence pointing to the critical contributions of individuals, groups, and organizations. In addition, an appendix in Part III reviews the technical evolution of a particular company with respect to chemical technology during the past half century. This review emphasizes the roles of technical and managerial leadership to various challenges and opportunities originating outside the organization. The analysis is far from exhaustive, however.

Recommendation 3: Chemical Productivity Studies

The chemical industry has long recorded favorable productivity and price performances, and it has done well in the recent years of inflation and of general deterioration of the nation's balance of payments. A better understanding of the sources of the chemical industry's unusual achievement should prove useful for planning and policy, even at the government level. It is therefore recommended that:

The Department of Labor, Department of Commerce, and National Commission on Productivity foster studies of the relation of chemical technology to recent productivity and price experience, preferably on a company basis.

A second group of three recommendations relates to the climate for competitive enterprise at home and abroad: technological-injury insurance, health and ecological risk research, and uniformity in international practices in such matters as product liability and environmental responsibility.

Recommendation 4: Insurance for Technological Injury

Increasing federal surveillance of erstwhile private decision-making with respect to new technology, however desirable, threatens slowdowns of innovation and investment that could well detract from national productivity growth and competitiveness in world trade. Formal programs of technological assessment, even if they succeed in identifying and evaluating the immediate and longer-range costs and benefits, are bound to hamper the introduction and use of new processes and products. Other monitoring devices and programs enacted in recent years also may deter various kinds of research and applications that would surely be welcomed if safety could be assumed or easily be proved. Since it is normally difficult to guarantee safety and since a guarantee becomes harder to formulate as more interests and as longer term and repercussive effects are taken into account, it is recommended that:

The desirability and feasibility be explored of an insurance program that would compensate for technological injury. Such a program might expedite at least tentative adoption of new processes and products that:

• Satisfy early screening and testing criteria.

• Are not thought in the light of current and authoritative knowledge to pose a latent threat of catastrophic, widespread, or illimitable damage.

Recommendation 5: Research on Health and Ecological Risks

Procedure along the lines just mentioned does not reduce the need for more research on understanding and correlating observations made in laboratories and clinics. With respect to personal health, for example, a more deliberate attack is required on the real significance of carcinogenic, mutagenic, and teratogenic effects produced in experimental animals by current and potential drugs, intentional and unintentional food additives, and other products that people may ingest in the course of their daily lives. More also needs to be known about the correlation of metabolism in animals with that in man, especially from the standpoints of efficacy and toxicity. To answer such questions, it is recommended that:

Both industry and government take leadership in the support of fundamental research in universities and other institutions on product efficacy and toxicity and on the relevance of laboratory and clinical findings as they pertain to humans.

Moreover, with respect to the environment there are also many unresolved problems, and their number and complexity grow daily. It is therefore further recommended that:

Consideration be given for expanded public and private support of research on ecological problems, for their solution is critically important to the economics of an intricate technological society as well as to personal well-being and safety.

Recommendation 6: Harmonizing International Practices

The time is ripe for greater international harmonization of practices relating to such matters as health, food, and other product liabilities and environmental responsibility. Accordingly, it is recommended that:

Appropriate agencies in the U.S. cooperate with their counterparts in other countries to reduce disparities in practices in chemically-related manufacturing industries, thus leading to international improvement of economic, social, and environmental standards as well as encouraging the advancement of U.S. chemical technology and productivity.

The remaining recommendations relate to the initial and continuing education of chemistry's professional manpower. Before turning to specific recommendations, however, a few introductory remarks appear in order about graduate policies and programs. The overexpansion of capacity for graduate education, the limited supply of able and interested students, particularly of U.S. citizenship, and the moderation of the growth rate for employment in the 1970's are circumstances and prospects that must be given serious attention. These factors imply the following propositions that should be of interest to faculties and governing boards of universities, to regional accrediting agencies, and to concerned committees of the professional societies:

• That new graduate programs be discouraged unless it is very clear that they will meet unique needs not otherwise satisfied.

• That all Ph.D.-granting institutions review carefully their admission and screening policies and consider whether their programs could gain in quality by a reduction in scale.

• That schools with marginal graduate programs realistically examine their prospects and weigh the relative merits and costs of continuation or abandonment.

• That all Ph.D.-granting institutions examine their dependence on foreign students and consider training only those of exceptional ability or those who have firm prospects of employment in their home countries and who preferably can supply their own support.

These suggestions may sound elitist or excessively restrictive in the face of liberalizing public attitudes on access to higher education, but they are intended to bring the scale of our graduate enterprise into better balance with plausible projections, to improve the cost effectiveness of this enterprise, and to upgrade the quality of its product.

Recommendation 7: Educational Statistics

In connection with education, it is useful at this point to recall Recommendation 1 for a comprehensive manpower accounting system and to extend the thoughts expressed there with the recommendation that:

Relevant government agencies and nonprofit organizations be encouraged in their efforts to compile and publish more detailed, more complete, and more timely information on enrollments, degrees granted, and other aspects of education in chemistry and chemical engineering.

Recommendation 8: Uncoupling Graduate Training and Research

At the same time that our universities engage in instruction, they must conduct a major part of the nation's fundamental research in chemistry. If this coordinate activity is to continue with success in the face of a decreasing output of Ph.D.'s, the quality of graduate training has to be kept high. The research role and the training of graduate students, however, may need to be uncoupled to some extent. It is recommended that:

Government and private funding agencies and academic leaders explore ways to:

• Maintain and increase the relative importance of fellowships and traineeships in the support of graduate education, administering these with a view to making most effective use of students in the teaching assistant pool.

• Provide alternatives to reliance on graduate students for research, as for example through the use of technicians and postdoctoral fellows and through the support of research at non-Ph.D.-granting institutions.

Recommendation 9: Chemical Education for Nonchemists

In contrast to the concern about the overexpansion of the graduate enterprise, it appears that training in chemistry at the undergraduate level is adequate in many areas. It also appears that no problem of oversupply at the bachelor level impends, since B.S. chemists often do not enter or remain in the chemical field but disperse widely throughout the occupational spectrum. Moreover, society surely benefits from an increase in the level of citizen understanding of the implications of chemistry for the ways we work and live and of the problems entailed by the practical use of chemical science. It is therefore recommended that:

• Counselors and school officials generally encourage training in chemistry for capable students, even if the career objectives of such students lie in other fields.

• Colleges and universities increase their efforts in offering courses in chemistry that are specially designed for persons majoring in nonscience fields, thus making an appreciation of chemistry a part of a liberal education.

Recommendation 10: Graduate Student Career Orientation

The present concept of graduate education as an apprenticeship in research culminating in the preparation of a thesis is believed to be basically sound. This plan has proved remarkably effective in training chemists for roles in teaching, academic research, and industry. On the other hand, the scheme also encourages a tendency toward early specialization and fosters an attitude that favors academic research over other forms of chemical endeavor. To counteract these trends and to prepare students better for the variety of careers actually accessible to them, it is recommended that:

• Qualified counselors, familiar with the requirements, conditions, and rewards of industrial as well as other positions, be available to help students make realistic career choices.

• A "core-type curriculum" be provided in the graduate program to help students avoid premature specialization, compensate for undergraduate deficiencies, and choose more meaningfully among careers in academic research, industrial research, and teaching.

• Graduate programs be kept flexible and courses be planned to allow exposure to a wide range of subject matter, including interdisciplinary topics.

(A student would thus be enabled to select a course of study consistent with his future plans, and the proliferation of specialized advanced degrees would also be avoided.)

• Greater contact with industry on the part of the teaching staff be encouraged—through consultantships, lectures, and visits—by personnel of both universities and industry.

• Industrial chemists play a more active part in graduate teaching by service, for example, as visiting professors for special courses or as advisers to students and faculty.

• Postdoctoral programs involving industry-university exchange be encouraged.

Recommendation 11: Company Cooperation with Schools

Implementing these proposals requires the collaboration of industry, which is, after all, the intended beneficiary. Accordingly, it is recommended that:

• Companies participate actively in curriculum planning and provide staff scientists to serve as adjunct professors for special courses and as departmental advisers, examiners, or instructors.

• Companies make wider use of university faculty members, particularly younger ones, as consultants, lecturers, and visitors.

• Companies bear a reasonable share of the cost of industrial fellowship programs in the chemistry departments of universities, programs that have eroded drastically during the past decade.

• Companies take into account the diminishing public support for training chemists and chemical engineers, so they may assure a supply of adequately trained manpower for their needs.

Recommendation 12: Improved Training in English and Economics

The frequency of two complaints concerning both B.S. and Ph.D. graduates is impressive in the survey of employer views; namely, the inability of graduates to communicate effectively and their deficient knowledge of economics. To overcome these limitations in training it is recommended that:

• Chemistry programs contain strong academic requirements for written and oral communication in English.

• Even more important, experience in oral reporting and in writing technical English be made a part of the chemistry curriculum itself.

• Curricula for chemistry students make adequate provision for instruction in economics and other social sciences, with special attention to such practical matters as the distinction between technical feasibility and profitability, the comparison of costs and benefits, and the difference between private and social accounting.

• Companies recognize the limitations of college educations and accept the responsibility to further the education of professional chemists and chemical engineers in economics, especially as economics may relate to company conditions and activities.

Recommendation 13: "Real-World" Research and Statistical Methods

Finally in considering curricula, it should be affirmed that chemistry remains, above all, a laboratory science. In approval of the efforts now being made to upgrade laboratory courses in universities throughout the country, it is recommended that:

- **Formats be developed that emphasize the kinds of situations and problems encountered in "real-world" research and production.**

- **Adequate training and practice in the statistical design of experiments and treatment of data be made an integral part of the chemistry curriculum.**

Recommendation 14: Continuing Education

Although good training at the bachelor, master, or Ph.D. level provides a start in the profession, it is certainly less than half the battle. The successful professional must look forward to a lifetime of continuing self-education. The ultimate responsibility for this sequel rests on the individual himself, but companies, universities, and professional societies all have a share of the responsibility. Companies, for example, have increasingly recognized the wisdom of assisting employees to keep current with new technical developments in or near their areas of education and work assignment. Universities, meantime, have a vital role in refurbishing skills, partly because they are primarily centers of education and partly because of the value of the feedback to their own primary programs. Finally, the importance of the professional society in the continuing education of chemists and chemical engineers is often underestimated or goes unrecognized. Such a society, however, can help professionals maintain technical competence by means of primary and abstract journals, meetings, local section forums, and packaged tutorial short courses.

All these means of assistance are critically important for avoiding or reducing the enormous cost to companies and to society of technological obsolescence. To minimize obsolescence even further, it is recommended that:

- **Companies make even greater use of postdoctoral arrangements, university extension classes, selective tuition assistance, sabbaticals, special and regular seminars, visiting speakers, in-house courses, closed-circuit television, book purchases, library services, and other means of maintaining, upgrading, and extending the skills of staff chemists and chemical engineers.**

- **Universities and colleges develop extension courses designed and scheduled to meet the needs of continuing education for mature scientists and engineers employed in neighboring companies.**

- **Universities and colleges develop and promote packaged courses that take advantage of all modern education media and that companies assist their employees in making off-campus use of such courses.**

- **Scientific and professional societies make a particular effort to meet the tutorial needs of isolated individuals or companies unable to take advantage of university extension courses.**

Part I

THE ACCOMPLISHMENTS OF CHEMISTRY

Much of the nation's economy in essence provides for only a relatively few simple needs. Among them are the traditional food, clothing, and shelter plus perhaps health services for well-being and longevity and transportation and communication to make an exchange of goods, services, and ideas easy. But while many of its activities may be thus simply categorized, the entire economy itself is exceedingly complex. It uses many paths and presents many options for choice in meeting those few and simple needs. In addition, in an affluent country such as the U.S., the economy provides many other goods and services that make life more varied and stimulating.

In the chapters of Part I we have sought to document in one place some of the major accomplishments of American chemists and chemical engineers that have helped meet such needs and provide such other goods and services. Only many volumes the size of this one, of course, could present fully all of their accomplishments. Accordingly, we have had to be selective in this study, and we have included only some of the more major chemical innovations that have changed products and services a reasonable amount. Others doubtless would have included some different events, and they possibly would have omitted some that are included here. The study has been designed to be exemplary, not comprehensive, however. To create the definitive work on the usefulness of chemistry in the U.S. would be a task beyond human and financial resources at the American Chemical Society's disposal in the present instance. Quite apart from resources, moreover, the definitive work would not have been a sensible goal for this maiden effort. In any event, substitutions and additions would not change the basic findings documented by this part of the study:

• That chemistry has given distinctive shape to the ways man works and lives.

• That the principles and techniques of chemistry pervade much of an industrial society and extend beyond the chemical industry itself.

• That chemistry is sure to remain a principal instrument in the continuing adaptation of nature to man's purposes.

Experts organized in 24 committees supplied the basic information for this part of the study. They came mostly from industry, but an infusion of participants from government agencies and educational institutions assured broadly based inputs. The information they supplied was then supplemented and adapted by the study's staff, but the committees collaborated and fully reviewed the resulting chapters.

The treatments in the chapters vary somewhat in breadth of coverage, in the handling of names of people and organizations, and in other respects. Some of this variability reflects differences in the nature of the sectors themselves, since standards of value and the intensity of chemical involvement naturally differ from sector to sector. Some of the variability, however, reflects the differing approaches of the contributing committees to their assigned tasks. The committees worked independently, and homogeneity was not accorded high priority in their efforts.

The chapters of Part I make their points largely in anecdotal fashion. Numbers are used often to provide perspective, but they do not comprise a comprehensive body of hard data. The essence of this part of the study, as a matter of fact, is not truly quantifiable. It is rather to show by example how chemistry actually has functioned in industry and a few of the ways it might function, in both instances in terms of the people and organizations involved and of the manifold forces that shape their actions.

Attempts to name individuals and organizations contributing in important ways to the advancement of chemical technology are also limited. Even were it possible to assign credit fairly and unambiguously, the resulting document at the very least would have been unwieldy and not very readable. Therefore, names of individuals are included only for those advances that clearly marked fairly sharp departures from earlier history. In other cases where many individuals contributed and built largely on the work of each other, we have attempted to name the industrial, governmental, and educational organizations that were among the pioneers in the developments concerned.

Part I makes it clear that no scientific discipline is really independent of all others. Any given scientific or engineering talent must be complemented by other skills, and many of the accomplishments recorded here have resulted from work performed by persons trained in many disciplines. While we have naturally concentrated on the work of chemists and chemical engineers, the contributions from other disciplines are noted from time to time. Even where they are not, however, we hope readers will remember other science and engineering skills are always called for, even if we do not remind them on every page.

Authorities have stressed for years in various contexts that geographical or political boundaries cannot for long restrict the spread of scientific and technological knowledge. The truth of that concept emerges clearly from this part of the study. One of the many examples is the basic British work on polyethylene, which drew particularly on the published research of Bridgeman and Conant in this country and on the high-pressure expertise of Michels in the Netherlands. Among other examples, penicillin emerged from a laboratory in Great Britain and the forerunner of today's general-purpose synthetic rubber from a laboratory in Germany. Although accomplishments in the U.S. are stressed, pioneering ones elsewhere on which Americans built are not ignored. In many instances, the foreign antecedents are acknowledged, but there are doubtless some that are not. Any such oversights are entirely unintentional.

Finally, it should be noted that the chemical accomplishments detailed in this study were judged by their social and economic impacts in the context of their times. Many persons today would quarrel with the view that such developments as lead antiknock compounds, DDT, or detergent phosphates represent "accomplishments." Regardless of the merits of such arguments that might be advanced, it can hardly be disputed that these and other products now questioned have exerted important social and economic effects. Nor can it be disputed that those effects also have been widely regarded as beneficial. That the lead compounds or DDT or detergent phosphates are falling from favor now, whether justifiably so or not, is not truly relevant to their status as chemical accomplishments in their own time.

This is not to imply that the practice of chemistry is inherently good. The point is that it is useful and, in a sense, sometimes ironically so. A few parts per million of organochlorine residue in human body fat or of sulfur dioxide in urban air would go undetected without the highly developed capabilities of analytical chemistry. Chemistry is also required to make some fabrics flame resistant, to devise emission control systems for vehicle exhausts, and to develop recycle systems that at once reduce waste and are economically attractive. As these chapters make clear, the utility of chemistry in solving such problems has important economic implications, too.

1 / Organic Chemicals

SOCIAL AND ECONOMIC IMPACTS

A distinctive characteristic of organic intermediate chemicals is their anonymity among the consumers they serve so diversely. Most automobile drivers know of "permanent antifreeze," and many know that it is mainly ethylene glycol, in this use an organic end chemical. Few consumers, however, will know that ethylene glycol is made from an organic intermediate, ethylene oxide, which is made in turn from a primary organic chemical, ethylene. Nor will many know that ethylene glycol itself is used as an organic intermediate to make polyester fibers for clothing and polyester tape for tape recorders and computers.

The anonymity of organic intermediates is due partly to the complexities of organic chemical nomenclature. The main reason, however, is the fact that, as a rule, these compounds reach the consumer in altered form. The intermediates acetic anhydride and salicylic acid are used to make aspirin, the most effective nonprescription pain killer known today. Hexamethylene diamine and adipic acid are used to make nylon, whose utility and economic effects alike have been worldwide.

The great versatility of organic intermediates is largely inherent in the chemistry of organic compounds. The discovery and use of the knowledge required to exploit that versatility, however, constitutes a classic example of the economic consequences of scientific research. Chemical research and development in roughly the past century has created thousands of uses for organic intermediates in a variety of industries, including agricultural chemicals, paints, pharmaceuticals, plastics, synthetic detergents, synthetic fibers, and synthetic rubber.

Whether an organic intermediates industry exists, as such, is open to considerable debate. The compounds are produced almost entirely by companies whose primary business is chemicals or petroleum processing. (The chemical operations of some petroleum companies are larger than those of most chemical companies.) Nearly all major chemical companies make such compounds and, overall, close to 400 companies produce them. The two leading producers of each of the top 100 organic intermediates total some 50 companies in all. Despite this diversity of source, however, the very high commercial standards of purity for the compounds make them largely interchangeable from one manufacturer to another.

The production value of organic intermediates in the U.S. was about $17 billion in 1970. This was roughly half of the production value of all industrial chemicals. Close to 90% of organic intermediates today are made from petroleum and natural gas (and thus are "petrochemicals"). Most of the remainder come from coal tar, produced by the steel industry's coke ovens, and a very small amount is made from animal fats, vegetable oils, rosin, and grains.

The chemistry and economics of organic chemicals are such that producers tend often to use continuous processes on a large scale. They tend also to integrate both backward toward raw materials and forward toward end products. A maker of ethylene glycol, for example, may make ethylene, ethylene oxide, and the glycol in successive steps and sell only the glycol. Because of such vertical integration, companies that make organic intermediates sell only about half their output, overall, and use the rest captively to make other products.

The economic logic of making and using organic intermediates in large, continuous, highly-integrated plants is one of several interrelated reasons for the huge economic impact of petrochemicals on the Gulf Coasts of Texas and Louisiana. The major reason, perhaps, is the nature of a basic organic chemical, ethylene, a precursor of many organic intermediates. The sources of ethylene, petroleum fractions and natural gas liquids, have been abundant in the Gulf Coast area. The compound itself can be shipped with comparative ease by pipeline and is used extensively to make other chemicals and plastics. Its production economics favor very large plants, which range today upward of 1 billion pounds of production capacity per year. These characteristics of ethylene account in large measure for the giant petrochemical complexes that have sprung up on the Gulf Coast since World War II.

By no means all organic intermediates, however, are produced in enormous volume. Among the top 100 in production value, annual output ranges from 3 billion pounds to 10 million pounds or less. Among compounds below the top 100, annual output may be only a few thousand pounds. Prices follow the same pattern, ranging from a few cents per pound for large-volume items to $15 or more for low-volume, highly specialized compounds. Many small-volume intermediates are made by batch processes. Such products tend to be labor-intensive, as opposed to the capital-intensive, high-volume intermediates made in big, continuous-process plants.

In addition to wide variations in volume and price, organic intermediates are marked by complex interrelationships. These are rooted largely in the fact that, chemically, two or more combinations of raw materials often can be used to make the same product. A relatively simple example is acrylonitrile. This intermediate is the main component of acrylic synthetic fibers and nitrile synthetic rubber and an important component of certain plastics, including the versatile, large-volume ABS (acrylonitrile-butadiene-styrene).

Since acrylonitrile was introduced commercially, in about 1940, it has been made from three different sets of starting materials. The most recent process, based on propylene and ammonia, was first used in 1960. It capitalized on the ready supply of cheap propylene that had become available by the late 1950's as a by-product or co-product of petroleum and petrochemical processes. The propylene-ammonia process by now has entirely displaced the earlier processes. And the price of acrylonitrile, partly as a result of continuous process research and development, has fallen from its high point of about 38 cents a pound to about 12 cents today.

The chemistry and versatility of organic intermediates involve them not only in many products and industries but in a variety of related issues. Not the least of these is the ubiquitous concern over environmental effects. Chemical producers have worked over the years to develop products useful to industry and individuals but in the process have outrun the general ability to evaluate the long-term effects of the progress made. By the same token, industrial chemists often are able to resolve long-term environmental problems, when they appear, without sacrificing the benefits involved.

A familiar example is the unsightly foam that was appearing with growing frequency in the nation's waters by the early 1960's as a result of heavy use of household synthetic detergents. Research disclosed that the solution was to replace a branched chain of carbon atoms in the molecule of the detergent intermediate, alkylbenzene, with a more biodegradable straight chain. At the time, chemists did not know how to produce the straight-chain part of the molecule economically in the required volume.

But in 1965, following extensive research, development, and testing for product safety, the U.S. detergent industry converted to the straight-chain alkylbenzene.

The economic utility of organic intermediate chemicals, overall, combines the individual utilities of thousands of end products whose impact often is not fully evident in terms of dollars or pounds. Synthetic glycerol, for example, has filled a world need that could not be supplied from natural sources. Synthetic rubbers offer a large measure of relief from dependence on natural rubber and often offer better properties than the natural product possesses. These and the many other economic ramifications of organic intermediates cannot feasibly be examined here in great detail. The nature of the economic impact of the compounds can be seen clearly enough, however, in a broad examination of certain exemplary intermediates and their end products.

CHEMICAL ACCOMPLISHMENTS

Intermediates for Rubber and Plastics

Synthetic rubbers and thermoplastics have revolutionized materials engineering in the past half century and in the process have grown into multibillion-dollar industries. All of these materials are polymers, giant molecules made by linking thousands of molecules of primary organic chemicals or intermediates end to end. A polymer may contain only one kind of repeating unit, or it may contain two or more kinds. Neoprene synthetic rubber, for example, is polychloroprene, a polymer of an organic intermediate, chloroprene. ABS plastic is a polymer of the intermediates acrylonitrile, butadiene, and styrene.

Much of the early interest in polymers arose from the desire to make synthetic rubber. The goal was to relieve the then-total dependence on natural rubber, which varied in quality, supply, and price. Research in this area received considerable impetus in the mid-1920's as a result of the Stevenson Act, a British law limiting the output of rubber plantations. One effect of the act was to force the price of rubber in New York from 11 cents per pound to $1.23 in 1925. The price returned to normal by 1926, but research continued, especially at du Pont in the U.S. and I. G. Farbenindustrie in Germany.

By the mid-1930's, du Pont was producing the first synthetic rubber, neoprene, on a commercial basis. The years since then have seen extensive research and development in polymer chemistry and physics in many industrial and university laboratories in this country and abroad. Out of this work have come the commercial developments that created the giant industries of today.

Both plastics and synthetic rubber are produced primarily from organic chemicals derived from petroleum. There are, of course, competing uses for petroleum, such as gasoline and fuel oil, and most of the intermediates made from it are basic building blocks not only for polymers but for many other industrial and consumer products. Process research and engineering improvements nevertheless have enabled chemical and petroleum companies to meet the steadily growing demand for organic intermediates. Often they have done so over the long term at decreasing prices in spite of inflationary trends.

Ethylene, Propylene. Ethylene and propylene are the two simplest building blocks for organic chemicals and polymers and are also among the most versatile. Both compounds, strictly speaking, are primary organic chemicals as opposed to intermediates. They are intimately related to many intermediates, however, and may be considered intermediates themselves when polymerized directly to make two widely-used plastics, polyethylene and polypropylene.

Refinery streams have contained ethylene and propylene since thermal cracking of petroleum began in 1913. Their use in petrochemicals was pioneered in the 1920's and 1930's by Union Carbide and Standard Oil (New Jersey). Production of ethylene in the U.S. exceeded 18 billion pounds in 1970 and is expected to double by 1980. Chemicals and polymers consume more than 8 billion pounds of propylene per year today, and the figure is expected to exceed 20 billion pounds per year by 1980.

Industrial chemists recognized at the outset that derivatives of ethylene and propylene could fill many industrial and consumer needs. This concept combined with several economic and technical developments to spur the growth of a chemical industry based on the two compounds.

In 1920, Standard Oil (New Jersey) began to make isopropyl alcohol from propylene. Isopropyl alcohol, familiar to consumers today in rubbing alcohol, was used mainly to make an important industrial solvent, acetone. The latter at the time was produced by the generally less precise fermentation process with butyl alcohol as a co-product. Isopropyl alcohol was the first commercial synthetic alcohol and is generally considered to have been the first petrochemical.

By 1923, equipment and know-how were available to make synthetic ethyl alcohol from ethylene. The early Prohibition era, however, was not considered a propitious time to introduce a new source of this alcohol, which then was made entirely by fermentation. Little ethylene, moreover, was present in the refinery gases of the day.

Much of the early research on ethylene-based industrial chemicals was supported by a predecessor company of Union Carbide. Out of this work came a number of commercial products, including ethylene glycol (1925) and synthetic ethyl alcohol (1930).

Starting about 1937, the growing use of tetraethyl lead as a gasoline additive spurred intensive study of processes for producing ethylene. The antiknock compound is made from lead-sodium alloy and ethyl chloride, which is made in turn from hydrochloric acid and ethylene. The consequent demand for ethylene was thus threatening to outrun the supply.

Improvements were made in petroleum cracking and gas separation processes, and the first naphtha steam cracking unit went on-stream in 1941 at the Baton Rouge refinery of Standard Oil (New Jersey). The key to the process was the use of steam in the feed to minimize the coking that occurs in the furnace tubes when petroleum vapors are cracked at high temperature and low pressure. The gases from the naphtha steam cracking process contained not only ethylene but butadiene and isobutylene as well. These two intermediates were soon to become important in making synthetic rubber.

World War II created two major new uses for ethylene. One was polyethylene, important to the war effort because of its electrical insulation properties. The other new use was styrene, an intermediate made from ethylene and benzene and used in enormous quantities in styrene-butadiene synthetic rubber.

Demand for a variety of ethylene derivatives has grown steadily since World War II. Growth and process and engineering developments have justified larger plants that produce at lower cost. The price of ethylene has declined steadily, reaching roughly 5 cents per pound in 1955–60 and 3 cents today. Cheap, plentiful ethylene spurred research on means of substituting the compound for other chemical starting materials. As a result of this evolutionary process, an ethylene plant built today may be designed to produce 1 billion pounds per year or more.

Of the 18 billion pounds of ethylene produced annually in the U.S., some 35% is used to make polyethylene. An additional 40% or so is accounted for by ethylene oxide and its derivatives, by ethyl alcohol, and by ethylbenzene, the precursor of styrene.

Despite the large and growing demand for propylene, little need has existed for better methods of making the compound. It is an abundant coproduct of ethylene production and other refinery operations. Catalytic cracking of petroleum to produce high-octane gasoline produces large amounts of propylene, and only about a quarter of it is separated for chemical use. Of the 8 billion pounds of propylene used annually to make chemicals and polymers, roughly 20% is used to make isopropyl alcohol, 18% for acrylonitrile, 16% for polypropylene, and 13% for propylene oxide. The remaining one third is used to make a number of other products, including cumene, an intermediate used to make phenol, and oxo alcohols, the basis of plasticizers for plastics. Among the smaller but more sharply growing uses of propylene is ethylene-propylene synthetic rubber.

Styrene. The organic intermediate styrene is an international commodity made in more than 30 countries. U.S. production alone exceeds 5 billion pounds per year and should reach 6 billion pounds by about 1973. Nearly all styrene is consumed in polymers: polystyrene, ABS, and related plastics; styrene-butadiene synthetic rubbers; and styrene-butadiene latices, such as those used in water-based paints and paper coatings.

Small-scale production of styrene began in the 1930's as a result of a technological breakthrough by Dow Chemical in the U.S. and, at about the same time, by I. G. Farbenindustrie in Germany. The method was catalytic dehydrogenation of another intermediate, ethylbenzene. The latter is made readily and cheaply by reacting ethylene with benzene separated from refinery streams.

This process made possible the rapid buildup of U.S. styrene capacity several years later for the World War II synthetic rubber program. More broadly, it led to the production of cheap, highly-pure styrene, the economic key to the styrene plastics industry. Today, in the U.S. the industry supplies more than 3 billion pounds per year of styrene-based plastics for a variety of uses: packaging, appliances, construction, automobiles, toys, radio and TV, furniture, housewares, and luggage.

All U.S. styrene is made, basically, by the original method. Process improvements have been made, however, and these and the growth of the market are reflected in the price of the intermediate. From a range of 15 to 20 cents per pound in 1948–57, it has declined to 7 to 8 cents today. As the price has declined, styrene producers have found it economically advantageous to build very large plants, some with production capacities exceeding 500 million pounds annually. They have tended also to integrate vertically to make both the styrene and its raw materials and polymers.

Uncertainties in styrene production costs have been created by the anticipated removal of lead from automotive gasoline. The problem revolves around benzene, which is reacted with ethylene to make more than 90% of the ethylbenzene used in styrene plants. The styrene process has been made so efficient that the single most important factor in the price of the intermediate is the cost of benzene. The major source of benzene is catalytic reforming of petroleum fractions that also can be used to raise the octane level of gasoline. The degree to which refiners use these fractions to counteract the effects of removing lead will thus directly affect the demand-supply-price structure for benzene. What the structure will be depends on how automotive emission standards evolve and how they are met.

Isobutylene. Polymers and chemicals in the U.S. consume today more than 500 million pounds per year of the organic intermediate isobutylene. The largest use by far is in polymers, the major one being butyl rubber. More than 200 million pounds of this specialty synthetic rubber are produced annually in the U.S., and production capacity worldwide is more than 750 million pounds per year.

The first commercial chemical use of isobutylene came in 1929, when Standard Oil (New Jersey) began using it to make diisobutylene. This compound goes into surface active agents, lube oil additives, and plasticizers. Another product, polyisobutylene, is used in lube oil additives, sealants, adhesives, and caulking compounds. Isobutylene also has a number of small-volume uses.

Early attempts to polymerize isobutylene produced only oily products of limited utility, but by the early 1930's research had produced the polyisobutylenes of reasonably high molecular weight that still are used today. A few years later, chemists at Standard Oil Development Co. learned to make butyl rubber, a copolymer of isobutylene with a few percent of another intermediate, isoprene. Standard Oil (New Jersey) was operating two commercial plants by 1943–44.

After World War II, butyl rubber captured the entire inner tube market, but this use began to decline in 1953 with the introduction of the tubeless tire. Other uses for butyl have since been developed. Chlorinated butyl, a modification discovered at Esso Research and Engineering and commercialized in 1960, has helped to expand the industrial uses of the elastomer. Butyl rubber by now is a mature product, although consumption is expected to grow at least 15% in 1970–75. Vigorous competition from newer specialty elastomers, such as ethylene-propylene rubber, will likely restrict butyl's growth in the future and thus the growth of isobutylene.

Over the long range, isobutylene could come into use in several newly developed processes for converting the intermediate to isoprene, the monomer for polyisoprene synthetic rubber. Over the short range, however, existing processes based on other raw materials should be able to support demand for the elastomer.

Diolefins: Butadiene, Chloroprene, Isoprene. Synthetic rubber is the main outlet for the three most commercially important diolefin compounds: butadiene, chloroprene, and isoprene. Of the three, only butadiene is truly a large-volume commodity chemical, although isoprene could achieve fairly large volume over the longer range.

Scientists knew more than a century ago that natural rubber is a polymer of isoprene. Not until the 1950's, however, did they learn to polymerize it into "synthetic natural rubber." In the interim, research chemists working on synthetic rubber turned their attention to other monomers.

Butadiene's structural relationship to isoprene made the intermediate an early candidate for experiments on synthetic rubber. Polybutadiene elastomers with inferior properties were made in Germany and Russia as early as the 1930's. More promising were copolymers of butadiene with styrene or acrylonitrile. These became the basis of this country's huge synthetic rubber program in World War II and made butadiene commercially important by the early 1940's. The next milestone for the intermediate was the development of stereospecific polymerization techniques in Europe in the 1950's which led to the production of polybutadiene rubber.

Butadiene production in the U.S. today is in the range of 3 to 3.5 billion pounds per year and is projected to reach 5 billion pounds by 1980. Styrene-butadiene rubber consumes about 60% of the butadiene produced and polybutadiene rubber about 17%. The remainder is used to make products that include styrene-based resins and nitrile rubber.

Almost 80% of the butadiene in the U.S. is made by dehydrogenating butene or butane and nearly all the remainder by steam cracking petroleum fractions. A growing supply of relatively cheap butene was assured by the discovery of catalytic cracking in the 1930's. Use of the process expanded rapidly after World War II to meet the demand for high-octane gasoline. Butane can be recovered from natural gas and refinery streams.

Standard Oil Development Co. developed the basic process for making butadiene from butene after earlier work at Universal Oil Products Co. Dow Chemical later developed the improved catalyst used today to make most butadiene in the U.S. The shortage of butene during World War II led to the development of butane dehydrogenation processes by Phillips Petroleum and Houdry Process Co. Both are still used, but the Houdry process predominates.

All commercial butadiene processes yield mixtures of the intermediate with other hydrocarbons that boil in the same range. The cost of separating the butadiene thus contributes significantly to total capital and operating costs. Extractive distillation processes for recovering butadiene have been developed in this country by Standard Oil (New Jersey), Phillips Petroleum, Shell Oil, and Union Carbide

The organic intermediate chloroprene is the monomer for neoprene, the specialty synthetic rubber introduced by du Pont in the early 1930's. Neoprene was an immediate success, even at a premium price, because it resists ozone and solvents much better than does natural rubber.

Chloroprene is still made by the original du Pont process, addition of hydrogen chloride to vinylacetylene. It can also be made, however, from butadiene by a process developed in England in the 1950's. In 1970, Petro-Tex Chemical Co. became the second U.S. producer of neoprene, using the butadiene route. Market growth for neoprene has slowed in recent years, partly because of inroads by newer specialty rubbers, especially the ethylene-propylene types. But in the longer range, neoprene could become more competitive because of price reduction catalyzed by increased competition and the more favorable economics of the butadiene route to chloroprene.

Until the 1960's, the only large use for isoprene was in butyl rubber, which contains 1 to 3% of the monomer. The advent of stereospecific polymerization, however, made it possible to produce a polyisoprene that approximates the properties of natural rubber. This product, commercialized in 1959, sharply increased the market for isoprene. Demand for polyisoprene already exceeds 100,000 long tons per year. If it grows as projected, it could be using up to 500 million pounds of isoprene per year by 1975.

This new market was seen at the outset to far exceed the supply of isoprene available from cracked refinery streams, the traditional source. Research emphasis thus turned to other sources of isoprene, whose cost is crucial to polyisoprene's ability to compete with natural, polybutadiene, and styrene-butadiene rubbers. Of the variety of isoprene processes studied, two are now used in this country. Shell and Goodrich make the intermediate by dehydrogenating isopentene extracted from cracked refinery streams. Goodyear uses a three-step process that starts with propylene.

Intermediates for Synthetic Fibers

Among the topmost achievements of industrial chemistry are the noncellulosic synthetic fibers, a worldwide, multibillion-pound industry. U.S. production of these fibers—nylons, acrylics, polyesters—totaled roughly

3 billion pounds in 1970. Their utility has long been self-evident. Less evident, perhaps, is the importance of the ability to make the organic intermediates for the fibers cheaply and in volume. Research and development on economic routes to the intermediates began in the 1930's with du Pont's decision to make the polyamide fiber, nylon. The work has not yet ended.

Nylon Intermediates. The first of the nylons, and still the most important in the U.S., is nylon 66, based on the pioneeering research of W. H. Carothers and his colleagues at du Pont in the 1930's. In 30 years, U.S. production of nylon 66 fibers has grown to roughly 1 billion pounds annually. Production capacity for the fiber intermediates has reached 1.8 billion pounds per year, with another 1.4 billion pounds in the rest of the Free World.

Of the two intermediates for nylon 66, adipic acid and hexamethylene diamine, only the former was available commercially when du Pont decided to make the fiber. It was made in Germany from phenol. Du Pont thus had to develop processes for making adipic acid from phenol, adiponitrile from adipic acid, and hexamethylene diamine from adiponitrile. The first commercial intermediates plant began operating in 1939 with a capacity equivalent to 3 million pounds per year of nylon fiber.

A major expansion in nylon came in 1941–42, just at the time the U.S. entered World War II. As part of this expansion, the raw material base for nylon was changed from phenol to benzene. The latter can be hydrogenated to cyclohexane, which in turn is oxidized to adipic acid. The expansion included a continuous process for hydrogenating adiponitrile to hexamethylene diamine and the world's first plant for oxidizing cyclohexane to cyclohexanone and cyclohexanol.

As World War II neared an end, it was evident that pent-up demand would require a second, massive expansion of nylon capacity. Du Pont faced undesirably large investment in adipic acid facilities unless it could find means of making hexamethylene diamine from other compounds. From the resulting effort came an adiponitrile process using furfural made from cellulosic waste materials such as oat and rice hulls and corncobs. A commercial plant went into operation in 1948. Only 13 years later came a still more economical process, starting with butadiene. Research again was responsible, but the process also involved technology developed years before in seeking an alternative route to neoprene. The first plant using the butadiene route started up in 1951. By 1953, du Pont had closed down its last domestic adiponitrile plant based on adipic acid and was making all adiponitrile from furfural or butadiene.

Competitive interest in nylon 66 increased in the mid-1950's when du Pont's basic composition-of-matter patents expired. The company licensed Chemstrand Corp. under its patents and provided the technology for both nylon and its intermediates. Other new and potential entrants into the nylon business developed new processes for the intermediates. The most important of these commercially are:

- The boric acid–assisted cyclohexane oxidation process developed by Halcon International. It provides high yields of the cyclohexanone-cyclohexanol mixture which is subsequently oxidized to adipic acid.
- The electrohydrodimerization of acrylonitrile to produce adiponitrile in high yield. The process was developed by Monsanto Co.
- Celanese Corp.'s hexamethylene diamine process, which bypasses the adiponitrile intermediate and is the basis of the company's intermediates manufacture. In the process, adipic acid and other six-carbon cyclohexane oxidation products are hydrogenated to 1,6-hexanediol which is then aminated to hexamethylene diamine.

Whereas the nylon fiber development involved nylon 66 in the U.S., Canada, the United Kingdom, and France, the base was nylon 6 in many other countries, particularly Germany, Italy, and Japan. In this country, nevertheless, production capacity for caprolactam, the monomer for nylon 6, by now has reached nearly 600 million pounds per year. Capacity is more than 2 billion pounds per year in the rest of the Free World, so that total capacity almost equals world capacity for nylon 66 intermediates. U.S. production of nylon 6, meanwhile, has reached the range of 350 million pounds annually.

Allied Chemical, the first U.S. maker of caprolactam, started large-scale production in 1955, using phenol as the raw material. Allied has since been joined by Dow-Badische and Columbia Nipro. Both companies use processes developed in Europe and based on cyclohexane.

Because nylon 6 has been the senior nylon in Germany, Benelux, Italy, and Japan, the most significant recent developments in caprolactam technology have appeared outside this country. Among them are those aimed at reducing the production of by-product ammonium sulfate, an increasing economic and disposal problem. The quantity produced has been halved by improved methods of manufacturing hydroxylamine or cyclohexanone oxime, the intermediates for caprolactam.

Polyester Fiber Intermediates. Polyester fiber, like the nylons, has grown explosively since it became a commercial product in the U.S. in the early 1950's. U.S. production of the fiber topped 1 billion pounds in 1970. Production of the intermediates, dimethyl terephthalate (DMT) and terephthalic acid (TPA), totaled roughly 1.8 billion pounds. DMT and TPA are used also to make polyester film as well as herbicides, adhesives, and other products.

Polyester fiber was invented in England in the early 1940's. It was commercialized there by Imperial Chemical Industries and in this country by du Pont. The polymer, polyethylene terephthalate, can be made in two ways: by reacting TPA and ethylene glycol or by converting TPA to DMT, reacting the latter with ethylene glycol, and polymerizing the reaction product. The latter method was used first, because the intermediates must be very pure, and DMT can be purified much more readily than can TPA.

Neither TPA nor DMT could be made by the methods at first available in the quantities and at the costs required for a fiber that would have to compete with cotton, wool, and rayon. Scientists at du Pont therefore developed a process using para-xylene and nitric acid to make TPA and reacting the latter with methyl alcohol to yield DMT. Variations of the para-xylene process apparently were used by both du Pont and Imperial Chemical Industries in the first large-scale production of TPA and DMT in 1949–51.

Several processes since have been developed for TPA and DMT. Eastman Chemical Products, Mobil Chemical Co., and Amoco Chemical Co. make TPA by methods that differ in detail but are based on oxidizing para-xylene with air or oxygen. Hercules produces DMT by a process that starts with para-xylene but does not involve TPA. Mobil Chemical and Amoco Chemical in addition have worked out undisclosed methods of making TPA pure enough to make the polyester without going through the DMT step. The use of polymer-grade TPA may now be close to 300 million pounds annually.

The effects of process improvement and market growth are evident in the cost of DMT. From a unit value of 32 cents per pound in 1962, the intermediate declined to 14 cents in 1970.

Acrylic Fiber Intermediates. The third of the three major noncellulosic fibers is the acrylics, polymers of the organic intermediate acrylonitrile. The acrylics include the modacrylics, which are copolymers of 35 to 85%

acrylonitrile with other intermediates such as vinyl chloride. The U.S. produced roughly 500 million pounds of acrylic fibers in 1970. Fibers, however, use only 60% of the acrylonitrile produced; the rest goes into plastics (12%), nitrile synthetic rubber (8%), miscellaneous uses (10%), and export (10%). Overall, U.S. output of acrylonitrile climbed from 1 million pounds in 1940 to more than 1 billion pounds in 1970.

Acrylonitrile was a laboratory curiosity from 1893, when it was discovered, until World War II, when it was first manufactured in quantity. The main use was oil-resistant nitrile rubber, a copolymer of acrylonitrile and butadiene. The rubber was discovered in Germany and used in self-sealing liners for aircraft fuel tanks. In 1940–51, the intermediate was made in this country only by American Cyanamid. The company made it first from ethylene oxide and hydrogen cyanide and, later, from acetylene and hydrogen cyanide. In the 1950's, Union Carbide entered the business, using the ethylene oxide process, and Monsanto, Goodrich, and du Pont came in with the acetylene process.

Then in 1960, Standard Oil (Ohio) began producing acrylonitrile by the propylene-ammonia process, discovered by Sohio chemists in the late 1950's. The process was intrinsically economic and gave relatively high yield of acrylonitrile. Its main raw material, propylene, is a by-product of refinery cracking and a co-product of ethylene manufacture. Rapid growth in both areas had made propylene cheap and plentiful by the late 1950's. The propylene-ammonia process as a result soon dominated the industry and now is the only one used in the U.S.

The advent of cheaper acrylonitrile from this process and the versatility of the products made from the intermediate sharply stimulated its growth. Production has risen fivefold in the decade since the Sohio process was first used. The price of acrylonitrile, meanwhile, has dropped from an historic high of 38 cents a pound to 12 cents today.

The ready availability of cheap acrylonitrile should contribute to steady growth in the intermediate's end uses. One that is growing rapidly is acrylonitrile-butadiene-styrene (ABS) resins for automotive parts, pipe and fittings, and major appliances. ABS probably will be the material used if automobile bodies are made of plastic in significant numbers. The acrylic fibers should take an increasing share of the total textile market. The Barex resin being introduced by Vistron Corp., a subsidiary of Sohio, in disposable soft drink bottles contains about 50% acrylonitrile. Overall, acrylonitrile production in 1975 is projected at 2.15 billion pounds and the price at 10 cents per pound.

Intermediates for Synthetic Detergents

Organic intermediates are used in very large volume in the U.S. and abroad to make the surfactants (surface active agents) that provide the basic cleansing power of modern synthetic detergents. The U.S. today consumes a total of more than 4 billion pounds annually of some 700 different types of surfactants. These are derived in turn from a variety of intermediates: alkylbenzene, linear alcohols, phenol derivatives, ethylene oxide and glycols, organic amines, and others.

While the performance of synthetic detergents depends most heavily on one or more surfactants, the products often are multicomponent formulations. Their overall performance is adjusted to the end use by builders such as carbonates, phosphates, silicates, and sulfates and by additives such as borates, carboxymethyl cellulose, optical brighteners, enzymes, and bleaches.

The surfactant market is characterized by nearly unlimited numbers of end uses, most of them highly specialized. The market nevertheless has two broad segments, household and nonhousehold. Each accounts for about 2 billion pounds of surfactant consumption annually. Relatively few basic chemical suppliers provide raw materials to surfactant manufacturers, but more than 200 companies produce synthetic organic intermediates or end products for the surfactant market.

The performance requirements of household synthetic detergents are concisely defined and well established, and more than 80% of the market is divided among three producers: Procter & Gamble, Colgate-Palmolive, and Lever Brothers. The industrial market, in contrast, has more than 25 categories. They range from asphalt through cosmetics and toiletries to textile processing and, overall, involve some 3,000 companies.

The evolution of modern synthetic detergents began more than 2,000 years ago with the discovery of the cleansing powers of soap, which is itself a surfactant. The date of the first (partly) synthetic detergent is uncertain. In 1834, however, the German chemist Runge made surfactants for use in the textile industry by treating sulfonated almond and olive oils with caustic. In 1916 in Germany came production of diisopropyl naphthalene sodium sulfonate, which many consider the first true synthetic detergent.

Whatever the origin of synthetic detergents, the goal was to make from organic intermediates materials that would do the work of soap without the attendant disadvantages. In hard water, soap forms insoluble metallic compounds. The resulting adherent scum creates among other problems the rings in bathtubs, the films that spoil the luster of hair, and the grayness and harshness of feel that build up in textiles upon repeated washing. Like many natural products, moreover, the fats and oils from which soap is made tend to vary in price and quality to a degree that can cause problems in manufacture.

Despite considerable effort to replace soap, significant progress did not come until the end of World War II. By then it had become technologically feasible to make the intermediate alkylbenzene from benzene and propylene tetramer, both derived from petroleum. This development opened the door to new ventures in surfactant chemistry.

Alkylbenzene could be treated with sulfuric acid and then with caustic to yield alkylbenzene sulfonate (ABS). This low-cost synthetic surfactant possessed the cleansing action of soap without its disadvantages. The use of petroleum as raw material, moreover, not only eased manufacturers' problems on that score. It had the desirable side effect of freeing large amounts of the fats and oils used in soap for potentially more important uses in food products.

U.S. production of synthetic detergents increased 20-fold in the decade after World War II. Today they command close to 90% of the total detergent market once held by soap. In the interim, however, the sharply rising use of synthetic detergents, bolstered by population growth, created a problem whose solution directly involved detergent intermediates.

The problem was excessive foaming in sewage treatment plants and natural waters. Much of it was found to be caused by alkylbenzene sulfonate, the surfactant then used widely in household detergents. The solution was to replace the branched chain of carbon atoms on the alkylbenzene intermediate with a straight chain, which is degraded much more readily by water microorganisms (excepting anaerobic microorganisms, the type that predominate in septic tanks and other waters that contain little free oxygen). In 1965, following extensive research, development, and product testing, the detergent industry converted entirely to the new surfactant, linear alkylate sulfonate (LAS). To make the change economically feasible, however, the industry had to develop means of obtaining the straight chain of carbon atoms (normal paraffins) cheaply and in volume.

The required breakthrough was the molecular sieve, a modified zeolite that can be used to separate normal paraffins from petroleum fractions. The biodegradation problem helped also to speed the commercialization of other breakthroughs. One was the use of Ziegler catalysis to make certain straight-chain olefins and alcohols from ethylene. A second development was production of the same types of olefins by cracking petroleum waxes. The intermediates thus made available have permitted economic production of surfactants that are not highly branched and do not contain the aromatic (benzene) nucleus found in ABS and LAS. Both features ease the biodegradation of these products completely to carbon dioxide and water.

The pattern of consumption of detergent intermediates will likely continue to change in the future, influenced by changes in consumer products and in ecological requirements. All sectors of the household, personal care, and industrial markets will be affected. The low cost of the alkylbenzene intermediates used in ABS and LAS, for example, has tended in the past to restrict competition. Foreseeable ecological requirements suggest, however, that the cheapest product may not always be the dominant product. The use of LAS as a result is expected to decline, while the use of alpha-olefin sulfonates, paraffin sulfonates, and detergent-type alcohol derivatives is expected to increase.

Organic intermediates also are directly involved in the phosphate-eutrophication question. Formaldehyde is the intermediate used with ammonia, hydrogen cyanide, and sodium hydroxide to make nitrilotriacetic acid (NTA), the compound that for several years seemed most likely to replace detergent phosphates. NTA was withdrawn from the market in 1971 because of research findings that, under certain conditions, it might be instrumental in causing birth defects. Later in 1971 the Federal Government revised its stand on phosphates, stating that, on balance, they were the best materials currently available for the job.

The situation could change again, however. Subsequent research might clear NTA, although the bad name acquired by the compound could make it difficult to sell to the consumer. Other phosphate substitutes have been proposed, including citrates, diglycolates, polymaleates, and sodium oxydiacetate, but it is too early to assess the future of these compounds.

Glycerol

The adaptability of organic intermediate chemistry to new industrial developments is clear in the relationship between synthetic detergents and synthetic glycerol (glycerine). Until 1948, almost all glycerol was recovered from soap plants, which produce roughly 60 pounds of it per ton of soap. Price and supply problems inherent in dependence on a by-product provided an incentive for making synthetic glycerol. The synthetic product became feasible in the 1930's, when Shell Development Co. discovered that the intermediate allyl chloride could be made by reacting propylene with chlorine at high temperature. Allyl chloride can be converted to another intermediate, epichlorohydrin, which is converted to glycerol. Shell made pilot plant quantities of synthetic glycerol available in 1937, but the ample supply of the natural product at the time discouraged a commercial venture.

With the end of World War II, however, came the synthetic detergents and the consequent decline in supply of by-product glycerol from soap. In 1948, as a result, Shell Chemical started production at the world's first synthetic glycerol plant. Dow and FMC Corp. also are producers today.

One outgrowth of the original glycerol process was the availability of epichlorohydrin, which later served as an intermediate in the commercialization of epoxy resins.

The availability of synthetic glycerol encouraged the development of new and existing uses. The U.S. consumes currently some 300 million pounds per year of the compound, both by-product and synthetic, in more than 1,500 applications. The synthetic product accounts for 40% of the Free World's supply of glycerol.

Intermediates for Fluorocarbons

The fluorocarbons are a series of compounds of carbon, fluorine, and chlorine that were developed initially in the 1920's in a purposeful search for improved refrigerants. The first commercial fluorocarbon was produced in 1931. By 1970, U.S. production of these compounds had grown to something more than 700 million pounds annually, and another 600 million pounds per year is produced elsewhere in the world. In this country, refrigerants account for about 30% of fluorocarbon consumption and aerosol propellants for more than 50%. Production of fluorocarbons is the largest use for each of two organic intermediates, carbon tetrachloride and chloroform.

Of the refrigerants in use in the 1920's, none was without serious drawbacks. Some, like ethylene, were flammable. Others, like sulfur dioxide, were corrosive and quite toxic. Still others, like ammonia, combined all three hazards. Carbon dioxide would have been ideal except that it had to be used at relatively high pressure. The equipment required as a result was subject to leaks and almost prohibitively bulky. The perils of other refrigerants were so great, however, that the U.S. Navy in that period specified that its vessels would use only carbon dioxide refrigeration systems.

The makers of refrigeration equipment in the 1920's recognized the shortcomings of the refrigerants then in use. Thus, chemist Thomas Midgley, working with Albert Henne and others at General Motors Corp., set out to find the ideal refrigerant. Their plan was to produce such a refrigerant by deliberate arrangement of chemical elements, and the approach in consequence was fundamental.

The group began by plotting trends of toxicity, flammability, and boiling point on a periodic table of the elements. After considerable work, Midgley and Henne selected as their first choice dichlorodifluoromethane. The compound proved to be all that they had hoped for, and Midgley announced the discovery at the National Meeting of the American Chemical Society in April 1930.

A joint company formed by General Motors and du Pont began to produce the refrigerant, Freon 12, in 1931, and the refrigeration industry began to modify its equipment to use the new product. By the end of 1931 the joint company had begun to manufacture anhydrous hydrofluoric acid, a basic raw material for Freon not previously available in the necessary volume. By 1935 the product line included five of the new fluorocarbons, three derivatives of methane and two of ethane. These five covered a boiling-temperature range of $-41°$ to $+48°$ C. They filled the needs of the refrigeration and air conditioning industries of the time and even today account for more than 95% of world fluorocarbon production.

A major new use for the Freons came in World War II with the development of aerosol insecticides, in which they served as propellants. Shortly after the war the du Pont patents on the fluorocarbons expired, and Allied Chemical entered the business. The du Pont–GM joint company was dissolved, and its Freon business passed entirely to du Pont. In addition to du Pont and Allied, fluorocarbon producers today include Pennwalt Corp., Union Carbide, Kaiser Aluminum & Chemical, and Racon, Inc.

Many different types of fluorocarbons are on the market by now, both in this country and abroad. Besides the two major uses, aerosol propellants and refrigerants, they serve as dry-cleaning agents, solvents, blowing agents in foamed plastics, electrical insulators, inert liquids, and fire extinguishants.

The fluorocarbons are manufactured today by reacting an organic intermediate with hydrofluoric acid in the presence of a catalyst, antimony chlorofluoride. The intermediate used depends on the product desired. Carbon tetrachloride, for example, is used to make Freon 11, trichloromonofluoromethane; chloroform to make Freon 22, monochlorodifluoromethane; and perchloroethylene to make Freon 113, trichlorotrifluoroethane. The gas stream leaving the reaction vessel contains crude Freon and acidic gases formed in the reaction. One of the latter is hydrogen chloride, which is recovered for other uses. Purification of the Freon accounts for about 70% of the cost of the manufacturing process.

Vinyl Acetate

U.S. production of vinyl acetate has climbed almost fourfold in the past decade and today exceeds 1 billion pounds per year. The intermediate is consumed entirely in its polymerized form, polyvinyl acetate, or as a derivative of the polymer. It is used in emulsion and solvent systems for adhesives, water-based paints, textiles, and paper chemicals; as a copolymer in resins, particularly with vinyl chloride; and as a raw material for polyvinyl alcohol. Projected growth rates for the various applications of the intermediate range from 5 to 20% per year.

Vinyl acetate has displaced other chemicals in these applications in part because of ease of handling and specific advantages in properties. A third important factor, however, is that industrial chemical research has more than halved the price of the intermediate in the past 10 years.

Until shortly after World War II, vinyl acetate was made by reacting acetic acid and acetylene in the liquid phase. In the mid-1920's, German chemists discovered a vapor-phase process using acetylene and acetic acid with metal catalysts at high temperature. During the 1920's and 1930's, Union Carbide, du Pont, and British Celanese searched vigorously for new catalysts for this reaction and obtained many patents. Henry Dreyfus, founder of British Celanese, assigned a 1940 patent in this area to Celanese Corp. of America. The vapor-phase process for vinyl acetate finally took hold about 1948.

During World War II, meanwhile, British Celanese had found a way to make vinyl acetate from acetaldehyde and acetic anhydride. The discovery was of no practical use at the time. Then, growing markets for the two intermediates in the U.S. made it possible for Celanese Corp. to develop the capability to make them at raw material values comparable to that of acetylene. The company then developed a commercial vinyl acetate process based on the British Celanese invention and commercialized it in 1952.

By the late 1950's, the sharp growth and falling price of ethylene were increasing the interest in that hydrocarbon as a chemical raw material. In 1959, a German research team described the oxidation of ethylene using noble metal catalysts. Soviet chemists proposed soon afterward that the reaction might yield vinyl acetate if the solvent used were acetic acid instead of water. The details of the Soviet theory proved not to be precisely accurate, but the idea was sufficiently close to set industrial research groups on the trail throughout the world. Imperial Chemical Industries obtained the basic patents in 1963, and a liquid-phase vinyl acetate process based on this technology was first commercialized in 1965 by Celanese Corp.

The still-declining price of ethylene prompted even further research. Chemists at Farbenfabriken Bayer, at the Knapsack, A.G., subsidiary of Farbwerke Hoechst, and at National Distillers in the U.S. did development work on vapor-phase catalytic oxidation processes that produce vinyl acetate from ethylene, acetic acid, and oxygen. This technology was further developed and commercialized in this country in 1970 by Celanese Corp. and by National Distillers. Vinyl acetate processes based on both acetylene and ethylene are used commercially today, with the vapor-phase ethylene process accounting for essentially all the new capacity.

Ethylene Oxide

The rise of ethylene oxide from a laboratory curiosity to a 4 billion pound-per-year intermediate used in thousands of consumer products has been one of the two or three most powerful forces in the establishment of the U.S. petrochemical industry. In 1970, the U.S. produced 3.9 billion pounds of ethylene oxide, valued at $285 millon.

The largest use of ethylene oxide is ethylene glycol (50%), whose major uses in turn are automotive antifreeze and polyester fibers and films. Besides ethylene glycol, ethylene oxide is the starting point for many other materials. Among them are ethanol amines, glycol esters and ethers, polyglycol ethers, polyurethanes, and polyethylene glycols. These derivatives and ethylene glycol find important uses in synthetic rubber, synthetic fibers, resins, paints, adhesives, plastic films, molded articles, plasticizers, solvents, synthetic detergents, brake fluids, and cosmetics.

When the U.S. entered World War I, chemical warfare was much in the plans and to a lesser extent a part of the weaponry of both sides. The world dominance of the German chemical industry found the Americans in 1918 with limited chemical resources. The demand for mustard gas, made of ethylene and sulfur chloride, underscored the dilemma of the war. The ethylene had to come from ethyl alcohol, already in critical supply. More alcohol could be made only by scaling up the fermentation of grains in a nation then feeling the pinch of reduced food supply. The Army handed the problem to R. F. Bacon, director of Mellon Institute and technical director of the U.S. Chemical Warfare Service.

Bacon turned for help to Mellon's G. O. Curme, who was working on ethylene with the support of a predecessor company of Union Carbide. Curme had developed the use of a high-frequency electric arc submerged in gas oil to crack that petroleum fraction to acetylene, with ethylene and propylene as by-products. The wartime shortage of electric power, however, caused him to select as an ethylene process the thermal cracking of natural gas. Union Carbide perfected the process and produced quantities of pure ethylene and propylene, a by-product of the process, for making chemical derivatives. This work in 1919 came a little late to influence hostilities, but it liberated ethylene from the constraints of grain-derived ethyl alcohol.

Curme's group continued to work on ethylene, propylene, and their derivatives, and Union Carbide began to produce them commercially. The key compound was ethylene oxide. It was made from ethylene chlorohydrin, which was made in turn from ethylene and hypochlorous acid. Ethylene oxide had been discovered by the French chemist Charles Wurtz in 1859, but it was left to Curme and his colleagues to develop the practical processes for making the compound and its relatives on an industrial scale. In the process they taught the growing petrochemical industry to make hundreds of derivatives that imparted useful properties to a broad range of products that still is growing after 50 years.

Ethylene glycol was among the first derivatives of ethylene oxide to be produced. The first major customers were dynamite manufacturers, who used the glycol with glycerol to lower the freezing point of their nitrate explosives. Dow Chemical studied production of ethylene glycol from ethylene dichloride during World War I, but the early exploitation of the compound for explosives and antifreeze was handled largely by Union Carbide. That company and Dow rank first and second among the many world producers of glycol.

As the output of ethylene oxide grew, the amount of by-product calcium chloride produced by Curme's ethylene chlorohydrin process began far to exceed the limited demand for that compound. Chemists thus faced a new economic challenge: to find a means of making ethylene oxide directly from ethylene and oxygen with no by-products.

The French chemist T. E. Lefort in 1931 reported the use of a silver catalyst to oxidize ethylene with air. Lefort assigned his U.S. patent rights to Union Carbide, which perfected the process and in 1937 started up the world's first direct oxidation plant for ethylene oxide. In 1953, Shell Development Co. announced a direct oxidation process using oxygen instead of air, and Scientific Design Co. also developed a direct oxidation process. Both processes have been widely used. Among the many ethylene oxide producers around the world today, direct oxidation is the process of choice. Whether air or oxygen is used depends on the circumstances of the individual producer.

Propylene oxide is similar to ethylene oxide in some respects and has grown steadily if at a lower level than the ethylene compound. U.S. output was just over 1 billion pounds in 1970, roughly 30% of the amount of ethylene oxide produced. The major uses of propylene oxide are in making the proplyene glycols used in polyurethane and polyester resins. The oxide also is used in surface active agents, lubricants, and hydraulic fluids.

Propylene oxide has long been made by the chlorohydrin process in essentially the same manner as ethylene oxide. The calcum chloride by-product was not so severe a problem for propylene oxide because of the lower volume produced, and direct oxidation of propylene was in any case a more difficult proposition than for ethylene. As direct oxidation became the dominant process for ethylene oxide, a number of chlorohydrin plants were converted to make propylene oxide instead of building new plants to meet the growing demand.

The economic incentive for a new process intensified with the growth of propylene oxide, however. In 1969, Oxirane Corp. began to produce the compound by an indirect oxidation method developed with Halcon International. The process uses tertiary-butyl hydroperoxide to oxidize propylene to the oxide and produces tertiary-butyl alcohol as a by-product. The alcohol can be dehydrated to isobutylene, which can be used to make high-octane gasoline and has potential uses as an organic intermediate. Alternatively, the process can oxidize the propylene with alpha-methylbenzyl hydroperoxide, producing alpha-methylbenzyl alcohol as a by-product. The latter can be dehydrated to styrene, which is used widely in polymers. The Oxirane-Halcon process or variants of it will certainly dominate propylene oxide production, but direct oxidation seems secure for ethylene oxide.

The versatility and broad spectrum of uses of ethylene oxide promise that demand for the compound will continue to grow at a healthy rate as living standards improve around the world. A significant factor here is the future of the liquid-cooled automotive engine. Antifreeze accounted for roughly 60% of the ethylene glycol made in the U.S. in 1970, which translates to about 30% of the ethylene oxide produced. Should the future see changes that downgrade the role of the engine coolant, consumption of ethylene oxide would be affected markedly. Even so, however, new and existing uses seem likely to ensure a strong market for the compound for many years to come.

Methanol, Formaldehyde

Methanol (methyl alcohol) is a major commodity chemical worldwide, and the U.S. produces more than 40% of the world supply. Current world production capacity for the alcohol exceeds 2.5 billion gallons annually. U.S. output in 1970 was 750 million gallons (just under 5 billion pounds).

Almost half the methanol produced is used to make formaldehyde, a basic building block of the chemical industry. Methanol also serves widely as a solvent and in making other organic intermediates for a variety of uses. It's used in volume today, for example, to make dimethyl terephthalate, the predominant intermediate for polyester fibers and film.

Methanol was first made commercially in the 17th Century by destructive distillation of wood. Hence the familiar name, wood alcohol. This process remained the sole source of the alcohol until 1913, when the German chemists Haber and Bosch at Badische Anilin- und Soda-Fabrik learned to make it synthetically. Their method was the high-pressure catalytic reaction of carbon monoxide with hydrogen. After a delay incurred by World War I, the company built a commercial plant and by 1924 was marketing a substantial amount of methanol throughout the world. The synthetic product was cheaper than wood alcohol and free of the latter's characteristic stench.

The basic problem in making methanol from carbon monoxide and hydrogen is that the two can react in many different ways. The reaction that yields methanol is the least favored thermodynamically. The alcohol is produced only when certain selective catalysts are used and the process is directed to avoid other reactions. These undesired reactions can be suppressed in part by operating at very high pressure, and until recently, methanol synthesis plants used pressures of 3,000 to 5,000 pounds per square inch.

During the 1920's, the manufacture of methanol directly from synthesis gas (carbon monoxide and hydrogen) seemed such a formidable problem that attempts were made to use a more direct route, the oxidation of methane with air. Cities Service Oil Co. built a plant in 1928 to make methanol and formaldehyde by catalytic oxidation of methane with air. Celanese Corp. later began making the same products by noncatalytic oxidation of higher hydrocarbons with air. Only Celanese still makes methanol by this route.

At any rate, the success of the original Haber-Bosch process stimulated worldwide research on catalysts, synthesis gas manufacture, plant equipment, and process controls. This work, particularly on catalysts, has continued to the present.

Haber and Bosch had identified copper–zinc oxide catalysts as very reactive, but the copper made them sensitive to inactivation by sulfur. Today's improved methods of removing sulfur compounds from the synthesis gas, however, have made it possible to use newly-perfected copper-zinc catalysts that function below 1,500 pounds per square inch, less than half the operating pressure of earlier methanol processes. Low-pressure technology based on these catalysts and developed by Imperial Chemical Industries affords significant savings in operating cost. This technology is used in several new methanol plants in the U.S. whose production capacity totals over 400 million gallons per year.

The original German process was designed to use methanol synthesis gas made from coke and steam. Modern processes, however, use synthesis gas made by the high-temperature catalytic reaction of steam and methane (natural gas) or light petroleum fractions. In this country, methanol production is based almost exclusively on natural gas or petroleum.

Formaldehyde. U.S. companies produce about 2.2 billion pounds per year of formaldehyde, or about half of world production. Formaldehyde was used in the phenol-formaldehyde polymer, Bakelite, the first commercial plastic. Today, formaldehyde-based resins consume about two thirds of domestic production of the compound.

A major step in fibers involved the use of formaldehyde resins to impart permanent-press, wash-and-wear characteristics to clothing. Du Pont and Celanese in the past decade or so have developed formaldehyde polymers that can be machined, remain rigid at high temperatures, and are insensitive to attack by organic chemicals. These engineering plastics are used to replace metal, glass, wood, rubber, and other plastics. They may one day surpass in importance all other end uses of formaldehyde.

Commercial production of formaldehyde started in Germany in 1890. The method was catalytic oxidation of methanol, using copper and silver catalysts. It had been discovered in 1868 by August von Hoffmann, who made formaldehyde by passing methanol vapor over a heated platinum wire.

Although many process improvements have been made since, much of the world's formaldehyde is still produced by oxidizing methanol over silver catalysts. A significant fraction is made by the Reichhold-Skanska process, which uses mixed metallic oxides as the catalyst, and a relatively small fraction is made by noncatalytic oxidation of butane and propane. Intensive research has been done on the direct synthesis of formaldehyde from methane, but no such process has been commercialized.

Phenol

U.S. companies produced 1.7 billion pounds of the organic intermediate phenol (carbolic acid) in 1970, and they account for roughly half the world supply. The most important use of phenol, its reaction with formaldehyde and other aldehydes to make resins, consumes more than half of the U.S. output. In 1970, this country produced about 1.2 billion pounds of phenolic resins, worth $240 million.

The phenolic resins have a variety of uses: molding and laminating materials, coatings, adhesives in plywood and other wood-based products, foundry cores, and others. About one quarter of the U.S. consumption of phenol is used in making epoxy resins, nylon 6, and surface active agents, and another quarter is used for a number of miscellaneous purposes.

Phenol provides still another example of the flexibility of organic chemical processes and the interdependence of their products. Coal tar was the only source of phenol until 1918, when the compound was first made synthetically in Germany (by sulfonating benzene and treating the reaction product with caustic). Four additional processes have reached commercial status since then and, in the U.S., nearly all phenol is based on petroleum.

The preferred process today is the cumene process, which accounts for about 70% of U.S. production. It was discovered in Germany in 1944 and developed in the U.S. by Allied Chemical and Hercules (with Distillers Co., Ltd., of England). For each pound of phenol produced, the process yields about 0.6 pound of acetone, which also is made from isopropyl alcohol. The success of the cumene process has thus depended not only on the low costs of its raw materials (benzene and propylene), but on the market for acetone as well.

The chlorobenzene process for phenol was commercialized in 1924 by Dow Chemical. Economically it depends in part on advanced chlorine-cell technology, since relatively cheap chlorine is required to make the chlorobenzene starting material. The economics of the process also depend on the development of uses for diphenyl oxide and phenylphenols, both being produced as by-products.

The Raschig-Hooker process originated in Germany in the 1930's and was first operated in the U.S. in 1940 by Hooker Chemical. The process yields very small amounts of by-products, but the reactants (initially benzene and hydrogen chloride) must be recycled a number of times to achieve economically high conversion of benzene to phenol.

The sulfonation process for phenol, the first to be used commercially, was developed in the U.S. by Monsanto and Reichhold. The process produced sodium sulfite as a by-product, and its economics depend in part on a nearby market for this compound, such as a paper mill. Most sulfonation plants by now have succumbed to competition from other phenol processes, notably the cumene process.

The toluene process for phenol was developed by Dow Chemical in 1961. The company built only one small plant, however, and since has sold it. The economics of the process depend on a low ratio between the prices of toluene and benzene and on the ability to minimize and dispose of useless tars produced by the process.

All five phenol processes have been able to operate competitively because of differences in conditions at various plant locations, even though the cumene process has risen to the dominant position. In addition, the picture has been influenced by the prices of sulfuric acid, caustic, and chlorine, and by the market for acetone and sodium sulfite.

Melamine

Melamine resins, made from an organic intermediate, melamine, have found wide consumer acceptance in the past 25 years. Two of the most familiar uses of the resins are break-resistant (Melmac) dinnerware and heat- and stain-resistant (Formica) laminates used as counter and table tops and as wall paneling. Other end uses of the resins include surface coatings, textile finishes, and wet-strength resins for paper. As of February 1971 the major producers of melamine were Allied Chemical, American Cyanamid, and Ashland Chemical. Total U.S. production of the intermediate in 1970 was 68.5 million pounds.

Melamine, which contains 66% nitrogen, was first prepared by the German chemist Liebig in 1834. The compound remained a curiosity for about 100 years, but then American and European chemists capitalized on its resin-forming properties and developed practical means of making it commercially. Melamine was introduced into commerce, as an intermediate and in resin form, about 1940, principally as the result of research and development by American Cyanamid.

Until the 1960's, all industrial melamine was prepared from dicyandiamide, derived in turn from calcium cyanamide, a chemical made from coal, air, and limestone. Despite melamine's relatively low price, however, and because of its widespread use, U.S. chemists continued to seek cheaper and simpler processes for making it. A particular goal was a method that did not require the costly high-pressure equipment of earlier processes.

Both goals were realized ultimately, starting with the discovery by

J. H. Paden and J. S. Mackay of American Cyanamid that melamine could be made by heating urea, a very cheap intermediate. A patent issued on the discovery in 1951, and a second patent issued on Mackay's later discovery that the reaction proceeds at atmospheric pressure in the presence of certain catalysts. Some of the melamine produced today is made in this fashion, using adaptations of the fluidized bed technology developed by the petroleum industry.

The particular properties of melamine resins should assure them of a significant position in the economy for many years to come. Any technological advances that further reduce the cost of melamine and its resins would certainly increase their consumption.

Other Intermediates

A variety of organic intermediates not mentioned here exert significant economic impact in the U.S. and abroad. Among them are vinyl chloride, the chlorinated solvents, a number of alcohols, acetic anhydride, and acetic acid. The economic and technical characteristics of these compounds and their uses, however, differ mainly in detail from those of the intermediates already covered and, for that reason, will not be examined here.

FUTURE DEVELOPMENTS

One of the most significant trends in petrochemical raw materials and intermediates is the continued need for additional hydrocarbon feedstocks to meet the ever-increasing demand for the compounds. The most widely used petrochemical building block in the world is ethylene, and in this country about 80% of it is made from the light refinery and petroleum gases ethane and propane. Supplies of these hydrocarbons are limited in Europe and Japan. The main feedstock there for ethylene is petroleum naphthas boiling in the gasoline range, and the trend has been toward higher-boiling feeds. In the U.S., as ethylene demand outstrips the availability of ethane and propane, the trend will also be toward higher-boiling feedstocks.

Relaxation of this country's oil import controls could stimulate the trend toward higher-boiling feedstocks. A second stimulant is the continued need to process the entire petroleum crude more effectively, that is, into higher-value products. (Processes designed to produce ethylene from feedstocks heavier than ethane and propane also produce a spectrum of other building blocks, such as propylene, butenes, butadiene, and the like.)

The worldwide demand for naphtha, at any rate, has been increasing, and so has its price. There has been an incentive, therefore, to develop an ethylene-manufacturing technology based on cheaper, higher-boiling feedstocks, including total crude. The cheaper feedstocks, however, are not easily adaptable, as a rule, to the conventional tubular furnace cracking processes used with naphthas. Thus, the cracking of crude oil to produce ethylene and propylene is not expected to become an important factor unless some major improvement in technology occurs. The patent literature reveals a considerable technical effort both in the U.S. and abroad to achieve this goal of low-cost ethylene from heavy oil feeds by an economical and technically viable process.

Petrochemical feedstocks account for only about 5 or 6% of the nation's consumption of petroleum hydrocarbons, whose major use by far is as a source of energy. Trends in feedstocks nevertheless are influenced strongly by developments in energy resources. Shale oil, for example, can be expected to become an important source of energy over the next several decades. At the same time it can be expected to become an important source of petrochemical raw materials, although new technology will be required to accommodate the differences in composition between shale oil and conventional petroleum. This means among other things that petrochemical complexes of the future may very well not congregate along the Gulf Coast but in the Mountain States, where the nation's shale resources are concentrated. (A significant increase in oil imports also could lessen the economic incentive to concentrate petrochemical complexes on the Gulf Coast.)

Energy problems in the U.S. also have stimulated a considerable technical effort on processes for making synthetic crude oil and pipeline gas from coal. Over the long range, a coal chemicals industry seems likely to emerge from this work, although it would differ significantly from the coal chemicals industry of the early 20th Century. One result might well be a relocation of the "petrochemicals" industry to areas closer to major coal deposits, consumers, and more favorable distribution locations.

Environmental Aspects

The broad spectrum of uses of organic intermediates seems likely to involve them increasingly in environmental matters. Familiar examples include the removal of cyclamate nonnutritive sweeteners from consumer markets and the questionable status of nitrilotriacetic acid (NTA) as a replacement for phosphates in synthetic detergents. The cyclamates caused tumors in laboratory animals; NTA, among other problems, may cause birth defects under certain conditions. These problems did not crop up until after extensive premarket testing of both materials and extensive consumer use of the cyclamates. Many scientists will argue, in fact, that the problems are unreal, that they were detected under experimental conditions that bear little resemblance to the actual conditions of use of the materials by humans. These arguments, regardless of their merits, suggest the types of environmental complexities that organic intermediates (and other chemicals) are likely to be facing in the future.

A related indication of the future is the Toxic Substances Control Act passed by the U.S. Senate in May 1972 and a similar bill then moving through the House. Among other requirements the Senate-passed bill called for premarket testing of all new commercial chemical substances and of existing ones where indicated. Such tests might be made for carcinogenesis, teratogenesis, mutagenesis, persistence, and other characteristics. While premarket testing of chemicals for safety and the like has been routine for years, considerable controversy exists over the lengths to which such tests should reasonably be pursued. Significant changes in current practice, at any rate, could exert substantial effects on makers of organic intermediates, particularly in the allocation of R&D resources.

Other Trends

Among other, more diverse trends in organic intermediates is the move toward larger chemical plants based on direct catalytic conversion of simple raw materials into complex chemical products (propylene to acrylonitrile, ethylene to ethylene oxide, etc.). A second trend concerns the distribution of large-volume chemicals, which is involving increasing environmental risks as well as increasing costs. This situation could lead to greater use of pipelines in distribution. A number of developments of this type might be singled out, but all have a common element: the utility of chemistry both in breaking new ground and in solving the ancillary problems that attend technological change.

2 / **Industrial Inorganic Chemicals**

SOCIAL AND ECONOMIC IMPACTS

Primary and intermediate inorganic chemicals and the end products [1] made from them are principally products of the chemical and allied products industry. In many respects these industrial inorganic chemicals resemble their industrial organic chemical counterparts, which that industry also produces and which we dealt with in the previous chapter. In their primary and intermediate forms, for example, they serve as the initial materials that the chemical industry itself uses to make end products. In addition, the chemical industry consumes many inorganic chemical end products internally, while those it sells externally are bought largely by companies in other manufacturing industries and in service industries. Finally, individual consumers seldom buy industrial inorganics as such.

As a result, industrial inorganic chemicals affect the economy mostly at the manufacturing level, and they thus affect consumers largely indirectly through goods and services bought at retail. Industrial inorganics do affect consumers more directly, however, when they appear as end products in formulations made by the chemical and other industries and are bought by consumers. In these products, they are important—and sometimes critical—commodities for carrying out daily activities associated with providing food, clothing, shelter, health care, transportation, and recreation.

Manufacturers recover or produce inorganic chemicals for the most part from naturally occurring mineral ores or brines. They produce some, however, from atmospheric gases or gases made in such basic manufacturing processes as primary metals production, petroleum refining, and coal processing. Manufacturers then react the primary chemicals with different chemicals or otherwise process them into many intermediate and end-product chemicals.

Data on tonnages, dollar values, and employment often accurately reflect the economic impact of an industry sector. For industrial inorganic chemicals, however, complete data are not readily available. The problem occurs in part because of the difficulties in classification in an industry where products are so numerous and are used in so many different ways. Some industrial inorganics may serve as either primary or end-product chemicals, for example, depending on the application. Most are intermediates, but many also function as intermediates in some uses and as end products in others. In addition, once the early transformations from original raw materials have taken place, inorganic and organic intermediates may be combined to form other intermediates and end products that contain both categories of chemicals. Apportioning value and manpower data in the face of such complexities clearly becomes impossible.

Such difficulties aside, some data do exist. Sales by the chemical and allied products industry in 1970 amounted to nearly $50 billion. Industrial

---

1. As with industrial organic chemicals, industrial inorganic chemicals may be classified as primary, intermediate, or end-product chemicals. They are termed primary at the point that they exist for the first time in reasonably pure form and in reasonably definite composition following production from raw materials. Intermediates are those derived by chemical reaction either from primary chemicals or from other intermediates. End products are made from intermediates, also by chemical reaction. They are not changed further chemically, but they may be modified physically or formulated with other materials for use.

inorganics accounted for an estimated $10 billion of that sum. Among individual inorganics for which data are gathered, the compounds of sulfur, nitrogen, sodium, chlorine, and phosphorus dominate data for both tonnages and dollar values. The chemical and allied products industry also makes some two dozen other groups of industrial inorganic chemicals as classified by their major metallic or nonmetallic elements, however. About 75 individual industrial inorganic chemicals have annual production values exceeding $10 million, and three have values exceeding $500 million. Of the chemicals produced in the largest tonnages in the U.S., inorganics occupy all of the first five places and nine of the first 10 places.

Once one moves beyond such basic data, the economic and social impacts stemming from the production and use of industrial inorganics are in some respects more readily inferred than demonstrated with hard numbers. Inorganics are so widely used by all manufacturing industries and most service industries, for example, that few activities are undertaken without their participation at some point. In addition, many companies have been formed expressly to make and sell inorganic chemicals. Du Pont began as a manufacturer of explosives in 1802. It is now the largest U.S. chemical company, with 1970 revenues of $3.64 billion. The second largest U.S. chemical company, Union Carbide Corp., traces its origin to a company of the same name incorporated in 1898 to make calcium carbide. Union Carbide's 1970 sales amounted to $3.03 billion. Formation of Solvay Process Co. in 1881 to produce sodium carbonate makes it the oldest predecessor company of Allied Chemical Corp., whose 1970 sales totaled $1.25 billion.

Among smaller chemical companies, Pennwalt Corp. began as Pennsylvania Salt Manufacturing Co. in 1850, manufacturing lye and alkaline salts. Its 1970 sales were $432 million. Philadelphia Quartz Co. was formed in 1864 as the successor to a firm organized in 1831 to make candles and soap. The firm had begun adding sodium silicate to its soap in 1861 and also began making silicates themselves in 1864. Still a privately held company, it does not publish sales figures, but it is well known in the industry. Foote Mineral Co. was incorporated in 1916, but it started with a business founded 40 years earlier by A. E. Foote. Professor of chemistry and mineralogy in the Midwest, he was invited to exhibit his extensive mineral collection at the country's centennial celebration in Philadelphia in 1876. So many visitors asked to buy specimens that he never left Philadelphia but remained to become a supplier of minerals. Foote's sales in 1970 were about $100 million. Victor Chemical Works, at one time a medium-sized chemical company and now part of Stauffer Chemical Co., dates from 1902 when its founder, August Kochs, acquired Peterson and Mansar, Flour Merchants, and improved its process for making monocalcium phosphate for use in baking. Stauffer itself began in California in 1885 as the partnership Stauffer & Co., which was formed to recover chalk ballast dumped by ships in San Francisco Bay and sold by Stauffer mainly as a whitening agent. Stauffer Chemical's 1970 sales exceeded $480 million.

These companies have all expanded their product lines considerably since their beginnings, and they are only a few of the hundreds originally organized to make inorganic chemicals in the U.S. They and the others have helped establish the country's manufacturing base, provided many employment opportunities, and otherwise contributed to the Gross National Product.

The transactions that occur among all of them and their customers comprise an intricate labyrinth that extends throughout all manufacturing and service industries. We cannot examine that labyrinth in detail, but two examples selected at random will indicate how widely inorganic chemicals range throughout the economy. One example involves chemicals containing a metallic element, manganese, and the other chemicals containing a nonmetallic element, fluorine.

Manganese ores, which are essentially manganese dioxides, are processed to a number of primary industrial chemicals, among them manganous sulfate. Manganous sulfate itself may serve as an end-product chemical by supplying manganese as a trace nutrient for growing crops. It is also used as an end-product chemical in paints, ceramics, and dyes and as a fungicide, food additive, and ore flotation agent.

As an intermediate chemical, manganous sulfate is used to make several other industrial chemicals, including manganous hydroxide. Manganous hydroxide is an intermediate for making several more chemicals, including manganese dioxide. As an end product, manganese dioxide serves as a decolorizer in glass manufacture, as a component of dry batteries, and as an alloying agent in the preparation of manganese steels. As an intermediate, it is used to make potassium manganate and in turn potassium permanganate.

Among its many important uses, potassium permanganate finally becomes an end-product manganese chemical for use as a disinfectant in agricultural, medicinal, and veterinary applications. It serves in bleaching and water treatment and as an oxidizing agent in chemical reactions. It also functions as a flux in soldering stainless and chrome steels and as a plant growth regulator in agriculture.

In the case of fluorine, mined fluorspar is refined to a primary chemical, calcium fluoride. Several grades of calcium fluoride are produced, among them a relatively pure one from which an important inorganic chemical, hydrogen fluoride (HF), is made. HF's most important single use is in the production of aluminum metal. It has many other important uses, however, as an intermediate for a number of fluorine chemicals. One such large-volume use is in making refrigerant gases and aerosol propellants. Among its many other uses as an intermediate chemical, it is converted to stannous fluoride for use as a dentifrice additive. Another use as an intermediate, and one consuming much larger tonnages than toothpastes, is in processing uranium ores to uranium hexafluoride for making atomic reactor fuels.

In an end-product application, HF is highly important as an alkylation catalyst in petroleum refining, especially in making dodecylbenzene for further processing into detergents and in alkylating olefins to make high-octane gasolines. In another use, this time as an intermediate, HF is the source of fluorine for making polyfluorohydrocarbon resins. Such resins are used in gaskets, cooking utensils, processing equipment, and other applications requiring nonstick properties and performance under corrosive, high temperature, or other severe conditions. Potassium fluoride made from HF is used for etching glass or serving as a flux in soldering, among other applications. Sodium fluoride, also made from HF, serves as an insecticide, water fluoridating agent, metallurgical flux, enamel opacifier, paint preservative, and beverage equipment sterilizer.

As examples, these brief process sequences for a few manganese and fluorine chemicals at best merely indicate some of the intricacies of making and using primary and intermediate inorganic chemicals. Many other sequences are often more intricate and deal with chemicals that are ultimately more important in quantity and dollar value. In addition, the sequences concentrate largely on purely inorganic chemicals, and they mostly ignore intermediates and end products made by combining inorganic and organic intermediates. Such combinations, however, add their own complexities to an already complex flow of products. While industrial chemicals that combine inorganic and organic intermediates do not often become groups of products as important as many groups of organics or inorganics, they nonetheless do become significant in many uses. Pharmaceuticals, pesticides, dyes, and soaps and detergents are only four groups of chemical products made from inorganic and organic intermediates whose uses range widely throughout the economy. Among miscellaneous examples, tetraethyl lead (TEL) has long been used to raise gasoline octane ratings, and ethylene dichloride and ethylene dibromide have paralleled that use in helping remove lead deposits that TEL leaves in engines. (The use of TEL is sharply controversial today, of course, but without doubt it has affected the economy in a major way through its impact on auto development.) Polymers based on vinyl chloride and related monomers containing chlorine comprise one of the largest classes of macromolecular products for use as plastics, fibers, and protective coatings. Alkyl aluminum compounds are very important components of catalysts used to polymerize olefins for use as plastics and elastomers. Fire extinguishers often contain products combining inorganic and organic intermediates, while other combinations act as flame retardants in many textiles, plastics, paints, and other products.

Such examples, of course, are only a few of the hundreds of industrial inorganic chemicals that affect society and the U.S. economy. In the following pages we shall discuss in more detail—although still briefly compared to the information available—some of the accomplishments of chemists and chemical engineers in the production and use of a number of industrial inorganic chemicals. In doing so, we shall concentrate on primary and intermediate inorganics from the top five groups of compounds of sulfur, nitrogen, sodium, chlorine, and phosphorus, but we shall also include developments involving other less broadly used chemicals. If space were available, similar descriptions could also be provided for each of the other groups of inorganic chemicals and for many other individual chemicals as well.

Although somewhat expanded, the following explanations still gloss over the considerable work needed for economic success on the part of chemists, chemical engineers, and persons trained in other scientific and engineering disciplines. The passing of many man-years and the expenditure of many dollars lie behind developments that we shall often mention in only a sentence or two.

CHEMICAL ACCOMPLISHMENTS

Primary and intermediate inorganic chemicals may function in the economy in ways very similar to the ways that comparable organic chemicals function, but their development and production differ in several ways. A number of them, for example, have fairly small molecules that are not structurally complex. In addition, their production processes in some cases tend to be relatively simple and usually require only a few processing steps. As a result, some are made with technology that has existed for several decades, and many have been produced in large industrial quantities for significantly longer times than industrial organics. In fact, the first organic chemical was not made synthetically until the German chemist Friedrich Wöhler succeeded in synthesizing urea in 1828. On the other hand, alchemists and others began making inorganic chemicals a millennium and more ago. Phoenician sailors even earlier may have been responsible for man's first synthetic inorganic chemical when they used lumps of natural soda (sodium carbonate) to support pots over a fire on a sandy beach and produced molten sodium silicate. Among other early reports are those of the 8th Century from the Middle East describing the distillation of niter (potassium nitrate in today's nomenclature) with green vitriol (today's hydrated ferrous sulfate) to make what later workers would name sulfuric acid. The Chinese centuries ago produced gunpowder from what today is called potassium nitrate, which they had made by leaching soil in which nitrogen from urine had combined with a potassium mineral.

Such efforts might be considered simply dabbling by today's standards. Inorganic chemistry began assuming a more definite industrial form, however, as formal experimentation brought about the evolution of alchemy to chemistry. Prominent among early experimenters was the German chemist Johann Glauber. In the mid-1600's, he first described ammonium nitrate, and his name is still associated with chemistry in Glauber's salt. He recovered this hydrated sodium sulfate from a reaction between sulfuric acid and sodium chloride, and he later identified it as the same compound he had isolated from mineral spring waters promoted as beneficial to health.

Others before and after Glauber also carried out experiments that led to the inorganic chemical industry that continued to emerge in Europe in the 18th Century. Among early industrial inorganic chemical plants was one built in England in about 1740 that made sulfuric acid by burning sulfur in the presence of saltpeter (potassium nitrate). Another pioneering plant was one built in France in 1791 to make sodium carbonate by reacting sodium chloride with sulfuric acid, heating the resulting sulfate with limestone (calcium carbonate) and coal, and leaching the sodium carbonate from the reaction mixture. Many other plants built both earlier and later attested to the growing importance of inorganic chemicals as industry expanded in western Europe. It was with such technology imported from Europe that the American chemical industry got its start.

Production of primary and intermediate inorganic chemicals is in many ways characterized by a relatively old technology that hasn't changed as dramatically as has the more recently developed technology for making primary and intermediate organics. For inorganics, especially those produced in the largest volumes, improvements in recent decades have mostly involved changes in details such as working out better material flows to make plants more efficient and reducing costs with automation. Using more resistant materials of construction, building plants with greater capacities, and employing better analytical techniques have also contributed. While such improvements often may have been individually small and unheralded, together they have often made the difference between successful competition and economic failure.

While some inorganics may be made with old and apparently undramatic technology, all primary and intermediate inorganics by no means are made that way. In fact, totally new processes based on new chemistry have broadened the production of inorganics in many ways in recent years. New inorganics resulting from these innovations are now being sold to compete with products used earlier or to meet needs not formerly satisfied.

In the pages that follow, we shall deal both with some of the more primary and old-time inorganics and also with some of the more recently developed new inorganics made with new technology. Unfortunately, because of the industry's size and the pervasiveness of its products, we shall be able to discuss only a few of the innovations that have occurred. They have been selected, however, as typical of developments in this industry sector.

Sulfuric Acid

Sulfuric acid ranks as one of the earliest intermediate inorganic chemicals made on an industrial scale, and in time it has become one of the most important industrial commodities. The U.S. chemical industry produces more of it than any other single chemical, and manufacturers and others use it directly or indirectly in nearly every industrial process. In fact, the use of sulfuric acid extends so widely throughout industry that many economists feel its consumption quite accurately gauges economic conditions in the U.S. as a whole.

Sulfuric acid has reached first place among all chemicals because of several important properties. Most importantly, it reacts with many other chemicals to make useful products. It also dehydrates, catalyzes, dissolves, absorbs, and otherwise promotes chemical reactions, adjusts process conditions, and modifies properties of materials. In addition, the economics of its use often may be improved by recovering and regenerating it for recycling or by recovering and selling it for uses in strengths and purities lower than those for which it was used initially.

The fertilizer industry uses most of the sulfuric acid in the U.S. Making phosphate fertilizers accounts for 40% of total consumption, while reaction with ammonia to make ammonium sulfate accounts for nearly another 10%. The chemical industry consumes the second largest amount (about 20%), and uses it in every conceivable way its properties permit. The petroleum industry at about 10% of total consumption ranks third, using about 5% in alkylation processes for making high-octane gasoline and the other 5% in many refining uses.

Production of sulfuric acid possibly embodies the most standardized process technology and raw material supply situation in the U.S. today. About 100 producers make it at more than 200 locations. They mainly use elemental sulfur (for 81% of production), but they also use smelter gas (7%), spent acid (6%), hydrogen sulfide (4%), and pyrites (2%). The U.S. produces nearly 30 million tons a year with a value of more than $500 million. Projections indicate consumption of about 43 million tons a year by 1980.

Most sulfuric acid technology was developed in Europe and was imported to the U.S. Production apparently began in the U.S. in Philadelphia in 1797, and by 1850 it had spread to half a dozen cities or more. Important consumers included manufacturers of pharmaceuticals, textiles, and steel plus the growing fertilizer, metallurgical, and chemical industries. After the Civil War, sulfuric acid production continued to expand, largely because of the demand of the fertilizer and petroleum refining industries.

At first, producers made sulfuric acid in what came to be known as the chamber process. In this process, they burned sulfur in the presence of potassium nitrate in a chamber lined with lead and then condensed the acid. The chamber process accounted for all production of sulfuric acid both here and in Europe until the invention in Europe of the contact process toward the end of the 19th Century.

In the contact process, sulfur dioxide reacts with atmospheric oxygen when heated over platinum wire or some other catalyst. The product, sulfuric anhydride (SO_3), is then hydrated to sulfuric acid in absorption towers. In 1898, Badische Anilin- & Soda-Fabrik built the first successful contact plant in the U.S., and General Chemical (now part of Allied Chemical) pioneered with the development in 1900 of the first contact plant in the U.S. based on burning pyrites.

Today, more than 95% of new acid is made by the contact process in the U.S. It permits direct production of pure acid of any strength from dilute solutions to oleum (100% sulfuric acid containing excess sulfur trioxide) and even stabilized liquid sulfuric anhydride.

Prior to the late 1950's, sulfuric acid plants in the U.S. normally had three catalyst stages. Many older plants operated at somewhat low conversion ratios, and no plant operated higher than about 97% conversion. Although these conversion ratios allowed 2,500 to 3,000 parts per million (p.p.m.) of sulfur dioxide to be emitted in stack effluents, manufacturers faced no economic incentive to design for higher conversions. Besides, strong concern over pollution had not as of then developed very widely.

Beginning in the early 1960's, however, air pollution control agencies began to function in some heavily populated areas, and several established an upper level of 2,000 p.p.m. for sulfuric acid plant effluents (equivalent to 98% conversion). Largely as a result of implementation of the Air Quality Act of 1967, sulfuric acid plant emissions as low as 500 p.p.m. are now a prospect as federal, state, and local authorities propose increasingly stringent regulations on sulfur dioxide emissions. While modern plants burning elemental sulfur can meet the 2,000 p.p.m. level, mainly by adding a fourth stage to the standard three-stage system, the more complex plants designed to process other raw materials will not easily meet the standard even with four conversion passes. Most acid producers, therefore, have invested or are investing large sums for abatement equipment with no expectation of normal economic return.

Ammonia and Other Nitrogen Chemicals

Sulfuric acid may still rank as the single most important industrial chemical, but ammonia in recent years has become a close competitor. In 1970, production of 13.1 million tons put it second to sulfuric acid among all chemicals made in the U.S. It ranks high largely because nitrogen is so vital to life processes. Ammonia serves as the source of most nitrogen in fertilizers today, and livestock feed supplements such as urea made from it and molasses fortified with it also consume large tonnages. Like sulfuric acid, however, it also ranks high because of its utility in making many chemicals and because its properties permit its use in many applications.

Nitric acid, ammonium nitrate, ammonium sulfate, and potassium nitrate are among the major intermediate and end-product chemicals made from ammonia. The sulfate and both nitrates are mainly used as fertilizers. Nitric acid is the second most important industrial acid and the eighth largest tonnage industrial chemical. It is also mainly of interest to the fertilizer industry, which uses it to make ammonium nitrate and to solubilize some phosphate rock in making phosphate fertilizers. Nitric acid is also important in many other ways, however. Ammonium nitrate made from it is an important explosive, and other military and industrial explosives contain nitrogen derived from nitric acid. Nitric acid is also used to convert cyclohexanone to adipic acid, a monomer for one of the most important nylons. Making toluene diisocyanate for production of polyurethanes is another important outlet, as is production of many chemicals for such end uses as dyes, pesticides, and detergents.

Ammonia is also used in many other ways. The alkali industry, for example, uses some to make sodium carbonate from sodium chloride, and sodium hydroxide producers use it to purify their product. It is also an important intermediate in making sulfa drugs, antimalarials, vitamins, and methionine and other amino acids. Ammonia and its derivatives are often used in making dentifrices, lotions, and other personal care products.

As one of the lowest cost alkaline agents available, ammonia is used by the petroleum industry to neutralize acid constituents in oil to protect equipment. Mining companies use it as a processing agent in extracting metals from ores, while the metal industry de-tins scrap with it. In the fermentation industry it is used as a source of nitrogen for microorganisms and also to adjust acidity. In the pulp and paper industry, using ammonium bisulfite in place of calcium bisulfite lessens costs, increases yields, and allows the industry to pulp some hardwoods it could not pulp otherwise.

The beverage industry ammoniates licorice with it and also extracts bitter principles from plants and other sources. It prevents slimes and molds from forming in tanning liquors and acts as a depilatory and curing agent in making leather. In solution it serves as a cleaning agent in homes and factories, and it is also used to compound dry cleaning and specialty soaps. It is the refrigerant used in the largest tonnages, where applications include making ice, processing frozen foods, providing cooling in industrial processes such as dewaxing lubricating oils, and air conditioning large buildings and plants.

Early attempts to fix atmospheric nitrogen date from the work in the 1780's of an English chemist and physicist, Henry Cavendish. No one succeeded in developing a commercial process until 1900, however, when workers in Norway did so with a process that used an electric arc. This process is now of historical significance only, having long been displaced by the Haber-Bosch process developed in Germany shortly before World War I. Since ammonia's history is so intimately associated with that of the fertilizer industry, the Haber-Bosch development and subsequent innovations in ammonia technology are covered in the chapter on fertilizers.

Chlor-Alkali Chemicals

Some 30 manufacturers in the U.S. electrolyze sodium chloride solution to make the important industrial inorganic chemical co-products of sodium hydroxide and chlorine. They produced about 10.1 million tons of sodium hydroxide in 1970 and about 9.7 million tons of chlorine, putting sodium hydroxide in fourth place and chlorine in fifth place among chemicals made in the U.S.

As with sulfuric acid and ammonia, both sodium hydroxide (caustic soda, or simply caustic, in trade terminology) and chlorine pervade all segments of the chemical and allied products industry and many other industries as well. Caustic, for example, is essential for making rayon, various sulfur compounds, cellophane, soap and detergents, pulp and paper, and such large-volume chemicals as phenol, resorcinol, naphthol, and oxalic acid. It is also widely used to make dyes, ceramics, cosmetics, pharmaceuticals, boiler water softening and cleaning compounds, and household bleach products. The food processing, metal and petroleum refining, engraving and printing, and many other industries use it extensively as well.

Roughly four fifths of the chlorine produced in the U.S. is used as raw material for other chemicals. In fact, more than a thousand major chemicals made today depend on chlorine, and they are used in many ways. During the past three decades, much food in the U.S. has been grown more efficiently through the use of herbicides and insecticides containing chlorine. Almost all of the 25 billion yards of textiles woven in the U.S. are treated with chlorine or chlorine derivatives, and almost all paper is bleached with chlorine. Most drinking water depends on chlorine for purity and safety. Polyvinyl chloride plastics are among the most important made today, and they are used in many industrial and consumer products, such as upholstery, wall coverings, floor tiles, packaging materials, cable insulation, and piping. The chlorinated hydrocarbon perchloroethylene is widely used in dry cleaning and trichloroethylene in metal degreasing. Carbon tetrachloride and chloroform are important precursors in organic syntheses leading to plastics, pharmaceuticals, fluorochemical refrigerants and aerosol propellants, and other chemicals.

At one time, the chemical industry made large quantities of caustic from sodium carbonate. With the rapid growth in demand for chlorine, however, production from sodium carbonate has practically ceased except in some captive operations such as in the pulp and paper industry. Today, the electrolysis of a solution of sodium chloride in water is used to produce caustic and chlorine (and also hydrogen). Other means exist for making chlorine besides electrolysis, but none has yet proved economically competitive on a large scale, although a new process called Kel Chlor is beginning to appear as a possibility.

Caustic and chlorine are made in the U.S. by two processes developed here in the 1890's. They are commonly referred to as the diaphragm-cell process and the mercury-cell process. LeSueur used a diaphragm cell to produce the first electrolytic caustic-chlorine in 1893 at Rumford Falls, Me., and Mathieson Alkali Works introduced the mercury cell in full-scale production in 1897 at Niagara Falls, N.Y.

The electrolysis of sodium chloride solution may appear relatively straightforward, but much development work was required for commercially successful processes. In each case, the key proved to be a practical way to keep the caustic and the chlorine separate once formed. In the diaphragm cell, a porous diaphragm, usually of asbestos, separates the anode and cathode chambers. The diaphragm permits the sodium chloride solution to flow from anode to cathode. Chlorine evolves at the anode and hydrogen at the cathode. Caustic forms at the cathode when sodium ions from the salt combine with the hydroxyl ions from the hydrolyzed water, and it is removed as cell effluent. In the mercury cell, the anode is suspended above the cathode, which consists of a flowing film of mercury. Chlorine evolves at the anode, while sodium forms at the cathode and amalgamates with the mercury. The sodium amalgam is removed from the electrolyzer and is decomposed with water to caustic, hydrogen, and mercury in a separate chamber.

The processes have remained basically the same since their inception, but they have been refined in many ways to increase efficiency and control costs. Diaphragm cells now produce nearly three fourths of the caustic and chlorine in the U.S., and mercury cells produce the balance. Both processes yield comparable grades of chlorine, but the mercury cell, in contrast to the diaphragm cell, produces concentrated, commercial strength caustic of high purity directly from the cell. Mercury cells, however, consume more electrical energy than diaphragm cells do. Moreover, discovery of mercury in fish early in 1970 triggered great concern about effluents from mercury-cell plants and the possible hazards of mercury and other heavy metals in the environment. As a result of this concern, plants have trimmed mercury emissions to receiving waters by 90 to 95%. Regulatory agencies are also scrutinizing emissions of mercury to the air. Producers using mercury cells are developing plans for meeting the standard proposed by the Environmental Protection Agency (EPA) limiting emissions to the air of not more than 5 pounds of mercury per day per plant. If EPA tightens its limits any further, the mercury-cell process may diminish in importance in the U.S. in the coming years.

Hydrochloric Acid. Of the many chemicals based on chlorine, we shall single out only hydrochloric acid, hypochlorites, chlorites, and chlorates for special coverage. Hydrochloric acid is an important industrial chemical in its own right, while the chlorites and chlorates illustrate some of the important impacts on typical consumer products that developments in industrial intermediate inorganic chemicals can have.

Hydrochloric acid is produced commercially by four different methods. In one, sodium chloride and sulfuric acid react at high temperature to produce sodium sulfate and hydrogen chloride, which is absorbed in water to make hydrochloric acid. In a second and somewhat similar one, sulfur, oxygen, and sodium chloride serve as raw materials. In a third process, burning hydrogen and chlorine produces hydrogen chloride. Finally, hydrochloric acid is also produced as a by-product from reactions between chlorine and organic chemicals. Typical examples of this last one include chlorination of methane or benzene and the pyrolysis of chlorinated organics such as ethylene dichloride to make vinyl chloride and tetrachloroethane to make trichloroethylene.

Nearly 90% of the hydrochloric acid produced in the U.S. comes from the chlorination of organics, while the sodium chloride–sulfuric acid/sulfur processes and direct burning share equally in producing the balance. Two major outlets dominate among its uses: to pickle steel, where it now has about 85% of the market once held by sulfuric acid, and to activate oil wells. It is also widely used elsewhere in industry, however. Among important uses are metal cleaning, regenerating activated charcoal for sugar refining, and hydrolysis of starch to make dextrose and corn syrup. Among applications in other manufacturing processes, its use in the recovery from seawater of the important light metal, magnesium, ranks as significant. Ore extraction with hydrochloric acid is also important for such metals as germanium, manganese, radium, tantalum, tungsten, and vanadium. In fact, the hydrochlorination of ores is under intensive development and should become a major consumer of hydrochloric acid (or of chlorine) in the foreseeable future.

Since most hydrochloric acid is produced as a by-product, manufacturers cannot control supply and demand as they normally do for products made for primary purposes. To prevent hydrochloric acid from becoming a severe disposal problem when production outstrips its markets, the industry has found it advantageous to develop processes in which hydrochloric acid can be used directly or in which elemental chlorine can be recovered. Typical and most important of these has been the oxychlorination process for making dichloroethane from ethylene, oxygen, and hydrogen chloride. Hydrogen chloride produced in the pyrolysis of dichloroethane to make vinyl chloride monomer is used to make more dichloroethane. Combining oxychlorination with direct chlorination has made it possible to make vinyl chloride monomer without making hydrogen chloride as a by-product.

Hypochlorites, Chlorites, and Chlorates. These reactive chemicals are used in many applications, including matches, pyrotechnics, explosives, rocket fuels, ore dressing, herbicides, defoliants, and photography. Their uses as disinfectants in food processing and in water and waste treatment and as bleaching agents in laundering and in pulp and paper and textile industries rank as the largest, however. Bleaching pulp and paper, for example, accounts for about 70% of the annual U.S. consumption of sodium chlorate (or about 140,000 tons of nearly 200,000 tons produced in 1970). About one third of U.S. hypochlorite consumption is accounted for by swimming pool sanitation, and much of the balance is used for other sanitation and bleaching applications. As for sodium chlorite, it is used primarily for making chlorine dioxide (itself important in water treatment and bleaching), bleaching tallow, treating water, and controlling algae and slime.

Before development of chlorinated bleaches, textile manufacturers spread their goods on grass fields to be bleached by the sun. Long and laborious, the process suffered from high costs for land rent and maintenance, and only a few consumers could afford the luxury of cloth bleached in that way. Then in the last quarter of the 18th Century, chemists found that a solution of chlorine in water bleached a variety of plants, flowers, and leaves. During the same period they also found that chlorine bubbled through a solution of potassium carbonate produced potassium hypochlorite, which proved to be a useful textile bleach. Chlorination of solid calcium hydroxide (slaked lime) also produced a useful bleach that was especially important to the textile industry. Among other advantages, it provided a means of shipping chlorine for use by simple dissolution in water at points distant from its manufacture.

While the production of chlorine and caustic occurred in the U.S. in the 1890's, and with them probably sodium hypochlorite solutions, the first plant-scale studies of sewage disinfection by hypochlorination did not occur until about 1907 at Red Bank, N.J., followed the next year with the use of chloride of lime (or bleaching powder) as a water-treating agent at Union Stock Yards Co. in Chicago.

In the first years of water chlorination, the only commercial sources of chlorine were bleaching powder (chlorinated lime) and sodium hypochlorite solution. The available chlorine content of these products was low, while transportation costs and product variability were serious problems. Manufacturers accordingly developed processes to make liquid chlorine, which could be used to purify water directly or converted to bleach at consuming locations by reactions with alkali.

Sodium Carbonate

Sodium carbonate, or soda ash, is another industrial inorganic chemical widely used in the chemical industry and in other industries as well. As with a number of other industrial inorganic chemicals, its market is dominated by a large single use, in this case one provided by the glass industry. Fully 95% of all glass made in the U.S. is soda-lime-silica glass, and the soda (or sodium) part of that formulation accounts for almost half of the nearly 7 million tons of sodium carbonate consumed annually in the U.S.

As with other industrial inorganics, sodium carbonate is also widely used as an intermediate and end-product chemical in many different ways. To select only several chemicals from among the many made from it, sodium acetate serves such markets as textile dyeing, meat preservation, and tanning. Sodium arsenate and arsenite are important pesticides and germicides. The pulp and paper, textile, and photographic products industries use sodium bisulfite and sodium sulfite extensively. Sodium carboxymethyl cellulose serves as an antisoil redeposition agent in detergents, while sodium perborate finds uses in dental compositions, cold wave neutralizers, and deodorants. Sodium linoleate functions as a drier in paints and varnishes, sodium thiosulfate as an antioxidant in soaps and a fixative in photography, and sodium hydrosulfide as a depilatory in leather manufacturing. Sulfite pulping processes provide another major market for sodium carbonate in the pulp and paper industry, while large tonnages are also used in sewage treatment processes and to soften both industrial process and municipal water.

U.S. producers either make sodium carbonate synthetically or recover it from naturally occurring deposits. Synthetic carbonate has long been the principal source, but economics of production and distribution and stricter antipollution laws have recently helped shift the balance in favor of processes in which it is recovered from the mineral trona. The expectation today as a result is that synthetic carbonate production will decline from having a 75% share in 1965 to having about 45% in 1975.

Synthetic sodium carbonate is made by processes that originated in Europe: the Leblanc process, introduced in France in 1791 (but no longer used in the U.S.), and the Solvay process, introduced in France and England in 1874 and in the U.S. in the early 1880's.

Natural sodium carbonate occurs mainly in the brines of dry lakes in Southern California and in underground deposits in Wyoming. Exploitation of each has resulted from the work of American chemists and chemical engineers, with successful production coming in Southern California immediately after World War I and in Wyoming in the 1950's.

Brine and ore processing both require extensive use of phase rule chemistry dealing with inorganic salts in solution. At the present time, natural brines are either carbonated and the sodium bicarbonate precipitate dried and calcined to sodium carbonate or the brines evaporated to precipitate sodium carbonate–sodium sulfate crystals that are then processed to sodium carbonate. The processes are expected to be joined late in 1972 by one that will produce sodium carbonate by solar evaporation.

In Wyoming, meantime, deep underground deposits of nearly pure sodium sesquicarbonate, contaminated with a little shale, are mined. The mined material is calcined to convert it to crude sodium carbonate and then dissolved in water. Impurities are removed and the liquor evaporated to produce sodium carbonate monohydrate crystals, which are converted to sodium carbonate by drying. In an alternate process, the mined ore is dissolved and the impurities removed. Pure sodium sesquicarbonate crystals are then produced by evaporative crystallization and calcined to sodium carbonate.

Sodium Silicates

The sodium silicates are an important family of sodium chemicals. Composed of sodium oxide and silicon dioxide, they are sold occasionally as anhydrous glasses but more often with some amount of water. Most commercial products are not true compounds but rather materials of indefinite composition that are made by the high-temperature fusion of high-purity sand (silicon dioxide) and sodium carbonate. By varying proportions of the two constituents, manufacturers and users obtain products with a wide range of properties for many industrial uses. Among the largest are applications as catalysts; soap and detergent builders; adhesives; pigments; water, paper, and ore treatment agents; and desiccants. Sodium silicates also serve in many smaller uses as well, including construction materials, metal fabrication, and petroleum recovery and refining.

Commercialization of sodium silicates in the U.S. began in the early 1860's. Some 20 years earlier, work in Europe had led to their commercialization there principally as builders in soap. When northern soapmakers in the U.S. lost access to southern rosin at the time of the Civil

War, they reformulated their products with sodium silicate to enhance soil removal, as Europeans had earlier. The use grew with time and eventually amounted to a substantial market. Unfortunately for silicate manufacturers, however, consumption of soap fell sharply with the introduction of synthetic surfactants, especially in the late 1940's when soap companies found sodium tripolyphosphate to be so effective in heavy-duty laundry formulations. Even so, silicates are still used both in soaps and synthetic detergents, and these products still rank as major silicate users, accounting for some 250,000 tons annually (41° Baumé liquid basis). In addition to their usual function of enhancing detergency, they help protect aluminum, zinc, and brass parts of washing machines that phosphate builders might otherwise corrode. In dishwashing compounds, they prevent alkali leaching of fine china, crystal, and porcelain enamel.

Building soaps and detergents is a large market for sodium silicate, and it may become larger as limits are placed on phosphates because of concern over environmental contamination. However, at the moment the use of sodium silicate in the form of silica gel constitutes the largest market. Silica gel has an immense internal pore area and therefore high absorptive capacity. Combined with a relatively small amount of alumina, it was developed as a petroleum cracking catalyst at the beginning of World War II. In that application, it allowed refiners to produce the very large volumes of aviation gasoline required during those years. Combined with different materials, silica gels are also used in formulating catalysts for other catalytic processes, such as oxidizing sulfur in making sulfuric acid and in oxidizing organic compounds such as alcohols to make aldehydes and acids.

Companies also use large volumes of silica gel to dry air and other gases. In some applications, silica gel is employed to process gas streams, while in others it is packed with instruments and other equipment to prevent corrosion during shipment and occasionally during use. One of its largest single uses as a desiccant is at the U.S. Air Force's engineering development center where about 600 tons (solids basis) are installed in the world's largest air dryer. The gel dries air for two tunnels that produce winds up to 3,500 miles an hour under conditions resembling those at extremely high altitudes.

Adhesives comprise the third largest market for sodium silicates, accounting for nearly 15% of production and close to the consumption in soaps and detergents. This market began developing in the early 1900's, when freight carriers agreed to handle boxboard shipping containers in spite of stiff resistance from the lumber industry. Sodium silicate's low cost, extremely fast set, and strength-reinforcing properties paved the way for today's continuous high-speed corrugating industry, which now makes more than 180 billion feet a year.

Corrugated board has largely displaced wooden crates. Such packaging still requires trees for production, to be sure, but it does so much more efficiently and thus uses forest resources more effectively. The corrugated industry has continued to grow rapidly, but its use of silicates peaked at about 400,000 tons (liquid basis) in the early 1950's as starch adhesives began taking over the market. Since starch requires heat to set, silicates remain the major adhesive in cold set uses such as fiber drums, tubes, and other laminations. The paper converting field, therefore, continues as a major outlet for sodium silicates, using about 250,000 tons (liquid basis) a year.

Acting as fillers and pigments is the fourth largest market for sodium silicates. The extremely fine amorphous silicas and calcium silicates precipitated from sodium silicates are used in large volumes as reinforcing fillers for elastomers. In many cases they are superior to other reinforcing fillers, and they must be used in place of carbon black in nonblack rubbers. They are also used in flattening agents in paints.

Sodium silicates have also been adapted to many other uses. Individually these uses consume from a few hundred to several thousand tons a year, but collectively they constitute a large market, amounting to some 600,000 tons (liquid basis). Among them is the use of activated silica sols as a coagulant aid in removing suspended impurities to purify water. These sols are particularly effective in a process developed in the 1930's for periods when water has a high color, low temperature, or high turbidity. While the use has been mostly for potable waters, it is now gaining attention for pollution control in waste water treatment. Meantime, sodium silicates added to water in very low concentrations inhibit corrosion of water lines and equipment and at the same time are nonpolluters.

The construction industry uses silicates in wallboard, fireproof insulations, and roofing granule coatings. Hardness of concrete is increased by coating fresh slabs with silicates, and silicates are also important ingredients in refractory and acid-proof cements. Injection of silicate solutions containing a gelling agent into weak, porous soils beneath existing buildings or in excavations for new buildings helps stabilize and markedly increase the strength of such soils. Injection also seals against encroachment of water into subterranean structures such as mines, tunnels, and cofferdams.

Metal industries use the almost instantaneous reaction between sodium silicates and carbon dioxide to form foundry cores and molds. Doing so eliminates odors and toxicity associated with organic binders. The method also is faster, requires less space, and gives castings that are more accurate. Mining companies use silicates in ore flotation because of their wetting and deflocculation properties.

The petroleum industry uses silicates in drilling muds, where they help control heaving shale and reduce corrosion of drill stems and well casings. Refiners also use the wetting and emulsifying power of sodium silicates in reclaiming spent lubricating oils. The paper and textile industries use them widely because of their detergency, buffering action, colloidal properties, and ability to stabilize bleach baths. The ceramics industry uses them in clay deflocculation and refining, in preparing slip castings, and in formulating frits, glazes, and enamels. Sodium metasilicate, one of the very few crystalline compounds of sodium silicate, is widely used in cleaning operations. Household laundry detergents and surface cleaners, industrial floor and wall cleaners, and cleaners for metals, bottles, and dairy equipment are among its many uses. Commercial and institutional laundries use it in large volumes for heavy-duty cleaning of work garments and shop towels.

Phosphorus Chemicals

The primary, intermediate, and end-product chemicals derived from phosphorus stand close in importance as a group to those derived from sulfur, nitrogen, sodium, and chlorine. The fertilizer industry dominates the market, but phosphorus chemicals are also used in many other industries. The soap and detergent industry, for example, often depends absolutely on phosphorus chemicals for product performance (although they are now being attacked as pollution problems, as mentioned earlier). In addition, phosphorus chemicals serve in important—if not always so critical—ways in the food and feed, metal, mining, petroleum, textile, and pyrotechnic product industries. Moreover, they are widely used in industries making such items as dyes, pharmaceuticals, plastics and resins, synthetic rubber, and glass and ceramics.

The U.S. derives all phosphorus chemicals today from phosphate rock that is mined mostly in Florida, North Carolina, Tennessee, Wyoming, Montana, and Idaho. By far the largest tonnages of rock (about 80%) are acidulated with sulfuric, phosphoric, or nitric acids to make phosphate fertilizers, and we have dealt with chemical developments in such processes in the chapter of this study devoted to the fertilizer industry. We shall accordingly concentrate in this chapter on what are known as industrial phosphorus chemicals (that is, those not used as fertilizers).

A very small amount of phosphates from acidulation of phosphate rock becomes the raw material for further processing into industrial phosphorus chemicals, but most industrial phosphorus chemicals are made from elemental phosphorus produced from rock smelted in electric arc furnaces. U.S. demand for elemental phosphorus amounts to more than 600,000 tons a year. Almost 90% is converted into phosphoric acid. Some acid (or simple acid salts) is used to clean and protect metals and metal products. Most acid, however, is used to make high purity phosphates and phosphorus chemicals. Making the sodium and potassium phosphates that the soap and detergent industry uses as builders to enhance surfactant performance consumes about two thirds of such acid every year. Phosphatic animal feed supplements and clear liquid mixed fertilizers are two other big uses for phosphoric acid.

About 5% of the elemental phosphorus is converted directly into intermediate phosphorus chemicals. Chief among them are phosphorus trichloride, phosphorus oxychloride, and phosphorus pentasulfide. These chemicals in turn are intermediates for making such products as gasoline and lubricating oil additives, plasticizers, flame retardants, and insecticides. About 1% of the elemental phosphorus is used directly to make munitions, matches, fireworks, and pyrotechnic products. Exports and miscellaneous minor uses account for the balance of slightly more than 5% of annual elemental phosphorus production.

Commercial manufacture of elemental phosphorus began in Europe in the 1830's. It was based on acidifying bone chars (and later phosphate rock) to make phosphoric acid and then distilling the acid mixed with carbon in a clay retort and collecting the phosphorus distillate. In 1888, a process patented in England opened the way to using electric arc energy as the source of heat, and all elemental phosphorus today is made by such processes. In them, the basic reaction is one among calcined calcium phosphate rock, carbon, and silicon dioxide to yield calcium silicate, carbon monoxide, and elemental phosphorus.

Commercial manufacture of elemental phosphorus began in the U.S. in 1870 using bone char technology imported from England. The first electric arc furnace began operating here in 1897 at the plant of Oldbury Electrochemical Co. (later acquired by Hooker Chemical Co., which is now part of Occidental Petroleum Corp.). Other manufacturers joined the pioneers in the early 1900's, but it was not until the 1930's and later that elemental phosphorus began to become a chemical of much industrial significance. Today, its production has a value of more than $210 million.

Many organizations have contributed to the evolution of electric arc smelting technology in the U.S. Among the early ones were Swann Chemical (subsequently acquired by Monsanto), American Agricultural

Chemical (now part of Conoco), Virginia Carolina Chemical (now part of Mobil Oil), and Victor Chemical Works (now part of Stauffer Chemical). Probably more than any other organization, however, Tennessee Valley Authority is responsible for much of the smelting technology used today and especially of that used for the large expansions in capacity that manufacturers undertook in the 1940's and 1950's under the impetus of a rapidly growing market in the soap and detergent industry. TVA's interest may have centered largely on developing phosphorus technology for fertilizer production, but its findings made possible a rapid expansion of phosphorus production for other industrial uses.

About 90% of all elemental phosphorus made in the U.S. is reacted with oxygen in air to make phosphorus pentoxide, which is absorbed in water to make phosphoric acid. Phosphoric acid is then reacted with other chemicals to produce many of the important industrial phosphorus chemicals. Reaction with sodium carbonate, for example, produces monosodium phosphate. Most monosodium phosphate becomes an intermediate for other chemicals, but it is also used directly as a buffer, a boiler water softener, and an additive in such food products as powdered drinks, puddings, and baby foods. Because of its acidity, it is used to control pH in electroplating and in textile manufacturing. It is also used as an acidic cleaner for metals and as an acid in medicinals such as effervescent laxative tablets.

Heating monosodium phosphate produces either sodium acid pyrophosphate or sodium metaphosphate. Bakers use the pyrophosphate, which reacts slowly with sodium bicarbonate and allows them to mix dough and batter in large batches and then bake at convenient times. Sodium acid pyrophosphate is also used to prevent discoloration in potatoes. Potatoes contain iron complexed with organic matter. During cooking, the iron is freed, and it combines with tannins in the potatoes. When exposed to air, these compounds oxidize to deeply colored ferric compounds. Sodium acid pyrophosphate sequesters the iron in a colorless iron pyrophosphate and prevents color formation.

Sodium metaphosphate, meantime, sequesters many metal ions in water. It permits soaps to be used in hard water without formation of dirty calcium soap scum and is most familiar to consumers in this use as a proprietary product, Calgon. In boiler waters, it interferes with growth of insoluble calcium and magnesium salt crystals and prevents scale formation in boiler tubes. The insoluble crystalline form of sodium metaphosphate is a polishing agent in toothpastes, where its compatibility with an anticaries agent, sodium fluorophosphate, is important.

The reaction of phosphoric acid with sodium carbonate and then with sodium hydroxide produces disodium and trisodium phosphates. The disodium form was first made in the U.S. before the turn of the century, and it was mostly used to weight silk. Today, it is mainly used as an intermediate in making sodium tripolyphosphate and tetrasodium pyrophosphate, of which more in a moment. It is also used in a number of ways in the food industry. It serves as an emulsifier in the production of processed cheeses, for example. In milk processing, it prevents gelation at temperatures reached in making evaporated milk. Added to brine pickling solutions, it helps produce hams that are much more tender and juicy when cured and smoked than they would be otherwise. Starch treated with it forms gels in cold water in instant puddings. Added to cereals, it raises the pH to about neutral, making a quick-cooking cereal.

Trisodium phosphate uses depend on its high alkalinity. Its ability to convert fats and greases into soaps makes it useful when combined with abrasives and with bleaches that release chlorine to make scouring powders to remove cooking greases and stains. It forms a crystalline complex with sodium hypochlorite. This complex has a combination of properties that makes it especially effective in cleaning dairy equipment, where its high alkalinity removes fatty deposits and the released chlorine sanitizes the equipment. It is also used in automatic dishwasher detergent formulations.

Trisodium phosphate has been used to soften water since 1920, when chemists found it precipitates calcium and magnesium ions. The precipitates are flocculent and do not form hard scales inside power-generating boilers, and they are easily purged from boiler tubes. Used alone, it also cleans hard surfaces such as painted walls, where its alkalinity removes a thin layer of paint from the oxidized surface and exposes a clean underlayer. It is also used to clean varnished surfaces, linoleum, vinyl tile, and other common floor and wall coverings. Finally, it is used as a calcium-free phosphate supplement to adjust the calcium/phosphate ratio in livestock feeds.

Tetrasodium pyrophosphate is made by heating disodium phosphate. Its use remained modest until 1935, when it was introduced as competition for sodium carbonate, sodium silicates, and some trisodium phosphate used as builders in soap. An excellent sequestrant for calcium, iron, and magnesium, it prevented formation of curds when soaps were used in hard water. Until the introduction of sodium tripolyphosphate in synthetic detergents in the late 1940's, soaps built with it were the mainstay among laundry products.

Sodium tripolyphosphate is the reaction product obtained by heating a combination of monosodium and disodium phosphate. The greatest impetus for expansion in industrial phosphorus chemicals came shortly after World War II with the introduction of sodium tripolyphosphate as a builder for synthetic detergents. From an insignificant 500 tons in 1941, production of sodium tripolyphosphate started rising rapidly in the late 1940's. By 1969, it had reached 1.2 million tons. The products responsible for this phenomenal growth are the synthetic detergents used for heavy-duty laundering. In 1947, Procter & Gamble (P&G) introduced the first commercial synthetic detergent (Tide) that combined the synthetic surfactant sodium alkylbenzene sulfonate with sodium tripolyphosphate as a builder. This basic formulation revolutionized the soap and detergent industry. It cleaned noticeably better than soap, and it cost less as well. Since it left no scum when used in hard water, it made possible the design of modern automatic washing machines. Other soap companies quickly followed with commercial products, and within a few years housewives had relegated soap, used since the dawn of history, to a minor share of the laundry product market.

The basic ingredient of all detergents today, whether built with phosphates or not, is a biodegradable surfactant. For heavy-duty laundry service, which is the most demanding general service and which comprises the largest single soap and detergent market, about 15 to 20% of the packaged product is surfactant. It is responsible for actual soil removal. However, surfactants work very poorly unless assisted by builders. Accordingly, sodium tripolyphosphate is used as a builder, and it accounts for 35 to 55% by weight of most heavy-duty laundry detergent formulations. It assists the surfactant chiefly by softening the water and keeping the dirt particles in suspension and preventing them from redepositing on the laundry or the machine. Phosphate-built detergents are also particularly effective in automatic dishwasher products and in some industrial and institutional cleaning products.

Because of the recent controversy over phosphates in detergents as a possible source of nutrients that support algal growth in streams and lakes, Lever Bros. and Colgate-Palmolive have elected to stabilize the phosphate content of all their heavy-duty laundry detergents at about 35% tripolyphosphate. This level equals 8.7% elemental phosphorus and is a reduction from a previous average of roughly 10% phosphorus. It is the level generally considered the minimum needed for effective heavy-duty cleaning. Higher concentrations, however, do increase cleaning power somewhat, particularly with heavily soiled clothes washed in hard water. P&G, where legally permitted, continues to market some detergents containing up to 14.6% phosphorus.

Because of the confusion about the role of detergent phosphate in the environment, some manufacturers have formulated and introduced nonphosphate detergents. They have captured about 10% of the market. These nonphosphate detergents use sodium carbonate and sodium silicate as builders. Unfortunately, nonphosphate detergents may leave a scaly deposit that can build up and interfere with washing machine pumps, agitators, and filters. Some nonphosphate detergents also tend to stiffen fabrics and leave white deposits on dark colors after repeated washing in hard water. Use of such nonphosphate detergents, however, reduced the production of sodium tripolyphosphate from 1,215,100 tons in 1969 to 1,190,100 tons in 1970.

Building soaps and detergents for laundering and dishwashing accounts for about 90% of U.S. consumption of sodium tripolyphosphate. Adding the productive capacity to meet the demand for tripolyphosphate provided by soaps and detergents obviously has had a major impact on the chemical industry. Should phosphates eventually be displaced from such widely used products, the chemical industry will obviously suffer economically. (While other uses exist for tripolyphosphate, they are relatively very small. Among them are ones in water treatment, as a deflocculent in reducing the viscosity of oil well drilling muds, and as an additive to pickling solutions for hams to increase the water-binding property of the meat protein. These and other uses could by no means replace the market in soaps and detergents.)

Reaction of phosphoric acid with calcium hydroxide (milk of lime) at different concentrations, ratios, and process conditions produces mono-, di-, or tricalcium phosphate. The tri form is used in only a few ways, such as a conditioning agent to improve the flow properties of salt, sugar, and other dry products and also as a whitening agent in making some ceramics, glass, and glazes. The mono and di forms, meantime, are used much more widely.

Of the two forms, dicalcium phosphate is used more extensively. In fact, it stands second (at 535,000 tons in 1970) to sodium tripolyphosphate (at 1.2 million tons in 1970) among industrial phosphate chemicals. Dicalcium phosphate's largest use is as a mineral supplement in animal feeds. Large tonnages (about 25,000 tons a year) are also used as a polishing agent in toothpastes, either directly or as an intermediate in making calcium pyrophosphate for the same use. Other outlets include glass manufacturing and plastic stabilizers.

Monocalcium phosphate is widely used as a leavening agent, and it has helped change both household and industrial baking. In earlier times, housewives and bakers used yeast or sodium bicarbonate and sour milk as a leavening agent. Then in the 1860's, Chas. Pfizer & Co. began making cream of tartar to replace sour milk, and at about the same time a Harvard scientist found that monocalcium phosphate could also replace it.

Following expiration of the original monocalcium phosphate patent in 1874, a number of companies began making it. Its use in baking over the years gradually has evolved to the point that about 85% of today's household baking powders consist of a mixture of sodium bicarbonate, monocalcium phosphate monohydrate, and sodium aluminum sulfate combined with starch.

Growth of monocalcium phosphate parallels that of self-rising flours. Producers introduced the high-quality monohydrate used in such flours in the early 1930's. These flours produced fine products, but moisture in the flour and moisture that diffused through the bag caused the monohydrate to react slowly with the sodium bicarbonate. In three to four months, the self-rising feature disappeared. Then in the mid-1930's, a chemist at Victor Chemical Works found that crystallizing monocalcium phosphate from a mother liquor containing traces of potassium, aluminum, magnesium, or sodium and then heating the product would form a very thin coating of mixed metal metaphosphates around anhydrous monocalcium phosphate crystals.

Anhydrous monocalcium phosphate would normally be too hygroscopic to use, but the coated form resisted attack by moisture. Introduced commercially in 1939, coated anhydrous monocalcium phosphate helped revolutionize baking. It extended the shelf life of self-rising flour considerably and also opened the way for subsequent development of prepared mixes for biscuits, cakes, pancakes, and waffles. Further improvements in prepared baking mixes were also made in the mid-1950's with the introduction of sodium aluminum acid phosphates as baking acids.

While most elemental phosphorus (about 90%) is used to make phosphoric acid, which is used to make other industrial phosphorus intermediates and end products, some (about 5%) is used to make other intermediates. In these uses, reactions of elemental phosphorus with chlorine or sulfur are most prevalent. They lead to chemicals that have important applications in a number of industries.

Chlorinating elemental phosphorus produces a very reactive chemical, phosphorus trichloride. Phosphorus trichloride reacts readily with phenols to form triaryl phosphites, and it reacts with alcohols to form trialkyl phosphites. These phosphites are used as stabilizers during processing of alkyds to make enamels and as stabilizers to protect styrene-butadiene rubbers and polyvinyl chloride and olefin plastics against discoloration by heat and light. Trialkyl phosphites also serve as intermediates in the synthesis of insecticides.

The reaction of phosphorus trichloride with alcohol also produces dialkyl phosphonate, an important intermediate in making a flame-retardant agent for rigid urethanes. Phosphorus trichloride can also be used to chlorinate alcohols to alkyl chlorides and fatty acids to fatty acid chlorides. The fatty acid chlorides in turn are intermediates for preparing such surfactants as sodium tallow taurate.

The largest use for phosphorus trichloride is for making phosphorus oxychloride by oxidizing it with oxygen. Phosphorus oxychloride is also made by chlorinating phosphorus pentoxide suspended in phosphorus trichloride. The major use for phosphorus oxychloride is as an intermediate in making triaryl phosphates, especially tricresyl phosphate. Tricresyl phosphate was first used as a plasticizer for celluloid billiard balls and more recently as a plasticizer for cellulose acetate and vinyl polymers. It has the advantage of plasticizing well and at the same time of imparting some flame resistance. It is also used as a gasoline additive, an application Shell Oil promoted under the trademark TCP beginning in 1953. A minor but important use is as an additive in lubricating oil to reduce wear on bearing surfaces and also to act as a load-carrying additive.

Phosphorus oxychloride is also an intermediate for making trialkyl phosphates. Tributyl phosphate is one important one, which is used in aircraft hydraulic fluid, as an antifoaming agent, and as an extractant for uranium salts for atomic reactors. Another is bromopropyl phosphate, which is used to help flameproof polymers.

Phosphorus pentasulfide is the other important intermediate made from elemental phosphorus. It is made by adding molten yellow phosphorus (one of several forms of elemental phosphorus) to molten sulfur in the correct molar ratio. Phosphorus pentasulfide's most important reaction is with alcohols to form dialkyl phosphorodithioic acids. One such reaction product from short-chain alcohols was introduced around the turn of the century as a collector in flotation processes for concentrating sulfide minerals. Absorbed on the surfaces of those minerals, it gives them specific wetting properties and permits them to be separated from worthless rock. Sulfide minerals of copper, lead, nickel, and zinc are the largest classes processed in this way.

Around 1920, dialkyl and diaryl phosphorothioates were introduced as lubricating oil additives to inhibit corrosion and wear. They also impart some extreme-pressure characteristics to the oils.

The greatest impetus to production of phosphorus pentasulfide came in the late 1940's, and growth has continued to the present. The event was the discovery of thiophosphorus insecticides. About half of the pentasulfide today goes to make such insecticides, which include the widely used malathion and methyl and ethyl parathion.

Phosphoric acid most often serves as an intermediate chemical for making other intermediate phosphorus chemicals and phosphorus chemical end products. However, it also serves as an important end product itself in metal treatment. Treating metal surfaces with phosphoric acid (or simple phosphoric acid salts) produces a phosphatized surface that is reasonably hard and electrically nonconductive. The insoluble phosphate surface coating adheres tightly to the underlying metal, and it is considerably more absorptive than the metal. It accordingly not only protects the metal but also serves to make paint and other coatings adhere better.

The modern phosphatizing industry dates from 1906 with the issuance of a patent in England. Many of the improvements had been recorded by the time of World War II and exploited commercially in Great Britain and the U.S. Phosphatizing was found useful for iron and steel and also for coating metals based on aluminum, cadmium, magnesium, and zinc. Refrigerators, washing machines, and other electrical appliances as well as a wide range of other products are all phosphatized today before they are painted or enameled.

Another phosphatizing process serves for surfaces that are not subsequently painted. One such process is Parkerizing, which Parker Rustproofing Co. developed in 1918. In it, a manganese salt in phosphoric acid produces a heavy and absorbent manganese phosphate coating. It provides good resistance to corrosion, especially after a thin film of oil or wax is applied. The black and slightly oily surface on nuts, bolts, screws, tools, and ordnance parts results from such treatment.

Phosphatized coatings from the Parkerize process are also applied to auto engine parts, such as rings, tappets, cams, and gears. When wetted with oil, the coatings resist wear on the friction surfaces. Burnishing by friction produces a smooth surface, and friction scoring of metal surfaces is reduced to a minimum.

Another important use of phosphoric acid in metal surface treatment is in the bright-dip process that polishes many items of aluminum and its alloys. The largest use for such chemical polishing is for trim for autos and other vehicles. Chemically polished and subsequently anodized aluminum has been used since the 1950's to replace much of the chrome plate formerly used as trim. Another large use is for handles and trim in home appliances and utensils.

Calcium Carbide

Calcium carbide is unique among industrial inorganics by being the intermediate chemical for making a major industrial organic chemical. That chemical, acetylene, is in turn the organic chemical largely responsible for the inception of what we know today as the petrochemical industry.

Acetylene eventually proved to be more important in Europe than in the U.S. as the route to volume production of many petrochemicals, to be sure. U.S. companies now primarily use natural gas and petroleum fractions for such chemicals, but they have done so to a considerable degree because of pioneering research in the U.S. that started with carbide acetylene and that helped show the potential for petrochemicals. In addition, American companies have also used carbide acetylene for making a number of important organics. Now, even though carbide acetylene has been sharply challenged by acetylene made from natural gas and petroleum, it still has a market in the U.S. Moreover, with any curtailment in natural gas or petroleum for chemical production, carbide acetylene might again become of increasing importance to the petrochemical industry.

Calcium carbide is made by combining quicklime (calcium oxide) with coke (carbon) at high temperature in electric furnaces. The process that brought acetylene to commercial significance has remained the same in principle since its discovery in the U.S. in 1892. Its first commercial preparation, incidentally, was inadvertent in that the developers sought to extract aluminum from ore rather than to make calcium carbide.

A simple reaction between water and calcium carbide produces acetylene. Illumination provided the initial outlet for acetylene, and carbide production grew rapidly in the early 1900's to meet the demand for lighting hotels, factories, schools, hospitals, and vehicles. Perfection of incandescent lamps soon made acetylene illumination obsolete, of course, but the market that illumination provided in the early years furnished an economic base from which the carbide acetylene industry could develop other applications that are very important today.

The first of these important developments came nearly concurrently at the turn of the century. The exceedingly high temperature of $5,600°$ F. reached by burning acetylene and oxygen had been pointed out in 1895, but commercialization required a source of low-cost oxygen. Acquisition of U.S. rights to the Linde process developed in Germany for recovering oxygen and development in 1901 of the blowpipe for producing the oxyacetylene flame opened the way to the new art of welding and cutting metals. As the industrial growth of the U.S. mushroomed during and after World War I, so did applications for the flame's high temperature and burning velocity. Today, despite strong competition from cheaper fuel gases and new processes, the oxyacetylene torch still rates as a primary tool in metal construction, fabrication, maintenance, and salvage.

The market for oxygen and acetylene in welding and cutting in part also led producers to improve methods for separating and handling gases. Some of the information was later of benefit in the petroleum industry as it processed an increasing volume of natural and cracked gases. The entire technology of cryogenics as it is known today ultimately grew around these refinements.

It was acetylene's use in chemical synthesis, however, that was destined to be most important for calcium carbide. Although development in 1905 of the cyanamide process for fixing nitrogen for use as a fertilizer preceded the use of calcium carbide for making acetylene for hydrocarbon syntheses, it was the latter that proved to be more important. This new role of acetylene as a chemical intermediate began to take form around the commercial production of trichloroethylene in Germany in 1910. Having no natural petroleum and short of carbohydrate material for making ethyl alcohol, Germany had to rely on acetylene from calcium carbide for making organic chemicals. In these uses, acetylene was one of the factors in the dominance of the German chemical industry at the onset of World War I.

In the U.S., meantime, the war also speeded up investigations of acetylene chemistry. By 1919, research by a predecessor company of Union Carbide at Mellon Institute on making acetylene from sources other than calcium carbide and on making chemicals from acetylene revealed what seemed to be limitless possibilities. Most significant was the development of chemicals from petroleum gases. Their production soon led to an entirely new chemical industry of open-chain aliphatic hydrocarbons comparable to that previously developed to use the cyclic aromatic hydrocarbons from coal.

Expansion of production and diversification of uses for acetylene in the 1930's and early 1940's was dominated by the development of polymer chemistry and technology, particularly emulsion polymerization that made possible the growth of the synthetic rubber and plastics industries. Proliferation of plastics and chemicals, many of which were based on polymers made from acetylene, turned acetylene into a major chemical intermediate and established calcium carbide as an industrial commodity.

Today, the chemical and allied products industry produces acetylene from about 70% of the some 750,000 tons of calcium carbide made in the U.S. From this acetylene, the industry makes such major products as neoprene, per- and trichloroethylene, and acrylates. Another 25% of the calcium carbide is used to make acetylene for oxyacetylene welding, cutting, and scarfing, and the iron and steel industry uses the balance in desulfurizing and other applications in making steel.

Until the early 1950's, U.S. companies derived nearly all their acetylene from calcium carbide. During the following several years, newer processes for obtaining acetylene by cracking natural gas and other hydrocarbons began growing more important, and acetylene made from hydrocarbons supplied about 20% of the market by 1961. Hydrocarbon acetylene's share of the market has now risen to more than 80%, and expansion is continuing. In addition, ethylene and propylene have become successful replacements for acetylene as intermediates in reactions once limited to acetylene and have reduced carbide acetylene's market even more. As a result of this competition from hydrocarbon acetylene and from ethylene and propylene, calcium carbide last appeared in 1970 among the 50 largest volume chemicals made in the U.S. Hydrocarbon acetylene's continued encroachment now seems assured for the next several years, but its long-range dominance depends on the price and availability of natural gas and other suitable feedstocks as well as its price relative to competing petrochemical intermediates.

Boron Chemicals

In 1970, U.S. companies produced nearly 570,000 tons of borates and related compounds (on a boric oxide, or B_2O_3 basis). This amount represented more than 60% of total world production of borates. The U.S. accordingly is a large exporter of borates, with slightly more than 50% of its production being marketed abroad.

The glass industry consumes by far the most borate in the U.S., accounting for about 40% (or more than 110,000 tons of the 280,000 tons used in the U.S.). Specific applications in that industry include the use of sodium borate in making borosilicate glasses and glass fiber insulation and the use of sodium-free borates such as calcium borate, boric acid, and boric oxide to make textile glass fibers. Among other major uses are soaps and detergents (16% in 1969), porcelain enamel (9%), and agricultural chemicals (7%). In contrast to the U.S., incidentally, the major use of borates overseas has been as raw materials for perborate bleaches, primarily in Europe.

U.S. producers recover all borates in the desert areas of California. The largest single producer, U.S. Borax & Chemical Corp. (a subsidiary of Rio Tinto Zinc Corp.), recovers borates principally as sodium tetraborate decahydrate (tincal or borax) in an open pit mining operation. The other long-time major producers, American Potash and Chemical Corp. (now part of Kerr-McGee Corp.) and Stauffer Chemical, recover sodium borates from complex lake brines. These three have recently been joined by Tenneco, Inc., which made its first shipments of calcium borate (colemanite) in 1971, and they are to be joined by Occidental Petroleum, which plans to produce sodium borate from lake brines beginning in 1973.

The basic minerals recovered in these operations are converted to a variety of boron products in a series of refining processes. The major commercial products today are borax decahydrate and pentahydrate, anhydrous borax, boric acid, and boric oxide, while calcium borate is expected to become a more significant commercial product soon.

Most of the major borate chemicals are made by fairly standard processes that have existed for some time. Borax decahydrate and pentahydrate, for example, are refined by crystallization of crude sodium borates, anhydrous borax is made from them by fusion, and boric oxide is made by fusion of boric acid. Manufacturers have introduced new and improved processes in recent years, however. In the 1960's, for example, American Potash and Chemical Corp. developed an alternative method for recovering boric acid from dilute lake brines by extracting the brines with a kerosine solution of a complex polyol chelating agent. It then recovers boric acid by treating the kerosine fraction with sulfuric acid followed by fractional crystallization. In 1967, U.S. Borax & Chemical developed a new process for boric oxide in which it mixes borax with sulfuric acid in a fusion furnace and then separates the boric oxide and sodium sulfate products in the molten state.

Vanadium Pentoxide

Though perhaps most widely recognized by chemists as an oxidation catalyst, vanadium pentoxide has earned commercial status as the primary vanadium chemical from which nearly all intermediate and end-use vanadium compounds are made. Discovered in 1830, vanadium in early commercial uses was confined to salts for coloring textiles, leather, glass, and pottery. A few years before the turn of the century, however, the French began using vanadium as an alloying agent for armor plate, and shortly thereafter a professor at the University of Sheffield in England found that vanadium added to steel improved its cutting qualities. In time, vanadium has become a major alloying agent for steel.

In the U.S., Henry Ford is credited with pioneering recognition of the importance of vanadium in making steel, and the first successful heat of vanadium steel was tapped in 1907 under contract to him. Ford first used chromium-vanadium steels as standard materials for gears, axles, springs, and many other components, and full-page ads heralded the superior qualities of vanadium steels in 1908 when Ford introduced what was to become the legendary Model T. The competitive advantage he gained by his extensive use of vanadium steel helped him keep prices low, increase sales, and thus create a mass market for autos.

Until the 1960's, tool and full alloy steels accounted for most of the vanadium used in steel. Today, however, the rapidly growing high-strength low-alloy (HSLA) steels have given the steel market a new direction. HSLA steels, many of which contain vanadium, registered a fivefold increase in production during the 1960's, and they are taking over many of the markets previously held by other types of steels. With a high strength to weight ratio and other desirable properties, they are preferred where greater strength can be used either to decrease weight or increase durability. As a result, buildings and bridges are becoming leaner and more graceful and also less costly and more durable than those built entirely of carbon steels. In addition, thousands of miles of large-diameter high-pressure HSLA steel pipe now link oil and gas fields with consuming areas, and HSLA steels are also being used to advantage in storage tanks, overhead roadways, vehicles, and heavy industrial equipment.

Vanadium is also used to make titanium-base alloys for the aircraft and aerospace industries. In addition, it may find uses in the atomic energy field, where it would offer low capture cross section for fast neutrons, a high melting point, and high creep strength. At first, it appeared vanadium would have an important market as a structural and cladding material for fast breeder reactors, but embrittlement caused by low concentrations of oxygen has deterred such use. Now, however, vanadium-titanium alloys are being actively considered for use in the prototype controlled thermonuclear reactor expected to be built in the years ahead.

A small but important part of vanadium production goes to the chemical industry, which uses vanadium compounds mainly as oxidation catalysts. One important use is that of vanadium pentoxide as the catalyst in making sulfuric acid by the contact process. In another important use, vanadium halides and oxyhalides serve as catalysts of the Ziegler-Natta type to copolymerize ethylene and propylene. Other catalyst uses include the production of phthalic and maleic anhydrides and of polyamides such as nylon, while a market may come with its use in some of the catalytic converters under development to meet the emission standards for vehicle exhausts.

Recovery of vanadium involves costly handling of large quantities of ore and difficult processing problems. Most processes employ salt roasting and leaching, some of which are modified by solvent extraction or ion exchange procedures. Vanadium minerals are widely distributed, but they rarely occur in sufficient quantities for economic mining and processing for vanadium content alone. Vanadium, therefore, is generally recovered as a co-product or by-product as ores are processed for other metals. In the U.S., for example, vanadium's production since about 1950 has been geared to the production of uranium from Colorado Plateau ores and of phosphorus from Idaho-Wyoming ores.

Cutbacks in uranium production in the 1960's plus unprecedented vanadium demand during the same years led to supply shortages. In the resulting scramble for new sources, geologists at Union Carbide Corp. found one of the few vanadium deposits in the world that can be mined exclusively for vanadium. Located in the Arkansas shales, this large deposit should have reserves for about 10 years, and vanadium supply patterns are now shifting to production from the Arkansas mine-mill complex that Carbide began operating in 1969.

Fluorine Chemicals

Fluorine chemicals, in common with the other inorganic chemicals discussed in this chapter, are widely used in industrial and consumer applications. They serve as aerosol propellants, refrigerants, and plastics, for example. As intermediate chemicals, they are essential in the production of such diverse materials as aluminum, a major material of construction, and uranium hexafluoride, an intermediate for atomic reactor fuels. In addition, pharmaceuticals, dyes, pesticides, surface active agents, solvents, protective coatings, and water and metal treatment chemicals are among the many end-product chemicals derived from them. In the petroleum industry they function as catalysts for olefin alkylation to make high-octane gasolines, and they also find important uses in processes in the primary metals and ceramics industries. The simplest fluorine compound, hydrogen fluoride, is one of the very few chemicals that will etch glass.

Of all the intermediate industrial fluorine chemicals now made, hydrogen fluoride (HF) is the most important. It is the intermediate from which the vast majority of other intermediate and end-product chemicals containing fluorine are derived. HF is made as aqueous hydrofluoric acid and anhydrous hydrogen fluoride, each of which is used as an intermediate and end-product chemical in different applications.

Total production of HF in 1970 amounted to nearly 320,000 tons that sold for nearly $102 million. HF is made from what is known as acid-grade fluorspar, which is 97+% calcium fluoride and which is one of three refined grades of fluorspar made industrially from mined fluorspar, an impure calcium fluoride mineral. (The other two grades of refined fluorspar are the less pure ceramic and metallurgical grades.)

Hydrogen fluoride's production results from an old and chemically simple process typical of those used for making many industrial inorganic chemicals, but the process has benefited from a number of improvements. One improvement involved chemical engineering changes to increase production from a somewhat recalcitrant mixture of reactants, and the other called for changes that would produce much more concentrated acid to supply developing markets for fluorocarbons, aluminum fluoride, and catalysts.

Treating refined fluorspar with sulfuric acid produces hydrogen fluoride gas and calcium sulfate. Little reaction occurs until the suspension of fluorspar and acid is heated, at which time the slurry thickens rapidly and builds up as a sticky paste on furnace walls. It was this buildup that led chemists and chemical engineers in the early 1930's at the then only two producing companies to devise two more or less satisfactory answers to the problem. Their solutions opened the way to economical increases in production that were important at the time and became more so as new uses for hydrogen fluoride in the late 1930's and early 1940's called for expanded production.

At Allied Chemical's former General Chemical division, the solution proved to be loose rails extending the length of rotating horizontal furnaces that scraped the paste loose as it formed. At Aluminum Co. of America, the furnaces evolved into stationary cast iron shells fitted with heavy scrapers that cleaned the shell walls. These two types of furnaces served as the basic types underlying all hydrogen fluoride technology in the U.S. until 1960, when newer developments occurred, primarily at du Pont and Allied Chemical.

Early major uses for hydrogen fluoride included its use as an intermediate for making sodium fluoride for insecticide and steel ingot rimming applications, sodium bifluoride for laundry sour applications, and aluminum fluoride and synthetic cryolite for primary aluminum production. These and other uses of the time called for hydrogen fluoride in aqueous solutions no stronger than about 60%. With the introduction of fluorocarbons by du Pont in the 1930's, however, a new high strength of about 99.5% hydrogen fluoride was needed. Du Pont built its own plant to supply its needs, and it was soon joined by Allied Chemical and, a few years later, by Pennsylvania Salt Manufacturing Co. (now Pennwalt).

Manufacturers producing such anhydrous hydrogen fluoride have for the most part developed their own technology and have not sold it to others. Earlier processes were based on redistilling aqueous hydrogen fluoride, but they were difficult to operate because of the severe corrosion encountered. Later processes produced an essentially anhydrous product directly without redistillation. Much of the know-how is unpatentable because of long commercial usage, but over the years an interchange of personnel has spread the knowledge within the industry of the general principles employed. Specific details, however, such as mechanical features and materials of construction at various points, are of great importance, and they are mostly individual company secrets. Nonetheless, a few innovations have reached the patent literature in the past decade or so, including an adiabatic reactor patented by workers at du Pont in 1963 and an Archimedes screw recycle furnace patented by Allied Chemical personnel in 1971.

Aluminum Fluoride and Synthetic Cryolite. Producing fluorocarbon aerosol propellants, refrigerants, plastics, blowing agents, and solvents provides the largest market for hydrogen fluoride, accounting for about 45% of the U.S. output. The aluminum industry ranks as a fairly close second, however, using about 35% of U.S. production. In making aluminum, bauxite ore is first refined to alumina (or aluminum oxide), which is then dissolved in a molten electrolyte that is largely a mixture of aluminum fluoride and cryolite. Application of an electric current to the cell causes aluminum metal to deposit at the cathode (and oxygen to evolve at the anode).

Aluminum fluoride is used in larger amounts in the electrolyte solution. It is made synthetically by one of two processes. In one, alumina reacts with anhydrous hydrogen fluoride in a fixed bed tower reactor, and alumina trihydrate reacts with anhydrous hydrogen fluoride in a fluid bed reactor in the other. U.S. production in 1970 amounted to about 290 million pounds.

Cryolite is a naturally-occurring sodium aluminum fluoride mineral. Minable deposits have in time proved to be not nearly sufficient to meet aluminum industry needs, especially those brought about by the rapid expansion during World War II. U.S. companies for some years, therefore, have had to build plants to make cryolite. U.S. production of synthetic cryolite amounted to about 205 million pounds in 1970. About half of this total was recovered at aluminum plants for recycle. The balance generally was made by reacting hydrofluoric acid with sodium aluminate produced during the recovery of alumina from bauxite.

Rising prices for hydrofluoric acid and a shortage of fluorspar are sparking new interest in processes to make both aluminum fluoride and synthetic cryolite from fluosilicic acid, a waste by-product from phosphate fertilizer plants. Containing fluorine as it does, fluosilicic acid can serve as the intermediate for the production of either anhydrous hydrogen fluoride or aluminum fluoride and synthetic cryolite. Some studies have indicated that making aluminum fluoride and cryolite offers a more practical goal.

Kaiser has been producing cryolite from fluosilicic acid since 1958, and Aluminum Co. of America in 1970 built the first U.S. plant to make aluminum fluoride and also synthetic cryolite from fluosilicic acid. Proponents of fluosilicic acid claim not only attractive economics but also point out the relief its use brings to a disposal problem that phosphate rock processors have long faced. Should pressures for fluoride waste control by the U.S. Environmental Protection Agency materialize and the projected shortage of fluorspar occur, the use of fluosilicic acid to make aluminum fluoride and synthetic cryolite could be given added impetus. As much as 10% of U.S. aluminum fluoride production may have come from fluosilicic acid in 1971, and industry observers are predicting an increase to 20% by 1974. As for synthetic cryolite, they expect existing plants to continue to operate but look for no expansions or new facilities to be based on hydrogen fluoride.

Boron Trifluoride. Among the many other intermediate and end-product fluorine chemicals, boron trifluoride ranks high in significance because of its versatility as a catalyst in many chemical reactions. Among the more important examples of such uses are alkylation of benzene with long-chain hydrocarbons to make alkylbenzene for synthetic detergents and production of vinyl ether polymers and copolymers that are used in such products as adhesives, surface active agents, and sizings. Isomerizing petroleum fractions to make gasoline and petrochemical products is likewise a significant use. In addition, the nuclear energy industry takes advantage of boron's absorption capacity for thermal neutrons by using boron trifluoride made from the appropriate boron isomer as a neutron-absorbing medium in proportional neutron counters.

Boron trifluoride's history dates from the early 1880's, but it was not until the 1920's and 1930's that commercial applications began to appear, such as the possibilities for its use as a catalyst in olefin polymerization and acetal preparation. It became available commercially in 1936, when Harshaw Chemical Co. began operating a plant using a process its chemists and chemical engineers had developed during the preceding four years.

Producers make boron trifluoride by reacting a boron compound with a fluorine compound in the presence of an acid. The Harshaw method used the classic approach employing fluorspar, boric acid, and sulfuric acid but overcame the original impractically low yield. In 1947, Allied Chemical developed a simplified and continuous process in which the reactants are fluosulfonic acid and boric acid or boric oxide. Allied also pioneered in the early 1940's with the commercial use of boron trifluoride in a variety of uses but primarily in making petrochemicals.

Chromium Chemicals

Chromium chemicals perform many vital and useful functions in industrial processes and in consumer products. Chromium pigments, for example, provide some of the brightest, most corrosion resistant, and best covering-power greens, yellows, and oranges for textiles, inks, and paints, enamels, and lacquers. Inhibiting corrosion and protecting metals offer wide applications for sodium bichromate (pickling and bright dipping) and chromic acid (chrome plating). In addition, sodium bichromate is used in process water treating. The leather industry uses chrome chemicals to tan hides from cattle (except very heavy hides), calves, goats, kids, sheep, and lambs. Added to water-soluble toxic salts such as those of copper and arsenic, chromium compounds improve the performance of wood preservatives. In textile processing, sodium bichromate and chromic salts function as a mordant to fix dyes on fibers such as wool, silk, and nylon, improving the fastness of colors. In dye manufacturing, sodium bichromate acts as an oxidizing agent to make methylene blue, thionine blue, safranine, aniline purples, and others, while it or chromic acid serves as an oxidizing agent in a large number of commercial processes for making organic chemicals, including saccharin, Vitamin K, and steroids such as cortisone and those used in oral contraceptives.

Sodium bichromate is the principal commercial chromium chemical, and all others are derived from it. Its production processes have not changed much during the past half century and are based on roasting chrome ore with sodium carbonate, leaching the roast and acidifying the leach liquor with sulfuric acid to produce bichromate, and removing the last remaining impurities by crystallization. Differences among producers mostly involve the ratio of reactants and the type of oxidation furnaces used (e.g., rotating kilns vs. rotating ring hearths). Successful production, however, has called for considerable chemical engineering ingenuity and careful attention to economic details. Increased emphasis is now being placed on recovery and recycle processes to minimize the effluents from operations.

FUTURE DEVELOPMENTS

The industry sector responsible for producing primary and intermediate inorganics was at one time characterized by large-scale production of commodity inorganics by fairly standard and unchanging technology. It may still have this characteristic in some ways, but many new primary and intermediate chemicals have been added to its production list in recent years. While no one can predict the discovery of startlingly new chemicals or chemical processes, no doubt chemical industry personnel will continue to make such discoveries. The only possible difference in the foreseeable future is that such discoveries may not be made with quite the rapidity to which everyone has become accustomed.

Meanwhile, many opportunities exist to improve products and processes and to reduce costs—or at least to hold increases to competitive levels. We shall not attempt to recite the many improvements being worked on but instead shall single out only a few examples as typical of improvements for chemicals discussed in the foregoing pages.

Anodes for both the diaphragm and mercury cells for making caustic and chlorine have traditionally been made from graphite, for example, but the graphite is slowly consumed as the cells operate. Mercury-cell design has been improved to permit adjustments to be made to electrode gaps to compensate for wear and minimize electrical energy consumption. Diaphragm-cell design is such that no practical way exists to adjust electrode gaps, so power consumption increases as anodes wear.

Manufacturers have sought more permanent anodes for many years. In the past few years, a significant advance has been made leading to what are known as dimensionally stable anodes. These anodes consist of a titanium substrate coated with oxides of ruthenium and closely related elements, and their production has been made possible in part because of the reduced price and improved availability of titanium. Further development of such stable electrodes will probably continue for the next several years, resulting in reduced production costs for both chlorine and caustic. Such work will also continue to stimulate other work, as it already has, in reducing the cost of graphite anodes and of the chlor-alkali plants using them.

Another area of improvement for caustic-chlorine processes involves ion-permeable membranes. Successful development promises to produce a caustic stream free of salt without the use of mercury. In addition, air- or oxygen-depolarized cathodes may be developed that will significantly reduce the electrical energy needed to make the co-products. Pressure electrolysis also offers potential major production economies.

Many similar improvements are being worked on throughout the chemical industry. Meantime, all manufacturers of primary and intermediate industrial inorganics face an uncertain future under the current alarm over actual and possible impacts on the environment of the production and use of chemicals and chemical products.

In the earlier part on chemical achievements with industrial inorganics, we have already alluded to some of the impacts that concern is causing. Thus, sulfuric acid manufacturers are being required to reduce emissions of sulfur oxides to meet new and more stringent controls. Some are able to do so by simply adding equipment to recover sulfur oxides more effectively. Others, meantime, face making less economically attractive changes or shutting down operations. Caustic-chlorine manufacturers who use mercury cells are likewise being required to clean up their plant effluents. All have reduced mercury discharges to streams and rivers, and some plants have stopped operating because of their inability to meet the regulations and still operate economically. Moreover, the problem occurs in plants as well as outside of plants, and one company has closed a mercury-cell caustic-chlorine plant because of labor problems involving unsatisfactory mercury levels in the plant.

Elsewhere among similar problems covered earlier, soap and detergent companies, companies supplying them, and government agencies are all seeking alternatives to phosphates for use in soaps and detergents. In addition, sodium carbonate producers are abandoning synthetic routes in favor of processing natural deposits, since they often cannot remain competitive and still meet pollution control regulations.

Related concerns over possible harm to the environment by end-product chemicals will also bring changes to producers of primary and intermediate industrial inorganics. Among chlorinated products, for example, there is a growing interest in limiting solvent emissions to the atmosphere that will in all likelihood slow the overall demand for chlorinated solvents (and simultaneously require development of effective substitutes). In addition, the problems associated with persistent chlorinated insecticides and herbicides will probably affect the future use of chlorine in these markets. The total impact on chlorine will not be great, however, since making pesticides consumes relatively minor amounts of chlorine compared to total production.

Manufacturers of primary and intermediate inorganics also expect major changes among some of their raw materials because of environmental problems. Today, for example, most sulfur in the U.S. originates with the recovery of elemental sulfur from deposits along the Gulf Coast of the U.S. and Mexico. A major potential source of sulfur in the next decade will be from sulfur dioxide emissions now vented by power plants and smelters. As a nondiscretionary source of sulfur, it should assume a dominant role as a raw material for sulfuric acid.

Power generation, manufacturing, and other activities were emitting about 30 million tons of sulfur oxides to the atmosphere in the U.S. annually as the 1960's ended. That total contains about one and a half times the sulfur content of all sulfuric acid made here. Electric power generation is the most prominent target for regulation of this air pollution source, since it contributes about 50% of all sulfur dioxide emissions. Industrial and commercial fuel uses account for somewhat more than 30%, followed by ore smelting, which contributes about 12%. Since smelter emissions are vastly more concentrated both in composition and in amounts at a limited number of discharging locations (fewer than 40 compared to more than 950 for power plants burning fossil fuel), it follows that control of smelter sulfur dioxide emissions appears to be the first target for control in sight of U.S. authorities. Other sources will also feel the effect of tightening regulations, however. Regardless of the source of future sulfur supplies, suitable solutions will be required for recovering and using the sulfur values.

Similarly, another challenge is the supply-demand situation for some products. Typical is one covered earlier in which uses are sought for hydrochloric acid. Otherwise, the by-product supply from chlorination would soon far outstrip the country's ability to absorb the production.

Another typical example occurs for caustic-chlorine producers. Electrolytic cells make an unvarying ratio of caustic to chlorine, but chlorine markets have grown faster than caustic markets in the past and are expected to continue to grow somewhat more rapidly in the future. If producers meet the demand for chlorine, the potential for oversupply in caustic is obvious. Although caustic is now in short supply, chlorine manufacturers are looking for processes that don't also produce caustic.

Despite the negative aspects of an uncertain future predicted for the industrial inorganic sector primarily in the face of the alarm currently being sounded by those concerned with the environment, these same liabilities present many positive and challenging opportunities for chemists and chemical engineers in the years ahead. Certainly no manufacturer has solved all of his production and distribution problems for today's products—either for old-line commodities or newer and often more exotic compounds. Even daily newspapers tell of diminishing supplies of raw materials and energy sources and point out that most pollution problems are far from being solved by many companies. But normal growth and steady demand for improvements, plus the present concern over environmental harm, are leading forces that will stimulate change. All these factors point up the need for chemically trained personnel to turn these changes into assets, to improve processes using different raw materials, and to develop new technologies for appropriate solutions to today's problems in an ecologically awakened world.

SOCIAL AND ECONOMIC IMPACTS

The economic power of industrial chemical research is shown clearly by the explosive growth of plastics and resins in the past three decades. Starting with the billiard ball and the celluloid collar before the turn of the century, these materials have mushroomed since the mid-1940's into a worldwide, multibillion-dollar industry. Companies in the U.S. alone in 1969 produced 19 billion pounds of plastics, valued at $3.8 billion, and the average annual growth in poundage is estimated at 12% in 1969–74. In 1971, U.S. output exceeded 21 billion pounds.

A few basic characteristics account largely for this phenomenal growth. The raw materials for plastics, derived mainly from petroleum, are generally cheap and plentiful. Chemists, moreover, have learned to manipulate these raw materials at the molecular level to produce plastics of widely differing properties. Within broad limits, they can tailor-make the plastic to the requirements of the end use. A third factor is that plastics often can be fabricated more cheaply than competing materials. They offer in addition a relatively low specific gravity compared to metals or glass and resist the normal forms of corrosion to an unusual degree.

Plastics as a result have displaced traditional materials in thousands of products, aided by rising production and favorable price-volume trends. They are used as well in certain new products and processes to which they are uniquely adapted. Plastics technology created the water-based (latex) paints that have revolutionized the coatings industry. Plastic films and foams have done the same in the packaging industry. Few homes, if any, are without products made at least partly of plastics: nonstick frying pans, appliances, dinnerware, coverings for floors and working surfaces. Industry, too, uses an almost endless list of plastic products: small gears and other engineered parts, electrical and thermal insulation, automobile parts and even bodies, water purification systems, corrosion-resistant piping.

CHEMICAL ACCOMPLISHMENTS

Plastics are among the high-molecular-weight substances classified more broadly as polymers. The first commercial plastic was cellulose nitrate (celluloid), patented by the Hyatt brothers in 1870. Four decades elapsed before a predecessor company of Union Carbide introduced the phenolics (Bakelite), in 1909, but by the 1930's more than a dozen plastics were on the market or well along in development. Then came World War II with its heavy demand for materials of all kinds. Both polymer research and production accelerated markedly, thus establishing the technological base for the major new industry that was to come.

Synthetic rubbers and fibers, like plastics, are classified broadly as polymers. The three are made, in fact, from many of the same organic intermediates or monomers, but they differ in certain important respects. Plastics are hard and "glassy" at room temperature, but synthetic rubbers tend to grow brittle and "glassy" only below about −50° C. In order to prevent plastic flow, rubbers are generally used in a vulcanized (crosslinked) state. Unlike plastics and rubbers, useful synthetic fibers tend to melt at relatively high temperatures, usually above 200° C.

Plastics themselves fall into two broad classes: thermoplastics, which are soluble and capable of repeated softening and hardening; and thermosets which, once heat-formed, are insoluble and infusible. The thermoplastics include materials such as polystyrene, polyvinyl chloride, polyethylene, polypropylene, and polyvinylidene chloride (saran). The thermosets include the phenolics, melamines, epoxies, and styrene-based polyesters and often are used with reinforcing fillers such as glass fibers.

Complexities of Polymerization. The conventional portrayal of the chemical reaction that forms the basic polymer chain of a plastic can be disarmingly simple. Polyvinyl chloride (PVC), for example, is made by reacting n molecules of vinyl chloride, the monomer, to form a polymer containing n units linked together:

$$nCH_2{=}CHCl \xrightarrow[\text{catalyst}]{\text{heat}} (CH_2{-}CHCl)_n$$

The role of the chemist, however, is much more complex than this portrayal suggests.

First, the polymerization may be carried out by any of several methods: suspension, emulsion, solution, or mass (loosely defined as no liquid medium in addition to the monomer). Suspending agents are needed for suspension polymerization, emulsifying agents for the emulsion method. The latter also requires coagulating aids to bring the polymer out of emulsion. The catalyst system usually is complex, in that two or more catalysts may be used.

Secondly, the polymerization is a chain reaction and thus extremely sensitive to impurities or additives that can terminate growing chains and thus reduce molecular weight undesirably. The polymer chains will average, say, 2,000 monomer units, with individual chains ranging from a few to as many as 10,000 units. The actual molecular weight distribution among the chains, which affects the properties of the plastic, depends on the catalyst, the polymerization method, the fraction of monomer polymerized, and the temperature schedule during polymerization.

In addition, a comonomer may be used in small amounts with the vinyl chloride to contribute some desirable property, such as better adhesion, greater strength at low temperature, and the like. The comonomer may enter the polymer chain randomly or in blocks, depending on how its chemical reactivity matches that of the major component and on the polymerization process.

These and other variables affect the fabricating characteristics and end-use properties of the PVC produced. In thermoplastics other than PVC, the structure of the base polymer chain may be affected by even further variables. It is the chemist's job to evaluate such variables and learn to control them on an industrial scale.

Polymer Additives. The base polymer is seldom used alone in plastics. PVC requires a thermal stabilizer to prevent loss of hydrogen chloride in any processing step at temperatures above 100° C. Flexible PVC films and foams require 20 to 40% of a plasticizer. Flow agents may be included to facilitate heat fabrication. Other additives may include colorants, solid fillers to improve properties and reduce costs, and rubbers to improve low-temperature properties. Plastics in general may contain such additional additives as antioxidants, agents that control surface properties, carbon black to stabilize against degradation by light, delusterants, self-extinguishing agents, and so on.

In the selection and use of additives, as in making the base polymer, the chemist plays a major role. He must select not only the best additives for the purpose but the best combination of additives. For new additives, he may have to develop manufacturing processes. If the plastic will contact foods or be used by children, he must do the required toxicity studies on the additives. He must develop the additive package, moreover, within the economic constraints imposed by the plastic's end uses.

Problems of Thermosets. Thermosets involve all of the problems characteristic of thermoplastics plus many additional ones associated with crosslinking. These plastics, moreover, are often used with fibrous or particulate fillers, which raise additional complexities. Furthermore, thermosets become insoluble and infusible above a certain percent conversion of the starting ingredients. Thus, off-grade product cannot be recycled as is done with thermoplastics. During the reaction, the product must be prevented from gelling prematurely. In addition, the reaction must be driven to completion while being rid of reaction by-products that otherwise might be trapped in the product.

Polymer Characterization. Implicit in the foregoing is the development of a whole new science: the characterization of macromolecules or polymers. The effort has been a collaborative one among chemists who synthesize new structures and chemists and physicists skilled in unraveling them. Physicists generally have discovered and chemists have then developed the more powerful characterization tools, such as infrared, mass spectrometry, nuclear magnetic resonance, and the ultracentrifuge. One of the most powerful tools is gel permeation chromatography, developed by an organic chemist who was studying the nature of polyglycols used to make polyurethanes.

Polyolefin Plastics

Commercially important polyolefins today include:

- Low-density polyethylene, made at 1,000 to 3,000 atmospheres of pressure and 80° to 300° C. Also called branched or high-pressure polyethylene.
- High-density polyethylene, made at 1 to 500 atmospheres and 0° to 300° C. Also called linear or low-pressure polyethylene.
- Copolymers of ethylene with minor amounts of a comonomer.
- Ethylene-propylene copolymers.
- Block copolymers of ethylene, such as "polyallomers."
- Chemically modified polyethylenes.
- Polypropylene and blends of that polymer with rubber or chopped glass fiber.
- Polybutene-1.
- Poly-4-methylpentene-1.

The polyolefins, mainly polyethylene and polypropylene, are the largest tonnage plastics in the world. Companies in the U.S. in 1969 produced some 6.7 billion pounds of polyethylene and polypropylene valued at $975 million. This one narrow field of chemistry thus accounted in that year for about 0.1% of the nation's Gross National Product. In 1971, this country produced 7.7 billion pounds of polyolefins.

Polyethylene is the largest-volume plastic in the U.S. and in the world. Of the 12 billion pounds of polyethylene made worldwide in 1970, the U.S. produced about 45%. Polypropylene ranked fourth in volume in the U.S. in 1969, behind the polystyrenes and polyvinyl chloride and at the same level as the phenolics.

So many gimcrack novelties have been made of polyethylene as to obscure the fact that the plastic's success has always stemmed in part from its very unusual properties. The same has been true of the much younger polypropylene. The growth of the polyolefins also can be interpreted in terms of two general rules of thumb for commercial activity in plastics:

• Annual sales of any plastic, in pounds, varies inversely with the cube of the selling price in cents per pound.

• A polymer generally costs two to three times the cost of its monomer.

Ethylene is a very cheap monomer, and propylene is a close second. These factors, price and properties, inevitably made the polyolefins the basis of an international development program that began to accelerate after World War II.

Polyethylene. Polyethylene has been an international development almost from the beginning. The polymer was discovered in 1932–35 at Imperial Chemical Industries, in England, in the course of a research program on the effects of very high pressure on chemical reactions. The program was directed by M. W. Perrin and John Swallow. It drew particularly on the published work of physicist P. W. Bridgman and chemist J. B. Conant, both of Harvard, and on the expertise of Prof. Michels, at Amsterdam University, in high-pressure equipment.

The initial discovery was accidental. By 1935, however, the ICI chemists had learned to polymerize ethylene to a solid whose properties indicated that it was a reasonably straight-chain polymer of fairly high molecular weight. (The use of "straight-chain" is relative. By today's standards, the polymer was a branched chain.) In the ensuing period it became evident that polyethylene (the British name is polythene) was the ideal material for shielding microwave cable in military radar equipment. The British work on radar was a top-priority project as World War II drew near and sharply spurred the development of polyethylene.

Besides its electrical insulating characteristics, the plastic had other unusual properties: ease of fabrication, toughness, good solvent and corrosion resistance, and the lowest specific gravity (0.915) of any plastic known. This combination of properties would stimulate in the two decades after World War II a succession of developments in polyolefins replete with international scientific, legal, and financial complexities. The economic implications of polyethylene are clear in the number of companies that became involved and in the magnitude of their efforts.

Low-Density Polyethylene. Low-density polyethylene accounts for roughly two thirds of U.S. output of polyethylene. About 60% of the low-density polymer is used in packaging, mainly in film and extrusion coating. Other important uses include wire and cable insulation, pipe, and a variety of injection-molded and blow-molded products. In price per unit volume, low-density polyethylene is the cheapest of all plastics. Among the polymer's other useful characteristics are flexibility, clarity, heat sealing properties, and ease of fabrication.

Toward the end of World War II, du Pont began to produce polyethylene commercially under a process (stirred autoclave) licensed from ICI. The product was the low-density or branched-chain polymer. Du Pont chemists were to discover later that polyethylene made at extremely high pressure (up to 20,000 atmospheres) differed substantially from the branched material. The company filed a patent application in 1951. In December 1957, after a long battle with the U.S. Patent Office, the Larcher-Pease patent on linear polyethylene issued and was assigned to du Pont. The patent was issued as a result of a powerful study in polymer characterization made by analytical and physical chemists as well as physicists.

In April 1941, Union Carbide independently started a development program for producing polyethylene. The resulting process, based on tubular reactors, yielded a low-density polyethylene that proved superior to both ICI and du Pont material for insulating radio-frequency cable. The U.S. military in consequence used the Union Carbide polymer almost exclusively for that purpose after the company started up its first plant (in April 1943).

In 1952, as the result of a civil antitrust suit, the U.S. Government compelled ICI to license its 1950 polyethylene technology to bona fide applicants in the U.S. Early in 1953, ICI licensed Spencer Chemical, Dow Chemical, Eastman Kodak, and National Petrochemical (later National Distillers). Rexall acquired an ICI license at some later date.

Monsanto and Koppers, meanwhile, licensed a tubular-reactor process developed by the German company Badische Anilin- & Soda-Fabrik. The U.S. thus jumped from the two original polyethylene producers, du Pont and Union Carbide, at the end of World War II to nine producers by 1960. Today, 12 companies make low-density polyethylene in the U.S.

Linear (High-Density) Polyethylene. Linear or high-density polyethylene is made in the U.S. by three different processes: Ziegler, Phillips, and Standard Oil (Ind.). The polymer's main use is in blow-molded containers such as bottles for household chemicals, milk, and other beverages. It is more rigid than the low-density material and thus can be used to make lighter, thinner-walled containers. Other important uses include a range of injection-molded and extruded products.

In 1954 in Germany, chemist Karl Ziegler revealed a process for making polyethylene at atmospheric pressure. The excitement that the announcement generated in the community of chemists was intensified by the properties of the Ziegler polymer. It was essentially linear and had a higher melting point and degree of crystallinity than the branched-chain polymer. In addition, it exhibited significantly higher stiffness and tensile strength.

Not long after the Ziegler revelation, an announcement by G. Natta in Italy transformed the excitement into a scientific and industrial frenzy. Natta had found that variants of the Ziegler catalysts could cause propylene, styrene, and alpha olefins in general to polymerize stereospecifically to crystalline polymers. Both Ziegler and Natta were later to receive the Nobel Prize in Chemistry.

The Ziegler catalysts, like the original ICI polyethylene, were discovered accidentally. Ziegler had been making linear compounds of molecular weights up to about 2,000 by reacting ethylene with aluminum alkyls. The formation of a truly high-molecular-weight (viscosity-average of around 100,000) polyethylene was unexpected and was traced to a metal salt present as an impurity in the reaction vessel. Out of this came the invention whose best-known example is the use of aluminum alkyls plus titanium tetrachloride as catalysts in polymerizing olefins.

Ziegler's research was supported by German industry, and he was thus required to license any German company that wished to practice his invention. Several began rather quickly to do so. Ziegler as an individual, however, had the right to license outside Germany. U.S. companies that acquired a license to his process were Union Carbide, Dow, du Pont, Standard Oil (New Jersey), Hercules, Goodrich Gulf, Koppers, and Monsanto. Hercules in addition had a know-how exchange agreement with the German firm, Hoechst, a Ziegler licensee, and thus was able to take an early lead in the U.S. competition in linear polyethylene.

Ziegler learned, of course, that his linear polyethylene was dominated by the composition-of-matter claims in the Larcher-Pease patent (application) that was to issue in 1957 and be assigned to du Pont. Thus, Ziegler had the process, and du Pont was soon to have a dominating patent. Some Ziegler licensees acquired a license also from du Pont, and Ziegler and du Pont worked out an agreement for sharing royalties.

Despite the excitement over the Ziegler process, it is clear that the German chemist was not the first to have a low-pressure process nor to be aware of linear polyethylene. In addition to Larcher and Pease, chemists at ICI had reported on a linear polymer in 1950. An Australian patent issued and was assigned to Phillips Petroleum in August 1954 on a process for making linear polyethylene at moderate temperatures and pressures. And Standard Oil Co. (Ind.) by 1950 had some promising results using heterogeneous catalysts which led to a series of patents that began to issue in 1954.

The Phillips polyolefins process has had a tremendous economic impact. In 1970 it accounted for 1.5 billion pounds of linear polyethylene produced worldwide. The process grew out of Phillips' discovery in 1950 that olefins could be converted to solid polymers over a chromium oxide–silica-alumina catalyst. The process was well in hand by 1954 when Ziegler announced his own method.

When Phillips decided to license its process, Ziegler had seven licensees in the U.S. but only a laboratory method that each licensee had to scale up. Phillips had a commercial process, however, and offered to license both its patent package and commercial know-how. In 1955 and 1956, the company licensed nine companies in seven countries and in late 1956 started up its own commercial plant. The Phillips-process polymer quickly became the dominant linear polyethylene in the U.S. and the material against which Ziegler product had to compete. The Ziegler polymer eventually closed the gap, but the Phillips process, by now much improved over the original, remains a very important factor.

The linear polyethylene process developed by Standard Oil (Ind.) is quite different from the Phillips process in terms of economic impact. The basic research that led to Standard Oil's patents was original and of high quality. Apparently, however, the company asked a large U.S. plastics manufacturer to evaluate some of its early linear polymers, and the report was sufficiently negative to discourage Standard Oil from a commercial venture until the Ziegler-Phillips explosion occurred. By then it was almost too late. The first plant to use the Standard Oil process was built in Japan, and as of early 1972 Standard Oil planned substantial U.S. production of polyethylene (and polypropylene).

Polypropylene. Polypropylene is made by nine companies in the U.S. and reached the billion pound-per-year level in less than 15 years from the time it was first produced in this country. U.S. output in 1971 was 1.3 billion pounds. About 28% of the polypropylene made is used in filament and fiber and 17% in packaging applications such as film, bottles, and closures. The remainder goes into a range of items including appliances, housewares, pipe, and wire and cable coating.

The polymer's advantages over linear polyethylene are related mainly to its higher (by 40° C.) crystalline melting point. They include strength and toughness, solvent and water resistance, dimensional stability, and gaseous barrier properties.

Propylene is a dominant co-product in the production of ethylene from petroleum fractions. As ethylene production grew, propylene became a drug on the market, and a widespread search began for ways to polymerize the olefin. Polypropylene was made by cationic polymerization at least 25 years ago but was a tacky, semirubbery material with no apparent uses.

The catalyst systems in the Phillips and Standard Oil (Ind.) polyethylene processes could make a crystalline polypropylene, but the polymer apparently was neglected because the dominant interest was linear polyethylene. Early visitors to Ziegler's laboratory brought back a sample of polypropylene that was nontacky and thus different from the cationic product. The difference was ascribed to higher molecular weight and not to crystallinity, because chemists were not accustomed to the concept of stereospecificity (tacticity) in polymers of higher olefins.

Then, Montecatini acquired a Ziegler license for Italy. Natta, a consultant to the company, started a research program on polymerizing propylene and found that polypropylene made by the Ziegler process was partly crystalline. He then discovered special variants of the Ziegler catalyst which would give highly crystalline polypropylene as well as other polyolefins. Ziegler's original invention appeared to cover polypropylene, but Montecatini obtained some significant improvement patents based on Natta's research. As a result, Ziegler and Montecatini set up the so-called Pool Agreement that U.S. licensees would have to reckon with.

Hercules was the first U.S. company to make polypropylene by the Ziegler-Natta slurry process and today is the leading domestic producer. Standard Oil (Ind.), Shell, Standard Oil (New Jersey), Dart Industries, Diamond Shamrock, Eastman Kodak, and Phillips Petroleum also make it.

Role of U.S. Chemists. The work of ICI, Ziegler, and Natta clearly contributed much to the polyolefins technology used in this country, but U.S. chemists have contributed significantly, too. One indication of the fact is the lively patent interference and infringement battle of the past 15 years over process and composition-of-matter claims. It centers mainly on polypropylene. Generally involved have been Ziegler, Montecatini, du Pont, Hercules, Phillips, and Standard Oil (Ind.). The U.S. Department of Justice has surveyed all patent and licensing agreements in polyolefins but thus far has taken no action.

American teams sent to England to learn the ICI know-how on low-density polyethylene were shown a plant where pressure on the reactor was controlled manually by operators working around the clock. Such labor-intensive processes were out of the question in this country, and U.S. chemists had to automate the process before it could be used.

Product competition in the U.S., moreover, was different from and more rigorous than that in Europe. Characterization techniques had to be adapted to the U.S. market. The purity of feed and recycle streams had to be controlled. Special copolymers had to be studied and patented. One source estimated that 440 U.S. patents had been issued in the polyolefins field by the end of 1960. The magnitude of the effort is evident in the fact that each patent usually requires up to three or four man-years of research. The same source estimated the U.S. industrial research and development budget for polyolefins at $30 million in 1960, which may have been a peak year. Major plastics producers who were marketing polyethylene products found a substantial part of their research and development budgets (10 to 30% at the peak) going into polyolefins research.

Since the Ziegler license was to a laboratory process, U.S. licensees had to scale up to commercial size. The Ziegler catalysts are sensitive to moisture and burst into flame in air and thus presented a large array of handling and safety problems to the laboratory chemist, the chemical engineer, and the production manager. A most significant U.S. invention was the use of hydrogen as a chain transfer agent to control molecular weight in Ziegler polymerization. The use of small amounts of comonomers, such as propylene or butene-1, to control chain branching proved good for some uses.

The story has been much the same with polypropylene. U.S. chemists had to learn how to make the polymer in high yield and to control its properties. They developed preferred polymerization catalysts, discovered how to control molecular weight by using hydrogen in the polymerization process, and developed suitable copolymer compositions. Polypropylene is less crystalline than linear polyethylene and much more susceptible to oxidative degradation. Thus, formulations had to be developed to control the polymer's stability on exposure to heat and light. Coloring techniques, using both pigmentation and dyeing, were important developments for polypropylene fibers.

Other Polyolefins. Chlorinated linear polyethylene (CPE) has certain unique properties that make it more desirable for many end uses than the chlorinated branched polymer. CPE was developed in Germany by Hoechst, which licensed the process patent to Hercules. Allied Chemical developed a CPE process under a sublicense from Hercules and put considerable effort into developing U.S. markets for the product. Dow Chemical independently did a process, product development, and market study and at the moment is the only U.S. producer of CPE. It is used as a blending aid between two or more polymers, as a toughening agent in polyvinyl chloride, and as a weather-resistant film in such uses as roofing membranes and liners for irrigation ponds and ditches.

Du Pont chemists in the 1960's developed a patented product line called "Surlyns." They are made by copolymerizing ethylene with 5 to 15% of methacrylic acid at high pressure and then preparing half-metal salts with, for example, lithium. The Surlyns are a new class of hydrocarbon polymers with ionic bonding forces, which improve high-temperature properties and clarity. Du Pont also developed the chlorosulfonated branched polyethylene sold commercially as Hypalon and used as a weather-resistant coating.

Union Carbide chemists studied a whole series of high-pressure ethylene copolymers, of which the most important appear to be those made with acrylic acid and also with alkyl acrylates. Those made with acrylic acid have adhesive bonding applications, and those made with alkyl acrylates are toughening agents for other plastics.

Dow chemists developed an improved process for making high-pressure copolymers of ethylene and acrylic acid for use in adhesive bonding. A new Dow product line, Zetabon, consists of metal foil clad on one or both sides with this acid-containing copolymer. It is used as a cable wrap in making improved communications and power cables.

Styrene Plastics

The styrene plastics, the second largest category of plastics in the U.S., include four types of polymers: polystyrene, impact polystyrene, acrylonitrile-butadiene-styrene (ABS), and styrene-acrylonitrile (SAN). Companies in the U.S. in 1969 produced about 3.4 billion pounds of these materials, valued at $645 million. Output in 1971 was 3.8 billion pounds.

The low cost, relative ease of fabrication, and other properties of styrene-based plastics have made them useful in a broad spectrum of industries and products, including packaging, appliances, construction, automobiles, toys, radio and TV, furniture, housewares, and luggage. The technological base for the styrene plastics was established initially during World War II as a result of the huge expansion in styrene-butadiene synthetic rubber. U.S. styrene production expanded from 2 million pounds per year in 1940 to more than 300 million pounds by the end of the war.

Polystyrene. Polystyrene has been known since 1839, but little was done with it until styrene became cheap and abundant. Of the thousands of scientists who have done basic work on the chemistry and properties of polystyrene, the most significant contributor was a German chemist and Nobel laureate, Hermann Staudinger. In the 1920's and early 1930's he played a major role in building the basic chemical knowledge of styrene and polystyrene.

Today in the U.S., the dominant method of making polystyrene is continuous mass polymerization, in which liquid styrene serves as both monomer and vehicle. Chemists at Dow Chemical and Union Carbide contributed significantly to the development of the process. A second common production method is suspension polymerization, in which drops of styrene suspended in hot water polymerize to polystyrene beads. Chemists at Koppers Co. developed a suspension stabilizing system, based on tricalcium phosphate, that led to the use of suspension polymerization on a large scale.

The phenomenal growth of polystyrene has depended in large measure on its cost, properties, and excellent moldability. High-speed injection molding has been particularly important.

In addition to molded products, polystyrene is used widely as a foam, whose structural and insulating properties give it important uses in construction and elsewhere. Polystyrene foam originated in Europe, but Dow Chemical later developed a continuous process for making it. The German company, BASF, developed a process for making expandable polystyrene beads and fabricating foamed objects from them. Many companies in this country now make polystyrene foams as beads, sheeting, and board. U.S. chemists have contributed to the technology of the blowing agents, fire retardants, and other additives used in these foams.

Polystyrene films and sheeting also have become important commercial products. Many scientists have contributed to the knowledge of antifog coatings, surface slip agents, and surface coatings used on them.

Impact Polystyrene. The brittleness and rigidity of unmodified general-purpose polystyrenes restricted their use in many applications and spurred the development of impact polystyrenes. These materials are toughened by incorporating a rubber in the polystyrene matrix. They were introduced in 1948, and U.S. consumption by now has reached 1.2 billion pounds annually. The growth potential of impact polystyrenes is evident in their use to replace wood in furniture. This application started in 1969. In 1971, it consumed some 90 million pounds of impact polystyrenes.

The first commercial impact polystyrenes in the U.S. were made by Dow Chemical by polymerizing a solution of styrene-butadiene rubber in styrene. This polymer had a significant effect on the plastics market but suffered from surface blemishes or "fish eyes." Dow chemists later improved the process to yield a product with excellent toughness whose extruded surfaces are free or substantially free of blemishes. This modified technology is used widely throughout the world. Impact polystyrene also is made extensively by suspension polymerization of a prepolymerized mass. Chemists at Monsanto Co. contributed to this development.

Styrene-butadiene rubber (SBR) is still used extensively in impact polystyrenes, but stereospecific rubbers, such as cis-polybutadiene, are coming into wide use. They tend to improve the toughness of the finished plastic, especially at low temperature.

Acrylonitrile-Butadiene-Styrene. The first acrylonitrile-butadiene-styrene (ABS) products in the U.S. were developed at U.S. Rubber Co. (now Uniroyal, Inc.) and introduced about 1950. Annual production exceeds 550 million pounds, about half of world production capacity.

The first ABS polymers were very tough, but they also were extremely difficult to fabricate and had poor appearance and stability. A number of companies made incremental improvements in the materials, whose toughness, hardness, and chemical resistance provided a combination of properties that led to use in such forms as pipe and sheet. The real breakthrough came in the mid-1950's at Marbon Chemical Division of Borg-Warner Corp. It was based on graft polymerization of styrene and acrylonitrile on a preformed polybutadiene elastomer substrate. The resulting polymer system had vastly improved moldability, better low-temperature impact strength, and an attractive high gloss on molded parts.

The new ABS polymers soon found outlets in applications such as radio housings and color telephones, and consumption began to expand rapidly. Uses for ABS resins are now numerous, and many special types and grades have been developed to meet specific market needs.

Styrene-Acrylonitrile Copolymer. Styrene-acrylonitrile (SAN) copolymer was developed for two main reasons: the availability of cheap acrylonitrile made by a process developed at Standard Oil Co. (Ohio) in the late 1950's and the need for a polymer with better chemical resistance than polystyrene. The U.S. today consumes more than 100 million pounds of SAN resins annually. The material is used in molded, extruded, and foamed products and as a blending stock in making ABS products.

Vinyl Chloride Polymers and Copolymers

The first vinyl chloride resin was discovered in 1872, but the material remained a laboratory curiosity until 1927. Commercialization of vinyl chloride polymers started after E. W. Reid and his co-workers discovered that copolymers of vinyl acetate and vinyl chloride were more soluble and more readily softened by heat and plasticizers than polyvinyl chloride itself. Reid was working on the Union Carbide fellowship at Mellon Institute, and Carbide opened its first commercial vinyl plant in 1936. Its capacity was 5 million pounds per year. U.S. production today exceeds 3 billion pounds per year, and world production exceeds 12 billion.

Union Carbide's Vinylite copolymer resins found their first industrial uses in surface coatings because of their outstanding resistance to alcohol, gasoline, acids, and alkalies. The use of vinyl resins in can coatings grew rapidly with the introduction of canned beer at the end of Prohibition in the U.S. The growth of these vinyl coatings led to a demand for certain ketone solvents to supplement the then-available acetone and spurred the growth of that part of the solvents industry.

Union Carbide chemists developed heat stabilizers, lubricants, and plasticizers which made possible the hot compounding of vinyl chloride copolymers and their fabrication by molding, extrusion, and calendering. Among the many articles made by these processes were dentures, phonograph records, tile, wall coverings, draperies, upholstery, insulation, and extruded wainscoting.

Chemists at B. F. Goodrich Co. learned to plasticize polyvinyl chloride (PVC) itself, and the company introduced the resulting Koroseal products in the form of tubing, sheet, molded shapes, and insulation. In 1940, Goodrich opened its own vinyl chloride resins plant.

Union Carbide developed dibutyl tin stabilizers, which made it possible to make clear, transparent vinyl film and sheeting. Demand grew rapidly, and the film displaced the flammable and short-lived nitrocellulose-coated fabric. Rigid sheeting was used widely. Flexible copolymer sheeting enjoyed wide popularity in belts, shoes, watchbands, and other novelties until the military demand of World War II cut off the supply of resin.

The U.S. Navy's interest in vinyl chloride resins was spurred by the discovery that experimental vinyl insulation was intact in a recovered sunken submarine, whereas the rubber-coated wiring was destroyed. Further attention was focused on flameproof polymers when the German warship, *Graf Spee,* was crippled by destruction of its electrical controls by burning of rubber insulation. Throughout the war, industry was pushed to supply enough vinyl chloride resin for electrical insulation and for coating of cloth.

The introduction of resins with acid and alcohol groups in the polymer during the 1940's broadened the uses of vinyl chloride coatings to maintenance, underwater finishes, antifouling coatings, and paper and foil coatings. In 1944, Union Carbide introduced organosols and plastisols, dispersions of finely-divided PVC in plasticizer. These materials could be used to coat cloth and paper, and the technique largely displaced solution coating methods. A new industry developed when plastisols were adapted to fabrication of flexible articles, such as dolls and toys, by low-pressure rotational molding.

Vinyl applications grew rapidly in the 1950's, when industry made low-cost resin available in quantity. Plasticized resins found outlets in coated cloth and film for upholstery, drapes, clothing, tarpaulins, and general industrial use. Vinyl chloride polymers captured the phonograph record and tile market almost entirely. Clear, rigid bottles are a growing market, and the nonflammability of vinyl chloride polymers has brought them a growing role in the construction industry.

Methyl Methacrylate Polymers

The first acrylic polymers date from 1901, when Otto Rohm prepared acrylate esters and observed their polymerization. Some 30 years later the first commercial production and polymerization of acrylic esters in the U.S. occurred at Rohm & Haas Co., which in 1936 produced the first cast sheets of methyl methacrylate (MMA) polymers in the U.S. Methyl methacrylate monomer capacity in this country by 1970 had reached 720 million pounds annually, divided among American Cyanamid, du Pont, and Rohm & Haas.

Actual U.S. production of MMA in 1970 was an estimated 480 million pounds, the predominant portion of it being converted to polymer. About 41% of the MMA went into cast sheet, 23% into coatings, 21% into molding powder, and the remaining 7% into oil additives and syrup. The most familiar MMA polymer is the cast sheet (Lucite, Plexiglas). More than half of it is estimated to be used in advertising signs and displays. In addition, cast and extruded MMA polymers are used extensively in breakage-resistant glazing, lighting fixture lenses, and automotive parts such as tail-light assemblies and decorative medallions.

In the late 1930's, du Pont joined Rohm & Haas as a producer of MMA polymers, but development was gradual until World War II. Then, improved manufacturing methods lowered the cost of the monomer. The result was an upsurge in the use of cast sheet, particularly for aircraft canopies and bomber noses.

All commercial production of MMA today involves preparation of acetone cyanohydrin from acetone, conversion to methacrylamide sulfate, and further conversion to MMA. MMA is used in polymers as the sole or dominant ingredient or as a minor component of a copolymer composition, depending on the desired contribution of MMA to the properties of the finished polymer. Homopolymers of MMA are characterized by optical clarity and brilliance, outdoor durability, and good handling properties.

Acrylic copolymers are tailored to particular applications by using MMA with acrylate esters, higher alkyl methacrylates, or other classes of co-monomers.

Many companies in the U.S. now fabricate MMA polymers to fill specific requirements. Inclusion of a phosphorus derivative gives the polymer fire-resistant properties. Elastomeric-based MMA polymers can be used to prepare impact modifiers for polyvinyl chloride resins. Heteropolymers involving a major amount of MMA can be used for free film applications.

Another major use for MMA polymers is in coatings, which can be cast from solution or emulsion. Coatings may be the best example of the chemist's ability to tailor copolymer composition to obtain the desired properties. Oil soluble methacrylate polymers also are used as oil additives for purposes such as improving the viscosity index of petroleum oils.

Domestic consumption of methyl methacrylate and its polymers currently is increasing at about 10% per year. Demand in 1975 will thus about equal 1970 production capacity of 720 million pounds of MMA. This steady growth is due largely to the fundamental properties of MMA polymers, chemists' ability to tailor the properties of the polymers to specific requirements, and continuing research to improve the capabilities of both the monomer and its polymers.

Phenolic Resins

The phenolic resins, a family of thermosetting materials made from phenols and aldehydes, offer a wider range of properties than any other resin. U.S. companies produced 1.1 billion pounds of phenolics in 1971, placing them fourth in production value among plastics in this country. Some 60% of the amount produced is used as bonding and adhesive resins in products that include plywood, thermal insulation, foundry molds, and coated and bonded abrasives. About 28% is used as molding materials and the remainder in coatings and other, smaller outlets.

The German chemist, Adolf Baeyer, reported the reaction of phenol and formaldehyde in 1872. In 1905, in the U.S., a Belgian-born chemist, Leo Baekeland, set to work to develop commercial products based on the phenol-formaldehyde reaction and by 1909 had received patents on the results of his work. In addition to working out the necessary chemistry, Baekeland learned that phenolic resins blended with wood flour produced strong, heat-resistant moldings that did not deteriorate with age. Baekeland's phenolic products, under the trade name Bakelite, found a ready market in molded parts, insulating varnishes, and laminated sheets, particularly in the automotive and electrical industries, which were growing rapidly.

The success of the phenolics, which are still growing after 50 years, is due partly to the fact that their properties can be modified widely by adding other components. The resins themselves are made today not only from phenol and formaldehyde but from substituted phenols such as ortho-cresol, butylphenol, and phenylphenol and aldehydes such as acetaldehyde and butyraldehyde. The reaction products may be liquid or solid and may be modified with fillers or polymeric compounds. One group of phenolics, the novolacs, are thermoplastic initially but are made thermosetting by adding a hardener.

Other Plastics Materials

Of the 21 billion pounds of plastics materials produced in the U.S. in 1971, some 75% was accounted for by the classes of polymers already examined here: the polyolefins, the styrenes, the vinyl chlorides, the phenolics, and the methacrylates. Annual production of the many other commercial plastics materials ranges from 700 million pounds for the unsaturated (nonfiber) polyesters to a few million pounds or even less for some of the more costly, more highly specialized polymers. The full range of these other materials embraces a very large variety of properties and uses and will not be treated exhaustively here. Their economic implications can be seen clearly enough in terms of brief, selected examples.

Unsaturated Polyesters. The unsaturated polyesters are thermosetting resins that harden at close to room temperature under little or no pressure. In general they are made of phthalic anhydride and a dialcohol, such as ethylene glycol, plus smaller amounts of maleic anhydride and styrene or other vinyl monomers. The maleic anhydride serves as a crosslinking agent. The styrene or other vinyls serve partly as crosslinking agents and partly as diluents. Polyesters increasingly are being made fire retardant by additives such as antimony oxide or by building chlorine or bromine atoms into the polyester structure. Often the polyester plastics are reinforced with glass cloth or fiber to which the liquid resin is applied before curing.

The first pilot-plant lot of unsaturated polyester crosslinked with styrene was produced in September 1942 by American Cyanamid as a result of a development agreement with U.S. Rubber and Owens Corning Fiberglas. By the end of World War II, Cyanamid had produced about 5 million pounds of this material, mostly for use in aircraft. U.S. production of polyesters in 1970 was 570 million pounds.

Polyesters are used today to make a variety of products: fishing rods, bathtubs, chemical process equipment, luminescent construction panels, and tank cars. The Chevrolet Corvette has a glass-reinforced polyester body, and the same material is used widely in boat hulls. These polyester products are fabricated in a number of ways, including hand layup, filament winding, and matched die molding.

Polyvinyl Acetate, Polyvinyl Alcohol, Polyvinyl Butyral. While polyvinyl acetate is produced in this country at a rate of more than 400 million pounds per year, none of it is used as a plastic per se. The reason is the polymer's glass temperature of 25° C., which is too high to make it a rubber and too low to permit its use as a rigid plastic. The major use is in latex paints.

Polyvinyl acetate also is a raw material for polyvinyl alcohol. This water-soluble resin (45 million pounds per year) is used in adhesives, sizes, binders, and the like. Also made from polyvinyl acetate is polyvinyl butyral (20 to 25 million pounds per year), the interlayer in automobile safety glass, in its own way one of the most significant inventions of this century.

During the 1930's, the increasing speeds at which automobiles were being driven, as well as the increasing popularity of enclosed cars, developed a demand for a safety glass much more effective than the cellulose acetate laminated glass then used. The solution to the problem required the cooperation of many specialists. Ceramists produced new formulas for glass; engineers devised means of casting, grinding, and polishing plate glass and reducing its thickness; and chemists developed the plastic that served as the interlayer between the glass sheets.

The polyvinyl butyral interlayer was developed by H. F. Robertson and his colleagues at Union Carbide Corp., working closely with the laboratories of Pittsburgh Plate Glass Co. The goal was an elastic material that, in case of accident, would act more like a fireman's net than a rigid plate. The interlayer in addition had to be relatively cheap, remain clear and colorless indefinitely, and adhere permanently to the glass to lessen the danger of flying fragments.

Robertson had become interested in the properties of polyvinyl butyral while collaborating with the National Bureau of Standards in improving the fabric of the gas bags used in dirigibles. The plasticized sheeting developed for safety glass required the invention of a plastic composition with a special plasticizer, triglycol 2-ethyl butyrate, and a vinyl polymer of suitable molecular weight with a definite ratio of butyral and hydroxyl groups as side chains. In 1936, polyvinyl butyral accounted for less than 2% of the interlayer material used in safety glass in this country; by 1941, it accounted for 98%.

Tetrafluoroethylene Polymers. The tetrafluoroethylene polymers, familiar as the Teflon nonstick finish in frying pans, were discovered accidentally in 1938 by a du Pont chemist who has doing research on refrigerants. Subsequent tests showed the material to have very unusual properties: extraordinary chemical resistance, excellent electrical properties, outstanding heat resistance, and unusual frictional properties.

Du Pont pursued the discovery and began small-scale production in 1941. The small amount of polymer produced was preempted at once by the Manhattan Project, where it was used to contain a highly-corrosive fluorine compound used to separate isotopes of uranium for the atomic bomb. Other wartime uses for polytetrafluoroethylene were in nitric acid service in explosive plants, in radar equipment, and in aircraft engines.

Du Pont built a commercial plant in 1948, but Teflon did not become widely known until 1961, when the nonstick frying pans were introduced in U.S. stores. Before that, the uses of the polymer were obscure, even invisible: gaskets in chemical plants, bearings for automobiles, nonstick equipment for candy manufacturers and commercial bakers, seals for rotating equipment, and a number of others.

Epoxy Resins. Among the unusual properties of the versatile epoxy resins are easy cure, low shrinkage, high adhesive and mechanical strength, excellent electrical and electronic insulating properties, and chemical resistance. From experimental amounts in 1948, U.S. production of these thermoset resins reached 165 million pounds in 1970. Their uses include protective coatings, reinforced plastics, adhesives, and electrical applications such as potting compounds.

The composition of the epoxies may be varied widely to achieve the desired properties, but the resins are based in general on the reaction of bisphenol A with epichlorohydrin. Early workers on epoxy resins were Pierre Castan in Switzerland in the mid-1930's and S. O. Greenlee of Devoe and Raynolds (now Celanese Coatings) in the U.S. starting early in 1939. Greenlee's basic concept was to use a wide range of unsaturated acids to make resins of controlled molecular weights for selective applications.

Many epoxy producers have contributed to the technology that grew out of the Castan-Greenlee patent position. Among them were the largest U.S. producer, Shell Chemical, and the second largest, Dow Chemical. Other U.S. manufacturers include Celanese, Ciba Corp., Reichhold Chemicals, Resyn Corp., and Union Carbide.

Polyvinylidene Chloride. Polyvinylidene chloride, best known as saran, was discovered in Europe in 1838. It was rediscovered in the early 1930's at Dow Chemical during the development of the salt-bath direct chlorination process for making tetrachloroethylene. The polymer's unusual resistance to solvents and chemicals made it a candidate for use in molded battery cases, an application for which it proved to be too costly. The concept did, however, precipitate the Dow research and development program that produced saran plastics.

U.S. production of vinylidene chloride–based polymers today is about 100 million pounds per year. The material's properties make it useful in certain types of packaging, especially for meats and poultry; in piping for chemical process operations; and in fibers for use in filter cloth, screens, upholstery, and the like.

Saran has been largely a Dow development, starting with a block of 18 U.S. patents issued to Dow inventors in 1939. One of these covered the production of molecularly oriented films and filaments by quenching saran from above its melting point into ice water and then stretching the polymer to induce crystallization. Another patent covered saran "shrink films," an extruded copolymer of vinylidene chloride and vinyl chloride that shrinks upon heating. Another Dow development was a multilayer packaging film with saran film in the center, polyethylene film on the two outer faces, and two adhesive layers of ethylene–vinyl acetate copolymer. The product combines the gas and moisture barrier properties of saran with the easy heat sealability of polyethylene.

Cellulose Plastics. The cellulosic plastics—cellulose acetate, cellulose acetate butyrate, and cellulose acetate propionate—have significantly affected the growth of the U.S. plastics industry. They offer an exceptional combination of properties: clarity, colorability, processability, hardness, strength, toughness, and (with ultraviolet stabilizers added) weatherability. The cellulosics are used in a variety of industries, including communications, automotive, electrical, home appliances, packaging, sports equipment, and toys. U.S. production of these plastics was about 182 million pounds in 1970.

Eastman Kodak Co. cast photographic film base from cellulose acetate as early as 1908, after G. W. Miles developed partially hydrolyzed cellulose acetate that was soluble in common, inexpensive solvents. Not until 1927, however, did Celanese Corp. introduce cellulose acetate sheets, rods, and tubes. Thermoplastic compositions based on cellulose acetate were marketed in granular form in 1929. These compositions and the injection-molding process, developed about 1920, initiated today's injection-molding industry. Cellulose acetate was the first thermoplastic to be injection molded with heat and pressure.

Chemists at Hercules Powder Co. and Eastman Kodak developed cellulose acetate butyrate, introduced in 1938 by Tennessee Eastman. Celanese Corp. introduced cellulose acetate propionate in 1945.

Ion Exchange Resins

The synthetic ion exchange resins, though not plastics in the ordinary sense, grew directly out of polymer research in the mid-1930's. Ion exchange resins are used today for a variety of purposes, including softening and deionizing water and as a method of separation in chemical analysis and many industrial chemical processes. They have special utility in chemical synthesis, medical research, food processing, and elsewhere.

Ion exchange is the reversible interchange of ions between a solid and a liquid phase with no permanent change in the structure of the solid or resin. The important point about the ion exchange resins is that they can be used and reused through many cycles. An ion exchange water softener, for example, exchanges its sodium ions for the calcium ions in the water. When the resin is loaded with calcium, it can be removed, regenerated with sodium chloride solution to its original sodium form, and used again.

The principles of ion exchange have been known for many years, and industry began to use natural ion exchangers (zeolites) in about 1910. In 1935, the British chemists B. A. Adams and E. L. Holmes discovered the first organic ion exchange resins, the phenolics, during their research on the condensation of polyhydric phenols and polyamines with formaldehyde. A few years later, Rohm & Haas introduced the resins into the U.S., and the early 1940's saw extensive work on the synthesis, properties, and uses of phenolic ion exchange resins.

Ion exchange was transformed from an art to a science with the discovery of styrene-divinylbenzene resins by G. F. D'Alelio at General Electric Co. These resins were introduced commercially by Dow Chemical in about 1947. The styrene-divinylbenzene resins had a uniform gel structure achieved by controlled crosslinking, a single active exchange group (SO_3H) of high acidity, and spherical particles. These characteristics permitted a quantitative measure of the equilibriums and kinetics of ion exchange and the identification of the mechanism of cation exchange as an equilibration of two electrolyte phases. Later, Dow and Rohm & Haas developed concurrently the use of quaternary amine groups in the styrene-divinylbenzene structure to produce a comparable anion exchange resin.

As a result of these and subsequent developments, ion exchange resins today are precision materials with porosity, particle size, and functional groups adapted selectively to the end use. They have displaced inorganic ion exchange materials almost completely.

Newer Commercial Polymers

Although the major classes of commercial plastics and resins date largely from World War II and earlier, important new polymers have been discovered in the U.S. and commercialized in the past decade or so:

- General Electric's Lexan polycarbonate in 1960.
- Du Pont's Delrin polyacetal (polyformaldehyde) in 1960.
- Celanese Corp.'s Celcon, a copolymer of formaldehyde with a small amount of ethylene oxide, in 1963.
- General Electric's PPO (polyphenylene oxide) in 1965.
- Union Carbide's parylenes [poly (p-xylylenes)] in 1965.
- Union Carbide's polysulfone in 1965.

The relatively high cost and unusual properties of these newer polymers seem likely to preclude their use in tremendous volume, although U.S. production capacity for the polyacetals (Delrin, Celcon) has reached 150 million pounds per year. These materials do, however, have great if somewhat specialized utility. Lexan's uses include structural parts, meter faceplates, and prosthetic devices. The polyacetals often substitute for metals, as in hardware and appliance parts. Polyphenylene oxide substitutes for die-cast metals and is used in dielectric components and hospital and laboratory equipment. Parylene coatings, among other uses, promote dropwise condensation on heat transfer surfaces, eliminating condensate films and thus markedly increasing heat transfer efficiency. Polysulfone has certain desirable mechanical properties and is used widely in the appliance, automotive, electrical, and other industries.

FUTURE DEVELOPMENTS

Trends exist in plastics materials and, more broadly, in polymer chemistry that can reasonably be extrapolated two or three decades hence. As in any field of science, however, developments unthought of now will almost certainly materialize in the next 20 to 30 years.

Pertinent here is the situation at the time of the basic discovery that led to the Lexan polycarbonate resins. The consensus of the experts in 1953 was that polycarbonates could be made but that they would fall apart in water. A General Electric chemist recalled later that, "If you read the chemical literature, and believed everything, we would never have made Lexan." There is no sound reason to believe that the same sort of thing will not happen again.

Whatever new polymers are discovered, however, seem unlikely to combine the broad utility and low cost required to rival plastics such as the polyolefins or polystyrenes in production volume. Much more likely is the advent of new, specialized materials such as the polycarbonates, the polyacetals, and others that have been commercialized in the past decade or so.

Although industrial chemists already have developed a very considerable body of technology in the commercial development of plastics materials, their role does not appear to be diminishing. This is so for a variety of reasons. End-use patterns for plastics are changing. Faster speeds of fabrication require better control of molecular architecture. Food and Drug Administration regulations are growing more stringent. Competition is becoming stiffer, and economic conditions are changing.

In addition, new materials and techniques are becoming available. They include:

- Electron beam curing of coatings on metals.
- Injection molding of thermoset resins.
- Methods of achieving wood-grain appearance in plastic furniture by using high-density plastic foams.
- Techniques for making multilayer plastic films.
- Plastics that bar passage of carbon dioxide for use in replacing glass bottles for carbonated beverages.

The Environment. It is much too early to predict in detail how plastics will be affected by the general concern over ecology. It seems certain, however, that all chemical processes involved must be revised either to avoid production of undesirable waste products or to eliminate such products if they are formed. In addition, chemists worked for many years to develop plastics that resist attack by bacteria, fungi, sunlight, ozone, and oxygen. Today, ironically enough, many talk of the need for biodegradable and/or sunlight-degradable polymer systems to solve the plastics litter problem.

It is apparent from the technical literature that chemists already have demonstrated ingenious schemes for solving some of these problems. Social pressures and legislative developments will have much to do with determining whether such schemes are commercially practical and with the course to be taken in research in the area.

Nonflammability. Emphasis on the flammability of plastics is increasing sharply for several interrelated reasons: local and national legislation, action by federal regulatory agencies, concern of the Fire Underwriters Association, and policy set by insurance companies. The Flammable Fabrics Act, which applies to wearing apparel, is an example of federal legislation in this area. Flammability also is a factor in local building codes, which govern the use of plastics in construction.

All major producers of plastics materials are working hard on the problem. The trend is focusing increased emphasis on nonburning plastics such as polycarbonates, PVC, saran, and fluorinated polymers. Use is growing of chlorine-, bromine-, and fluorine-containing co-reactants to make polyester and epoxy resins nonburning. Companies are searching intensively for additives to make the pure hydrocarbon polymers—polyethylene, polypropylene, polystyrene—either self-extinguishing or nonburning. Research on these new additives or co-reactants, of course, involves a complex interplay of factors such as effects on physical properties, requirements of the Food and Drug Administration, ecological matters, and economics.

Living Polymers. A concept that holds important implications for industrial polymer chemistry is the so-called "living polymer," discovered in the mid-1950's by Michael Szwarc at Syracuse University. The use of the concept in industry is widespread already and seems likely to become much more so.

The living polymer concept has to do with the manner in which the growth of a polymer chain is terminated. In conventional polymerization processes, the growing chains are terminated in random fashion, and unreacted monomer continues to add to chains that are still growing. The molecular weights of the chains in the finished polymer thus vary within a range determined by the monomer(s) and the conditions of polymerization. With the living polymer method, an electron donor is used to maintain the ends of the polymer chain in an active condition. As long as unreacted monomer is present, the chains continue to grow. It is possible also to feed a living polymer of monomer A with monomer B to form a block copolymer, in which a string of B units follows a string of A units.

The use of living polymers thus gives the industrial chemist much greater control of the polymerization process than in the past. The polymer can be grown to any size by feeding fresh monomer, and the rate of growth can be regulated by the rate of feeding. The chemist can produce polymers whose chains are of more nearly uniform molecular weight instead of the conventional types in which the chains are distributed over a range of molecular weights. He can make polymers with functional end groups which will undergo some further desired chemical reaction. Different monomers can be distributed at will along the chain to form block polymers of any desired pattern. The incompatibility of the blocks linked by chemical bonds creates well-defined and regularly spaced domains in otherwise homogeneous materials. The organized structure thus acquired provides important new properties.

Chemists at Shell Development Co. have used the living polymer concept to make tri-block copolymers of styrene-butadiene-styrene. These give the so-called thermoplastic rubbers, which mold like thermoplastics and "cure" at lower temperatures without the need of chemical crosslinkers. Chemists at Dow Chemical have used the concept to develop several materials and processes: a carboxyl-terminated polybutadiene initiated by methylstyrene dimer and a process for making the material; a method of obtaining polystyrenes with molecular weights in excess of 500,000, as opposed to the normal range of 100,000 to 200,000; and a method of making soluble (noncrosslinked) styrene polymers containing halogens such as chlorine or bromine in the ortho position, as in 2-bromostyrene or 2,6-dichlorostyrene.

Phillips Petroleum has made living polymers of butadiene capped (terminated) with carboxyl groups to give long-chain dicarboxylic acids. These materials are useful as binders for rocket fuels and as toughening agents for epoxy resins.

Polymeric Oil-in-Oil Emulsions. Polymeric oil-in-oil emulsions are a new class of emulsions. They consist of immiscible solutions of polymers in organic solvents or monomers, with graft or block copolymers as emulsifying agents. The emulsions were discovered in the early 1960's by G. E. Molau, who found them first in high-impact polystyrenes.

The polymeric oil-in-oil emulsions have two key aspects: the finding that graft and block copolymers have colloidal properties that enable them to stabilize organic dispersions and the novelty of emulsions without water. The discovery of these new colloidal systems led to the understanding of the formation of high-impact polystyrenes and to new ways of making such polymers and improvement of existing methods.

The emulsifying action of block or graft copolymers can be visualized in general in terms of a system comprising two homopolymers, poly-A and poly-B, a mutual solvent, and a block copolymer, AAAA-BBBB. (The actions of graft and block copolymers are analogous.) When the two homopolymers are dissolved in the mutual solvent, two distinct layers of polymer solution are formed, one containing poly-A, the other, poly-B. If a very small amount of the block copolymer, say, 0.5 to 1%, is added to this system, and the layers are mixed by agitation, one layer of solution disperses into the other in the form of tiny discrete droplets. The resulting emulsion is very stable. Its stability depends on the molecular weights of the homopolymers and the nature of the chain ends of the block copolymer.

The use of the novel colloidal properties of graft and block copolymers has been extended from plastics into the dispersion of solid particles in organic media. Commercial use can be made of the phenomenon wherever stable dispersions in organic media are required. A typical application is the stabilization of pigment dispersions in oil-based paints and stains.

Many new uses of polymeric oil-in-oil emulsions can be visualized in the future, when a greater variety of graft and block copolymers will become more readily available. It appears that the potential of these new colloidal systems is just beginning to be tapped today.

Synthetic Polymers in Catalysis. Homogeneous catalysis offers certain advantages in chemical processes, but it has the disadvantage that the catalyst can contaminate the product. Such contamination can be avoided, while retaining the advantages of homogeneous catalysis, by passing solutions of chemical reactants through columns packed with swollen, catalytically active polymer beads.

The first studies of this approach were made in the U.S. and involved acid-catalyzed solvolysis (reaction of solute with solvent) in the presence of ion exchange resins. The work demonstrated that the reactive substance can be partitioned between the solution and resin in a manner that may make acid catalysis in the resin more efficient than catalysis by the same concentration of acid in homogeneous solution. It showed also that considerable selectivity in reaction may be achieved, particularly as between reactants of different charge. Similar results may be achieved with nonionic resin beads (which are not ion exchangers) carrying catalytically active nucleophilic (electron-sharing or transferring) substituents such as pyridine residues.

The results of studies in which soluble, catalytically active polymers are used may lay the groundwork for the design of new catalytic resins. Such polymers exhibit high catalytic activity if they carry catalytic substituents or a chelated catalytically active metal ion. Extremely large catalytic effects have been observed for reactions involving two cationic or positively charged species in dilute solutions of polyanions or negatively charged polymers. It is also possible to enhance the catalytic effect by concerted action of two catalytic groups carried by a polymer molecule.

Particularly strong potential lies in resins with enzymes or natural catalysts bound to them. Such enzymes are highly stable, and their optimum operating pH may be shifted from their optimum when in solution. An enzyme-resin column has been used to separate the two isomers of a synthetic amino acid.

Polymers of Amino Acids as Foods. The total synthesis of ribonuclease in 1969 indicates that it would be possible to synthesize proteins that occur naturally in food. To do so would be difficult and expensive, however, because of the great number of steps involved in building up molecules from 21 different amino acids. It might be more feasible to make simpler polymers of amino acids as food. The eight essential amino acids (10 for some purposes) could be used with only one or two instead of 11 or 13 of the nonspecific or dispensable amino acids, since the body can use the latter interchangeably.

Such a polymer, to serve as a nutritionally complete protein, should meet three requirements:

- It should contain and yield on digestion the essential amino acids in the same proportion as in high quality animal protein such as whole egg.
- It should contain and yield on digestion nonspecific amino acids in the same ratio to the essential amino acids as in the high quality protein.
- It should be capable of being produced in a form or forms of acceptable taste, consistency, and texture.

In addition to polymers serving as complete proteins, there might also be nutritional uses for acceptable polymers of one, two, or three different amino acids. Such polymers, which probably could be made by relatively simple means, might be used to supplement incomplete natural proteins. Logistically, for example, the protein of rice might be supplemented more practically by a suitably proportioned copolymer of lysine and threonine than by the pure amino acids. The nutritional value of simulated meat made from spun soy protein might be increased very significantly by adding a polymer of methionine of sufficiently high molecular weight to be tasteless. Methionine is used regularly now to supplement soybean meal and other oilseed products for animal feeding, but its use in human food is limited by its taste, which some people find objectionable.

Any consideration of the use of amino acids or their polymers as food must take into account the problems of synthesizing the individual acids. In addition, the amino acids occur in two structural forms, the D-isomers and the L-isomers. Except for methionine and phenylalanine, only the L-form is physiologically active, and it must thus be separated from the D-form.

Studies of the origin of life on the primitive earth suggest that it might be possible to bypass the production of individual amino acids. A nutritionally adequate polymer might be made instead directly from simple, abundant raw materials in a single operation. In research on reactions that might have occurred on the primitive earth, chemists passed mixtures of methane, ammonia, and water vapor over silica at about 1,000° C. They obtained a product that contained all of the amino acids common to protein except those containing sulfur, which was not among the reactants. Other U.S. chemists have studied the polymerization of hydrogen cyanide in the presence of alkali. From the resulting polymers they obtained 12 of the amino acids that occur in natural proteins.

The yields obtained in these investigations were small, and the individual amino acids in the products were not in the proper proportions for use as food. The results suggest, however, that further research might establish conditions in which simple raw materials might be converted directly to polymers nutritionally equivalent to high quality natural proteins.

Biological Activity of Biopolymers. "Biopolymers" are defined here as polymers derived from biological systems or synthetic polymers that show the biological activity of biologically-derived polymers. Key patents in this field are still in review, and direct widespread utility has not yet been demonstrated for biopolymers. These materials nonetheless display an exceptional range of new physical and biological properties.

One area of research on biopolymers involves interferon, a protein in higher animals that constitutes an important part of the natural protective mechanism against viruses. Interferon has been isolated and characterized, but the manner in which it inhibits the replication cycle of a large number of viruses is not well understood. It was shown in the mid-1960's, at any rate, that synthetic pyran–maleic anhydride polymers administered to animals can stimulate the release of interferon with resultant antiviral protection. Since then, a number of both natural and synthetic polymers have been found that release or induce interferon. They function, presumably, by mimicking the entry into the cell of "foreign" ribonucleic acid, triggering the interferon response. The toxicity and antiviral effects of these biopolymers are sufficiently promising that clinical studies in man are anticipated.

Considerable research also is under way on synthetic polypeptides or polymers of amino acids smaller than proteins. This work falls into two general areas: homo- and copolypeptides and fixed-sequence polypeptides.

The work on homo- and copolypeptides has emphasized mainly the correlation of the physical and biological properties of polypeptides with particular side chains. The hope is that the side chain behaves similarly when it is part of a protein sequence. The use of only one or a few amino acids in a synthetic polypeptide greatly simplifies these correlations. It results, however, in loss of the subtle and specific interactions between the various parts of a fixed-sequence polypeptide or protein that contribute heavily to the structure and function of a polymer. Nonetheless, hydrodynamic and spectral studies of synthetic copolypeptides have yielded important information.

Biological properties also have been correlated with side chains using synthetic homopolypeptides, particularly in studies of tryptic digestion and generation of antibodies. Some synthetic homopolypeptides in addition have antiviral and antibacterial properties. The utility of these materials appears to be limited, however, by problems of toxicity and administration.

Peptide hormones, which are fixed-sequence polypeptides, are essential and extremely active biological substances. For this reason, and because they are relatively small compared to proteins, they have become the first targets for chemical synthesis of fixed-sequence polypeptides. Many of the naturally-occurring hormones, as well as analogs and derivatives, have now been synthesized. Of particular importance, medically, is the synthesis of a derivative of the hormone ACTH that has much of the biological activity of the hormone and a greatly decreased tendency to stimulate antibodies. Also medically important is what appears to be the correct synthesis of human growth hormone.

4 / **Textile Fibers**

SOCIAL AND ECONOMIC IMPACTS

One day in 1935, a small group of chemists gathered around a simple hypodermic syringe in the laboratory of E. I. du Pont de Nemours & Co., in Wilmington, Del. Led by Wallace Carothers for the past seven years, they had sought to make large organic molecules from smaller, simpler ones and had investigated their properties. The syringe held a molten polymer of hexamethylene diamine and adipic acid. One of them pushed the plunger, and a thin, fiber-like stream extruded into the air, cooled and solidified, and was wound on a small spool. The fiber and textile industries throughout the world have never been the same since.

The du Pont chemists had spun what others in the company would shortly name nylon. At first, nylon served as a plastic, not as a fiber, replacing hog bristles in toothbrushes in 1937, and another two years passed before the company sold the first nylon fibers (in 1939 in women's stockings). While nylon eventually became an important engineering plastic, it is as a fiber that it has caused the greatest change in the U.S. and elsewhere.

Other synthetic fibers followed nylon in short order from chemical companies in the U.S. and overseas. I. G. Farbenindustrie in Germany sold a slightly different nylon commercially in 1939 (Perlon, based on caprolactam instead of diamine and dibasic acid). In England, Calico Printers Association produced the first polyester fiber (Terylene) on a pilot-plant scale in 1948. Union Carbide began selling an acrylonitrile–vinyl chloride fiber (Dynel) in 1949, and du Pont followed with an acrylonitrile fiber (Orlon) in 1950. Du Pont bought American rights to Calico's polyester in 1946, produced experimental quantities in 1950, and began full-scale commercial production (of Dacron) in the U.S. in 1953. Du Pont added a urethane fiber (spandex) in 1959, and Hercules produced textile-grade polypropylene fibers (Herculon) in 1961.

Today there are 17 man-made fibers[1] available to the world's textile industry. That number, however, does not tell the whole story of the man-made fiber revolution, for within each of the 17 main types are up to several thousand variations, each individually tailored to meet the requirements of a specific end use.

Many innovations have helped revolutionize per capita consumption of fibers in the U.S. A century ago (in 1870) per capita fiber consumption was just over 13 pounds per year. By 1920, a half century later, improvements in cotton and wool and growing affluence of the country had raised per capita consumption to nearly 30 pounds a year. Today, following the post–World War II boom in new fibers and fiber treatments, per capita consumption lies just shy of 45 pounds per year.

Per capita growth may be dramatic, but the changing pattern of fiber consumption is even more so. In 1870, some 40 million Americans used nearly 515 million pounds of cotton and perhaps 190 million pounds of wool. By 1920, the nearly 106 million Americans used about 3.1 billion pounds of fiber—2.7 billion pounds of cotton, about 400 million pounds of wool, and about 15 million pounds of rayon.

---

1. As defined by the Federal Trade Commission—see table on page 88. Four of the 17 are not at the moment produced commercially in the U.S.

Man-Made Fibers Today

Just over a decade ago, Congress passed the Textile Products Identification Act to bring order to a business that had become somewhat chaotic, with a proliferation of fibers, a jumble of trademarks, and sharp claims in the marketplace. Under terms of this act, which took effect in March 1960, the Federal Trade Commission named 17 "manufactured" fibers generically, and manufacturers must use these generic names on textiles. The names, monomers or polymers, the first commercial U.S. producer, and the year introduced in the U.S. follow in the table.

Generic Name	Monomer or Polymer*	First Commercial U.S. Producer	Year
Rayon	Regenerated cellulose	The American Viscose Corp. (now FMC Corp.)	1910
Acetate	Cellulose acetate (80% acetylated)	Celanese Corp. of America (now Celanese Corp.)	1924
Rubber	Natural or synthetic	U.S. Rubber Co. (now Uniroyal, Inc.)	1930
Glass	Glass	Owens-Corning Fiberglas Corp.	1936
Nylon	Polyamide	E. I. du Pont de Nemours & Co.	1939
Vinyon	> 85% Vinyl chloride	The American Viscose Corp. (now FMC Corp.) and Union Carbide Corp.	1939
Saran	Vinylidene chloride	Firestone Plastics Co. (now Firestone Synthetic Fibers and Textiles Co.)	1941
Metallic	Metal, coated metal, and metallic coated	Dobeckmun Co. (production since absorbed by Dow Badische Co.)	1946
Modacrylic	35-85% Acrylonitrile	Union Carbide Corp.	1949
Acrylic	> 85% Acrylonitrile	E. I. du Pont de Nemours & Co.	1950
Polyester	Ester of dihydric alcohol and terephthalic acid	E. I. du Pont de Nemours & Co.	1953
Triacetate**	Cellulose acetate (97% acetylated)	Celanese Corp. of America (now Celanese Corp.)	1954
Spandex	Urethane	E. I. du Pont de Nemours & Co.	1959
Olefin***	Ethylene, propylene, or other olefin	Hercules Powder Co. (now Hercules, Inc.)	1961
Anidex	Monohydric alcohol/ acrylic acid ester	Rohm & Haas Co.	1969

---

* Indicative only, as some definitions are more complex than shown.
** Actually a subclassification of acetate, but placed here chronologically.
*** Hercules produced olefin monofilaments for specialized uses in 1949, but the first production of textile-grade monofilament (of polypropylene) occurred in 1961.
In addition to the above man-made fibers there are three others—azlon, nytrile, and vinal—that are no longer produced commercially in the U.S. and one—lastrile—that has not been produced commercially to the present.

By 1970, total consumption by the then 205 million Americans had risen to nearly 9.2 billion pounds. Man-made fibers at just under 5 billion pounds outranked cotton and wool at about 4.2 billion pounds (for the third year, 1968 having been the go-ahead year). Cotton, of course, was still the No. 1 single fiber (at 3.9 billion pounds), but its per capita position had slipped from a high of nearly 31 pounds in 1950 to just over 19 pounds in 1970. The noncellulosic man-made fibers in the same 20 years, meantime, had risen to nearly 3.6 billion pounds (17.5 pounds per capita), and they will probably pass cotton within a year or two.

Americans have also used silk during all these years, consuming 3.3 million pounds, as a matter of fact, in 1881, the first year for which statistics are available. However, since little chemistry is involved in silk's use and since silk has virtually disappeared from the U.S. market, we shall not cover it, except to note here in passing that in earlier times it was a luxury fiber for the rich. In more recent times—between the two world wars, say —it achieved some popularity in the U.S. despite the fact it is hard to sew and hard to care for. Use peaked in 1930 at just under 90 million pounds— about 0.7 pound per capita—and it has steadily declined since then, amounting to a minuscule 900,000 pounds in 1970.

Such numbers show economic impact, of course, but what probably matters more to the individual is how he meets his need for clothing and shelter today compared with yesterday. For example, most people now have diverse summer and winter wardrobes instead of the limited ones of earlier times. Carpets, draperies, and other fabric furnishings surround us. Colors seldom fade, garments retain their sizes and shapes, and moths and mildew cause little damage. People now often discard clothing and furnishings because fashions change more than they do because of simple wear. Equally important, housewives may save as much as a day per week caring for clothing compared to half a century ago, and their physical labor certainly is reduced sharply. Fabrics come clean more easily than ever before, and many clothes and some household furnishings dry nearly wrinkle free, requiring touch-up ironing at most compared to much more laborious ironing some years ago. Yet with all these improvements, clothing (including shoes) accounts for less than 9% of spending for personal consumption compared to 12% nearly a half century ago.

Consumer goods lead industrial goods as a market for fibers in the U.S. about four to one, with some 7.8 billion pounds used for consumer goods and 1.8 billion pounds for industrial goods (in 1968, latest year for which such figures are available). Within consumer goods, apparel ranks first at 3.8 billion pounds, with home furnishings (mainly bedroom and bathroom supplies, floor coverings, upholstery, and draperies) second at 2.9 billion pounds. Other consumer goods (such as toys, luggage, hospital supplies, and shoes) account for the 1.1 billion-pound balance.

Among major industrial markets, tires take the most—580 million pounds. Reinforced plastics rank second at nearly 260 million pounds, and the balance of nearly 1 billion pounds goes into a diverse array of products such as hose, rope, belting, tarps, bags, filter cloths, and the like.

CHEMICAL ACCOMPLISHMENTS

Few consumer industries deal with materials quite so complex as do the natural and man-made fibers and textile industries. Not only do these industries have an array of fibers and fiber processes to choose among, but they must also contend with a wide variety of uses for fibers. Fabrics must be durable yet light and flexible. They must be permeable in some cases and not in others. They mustn't burn quickly. They should shed wrinkles readily and wash clean easily. Their dyes shouldn't fade, and they should resist oxidation, water, bacteria, fungi, insects, and light. They have to be aesthetically appealing, and they ought to help mask or reshape what are unsightly bulges in their wearers' eyes. Industrial markets impose many of the same requirements and add a few of their own.

Man used furs as clothing at first in cooler climates, while vegetable fibers and leaves served the same purpose in warmer ones. No record exists of who spun the first yarn or wove the first cloth, but textiles undoubtedly existed long before the earliest written records found so far. Archeologists working in India appear to have found fair evidence of cotton fabric and string at sites dating to about 3000 B.C., and some speculation has been published that Middle Egypt (about 2000 B.C. to 1600 B.C.) grew and used cotton. Certainly as long ago as 2,400 years a cotton industry flourished in India. At that time, for example, Herodotus wrote about "trees which grow wild (in India), the fruit of which is a wool exceeding in beauty and goodness that of sheep . . ." and noted that Indians made their clothes of "this tree wool." People in the Western Hemisphere also used cotton long ago. Investigations have unearthed cotton specimens in Peru dated to 2500 B.C., for example, and evidence also indicates very early use in Mexico.

Wool and cotton have been the most widely used fibers up to recent times, although man has used other fibers as well to some extent. Regardless of the fibers used, however, the fiber and textile industries were for much of history mechanically oriented industries. Of course, they achieved remarkable results in earlier years (and still do, for that matter) by balancing friction, tension, and the like as they twisted fibers into yarns and wove yarns into fabrics. In these earlier years, the fiber and textile industries depended on chemistry only very little, at least in the ways chemistry is thought of today. Then toward the end of the 18th Century and the beginning of the 19th, chemistry in its more modern form began to assume a more important role. In those years, dilute sulfuric acid began being used to neutralize scouring mixtures, and synthetic alkali replaced vegetable ashes. Chlorine's discovery in 1774 led to its use for bleaching, and introduction of bleaching powder in 1798 provided a more effective process. Also during the first half of the 19th Century, dyeing calico printing and mercerization were introduced. Then in 1856 William Henry Perkin produced the first synthetic dye (aniline purple or mauve) and introduced additional chemical technology to the fiber and textile industries.

While such advances were important in their times, it was mainly when the chemical and fiber industries began making fibers themselves to substitute for those growing naturally and then to modify chemically both the man-made and natural fibers that chemistry began profoundly influencing the way people clothe themselves, furnish their homes, and use textile products industrially.

The new chemical fibers succeeded so dramatically and changed clothing, household furnishings, and industrial textiles for two very simple reasons: They last longer, and they are easier to care for. Nylon killed silk for women's stockings, and it and the polyesters and acrylics that followed soon successfully invaded other apparel markets as well as household furnishing and industrial markets. To be sure, the new fibers didn't succeed completely in every use in their early growth years of the late 1940's and early 1950's. Decidedly hydrophobic and lacking reasonable moisture transport, they felt uncomfortable in underwear and shirts. Strong beyond previous fiber experience, they sometimes formed unsightly pills at abrasion points in stockings, sweaters, and suits. While they washed easily and dried with few wrinkles, they often turned yellow or gray and lost their desired bright whiteness after several washings.

As these and other shortcomings cropped up, fiber chemists began modifying polymers, finding new treatments for fibers, and blending different fibers together in new ways. In time they largely overcame many of the difficulties. U.S. fiber consumption continued to change markedly as a result, both in volume and in the relative amounts of different fibers used.

The story of the fiber revolution in the U.S. is a chemical story not only for the man-made fibers but for the natural fibers as well. The new man-made fibers from nylon's day on exist largely because of what chemists have done, and much of the glamour that fibers have in consumer eyes focuses on nylon and its chemical successors. However, the natural fibers of cotton and wool, used for centuries and long dependent on the mechanics of spinning and weaving for performance innovation, and the first man-made fibers of rayon and acetate, used for less than a century, deserve to share the glamour. Cotton, wool, and the cellulosics today in many ways hardly resemble their forerunners, thanks to chemistry's having made essentially new fibers of them.

Chemical innovations for the natural fibers, however, tended to come slowly. In cotton's case, for example, they occurred with few exceptions largely because the chemical fibers threatened—and often took much of—cotton's supposedly secure markets. The new fibers not only outlasted cotton but most brought a new concept to textiles—easy care. They washed easily and dried quickly and largely free of wrinkles. As it happened, the chemical innovation that would let cotton compete successfully in a market where consumers eventually clamored for easy care had its inception before the first chemical fibers were spun. The innovation made its greatest impact, in the U.S. at least, well after the chemical fibers did, however. Interestingly enough, a textile company (Tootal Broadhurst Lee in England) did the pioneering work, one of not very many examples of a major chemical innovation by a textile company and not by a chemical company seeking markets in the textile industry.

The U.S. textile industry in the mid-1930's began treating cellulosic textiles to make them resist wrinkles. It was not until 1948, however, that the concept received much public attention here. In that year the U.S. textile industry claimed production of no-iron nylon. As consumers embraced this new (to America) sales feature, the textile industry gradually extended the idea to cotton. Textile finishers put no-iron cotton goods on the market in 1954, and by the following year they were selling at least 50 no-iron fabrics with such claims as wash-wear, drip-dry, and the like. By the end of the decade the industry was treating nearly 60% of the woven apparel fabric in the U.S. to give it no-iron properties.

Development of easy care treatments for cotton, rayon, and their blends and invention of man-made fibers are without doubt the two most significant contributions of chemists and chemical engineers to the fiber and textile industries and thus to helping meet needs for clothing and shelter. There have been many other developments, however, that while not quite as significant, have nonetheless made an impact. A host of specialty fibers followed nylon, the polyesters, and the acrylics to offer greater choice to consumers and industry. Fabrics tend to burn, often distressingly easily, and many scientists have worked on treatments to prevent their doing so. Much effort has gone into dyes and dyeing, and today fabrics woven from one or more fibers can be dyed one or more colors in a single dyebath. Dirt, oil, and water repellents abound, as do treatments to improve hand.

to make fabrics soft, and to prevent static buildup. Chemists have also found ways to make fabrics more resistant to degradation by light, microorganisms, and oxidation.

Chemists, chemical engineers, and physicists have found ways to give man-made fibers a built-in crimp to make fabrics woven from them bulkier and more resilient. They have also given the man-made fibers new, nonround cross sections in the battle against soiling, undesirable luster, and limp hand. Bonded and laminated fabrics produce a composite structure that can be handled as if it were a single fabric. Such fabrics are self-lining, warm without weight, and strong enough for easy care although the outer layer might have delicate attractiveness. Nonwovens are making an impact on diapers and in hospitals and other institutions where disposal replaces simple inconvenience or labor-intensive laundering.

Rayon

Fibers made from regenerated and chemically modified cellulose were the first of the man-made fibers to become and remain commercially important. Rayon, which is regenerated cellulose, was made commercially in Europe in 1891 and in the U.S. in 1910. Cellulose acetate, which is cellulose acetylated to the extent of 80%, became commercial in Europe right after World War I and in the U.S. in 1924. Cellulose triacetate, which is cellulose with 97% of its hydroxyl groups acetylated, became commercial in the U.S. in 1954.

Chemistry's impact on fibers and textiles generally is considered to date from the end of the 19th Century when French chemist Count Hilaire de Chardonnet invented rayon as a textile fiber and produced it commercially. De Chardonnet began experimenting in 1878 and within half a dozen years had made his first fiber by coagulating nitrocellulose in heated air. He showed articles woven from the fibers at a Paris exhibition in 1889, got financial backing for a factory, and began commercial production in France in 1891. The chemical revolution in fibers was under way.

Across the channel, meantime, English chemists C. F. Cross, E. J. Bevan, and C. Beadle improved on de Chardonnet's original nitrocellulose process in 1892 by discovering the viscose process for making rayon. Courtaulds bought the U.K. patent rights to the process in 1904 and began using it commercially there the same year.

Fiber scientists have made a number of significant innovations in rayon down through the years. Among the more important are permanently crimped rayon staple, rayon tire cord, and flame-retardant rayon. Crimped rayon has become a multimillion-dollar business in the U.S., while rayon tire cord through a continuing series of improvements has retained a significant share of the competitive tire cord market. Although still under development, flame-retardant rayon is becoming increasingly important with the growth of federal regulations controlling flammability of fabrics.

American Viscose (now a division of FMC Corp.) developed permanently crimped rayon and offered it for sale in the U.S. in 1943. For the first time it gave rayon a wool-like helical crimp that stretching or washing and drying did not destroy. In addition to adding greater bulk to rayon fabrics, it allowed rayon to be carded at a higher speed, a significant production saving. Today, crimped rayon staple is supplied by most producers throughout the world.

It is made from a filament whose cross section is unbalanced, with one side having a thin skin or no skin. Because the skin is more elastic than the core, the thick-skinned side shrinks more after stretching than does the other side, and it thus takes the inside bend of the crimp. The crimp forms instantly during production when the stretched rayon is cut into staple.

Tires have long represented a major fiber market. Until the early 1930's, cotton had a monopoly. Then in 1933, du Pont introduced Cordura, a high-tenacity rayon filament, and cotton's place in the market was destined to disappear. Tires made from Cordura and from similar rayons introduced by other producers were stronger, lighter, and lasted longer. By 1946, tire-makers were using as much rayon cord as cotton cord, and cotton soon was phased out altogether.

Rayon's preeminence in this market was itself destined for attack, and rayon cord sales peaked in 1954 at 440 million pounds. The advent of nylon tire cord during the early 1940's and of polyester cord during the 1960's gradually eroded rayon's position. Only the continual development of stronger, more fatigue-resistant rayon and its inherent economic advantage have enabled rayon cord to compete with the newer fibers. Development of new ways of coagulating and stretching the rayon filaments as they are formed and the use of special additives to slow the coagulation of the viscose have given super-high-tenacity rayon tire cords that are stronger, resist abrasion better, and flex longer without breaking. Now the trend toward bias-belted and radial tires is expected to improve the competitive position of rayon tire cord, for a time at least.

The need for flame-retardant cellulosic fibers in apparel, especially in children's nightwear, and in bedding, upholstery, draperies, and carpets has long been recognized. Cotton was the first beneficiary of research in the area, and work in the 1950's at the U.S. Department of Agriculture's Southern Regional Research Laboratory led to the development of flame-retardant finishes that could be applied to cotton or rayon.

Because rayon is a regenerated fiber, fiber chemists have turned to the possibility of incorporating durable flame retardants into the fiber itself rather than of applying them topically. Accordingly, research at American Viscose beginning in 1948 aimed at finding such materials that would be compatible with the viscose process and that would produce fibers that dyed well, had good hand, withstood laundering and dry cleaning, didn't deteriorate in light, had low toxicity, and did not cause dermatitis. By the mid-1960's, American Viscose had a fire-retardant rayon that contained tris(2,3-dibromopropyl)phosphate. It satisfied some of the criteria, but it degraded in ultraviolet light and wasn't stable to alkaline processing and laundering.

Research at FMC Corp.'s American Viscose Division later led to a selection of certain alkoxyphosphazenes as flame retardants for rayon. They are well suited for use in the viscose process and give excellent fire-retardant rayon fibers. Potential markets for these rayons are in bedding, children's nightwear, upholstery, draperies, carpeting, and a host of other applications where flame retardancy is needed.

Cellulose Acetate and Triacetate

Acetate fiber ranks fourth in production of all man-made fibers in the U.S., being surpassed only by rayon, polyester, and nylon. Because of its uniform quality, color, versatility, drape, hand, and other desirable aesthetic properties, acetate goes mostly into women's apparel and home furnishings. Its most important nontextile use is as filter material in cigarettes.

Cellulose acetate as a chemical material was known as early as 1865, but half a century was to pass before two Swiss chemists, Camille and Henri Dreyfus, recognized the industrial importance of a fiber less hygroscopic and more water-resistant than rayon. The Dreyfuses successfully spun acetate fibers in England shortly after World War I ended, and a company founded by them in the U.S., predecessor of today's Celanese Corp., produced the first acetate fiber here on Christmas Day in 1924.

Fiber and textile chemists and engineers have since modified the original manufacturing techniques for acetate considerably. These modifications especially involve pretreating the cellulose for esterification, precipitation, washing, drying, and solutioning of the polymer. U.S. production of acetate and triacetate (discussed in the following paragraph) totaled nearly 670 million pounds in 1970.

Cellulose triacetate, trademarked Arnel, was an American innovation by Celanese in 1954. Interest in easy care fabrics was the spur to its success. The properties of hydrophobicity, resistance to wrinkling, dimensional stability, high crystallinity after heat setting, and rapid drying allow triacetate fibers to impart to fabrics the desirable properties of easy care.

Triacetate finds wide acceptance in pleated dresses and skirts. Because of its chemical nature and the properties induced through heat treatment, triacetate is more closely allied to, and competitive with, the noncellulosic man-made fibers. It is most often used in cross-dyed blends with rayon.

Contributions by chemists and chemical engineers that were necessary for successful commercialization of triacetate include the acetylation procedure, the mode of precipitating the polymer for efficient washing and drying, solutioning of the polymer, dyeing procedures, and heat treating fabrics.

Polyamides

The research that led to nylon began in 1928 at du Pont's Experimental Station in Delaware. After examining polymers made from dibasic acids and dihydric alcohols, the du Pont scientists turned to hexamethylene diamine and adipic acid. They spun their first fiber in 1935, and by early 1937 they had produced enough fiber to knit the first experimental stockings. The company coined the name nylon for the polymer and introduced it later that year in toothbrushes, as mentioned earlier.

It was as a fiber, however, that nylon was destined to make its greatest impact. Nylon stockings went on sale in Wilmington, Del., in December 1939, and du Pont began full-scale fiber production the following month. Nylons, as they were soon called, were on sale in all principal cities by May, and retailers sold some 15 million pairs that first year.

Nylon promptly entered markets provided by foundation garments, underwear, and dress lace, but within a few months the war cut off civilian sales. Nylon's high tensile strength and toughness suited it well for such outlets as parachutes, tires, and tents, and all nylon went to military uses until war's end.

After the war, sales of nylon hose boomed, thanks to the sheerness, strength, and durability that nylon provided well beyond that attainable with silk or rayon. Nylon's properties, however, suited it for more than hosiery. Its success in other markets was particularly rapid as manufacturers were able to apply quickly the experience they had gained during 30 years of producing the first man-made fibers, rayon and cellulose acetate. Chemists began tailoring nylon to fit other end uses. They modified the polymer's molecular weight and changed the concentration of end groups on the individual polymers. They also improved its morphology with new drawing processes and crystallization sequences, developed new nonround cross sections and other methods to control luster and cover, and introduced mechanical bulking and tufting processes. Thus modified, nylon fibers with widely varying properties successfully invaded other markets, including knit and woven apparel, tires, carpets and upholstery.

From a mere 2.7 million pounds in 1940, nylon production in the U.S. rose to nearly 1.4 billion pounds in 1970. For three decades, nylon led among noncellulosic man-made fibers, but in 1970 polyester fibers finally passed it with production of just slightly more than 1.4 billion pounds.

Polyesters

The fastest growing man-made fibers today are the polyesters. First introduced in England on a pilot-plant scale in 1948, polyesters are now being made in more than 30 countries around the world at a rate of more than 3 billion pounds a year. In the U.S., polyesters rank first in production among the noncellulosic man-made fibers, having passed nylon for the first time in 1970. Some in the industry are now estimating that U.S. production of polyesters will reach 4 billion pounds by 1979, not only surpassing all other man-made fibers but also cotton as well.

Work at du Pont provided the basis for polyester fibers when investigations of macromolecules in the 1930's showed dibasic acids and polyhydric alcohols could be polymerized. The du Pont scientists turned to the polyamides, however, as more likely to have the desirable properties for forming fibers. Meantime, chemists J. R. Whinfield and J. T. Dickson at Calico Printers Association in England, having studied the du Pont results in detail, took up where du Pont left off. In the early 1940's, they succeeded in showing that the polyester from terephthalic acid and ethylene glycol could be aligned and crystallized by forming fibers from the molten polymer.

Calico, not in a position to develop a fiber process, sold its rights to Imperial Chemical Industries, which introduced the fiber in England as Terylene. Du Pont bought U.S. rights in 1946, produced experimental quantities in 1950, and began full-scale commercial production in 1953 under the trademark Dacron.

As with the other man-made fibers, a number of modifications have been introduced. For example, scientists at Tennessee Eastman Co. (division of Eastman Kodak Co.) produced a polyester of basically different chemical composition. Although not successful in apparel, it is particularly suited for carpets. The polyester from 1,4-cyclohexane dimethanol has a bulkier molecule than polyethylene terephthalate. Carpet fiber made from it has a decided increase in bulk and covering power. Trademarked Kodel and introduced in 1958 by Tennessee Eastman, it is somewhat easier to dye than polyethylene terephthalate. It also has good abrasion resistance, a soft and luxurious hand, and easier cleanability.

Of all the advances scored by polyesters, one of the most striking ones in recent years has been their invasion of the lucrative tire cord market. Of the individual markets for man-made fibers, apparel ranks a strong first (at about 1.8 billion pounds in 1968) and carpets second (at 950 million pounds in 1968). Tires, however, are third (at 575 million pounds in 1968 and ahead of drapery, upholstery, and slip covers at 357 million pounds). Unlike the other uses, however, tires represent a very cohesive market, production being dominated by the Big Five of Goodyear, Firestone, Goodrich, Uniroyal, and General. Whatever fiber captures this market captures an enviable (if highly competitive) one.

Goodyear Tire & Rubber Co. introduced the first polyester-reinforced passenger car tire in 1962. Before 1962, tires were either reinforced by cellulosic fibers or by nylon, neither of which was completely free of problems. Early research indicated that the chemical structure of polyethylene terephthalate could combine strength, fatigue resistance, and a good ride for general use as a tire cord. However, two important deficiencies of poor adhesion to rubber and low strength-to-cost ratio retarded penetration into the tire yarn field. Development of new chemical adhesive systems by fiber producers and tire manufacturers solved the adhesion problem, and improved processes and greater production efficiencies eventually led to competitive prices.

Early in 1962, du Pont introduced a new polyester yarn especially for tires (Type 68 Dacron). Its particular combination of fiber structural parameters that gave high strength, fatigue resistance, and excellent ride aesthetics led to its early acceptance in conventional bias-ply automobile tires. Other manufacturers have now also introduced high-strength polyethylene terephthalate tire yarns.

In the mid-1960's, American tire manufacturers introduced a new type of tire construction that offered 40 to 50% longer tread life. Named bias-belted tires, they were made with carcasses reinforced with polyester cords and treads reinforced with belts of glass fibers. They soon became standard as original equipment on most American automobiles. Polyester's greater overall dimensional stability compared to that of rayon or nylon and its lack of nylon's annoying flatspotting spurred rapid expansion of sales of polyester tire yarns to the 200 million pound-per-year level by 1970. Half of all passenger car tires made in the U.S. and 85% of all original equipment tires in that year were made with polyester cord.

Acrylics and Modacrylics

Acrylic fibers were first produced commercially in 1949 by du Pont, followed by Chemstrand (now Monsanto Textile Co.) in 1952, Dow in 1958, and American Cyanamid in 1959. These fibers are now used extensively in carpets, sweaters, and outerwear. They offer warmth, softness, and luxuriousness. Resistance to degradation by sunlight, weather, chemicals, and microorganisms also makes acrylics particularly suited for such applications as indoor-outdoor carpets, awnings, filter fabrics, and sandbags.

The potential of fibers based on polyacrylonitrile was recognized in the 1930's. Fiber chemists were thwarted in their attempts to make fibers, however, because they could neither melt nor dissolve the polymer. Then chemists at I. G. Farbenindustrie spun the first acrylic in 1938, but their solvent was not practical for commercial production. Finally, chemists at du Pont succeeded in 1942 when they prepared experimental fibers by dry-spinning from dimethyl formamide solutions.

In the two decades since its introduction, polyacrylonitrile has benefited from many new developments. Among the most important of these are the wet-spinning process, which offered new flexibility in fiber production, polymer compositions for improved dyeability with basic and acid dyes, bicomponent fibers with superior bulk and resilience, and pigmented products for maximum color stability.

The present U.S. capacity for acrylic fibers is about 600 million pounds per year, about one third of worldwide capacity. Domestic shipments totaled 475 million pounds in 1969, of which the carpet industry consumed about 150 million pounds and apparel and industrial markets the rest.

Modacrylic is the generic name for fibers composed of 35 to 85% acrylonitrile. Recognizing the limitations imposed by the heat sensitivity of earlier man-made fibers based on vinyl chloride copolymers, scientists at Union Carbide turned their attention to ways of overcoming the deficiency. Copolymerization of vinyl chloride with acrylonitrile proved to be an effective way to increase flame resistance, and the first modacrylic fiber was developed on an experimental basis during the 1940's. Union Carbide introduced continuous filament yarns (Vinyon N) in 1948 and staple fiber (Dynel) in 1949. Dynel has subsequently replaced the filament yarn.

Since the introduction of modacrylics in 1948, use by the textile industry has increased steadily. Use was 10 million pounds in 1960 and rose to an estimated 45 million pounds in 1969. They are not used alone but are blended with other fibers. Blending lowers cost, but probably more importantly, modacrylic fibers resist burning to such a degree that blends with other fibers meet the recently imposed safety requirements specified by federal regulations for carpeting.

Polypropylene

An Italian chemist, Giulio Natta, announced in 1954 that he had made an isotactic polymer of propylene that was highly crystalline and could be transformed into fiber. Hercules Powder Co. (now Hercules, Inc.) produced the first textile-grade monofilaments in the U.S. in 1961. Polypropylene proved difficult to dye at first because of its lack of dye receptors and general impermeability to dyes. Chemists accordingly had to find ways to modify it before it could become successful commercially. Three companies that did much to overcome the problem were Uniroyal, Union Carbide, and Hercules.

In June 1963, Uniroyal scientists found a propylene fiber that was completely dyeable in conventional equipment by conventional dyes. Their simple way of making the fiber more permeable to dyes consisted of using a fiber that contains a receptor for a basic dye and treating it with a Lewis acid after spinning. Five months later, chemists at Union Carbide announced they had made a dyeable polypropylene fiber by adding a dye receptor before melt spinning. In this same period, Hercules and Reeves also attained some success by incorporating nickel or aluminum salts to provide dye sites for metallizable disperse dyes. Finally, Hercules in 1966 found a polypropylene fiber that was dyeable with disperse dyes and in 1970 announced an acid-dyeable propylene fiber.

Polypropylene is the major polyolefin fiber in the U.S. It is widely used in carpets, carpet backing, and cordage. Its production is not listed separately but is included with polyethylene and listed as olefin fiber production. In 1960, U.S. companies made nearly 14 million pounds of olefin yarn and by 1969 had increased their output to just over 200 million pounds.

Durable Press

To chemists, the development of easy-care treatments for cotton has meant modification of natural polymers to make virtually new fibers of them. To economists, the development has meant the revival of the hard-pressed cotton industry. But to consumers, on whose acceptance success depended, the development has meant a dramatic decrease in the need for home ironing.

One of the faults of cotton (and rayon) fabrics is that they wrinkle when washed and dried. Relatively weak forces in individual fibers hold the many small fibrils of cellulose together. The fibers swell when wet, and the fibrils and/or adjacent cellulosic chains slip into new alignments. Then when the fiber dries it assumes a different shape, and the fabric wrinkles.

The textile industry around the turn of the century first used formaldehyde to control shrinkage in rayon fabrics. From that beginning, it gradually extended the concept of using chemicals to modify cellulosic textiles to make them dimensionally stable and easier to launder. Making garments that don't wrinkle has evolved from simple treatments improving wrinkle resistance only slightly to ever more complex ones that have led in successive stages to the labels easy-care, wash-and-wear, and now today to durable-press garments.

Many chemists in government, university, and private industrial laboratories have contributed, and it is difficult to recognize them all. However, R. P. Foulds, J. T. Marsh, and F. E. Wood at Tootal Broadhurst Lee Co. in England during the late 1920's probably developed the first commercially applicable finishes for wrinkle-resistant cellulosic fabrics. Working for this medium-size textile company, Foulds had set out after World War I specifically to improve cotton's resistance to creasing. Within half a dozen years he and his associates found they could make cotton fibers shed wrinkles by first impregnating them with certain monomers (they used urea-formaldehyde) and then polymerizing the monomers to resin.

Early workers assumed the monomers polymerized within the cotton fibers to provide wrinkle recovery. More recent work has indicated instead that the resins crosslink adjacent fibrils or cellulosic chains. Regardless of the mechanism, fabrics dried smooth after washing. Moreover, the method gave cloth that retained much of the desirable hand and comfort of untreated cotton, unlike earlier methods to produce wrinkle resistance. They had coated the fibers and made the fabric stiff and uncomfortable.

Tootal Broadhurst Lee introduced its process commercially in England in the early 1930's, and U.S. textile companies began similar treatments soon thereafter. Nearly two decades passed, however, before resin treatment of cotton received much public attention, and it was not until 1954 that U.S. textile finishers promoted no-iron cotton goods on the market.

Following Foulds' innovation, the textile industry first treated cotton fabrics but later extended the treatment to rayon and then as they appeared to blends of cotton and rayon with other man-made fibers, especially the polyesters. The easy-care fabrics of the 1970's differ markedly from those of the mid-1950's, of course. It was soon clear that the early treatments led to fabrics that tore and abraded more easily, soiled more readily, and soon lost whatever no-iron properties they had after not too many washings. For the past decade and a half, therefore, chemists have sought to overcome such difficulties. Developing different treating agents and curing methods and also blending cotton and rayon with polyester fibers in time increased strength and abrasion resistance. In addition, treating fibers with still different chemicals made them soil less easily and release dirt more readily when washed. Finally, new resin systems eventually put much more permanence into permanent press.

Of the original failings of resin-treated cellulosic fabrics, loss of no-iron properties and reduced resistance to soiling were doubtless the most serious. Smooth-drying properties greatly outweighed loss of wear life to many a housewife, and she willingly traded a little durability for freedom from ironing. However, if that freedom from ironing soon disappeared in the wash water she was none too happy, and chemists set about improving launderability. They eventually found that no one treatment worked in all cases, and today some half a dozen leading contenders divide the market among them.

The early crosslinking agents used in England, such as dimethylol urea or dimethylol ethylene urea, efficiently crosslinked the fibrils and/or adjacent cellulosic chains, and the garments dried satisfactorily smooth (provided finisher and garment manufacturer did a good job). In the rush to get into production, Americans adopted these agents, and they shortly fell afoul of American laundering practices. American dedication to chlorine bleaching caused the urea-based resins to pick up chlorine. If the chlorinated fabric were scorched during drying, it released hydrochloric acid, not the most desirable chemical to have in a fabric. Moreover, although the advantage touted for wash-wear garments was home laundering, the textile industry soon found that many garments did go to commercial laundries occasionally or frequently. Commercial laundries end with an acid extraction (or "acid sour") to remove metal stains and residual soap. Such a step, however, also efficiently removed the resins. Without resins, the cotton fibers returned to their old ways and accumulated wrinkles or shrank like ordinary, untreated cotton fibers.

Changing the laundry habits of a nation was obviously a hopeless task, so chemists turned to modifying the treatment so it could resist abusive laundering. In 1961, chemists at the Department of Agriculture's Southern Regional Research Laboratory in New Orleans revealed their success with dimethylol alkyl carbamate crosslinking agents. They resisted alkaline laundering, chlorine damage, and acid souring of commercial laundering. Furthermore, the new agents were not expensive, they were not difficult to use, and white fabrics could be treated without obtaining off-shades.

While some chemists were working at improving strength and resistance to abrasion or at devising finishes that did not wash out soon, others looked for still new advantages to be developed from the original concept of resin finishes. Their work resulted in finding that if they wanted creases (as in trousers) or pleats (as in skirts) they could dry certain monomers in the cloth, make and press the garment into shape, and then crosslink the monomers with cellulose. After washing and drying, the creases and pleats reappeared where set and the flat areas resumed their smooth appearance (with brief touch-up ironing at most). This development occurred in the mid-1960's when improved cellulose reactants, polyester/cellulosic fiber blends, and radical changes in textile processing combined to make possible what we now call durable-press garments. Major contributors to the development were Koret, Levi Strauss, and Sun Chemical.

While carbamates are the most recently developed crosslinking agents, the largest share of the cellulose reactant market is held by dimethyloldihydroxyethylene urea. DMDHEU is used now in both the precure and the delayed cure durable-press market. It shows little or no chlorine damage when properly applied and has a very low free-formaldehyde content. The latter is important in overcoming the odor problem.

Wool

Wool is obtained primarily from sheep and lambs, with smaller quantities being obtained from Angora and Cashmere goats (and still smaller quantities from camels, alpacas, llamas, and vicunas). Its use as a textile fiber occurred before that of cotton, and it may well have been the first fiber to be woven by man into textiles (although there is some possibility that flax had that distinction). Highly suitable in its natural state as a textile fiber, wool is still considered by many to have the best combination of properties for comfort in apparel, at least in temperate and cold climates, despite strong competition from man-made fibers. Naturally resilient, resistant to wrinkling and dry soiling, having good moisture transport properties, and least flammable among the natural and man-made organic fibers, wool is used largely in apparel, with household furnishings (carpets, blankets, upholstery) a close second.

Wool has long been an important fiber for the U.S. textile industry, although its position today has dropped rather sharply from former ones. The first wool artisans arrived in the American colonies (in Massachusetts) in 1643, and in time wool production and weaving became an important American industry. Consumption a century ago (in 1870) had reached about 190 million pounds (4.9 pounds per capita), and growth more or less paralleled population growth in the U.S. (except for abnormalities during war and depression years) until the peak year of 1946. At that time, consumption reached nearly 740 million pounds (about 5.3 pounds per capita), but it has declined steadily since then, falling to 240 million pounds by 1970 (1.2 pounds per capita).

The decline in wool consumption in the U.S. stems from a number of reasons, but wool's price relative to man-made fibers and to combinations of man-made fibers with cotton is probably among the most important. The decline, however, also stems from the fact that other fibers and combinations of fibers, especially those modified with chemicals, offer easier care in the home. In addition, the U.S. climate is also a factor, since wool is not suited for use in the heat of American summers.

As with the other natural textile fibers, the history of innovations in the production and use of wool has been largely associated with mechanics and deals mostly with such processing methods as carding, combing, topmaking, worsted drawing, felting, napping, and shearing. As with the other natural fibers, however, the textile industry in earlier times did make some use of chemical technology, as in scouring and dyeing, for example. In addition, in the 1800's and earlier a process known as fulling combined a soaking in warm soapy water with mechanical beating to help improve the appearance and properties of just-woven cloth. Also in the 1800's, a process invented in Germany made use of carbonization to remove vegetable matter from raw wool. Despite these and other applications of chemical technology, however, it has only been roughly in the past four decades or so that chemists have investigated wool in detail, and it is only within relatively recent years that the information they have developed has indicated the considerable potential that chemical technology has for the wool industry. To be sure, the improvements in wool's properties that application of chemical technology has made and will make possible will not lead wool to surpass its man-made competitors in the total textile market, but appropriate chemical modifications should assure wool a better market in the future. In this connection, incidentally, it is perhaps not inappropriate to note that the situation in wool research and development is somewhat typical of the paradoxical position that chemists often face. Thus, the decline in wool consumption has in part been caused by the development by chemists of new synthetic polymers. Now, it is this same technology, when applied intelligently and expanded by chemists, that offers much of the hope for wool to remain competitive with the synthetics.

John B. Speakman and his associates at Leeds University in England are generally credited with pioneering in studying the physical and chemical properties of wool fibers. In the years since they began in the early 1930's, they and many others primarily in England and later in Australia and the U.S. have done much to elucidate the chemistry of wool keratin, with particular attention to the existence and extent of covalent crosslinking and hydrogen bonding, the amphoteric nature of the molecules, and the reactions of various chemicals with the great variety of functional groups present in wool fiber molecules. Since wool is a protein, scientists in the wool industry have also profited from studies by others on the nature and structure of proteins in general, and the literature on wool contains many references to the findings of such Nobel Prize winners as Linus Pauling, Frederick Sanger, and Francis H. C. Crick.

Chemically, wool is a complex protein consisting of an extremely large number of about 20 different amino acids. The exact configuration of the protein molecule's structure has yet to be determined, although X-ray analyses have presented many clues. Also still to be fully determined are the exact sequences of amino acids in the protein molecules. In connection with the latter determination, incidentally, chemists have faced considerable difficulty in isolating definite protein fractions from the parent material. What progress they have made is due in great part to the development of partition chromatography in 1941 by Archer J. P. Martin and Richard L. M. Synge at the Wool Industries Research Association in England. (Martin and Synge received the Nobel Prize in Chemistry in 1952 for their work. Subsequent modifications to chromatography have made it one of the most potent and useful analytical methods available to chemists in all fields.)

While wool in an unmodified state has many outstanding attributes as a textile fiber, it does have a few drawbacks. Of these, its tendency to felt and shrink when washed and agitated in water and its inability to retain permanently a pressed and creased appearance have been the most serious for apparel applications. Studies on the fundamental structure and chemical nature of wool fibers, therefore, have had some important practical implications for the wool industry, since they have led to chemical treatments that provide both permanent set and resistance to shrinking.

Speakman developed the first commercial process for permanent set wool, and it became available for commercial use in 1936. More recently, work in Australia has resulted in a process in which solutions of such chemicals as ammonium thioglycolate or monoethanolamine sulfite are sprayed on a garment and the creases set by steam pressing with an iron or in a steambox or autoclave.

Shrinkproofing is also a matter of extreme concern to the wool industry. The scales on individual wool fibers form during growth with the scale edges pointing toward the tip and away from the root. This structure results in what is called the directional friction effect, in which the coefficient of friction in the tip-to-root direction is greater than in the root-to-tip direction. Under the influences of heat, moisture, and mechanical agitation (such as in washing), the wool fibers slip irreversibly past one another, and the fabric shrinks.

The textile industry uses such felting shrinking to advantage when it produces highly fulled flannel fabric and felt hats. Such shrinkage is highly undesirable in machine-washable garments made from knitted or woven fabrics, however, and chemists have undertaken much research to solve the problem.

Two general ways have emerged from such research to minimize the felting tendency brought about by the directional friction effect. In one, the irreversible slippage is made considerably more reversible by modifying the fiber surfaces with chemical treatments. In the other, slippage is prevented in the first place by fixing the fibers more or less in place in the fabric. Treatment with chlorine, developed in the 1920's, was one of the early approaches to reversing the directional friction effect. The process, however, was difficult to control, and more recently the textile industry has modified its approach by providing controlled chlorine generation with the use of salts of dichlorocyanuric acid. A combination of chlorination with a surface deposition of synthetic resins is now frequently used on an industrial scale to provide machine-washable wool garments. Work is now well under way to utilize the polymer deposition alone for this purpose.

Among other chemical treatments applied to wool, formaldehyde has been used to make wool resist alkaline solutions. Chlorine treatment not only has been used for shrinkproofing, as just mentioned. It also gives wool a desirable luster in some uses and makes it faster dyeing. Treatment with substituted sulfonic acids containing chlorine has been one fairly satisfactory approach to retarding damage by microorganisms (such as moths and carpet beetles). Even more effective, however, has been the use of an insecticide, dieldrin. Dieldrin, unfortunately, is one of the persistent chlorinated hydrocarbons that have become serious concerns to an environment-conscious nation. Its use in mothproofing apparel and carpets may well be banned in the near future. If it is, those who benefit from mothproofing treatments appear to face a serious problem, for no effective and permissible substitute is now available. Wool may be particularly susceptible to moths and carpet beetles because of the presence of cystine-disulfide crosslinkages. Therefore, modifications of those linkages offer one possible solution, such as conversion of the linkage to such forms as a bis-thioether crosslink reported by U.S. chemists some years ago as a possible way to mothproof wool.

Chemists have also been concerned with processes for preparing raw wool for spinning and weaving and for dyeing. Among their current concerns are the possibilities for shifting from detergent systems to organic solvent systems for removing wool fat and other contaminants to provide clean, uniform fibers. A factor in their concern will obviously be the possible impacts on the environment of residues remaining for disposal. In dyeing, a matter of much interest is producing fabrics with brighter colors to allow wool fabrics to compete with the brightly colored fabrics now woven from other fibers. Wool is not normally susceptible to such dyeing, while both cotton and man-made fibers are with the use of fluorescent dyes. Chemists have now developed a dyeing system that permits the use of fluorescent colors on wool that are not harmed by moisture and do not yellow, as was the case earlier.

Coloration

Developments in dyes and dyeing technology have played a major part in increasing interest in apparel and household textiles and, hence, in the growth of the textile markets. In cotton textiles, key innovations have been synthesis of dyes of improved washfastness and developing process technology by which these dyes can be applied continuously. The development of anthraquinoid vat dyes starting about 1900 opened the door. As marketed, these dyes have no affinity for cotton, but on reduction to the leuco form are substantively absorbed by the fiber. After penetration, the dye is oxidized to the insoluble pigment form, yielding high wash- and lightfastness.

This process was initially a batch process. However, under the impetus of World War II and the attendant needs for large quantities of cotton fabrics dyed durably to a single color, du Pont developed the continuous Pad-Steam Process. This process enabled mills to dye all-cotton fabrics more uniformly and with high productivity at rates of up to 150 yards per minute. The key was the use of steam to promote rapid and efficient reduction of the vat dye into its leuco form and rapid diffusion of the leuco form into the cellulosic fibers.

Penetration of the cotton market by polyester/cotton blends required competitive continuous processes for imparting color. The development of the du Pont Thermosol Process opened this market for polyester fibers. In this process, dry heat is used to transfer dye from the surface of the cotton fibers to the polyester fibers and to diffuse in them. Aligning the Pad-Steam Process to follow the Thermosol Process permitted continuous application of disperse and vat dyes and thus a one-pass dyeing of both blend components at high speeds.

Disperse Dyes. In the mid-1920's, the advent of cellulose acetate, which the water-soluble dyes then available would not dye, resulted in the development by European chemists of low-molecular-weight water-insoluble compounds. These compounds dyed the fiber by forming a solid solution in it. They were expensive, however, and they had only mediocre lightfastness. Moreover, the lightfast blues resisted the oxides of nitrogen very poorly.

In the late 1930's, American companies succeeded in preparing less expensive but adequate dyes and in the early 1940's improved considerably the lightfastness of many of the shades. These advances resulted in increased acceptance of cellulose acetate in many fabrics, including apparel.

The poor resistance of lightfast blues to oxides of nitrogen (gasfastness) remained a problem until solved by American chemists about 1950. This time they imparted fiber affinity to some chemicals that until then had been without uses in the textile industry and imparted resistance to gas fading.

These were also the first days of the struggle to introduce polyesters to the textile industry, and the gasfast blue dyes for acetate also had satisfactory gasfastness on polyesters as well. Polyesters also required the invention of many new disperse dyes to meet new requirements, such as resistance to sublimation, to the effects of permanent-press finishes, and to harsher laundry conditions.

All of the new man-made fibers use disperse dyes to some degree. Polyesters and cellulose acetate are dyed almost exclusively with disperse dyes. These dyes are also used extensively on nylon, where they have the unique ability to cover imperfections introduced during production and processing. They are used on acrylics and modacrylics in light and medium shades. They are also used for some types of modified polypropylene.

Dye Variances and Multifiber Piece Dyeing. In the steps from fiber to fabric, the most critical one is dyeing. In most mills, profit or loss depends in good measure on costs in the dyehouse. Producers of man-made fibers therefore continuously seek ways to modify fibers and reduce dyeing costs. One way in the recent past that has been successful is to produce a series of fibers having the same physical characteristics but differing in the way they accept dyes. The mill can thus combine fibers in weaving, so that a single dyeing step colors them. Using this method, the mill may produce multiple-color styling from the same piece of fabric merely by modifying dyebath formulations. Doing so saves the mill both time and the need to carry large inventories.

The first multifiber fabrics to be piece dyed were woven from cotton and either wool or silk. By selecting the correct dyes and dyeing conditions, mills could obtain three different effects: union dye both fibers to the same color; reserve one fiber by leaving it undyed while dyeing the other; and dyeing each fiber to a different color.

The first step came after World War I when cellulose acetate fiber was introduced and proved to be undyeable with the dyes then known. In 1924, the simultaneous development of suspended, water-insoluble, dispersed azoic pigments and of various insoluble amino derivatives of anthraquinone brought about an entirely new class of dyes, the disperse dyes.

These disperse dyes are now extremely important in piece dyeing combinations of cellulose acetate, polyamides, polyesters, and polyacrylonitriles. Fortunately, many of the disperse dyes did not color ordinary cellulose, whether cotton or rayon. This difference enabled stylists to design fabrics that were easily dyed into almost any color combination thinkable.

The possibilities for multicolor dyeing multifiber fabrics expanded enormously as each new synthetic fiber came onto the market. Nylon, for example, exhibits affinity for acid dyes (the kind normally used to dye wool and silk), for the disperse dyes developed for dyeing cellulose acetate, and for some direct dyes used to dye cotton and rayon. The facts that some direct dyes can be prevented from dyeing nylon and that some acid dyes do not dye normal cellulose or cellulose acetate created colorful opportunities for the fabric stylists. Similar variations arose with the introduction of the acrylics, polyesters, and cellulose triacetate.

Along with new fibers, new dyes extended the range of older dye types. Among them was a new class of dyes, the reactive dyes, so called because of their ability to form a covalent bond by reaction with the hydroxyl groups on cellulose. Thus, previously unattainable bright, washfast shades with acid dyes can now be obtained on cotton by incorporating a cellulose reactive group in the dye molecule.

The first opportunities thus exploited were in the dyes themselves and in the dyeing procedures. In the past decade, fiber manufacturers have begun deliberately modifying existing synthetic fibers to permit mills to dye fabrics in three and even four colors at the same time. Typical of such modifications is development of differential dye nylon carpet fibers by Monsanto in early 1960. These fibers resulted from a modification of the chemical structure of regular anionic-dyeable nylon. The new fibers were made by changing the number and type of basic groups within the fiber that are able to react with anionic dyes used in normal nylon dyeing. The resultant fibers, which differ in amine end group analysis, can by careful dye selection and modification of dye methods produce three separate colors in a single dye bath. This development allowed mills to weave carpets of continuous filament, textured nylon and dye them in a broad spectrum of color combinations.

Later (in 1966), polymer chemists at du Pont took an additional step by neutralizing amine end groups in the fiber with strong sulfonic acid additives and then adding sufficient excess to make the fiber dyeable with cationic dyes, which were a different class of dyes than previously used on nylon. Through this change, fibers were produced that would give multicolor combinations by use of specific anionic dyes, disperse dyes, and cationic dyes in one dyebath.

To make acrylics dyeable with acid dyes, chemists at Chemstrand (now Monsanto Textile Co.) copolymerized basic monomers such as vinylpyridine with acrylonitrile. Acrilan 1656 made this way was introduced in 1952. Dyeability with acid dyes was designed to match wool, with which much of the early acrylic fiber was blended. In yarn form, it was also blended with Acrilan 16, a basic dyeable acrylic, to achieve two-color effects by dyeing with acid and basic dyes.

With the advent of piece dyeing of acrylic carpets, the ability to achieve multicolor effects in the same fabric became even more important. In 1969, American Cyanamid made two-dye variant acrylics available as companions to basic dyeable Creslan 83. One dyes lighter with basic dyes, and the other is dyeable with acid colors, making three-color effects possible.

Originally, polyester fibers were primarily dyed only with disperse dyes, but modified polyesters containing sulfonate dye sites are now available that can be dyed with cationic dyes. Also available is a modification that yields a deeper dyeing with the disperse dyes than with the normal yarn. This permits two depths of the same color and also a different color to be dyed at the same time on a 100% polyester fiber.

The resultant tremendous variety of colors and fibers that can be dyed at the same time from the same dyebath has brought a choice of colors to textiles at a cost hardly different from that of white goods or of solid colored materials. It has permitted the fabricmaker to enjoy the economy of weaving or knitting large yardages of a given style but later to deliver small lots especially dyed for a particular customer in colors that are timely. If it were not for this achievement, fibers or yarns would necessarily be dyed before spinning the fibers into yarn or before weaving or knitting yarn into fabric. Thus, orders for special colors would have to be placed months in advance in hopes that colors selected would be salable when delivered.

Polymers in Textile Processing

While the average consumer knows of the revolution brought about in fibers by polymer chemistry in the past 30 years or so, few know that other, relatively unknown polymeric chemicals have played an extensive role in the development and production of modern textiles. Not only have the synthetic fibers brought into use a host of polymeric auxiliaries, but the established natural fibers have benefited from a growing list of such chemicals as well. Included are chemicals to make fibers easier to process into yarns and yarns easier to weave into fabrics as well as a vast array of chemicals to modify the properties of fabrics once they are woven.

Yarns, for example, particularly those made from staple, contain many loose ends and loops that must be tied down before weaving. Chemical sizes are used to coat yarns to protect them in the loom. After weaving, fabrics are treated to remove the size and then are usually treated with other chemicals to make them softer, improve their hand, keep them from becoming dirty, and make them easier to clean once they are dirty.

Such treatments begin with yarns. For years, starch has been the preferred size because it is cheap and largely effective. It still dominates the market for sizes. Price and effectiveness notwithstanding, however, starch sizes do have a number of drawbacks. They tend to gel, produce sizing baths with high viscosity, and often require large amounts for effective sizing. They also often do not work well on synthetic fibers. Finally, treatment with enzymes or other products is needed to remove them after weaving, and disposal poses pollution control problems.

To overcome such difficulties, chemists turned first to modifying natural starches. They found that esters, such as starch acetates, and ethers, such as hydroxyethyl starch, form solutions that have stable viscosities, lower gelling temperatures, and good cohesion. They do not form hard gels as unmodified starches do, and they adhere better to fibers. They also permit weaving at lower humidities.

Chemists also turned to synthetic polymers to solve pollution problems associated with natural, and even modified, starches. Polyvinyl alcohol and carboxymethyl cellulose have been particularly successful among those tried. While they cost more than starch, water dissolves them from the fabric, they don't set up or gel on cooling, and smaller amounts give the same protection. Moreover, it is possible to reduce weave room temperatures and humidities below those required for starch-sized yarns.

While polyvinyl alcohol has been around since the 1920's and was used to size nylon hosiery yarns in the 1940's, its widespread use as a warp size did not become commonplace until the mid-1960's. Price reductions and the growing use of synthetic or blended yarns spurred the demand for polyvinyl alcohol (PVA) and also carboxymethyl cellulose (CMC). In recent years, the low biological oxygen demand of these sizes relative to starch has also been an important factor. Stricter enforcement of stream pollution laws has required weavers either to use PVA or CMC or to continue to use starch and buy the necessary pollution control equipment to dispose of the starch. Under these conditions, many textile companies have chosen PVA and CMC. Meantime, a developing trend in effluent control specifies total carbon content rather than biological oxygen demand. PVA and CMC may thus become as much of a problem as starch.

Once fabrics have been produced, they are treated in many cases with polymers to finish them. In ordinary finishing of apparel fabrics, for example, polyethylene is used to improve tear strength while polyacrylates change fabric hand, enhance wrinkle recovery and abrasion resistance, improve dye retention, and reduce pilling. Quaternary ammonium compounds, substituted imidazolines, polyolefins, and silicones serve as fabric softeners. Silicones allow fabrics to be sewn easier. Acrylic polymers and fluorocarbons find use in soil release and antisoiling treatments.

In addition to the preceding general uses of sizing, building hand, and softening, other uses have been developed for polymers in specialty fabrics that represent growing markets in today's fashion-conscious apparel industry. Thus, polyacrylates, butadiene-styrene copolymers, and other polymer systems backcoat corduroy fabrics to improve pile retention and filling strength of this fuzzy fabric. Bonded fabrics owe their rapid growth to the development of polyurethanes, polyacrylates, and other polymeric systems that bond the component fabrics together. The rapid development of nonwoven fabrics in today's textile markets likewise owes much to the effectiveness of polymeric additives in holding the fiber mat together. Polyvinyl acetate and also the polyacrylates are widely used for this purpose.

The trend to wash-wear and durable-press garments has extended the requirements of shape retention and dimensional stabilization to the older natural fibers. While chemical crosslinking agents have made cotton dimensionally stable, an interfacial polymerization method has been developed to make wool fibers dimensionally stable. This is another case of making essentially a new product by taking an old natural fiber and improving it with a polymeric finish.

On the other hand, it was the loss of strength caused by a polymer additive that brought about the durable-press garments of fabrics of polyester blended with cotton. The story started in 1963 when the National Cotton Council of America encouraged Levi Strauss, one of the nation's leading maker of casual and work clothes, to market durable-press men's trousers. The garments were crosslinked with dimethyloldihydroxyethylene urea, with some polyvinyl acetate added to the finish. Unfortunately, the stiffness of these fabrics from the polyvinyl acetate and the weakening of the cotton from the crosslinking agent caused the trousers to wear out rapidly from abrasion. Fortunately, producers of polyester fibers had the answer ready in blends of cotton with polyester. Garments made from the blended fabrics didn't abrade readily, and durable press soon found its place on the nation's clothes racks.

While the most important and best known form of durable press stemmed from the National Cotton Council–Levi Strauss work, it is important to note in closing that making durable-press garments by treating the garment instead of the fabric had occurred earlier. In 1954, for example, chemists at USDA's Southern Regional Research Laboratory (SRRL) had developed such a process, and a Texas firm began using it shortly thereafter. SRRL chemists also used vapor-phase methods to treat whole garments to produce durable creases and pleats, and the principle is practiced commercially today.

Repellency and Antisoiling

Fabrics, be they in clothes or household furnishings, often get wet and become stained and dirty. Because people like to be dry and clean and like to keep their furnishings that way, textile finishers and chemical companies that sell products to them have long sought to develop treatments that will repel water, stains, and dirt. In the past three to four decades they have achieved remarkable degrees of success. Today, few upholstery fabrics are produced that do not have one or more chemicals applied to help them repel water, stains, and dirt and to make them easier to wash or dry-clean once they do become soiled. Many apparel fabrics are similarly treated as well.

Water repellency has always been of much concern to the textile industry. In many cases, a textile is worth more and performs better when it resists wetting. In addition, in many cases the entire value of fabrics depends directly on water resistance, as in raincoats, tarpaulins, and awnings.

The earliest treatments for water repellency consisted of coating the fabric with various substances impervious to water, such as rubber, derivatives of asphalts and other tarry materials, and oxidizable oils. Fabrics produced by these early processes often felt uncomfortable, turned tacky when hot, and aged poorly.

Early attempts to produce apparel that repelled water yet remained permeable to air used waxes and metallic salts or soaps of aluminum and, later, zirconium. These wax-salt repellents have been available in a variety of single- and multiple-treatment forms for many years and still probably account for about 10% of the value and more than 25% of the poundage of water repellent finishes.

They may repel water well, but they aren't durable to repeated laundering and dry cleaning. A major step occurred in 1937 when Imperial Chemical Industries in England marketed the first water repellent that could be classified as durable (that is, resistant to laundering and dry cleaning). Du Pont introduced the same product in the U.S. the following year. Trade-named Velan PF by ICI and Zelan A by du Pont, the product contained stearamidomethyl pyridinium chloride. It had evolved from ICI research leading to the development of long-chain quaternary ammonium (pyridinium) compounds. It formed a chemical link with the hydroxyl group in cellulose fibers to provide water repellency and resisted laundering even at boiling temperatures.

Products of the stearamidomethyl pyridinium chloride type have been widely used on both civilian and military rainwear, and they predominated in the durable repellent market during and immediately after World War II. After the late 1950's, however, their use began to decline as other types of repellents became available. Today, they probably account for only about 5% of the value of water repellents, and they are used mostly with other types of repellents.

An outstanding advance involving quaternary (pyridinium) water repellents occurred in 1959. The U.S. Army's Natick Laboratories announced the development of its Quarpel water- and oil-resistant treatment for textiles. Chemists at Natick had found that applying a pyridinium and a fluorochemical repellent at the same time gave an improved finish. Specifically, they found that the combination gave initial repellency and laundering durability not obtainable with either component alone. The Quarpel combination finish gradually replaced single component water repellent finishes on many military combat garments.

Another type of water repellent finish that became popular in the 1950's was the resin-based finish produced by condensing stearamide with methylolated melamines and emulsifying the resulting product. Typified by the Permels marketed by American Cyanamid, such resin-based repellents produced durable water repellency on a broad range of fibers. By 1950 they probably accounted for about 25% of the value of water repellents.

Since then, use of resin-based products has declined, so that now they probably represent only about 5% of the value of repellents used.

Silicone water repellents were introduced in the early 1950's. These products are hydrogen methyl and dimethyl polysiloxane copolymers that react in the presence of metal salt catalysts to form crosslinked water repellent films. Many proprietary products have been marketed by companies that buy the silicone oils from basic silicone producers and emulsify them.

The silicone repellents are unique in the soft silky hand they produce, and they have been found especially useful on fabrics of all-synthetic fibers. They provide outstanding water repellency that is permanent to repeated dry cleaning and reasonably permanent to repeated laundering. In 1959, they probably accounted for 25% of the value of repellents sold. Although their share of the total market has declined, they nevertheless continue to grow and still hold an important segment of the market.

Durable oil- and water-repellent finishes based on fluorochemicals emerged at the end of the 1950's, and they are probably the type most consumers are familiar with today. They offer several advantages over durable water repellents (pyridinium- or silicone-based) by combining oil repellency with water repellency and by greatly improving durability to dry cleaning. In addition, being fluoropolymers they present the potential for combining repellency with new properties such as resistance to dry (particulate) soiling and to release of oily stains by washing.

Minnesota Mining & Manufacturing (now 3M) offered the first fluorochemical finishes for textiles to repel oils and stains in 1956 under the trade name Scotchgard, and du Pont followed six years later with its Zepel brand. Apparel, upholstery, and carpets are the main beneficiaries today.

Dirt is a recognized hazard for clothing and household textiles, but the effort required to remove it was not considered a major problem before the advent of wash-wear and durable-press treatments. Minor soiling and cleaning problems then arose when cotton fabrics began to be treated with nitrogenous crosslinking agents to produce wash-wear fabrics. Soiling and soil removal became really serious in the mid-1960's, however, when hydrophobic fibers, particularly the polyesters, were blended with cotton to make resin-finished fabrics with durable-press properties stronger and resist abrasion better. Consumers quickly became aware that their clothes and household furnishings with the durable-press finishes soiled more easily and were harder to clean than previous products.

Deering Milliken Research developed the first soil release finish, an acrylic polymer, in the early 1960's and introduced it in 1966 (as Visa). During the next two years, a number of soil release products came on the market, including Celanese's Fybrite, Burlington's Come Clean, McCampbell-Graniteville's X-It, and 3M's Dual-action Scotchgard.

The various soil-release compositions all make the fabric more hydrophilic, allowing it to be wetted more readily by detergent solutions and giving it soil release character with low soil redeposition. Such finishes work best with oily stains. Newer generation fluorocarbon finishes such as Scotchgard combine resistance to staining with resistance to dry soiling and improved wear life as well as wash release of oily soils.

Fiber Shape vs. Soiling

Chemists found not only that applying a variety of treatments permitted them to improve the antisoiling properties of fibers but that fiber shape also played an important role. It not only affects soiling, or at least the appearance of soiling, but also other properties as well, including luster and transparency.

When nylon was first introduced, it was used mainly for women's hosiery, because it was strong and durable yet could be woven as sheer as silk. Then within a few years it was being used broadly in woven fabrics, knit garments, carpets, and elsewhere. However, the early nylon yarns had some undesirable features. Fabrics were too shiny, did not cover adequately, and appeared dirty sooner than did those woven from natural fibers.

Although delusterants helped correct shininess and cover, their effectiveness was limited. Then about a decade ago an important improvement evolved from studying the optical properties of the filaments. It turns out that at critical angles the round surfaces of filaments in fabrics act as innumerable synchronized mirrors and reflect a high concentration of white light. This results in the fabric's undesirable shininess. Furthermore, the light that does penetrate the fabric is partly reflected out again to show and perhaps accentuate the soil on the fiber surfaces away from the viewer.

Scientists at du Pont in 1960 brought about a major change in such behavior by changing the cross section of the nylon filament from round to trilobal. Instead of having high glare, fabrics appeared with subdued luster highlights, because the light was scattered instead of being reflected mainly in one direction. In addition, the outside filaments reflected more of the incident light. Viewers saw less imbedded soil, and the fabric also covered better. Finally, the trilobal cross sections not only gave desirable luster but also made fabrics resist bending more, giving them a firmer hand.

These attributes enabled more attractive and useful woven and knit fabrics to be made for apparel from nylon and, later, polyester. In addition, improvements brought about by the trilobal cross-sectional filaments have resulted in the development of major outlets in carpets for such fibers.

Carpets are made to be walked on, of course. As a result, they have been subject to much research on ways to help keep them clean. Before the trilobal cross section made its mark in carpets, however, chemists were to use a variety of other approaches to control soiling. Early techniques included spinning fibers with smooth or specially hardened surfaces, "presoiling" with white particles to fill soiling sites on fibers and deny them to "dirty" soil, and chemical treatments.

One of the earliest studies on soiling of carpet fibers came in the late 1930's with the demonstration that using a smooth fiber surface minimized soiling. Scientists at C. H. Masland and Sons showed that soil gathered in the irregularities of the crenulated rayon surface and that a smooth, round fiber substantially reduced soil accumulation. The technique became particularly important for wet spun fibers, where special methods are needed to achieve the smooth, soil-resistant surface.

In the 1950's and early 1960's, numerous patents issued covering the principle of presoiling, another approach to keeping carpets clean. The theory was that potential soiling sites on fiber surfaces would be occupied by "presoiling" material and the fibers wouldn't accumulate "dirty" soil. Fibers, yarns, and carpets were treated with white or colorless solids such as bentonite, silica, alumina, clay, or a variety of metallic salts or oxides. Because of processing difficulties, dusting, and perhaps lack of demand, however, use of presoiling never became very widespread.

During this same period, chemists tried hardening fiber surfaces by a variety of ways to reduce soiling, but in general the processes were not particularly successful. Then in 1960, the principle of soil hiding began to develop, resulting first in the trilobal cross-sectional fiber, as already described. The following year chemists at du Pont put polyethylene oxide derivatives in nylon. They mainly wanted to improve antistatic properties, but they also made fibers with antisoiling properties as well. The additives existed as a separate phase in the fiber to alter its optical properties, and soil was hidden more effectively than with the trilobal fiber.

First exploitation of such an improved soil hiding fiber came in 1968 when du Pont introduced an Antron carpet fiber to the carpet trade. Other fiber producers that had been working on the same principle soon followed suit—Monsanto with Cadon, Allied with Anso, and American Enka with Enkalure, for example. The following year du Pont introduced Antron II, which also used the principle of soil hiding but altered optical properties by building voids into the fibers.

So far, the principle of soil hiding has been applied almost exclusively to nylon. In the coming years, however, the principle is certain to be used in many, if not all, of the synthetic fibers used in carpets.

Soiling upholstery presents a somewhat different problem from that of soiling carpets in that substances that soil it contain much more oil than those that soil carpets. Therefore, the approach has been to modify the fiber surface by reducing its surface energy to such a level that it repels not only water but oils as well.

Applying proprietary polymeric fluorocarbon surface finishes is the solution, as it has been with apparel. A difficulty with these finishes, however, is that they resist soiling, but they do not prevent it. Also, cleaning after the fabric has become soiled is difficult, because the finish repels the cleaning liquid as well.

To overcome these drawbacks, chemists have built hydrophilic blocks into the polymers with the oleophobic fluorocarbons. 3M, for example, included polyethylene oxide as part of the fluorocarbon polymer. The company says that when the fabric is in use the fluorocarbon part is on the outer surface and repels oily soils. When the fabric is being cleaned, however, as with a detergent system, the hydrophilic polyethylene oxide comes out and the fluorocarbon goes in. This inversion allows easy cleaning.

Antistatic Finishes and Fibers

Static buildup on synthetic fibers annoyed consumers from the beginning. Nylon slips that rode up, trouser legs that stuck to nylon socks, and nylon carpet that triggered shocks at doorknobs were less than endearing and may well have more than once been the difference between sale and no sale. Fiber and textile producers, therefore, were soon seeking ways to reduce or eliminate the problem. The first approach was to apply various finishes that would speed discharge of any static electricity that might accumulate, followed by attempts to build antistatic properties into the fibers themselves at the time of spinning.

Substances with high dielectric constants are good antistatic agents. Water is particularly effective. It is volatile, of course, but fabrics can replenish their supply from the humidity of the air. Thus, compounds that are hygroscopic and electrolytic are good antistats. One such agent is the condensation product of stearic acid, glycerin, and ethylene oxide.

Such chemicals unfortunately don't last through more than a washing or two, and it was not until 1957 that the textile industry had its first durable antistatic finishes. Chemists at Onyx Oil and Chemical Co. showed that a single compound having the proper hygroscopic and ionic properties could be deposited on a fabric and an insoluble lattice structure formed. The reaction product of diethylenetriamine and the diiodide of polyethylene glycol further reacted in place with an organic dihalide is one such combination, while polyhydroxypolyamines and tris-methoxypropionyl-s-perhydrotriazine is another.

While some in the industry worked on such finishes and made many antistatic agents to apply to the hydrophobic fibers, others sought approaches that would be even more durable. Monsanto, for example, launched a major research and development program directed toward this end a decade ago. As a result, in 1969 the company introduced the first permanently altered antistatic nylon fiber, 22N. Later in the year, du Pont introduced its textile fiber, Antron.

Production of 22N involves dispersing an ethoxylated hydrocarbon as a second phase in the polymer matrix before the fiber is formed. In the finished yarn this dispersed phase reduces the fiber's electrical resistivity, which allows static charges to dissipate rapidly. This virtually eliminates shocks normally experienced after walking on nylon carpets.

Producing a nylon with such a dispersed phase also provides some added benefits. Carpets, for example, don't attract as many airborne particles, since they don't retain a static charge. The dispersed phase also makes the filaments less transparent by increasing internal refractions. As a result, dirt behind the fibers isn't magnified, thus reducing apparent soiling. In textile fibers, the improved moisture transport properties are important for comfort. In addition, 22N retains all of the good properties of nylon as a class—strength, toughness, resistance to abrasion, and easy washing. It represents one of nylon's most significant functional advances since its introduction more than 30 years ago.

Flame-Resistant Textiles

Fire is the third leading cause of accidental deaths in the U.S. (after motor vehicles and falls), and burning textiles are involved in many of them. Among the natural and regenerated fibers, cotton and rayon may present serious fire hazards. Wool, on the other hand, is difficult to ignite and burns very slowly. Among the synthetic fibers, nylon, polyester, and polyolefin fibers are generally less serious fire hazards than the cellulosics, but they do present the hazard of molten drip.

Efforts to overcome flammability of textiles date back almost as far as textiles themselves. Egyptians, for example, treated woven cotton with various inorganic chemicals to make them nonflammable. In more recent times, the textile industry for more than 200 years has used water-soluble, nondurable chemicals to make fibrous materials resist fire. It was not until the early 1900's, however, that the textile and chemical industries began to develop semidurable and durable fire retardants for cellulosic fabrics, and it was not until the early 1950's that they had available truly durable finishes (that is, finishes that resisted ordinary washing and dry cleaning). Even then, these finishes at first were suitable for use only on heavier fabrics.

The first semidurable finishes were based on precipitating metal oxides within the fibers. Later developments included additive coatings of halogenated compounds. In the early 1930's, the first observations were made on the value of chlorinated compounds with inorganic oxides. However, the first major practical advance was patented by scientists at Wm. E. Hooper & Sons Co., who first combined chlorinated organic compounds and antimony oxide. In addition, other agents were included so that the treated fabrics complied with the federal specification for cotton duck that resisted fire, water, and weather. The early formulations, which contained chlorowax as the source of halogen, required relatively high add-ons to impart the necessary properties. Treated fabrics had poor drape, and use was therefore limited to tentage, tarpaulins, awnings, and other fabrics for outdoor use. The largest single usage has been by the military.

Modification of the synergistic antimony oxide–chlorowax formulations with urea-formaldehyde resins by USDA scientists broadened the scope of potential use to apparel goods and increased durability to laundering. Substitution of plasticized polyvinyl chloride for part or all of the chloro-wax produced even further improvements in aesthetic and laundry durability properties. However, treated fabrics still exhibited considerable afterglow. In the early 1950's, the afterglow was remedied by incorporating the organophosphorus compound THPC (of which more in a moment). The Lynn-Russ Finishing Co. used this process to produce military and industrial fire-retardant cotton apparel.

The major breakthrough in the discovery of a flame-retardant system for cotton that for the first time could be used on wearing apparel and household fabrics came in 1953 when scientists at the U.S. Department of Agriculture's Southern Regional Research Laboratory found a durable flame-retardant system for cotton based on tetrakis(hydroxymethyl) phosphonium chloride (THPC), a chemical first manufactured by Hooker Chemical Co.

They showed that THPC reacted with almost any compound containing an active hydrogen and that in the presence of nitrogen-containing compounds, such as melamine and urea, polymers were produced that retarded flames excellently. The finish, referred to as the THPC-amide finish and containing THPC, methylolmelamine, and urea, is the most widely used durable system today for making cotton apparel and household fabrics flame resistant. It and other systems now based on THPC account for most of the production of durable fire-retardant cellulosic textiles in the United States, England, and Western Europe. All these processes provide durable fire retardancy for some cotton and rayon fabrics, along with satisfactory hand, strength, abrasion resistance, sewability, and durability to laundering. Rot resistance and wrinkle resistance are also imparted to fabrics by this finish. These finishes have limited use on white goods, however, because of yellowing caused by bleaching. They are also susceptible to degradation by sunlight, chlorine, and carbonate detergents.

FUTURE DEVELOPMENTS

The textile fiber industries in many ways may appear to have matured. Chemically trained personnel working on polymer, fiber, and textile technology face mostly the same situation their counterparts face in many other industries represented in this study. They have a greater body of theoretical literature and many more tools available to them than their predecessors had. Any new products or processes they may develop and try to introduce must compete against those introduced by several generations of chemical technologists. Furthermore, any success they may now have will come only when they surpass much more stringent requirements than if the work had been done 20, 10, or even five years ago.

Requirements may now be more stringent, but there's no aura of pessimism in the industry about prospects for developing better products and processes in the future. The optimism of the 1970's, however, does differ in some respects from the optimism of, say, the 1950's. In those first postwar years, opportunities were quite good for inventing a totally new synthetic that would supply significantly large markets, and events eventually proved the optimism to be justified. Today, the chances are not too high of developing and producing a new fiber that differs radically from those now made and that would compete successfully with today's major fibers. Not only would finding such a new fiber be difficult in this well-researched field. The cost of building a plant with a large capacity and then of establishing that new fiber in fiercely competitive markets would probably lie beyond the resources—or at least the will to risk—of even the largest industrial organizations.

No one who has observed how suddenly an unanticipated discovery can occur would flatly predict no major new synthetic fiber. Even if the chances do not appear to be great, however, there is no reason for less than optimism about the development of new fiber industry technology. In fact, optimism for improvements should at least equal—and may even be greater—than the admittedly high optimism of the recent past. Continued research and development at a high level in the U.S., major research efforts in other countries, cooperation and exchange of information, and increasing affluence all provide a basis for such optimism.

For the foreseeable future, much of the research and development in textile fibers will continue along trends already established. Chemists will continue to seek to develop fabrics that weigh less and are more comfortable. They will also continue to develop chemical additives and finishes to improve resistance to soiling and to soil deposition during laundering. Resistance to static buildup, wrinkling, and flammability also continue to be numbered among their goals. The possibilities for developing new dyes and multicolor dyeing systems are far from fully exploited.

Chemists working with textile fibers are also trying to develop better moisture transport properties for some fibers. Some double-knit garments of 100% polyester already rate high in comfort, but some apparel woven of noncellulosic fibers still generally rates lower in comfort. Cotton, rayon, and wool and blends of them with noncellulosics provide much of the desirable comfort now, but companies making noncellulosic fibers would like to increase their competitive advantage. A hydrophilic synthetic fiber could be the answer, and nylon-4 may prove to be such a fiber.

New finishes beyond those for today's durable-press garments still remain to be developed, and proponents of durable press will continue to seek new markets. Bed linen is one of their markets that is now experiencing a sales boom. Households are a major outlet, but operators of hotels and motels are particularly enthusiastic about resin-treated blends of cotton and polyester fibers because of the potential saving in labor for ironing.

With the advent of bright and colorful patterns for apparel, garment makers are increasingly demanding accurate shade matching and freedom from shade changing. Scientists will be called on to meet this need, as well as to make dye and resin systems and other finishes more compatible. New dyeing methods, such as solvent dyeing with reusable solvents, are needed to minimize pollution. Development of simple print dyeing methods may eventually lead to custom printing of fabrics at the retail level to suit style-conscious customers with design ideas.

Innovations in weaving and knitting will continue, although chemistry will not often be heavily involved. One fabrication method in which chemistry is important is in nonwoven fabrics. Interest in nonwoven disposable fabrics on the part of some in the fiber and textile industries peaked in the mid-1960's only to fade as the decade ended. Now, interest appears to be reviving again. The concept of general-use throwaways, however—the one-wear underwear, for example—is now viewed as very limited at best, in contrast to some earlier popular predictions. Not only are manufacturers and consumers more conscious of the potential waste involved for limited raw materials, but they are also much more critical of overwhelming an already overburdened disposal system for solid wastes.

Suitably developed nonwoven disposables do appear to have a market in high-service-cost uses, however. The hospital supply field—bed linen, gowns for professional and nonprofessional personnel—is the best bet for now. Diapers are another good market, while restaurants, hotels, motels, and similar institutions are more distant possibilities.

The foregoing developments stem largely from extrapolating present trends. The fiber and textile industries and those supplying them face a major new challenge, however, in the growing battle over textile flammability. Methods exist now for making textiles resist burning, but they are a growing distance from meeting new and increasingly stringent regulations that now exist or are pending.

That people die or are permanently and badly disfigured because of burning textiles is incontrovertible. Beyond that statement, however, there is not too much agreement. Some persons argue that undue hazards exist and much more stringent regulations are needed now. Others view the problem as being much less serious and counsel a less hurried approach. Regardless of what the actual losses in deaths and disfigurements may be, it does seem clear that what concern there is plus requirements imposed by regulations will increase consumption of flame-retardant chemicals in textiles. Use of flame-retardant textiles (and plastics) may well grow 35% to 45% a year from 1970 to 1975, rising from 800 million pounds in 1970 to 3.5 to 4.5 billion pounds in 1975, according to some estimates.

Meantime, no little work will be required by chemical technologists in industry and government. Not only have they much to learn about burning and how to control it. They have yet to reach agreement on defining the risks, devising the specifications, and then developing tests that will assure specifications are met. In addition, properties of new products must finally be reconciled with those that consumers will spend money for in terms of comfort, easy care, and fashion.

The final realities of costs, prices, and competition will largely govern the future of every industry, as they have in the past. The truism, of course, applies to the fiber and textile industries and to those supplying them. The realities of costs, prices, and competition squeeze the fiber and textile industries to a different degree than they squeeze some others, however. The manufacturing spectrum from monomer to polymer to fiber to yarn to textile to finished product generally becomes more labor-intensive. One result in the past has been that high labor costs have driven the textile industry generally from northeastern states to southeastern ones. Now, competition from low-labor-cost countries such as Japan, the Republic of Korea, Taiwan, and even India and other less developed Southeast Asian countries may bring about an exodus of U.S. companies to those regions.

At first glance, such an exodus, if it does occur, may not concern chemists and chemical engineers in the U.S. A longer view, however, suggests it might eventually make sense for relocated textile and garment factories to make or buy polymers and fibers locally rather than to import them from the U.S. Moreover, lower costs for research and development in these other regions is helping erode U.S. leadership in fiber and textile R&D as well as in manufacturing. R&D might be more profitably done elsewhere, too.

Chemistry unfortunately does not often exert great economic leverage on all parts of the sequence from monomer to product on a department store shelf. The highly chemical processes of making fibers and then dyeing and finishing them, for example, may represent less than 10% of the retail price of many garments, whereas spinning and weaving fabric and designing and manufacturing garments may represent more than 50% of the retail price. Consumers may hardly notice a 10% drop in chemical process costs, whereas a 10% drop in mechanical costs may be significant. What the U.S. textile and garment industries see themselves as badly needing, therefore, are new and much less expensive fabrication methods.

Mechanical engineering will doubtless provide many of the solutions— higher speed shuttleless looms, more efficient knitting machines, more automatic or even automated (or computerized) garment manufacturing, for example. Successful processes involving adhesives or heat-bonding to replace sewing, however, offer an opportunity for chemical innovation.

Finally, while the fiber and textile industries will put most effort on the more immediate practical problems of short-range production improvement, they are not beyond speculating about longer-range possibilities. Included among such items are:

• Fabrics that heat or cool the way space suits do.
• Fabrics that change color depending on where the wearer is.
• Truly no-care clothing and household fabrics.
• Textiles that can be readily and practically recycled on a short-term basis.

SOCIAL AND ECONOMIC IMPACTS

A prominent Spanish historian, Antonio de Herrera y Tordesillas, first described it in western literature in 1725. An English chemist, Joseph Priestley, gave it the common English name in 1770. An English physicist and chemist, Michael Faraday, made one of the first reinforced hoses from it in 1824. But it was not until an American inventor, Charles Goodyear, developed a process of compounding and heating sulfur with it in the early 1840's that it could be made into useful commercial products and become more than a tantalizing curiosity.

The product was rubber, at the time the juice from the tropical plant *Hevea brasiliensis* and today increasingly the product of chemical plants. In the approximately 130 years since the invention of vulcanization, the rubber industry in the U.S. has grown to the point that nearly 500,000 Americans earn their livings working in it, its annual volume of business is about $17 billion, and U.S. elastomer consumption totals some 2.7 million long tons a year.

As with the other industries this study deals with, the rubber industry in recent years has both grown sharply and changed its raw materials base. As a result, the rubber industry of today hardly resembles the rubber industry of even a short three decades ago. In the 20 years 1940–59, U.S. consumption of rubber rose from 700,000 long tons a year to 1.6 million long tons, an increase of nearly 130%. In the past 10 years, growth has continued strong, and in the decade ending with 1969, U.S. consumption increased nearly another 70% to about 2.7 million long tons.[1] In no other 30 years during the rubber industry's history has it grown at a comparable rate.

Growth in the past 30 years may be striking, but the change in elastomers that the industry uses is even more so. From an industry that once depended on nature for its primary raw materials and on chemical technology only for its auxiliaries, it has become an industry that depends largely on chemical technology for its primary raw materials as well as entirely on chemical technology for its auxiliaries.

Three decades ago (in 1940), natural rubber (new or reclaimed) held 99.6% of the market in the U.S. and synthetic rubber a minuscule 0.4%. In little more than a decade, however, synthetic rubber overtook natural, capturing more than half (actually 53%) of the market in the U.S. by 1951. Synthetic rubber's share continued to rise to more than three quarters of the U.S. market by 1969, leaving natural rubber with less than a quarter (actually 22.8%).

The use of natural rubber in the U.S. has suffered during these years, of course, but some chemical innovations have helped it improve its position. Nevertheless, most of the U.S. rubber industry's growth during the past 30 years has occurred with the synthetics, and U.S. consumption of natural rubber in 1969 amounted to about 590,000 long tons,

---

1. All production and consumption figures in this chapter come from "Rubber Industry Facts, 1971," published by the Rubber Manufacturers Association. For the rubber industry, we are using 1969 data for the most part in contrast to our use of 1970 and 1971 data for other industries dealt with in this study. We might have used data for 1970 for the rubber industry, but production sagged badly that year because of a long strike. Results for 1970, therefore, fall below long-term trends and would give a false impression of the industry's position.

down about 10% from 1940's consumption (but up nearly 25% from 1960's consumption, incidentally).

Before the discovery of vulcanization that improved rubber's physical properties so much, products made from it disappointed manufacturers and users alike. Rubber's stiffness when cold and its stickiness when hot became common knowledge, and early attempts at making rubber goods too often proved disastrous.

Nevertheless, rubber's unique property of elasticity helped it remain popular despite many drawbacks, and entrepreneurs again and again tried to make serviceable products from it. The first U.S. patent on rubber goods issued in 1813 (on a rubber varnish to waterproof leather shoes), and crude shoes of native natural rubber made in Brazil were imported in 1820. Roxbury India-Rubber Co. made the first rubber goods in the U.S. in 1835 (and incidentally became the company where the forerunner of the present-day calender was invented, one of the most important machines now used by the rubber industry).

It was during the 1830's that Charles Goodyear became associated with Roxbury India-Rubber and turned his attention to improving the properties of rubber. In 1839, he noted that rubber did not melt in boiling sulfur but seemingly charred. Within a short time he had perfected vulcanization (as the process was later named after the Roman god of fire), and the real beginning of the rubber industry, in the U.S. as well as elsewhere, dates from his success.

One of the first products to be made in large quantities following the invention of vulcanization was elastic webbing. In 1841, Goodyear himself invented what is now known as proofed goods. The material was used in suspenders at first and then in corsets and boot linings. Vulcanized rubber products in large quantities soon began meeting other needs, however, and proofed goods were eventually joined—and often surpassed—by many other rubber products, including vehicle tires, molded goods of many types, hoses, belting, sponge and foam, shoe soles and heels, sports equipment, adhesives and sealants, and medical supplies.

Making rubber products involves the activities of companies that comprise the rubber industry itself as well as the activities of companies in the chemical and petroleum refining industries and of companies that operate natural rubber plantations. Many of their operations are highly or entirely chemical. Companies in the rubber industry mix a variety of chemicals with natural and synthetic rubbers. Using a combination of chemical reactions and mechanical processing steps, they cure the raw rubbers and fabricate them into finished products. Their main raw materials are natural and synthetic rubbers, but they also use many auxiliary chemicals in both large and small volumes to vulcanize and otherwise modify raw rubber properties to meet service conditions for the many types of rubber goods they produce.

All natural rubber used in the U.S. is imported, mainly from the Far East. The largest rubber industry companies operate their own plantations to meet all or part of their needs for natural rubber, while smaller rubber companies buy natural rubber from them or from others that operate plantations. Although plantation operations are primarily agricultural pursuits, they do make use of chemistry in such aspects of their operations as using fertilizers and, especially recently, applying chemicals to tapped trees to increase yields.

The synthetic rubbers used by the industry are made from ethylene, propylene, butadiene, styrene, and many other monomers. These chemicals are products of the chemical and petroleum refining industries.

Using monomers bought from chemical and petroleum refining companies, the largest U.S. rubber companies polymerize them in their own plants to meet all or part of their needs for synthetic rubbers. Smaller rubber companies, plus larger companies with insufficient polymerization capacity, buy synthetic rubber from other large rubber companies or from chemical and petroleum refining companies that own and operate polymerization plants.

Finally, chemical, petroleum refining, and the larger rubber companies all are involved in making the many auxiliary chemicals used to process raw natural and synthetic rubbers into finished goods. Companies in each industry have helped develop the auxiliary chemicals. For rubber companies, their development and production have on occasion resulted in the formation of sizable chemical operations.

Rubber products of many types have in time become highly useful in the economy and have helped improve the production of goods and services in the U.S. Of all rubber industry products, tires for vehicles have without doubt made the greatest single impact. The pneumatic tire gave automobiles a soft cushion of air to ride on, and they made practical the long-distance transport of goods by truck and of people by bus. The simultaneous improvements in vehicles, tires, and the highway system that began in the early 1900's and accelerated in the post-World War I years increased mobility in the U.S. tremendously and contributed heavily to the shift from a rural to an urban society. In addition, the market for vehicles caused sharp growth not only for auto, truck, and bus manufacturers but also for industries for which vehicles represent a major market, such as petroleum refining, steel, protective coatings, fibers and textiles, and plastics.

Tires for highway vehicles, of course, have not been the only tires to help change the economy. Goodyear Tire & Rubber Co. pioneered with the modern airplane tire in 1928, for example. Its design took the jolt out of landing and also kept the tire from collapsing should it rupture. A design similar to it is still used even though plane loads and landing speeds have increased tremendously.

The rubber industry also experimented with tires for farm equipment, and it put the first oversized, low pressure tires on tractors in Florida in 1930. These tires gave traction on a variety of surfaces in terrain previously considered too difficult to cultivate, and farmers were soon using them on tractors in large numbers to cultivate larger areas more efficiently and economically.

The rubber industry also improved and adapted the basic tractor tire design to off-the-road tires for big earth-moving equipment. Large graders and haulers have been used to expand the highway system for better transportation, build dams for reclamation and waterpower, recover fuel and minerals for an expanding economy, and undertake other similar projects. While some people may now question the wisdom of some individual projects of these types and the pace with which some of them are still occurring in particular cases, the availability of such equipment without question has eased the production of some goods and services in the U.S.

Impressive though these other impacts of developments in tires may be, the most direct impact on the individual consumer still comes from the auto tire. As is true with many other products of chemical technology, today's tire gives more service for a smaller part of a person's income than ever before. A 1927 Model A Ford, for example, was equipped with tires constructed of natural rubber and a four-ply cotton carcass. They listed for $23.95 each. With an average life of about 14,000 miles, a set of four cost about $6.80 per 1,000 miles (and not counting hazards to life from blowouts or drain on energy from flats). Today's comparable tire has a tread of oil-extended styrene-butadiene synthetic rubber, a carcass of either synthetic polyester or nylon cords coated with vinyl pyridine–resorcinol formaldehyde dip, and a similarly coated belt of glass fibers between the carcass and the tread. It lists for $43.65. With an average life of at least 30,000 miles, a set of four costs about $5.80 per 1,000 miles (in 1972 dollars and with blowouts and flats highly unlikely when tires are reasonably driven and properly maintained). Moreover, today's tires may travel over highways at speeds of 70 m.p.h. and higher for extended periods in all types of weather and carry a car that may weigh twice what Ford's 1927 model did. Now, tires made with steel instead of fabric cords and with a radial instead of a bias structure promise to provide even better service. They are becoming more popular and account for about 5% of the use on autos at this time.

Tires thus are undoubtedly the most important rubber product, but many thousands of other rubber products are also used that chemists and chemical engineers have helped develop and produce. Nontire rubber goods in the past decade, in fact, have accounted for about 50% of the dollar value and about 35% of the tonnage of the rubber industry's output. The thousands of items into which this output has been processed come in many sizes and shapes and meet many different needs. Only a catalog of many hundreds of pages could list them all.

One of the most useful and versatile of nontire rubber products is the endless belt that either transmits power or transports a great variety of products. Some $320 million worth of such belts were sold in the U.S. in 1969. All vehicles depend on them to power auxiliary equipment, and home appliances and office machines use them in several sizes. Industry uses thousands of different machines in which belts transmit power. Conveyor belts, meantime, are a prime tool in mines and factories. They have also been widely installed in airport terminals to move baggage and airfreight.

Grain was moved by conveyor belt in 1863 in one of the first such uses, and iron ore followed in 1892. For the next half century, growth areas were coal and other mines, grain warehouses, and flour mills, where large quantities of materials must be moved long distances at very low costs.

Additional expansion followed World War II when new polymers and fibers made it possible for conveyor belts to be useful where resistance to heat, oil, and ozone are important. Now, just about every product imaginable moves by conveyor, and the industry is looking forward to increasing the market in moving people. The concept isn't new, since people traveled on conveyor belts at the Paris Exposition in 1900. However, growing congestion in such high-density locations as sports arenas, shopping centers, airports, and even city centers appears tempting for expansion.

Hoses are another major rubber industry product. Virtually every industry depends on large and small ones to carry out its activities efficiently, and annual hose sales in the U.S. today total more than $650 million. Hosemakers had established the basic processes for extrusion, spiral wrapping, and braiding before 1900, and from then until roughly World War II they primarily refined and automated these basic techniques. Then from shortly before the war to the present, they have been able to serve a growing list of new markets as new synthetic polymers and fibers have permitted hoses to carry petroleum products, corrosive liquids, and other solutions over an expanding range of temperatures and constantly rising pressures.

About a decade before the war, polysulfide rubbers developed by Thiokol Corp. and a polychloroprene rubber developed by du Pont resisted solvents, hydrocarbon fuels, and hydraulic fluids, for example. Hypalon, a chlorosulfonated polyethylene commercialized by du Pont in the 1950's, began carrying sulfuric acid satisfactorily. More recently, ethylene-propylene polymers have become important materials of construction for hoses, as they resist low and high temperatures, some acids, and weather better than more unsaturated polymers.

Development of man-made organic and glass fibers has made it possible to construct hoses that may be used at pressures up to 4,000 p.s.i. and at temperatures up to 350° F. Such hoses are required for the hydraulic pressures that operate earth-moving equipment. Moreover, developments in elastomers, fibers, and hose fabricating technology have led to hoses with greater diameters and lengths. It is not uncommon to see dredge hoses in 4-foot diameters, and smaller diameters can be made in continuous lengths up to 5,000 feet. Among the many applications of such developments is a light, flexible hose that traveling sprinklers can pull to irrigate farms, a use that is especially important in less developed countries trying to grow more food.

Tanks made of metal or concrete and lined with rubber have for some time been used to store and transport materials, especially corrosive liquids. More recently, shippers have had containers made of rubber and fabric to transport liquids by ship, rail, and truck. The armed services use such containers to store fuels in strategic areas. They have the added advantage of being collapsible for storage when not in use. Airfilled bags of rubberized fabric, known as dunnage bags, are used in railroad cars to keep loads from shifting.

Rubber products play an increasingly important role in the construction industry. Not only are they often critically important in tools and equipment, but they find growing outlets in such structural applications as rubber sheeting for waterproofing, foamed-in-place polyurethane for insulation and roofing, expansion joints for air conditioning and heating ducts, and glazing and gaskets for structural panels.

Ethylene-propylene and butyl rubbers resist aging and water extremely well. As a result, sheeting made from them encloses entire foundations completely to protect new buildings from hydrostatic penetration and ground moisture. Liners between concrete decks bar water from dripping in parking garages, plazas, and promenades. Sheets of them line ponds, reservoirs, and irrigation ditches to prevent water loss.

Various rubber products are also used to prevent pollution or to control it. Sheets of ethylene-propylene rubber line holding ponds for brine from oil wells and effluents from dye and other chemical plants. Foam rubber can soak up oil spilled on inland waters, and booms of fabric coated with oil-resistant rubber and ranging up to 13 feet high have recently been developed to contain oil spills in the ocean.

Some of the most important, if small volume, uses of rubber occur in medical applications. Plastic surgeons often implant silicone rubber for ladies who feel nature may have shortchanged them, and molded parts are available for sculpturing a more attractive nose or chin. Such uses often solve psychological problems, but others relate more directly to physical health. Physicians implant silicone rubber devices entirely below the skin to drain excess fluid from the brain cavity into the heart to relieve "water on the brain," and many thousands of artificial valves have been implanted in persons with defective hearts.

The space program of the past decade has brought new demands for elastomeric materials. Among them are ablative compounds to protect launching equipment, to safeguard parts of vehicles during flight, and to shield reentering space crafts. In space itself, elastomers serve as seals and thermal control coatings where they must perform over a wide temperature range and resist high vacuum and ultraviolet radiation.

Solid fuels for rocket motors, as well as handling systems for high energy liquid propellants, have presented many challenges to chemists in recent years. Large amounts of elastomers have also been used as the binder in solid fuels, while elastomers filled with asbestos or silica serve to insulate rocket engines.

In recreation, generations of Americans have grown up wearing the ubiquitous tennis shoe, and other rubber products have shown up in much sports equipment ever since early Indians bounced crude rubber balls. Now, athletic events are increasingly taking place on synthetic turfs that lie on impact-resistant elastomers.

CHEMICAL ACCOMPLISHMENTS

The rubber industry was founded on useful products made from the juice of a tropical plant, and few products can be made with such a range of properties as can rubber, both the original natural and the more recent synthetic. Some of its products are soft, such as foam rubber, and others stretch and recover easily, such as elastic thread. However, some products are tough and horny, such as combs, bowling balls, and battery cases. Rubber puts the bounce in golf balls, but it can be compounded to absorb shocks on marine docks, bowling alley backstops, and auto bumpers. It can be made to resist electricity in wire coverings and electrician's gloves, but it can also be made to conduct electricity in sheets and floors in hospital rooms. Sandblast hose made from it resists abrasion, but erasers use the opposite property.

To provide such a wide spectrum of properties, chemists in the rubber industry use several different kinds of rubbers and then compound them with chemicals and fillers to obtain the required properties. Almost all rubbers must be vulcanized (or crosslinked) to form large elastic networks of polymers if they are to perform properly. Sulfur is the most common vulcanizing agent, but many specialty rubbers use different crosslinkers such as peroxides, resins, and other chemicals. In most cases the primary agent, such as sulfur, is not used alone but is combined with other chemicals such as metallic oxides, organic accelerators, and the like that control the speed and type of crosslinking reaction.

Even when vulcanized, some rubbers are weak, particularly at higher temperatures, and they need reinforcing for strength. Fine colloidal fillers, particularly carbon blacks and silicas developed by chemists in other industries, do the job. Sometimes, softeners and plasticizers are added to make rubber easier to process or to make it more flexible at low temperatures.

Rubber products, each in its own way, deteriorate with age. Oils, oxygen and ozone, light, heat, and mechanical action all make them age faster. Some stiffen while others soften to the point that they are no longer serviceable. Again, chemists have contributed by developing compounds that slow aging and give products longer useful lives.

Thus, rubber is seldom used alone but instead is compounded with many auxiliary chemicals. Furthermore, compounded rubber is often part of a composite structure containing other materials such as textiles, metals, and plastics. Producing such composites that are dimensionally stable and strong involves complex chemical modifications usually carried out after the article is molded or otherwise shaped.

Since rubber products are often composites, the dissimilar materials must develop good adhesive bonds to each other. Special chemicals must generally be used to make rubber adhere to textiles, metals, and plastics. In many cases, these adhesive systems are quite specific to the materials involved, and chemists again are called on to develop them to make serviceable products.

Vulcanization

People played games with rubber balls and eventually they waterproofed cloth with thin films of rubber during some three centuries following rubber's discovery in the New World. Unfortunately, these and other products never worked very well, as the rubber melted in the summer, froze in the winter, and turned sticky when exposed to solvents.

Rubber's apparent versatility, however, led manufacturers to keep trying, and by the early 1800's companies in the U.S. were turning out a variety of rubber items, including shoes and boots, aprons, and life preservers. They tried to remove the gum's natural adhesiveness by mixing it with lampblack or by dissolving it in solvents and then evaporating the solvents. Their successes could only be described as indifferent, however, until Charles Goodyear applied himself to the problem.

In poor health and with several business failures behind him, Goodyear had a contract with Roxbury India-Rubber Co. in Massachusetts to improve life preservers. Pleased with his results, company officials suggested that he try to improve rubber itself. He accepted and from that time on put all his effort into solving the problems inherent in the properties of natural rubber.

He first tried compounding rubber with magnesia and boiling it in quicklime and water. The process appeared to destroy surface tack, but the effect proved only temporary. Shortly thereafter, he developed what was termed an acid gas process, and he made the first all-rubber overshoes by the process in the winter of 1837–38.

At about this time, Goodyear met Nathaniel Hayward, a foreman at Eagle Co., a manufacturer of rubber goods also located in Massachusetts. Hayward had patented a process to remove rubber's adhesiveness by spreading sulfur dust on its surface and drying it in the sun. Learning of Hayward's work, Goodyear hired him and began making rubber products using his acid gas process and Hayward's "solarization" process.

Their products failed in warm weather, however. While trying to find out why, Goodyear noted that rubber previously treated with sulfur charred like leather when carelessly brought into contact with a hot stove. Since rubber had always melted when heated, Goodyear inferred that if the process of charring was stopped at the right point it might remove the gum's adhesiveness. He became convinced when he found that rubber charred but would not melt in boiling sulfur.

He soon reduced his ideas to a practical process, and his patent on vulcanization issued June 15, 1844. It covered combining rubber with sulfur and white lead, alternating layers of such rubber sheets with cotton batting, and "exposing this india rubber fabric to a high degree of heat."

Until his death in 1860, Goodyear obtained numerous patents on various uses of rubber. Elastic goods used in suspenders, corsets, and as goring in boots were among his early major contributions. Rubber goods in general also markedly improved following the invention of vulcanization. They had greater elasticity, and cold did not affect their pliability. The usual solvents did not dissolve them (although they did swell), and they no longer stuck to nearly everything, as they had in earlier times. They were impermeable to liquids, and they also served as excellent insulators for electricity.

Organic Accelerators

The next big step in the rubber industry following vulcanization occurred in 1906 when scientists in the rubber industry found that some chemicals would cause rubber to vulcanize faster. In that year, George Oenslager and A. H. Marks at Diamond Rubber Co. (which later became part of B. F. Goodrich Co.) compounded rubber with sulfur and lead oxide, as was customary, but they also added aniline. They discovered that the rubber cured much more rapidly than with sulfur and lead oxide alone. In addition, the rubber aged much more slowly than previously. Since workers could handle aniline only with difficulty because of its highly poisonous nature, the industry's first commercial organic accelerator was mainly thiocarbanilide, an aniline derivative. As it happened, thiocarbanilide was both safer and a faster accelerator than aniline.

The more efficient use of organic accelerators dates from about 1920. At that time, chemists learned that organic acids such as oleic and stearic in the presence of zinc oxide or the zinc salts of such acids were extremely helpful if not absolutely essential to make many organic accelerators function. In the early 1920's, chemists also developed a series of new accelerators, including diphenylguanidine (1921) and guanidine (1925).

Perhaps of more scientific importance, however, was the discovery (in 1921) of the accelerating properties of 2-mercaptobenzothiazole and its derivatives at Goodyear Tire & Rubber Co. Even with many new types of accelerators on the market today, thiazole derivatives account for 65% of total accelerator production. They speed curing but do not activate curing at the lower temperature used for mixing and processing. Another major advantage is the improved resistance to aging that they give vulcanized rubber. Their improvements combined with those supplied by newly developed antioxidants accounted for signal improvements in rubber products in the late 1920's.

Reinforcing Agents

The chemicals that vulcanize rubber are by far the most important in giving rubber articles the desired properties. Even when vulcanized, however, rubber still lacks the balance of properties required for good wear. The discovery of rubber reinforcement, therefore, especially with carbon black, ranks close in significance to vulcanization in improving rubber product properties.

Goodyear himself in 1855 wrote about lampblack that was "often used to cause the gum to endure the effects of sun and weather," and shortly after the turn of the century an English chemist noted for the first time that carbon black reinforced rubber. It was not until about 1912 in work at Diamond Rubber Co. to perfect a tread compound to outlast the fabric used in tires, however, that the most significant advance occurred. Results there showed that batches containing carbon black lasted 10 times longer than those containing clay or zinc oxide, popular rubber extenders of the time. The discovery made rubber goods more practical than ever before for such articles as tires for vehicles and belts for conveyors.

The chemical industry today produces some 3 billion pounds of carbon black a year, up from the 10 million pounds made in 1911. While the dollar cost of about $210 million for carbon black may not be considered high in a nation accustomed to counting in the billions, it is safe to say that the rubber industry would be very handicapped if carbon black were not available.

Such growth in carbon black production would not have occurred without several other significant discoveries in succeeding years, of course. For example, manufacturers could not mill the originally fluffy, low-density carbon black into rubber without much labor and considerable cleanup. Then in 1927, scientists at Binney and Smith patented a wet pellet process and at about the same time others at Cabot and also General Atlas and Powder perfected a dry pelletizing process. The new form from these processes reduced the labor and mess and let the industry develop more efficient mixing, automatic weighing, and conveyorized systems.

In time, new blacks also replaced the traditional channel black. Thermal blacks were produced commercially in 1922, and the first furnace blacks followed in 1928. These latter were especially responsible for how rapidly the synthetic rubber industry grew. Growth was particularly assured when tiremakers found that the most widely used synthetic rubber reinforced with furnace black let them build a tire that would give more mileage at lower cost than did channel black in natural rubber.

Age Resisters

A rubber may be well vulcanized and well reinforced for its intended uses, but it may soon age into a useless state unless it is protected. Industry personnel now know that rubber ages because of attack by oxygen and ozone in the air, ultraviolet radiation from the sun, heat in the environment, and metallic contaminants such as copper, iron, and manganese in rubber articles themselves. Chemists have long sought compounds that would slow aging, and rubber manufacturers today can buy any of more than 100 chemicals called age resisters to keep products from becoming hard and brittle, turning sticky, or developing cracks.

Developing and producing age resisters have put most of the large rubber companies into the chemical business and not a few chemical companies in effect in the rubber business. Goodrich, Goodyear, and Uniroyal led among rubber processors in developing age resisters. They were joined later by Firestone. Du Pont led among chemical companies, followed by Monsanto, American Cyanamid, Eastman Kodak, Ethyl Corp., Universal Oil Products, Pennwalt, and others.

Many of the chemicals used to retard aging in rubber are also used in other materials for the same reason. Among such other materials are gasoline, oils, greases, plastics, and even foods. U.S. companies yearly make about 240 million pounds of age resisters of all types, and the rubber and other industries spend about $150 million a year for them.

The some 170 million pounds of age resisters used in rubber products each year have a value far greater than their direct dollar cost (as is true of age resisters used by other industries, of course). If a rubber industry estimate is correct that rubber articles with age resisters on the average last at least four times longer than they would without age resisters, then the impact on the economy and on society is considerable. Not only must the cost of much more frequent product replacement be accounted for but also the cost of idle machinery, unemployed people, possible accidents, and even stalled autos on the expressway.

Of the chemicals and environmental conditions that age rubber, oxygen and ozone are probably the two most important. Building resistance to oxidation has accordingly concerned chemists most, especially at the beginning. Most oxidation inhibitors today are amines, phenols, phosphites, or compounds of sulfur. Phenols were suggested as early as 1870 to combat aging. Rubber industry chemists between then and the early 1900's used a variety of naturally occurring materials to slow aging, including creosote, naphthalene, asphalt, and coal tar pitch. Then in 1906 the work previously discussed on aniline acceleration of vulcanization also gave products that aged less rapidly, and chemists during the next two decades or so found a number of organic chemicals that protect rubber against oxidation. Among them in addition to phenol and aniline were such compounds as the cresols, hydroquinone, and many other hydroxy and amine derivatives.

In 1924, chemists found the first of the nonaccelerating antioxidants, a group of aldehyde-amine reaction compounds. These compounds made rubber products last two to three times longer than earlier age resisters did, and they also allowed products to be used at higher temperatures.

From the mid-1920's until about 1940, hundreds of patents issued on antioxidants, most of which dealt with secondary aromatic amine products. Amines generally proved to be the most effective oxidation inhibitors, but they tended to discolor light-colored rubber products and even stained surfaces the products contacted. As new uses of synthetic rubber developed after World War II, the rubber industry looked more and more for nondiscoloring and nonstaining antioxidants. During the 15-year period ending in about 1960, a whole new class of antioxidants became commercially available. Commonly called nondiscoloring antioxidants, they opened new markets for rubber products where light colors were aesthetically desirable.

At first, the most promising alternatives to the amines appeared to be phenol, p-cresol, and hydroquinone, but they unfortunately did not retard aging too effectively. After World War II, U.S. chemists learned that the Germans had used reaction products of cresols with olefins to stabilize their synthetic rubbers. At about the same time, work at Gulf Oil Co. showed that phenols with tertiary butyl groups ortho to the hydroxyl deterred oxidation. With these two clues, chemists made many hindered phenols (as they were labeled) as candidates for nondiscoloring antioxidants. The most effective was a phenol with tertiary butyl groups ortho to the hydroxyl and a methyl group para to it. This chemical evaporated too quickly for many uses, but chemists reacted a phenol closely related to it with formaldehyde and opened up a whole new field of antioxidants, the hindered bisphenols.

Up until World War II, most work on aging dealt with oxygen as the cause. Then during the war when unsaturated rubber cracked severely under stress, it became apparent that ozone also aged products. The problem became really critical, however, when the Korean War began. The army discovered that tires stored since 1945 had cracked severely and failed after running only a few hundred miles.

Chemists soon found that a chemical used as a gasoline antioxidant (N,N'disec.butyl-p-phenylenediamine) effectively protected rubber against ozone. Unfortunately, it was both highly toxic and quite volatile. Since considerably more had to be used to protect rubber goods against ozone than against oxygen, its use was both unsafe and expensive.

In 1954, chemists at Rock Island Arsenal laboratory found the first effective and safe substitutes and coined the term antiozidants for them. In the decade of the 1950's, chemists in many rubber and chemical company laboratories made and tested hundreds of antiozonants (as they came to be called). The successful ones not only solved the army's ozone cracking problem but aided civilians as well. A product known as oil-extended rubber had been developed in the early 1950's, and manufacturers soon adopted it for making auto tires. Ozone attacked such extended rubber severely, however, and disconcertingly deep cracks formed in sidewalls. Development of effective antiozonants occurred just in time to eliminate the problem. In addition, ozone concentrations at about that time began to rise in large cities, especially in Los Angeles, and antiozonants were available to make ozone-resistant goods for use in such regions.

The rubber industry now uses some 40 million pounds a year of antiozonants, and their development is one of the most important recent events in the industry. They are effective against both ozone and oxygen, they help decrease cracking caused by flexing, and they deactivate such contaminants as copper, iron, and manganese. To a large extent they have replaced the conventional amine antioxidants developed earlier for stabilizing synthetic polymers, and they are generally regarded as universal age resisters for black stocks today. They unfortunately aren't suitable for light-colored articles. While nondiscoloring antiozonants have been introduced recently, they generally lack the effectiveness of the phenylenediamines and are used in specific applications only.

The First Synthetic Rubbers

The unusual properties of rubber intrigued scientists for years, but only after some 100 years of research were they able to deduce the molecular structure needed for rubber-like elasticity and to develop a theory to explain rubber's elastic behavior. Today, because of their understanding and because they can make many molecules with elastic properties, chemists have provided a wide range of rubbers to meet many needs.

Until the early 1930's, natural rubber was the only elastomer available. While the rubber industry could process it to meet many service conditions, it still had certain limitations, including mainly relatively poor resistance to oils and limited resistance to heat and aging. Moreover, the U.S. had to import every pound, not a situation in which any nation feels secure.

By the 1920's, the research staffs of many organizations in the U.S. and in other countries had investigated natural rubber and had explored many possibilities for making synthetic rubber. The empirical formula of the main constituent of natural rubber had been established as C_5H_8 in the early 1800's. In succeeding years, chemists determined that the monomer had one double bond and that pyrolysis with heat and ozone showed it to contain many isoprene units. Chemists isolated natural rubber's basic unit, isoprene, in 1860 and its close relative, butadiene, in 1863. They also polymerized isoprene to a rubber-like substance in 1879.

By 1910, scientists had concluded that rubber was a polymer of *cis*-1,4-polyisoprene, but they neither agreed on its molecular weight nor understood why such a molecule should be elastic. By the late 1920's and early 1930's, however, the long accumulation of information culminated in two events that marked the beginning of the change in the rubber industry from natural to synthetic elastomers. One of these events was the genesis in Germany of the science of macromolecular chemistry and the other the announcement in the U.S. of the first two commercially available synthetic rubbers.

In Germany in about 1930, Hermann Staudinger proved that rubber, cellulose, and some plastics are made up of giant molecules. Professor of chemistry at University of Freiburg, Staudinger laid the scientific basis for macromolecular chemistry, as it was to become known. Today's rubber, plastics and resins, and man-made fiber industries rest largely on that science, and Staudinger received the Nobel Prize in Chemistry in 1953.

In the U.S., meantime, two companies had synthetic rubbers ready for commercialization. In 1931, du Pont introduced the first synthetic elastomer of great importance, polychloroprene, and Thiokol Corp. at about the same time announced a polysulfide rubber it called Thiokol. Articles made from both rubbers soon began supplying markets for specialty products. Although it had lower stress/strain values than natural rubber, Thiokol resisted solvents extremely well. Neoprene (as du Pont named its polychloroprene in 1936) also resisted solvents extremely well. In addition, its stress/strain values equaled natural rubber's, and it also weathered much better and did not become soft and sticky after prolonged aging.

Thiokol's production has increased appreciably in the 40 years since its introduction, but its production has never approached the large scales reached by synthetics developed later. It still finds uses in specialty applications, however, such as in flexible hoses for gasoline, paint thinners, and aromatic fuels. It is also used to make inking rollers and other products where resistance to hydrocarbons is required. A liquid form of it was one of the first elastomers used in making solid propellants for rockets during and after World War II, although such use is very small now.

Neoprene was introduced at $1.05 a pound at a time when natural rubber sold for less than 5 cents a pound. Nevertheless, it established itself within half a dozen years as a premium specialty rubber, and annual consumption reached 2 million pounds by 1938. Du Pont now operates two neoprene plants in the U.S. and has one competitor here, while other companies produce neoprene in France, Germany, and Japan. Total world production capacity is about 300,000 long tons a year, which contrasts rather well with the 12,000 pounds made in 1931, the introduction year.

No rubber has yet been found to equal neoprene's all-round balance of properties. Its vulcanizates resist most oils and greases and many chemicals. They withstand temperatures in the 250° F. range and intermittently higher temperatures, and they can be compounded for low temperature service as well. Oxygen, ozone, and ultraviolet radiation have virtually no effect, and neoprene does not propagate flame. Such attributes have led to its use in a wider variety of rubber products than any other elastomer.

The First General-Purpose Synthetic

Despite all their good properties, neither Thiokol nor neoprene proved to be a truly general-purpose synthetic rubber that could compete with natural rubber in its many uses. Such a rubber was not long in following the first synthetics, however.

Standard Oil Co. (New Jersey) is the company in the U.S. that pioneered with what was eventually to evolve after much work by it and many other companies into what is known today as styrene-butadiene rubber (SBR). The company first became interested in synthetic rubber in the post-World War I decade. The U.S. was using gasoline at a rapidly rising rate then as auto use grew, but the petroleum industry was not finding new reserves fast enough. During World War I, Germany had tried hard to make oil from coal, and in 1927 Standard Oil (New Jersey) sought to benefit from its experience by signing an agreement with Germany's I. G. Farbenindustrie to exchange information on making oil from coal. Then the next year, drillers found the East Texas oil field, and the U.S. oil industry shifted from scarcity to surplus.

Work on making oil from coal was promptly suspended, but Standard Oil and I. G. extended their agreement to include making chemical raw materials from oil. They also eventually included synthetic rubber in the agreement, since it appeared that the key monomer, butadiene, might best be made from oil or natural gas.

The Germans developed two synthetic rubbers and shared information with Standard Oil on them under terms of the agreement. One they called Buna S, a butadiene-styrene elastomer that was viewed as a general-purpose rubber. The other they called Buna N, a butadiene-acrylonitrile elastomer that was viewed as an oil-resistant rubber.

By the late 1930's, some Buna S had been imported to the U.S. and experimental tires built with it. When the war started in 1939, chemists and chemical engineers at a Standard Oil subsidiary, Esso, were working on making butadiene from oil and on catalytic cracking, which was to be a route to making aviation gasoline as well as to making butylene, the principal precursor of butadiene. With a rubber crisis obviously impending, the Federal Government was soon involved in helping get a pilot synthetic rubber program started in the U.S. A Standard Oil (New Jersey) subsidiary began building a plant to make Buna N to help supply the demand for an oil-resistant rubber and at the same time act as a pilot plant for a Buna S plant. (This plant was the first commercial size emulsion polymerization plant built in North America. It is still operating at Baton Rouge and is now owned by Uniroyal.) The subsidiary also began building a butadiene plant and agreed to license others to make Buna N provided they also experimented with Buna S at the Government's request.

Independent of this activity, some of the largest U.S. rubber companies were also working on practical ways of making synthetic rubber that would replace natural rubber, especially in all-important tires. Buna S imported from Germany before World War II processed poorly, for example, and tires made from it were inferior to those containing natural rubber. In addition, it was produced in a batch process, and its physical properties varied considerably. It also had to be heated in ovens at times to produce workable raw rubber.

The Government, the four major rubber companies, and Standard Oil Co. (New Jersey) agreed in December 1941 that information then existing on styrene-butadiene rubbers and to be developed through subsequent research on them would be pooled. Other companies later joined in the agreement. Within a short time, chemists soon developed a new all-purpose rubber that could be processed on conventional equipment, and tires made from it were greatly improved. Chemical engineers, meantime, also contributed by developing such processes as continuous polymerization, new types of agitation, efficient monomer stripping, faster coagulation, and continuous drying. Whereas the German product was made by a single company to suit its own needs, U.S. manufacturers needed—and got—a more uniform, workable, and high-grade product.

The shift that occurred during the early war years from natural rubber to what became known as GR–S (for Government Rubber–Styrene) ranks as one of the major chemical accomplishments contributing to the Allied war effort. In the years since then, what is now known as SBR (for styrene-butadiene rubber) has become a vastly different and improved product. Better polymerization techniques, new types of modifiers, and improved compounding materials such as high reinforcing blacks and better antioxidants and antiozonants have combined to give the high quality SBR produced today. Annual U.S. capacity for SBR now amounts to more than 1.9 million long tons. Moreover, capacity for the other major synthetics developed both before and after the war is now close to 900,000 long tons. In addition, the country's synthetic rubber plants produce in excess of 500,000 long tons a year of latex, the liquid form of rubber that goes into such products as foam, paints, adhesives, nonwoven fabrics, and asphalt.

Consumers and industry have benefited from products that are more serviceable when compounded from such rubbers. In addition, they have been freed from the vagaries of price so characteristic of natural rubber. Finally, benefits of synthetic rubber developments have been extended to other countries as well. New factories built in Europe and elsewhere after the war, often with U.S. financial aid, used American technology, and in 1969 total world production of all types of synthetic rubber amounted to more than 4.5 million long tons.

Butyl, the Third Prewar Synthetic

Another synthetic rubber had been developed before the war besides polysulfide, polychloroprene, butadiene-styrene, and butadiene-acryloni-trile rubbers. That rubber, butyl, supposedly had little or no future when first developed in 1937, but it has since become a most successful specialty rubber. It appears in many products today where its properties make it nearly unique.

During the period when Standard Oil Co. (New Jersey) and I. G. chemists were pursuing research on the butadiene-styrene and butadiene-acrylo-nitrile rubbers, chemists in Germany succeeded (in 1929) in preparing a gum-like homopolymer of isobutylene, and Standard Oil continued with the work. Polymers of isobutylene would not vulcanize, however, and it was not until 1937 that chemists William J. Sparks and Robert M. Thomas at what was then Standard Oil Development Co. (now Esso Research & Engineering Co.) copolymerized a small amount of butadiene with iso-butylene. In doing so they produced an elastomer that would vulcanize, had good initial tack, and aged well. Sparks and Thomas eventually selected isoprene as a more desirable diene for the copolymerization, and today isobutylene-isoprene copolymer, or butyl rubber, is the most important of the isobutylene family of elastomers.

The process for making butyl is unique in that it requires a plant that operates continuously at −140° F., and it must maintain this low tem-perature despite the large amount of heat the polymerization evolves. Also, all reactants and solvents in the polymerization must be very pure.

Despite some intriguing properties, butyl could find no markets before the war, and interest lagged badly until the clear possibility of the coun-try's being cut off from natural rubber in the Far East stimulated Standard Oil Co. (New Jersey) to revive its research program. A subsidiary was only six days into construction of a commercial plant at Baton Rouge on Dec. 7, 1941, and the Government was soon clamoring for production. Military and essential civilian uses got all butyl until 1945. After the war, civilian auto and truck tubes became butyl's major outlet, and the nearly daily stop by motorists for air gradually faded into history. Butyl producer Enjay, a subsidiary of Standard Oil Co. (New Jersey), purchased the butyl plants from the Government in 1955, but it received a rude shock before too long. Scientists at Goodrich had patented a tubeless tire in 1946, and Detroit finally put tubeless tires on as original equipment with the 1955 models. Within two short years of that event, butyl lost 40% of its market (or about 50 million pounds a year).

Undeterred, Enjay set about finding new uses. It developed a new butyl grade for tubeless tire innerliners. Since butyl isolates noise and vibration so well, it was also developed for body and engine mounts and suspension bumpers. Its excellent weather resistance opened markets for it in window sealing tapes for homes, commercial buildings, and autos.

Today, American autos contain more than 4 pounds of butyl on the average. In addition, the tire tube is still a growing market, especially in truck service. In fact, tubes now use about 75% of the butyl produced in the U.S. and about 70% of that produced worldwide.

Butyl never became the general-purpose rubber it was originally intended to be. There was a short time some years after the war when the butyl tire entered the market and it appeared butyl would succeed. Butyl tires gave a soft and quiet ride. They unfortunately often gave only a short one before wearing out (partly because butyl doesn't resist abrasion very well, but in butyl's defense also partly because some users tended to drive such tires harder since they did not *sound* as though they were wearing).

Lacking a significant market in tires, butyl (or any other rubber, for that matter) could not become a really high-volume product. Nevertheless, it has become one of the most important specialty rubbers, where its blend of properties is very important. These include very low permeability to gases and water, very high shock absorption capacity, good flexibility at low temperature, outstanding resistance to heat and aging, good electrical properties, high tear strength, fair to good resistance to ozone, and excellent resistance to such chemicals as acids, bases, oxygenated solvents, and fatty acids.

Such properties put butyl in countless articles in manufacturing processes, in buildings, and in consumer products. Of all its applications, however, those involving butyl's impermeability to gases and water are probably the most apparent to the average consumer. Before the butyl tube (and now butyl innerliner) became commercial, tires needed air at every stop for gas—and often in between. Now, drivers add air only every month or two (although they still should check tires more often as a safety precaution). The convenience of having to make fewer stops for air is known to everyone, but more important is the longer mileage people get from tires simply because they don't run underinflated so much.

Butyl's importance as a water barrier is also readily apparent. A square yard of butyl $1/32$ inch thick allows about a pint of water to seep through in 10 years, while a square yard of concrete 1 inch thick would let about 70 gallons through in the same time. Because of this property, and because sunlight, heat and cold, ozone, fertilizers, farm chemicals, molds, and bacteria do not affect it, butyl is used to line many irrigation ditches, ponds, and reservoirs. An outstanding example of such uses is the butyl lining of the 1.4 billion gallon Kualapuu Reservoir on Molokai that will irrigate some 17,000 acres of farmland in the dry central and western portions of the island.

Oil-Extended Rubber

In 1951, the Government in the form of its Reconstruction Finance Corp. (RFC) owned the synthetic rubber plants it had built in the U.S. during the war. On Feb. 21, RFC announced that its plants were making a new product described as an oil-extended GR–S masterbatch. The rubber industry probably never accepted an innovation quite so rapidly. From hardly visible production at 1951's beginning, output rose to 14 million pounds a month by December and totaled more than 70 million pounds for the year.

Of all the improvements made to GR–S (or SBR as it is now called), this one coming when the elastomer was barely a decade old has had more of an impact than any other one, and it comes close to rivaling the development of GR–S itself. In 1969, the U.S. rubber industry made nearly 1.2 million long tons of SBR, about 60% of which was oil-extended rubber. Americans, and to a lesser extent people in other countries, ride on tire treads consisting for the most part of SBR extended about 35% with petroleum hydrocarbons. Besides being considerably cheaper than SBR alone, the oil-extended polymer gives better traction and a softer ride. Moreover, without oil extension, the industry in the U.S. would have had to expand its capacity about 35% in the early 1950's to produce the SBR needed. Doing so would have decreased the effort the industry was then putting on the new cold rubbers and stereo rubbers.

To understand the significance of the accomplishment that oil-extended SBR represents, one must understand at least a few aspects of tire manufacturing, since oil extension affected tires mostly and tires in turn so dominate the market for rubber. Of the various parts of the tire, the tread rates among the most important. For a tire to perform properly, it must among other things have a tread with a uniform cross section that weighs the same around the tire's entire circumference. In addition, tread rubber must be tough, strong, and resilient, and it must resist cracking caused either by flexing or light.

Tire manufacturers may make treads either by molding or extrusion, but they have found that only extrusion allows them to obtain the required uniformity at competitive costs and in the volumes needed. For extrusion, rubber must be reasonably soft. Unfortunately, the rubbers that combine toughness, strength, and resilience best also extrude least easily. In fact, processors cannot extrude them without adding plasticizers and masticating them mechanically for some time to make them soft enough to process. Such processing raises costs, and it also produces vulcanizates with lower properties.

With the advent of the general-purpose synthetic, GR–S, the rubber industry grappled with the conflict between properties and processability by controlling the degree of polymerization and producing a suitably soft rubber in the first place. Tires made from such softer rubbers, unfortunately, did not equal those made from tougher rubbers carefully processed to a similar degree of breakdown.

Then scientists found that they could take a tough synthetic rubber not suitable for extrusion into tread stock, mix it with any of a number of high-boiling petroleum hydrocarbons in ratios as high as pound for pound, and make a tread stock that not only extruded readily with the required accuracy but also vulcanized to a tread superior to those obtained from the then available GR–S stocks. In addition, tire manufacturers also found oil-extended GR–S worked well as a carcass rubber. Among its advantages in this use, it developed less heat when flexed and helped overcome a defect from which the first synthetic rubber tires failed. (Oil extension was not without its disadvantages, however. The first oil-extended tires cracked severely in the sidewalls, but development of anti-ozonants overcame the problem, as recounted earlier.)

Oil extension allowed the rubber industry to use tough rubbers it had not been able to process earlier. It also reduced elastomer costs significantly, since the petroleum hydrocarbons sold for 2 to 3 cents a pound compared to 19 to 20 cents a pound for unextended (but otherwise compounded) rubber. Over the years, this cost reduction has saved motorists (and others, since oil-extended rubber now is used in products other than tires) millions of dollars. Meantime, of more immediate concern at the time, oil-extended GR–S overcame a severe rubber shortage in the U.S. At 1951's beginning, the country was six months into the Korean War and among other things exceedingly short on rubber. By extending available supplies with oil, rubber industry scientists lessened the scramble for more capacity both for monomer and polymer plants. The U.S. patent on oil-extended SBR was issued in 1960 to a team of scientists working at General Tire & Rubber Co.

Chlorosulfonated Polyethylene Rubber

In April 1952, du Pont formally introduced the second of its synthetic rubbers to American industry. Hypalon, as du Pont trademarked it, turned out rather different from what the company intended when it began research a decade and a half earlier, however.

The story started in 1936 when chemists at Western Cartridge Co. received a patent covering direct chlorosulfonation of hydrocarbons. Du Pont bought a license, and its chemists at first used the process in detergent research. They soon extended the Western Cartridge patent with one covering production of sulfonyl chlorides of olefin polymers. These products showed considerable promise as fabric coatings, but the war intervened to interrupt further development.

After the war, du Pont resumed work, with fabric coatings seen as a major use. The hoped-for market didn't materialize, however. Upholstery fabrics offered the greatest potential, but the chlorosulfonated polyethylene had every property wanted except the important one—it did not allow vapors to pass through and thus felt uncomfortable.

Du Pont dropped the fabric coating use and instead offered the polymer on the open market. Manufacturers showed considerable interest, because it resisted ozone and strong oxidizing chemicals outstandingly well. Also, it did not require carbon black to produce vulcanizates of good quality, and it could thus be used in a wide range of colors. Today, du Pont produces Hypalon in the country's only commercial plant (at Beaumont, Tex.). While annual consumption is small compared to that of other synthetic rubbers, Hypalon has met needs other elastomers cannot meet in a variety of uses ranging from protective coatings to chemical hose to special wire coverings.

The Stereo Rubbers

A new era in synthetic rubber began in the mid-1950's when Goodrich, Firestone, and Goodyear each revealed it had made a rubber from isoprene that matched natural rubber molecule for molecule. The discovery of how to duplicate in a reactor what nature does in the *Hevea brasiliensis* tree marked the beginning of the age of the so-called stereo rubbers. Stereo polybutadiene, stereo styrene-butadiene, and ethylene-propylene rubbers soon followed the stereo polyisoprene, and together they have provided a new and almost unlimited range of tailor-made rubbers.

In the past 15 years, the stereo rubbers have become an entirely new industry. In 1970, production totaled some 280,000 long tons of stereo polybutadiene, 120,000 long tons of stereo polyisoprene, 63,000 long tons of ethylene-propylene-diene rubbers, and 85,000 long tons of solution polymerized SBR using similar catalysts. New capacity is still being announced, and growth promises to continue for years to come.

It is still rather early to assess fully the total social and economic impact of the new rubbers. Clearly they have provided superior consumer and industrial products. In addition, they have completed the trend started by GR–S to stabilize and put a ceiling on the price of natural rubber.

Development of stereo rubbers has also expanded manufacturing opportunities for related products. New sources of isoprene had to be developed to keep pace with stereo rubber growth, for example, and production of butadiene has been expanded many times in the past 15 years. Making the polymerization catalysts, mainly organoaluminum compounds, has also led to an entirely new industry, since such chemicals were laboratory curiosities 15 years ago.

But probably the greatest benefit from the stereo rubbers has been to national security. The U.S. no longer must depend on imported natural rubber, and it can now meet all of its urgent emergency needs for elastomers with a domestic synthetic rubber industry should it have to.

The significance of what chemists accomplished in making the stereo rubbers lies in the spatial relations that exist around the double bond in polyisoprene and polybutadiene. The distinguishing feature of stereo rubbers is the orderly arrangement of their atoms in three dimensions with respect to the plane of the carbon-carbon double bond. In 1,4-polyisoprene, the elastomer is termed cis-1,4-polyisoprene when the hydrogen atoms and methyl groups are always on the same side of the double bond in the repeating isoprene units. When they are on the opposite sides of the double bond, the elastomer is termed trans-1,4-polyisoprene. The cis structure is identical to that of Hevea natural rubber, while the trans structure is identical to the natural resins known as gutta-percha and balata.

[Isoprene (and butadiene) can also be polymerized catalytically across the 1,2- or 3,4- double bonds in addition to across the 1,4- bonds. In these polymerizations, the elastomers have no unsaturated double bonds in the continuous chain but instead have pendant unsaturated vinyl groups along the chain. At present, only 1,4- polymers are produced commercially, while the 1,2- and 3,4- polymers are still experimental.]

The search for ways to make Hevea rubber has a long history, but success in controlling structural symmetry of molecules remained insignificant for years. Then when success did come in duplicating in the chemical plant what nature does in a rubber tree, it turned out to be in reality a chance discovery.

In the early 1950's, chemists Karl Ziegler in Germany and Giulio Natta in Italy in work with metal alkyl catalysts showed how to polymerize olefins with regular configurations. Goodrich-Gulf Chemical Co. obtained a license from Ziegler to work with the organometallic catalysts he had used for polymerizing α-olefins at low pressure. During their work, scientists at Goodrich, one of Goodrich-Gulf's co-owners, were attempting to use Ziegler's catalysts to make an unsaturated (and therefore vulcanizable) polyethylene. Since using isoprene earlier had proved successful as a way to introduce unsaturation in olefin polymers (as in making butyl rubber, for example), they copolymerized isoprene with ethylene using a trialkyl aluminum/titanium tetrachloride catalyst. Infrared examination of their first copolymer showed isoprene present in the cis-1,4 configuration, as it is in natural rubber. In short order they fractionated the polymer and separated pure polyethylene and pure cis-1,4-polyisoprene. Needless to say, they immediately forgot about making an unsaturated polyethylene and concentrated on making cis-1,4-polyisoprene. They obviously wasted little time, for within three months they had made ton lots of pure cis-1,4-polyisoprene and had polyisoprene truck tires running on their test tracks.

The Goodrich scientists worked at first with trialkyl aluminum/titanium tetrachloride catalysts, as just mentioned. They next found that the isoprene polymerized in the pure *trans*-1,4- configuration if they modified slightly the ratio of trialkyl aluminum to titanium tetrachloride. Thus, by proper manipulation they could use a single catalyst to prepare either *cis*-1,4- or *trans*-1,4-polyisoprene. They had succeeded in duplicating two natural materials within the space of a very few days.

While the Goodrich scientists were carrying out the foregoing work in 1954, chemists at Firestone Tire & Rubber were also successfully duplicating the Hevea rubber molecule. They used lithium metal in the polymerization, however. It is ironic that chemists had known since 1910 that lithium polymerized isoprene. The Firestone scientists, however, were the first to apply modern methods of determining polymer structures and to recognize the *cis*-1,4 structure in the polymer molecule.

Goodrich announced its polyisoprene (Ameripol SN) in December 1954, Firestone followed with its announcement (of Coral rubber) in August 1955, and Goodyear revealed its product (Natsyn) in October 1955. The first commercial production of a synthetic natural rubber using a lithium catalyst occurred in 1960 (by Shell Chemical Co.) and of a synthetic natural rubber using a catalyst system of aluminum triethyl and a co-catalyst in 1962 (by Goodyear Tire & Rubber Co.). Stereo-regular *cis* and *trans* polybutadienes followed in time from other companies, and with the stereo-regular polyisoprenes they launched an entirely new era in synthetic rubber. Production today amounts to about 400,000 long tons, and the end of plant expansions is not yet in sight.

Cis-1,4-polyisoprene duplicates Hevea natural rubber at the molecular level, and it can thus replace natural rubber in all uses. Moreover, manufacturers can compound and handle it almost exactly as they do Hevea, making change-over extremely simple. Auto and truck tires provide its largest outlet, with industrial products and miscellaneous items such as elastic bands in second place.

Cis-1,4-polybutadiene is used as a tire rubber in blends with SBR and *cis*-1,4-polyisoprene to give outstanding abrasion resistance, low heat build-up, high gum strength, high cured strength, and a degree of building tack. Its cure compatibility with other elastomers makes it very versatile in blends. *Cis*-1,4-polybutadiene also shows the highest bounce of any rubber and is flexible at very low temperatures.

Trans-1,4-polybutadiene ranges from elastic to nonextensible at room temperature, depending on the relative amounts of amorphous and crystalline structures chemists build into it with their choice of catalyst. It is only a small item of commerce to date, however, being used almost exclusively in synthetic leather applications.

Recently a stereo styrene-butadiene rubber has begun to make its presence felt in the tire industry. This material, called solution SBR to distinguish it from the standard general-purpose rubber, emulsion SBR, shows superior resistance to abrasion and to groove cracking under severe conditions. Solution SBR also has a higher cured modulus than that of emulsion SBR. Production is small now, but solution SBR could totally replace emulsion SBR if predictions for it come true.

Ethylene-Propylene Rubbers

With butadiene and isoprene so firmly established during and immediately after the war as monomers that polymerized to elastomeric products, it was only natural that chemists should examine the possibilities with other olefins. Ethylene was one of the obviously leading candidates.

By the early 1950's, chemists knew that polyethylene exhibited rubber-like properties at temperatures above its melting point, although it was a hard plastic at room temperature. They also knew that chemically modifying some 25% of the polyethylene molecule would destroy its crystallinity and change it from a plastic to a rubber. In fact, chemists at du Pont had already done so successfully in developing a specialty rubber, Hypalon.

A second leading elastomer candidate was propylene, but efforts to polymerize it had produced only low-molecular-weight oils up to the time of the discovery of the Ziegler catalysts. Then in 1955, Natta and his associates in Italy first demonstrated the possibility of making an elastomer of ethylene and propylene. Using modified Ziegler catalysts, Natta found he could make crystalline polymers of α-olefins, such as propylene. Once a way to polymerize propylene to high molecular weight became available, the concept of combining ethylene and propylene to make a rubber soon developed. Natta's approach was to copolymerize ethylene with other olefins, notably propylene, by using enough of the latter (about 30 to 50% by weight) to prevent crystallization. Modified Ziegler catalysts would not polymerize the monomers, since they were generally heterogeneous catalysts, but Natta and his co-workers solved the problem by finding suitable homogeneous catalysts.

Natta's work dealt with the synthesis of ethylene-propylene rubber, but teams of chemists in the U.S., Germany, Canada, and England were also all conducting similar explorations in ionic coordination polymerization during the late 1950's, all stimulated by the basic Ziegler discovery. The result of their efforts was widespread recognition of the potential value of the new saturated rubber, followed by engineering developments to prepare for its commercial production. Chemists and chemical engineers in several large U.S. companies made a number of significant contributions, such as finding optimum conditions of polymerizing and designing equipment to cope with the problems of agitation of viscous cements, heat transfer, maintenance of equilibrium between the gas and liquid phases, accurate metering of raw materials and polymerization catalyst, solvent and polymer recovery, drying, and finishing.

Enjay Chemical Co., a subsidiary of Standard Oil Co. (New Jersey), introduced ethylene-propylene rubber commercially in the U.S. in 1961. As a polymer only of ethylene and propylene, it contained no unsaturation. Conventional accelerated sulfur recipes would not vulcanize it, but organic peroxides would. In addition to good resistance to heat, chemicals, low temperatures, and weathering, it also has good electrical properties. It is used for wire and cable insulation.

Peroxide-cured ethylene-propylene crosslinks slowly, and it also has by-product odors that preclude its use in many applications. The rubber industry improved peroxide cures, but a more significant improvement came with the development of terpolymers containing ethylene and propylene plus about 5% of a nonconjugated diene that increased the polymer's propensity to be crosslinked with sulfur. This improved polymer is now commonly called EPDM to distinguish it from the first ethylene-propylene polymer, which is called EPM. Enjay Chemical and du Pont made EPDM commercially available in 1963, and Uniroyal followed in 1964. Copolymer Rubber and Chemical became an EPDM producer in 1967 and Goodrich in 1971.

EPDM is still a specialty rubber, but it is rapidly becoming a contender for the label of general-purpose rubber. In less than a decade its production has passed the annual rate of 58,000 long tons, and its production is still growing at a rate of 15 to 20% a year. EPDM enjoys such growth in part because it has certain properties that other polymers do not possess, and it can be made at a reasonable cost. Moreover, supplies of ethylene and propylene are practically unlimited. But probably most important, EPDM not only resists aging and chemicals well, but it also can be greatly extended with inexpensive hydrocarbon oil and fillers to lower its cost without sacrificing its properties.

Makers of rubber products find compounded EPDM particularly intriguing because of its excellent resistance to oxygen and ozone aging. Nontire auto parts consume nearly half of today's EPDM production, with tires consuming a relatively small amount. For EPDM to become a general-purpose rubber, it must be used more in tire applications. Nearly all auto tires contain a small amount of EPDM blended with other rubbers in white sidewalls and adjacent black overstripes, where it helps resist weather checking. Tire manufacturers have built all-EPDM tires and report very good performance. Significantly, the EPDM tread stocks contain 75% more extender oil than now used commercially, yet the tires wear quite well.

Nevertheless, many technical problems remain for chemists to solve before EPDM tires can be made to perform safely at high speeds. Perhaps the most serious obstacle is getting proper adhesion of the rubber to the carcass fabric. EPDM's price is also a problem when compared to the prices of SBR and natural rubber.

Problems aside, one industry expert predicts use of about 240,000 long tons of EPDM by 1975, nearly 90,000 long tons of which would be in tires and nearly 70,000 long tons of which would be in other vehicle uses. Whether EPDM will grow that much remains to be seen, of course, and even if it does, EPDM will still lie several levels below the some 1.3 million long tons of SBR used in the U.S. in 1969.

Polyurethanes

Urethanes today are a 600 million pound business in the U.S. They are produced as both rigid and flexible foams as well as elastomers, and they are used in a wide range of applications, including off-the-road tires, fabric coatings, and shoe soles.

Their origins lie in chemical accomplishments recorded somewhat more than 100 years ago in Germany with the first synthesis of aliphatic and aromatic isocyanates in both laboratory and commercial processes. It was not until the mid-1930's, however, that Otto Bayer and his co-workers at Farbenfabriken Bayer Co. in Germany discovered the diisocyanate addition polymerization that led to their production. Bayer was attempting to improve on du Pont patents on polyamides, and his work resulted in the preparation of many different types of polyurethanes and polyureas.

Du Pont chemists undertook the first work in the U.S., which resulted in the finding that hexamethylene diisocyanate would dry alkyd resins more rapidly than processes used up to that time. Their patent (since expired) is considered the basic urethane patent in the U.S.

Most of the early work on urethanes dealt with products for the textile and adhesives industries. After the war, chemists at Goodyear developed methods for making rigid polyurethane foams while working under a government contract. In 1952, arrival of flexible foams with high strength and very low densities assured polyurethanes' commercial success.

The first flexible foams were based on polyesters. They had several shortcomings, such as poor resistance to hydrolysis, high cost, and a boardy feel. Polyether foams introduced in the late 1950's replaced them. They were based on ethylene and propylene oxide. Since that time several improvements have occurred in process technology, and the U.S. market for flexible foam in 1969 was about 450 million pounds.

Most notable among improvements in the past decade and a half is one patented in 1958 by chemists at Goodyear Tire & Rubber Co. They invented foaming on an inclined plane, making it possible to foam larger, continuous slabs without the shears and splits that form when the foam flows back over itself. Another notable improvement is covered by a patent owned by General Tire & Rubber that deals with the use of auxiliary blowing agents such as trichloromonofluoromethane to make. lightweight foam with excellent physical properties.

While urethane foams were being developed, so was urethane elastomer technology. The first millable elastomers in the U.S. were developed by chemists at Goodyear in 1952. Isocyanate cured, they had excellent physical properties but poor shelf stability. To overcome this problem, Goodyear developed a new series of sulfur curable urethane elastomers that had all the processing characteristics of natural rubber combined with the good physical properties of polyurethanes. Several companies have licensed the process and are now making both types of millable urethanes.

Another notable event in urethane technology was Goodrich's introduction of thermoplastic urethanes in the late 1950's. These polymers were unique in that they displayed the properties of a rubber vulcanizate without going through the vulcanization process. Thermoplastic urethanes were improved in the early 1960's by Mobay with the development of a material that could be injection molded and also had the added advantage of crosslinking after molding.

The Germans first developed liquid cast urethane elastomers and casting methods for them. The early systems were based on polyesters, 1,5-naphthalene diisocyanate, and glycol curatives. Farbenfabriken Bayer chemists modified the system to a polyester diphenylmethane diisocyanate system that Mobay introduced to the U.S.

The first cast urethanes had excellent physical properties, but they resisted hydrolysis somewhat poorly. Chemists at Mobay corrected the hydrolysis problem and produced several improved elastomers. The greatest contribution to cast elastomer technology in the U.S. probably came from chemists at du Pont when they developed a polyether toluene diisocyanate elastomer that both diols and diamines would cure. The system, developed in 1956, yields a tough, durable material that is used for a wide variety of cast applications.

Natural Rubber

World production of natural rubber in 1970 amounted to 2.9 million long tons with a value of about $1.1 billion. Virtually all of it came from *Hevea brasiliensis* trees that grow only in tropical climates. At one time, natural rubber was recovered from trees growing wild in tropical forests, but today practically all trees are cultivated on plantations. About 14.5 million acres are devoted to rubber tree cultivation, about 90% of which are in the Far East. There, growing rubber earns a large part of the region's hard currency and is important to the balance of trade. Small farmers cultivate about two thirds of the acreage, with large estates accounting for the balance. On the average, an acre yields about 600 pounds of rubber a year, although well-managed estates consistently produce more than 1,100 pounds per acre.

Natural rubber producers have continuously tried to increase yields and make plantations more profitable ever since plantations were established. Agriculturists and botanists have been primarily responsible for much of the improvement with their programs of selectively breeding genetically superior trees. In the past 30 years they have succeeded in tripling potential yields, and today more than 40% of the acreage is planted in higher yielding strains of trees that produce more than 800 pounds per acre per year.

Natural rubber producers can also increase production by planting more trees, of course, while selective breeding will doubtless produce still higher yielding strains. Chemists, meantime, have and will continue to make their contributions to increased production. Some years ago, chemists showed that applying nitrogen fertilizers increased yields markedly. Later experiments also showed the economic advantages of using phosphorus and potassium as well as nitrogen on some soils. Still later experiments also showed the need for applying other elements such as magnesium, iron, and manganese at relatively high levels, calcium and sulfur at low levels, and zinc, boron, copper, and molybdenum at very low levels.

In more recent times, chemists have centered their attention on increasing latex flow from tapped trees by applying a variety of chemical stimulants. Shortly before World War II, scientists in the field had observed that scratching the bark below the tap and applying oils or irritating and toxic substances stimulated latex flow and increased overall yields. This observation led chemists to search for more effective chemicals. In the early 1950's, they found that growth regulators such as 2,4-D and 2,4,5-T increased annual yields significantly when applied as pastes or in oil solutions below the tap. Commercial use of these chemicals has increased yields by 20 to 30% even with older trees.

Meantime, plant physiologists at Rubber Research Institute of Malaya (RRIM) were trying to determine why latex flows rapidly during the initial period of tapping but then falls off and eventually stops while the tree still has an abundance of latex. In 1966, they found that the impediment is located within a few millimeters of the cut. A growth regulator such as 2,4,5-T decreases the attack by microorganisms on the latex and delays coagulation at or near the cut surface. More recently RRIM workers have found that the tendency to plug is a genetic characteristic of the tree and that yields are stimulated most effectively on strains of trees with a high tendency to plug.

In their search for more effective chemical stimulants, chemists also have found that they improve latex yields greatly if they release ethylene and contain it in close proximity to the tapping area. Such a procedure would obviously not be commercially practical, but chemists have now found compounds that release ethylene at the rate required for stimulation. One such chemical, 2-chloroethyl phosphonic acid, sold under the trade name Ethrel, is now being used commercially. Responses have exceeded 2,000 pounds per acre per year, double the previous yield of the hybrids involved. Indications are that some extrahigh-yielding trees may be stimulated to produce as much as 5,000 pounds. Moreover, a surprising observation has been that supposedly dry trees have started to yield rubber again after treatment with Ethrel.

Besides helping increase production of natural rubber, chemists have also helped develop processes to make natural rubber more suitable for its customers. An enzymatic reaction polymerizes natural rubber, and for this reason its structure is more regular than its man-made counterparts. With a regular molecular structure, natural rubber crystallizes more easily when stretched, giving it a greater breaking strength than any synthetic rubber when cured without reinforcement. The regular structure also gives raw rubber good building tack, a property necessary in the assembly of many complex items such as tires.

These are strong pluses for users. Unfortunately, natural rubber hardens on storage, making it impossible to guarantee a customer will have a particular viscosity at the time of use. In 1960, chemists at RRIM determined that a small number of aldehyde groups on the rubber molecules allowed them to crosslink during storage. They suggested this reaction be blocked chemically. As a result, the industry today markets natural rubbers with specified viscosities, just as does the synthetic rubber industry. These natural rubbers still retain the higher stereo-specificity and greater building tack of the natural polymer compared to the synthetic variety, and they have superior properties that allow them to compete with their synthetic counterparts. The natural rubber industry produced about 250,000 long tons of the new constant viscosity rubber in 1970, and it expects production will be at least 750,000 long tons by 1975.

Before chemists invented general-purpose rubber in the late 1930's to replace natural rubber, the price for natural rubber depended on the relationship of supply to demand, production and transportation costs, and the prices for whatever competing materials there might have been for particular uses. When general-purpose styrene-butadiene rubber began to be produced in volume for civilian markets after World War II, however, its price helped dampen fluctuations in the price of natural rubber and put a ceiling on it. From 1900 to 1940, the price of natural rubber ranged from 3 cents to more than $2 a pound. After the war, however, the price of natural rubber has been under 30 cents a pound (except briefly at the beginning of the Korean War) as natural rubber producers have kept the price down to meet the competition from synthetic rubber. Clearly the improvements in yields and in properties of natural rubber have been very important to the economic health of natural rubber producers. They have also allowed consumers to continue to benefit from products containing natural rubber at consistently lower prices.

Natural Rubber Derivatives

The commercially significant chemical derivatives of natural rubber are hard rubber, halogenated rubbers, isomerized rubbers, and polymerically modified rubbers. Of these, hard rubber and halogenated rubbers have been the most important.

Charles Goodyear in the U.S. and Thomas Hancock in England at about the same time in the 1840's discovered hard rubber, or ebonite. Ebonite, the oldest man-made rigid polymer, is made by hot vulcanization of natural rubber to give a product containing more than 15% combined sulfur. Its major uses depend on chemical resistance and mechanical properties. Battery boxes and linings for chemical process equipment take advantage of its resistance to corrosive liquids. Applications depending on its strength and toughness, meantime, are many although each is individually small. They include combs, surgical appliances, pipe stems, water meter parts, bowling balls, and woodwind instruments. Until about 1950, ebonite was also used extensively as an intermediate bonding layer to attach soft, vulcanized rubber to metal to line tanks and cover printing press rolls. This application has since been almost eliminated by chemical bonding agents, while the prevalence in recent years of cheaper insulating plastics has reduced ebonite's use as insulation against electricity.

Natural rubber behaves chemically as a trialkylethylene, and it undergoes addition reactions at the double bond or substitution reactions at the activated methylene or methyl groups. Of the many reactions investigated, however, only those involving the halogens and the halogen acids have been of significant commercial importance.

Treating rubber sheet, latex, or solutions with gaseous chlorine produces chlorinated rubber. The fully chlorinated rubber is not flammable, and it resists many chemicals very well. Such properties make it important as a base for protective paints. In fact, anticorrosive paints used in chemical plants and in marine and other construction applications account for about 65% (about 20,000 metric tons a year) of all chlorinated-rubber paints. The use is expected to grow about 16% a year, especially as specifications and codes involving product flammability become stricter. Other volume uses are in adhesives and traffic paints.

Hydrogen chloride readily adds to natural rubber to give a highly crystalline, flexible, tough, film-forming material. Its main use has been as a packaging material (Goodyear's Pliofilm). Partially halogenated rubbers, meantime, using either halogens or the haolgen acids, have been used extensively as thermally unstable constituents of rubber-to-metal adhesives. Such rubbers are stable at room temperature. When they are raised to rubber-curing temperatures, however, they decompose and supply free radicals that provide primary chemical bonding between adjacent rubber and a metal adhesive primer such as fully chlorinated rubber.

Treatment of natural rubber with sulfuric acid, aromatic sulfonic acids, or Friedel-Crafts catalysts forms resinous solids known as cyclized (or isomerized) rubber. These rubbers have the same empirical formula as polyisoprene. Their densities, refractive indexes, and softening points are higher than those of natural rubber, but their intrinsic viscosity and degree of unsaturation are lower. Cyclized rubbers have been used commercially in compounds for shoe soles, paper coatings, hard moldings, industrial rollers, and reinforcing resins. A cyclized rubber (Pliolite NR) made by Goodyear by reacting tin tetrachloride with natural rubber in solution is used for coatings for moisture-resistant paper and for adhesives. For many years a commercial process (Goodrich's Vulcalock) was used to bond soft rubber to metal, but more heat-resistant bonding agents have gradually replaced it. One use that is still commercially important, however, is in anticorrosion paints for marine uses, because it adheres to metal extremely well and is inherently tough.

Polymerically modified rubbers include those in which vinyl monomers have been polymerized or to which monomers such as methyl methacrylate, styrene, and vinyl pyridine have been grafted. Of these, the methyl methacrylate modification is the most important. Swelling rubber in the latex form with methyl methacrylate in the presence of a peroxide-polyamine catalyst grafts half or more of the rubber to make a product called Heveaplus.

Heveaplus can be used in latex form or in solution. It blends readily with natural rubber, and it can be processed and vulcanized in the regular manner. Vulcanizates have excellent physical properties, including high hardness. They are used to make rigid moldings requiring high impact strength.

Rubber Latices

Rubber latices are rubbery polymers stably dispersed in water. They account for about 8% of all rubber consumed. Foam rubber, dipped goods, nonwoven fabrics, carpets, paper, paints, adhesives, and many other products use increasingly large quantities. The highway industry uses latex in bitumen road surfaces, and if such surfaces were adopted universally they would use more rubber than any other application.

Particles in natural latex are mainly cis-1,4-polyisoprene. Consumption in 1969 amounted to just over 61,000 long tons (dry rubber basis).

Particles of synthetic latices consist of a variety of polymers having many different properties. Styrene-butadiene latex is the largest volume synthetic, accounting for nearly 160,000 long tons (dry basis) in 1969. Other synthetics include nitrile-butadiene, neoprene, and carboxylated versions of styrene-butadiene and nitrile-butadiene types.

Latex produced by the tree has a low solids content, and it isn't stable. Accordingly, rubber manufacturers had little use for it until well into the 19th Century. Then, rubber scientists began preserving latex by treating it with ammonia, and the industry started making dipped goods, rubber thread, and treated fabrics.

Even with suitable methods for preservation, however, the industry adopted latex slowly, largely because of its relatively low solids content. Plantations in Ceylon, Malaya, and Java in about 1920 developed regular supplies of preserved and stable latex, and in the next few years scientists found how to concentrate it to raise the solids content. Interest in using latex revived significantly. Rubber thread made by extruding latex into an acetic acid bath, tire cord impregnated with latex, and cans sealed with latex were some of the uses that resulted.

In the late 1920's and early 1930's, the industry developed the first foamed products. One process patented in 1927 in the U.S. called for beating air into a natural latex stabilized with soap, refining the foam to a fine-celled froth, and drying the froth in an oven. Dunlop Rubber, an English company, bought the patent and improved the process somewhat. It is the basis for the Dunlop latex foam process still used today.

In a second process developed in the mid-1930's, catalytic decomposition of hydrogen peroxide formed the foam. This process has been modified to its present form in which expansion of the foam takes place under vacuum. The foam is then frozen, gelled by permeation with carbon dioxide, and heated to vulcanize the rubber.

In the days since those developments gave birth to natural rubber latex products, the industry's major innovations have involved better ways to preserve the latex, to remove the preservatives before processing, and to increase particle sizes to hasten creaming rates.

Synthetic rubber latex dates from a German patent issued in 1912, but it was not commercially important until the advent of neoprene in 1931. In the following years, neoprene latex found its way into a number of commercial uses, such as dipped goods, adhesives, and foam rubber.

After World War II, the industry turned to the possibilities offered by commercial products made from styrene-butadiene latices. The first SBR latices had a low solids content, however. Moreover, the particles were much smaller than those in natural latex. Concentrating such particles beyond about 50% solids made the viscosity too high for practical use. Rubber industry scientists thus had both to make larger particles and increase solids content.

Chemists at Goodyear in the late 1940's succeeded in making the rubber particles in SBR latices larger by seeding the polymerization with latex particles prepared in a previous polymerization. Goodyear during the same period also introduced an SBR latex with 60% total solids that it prepared by increment soap addition and that had a lower nonrubber content than latices agglomerated by chemical additives. Uniroyal scientists during the same years also produced latices with 60% total solids.

Besides increasing particle sizes and total solids content, rubber industry chemists also worked on improving the stress/strain properties of early SBR latices. Putting more styrene in the copolymer was one way to reach their goal, but more styrene gave the latex products poor properties at low temperatures. They then found that polymerizations carried out at 40° to 50° F. to about 60% conversion instead of the 122° F. and 95% conversion used earlier produced a polymer with both better stress/strain and low temperature properties.

Even with these improvements, however, the rubber industry was forced to do better to compete with the new foams made from polyurethane when they became commercial in the early 1950's. In 1964, chemists at Goodyear applied for a patent on a latex whose particles consisted of a tough, high-styrene polymer encapsulated within a rubbery, low-styrene rubber. This latex had the advantage of permitting a firmer, flexible foam rubber to be made with an overall lower density, thus saving on cost of materials. Practically all latex foam rubber being made today uses an encapsulated latex that is made by this or other processes.

Tires

Geometrically, a tire is a torus; structurally, a flexible membrane containing a fluid under pressure; and chemically, a composite of macromolecules. Many of the innovations leading to today's high performance tires for autos, trucks, airplanes, and off-the-road equipment have occurred because of the way tire components are arranged physically in the tire and how the tire is shaped. Chemical innovations have also led to such tires, however, by providing the rubbers, fillers, fibers, and adhesives needed to build successful tires.

The first patent on an "air tube device" was issued in England in 1845. Producers soon abandoned this forerunner of the modern pneumatic tire, and it was not until 1865 that the first successful rubber tire was developed. It was a solid tire for bicycles.

Tires inflated with air were not to be ignored, however. The discovery of vulcanization had made practical the use of thin layers of fabric impregnated with rubber, and John B. Dunlop reinvented the pneumatic tire in 1884. This Scottish inventor, who had a veterinary practice in Belfast at the time, put his tires on bicycles at first, but the concept soon was adopted for automobiles.

The first tires containing fabric performed extremely poorly, and manufacturers put solid tires on many of the early cars and carriages. Solid tires outwore those containing fabric, but they rode extremely harshly. Holes drilled across the width of the tread helped improve the ride some, but solid tires were not to be the final solution (although they were to be used on intracity trucks until the 1920's and 1930's).

One of the early problems holding back the pneumatic tire was the difficulty of keeping it on the wheel. The Clincher Tire with an edge shaped like a hook engaged in a modified rim flange was an early solution. Clincher Tires were hardly the most easily put on and taken off, however, and within a few years they were displaced by tires having a flat tape of wire in the bead. Such a reinforced bead held an inflated tire securely to the wheel yet was flexible enough to stretch over the rim's edge for mounting and demounting. Tires using this principle of construction are the ones used today.

While some rubber technologists were concerned with tire shape and construction, others improved its materials of construction. Organic accelerators discovered in 1906 gave faster and better cures than sulfur alone, and carbon black reinforcement in 1912 improved tread life significantly. Fabric impregnated with rubber gave way to multiple layers of cotton cord molded between sheets of rubber to give tires a dimensional stability well beyond that of the first tires.

The four-ply cotton cord tire of the early 1930's gave immeasurably better service than its counterpart of the turn of the century, but better roads, higher speeds, and heavier loads stressed it severely. The demand for stronger tires was eventually met by a rayon fiber developed by chemists at du Pont, with an assist from chemists at Goodyear, who developed a casein-latex adhesive system to overcome the rayon's poor adhesion to rubber. Chemists at du Pont also contributed a resorcinol-formaldehyde latex adhesive system, and rayon auto tires were introduced to the general market in the U.S. in 1936.

In construction innovations, Goodrich pioneered with tubeless tires in the mid-1940's, and Michelin began promoting radial tires in France at about the same time after they had been made in the U.S. but not accepted. Armstrong Rubber, one of the smaller U.S. tire producers, in 1964 announced the first tire with bias-ply cords and a glass fiber belt. Shortly after Goodyear Tire & Rubber Co. introduced tires with bias plies of polyester cords and a belt of glass fibers, this type of construction was accepted by U.S. automakers as original equipment. Radial tires are now making inroads in the original equipment market and are becoming more popular in the aftermarket.

In the most recent construction innovation, Firestone in 1971 announced a cast pneumatic tire that does not require cord reinforcement of the carcass. It has suggested that production quantities would be available in 1975. Goodyear in the early 1960's had produced modest quantities of cordless tires made from cast urethane. Cost and some technical difficulties have kept them out of the auto market, although Goodyear today makes off-the-road urethane tires for use under severe conditions, such as in steel plants and foundries.

FUTURE DEVELOPMENTS

The rubber products manufacturing industry continues to be one of expansion and improvement in its many product lines. Competition runs throughout it strongly, however. Many risks exist, and profitability has been none too good lately. In 1969, for example, all manufacturing industries in the U.S. had an average profit of 4.6% of sales, while rubber manufacturers averaged only 3.7%. As return on investment in the same year, all manufacturing scored 7.8%, while rubber manufacturers averaged only 6.8%. Because of both national and international competition, the industry doesn't expect much improvement soon in such results.

Be that as it may, the industry expects to continue to grow and to improve its products. The effort being put on tire research and development is bound to result in even better tires with lower relative costs. Automation and computerization of tire lines will certainly be speeded up, and a completely cast tire has been forecast for commercial production by 1975. Industrial rubber products will be engineered for greater use in all transportation equipment, including impact-resistant bumpers and interiors, crash-resistant gasoline tanks, and even flexible front ends. Farmers will get new types of hose systems for irrigation and more efficient hay crusher rolls. Flame-resistant foam cushioning will reduce hazards from burning mattresses and upholstered furniture. Strategically located reservoirs lined with new types of rubberized fabric will conserve the nation's water supply more efficiently. Rubber flotation gear will be used to contain oil spilled in oceans and bays until it can be recovered. Sewage treatment plants will make greater use of rubber storage tanks to hold refuse from peak times for processing later.

The Future in Elastomers.   To exploit these and other opportunities, chemical and rubber industry scientists and engineers possibly will not have great chances to develop and produce on a large commercial scale a totally new elastomer to rival any of today's major ones. Polymer chemists have investigated rather thoroughly the possibilities using inexpensive elastomer monomers, and process improvements have lowered prices to the minimum. Moreover, the industry has depreciated much of its plant capacity. A novel elastomer that would challenge the established ones in more than small specialty markets faces stiff competition. Polymer chemists will continue to improve elastomers incrementally, however, and rivalry will continue among established synthetics and between synthetics and natural rubber for growth into new markets and for growth at the expense of each other in present markets.

Competition among elastomers in fact has existed for some time. SBR's position as the general-purpose synthetic rubber remains a strong first, for example, but it does not hold as great a part of the market as it once did. In 1960, nearly 85% of the U.S. consumption of synthetic was SBR. By 1969, SBR's share had dropped to 65% (although its total consumption had increased, of course). In the same period, the specialty rubbers of butyl, neoprene, and nitrile continued to share the market about the same, ranging from about 3% for nitrile to more than 6% for butyl. Much of the change came from the growth of the stereo rubbers of polybutadiene, polyisoprene, and EPDM. By 1969, they had 20% of the U.S. market for synthetic rubbers.

As of today, the industry is well endowed with capacity to meet whatever needs arise. In fact, capacity for synthetic rubbers amounts to nearly 3.0 million long tons (see table), and producers of natural rubber can grow just about whatever they can sell. (The gap between synthetic capacity of 3.0 million long tons and U.S. consumption of about 2.0 million long

U.S. Synthetic Rubber Capacity

Rubber	Capacity* (Long Tons)
SBR	1,945,000**
Polybutadiene	310,000
Neoprene	200,000
Butyl	165,000
Polyisoprene	150,000
Nitrile	150,000
EPDM	115,000

* Rounded to nearest 5,000 long tons.
** Consists of about 1,830,000 long tons of emulsion SBR and 115,000 long tons of solution SBR.

tons isn't as great as it might at first appear, incidentally. The industry has excess capacity, to be sure, but it also has a profitable export business of nearly a quarter of a million long tons.)

Tires Dictate the Future. What an industry capacity table will show, say, 10 years from now depends largely on what happens in tires. Developments in power transmission belts, conveyor belts, hoses, wire and cable sheathing, building sheeting, and the countless other rubber industry products will influence that table, to be sure, but they won't influence it to the extent tires will.

One product line controls the destiny of few major industries as tires control the rubber industry. About 65% of all elastomers consumed in the U.S. in 1969 went into tires (or about 1.8 million long tons out of about 2.7 million long tons). Within elastomer types the dominance is often even more striking:

Elastomer	Percent In Tires
Polybutadiene	95
Solution SBR	95
Butyl	75
SBR	70
Natural	70
Polyisoprene	65
EPDM	20

Only for EPDM are tires and tubes relatively minor. Even there, however, producers would dearly like to have a greater share of their output absorbed by tires, for tires represent the truly big market for any elastomer.

At the moment, tiremakers aren't giving elastomers top research and development priority. They are too busy absorbing the rash of changes that chemical technology has put before them in recent years in the form of new fibers, new elastomers, and new tire designs. World War II had hardly ended and manufacturers had hardly absorbed the substitution of rayon fiber for cotton fiber before first nylon in the late 1940's and early 1950's and then polyester in the early 1960's began battling for selection. Then while this battle still raged, a new one erupted in mechanical design with the advent in the mid-1960's of radial and bias-belted tires.

As of now, the bias-belted tire consisting of polyester body plies and a glass fiber belt holds the auto original equipment market. Nylon in conventional bias-ply tires dominates the auto replacement market and both segments of the truck and bus market, but the polyester-glass combination is increasing its share of the auto replacement market. Rayon, meantime, is hurting in all markets and counts on the radial tire for its future.

Nylon in tires grew from about 125 million pounds in 1960 to 310 million pounds in 1968 and then dropped to 305 million pounds in 1969. Polyester, meantime, grew from nothing in 1960 to 125 million pounds in 1969. Both grew at the expense of rayon, which dropped from 220 million pounds to 120 million pounds.

Rayon is by no means dead, however. Counted out first when nylon arrived, then again when polyester arrived, and finally when the belted tires arrived, it still hangs in and even shows signs of having a good future. The rayon tire cord originally introduced in 1936 wouldn't recognize its 1970 successor, however, as chemists have made it in many respects a new fiber.

The bias-ply tire appears destined for service where performance demands aren't high, while bias-belted and radial tires will battle for the high performance market. Tire manufacturers talk confidently today of 40,000 and more miles of service on autos (the biggest single tire market), and only bias-belted and radial tires perform this well. Nylon apparently works well in neither because of its flatspotting characteristic, so its future appears darkened. Polyester-glass has the edge in bias-belted. Rayon plies with a steel fiber belt appear to make a satisfactory radial, although rayon companies are pushing rayon plies with a rayon belt as giving a ride more satisfactory to Americans. Their push, plus the fact the U.S. presently has little steel fiber capacity and U.S. tiremakers little experience in building tires with it (except for some truck and off-the-road tires), may well give rayon the edge, at least for a while.

Be that as it may, the likely outcome is far from clear, and many fiber and rubber industry chemists in the next several years will be striving to score advantages with their products over those of their competitors. Typical of the speed with which developments can occur to make predicting a hazardous calling are two that occurred in the interval between accumulation of information and preparation of the final draft of this chapter on the rubber industry. Within a short period du Pont announced a plant to produce what it calls Fiber B, and each of the Big Five tiremakers and others began heavy promotion for auto tires built with steel fiber belts. Du Pont and several of its tire producer customers say Fiber B provides much better strength for tires than does steel, while the position of the largest tiremakers gives obvious impetus to steel fibers for tires. The assurance in some circles with which earlier statements downgraded the possibilities for introduction of a major new organic fiber and the use of steel belts in American-built tires clearly has been eroded.

The tire industry may be putting most effort into selecting options before it in fibers and tire design, but it isn't ignoring elastomers totally. Tires contain several different elastomers, and each provides certain performance pluses. To oversimplify somewhat, SBR is strong, resists flex cracking, and can be extended appreciably with oil and carbon black to reduce elastomer costs. Polybutadiene provides excellent abrasion resistance. Natural rubber and polyisoprene resist heat buildup very well, while EP-DM resists weather checking.

Polybutadiene is added to SBR or natural rubber in tire treads, where its chief function is to improve abrasion resistance and thus improve tire wear. However, when polybutadiene is added to too great an extent (above about 40% total rubber content), it improves abrasion resistance so much that traction is reduced. Such tires then tend to skid more, especially on wet roads.

The industry's chemists and engineers are now trying to improve skid resistance but at the same time maintain high mileage performance. They have reached their goal in part by development of bias-belted and radial tires, whose belts permit the tires to roll smoothly and maintain an even and constant contact with the road. Such contact improves skid resistance, and the tires also wear well. Meantime, they are also examining new rubbers for use in tire treads. Both solution SBR and emulsion SBR polymerized with lower soap content are among possibilities. Both rubbers have improved skid resistance and with a normal quantity of polybutadiene incorporated in them produce tires with both improved mileage performance and better skid resistance.

U.S. capacity for solution SBR is now about 115,000 long tons a year, with Firestone and Phillips accounting for most U.S. production. Solution SBR may have a cost advantage over emulsion SBR, and as older SBR plants become obsolete, solution SBR may take over an increasing share of the tire market. This is a dynamic field, however, and several of the older rubbers, such as natural rubber, are not to be overlooked. Polybutadiene added to treads of natural rubber greatly improves wear, and tire producers are also examining a recently available tread consisting of a blend of three types of natural rubber plus a small quantity of oil added at the plantations. In addition, rubber compounding also is quite important to tread performance, and the industry's chemists are continuously experimenting with new chemicals to improve tire performance.

Specialty Rubber Outlook. Not all of the rubber industry attention centers on elastomers for tires, of course, and other elastomer innovations appear in the offing. One application that could develop into a sizable market is the use of rubber in roads. Used for the past decade or so but still relatively small in total business, paving with rubberized surfaces involves the addition of small quantities of SBR latex to a heated and liquid bitumen. As little as 3% rubber improves the bitumen's properties in both summer and winter. The International Synthetic Rubber Organization is presently sponsoring a large-scale road experiment in Utah. If results are favorable, building asphalt and bitumen roads could consume hundreds of thousands of tons of rubber worldwide.

Polyurethanes are also finding greater acceptance each year. Their use is particularly advantageous in that production involves mixing liquids and pouring them into molds instead of the arduous mixing in mills required of regular dry rubbers. Practically every major rubber products company now makes one or more items of urethane. Flexible foam production has passed 500 million pounds a year, and its growth continues at 12% or more per year. Rigid foams for insulation and for structural uses when sandwiched between steel or aluminum represent two other fast-growing uses, while shoe soles of microcellular urethane represent a new application of much promise.

Several companies have introduced what are known as block polymers. They are being used in such applications as adhesives, footwear, and fabric coatings. Chemically, most consist of styrene and butadiene. In contrast to SBR, however, where the monomers occur randomly in the polymer, block polymers have systematic numbers of styrene molecules alternating with systematic numbers of butadiene molecules. Their advantage is that the polymers have tough rubbery properties up to certain temperatures without having to be vulcanized.

Cleaning the Environment.   The rubber industry no less than many another American manufacturing industry will find its future profoundly influenced in the name of the environment. The industry may feel harassed by demands that it clean itself up, and it may appear to some to do so only under great duress. It may not always move at the high rate of speed desirable for its own plants, but it isn't ignoring the problem, either. Moreover, it isn't hesitating to find new business opportunities that public concern about the environment is opening to its products. Old technology will serve in some cases, but new technology based on chemistry and chemical engineering will also be called for.

Manufacturing rubber products is not the cleanest of activities. Rubber and the chemicals used to process it often smell. Dusty products must be handled in large volumes. Liquid effluents tend to be dirty, too. Many of the industry's plants date from days when neighbors tolerated them for the jobs they provided and complained less about smells, dusts, and other exuda.

Today, many manufacturers are trying to improve. Goodyear says it committed nearly $3 million a year to control pollution from its plants in its first five-year program that began in 1966 and expects to spend at least $32 million in the effort by 1975. Uniroyal is currently spending about $2.5 million a year for the same purpose. The other rubber companies are also rapidly tackling the problems, and great improvements will occur during the next few years.

Applying known technology will largely solve rubber factory pollution problems. When the industry begins looking beyond its fences, however, solutions come less easily. For years, industry scientists have sought to make ever more durable products. In some respects they may have succeeded too well, for disposing of worn-out rubber goods plagues more than one community.

Take tires, for example. Nearly 225 million new tires now come from tire molds in the U.S. annually—50 million original equipment tires for autos, trucks, and buses; 130 million replacement tires for autos; nearly 30 million replacement tires for trucks and buses; and 15 million tires for farm, aircraft, and industrial markets. Every one of them turns partly to elastomer and carbon black dust as they wear, and what's left—which is most of the tire—becomes a disposal problem when the service life ends.

Neither dumping nor incinerating attracts any community as an end for its discarded tires—or for the tons of other rubber products that wear out and are discarded every year, for that matter. Since the rubber industry would find it an impossible mission to compound rubbers that self-destruct after their useful life ends, it must seek alternatives. At least three show considerable promise.

In one, wire beads and fabric are removed from tires, the remaining elastomer is ground up, and the particles are mixed with oil. Subjected to incomplete combustion, the oil mixture yields carbon black (much as incomplete combustion of oil now yields carbon black). The rubber from one auto tire produces enough carbon black for a new tire.

In the other, scrap rubber is hydrogenated under high temperature and pressure over a catalyst. The resultant primary product is a re-hydrogenated synthetic oil that might be combined with virgin crude for refining into the usual petroleum products. By-products include sulfur, carbon black, and off gases suitable for fuel.

Either process might help solve the growing solid waste disposal problem posed by the millions of old tires that Americans discard every year. Another pilot program to the same end that tire manufacturers are involved in would use worn-out but intact tires to build artificial reefs on ocean floors. Many marine animals need solid substrates to complete their life cycles on, and tires might help regenerate the ocean floor for aquatic plant and fish life. The Interior Department's Bureau of Sport Fisheries and Wildlife, sponsor of the project, has estimated that more than 1 billion old tires could be used advantageously and safely off the East Coast.

All the major rubber companies are working in other ways to help clean the environment. Recent mishaps with oil spills and seeps have focused world attention on the problems of transferring oil cargoes and on containing massive quantities of oil that might escape. Booms of nylon coated with abrasion- and oil-resistant elastomer have been developed for use in harbors and even on the oceans in seas up to 8 feet and in winds up to 30 knots. Once the spills are confined, they can be skimmed from the water with endless moving belts of oil-resistant rubber, or they may be pumped into flexible rubber tanks as large as 140,000 gallons that can be air-dropped at the point of the disaster.

Helping dispose of sewer wastes also offers opportunities to the rubber industry. Two have been tested in the prototype stage in the Anacostia River in Washington, D.C. In one, huge Pillow tanks stored storm sewer wastes temporarily during overflow periods for later processing. In the other, smaller tanks held wastes that boats discharged at marinas for later processing.

In other applications, rubber sheeting lines pollution control reservoirs at a major nitrogen fertilizer plant in the East. The linings keep nitric acid and ammonia contaminants from escaping to the nearby Allegheny River. Similar materials are also used in brine pits, water treatment lagoons, and refinery sumps.

Acidic effluents from abandoned coal mines are a particularly trouble-some source of pollution in mining areas. One way to control such pollu-tion is to keep water from entering the mine in the first place. In one experimental study, latex has been applied over the mined area about 2 feet below the surface (to prevent interference with surface vegeta-tion). Minerals in the soil interact with the latex to coagulate it, clogging the pores of the soil and forming a water-impermeable barrier.

An elastomer diaphragm to be used on auto gasoline tanks to reduce vapor emission has been developed. Recent government steps to upgrade safety in rail shipments of hazardous materials involves development of an energy absorption and puncture-sealing system to reduce danger to surroundings in rail mishaps. A hazardous waste container has also been developed for low-cost protection against crashes and fires during shipment and transfer of radioactive wastes from nuclear generating plants or other atomic energy facilities to a land or sea disposal site. Such containers could spawn a new generation of protective cocoons enclosing entire rail car shipments of dangerous materials.

SOCIAL AND ECONOMIC IMPACTS

Not quite half-a-century ago, General Motors lacquered the 1923 Oakland with nitrocellulose. GM made Oaklands for only a few more years, but putting lacquer on the 1923 model probably more than any other single event signaled the beginning of today's protective coatings industry. Up to that time, the coatings industry had made its products mostly from a limited number of naturally occurring oils and minerals, such as linseed oil and lead carbonate. In the years since, many new binders and other coating components have joined nitrocellulose, and the coatings industry today largely uses ingredients made by chemical processes. Coatings as a result perform significantly better in a wider range of uses. They protect substrates better from weathering, moisture intake, and staining, and they decorate many types of products with a great variety of durable colors and textures. In addition, they have been formulated to be applied by many different methods. Their use in all applications saves millions of dollars annually from losses caused by deterioration and corrosion.

The industry that makes these products is known as the paint, varnish, lacquer, enamel, and allied products industry (which we shall shorten to the coatings industry in this study). For statistical purposes, the Government places it in the chemical and allied products industry. In 1970, it sold nearly 830 million gallons of coatings for about $2.74 billion, about 5.5% of sales of all chemicals and allied products ($49.6 billion).

The coatings industry now employs about 71,000 people, of whom perhaps 4,600 work as chemists and chemical engineers. Such figures do not reflect the industry's size, however, for they include only those engaged directly in manufacturing coatings. They do not include the many persons, including chemists and chemical engineers, engaged in other industries in making coating precursors such as monomers and polymers, oils, solvents, pigments, and auxiliaries used in coatings.

About 50 companies account for nearly two thirds of the total value of coatings shipments in the U.S. The largest of these 50 companies also are the ones that perform the required research and development necessary for product improvement. Manufacturing and distributing coatings requires little capital, however, when companies operate on small and local scales, as most do. Moreover, technology is readily available. As a result, perhaps as many as 1,400 smaller companies also make protective coatings and add to the country's manufacturing base.

Coatings protect and decorate, and they have been used for both purposes from the earliest times. Americans in Colonial and post-Revolutionary days painted their homes, buildings, furniture, and equipment at first with imported products and later with ones made here. White lead, an important pigment, was first made in the U.S. in 1804 and the first varnish in 1815. The first coatings patent issued in 1855. The first ready-mixed paint became available in the U.S. in 1867 (a move that among other things transferred the art of making paints from painters to manufacturing plants). The man credited by some industry historians with being the first varnish chemist was graduated in 1878. (He was Henry M. Murphy, who studied under Charles F. Chandler, a founder of the American Chemical Society. After graduation, Murphy joined his brother, Franklin, at Murphy Varnish Co. in Newark, N.J.)

Manufacturers in the coatings industry and the industries that supply raw materials to it improved their products slowly during the early years, but the innovations introduced in the first decade or so after World War I have been the ones that have led to today's industry. The quick-drying nitrocellulose lacquer of 1923 was followed commercially in the U.S. as the decade ended by coatings based on rosin-modified and oil-soluble pure phenolics and on alkyds. The speed with which nitrocellulose lacquers dried cut the time for finishing autos from days to hours and helped to open the way for the mass production rates that made autos common commodities. The phenolics added a quick-drying feature to varnishes and allowed them to compete with nitrocellulose lacquers. The alkyds, meantime, outperformed them both, lasting longer, retaining their luster better, and being less expensive as well. The alkyds also contained more solids at application viscosities. Fewer coats covered and protected as well as resins used earlier, and finishing costs were reduced.

These early resins that chemical technologists developed and made bypassed natural agricultural products. They were quickly followed by other products of chemical technology on introductory or commercial scales. Urea-formaldehyde modifications were introduced in 1929 and vinyl copolymer resins in 1933. Melamine-formaldehyde and polyurethane resins joined the list in 1939 and silicone resins and styrenated and acrylated alkyds in 1944. Wall paints based on styrene-butadiene latices arrived in 1948. Epoxy resins and polyvinyl acetate latices in 1950, acrylic latices in 1953, and acrylic lacquers and thermosetting resins in 1958 were also some of the major innovations during those years.

These and other new binders immeasurably improved the properties of coatings. So did improvements in other coating components. Titanium dioxide (TiO_2) was introduced for use in coatings in 1924, for example. Produced by chemical processes from mined ores, it had the highest hiding power per pound of any white pigment. Improvements after its introduction raised its hiding power and performance even more, particularly when the rutile crystalline form became commercial.

Coatings pigmented principally with TiO_2 weather by slow erosion, whereas those containing previously popular lead and zinc pigments weather by checking and cracking. Erosion gives a surface that continues to appear good with time, since rain easily washes off the dirt. Erosion also provides a better surface for repainting than checking and cracking do. TiO_2 is now the most widely used pigment in coatings.

Among colored pigments, molybdate orange introduced in 1930 and phthalocyanine blue introduced in 1937 brought much better colorfastness to coatings in those hues. Introduction of lightfast yellow, orange, red, maroon, and violet in the mid-1950's extended the range of lightfast pale shades available to manufacturers.

As a result of these and other improvements, coatings have made wooden homes, steel bridges, and home appliances and furniture much more practical, to name only some of the more familiar applications. Improvements have also provided coatings that serve in more specialized uses. Some coatings, for example, control transmission of water, oxygen, and light in special ways. Others add or reduce friction, prevent static buildup, retard combustion, indicate temperature, or change a surface's wettability. Some coatings stretch with elastomeric substrates during compression and elongation, while others deter growth of microorganisms and other forms of life. Thixotropic paints become less viscous under shear stress. They apply easily, but they do not drip. Some paints readily accept an electrical charge and are applied by electrostatic attraction.

Others that are compatible with water ionize and are applied electrophoretically. To make products with such variety, manufacturers can select among thousands of resins, oils, solvents, plasticizers, extenders, pigments, dyes, driers, antiskinning agents, and the like.

CHEMICAL ACCOMPLISHMENTS

The coatings industry classifies its products as trade sales or industrial finishes. Trade sales are off-the-shelf products that move through extensive wholesale-retail distribution channels. Contractors, builders, repair shops, and the general public buy such coatings to apply to new construction and also to repaint old houses and buildings and to refinish a broad range of products such as furniture, autos and other vehicles, equipment of many types, and highway signs and markings. Manufacturers in other industries buy industrial finishes and use them for factory-applied coatings on vehicles, machinery, appliances, furniture, containers, and many other products fabricated from metal, wood, plastics, rubber, textiles, and paper. In 1970, trade sales amounted to about $1.50 billion and industrial finishes to about $1.24 billion.

Protective coatings consist of either two or three main components and a number of auxiliary ones. One main component is the binder (or vehicle), which consists of resins and/or drying oils or latices. A second main component is a volatile solvent or dispersing medium, which consists of an organic chemical or water. The third component is coloring and opacifying pigments, which consist of organic and inorganic chemicals and minerals. Auxiliary components, finally, are any of many chemicals that are added to control or adjust product properties. They include stabilizers, emulsifiers, defoamers, antiskinning and antisettling agents, driers, ultraviolet radiation absorbers, mildewcides and fungicides, thickeners, and flatting agents.

Of the four main types of protective coatings, paints and enamels have all three major components, and so do most lacquers. Varnishes, however, consist only of solvents and resins and/or oils and have no pigments (although they may contain flatting agents or colorants). Of the three components, the vehicle ranks high in determining how a coating performs, but pigments rank high as well. Pigments provide the decorating color. More importantly, they also hide the surface and also make many binders more durable by absorbing ultraviolet radiation, which so readily degrades some types of coatings. Pigments also have a major role in decreasing water permeability and, in specific instances, of adding anticorrosive properties. Solvents or water, meantime, serve as the dispersing medium for other constituents and influence how the coating applies and dries.

Paints, varnishes, lacquers, and enamels today impinge on our lives in many ways to protect possessions, decorate surroundings, and provide information. It would be impossible to examine in detail how they have done so in the space available in this study. Automotive finishes, however, can best serve as typical of the way coatings function and of the work chemists and chemical engineers have done to improve them. We shall deal first with innovations in this big coating market, therefore, before turning to more specific innovations in resins and pigments.

Automotive Finishes

The coating system for an automobile consists of metal pretreatment, undercoat or primer, and topcoat. This composite lamination of films is largely organic and totals less than 0.004 inch thick. The composite's manufacturers, the automakers who apply it, and the ultimate owners who live with it expect it to preserve the integrity and appearance of a product in which a substantial sum is invested. Moreover, owners expect the preservation to last with virtually no maintenance costs for half a dozen years or longer of exposure to antagonistic weather.

Coatings for automobiles at the beginning of the century were simply an extension of practices used in making carriages. The production rates the auto industry eventually reached, however, became possible only with modern production lines, and the first major technical development in auto coatings was a response to the production problem posed by finishes that dried slowly.

The solution proved to be a topcoat based on plasticized nitrocellulose that du Pont developed. Introduced on the 1923 Oakland, as mentioned earlier, nitrocellulose lacquer reduced painting time from several days to several hours. By 1926, all manufacturers except Ford had joined GM's lead in switching to nitrocellulose lacquer, and even Ford shifted in 1927 from varnish to a special oil-soluble resin lacquer that Glidden developed. From less than 1 million pounds in 1922, nitrocellulose lacquer sales grew to 10 million pounds by 1926. In addition, new markets also developed for pigments and solvents.

Nitrocellulose lacquers at first lacked the original depth and richness of the earlier oleoresinous finishes, but hard films formed by simple evaporation, speed of drying, ease of repair, and durability were attributes that served auto manufacturers well. Technical improvements to control molecular weight and hence nitrocellulose solubility, careful balancing of solvent-diluent ratios for solvency and evaporation rate, and plasticizing with mixtures of chemicals and saturated oil-modified alkyds in time made nitrocellulose coatings both appear and perform better. Although pressed by better products and eventually virtually eliminated from the auto market in the U.S., they had a market until well into the 1950's and are still used extensively in Europe.

Despite advantages, nitrocellulose lacquers did have disadvantages. Low content of solids meant multiple applications were needed. Films had a low gloss and had to be polished to develop acceptable appearance. Sunlight in the presence of atmospheric oxygen and moisture degraded them. Competition accordingly arrived before too long. Since nitrocellulose films formed by evaporation of solvent, the auto industry had added baking tunnels to hasten evaporation and shorten finishing schedules even more. This baking practice opened the way to heat-assisted cures of auto enamels formulated at first with oxidizable varnishes and later with alkyd resins modified with linseed, tung, and soybean oils.

Baked enamels became entrenched competitors of nitrocellulose lacquers during the early 1930's. They appealed strongly to carmakers, because their higher solids content reduced the number of coats needed for protection and opacity. Furthermore, they emerged glossy from the baking tunnel and eliminated laborious polishing.

With resumption of auto production following World War II, the coatings industry introduced baking enamels containing melamine-formaldehyde resins as crosslinking agents. The alkyds in such formulations crosslinked in a condensation reaction and mechanisms depended much less on oxidative polymerization. Since the mechanisms usually used for oxidation curing operate to degrade polymers, the new modified alkyds were more durable.

Auto stylists of the 1950's put a new demand on the coatings industry when they abandoned the traditionally dark, conservative colors of early auto days for a host of pastel shades. The new pastel tints made reproducible color matching difficult, as they could not obscure color substances formed in baking enamels that crosslinked through oxidation.

New Super Enamels introduced commercially in 1956 used oils having no unsaturation that were crosslinked with higher (20 to 30% in contrast to the former 6 to 10%) concentrations of melamine resins. Super Enamels led to trouble-free production of baked enamels in pastel tints. They also resisted marring and weathering better. In fact, automakers for the first time could suggest in their ads that motorists might be freed from periodic polishing to maintain exterior gloss, a goal they had long sought.

While the auto industry was testing and adopting Super Enamels, automakers that preferred the properties of lacquer had also been vigorously promoting coating development. As the 1950's ended, General Motors ushered in what many now view as the acrylic revolution. At that time, GM rapidly replaced all nitrocellulose lacquers with acrylic lacquers. Ford, Chrysler, and American Motors followed by 1963 with acrylic enamels, and all mass-produced autos in the U.S. have now been top-coated with acrylic lacquers or thermosetting enamels for several years.

Acrylics became an instant commercial success in the auto industry because of three important properties: They can be made to reflow at oven temperatures at appropriate times during film formation; they have a very low rate of degradation in sunlight; and they hydrolyze very slowly. Heat reflow makes acrylic lacquers the first auto thermoplastic coating to have full gloss without polishing (although nitrocellulose baking formulations had approached that goal). Slow degradation by sunlight and slow hydrolysis, meantime, make acrylic topcoats superbly durable.

Stylists took advantage of another acrylic property and began using coatings with pigmentations high in aluminum flakes. Acrylics have proved especially effective in protecting metallic flake pigments from environmental attack. Acrylic lacquers and enamels with metallic pigmentation also retain a high gloss for more than twice the exposure time of the best performing predecessors. As a result, more than 75% of the autos built in the U.S. in the late 1960's had such pigmentations.

As long as automakers continue to stamp bodies from mild steel, chemists in the coatings industry will have to help protect such a reactive alloy from corrosion induced by heavily salted roads in winter. The beneficial economic impact on owners of inhibiting corrosion is obvious.

Successfully resisting corrosion depends on interrupting electron flow from one pole of an electrochemical cell to another. If dissimilar metals serve as electrodes, interposing a nonconductor insulates them from each other. If adjacent crystals in the same piece of metal happen to become the two poles of a cell, however, then intimacy of contact by a protective coating with the metal surface and the coating's own anodic passivation must provide the insulation.

Epoxy esters crosslinked with melamine or urea resins insulate against corrosion extremely well. Hydrogen bonding of hydroxyls along the epoxy ester chain holds coatings tightly to the metal, while ether linkages provide chemical passivity. Introduction of epoxy spray primers in the mid-1950's, therefore, has done much to protect auto bodies against corrosion. So have the use of inhibitive pigmentations and the sacrificial protection provided by galvanizing.

Now, electrodeposition of priming coats provides even better protection. In this process, metal immersed in a tank of primer becomes the anode of a circuit, and coatings are deposited on it. The process offers a number of advantages. Most importantly, it coats interior surfaces of box sections and other inaccessible places to inhibit the insidious rusting of bodies from the inside out. In addition, it operates in aqueous media and eliminates fire hazards and hydrocarbon emissions. Finally, it avoids waste inherent in spraying and reduces the need for recovery and recycling.

A new method about to become commercial will produce cationic electrophoretic movement instead of anodic. This method deposits an even better inhibitive coating and also eliminates any chance for the process to disturb the crystalline inorganic layer produced on the metal by phosphate pretreatment etching.

Among the important trends in the auto industry of the past half dozen years or so has been increasing use of plastics. The engineering plastics, such as nylon, acetal, polycarbonate, polypropylene, acrylonitrile-butadiene-styrene, polyurethane, and polyester resins, are increasingly used for making grilles, bumpers, molding strips, instrument panels, body fairings, and even fairly large exterior and interior body panels. Chemists as a result have had to develop coatings that protect them, that match colors on metal panels, and that may be applied and repaired in an industry oriented to the mass production assembly line.

The acrylics developed for metals often protect plastics well. New elastomeric coatings have been required, however, where rubbery plastics or high density foams are used for bumpers or moldings to dissipate impact energy. Single package urethane enamels and lacquers have been used commercially in such applications since 1968. They are acceptably durable and are elastomeric in that elongation at break reaches or exceeds 100%. They also have high tensile strength and recover their original dimensions almost completely after high energy impacts distort them. They and coatings yet to be developed should contribute to the success of auto designs that decrease damage in low speed collisions and increase occupant safety in highway accidents.

Nitrocellulose Lacquers

Cellulose nitrate is the oldest cellulose derivative. Popularly known as nitrocellulose, it was made in Europe for the first time in the first half of the 1800's. In various forms it eventually became the basis for large textile, explosives, and plastics industries. It also became a significant raw material for the coatings industry.

In the 1840's, a French artist mixed ethyl alcohol and diethyl ether to make the first practical solvent for nitrocellulose, and he varnished his canvases with what became known as collodion. The medical profession introduced collodion as a dressing in 1847, and nitrocellulose eventually found another practical use as a carrier of light-sensitive chemicals in photography. Collodion proved too volatile and hygroscopic for coating uses, however, until a chemist at Celluloid Co. overcame the problem in the early 1880's. He found that amyl acetate worked as a solvent. Companies began selling cellulose lacquers in the U.S. shortly thereafter, but uses remained small until developments during World War I and immediately after led to significantly changed products.

Up to World War I, nitrocellulose had been used mainly to coat brass bedsteads, chandeliers, and other articles to prevent tarnishing. Only dilute solutions could be applied because of nitrocellulose's high molecular weight, and they produced only thin films that generally shrank and adhered poorly to substrates. Moreover, both nitrocellulose and its solvents were expensive, further limiting use.

World War I brought the first important change. Nitrocellulose dissolved in acetone, amyl acetate, or any of several other solvents proved to be an excellent dope for aircraft fabric. By war's end, large production facilities existed for nitrocellulose and its solvents and so did large stocks left over when the war ended.

The stockpiles and productive capacity eliminated cost and supply problems, but the high molecular weight of nitrocellulose still inhibited fully satisfactory coating use other than in small-scale decorative lacquers. At the time, du Pont had already established a position in nitrocellulose by using it as a base sheet in photographic product production. Stimulated in part by prodding from General Motors for faster-drying coatings for autos, du Pont began research in about 1920 on the problem. Within a short time, its chemists found that treating nitrocellulose with sodium acetate depolymerized it to a lower average molecular weight and that solutions with higher solids contents at application viscosities could then readily be made. The original batch digestion process for viscosity reduction of nitrocellulose that du Pont used is no longer employed in the U.S. In the early 1930's, Hercules Powder Co. (now Hercules, Inc.) achieved a significant advance when it developed a continuous digester, which eliminated some hazards encountered in batch digestion.

Du Pont introduced Duco nitrocellulose lacquers in 1923, and they were first used in the auto industry by GM that year, as mentioned earlier. Some industry historians, incidentally, have since credited Duco lacquers with being the first major accomplishment of organized research at du Pont.

Chemists at du Pont, Hercules, and elsewhere also contributed to nitrocellulose lacquer's success by developing suitable solvent systems and the commercial methods necessary to make the solvents in the volumes needed. The original solvents for nitrocellulose tended to evaporate too rapidly, and films could not level before the coatings became hard. In the mid-1920's, chemists evolved a number of new solvent systems with lower costs and greater ranges of evaporation rates, and large markets developed as a result. Butyl alcohol, for example, had originally been intended as a possible starting material for synthetic rubber. Chaim Weizmann, who later was to become one of the political leaders of Israel, had developed a fermentation process for butyl alcohol production in England before the war. The resulting availability of butyl alcohol and its esters at reasonable prices and in quantity helped make nitrocellulose lacquers industrially practical on a large scale. Butyl acetate production as a result rose from a few hundred thousand pounds in 1923 to nearly 31 million pounds by 1929.

The large-scale conversion to nitrocellulose lacquer by the auto industry also led to other developments in solvents. Amyl acetate, for example, had been used as a solvent since the 1880's. The whiskey distilling industry produced its precursor, amyl alcohol, as a by-product, and the arrival of Prohibition effectively stopped what cyclic production there had been. Chemists at Sharples Solvents Corp. developed a process for making amyl alcohol based on chlorination of pentane followed by hydrolysis with sodium hydroxide. Sharples marketed mixtures of pentyl alcohols in 1926 and acetates in 1927 from a commercial-scale plant that was one of the earliest in the U.S. to use petroleum hydrocarbons in organic synthesis.

Finishing wood products now provides the largest market for nitrocellulose lacquers. They are used in stains, wood fillers, sealers, toners, and topcoats for household and industrial furniture, radio and television cabinets, flooring, interior wood paneling, bowling pins and alleys, and many other products. Other important uses include decorative and functional lacquers for paper, cellophane, leather, cloth, and aluminum foil. Nitrocellulose is also used as a film former in rotogravure and flexographic inks.

The properties that have led to widespread use of nitrocellulose in industrial finishes include solubility in a variety of solvents and solvent systems, rapid solvent release, good mechanical properties of its films, and compatibility with an exceptionally large variety of resins, plasticizers, and other modifiers. Nitrocellulose's wide range of compatibilities with other resins permits manufacturers to formulate lacquers that meet many specialized performance requirements. By formulating nitrocellulose with thermosetting resins, for example, it is possible to combine the fast air-drying property of conventional lacquer with the solvent resistance of thermosetting resins. A recent development involves the crosslinking of nitrocellulose with certain urethane prepolymer resins to combine the desirable properties of both materials. It has also been possible to formulate nitrocellulose lacquers to meet regulations controlling emissions of photochemically reactive solvents, which are now assuming greater significance to the coatings industry.

Oil-Soluble Phenolic Resins

When Leo Baekeland announced the first synthetic resins in 1909, he revealed one of the classic episodes in a long series in which synthetic chemicals have replaced natural products. While Baekeland's resins succeeded immediately in a variety of molding, laminating, and casting applications, some two decades were to pass before they were successfully incorporated into significant commercial coatings. When this did happen (in the 1920's), varnishmakers got both a much faster drying product and also an alternative to combat the fast-drying nitrocellulose lacquers, which themselves had just begun changing the coatings industry radically.

Baekeland was a Belgian chemist who had emigrated to the U.S. in 1889. Financially secure following the sale to Eastman Kodak of a company he founded, he set out in independent research shortly after the turn of the century to develop commercially valuable products based on the reaction of phenol with formaldehyde. Adolf von Baeyer in Germany some 30 years earlier had found they produced only an "unmanageable resinous mass," but Baekeland believed such thermosetting resins offered possibilities. Among his targets was shellac, a natural product then used to mold a variety of items such as parts for the growing electrical equipment and telephone industries. Baekeland succeeded where Baeyer didn't. His alkali-catalyzed reaction product of phenol and formaldehyde, trademarked Bakelite, and co-worker Nathaniel Thurlow's acid-catalyzed reaction product, trademarked Novolac, soon supplanted shellac and other products and also opened new markets because of superior strength and electrical resistance.

Bakelite and Novolac may have introduced the age of synthetic resins but not an age of synthetic coatings. Chemists in the coatings industry at the time recognized that Baekeland's and Thurlow's phenolic resins resembled shellac, which was widely used in coatings, but they could not dissolve them in either drying oils or inexpensive hydrocarbons. Modifications were obviously called for.

Varnishes had historically consisted of turpentine, natural oils such as linseed, and natural resins called copals. Producers heated the copals to depolymerize them sufficiently to allow them to dissolve in the drying oils. They then added the oils and cooked the mixture to the desired viscosity. Turpentine thinned the varnish to application viscosities, and users produced somewhat dark coatings that hardened in three to four days to a rich appearance.

Such varnishes performed well, of course, but incentives for improvement did exist. Manufacturers, for example, had to contend with the usual drawbacks of natural products, such as variability in properties, supply, and price. Moreover, the cooking steps resulted in raw material losses of up to 30%. In working with Baekeland's resins, chemists at Kurt Albert Co. in Germany eventually found that adding alkyl or phenyl groups in the para position of phenol led to reaction products with formaldehyde that did dissolve in drying oils.

Kurt Albert introduced the first of the "four-hour varnishes" in Germany in 1919. Seven years later, an associate firm of Rohm & Haas in the U.S. (The Resinous Products and Chemical Co.) was formed, obtained rights to Kurt Albert's products, and introduced them in 1927 as Amberol resins. Several manufacturers now sell a variety of oil-soluble phenolic resins. The resins vary in the ratio of formaldehyde to phenol, in the catalyst used to promote the condensation (which affects the rate of polymerization when cooked with oil), and in substituent groups.

Most of the developments in phenolics occurred in the 1930's and 1940's. While the alkyds introduced shortly after them have largely replaced phenolics in uses where better color is needed, the phenolics still retain an important place in high traffic and severe environment uses such as spar and other marine varnishes, floor varnishes, and porch and deck enamels. They are also used in electrical insulating varnishes, food can coatings, and drum and pail enamels. They are particularly useful in printing inks.

Development of the phenolics also brought about several important economic changes. For one, they freed the U.S. from its dependency on imported resins. For another, they opened a market in varnishes for other oils besides linseed. Tung oil at the time was a particularly desirable oil to use, for example, as it dried rapidly and its films resisted water well. However, its use was limited because surfaces appeared frosted and wrinkled. Phenolic resins permitted manufacturers to use tung oil to prepare varnishes with excellent appearance and extended durability, especially for marine environments. Although imported tung oil served at first, growers soon established tung tree groves in the Gulf states and added another agricultural crop to the U.S. economy. Varnish manufacturers also found they could use phenolics with other oils such as soybean, castor, and sunflower, and these agricultural products thus had new nonfood outlets.

Alkyd Resins

In the space of less than a decade in the 1920's, three major new products changed the U.S. coatings industry as it had never been changed before. Quick-drying nitrocellulose lacquers began the shift from coatings based largely on natural oils to ones based largely on products made by chemical processes, and the oil-soluble phenolics continued the trend. Then as the 1920's ended, coatings based on alkyd resins became commercially available. In many respects they completed the trend.

Alkyds today rank as the single most important resin class in the U.S. coatings industry. Manufacturers use more pounds of alkyds than they do of all other resins combined. Alkyds appear in both industrial finishes and trade sales products, and they are used to make interior and exterior paints, enamels, lacquers, and varnishes for many uses.

When first introduced, alkyds fell short of some competing products, such as being dark and not too stable, resisting water poorly, and not performing adequately with some commonly used pigments. As chemists improved them to make them resist weather, heat, light, and many chemicals much better, alkyds soon began overtaking other resins. In time, formulations had high initial gloss and outstanding flow, leveling, and package stability. They also dried rapidly to hard films. Moreover, alkyd raw materials were (or soon became) readily available, inexpensive, and easily adaptable to existing production and application equipment and methods.

Many new resins with outstanding properties have been introduced in the intervening years, but extensive modifications to the alkyds during the same period have helped them retain their advantages. Today, other resins may outperform the best alkyds in one or two important properties and thus be used in special applications despite their shortcomings in other properties. No other resin class, however, can match the overall balance of properties the alkyds can be formulated to provide, and about half of today's industrial finishes and a quarter of trade sales products contain alkyds as the major resin.

Alkyds for coatings are made with a polyol, a dibasic acid, and an oil or fatty acid. The most common alcohol is glycerol, with pentaerythritol second. Phthalic anhydride is the most common acid, while isophthalic and maleic acids are also used. Manufacturers also use significant amounts of other polyols, such as trimethylolethane, trimethylolpropane, and sorbitol. Other acids used include adipic, azelaic, sebacic, succinic, and terephthalic. Castor, coconut, cottonseed, fish, linseed, oiticica, safflower, soya, tall, and tung oils or fatty acids modify the resins to a wide range of properties.

Alkyds have in time proved to be one of the few developments in coatings in the U.S. where the original research objective was primarily coating applications. Most developments in coatings have come from work in chemical and other companies where chemists mainly sought products for other uses and then somewhat secondarily adapted them to coatings. Chemists at both the Schenectady (New York) and Pittsfield (Massachusetts) laboratories of General Electric, however, deliberately set out a few years before World War I to make better coatings. They looked at first for coatings for insulating uses, but they eventually turned to more general applications.

Chemists had made resins from polyhydric alcohols and polybasic acids for some years before GE began its research. One of the simplest resins, glyceryl phthalate, had been prepared for the first time in 1901, for example, but it proved too intractable for use. Then in the 1912–14 period, chemists at GE turned to trying to make flexible films from glyceryl phthalate by modifying it with monobasic acids. Their first important patents issued in 1914, and by 1920 products that GE called glyptals were being used commercially as mica stickers (although admittedly not particularly successfully). Then in 1921, R. H. Kienle at GE modified alkyds with drying oil acids and in doing so produced the first really successful coating alkyd.

GE at first concentrated on wire enamels and baking varnishes, and it was not until 1928 that the air-drying characteristics of alkyds modified with unsaturated drying oils began to indicate the real potential in coatings for alkyds. Introduced commercially soon thereafter, these alkyds and their many modifications before long earned a commanding position in the coatings industry.

No other resin class in the coatings industry likely has been modified to the extent alkyds have, and the result is the widest range of properties for any resin class for manufacturers and users to choose among. Amino resins have improved curing rates, hardness, and durability, for example. Rosin has increased compatibility with other film formers. Phenol modifications have harder films, better initial gloss, and superior chemical resistance. Styrenated alkyds air-dry in minutes and become hard enough to handle in an hour or so, matching the drying performance of fast-drying lacquers. Alkyds have also been modified with acrylics, silicones, isocyanates, vinyltoluene, and chlorinated rubber to meet service applications.

Early applications for alkyds included electrical insulation and appliance finishes. Then in the early 1930's, alkyds modified with drying oils led to enamels that the auto industry found particularly advantageous. When baked, they dried about as fast as the nitrocellulose lacquers did, but at 50 to 60% solids they required fewer coats than nitrocellulose lacquers at about 20% solids. Ford switched to alkyds in 1932, Plymouth and Dodge in 1934, and alkyds (and nitrocellulose lacquers) dominated the auto finishes market until displaced by the acrylics in the late 1950's. Appliance manufacturers, meantime, also adopted alkyds quickly. By 1935, for example, some 90% of all refrigerators were finished with alkyds. Later, the melamine resin modifications provided better hardness, color retention, and stain resistance. Now, however, alkyds are being replaced with other coatings that provide better resistance to detergents and stains in such appliances as washing machines and refrigerators.

The Government listed alkyds in production statistics for 1933 for the first time. Six producers made 10 million pounds. By 1937, production had risen to more than 16 million pounds from 16 producers. Today, there are about 50 producers, and their output amounts to about 800 million pounds. About 95% of all alkyd production goes into coatings.

Latex Paints

Of all the many technical developments in the coatings industry, few can match the direct impact on the average consumer of the development of water-based latex paints. When Glidden introduced its Spred Satin latex emulsion paint in 1948, home owners and others for the first time had a paint that performed as well in many ways as did good quality oil-based flat paints. Ease of application enabled amateurs to achieve a professionally-applied appearance, and ease of cleanup encouraged them to try. By allowing home owners to change wall colors quickly and inexpensively, water-based latex paints unquestionably contributed much to the boom in home decorating and other do-it-yourself projects.

Sherwin-Williams, National Lead, Pittsburgh Plate Glass (now PPG Industries), du Pont, and others soon joined Glidden with their own brands of latex paints. From 100,000 gallons in 1948, production rose to 30 million gallons within three years. Today, the market for latex paints and semigloss enamels has grown to well above 40% of the $1.5 billion trade sales paint business. All indications point to water-based coatings' becoming even more commonplace with the development of enamels, floor paints, and appliance coatings.

In retrospect, the speed with which water-based paints grew may appear simply explained, but such paints were neither easily developed nor easily marketed, at least at first. Actually, the concept of water-based paints was not new in 1948. The first U.S. patent on water-based paint issued in 1865. Casein powder paints came on the U.S. market in 1910 and casein paste paints in 1930. Wartime shortages in the early 1940's forced coatings manufacturers to use casein vehicles for nonessential products, adding to consumer familiarity with them.

Casein paints may have been easy to apply, and they may have been inexpensive. They were technically inferior to oil-based paints, however, and the reputation for poor performance made them definitely second choice when there was a choice. With the war's end and the availability once more of quality oil-based paints, consumers quickly turned to them in preference to water paints.

Meantime, during the war Glidden had expanded its research to develop emulsion paints that would have the wear and wash resistance of solvent paints but at the same time apply as easily and dry as quickly as emulsion paints. Glidden's chemists investigated synthetic polymers based on vinyl chloride, vinylidene chloride, vinyl acetate (popular in Europe at the time), and acrylics, but they finally settled on butadiene-styrene.

Glidden introduced the first product from its research immediately after the war. This gloss emulsion paint was the first washable enamel on the market, but Glidden withdrew it because of poor gloss and adhesion. Glidden then introduced the first flat emulsion paint in 1948, and other manufacturers soon were selling similar products, as noted earlier.

Other polymers eventually joined butadiene-styrene as the main film-former in water-based paints, and improvements in resistance to weathering in time led to water-based paints that could compete quite satisfactorily with oil-based paints in outdoor as well as indoor applications. Besides being easy to apply and also equaling or surpassing performance standards set by other coating systems, water-based paints also have other advantages. They have eliminated fire hazards risked when oil-based coatings are used in many industrial finishing processes. More recently, they have been recognized as important in reducing the water and air pollution normally associated with using solvent coatings.

Urea and Melamine Resins

Urea resins were formally introduced to the coatings industry in the early 1930's and melamine resins in the late 1930's. Today, they are firmly established in coatings industry technology. They not only permitted production of better and more economical finishes in established fields, such as automobiles and appliances. Their speed of cure and excellent films also allowed the coatings industry to expand into new fields, such as formable aluminum strip coatings. Moreover, they made fast, automatic production lines possible, including those applying coatings with electrostatic sprays and electrodeposition and curing them with infrared radiation.

Urea was the first organic chemical made in the laboratory (by Friedrich Wöhler in 1828 in Germany), and melamine followed half a dozen years later (by Justus von Liebig in 1834, also in Germany). It was only when chemists began examining their reaction products with formaldehyde, however, that their potential in protective coatings became assured. The foundation was laid in about 1920 in a patent issued in the U.S. on a urea-formaldehyde syrup for use as an adhesive, a lacquer, and a casting compound in dental work. From that point on, many chemists here and abroad intensively investigated first the reaction of urea and formaldehyde and later melamine and formaldehyde, and scores of patents issued on such amino resins.

As with most combinations of monomers leading to polymers from which fibers, molded goods, and films can be formed, both urea and melamine react to form a wide range of resinous materials. By far the majority of the amino resins in coatings today are used with either an unsaturated or saturated alkyd. In all but a very few formulations using amino resins, alkyds are present in amounts ranging from equal parts to up to 10 times the amount of amino resin.

Both urea-formaldehyde and melamine-formaldehyde resins enhance properties of coatings based on alkyds. They make films harder; increase resistance to water, solvents, and other chemicals; and improve color retention and outside durability. These improved properties plus their outstanding resistance to marring have led to extensive use of such coatings on auto steering wheels and hardware. Resistance to humidity, grease, abrasion, and mild alkalies has made amino-modified alkyd coatings especially functional on refrigerators and metal furniture and ornamental hardware. Potentially large applications also apparently exist for them as overprint finishes on drink cases, food cartons, and similar packagings and also in floor coverings.

Acrylic Resins

As a class, the acrylic resins come closer than any other resin to rivaling the first-place alkyds in performance and in breadth of uses. Manufacturers produce them in either thermoplastic or thermosetting forms and supply them for coating uses as emulsions, as water-soluble and water-reducible polymers, and as organic-soluble polymers in solid or solution form. Formulators use emulsion and water-soluble and water-reducible polymers in water-based latex paints and enamels (and also in many noncoating applications such as floor polishes and finishes for textiles, paper, and leather). They use organic-soluble polymers widely in solvent-based industrial finishes such as auto topcoats, coil and strip coatings, and appliance finishes.

Sales of both latex and solution acrylic coatings in 1970 amounted to nearly 100 million gallons. The acrylics rank high for many reasons. Acrylic polymers, for example, offer excellent color retention, adhesion, flexibility, and durability. In water-based paints and enamels they share the advantages pioneered by the earlier water-based coatings containing butadiene-styrene. As a matter of fact, latex paints (which now include polyvinyl acetate as well as butadiene-styrene and acrylic types) have largely replaced flat oil paints in interior applications. In addition, acrylics finally allowed formulators to make latex paints that could compete in the exterior gloss market that oil-based coatings traditionally monopolized. Moreover, the suitability of acrylic latex coatings for use on new or sandblasted metal surfaces opened the industrial maintenance market to latex paints. In these uses, they adhere well and control corrosion. They are also easily applied and do not expose users to fire and pollution hazards characteristic of organic-soluble coatings.

As a result of such properties, acrylic latex polymers have contributed importantly to the trend to water-based coatings. Sales of all types of latex paints (including butadiene-styrene and polyvinyl acetate types) for the first time probably equaled those of oil paints in trade sales in 1968 or 1969.

Organic-soluble coatings containing acrylics have also been important in establishing acrylics in leading positions in parts of the coatings industry. Formulators use thermoplastic acrylics for clear and pigmented air-dry and bake-dry lacquers. In these uses, acrylics impart hardness, outstanding exterior durability, and chemical resistance. They also have high gloss, water-white color, and good dielectric properties. Formulators use thermosetting acrylic enamels widely in auto topcoats, appliance finishes, and coil coatings. In auto topcoats, the acrylics have replaced alkyd-melamine resins because they have better gloss and are more easily repaired on production lines. Equally important, they can be loaded with more aluminum flake pigment for today's popular metallized coatings. Finally, they have reduced the need for periodic polishing to protect autos and enhance their appearance. As a result, auto manufacturers finish about half the cars made in the U.S. today with thermosetting acrylics and the rest with thermoplastic acrylics.

In other major uses, acrylics probably hold more than half the market for appliance finishes, where they offer excellent appearance, hardness, and resistance to staining and to detergents. Metal coils coated with acrylics are fabricated into residential and industrial siding, truck and trailer bodies, metal awnings, electrical equipment housings, light fixtures, and containers. For exterior aluminum siding for houses, mobile homes, and other buildings, acrylic enamels may well have more than half the business. Their high degrees of durability and colorfastness relieve householder and plant manager alike of the tedious and expensive task of repainting every few years.

The acrylics are about 70 years old, dating from Otto Rohm's description of the synthesis of acrylic polymers from monomeric acrylic esters in his doctoral thesis in Germany in 1901. Rohm and Otto Haas founded Rohm & Haas Co. in Germany in 1909. Haas emigrated to the U.S. shortly thereafter and organized Rohm & Haas Co. in the U.S. The company made other products for nearly three decades, however, before research in acrylics led to the first production in the U.S. (in 1936) of its acrylic plastic. Since that time, the Rohm & Haas trademark, Plexiglas, has found its way into American vocabularies.

In 1936, Rohm & Haas also introduced the first acrylic and methacrylic solution resins for coatings. They were used for clear brass lacquers and heat-resistant baking enamels. Acrylate chemistry, however, continued largely restricted to producing polymethyl methacrylate plastic sheets, which were especially useful in aircraft during World War II. It was not until 1953 with the introduction of the first latex paint based on acrylic emulsion polymers that acrylic coatings began to boom. At first they were used for interior flat paints. By 1956, outdoor use on masonry became possible, and in 1960 a special polymer was introduced for use on wood surfaces to resist grain cracking and peeling.

In the auto industry, meantime, nitrocellulose lacquers had left their mark, but they suffered from some critical drawbacks. Not highly durable, they eventually wore away under constant and necessary (or at least desirable) polishing. Also, some industry personnel felt they could be formulated in only a limited range of colors, and color figures prominently in competition among automakers. Du Pont accordingly began research in the late 1940's to develop a superior auto finish based on acrylics.

Du Pont had originally become a producer of acrylic plastics with rights and technology bought in 1936 from Great Britain's Imperial Chemical Industries. Chemists at du Pont eventually found molecular weights and average molecular weights somewhat above 100,000 and up led only to fully plastic products. They also found changes in molecular weights and average molecular weights much below 100,000 led to unusable products. The key proved to be a process that produced a polymer closely controlled in the 100,000-molecular-weight range. Such a polymer could be formulated in a coating system at about 15% solids that could be sprayed. General Motors introduced it commercially on the 1956 Oldsmobile, and du Pont was in full production by 1959.

Cook Paint and Varnish, Jones Dabney (now part of Celanese Corp.), Pittsburgh Plate Glass (now PPG Industries), and Rinshed-Mason (now Inmont Corp.) were also active in addition to du Pont in developing acrylics for the auto industry. Thermosetting acrylic baking enamels began competing in 1962 with alkyd-melamine enamels, and today all mass-produced autos in the U.S. are finished with acrylics.

Another important commercial thermosetting acrylic system was introduced by PPG Industries. It is based on polymers containing acrylamide. In this system, the amide group is methylolated with formaldehyde and the methylol group etherified with an alcohol for stability. The acrylamide-based system fitted well in the coil coating industry. Its good adhesion and forming characteristics, high application solids, and lack of discoloration on baking led to quick acceptance in the growing coil coating and siding markets. More resistant varieties gave good one-coat appliance finishes, which also have been well accepted. Other uses have been container coatings, general industrial applications, and automotive primers.

Elsewhere in acrylic coatings, an acrylic latex semigloss system became available in 1964 to join the earlier flat acrylic latex coatings. Today, more than 100 companies make such coatings. In 1966, a new acrylic emulsion opened the industrial maintenance market to latex paints. About the same time, several companies introduced semigloss acrylic latex paints for exterior use, and they first became commercially important in the spring of 1969 with the availability of a new higher-gloss, excellent-flowing acrylic.

Epoxy Resins

Epoxy resins are products of polyhydric phenols and epihalohydrins. Those of commercial significance are produced from bisphenol A and epichlorohydrin. In coating applications they are most useful where they appear either as esters of fatty acids and/or are cured with a variety of resins. Among such resins are polyamine, polyamide, urea-formaldehyde, melamine-formaldehyde, and phenol-formaldehyde. Epoxies are widely used in industrial finishes, where many agents cure them without oxidation or solvent loss at most normally encountered temperatures. Their films are tough, flexible, and highly resistant to chemical attack. They are used as vehicles for primers and enamels in auto, appliance, marine, maintenance, container, wire, and similar markets. In addition, one of their very strong points is outstanding adhesion to hard-to-coat metals such as aluminum, brass, bronze, cadmium, magnesium, and zinc.

Research that led directly to epoxy resin technology began in Europe and the U.S. just before World War II. The Europeans sought improved resins for making dentures. In the U.S., however, epoxies proved to be another of less than a handful of examples where properties as coatings were sketched out first and then a product sought to meet those properties.

The prime objective at the time in the U.S. was to produce polyhydroxy compounds that could be esterified with fatty acids to obtain coatings with improved resistance to chemicals, mainly alkalies, and also much improved adhesion, particularly to metals. S. O. Greenlee, working under J. S. Long, research director at Devoe & Raynolds Co. (now Celanese

Coatings Co.), prepared the first epoxy resin in the U.S. in the early 1940's. His reaction product of epichlorohydrin and bisphenol A was a liquid resin substantially identical to the diglycidyl ethers of bisphenol A that are now the main products of the epoxy resin industry.

Commercial introduction came in 1947. The market for epoxies grew about 25% a year during the 1950's and 1960's, and U.S. production in 1970 amounted to 150 million pounds of more than 50 distinct types of epoxy resins. The coatings industry uses about 50% of total epoxy production, while the balance is used as castings, moldings, and adhesives in electrical, construction, vehicle, container, medical, tooling, and other applications.

Epoxy resins have been modified in many ways following Greenlee's original work. Scientists at Shell Development and General Mills in the early 1950's, for example, helped develop amine-, ketimine-, and poly-amide-cured epoxy coatings. Such two-package catalyzed epoxy systems allowed formulators for the first time to prepare air-dry coatings that exhibited the properties normally associated only with high performance baking systems. In the late 1950's, research at Dow Chemical on phenol-formaldehyde systems for high-temperature performance resulted in the first epoxy Novolac resins for higher chemical and thermal resistance uses. In 1960, Union Carbide introduced a method to process epoxy resin esters from liquid epoxy resins and bisphenol A to give chemists wide latitude in formulating coatings. In 1968, Shell Chemical introduced a precatalyzed liquid epoxy resin that allowed paint formulators to make both solid epoxy resins and epoxy esters with much formulating freedom. Several companies in recent years have introduced glycidyl esters of polycarboxylic acids that they promote as having improved electrical properties and outstanding resistance to chalking in exterior applications. Finally, Dow Chemical in 1972 introduced the first epoxy resin having 100% solids but a low viscosity for formulation in coatings having high solids.

Unsaturated Polyester Resins

The properties of unsaturated polyester systems have long intrigued chemists in the coatings industry, but the use of polyester resins as coatings gained status very slowly. Scientists and engineers in the chemical and plastics industries had fairly well established the technical aspects of making polyester resins in the 1930's, but real commercial production did not begin until 1940. Even then, however, producers emphasized castings and reinforced plastics, and it was not until later that polyesters were introduced as coating resins.

Polyesters developed slowly in the U.S. primarily because of higher labor costs associated with their use. In addition, wide use in the large wood finishing market has largely been closed to them, since Americans prefer dull finishes on furniture instead of the high-gloss finishes of polyesters. In Europe, by way of contrast, polyesters have about 10% of the total coatings market. In countries there, labor costs are less compared to raw material costs, and consumers prefer—and actually will pay a high premium for—high-gloss, glass-like finishes on furniture.

Despite labor costs and appearance shortcomings, polyesters do have advantages. For one, the monomer adjusts the viscosity of the coating following application, serving first as a self-solvent and thinner and then building into a polymer film by chemical reaction. For another, formulators can produce coatings with very high chemical and mar resistance, flexibility, and durability by balancing constituents in the very easy way the basic polymer permits. In addition, film thicknesses up to 5 mils from a single spray application are readily achieved, compared to the usual 1 mil for other coatings.

Fluorocarbon Resins

Coatings based on fluorocarbon resins began appearing in the early 1950's, and their uses have grown both in volume and diversity since that time. Today, they have a strong position in the market, and their potential for further growth appears large.

Du Pont introduced aqueous dispersions of polytetrafluoroethylene (PTFE) in 1950 for such industrial applications as saturants for packings of braided asbestos and as finishes for glass cloth used mainly for electrical tape and for gaskets and other sealing devices. Along with high service temperatures, chemical resistance, and excellent dielectric properties, PTFE coatings provided unique surface characteristics, such as low coefficients of friction, dry lubricity, and excellent release properties. Use of the coatings has allowed electrical motors to be miniaturized, jet engines to operate at higher temperatures, and chemical processing equipment such as valves, pumps, and conveyor belts to be self-lubricating and resistant to adhesions.

While PTFE coatings were becoming established in industrial markets in the 1950's, they had little impact on consumers. Then in the 1960's, du Pont introduced nonstick enamels based on PTFE aqueous dispersions for use on industrial bakeware. Consumer acceptance grew rapidly following the first use of PTFE enamels on frying pans, and today the nonstick finish is as well-known in the kitchen as it is in the factory. Other consumer items, primarily hardware products such as saws, shears, and trowels, are now available with such finishes. All provide release and self-lubrication vastly superior to untreated metals.

Another advance in fluorochemical finishes occurred in 1965 when Pennsalt (now Pennwalt) introduced polyvinylidene fluoride (PVDF). This resin is virtually unaffected by ultraviolet radiation. It also forms alloys with selective acrylic polymers, which lowers material costs while enhancing rheology, fusion, and film adhesion. A number of leading paint manufacturers such as du Pont, DeSoto, PPG Industries, and Glidden now use PVDF dispersions as a base for durable and decorative architectural finishes for coil and spray coated aluminum and galvanized steel. Such finishes are projected to last up to 30 years in exterior applications. This long-term protection has been the major factor in their increasing popularity for metal building components such as wall panels, window frames, and roofing.

By far the largest end use for these finishes has been on coil coated metal that is postformed into siding for architecturally designed industrial and commercial buildings. The building boom in the power plant segment represents a large user, since public utilities wanted the added protection to their long-term investments and at the same time became more conscious of color and color stability. More recently, manufacturers of pre-engineered buildings such as Armco Steel have upgraded lines to include exterior finishes containing PVDF. Although still a small factor in housing, these finishes have been adopted by primary aluminum manufacturers for their proprietary lines of aluminum roofing shingles.

Pennwalt followed the earlier success of PVDF with the introduction in 1970 of a vinylidene fluoride/tetrafluoroethylene copolymer as a base for exterior wood finishes. Unlike PVDF, the copolymer's lower fusion temperature and increased solubility in common polar solvents provide film forming characteristics at near ambient temperatures. Alloyed with acrylics, solutions and organosols have been formulated as decorative and durable finishes for mill work (windows and doors) and for prefabricated hardboard and plywood panels. Aging tests show these finishes over primed wood matched the durability of PVDF-coated metal. Similar finishes are being formulated for concrete and other building materials.

Responding to the federal regulations on the use of solvents and their influence on the ecology, manufacturers have developed aqueous based finishes of PVDF and vinylidene fluoride/tetrafluoroethylene copolymers. Latices of 50% solids blended with acrylic latex show promise for coil coating and spray applications in which no sacrifice is made on long-term film integrity and color stability. Another approach to the same problem is the use of solventless powder coatings, and work is in progress to develop fluorocarbon-based powders having the proper molecular weight and particle size for fluid bed and electrostatic spray applications.

Pigments

Basic lead carbonate (white lead) served for many years as the leading white pigment (white being both the most popular color and also the base from which many other colors are produced). While basic lead carbonate produces a coating that adheres well and resists transmission of water vapor, it does suffer from several drawbacks. Oil paints containing it chalk, check, and crack. They also darken in atmospheres that contain sulfur compounds. As a result, the protective coatings industry toward the end of the 1800's and beginning of the 1900's introduced a number of different white pigments to overcome the problems. Lithopone (a mixture of barium sulfate and zinc sulfide), zinc sulfide, and zinc oxide proved among the more successful replacements or supplements. None, however, rivaled titanium dioxide, which began to assume commercial importance as a coating pigment following World War I. (In more recent times, toxicity of lead pigments has become of great concern, and white lead is no longer used for interior architectural finishes. In addition, many states are now putting tight restrictions on the lead content of all paints.)

The coatings industry has produced improved forms of titanium dioxide since its introduction, and manufacturers consume more of it now in making coatings than of any other pigment by a wide margin and, as a matter of fact, about as much as all other pigments combined. Modern rutile titanium dioxide (one of its two forms) per unit of pigment provides opacity between five and 10 times greater than that of white lead. It also provides up to five times the opacity of zinc oxide, lithopone, and antimony trioxide and about two times that of zinc sulfide. To users, such hiding power means truly one-coat interior finishes, exterior finishes that last much longer, and painted products that remain bright and glossy throughout their useful life.

Titanium dioxide's position as the leading pigment also seems assured for the future. Since refractive index relates directly to covering power, both theoretical and experimental studies can be used to predict compounds of possible interest as white pigments. With the theoretical exception of silicon carbide, none in such studies matches titanium dioxide in hiding power and at the same time provides its degree of hydrolytic and thermal stability, lack of toxicity, and low cost.

The post-World War I titanium dioxide pigments were composites (with barium and calcium sulfates). The first noncomposite form, anatase titanium dioxide, followed in 1925. Noncomposited rutile titanium dioxide, which has a higher refractive index and thus covers better than the composite anatase form, became commercial in 1940.

Pigment producers at first made each of these titanium dioxide pigments by the sulfate process. In it, titaniferous ores are dissolved in hot sulfuric acid, the iron contaminants removed by fractional crystallization, and the product hydrolyzed to amorphous titanium dioxide that is then calcined to the proper particle size and crystal form.

In the late 1940's and early 1950's, du Pont pioneered with a significant new chloride process, and in the past decade or so all new capacity in the U.S. and much of the capacity elsewhere has used the chloride process as well. In it, ore is chlorinated in a fluid bed reactor, pure titanium tetrachloride separated by fractional crystallization from iron and other chlorides, and the titanium tetrachloride oxidized by an air or oxygen-enriched gas stream at high temperatures. Oxidation products are titanium dioxide and chlorine.

Control of purity and particle size is superior for the chloride process compared to the sulfate process. Waste disposal problems, while still formidable, also are less severe. Moreover, the close control provided by the chloride process permitted production of products with narrower size distributions, which in turn permitted experimentation leading to a deeper understanding of the influence of particle size and pigment concentration on the opacity of pigmented materials. By 1958, a new family of rutile titanium dioxide products emerged that had been optimized for specific end uses. Further refinements in surface treatment technology have provided improved optical characteristics, handling properties, and durability, and such is the direction of work now going on.

FUTURE DEVELOPMENTS

Buyers of trade sales paints and industrial finishes today all assume adequate and often outstanding durability, adhesion, corrosion resistance, and other performance characteristics in the products they buy. Even so, the coatings industry can still improve its products, so that articles they are used on suffer even less from early obsolescence. To the degree it succeeds, it will help minimize waste and conserve resources.

Many of the improvements will occur along lines of trends already established. The coatings industry and the industries supplying it may not in a similar time produce nearly the number of new coating components as they did in the past three to four decades, but research on new components has by no means ended.

Alkoxymethyl isocyanates, for example, are conceptually simple yet ingenious new materials of interest to the coatings industry. They react with compounds containing active hydrogens to produce groups that are stable under ordinary conditions. These groups crosslink, and thus originally thermoplastic polymers can be modified to thermosetting compositions.

Among other monomers of interest in coatings that companies are now working on, an acrylic monomer containing an activated chloromethyl group and copolymerized with a hydroxyl-containing monomer leads to a polymer that crosslinks at room temperature or after a short time at elevated temperatures. Acetoacetates containing a polymerizable group produce polymers or copolymers with pendant acetoacetoxy groups. The resulting resins are crosslinked by polyvalent cations, amines, aldehydes, ketones, and air oxidation. Finally, liquid coating compositions based on polyanhydrides and phthalic anhydride dissolved in a liquid epoxide will form polymers for use in coatings when heated in the presence of a tertiary amine catalyst.

The coatings industry will also change as older products establish themselves in new markets. Epoxy producers expect new markets, for example, especially in the auto industry and in highway and airfield construction. Epoxy resins have long been used in powder coatings applied in fluidized beds. Auto companies have now expressed interest in using such coatings to help eliminate pollution caused by paint solvents. A change to powder coatings could result in considerable growth for epoxies, and we shall return to the subject in a moment.

On highways, concrete fails rapidly on the thousands of bridges and overpasses in northern states because of freeze-thaw cycles. Manufacturers have largely completed initial testing and have recently introduced several epoxy topping systems that protect these surfaces satisfactorily. They are now ready to realize the growth potential the market offers, especially on the extensive federal interstate highway system. Meantime, in a development that is not entirely a coating application but one the coatings industry is participating in, epoxy resins (as well as latex resins) added to concrete increase strength and durability substantially. Because of high labor costs for repairs on concrete roadways and aircraft runways, the market appears ready for much growth in the 1970's.

Other improvements in coatings technology will come because of newly emerging needs. The possibility of broad adoption of stringent air pollution regulations similar to those pioneered in California, for example, is leading all large coatings companies into research and development on water-based industrial coatings and powder coatings as ways to reduce solvent pollution (as well as reduce the dangers of explosion and fire). The demand for such coatings is just beginning, and many companies expect acrylic resins especially to profit from it. With the current interest in consumer protection, meanwhile, manufacturers of polyesters are concentrating on improving fire retardance of coatings made from them. Incorporating chlorinated and brominated acids, anhydrides, and glycols now produces durable polyester coatings that retard flames, and further progress appears indicated.

The future coating that currently generates most speculation is undoubtedly powder that can be sprayed electrostatically, discussed briefly earlier in connection with epoxies. Such coatings are now used successfully on pipes, but much work is still required on polymer preparation, powdering techniques, coatings formulation, and film deposition and appearance to adapt the method to more complex shapes such as auto bodies. The prospective return from commercial success remains high, however.

Another potential applications development involves polymerization initiated by an electric current. In such electropolymerization, current passing through a conductive solution of a vinyl monomer produces a polymer, either in solution or as a precipitate depending on its solubility in the reaction medium. A recent innovation of considerable interest to the coatings industry is one in which the monomer is polymerized and at the same time deposited on a metallic object that is one electrode in the cell. The process is very similar to electrodeposition that is used with considerable commercial success for finishes applied in the auto and other industries, but it differs in one important respect. In electropolymerization, both polymerization and deposition occur in place in contrast to electrodeposition, where a preformed polymer deposits on the item serving as the electrode.

The coatings industry finds electropolymerization interesting and potentially useful by allowing a mixture of monomers to be polymerized on metal surfaces without the polymers having to be prepared first. Among potential advantages are coatings at less cost but with improved performance. Much development work is still needed, however, before the technique becomes important.

Radiation in various forms also offers possibilities for improving protective coatings technology. Microwaves generated by radio-frequency power tubes, for example, cause molecules in matter they pass through to align themselves with the rapidly changing electromagnetic field. This movement produces heat. Unlike thermal heating, however, where heat spreads inward from the surface, microwaves generate heat at every point in the material at the same time. Films cure rapidly and uniformly as a result. Shorter curing times normally reduce costs, and more uniform films normally fail less readily. Manufacturers must still design and synthesize suitable polymers and formulate them into compositions that respond to microwave heating before the method becomes commercially competitive, however.

Meantime, radiation-cured polyesters may soon challenge traditional lacquers in the wood finishing industry. Recent research in curing polyesters with electron beams has shown they cure in fractions of a second and still retain properties comparable to polyesters cured with peroxide catalysts. The method requires formulation changes, but manufacturers expect to work them out economically. The initial high cost of equipment for electron curing may deter use of the method, but high output may help offset the disadvantage.

Meantime, ultraviolet radiation may cure polyesters successfully commercially before electron beams do. UV cures as effectively as does an electron beam, and it requires less expensive equipment. While it doesn't cure quite as rapidly, the curing time is still short compared to the normal cure times with peroxide catalysts.

These radiation curing techniques are also being considered for polyesters in high-speed coil coating applications. Their excellent flexibility, color retention, and corrosion and abrasion resistance coupled with fast curing would make polyesters particularly attractive for many products fabricated from precoated steel and aluminum.

SOCIAL AND ECONOMIC IMPACTS

Drugs are at once the savior and the curse of mankind. Morphine, an ancient drug but still vitally important to hospital and physician, relieves pain in the seriously injured and terminally ill. Penicillin has saved thousands from formerly fatal infections. Aspirin more than any other drug has alleviated aches and pains that would otherwise have resulted in billions of dollars of lost working time. Yet heroin, a simple derivative of morphine, ravages thousands. Penicillin in the sensitized can cause rashes, shock, and—in rare instances—even death. Aspirin may lead to gastrointestinal bleeding and severe asthma in some patients. Probably only atomic energy among modern technologies rivals drugs in the duality of effect for good and evil.

Some drugs, to be sure, may be abused. Others may expose some people to risks more serious than their illnesses. Yet most Americans without doubt have very high levels of health care today even when the abuses and risks are considered. Many factors determine levels of health care, of course, such as education, sanitation, diet, living and working conditions, hospital facilities, and drugs. No one can determine exactly the share that each contributes. Nonetheless, certain trends are unmistakable. A child born in 1900 could expect to live about 47 years. By 1920, life expectancy for the newborn had risen almost a decade to 56 years and by 1970 another decade and a half to about 70 years. The baby of 1970 can thus expect to live about 50% longer than his great grandparents, and a person can accomplish a lot in an extra 20 to 25 years. All U.S. presidents, for example, entered office after 40, and many people achieve much in their later years.

The increase in life expectancy in the two decades from 1900 to 1920 must be credited in substantial measure to improved sanitation, better living conditions, and better hospital care. Nonetheless, the prophylactic vaccines for some diseases and the arsenicals for syphilis contributed significantly. In the 1930's, the synthetic vitamins, insulin, and sulfa drugs largely accounted for the decline in death rates. Then in the 1940's, 1950's, and 1960's, the most dramatic drop in death rates occurred, largely because of the discovery and use of antibiotics. Better drugs in the past 70 years may not have made the only contribution to better health and longevity, but they have made a major one.

As a result of these and other advances, the infectious diseases that killed some 500 of every 100,000 Americans in 1900 now kill only about 50. Deaths each year from influenza and pneumonia at about 30 per 100,000 people are about one seventh their 1900 rate, while those from tuberculosis and syphilis are about one tenth their former rates. Diphtheria, typhoid and paratyphoid, and whooping cough yearly killed fewer than one person each in every 100,000 beginning in the late 1940's and early 1950's compared to rates of 40, 30, and 10 per 100,000 in 1900. Among noninfectious ailments, fewer than 10 in 100,000 die annually from gastrointestinal problems today compared to about 150 in 1900. At the rates for 1900, some 1.3 million Americans could have expected to die in 1970 from these afflictions instead of the 107,000 that did die from them.

Unfortunately, these statistics on medical and drug successes against infectious and noninfectious illnesses are accompanied by less comforting ones. By 1970, for example, the death rate from cardiovascular diseases had risen 50% from its 1900 level and that from cancer had tripled. Sufferers from arthritis number in the millions. Americans may live longer, but they now become victims of other causes of illness and death.

The impact on the nation's health of any innovation involves more than changes in mortality rates and life expectancies, of course. Many people live more productively and enjoyably directly because of drug therapy. In few categories are the results more telling than in the broad range of problems termed mental illnesses. Less than 30 years ago, society had no other recourse than to confine numerous psychotics such as schizophrenics, paranoics, and others in sanitariums. Manic-depressives killed themselves too often unless hospitalized and treated by psychiatrists, who had only straitjackets, solitary confinement, and electroshock to work with.

While the direct cost to society for confining the mentally ill might have been calculated, who could measure the effects the untreated in society had on families, on communities, and on the economy? Then the discovery of chlorpromazine in 1950 opened the way to a chemical approach to the problems of mental illness. Today, physicians use many psychotropic drugs to relieve the symptoms of slightly disturbed people and help carry them through mental crises of one type or another. Even more importantly, these drugs help treat and correct more deep-seated mental illnesses. As a result, the number of patients now in state and county mental hospitals (where most mentally ill people are confined) has been reduced by one third from the peak of 1955 (of 560,000 patients) during a period when the country's total population has increased some 20 to 25%.

Accomplishments similar to those in treating mental illnesses have been recorded for drugs for many other major and minor afflictions, although direct statistical evidence is often less readily available. Such improvements in people's health are not surprising. The body is a chemical system, and chemical findings logically help medical scientists and physicians diagnose, treat, and often cure illnesses. As a matter of fact, men sought and used drugs long before recorded history. As man became more civilized, the tempo of his drug discoveries increased. The use of drugs in medicine in time paralleled the degree of civilization attained in different regions of the world, emerging largely at first in China, then in the Middle East, in Greece, and finally in western civilization.

Physicians still use drugs originating centuries ago, such as morphine, strychnine, digitalis, iodine, and alcohol. These drugs were discovered empirically, and the search for drugs became scientific hardly more than two centuries ago with the advent of chemistry and biology as rigorous sciences. Among readily remembered landmarks for biology and chemistry are:

- The discovery of the microscope by Anton van Leeuwenhoek (1632–1723).
- The research of Robert Boyle (1627–91) and Antoine Lavoisier (1743–94).
- The founding of organic chemistry by Friedrich Wöhler (1800–82), Justus von Liebig (1803-73), August Kekule (1829-96), and Jacobus van't Hoff (1852–1911) and others of their era.
- The concepts and experimentation of Louis Pasteur (1822–95).

Scientists have determined much in the past two centuries about normal and abnormal body functions. During the same time, they have accumulated detailed information on the structure of matter and developed the ability to make chemicals almost routinely. In their early research, they primarily reexamined ancient remedies by the then new scientific methods of chemistry and biology. Then in the mid-19th Century, scientists began systematic laboratory experiments on drug actions and established pharmacology as an independent science. In the last part of the 19th Century and the first third of the 20th, they made some progress as they continued to work on chemotherapeutic approaches to curing the body's ills. Then in the three decades from roughly the mid-1930's to the mid-1960's, countless new drugs flooded onto the market. Probably not even the long-time seekers of magic bullets, such as Paul Ehrlich, dreamed of such success with so many magic bullets:

- A variety of sulfa drugs, such as sulfanilamide, sulfathiazole, and sulfadiazine, and dozens of antibiotics, such as the penicillins, tetracyclines, streptomycin, and chloramphenicol, to treat infections successfully in many places throughout the body.
- Streptomycin and isoniazid to treat tuberculosis.
- Adrenocorticosteroids to treat rheumatoid arthritis and allergies.
- Rauwolfia and veratrum alkaloids, quaternary ammonium compounds, and other drugs to treat hypertension.
- Organomercurials and thiazides to remove excess fluids in treating heart, kidney, and liver ailments.
- Chlorpromazine and other tranquilizers for dealing with mental illnesses and anxiety states.
- Diphenhydramine and numerous other antihistamines to alleviate allergic conditions.
- Oral antidiabetic drugs to replace insulin in some treatments.
- Estrogen-progestogen chemicals to serve as oral contraceptives.
- Methantheline and other anticholinergics to treat peptic ulcers and other gastrointestinal tract ills.
- Chloroquine compounds as antimalarial and antiamebic drugs.
- Halogenated hydrocarbon anesthetics, lidocaine and other local anesthetics, and skeletal muscle relaxants for use during surgery.

As a result of these and similar contributions by medicinal chemists and others, today's physician in the U.S. can prescribe a great variety of drugs. These products, available in many forms for administration and often in several different dose levels, can be bought just about anywhere in the U.S. within a few hours. They are made by companies employing some 140,000 people in about 1,300 establishments.

Health services have become one of man's four basic needs (the other three being the traditional food, clothing, and shelter). In 1970, Americans spent about $67 billion for all facets of health care. Their payments went to a diverse array of individuals and organizations, ranging from family physicians to drugstores to hospitals to insurance companies. Of the total, they spent about $4.0 billion for ethical drugs and about $1.7 billion for proprietary products.[1] Making and selling drugs and other health products, providing health services, and performing medical research gave employment in 1970 to about 11,000 chemists and chemical engineers, according to Bureau of Labor Statistics estimates (and not counting persons teaching in colleges and universities whose research or other work may involve medical products and services). That number is exceeded only by the approximately 60,000 chemists and chemical

---

1. Ethical drugs are prescription and other drugs promoted only to physicians. Proprietary products are promoted directly to the public and sold over-the-counter. Figures are for retail sales in drugstores and do not include sales to and in hospitals.

engineers that the bureau estimated worked in 1970 in companies making chemicals (except drugs) and closely related products (such as plastics and resins, synthetic fibers, synthetic rubber, soaps and detergents, cosmetics and toiletries, pesticides, fertilizers, and paints).

CHEMICAL ACCOMPLISHMENTS

Many people working in health-related activities use the principles and techniques of chemistry in their work. In some cases, chemistry is peripheral. In others, it is basic. Since this study must have some limits, we shall confine the discussion on achievements in health-related activities to those pertaining to prescription drugs. These drugs aren't the only health-related products and services that depend strongly on chemistry, but their development is probably the single most important activity involving chemistry in the entire health field.

In singling out developments in prescription drugs, incidentally, we shall naturally concentrate on the accomplishments of chemists. Readers must not conclude, however, that chemists and chemistry are solely responsible. Much scientific work requires more than ordinary cooperation among people trained in different disciplines, but in no field is an empathetic cooperation more essential among people trained in a greater variety of disciplines than in drug research. No one, for example, is more frustrated than a chemist with a fascinating molecule without a biological test in glassware with enzymes (biochemistry), in animals (pharmacology), or in human patients (clinical research). Equally frustrated is the biologist with a great theory and no chemical with which to conduct an experiment. Thus, drug development from the earliest scientific days has involved a plurality of disciplines, including chemistry, biology, medicine, physics, and mathematics. Moreover, it has taken organizing forces in industry and government to bring benefits to bear. Otherwise, a unifying concept born in a university, say, without some agency to test its practical use would be of little ultimate good.

Unfortunately, we have not found it possible within the limits of this study to trace all contributions from many different disciplines and from many kinds of organizations. Even when we restrict ourselves to chemistry, we have not found it possible to mention much of what chemists and chemical engineers have done. Thus, what follows at best summarizes only some of the chemical highlights from an American point of view.

The Sulfa Drugs

While drug development is admittedly complex, its history does fall into several reasonably-well-defined periods. Until about 1900, drugs for the most part were natural products. They dealt primarily with illnesses caused by protozoans and metazoans (quinine for malaria, ipecacuanha for amebiasis, for example). Then in the first quarter of this century, medical scientists turned from extracting drugs from natural sources and launched the era of synthesized drugs (such as Ehrlich's arsphenamine for syphilis) and then developed a more extensive list of antiprotozoal and antimetazoal agents from 1925 to the present (chloroquine for malaria, piperazine for roundworms, for example).

What we now generally recognize as the modern era of chemotherapy dates from the early 1930's. The sulfa drugs launched that era, and they were followed by the antibiotics, the psychotropics, the corticosteroids, and many other types of chemicals for treating a wide range of illnesses.

In 1933, Gerhard Domagk in Germany found that a red dye made by condensing diazotized sulfanilamide and a diamine killed bacteria. Shortly thereafter, scientists at the Pasteur Institute in France showed that sulfanilamide itself was the component responsible for the antibacterial effect of what Domagk had named Prontosil. This finding spurred laboratories throughout the world to synthesize thousands of substituted aminobenzenesulfonamides and screen them for antibacterial action. As a result, physicians soon had a number of new drugs for treating a range of infectious diseases.

The sulfas were immediately labeled miracle drugs, for they dramatically helped sufferers of pneumonia and other infectious diseases. Before the sulfas, deaths from pneumonia, for instance, averaged two to four out of every 10 cases. After the sulfas, the average dropped to less than one in 10.

Although dramatic, the sulfas were not as dramatic as the antibiotics, which were developed soon after. Relatively few sulfas are still in use because of the antibiotics, which cover broader spectra of activities and are, on the whole, better tolerated by patients. While they may not rank high medically today, the sulfas did for a time rank as the best antibacterial agents available. Moreover, research on them led to a number of derivatives that now have considerable value. Scientists in England in 1940, for example, suggested that the sulfa drugs work as they do because they compete with p-aminobenzoic acid in some natural metabolic process. p-Aminobenzoic acid is a building block of folic acid, and cells use folic acid to make several components of the nucleic acids present in all living cells. When the sulfa drug competes successfully, the natural process does not occur, and the cells die.

Research workers at Burroughs Wellcome U.S.A. exploited this antimetabolite theory, and one practical result of their work was trimethoprim, an antibacterial agent. This diaminopyrimidine specifically inhibits dihydrofolate reductase, an enzyme required to keep the folic acid system running. Dihydrofolate reductase produced by bacteria is far more sensitive to trimethoprim than the corresponding enzymes produced by mammals, thus accounting for trimethoprim's effectiveness against bacterial infections in people. Trimethoprim and sulfisoxazole (one of the few sulfas still in use today) inhibit the folic acid cycle at two different points, and use of both results in true synergism. In Europe, the combination has been recently introduced to combat urinary tract infections.

The Antibiotics

In 1929, Scottish bacteriologist Alexander Fleming, working at St. Mary's Hospital in London, reported that a common *Penicillium* mold isolated as an accidental contaminant of a bacterial culture produced a potent antibacterial substance. Fleming called the substance penicillin.

A decade was to pass before the word penicillin was to make much impact on medicinal chemistry. By that time, three groups working independently, one in England and two in the U.S., made observations that put antibiotics in the history books.

The English group, working at Oxford University on a $5,000 grant from the Rockefeller Foundation, isolated penicillin in crude form and in 1939 showed that it spectacularly protected mice against experimental streptococcal infections. A year later the Oxford group reported the first cures of infections in people.

Meanwhile, at the Rockefeller Institute in New York, bacteriologist René Dubos isolated tyrothricin and demonstrated that this mixture of polypeptides protected animals against experimental pneumococcal infections. At Rutgers University, soil microbiologist Selman Waksman was studying the antagonistic effects of soil microorganisms. He and his co-workers made a series of discoveries (actinomycin in 1940, streptothricin in 1942, and streptomycin in 1944) that focused worldwide attention on the soil as a source of antibiotics.

From 1940 on, antibiotic discovery, development, and manufacture occupied many American chemists, chemical engineers, microbiologists, pharmacologists, and physicians. Their work revolutionized the treatment of virtually all infectious diseases except those of viral origin. Scarlet fever, streptococcal sore throat, mastoiditis, pneumonia, meningitis, plague, blood poisoning, syphilis, tuberculosis, and Rocky Mountain spotted fever all succumbed to antibiotics. Before the discovery and development of antibiotics, these diseases usually meant either death or serious illness.

After World War II, penicillin became the focal point for drug research in industry, the universities, medical schools, and government laboratories in many countries. In the 1950's and 1960's, chemists and other scientists introduced many new antibiotics as a result, and the penicillin of World War II is now only one of a series of drugs used to treat serious infectious diseases. The first orally effective penicillin was introduced in 1954 (by Eli Lilly). Chemists (at Massachusetts Institute of Technology) synthesized the basic penicillin nucleus (in 1958), and their work eventually led to the synthesis of a variety of penicillins by a simple acylation reaction. Several companies introduced a number of semisynthetic penicillins that had such advantages as more effective absorption in the body and treatment of *Staphylococcus aureus,* which resisted penicillin-G and was a serious problem in hospitals. Ampicillin (discovered by Beecham and available in the U.S. beginning in 1963) extended penicillin therapy to Gram-negative organisms not normally affected by other penicillin forms. Discovery (at Oxford) of cephalosporin, a new penicillin-related nucleus, led to the introduction (by Lilly) of four cephalosporin antibiotics (beginning in 1964).

The major goal of antibiotic research following the discovery of penicillin and streptomycin became finding safe agents to cure the broadest range of infectious diseases, both bacterial and viral. The goal hasn't been fully reached yet, but scientists have made remarkable progress in the past quarter of a century. The first success in finding broad-spectrum activity came in 1947 at Yale and Parke, Davis, whose scientists had collaborated in the discovery of chloramphenicol (also found independently at the University of Illinois). Chloramphenicol is now the best drug for treating typhoid and typhus and certain stubborn urinary tract infections. Like many really effective drugs, including penicillin and streptomycin, chloramphenicol exhibits dangerous side effects, but in the hands of competent physicians it will cure infections that resist other antibiotics.

What are now known as the tetracycline antibiotics represent another far-reaching advance in the quest for broad-spectrum but safe and generally useful drugs to treat infections. Scientists at Lederle discovered the first of them (chlortetracycline in 1948), but it turned out to be less safe than one found the same year at Pfizer (oxytetracycline). In 1954, scientists at Pfizer working with Nobel laureate Robert Woodward at Harvard reported the exceedingly complex structures of these drugs and of the parent tetracycline. Their results became the basis for the discovery and introduction of several antibiotics in the group, among which tetracycline itself, demethylchlortetracycline, and doxycycline are the most important.

Although pharmaceutical scientists continue to strive to find broadly useful antibiotics, they also search diligently for drugs to cure stubborn infectious diseases even though some of them are socially and economically minor. Some microbes, for example, have become resistant to the major antibiotics. They have posed special problems, but they have now largely been controlled by having many different antibiotics for physicians to use when they find resistance to a major antibiotic. Staphylococci resistant to penicillins and broad-spectrum antibiotics are now usually controlled by the macrolide antibiotics such as erythromycin (discovered at Eli Lilly in 1952) or triacetyloleandomycin (Pfizer, 1957). Others that can be used against Staphylococci are vancomycin (Lilly, 1955), lincomycin (Upjohn, 1962), and clindamycin (Upjohn, 1968). Problem Gram-negative pathogens are especially susceptible to gentamicin (Schering, 1963).

Effective though the sulfas and penicillin were in the late 1930's and early 1940's against many infectious diseases, they had no effect against tuberculosis. Waksman's streptomycin did, however, and Merck & Co. developed and marketed it in 1945, a year following its discovery. The number of people confined to TB sanatariums immediately began dropping following its use, and it is still one of the basic components of the multidrug approach to treating TB.

Two other drugs have also been important in controlling TB. One is isoniazid and the other p-aminosalicylic acid (PAS). Still working in Germany after World War II, Domagk followed his work on Prontosil with a report in 1946 that p-acetamidobenzaldehyde thiosemicarbazone was an antitubercular agent. This report led to almost simultaneous publications in 1952 by investigators at Hoffmann-La Roche and Squibb as well as by Domagk reporting isonicotinic acid hydrazide (or isoniazid) to be a potent antituberculosis drug. Highly effective and inexpensive, isoniazid, too, helped revolutionize the treatment of TB throughout the world.

Streptomycin, isoniazid, and PAS together have placed TB among the list of largely curable diseases. In 1900, for example, 200 people out of every 100,000 in the U.S. died from the disease. In 1954, however, TB claimed fewer than 10 per 100,000 and in 1969 fewer than three per 100,000. Unfortunately, TB still remains a stubborn disease among the very poor, but new agents such as ethambutol (from Lederle in 1964) and rifampicin (from CIBA and Dow in 1971) are still being introduced by pharmaceutical companies. Rifampicin combined with isoniazid is now considered the most advanced and useful therapy for treating TB.

Cardiovascular Agents

Diseases and malfunctions of the heart, arteries, and veins now kill more Americans than any other single class of ailments. They have accordingly attracted some of the most active research work in medicinal chemistry. The history of successes in this subject dealing with such problems as high blood pressure, improper rhythms in heart beats, and excess fluid in the body span only a brief quarter century or so, however.

An exceedingly complex series of chemical reactions controls the body's circulatory system. Intimately involved are the sympathetic nerve system, which raises blood pressure, and the parasympathetic nerve system, which lowers it. Biochemicals associated with nerve functions in these activities include:

• Acetylcholine, which carries nerve impulses across the synapses.
• Norepinephrine, which is liberated by nerve impulses at nerve endings and which activates muscles to contract.

- Monoamine oxidase and catechol-O-methyl transferase, which destroy or inactivate norepinephrine when its job has been accomplished.
- Dopamine, which is converted into norepinephrine in cells at nerve endings.

Chemical treatment of high blood pressure, which is one of the most widespread and serious circulatory system disorders, relies on controlling the production of these biochemicals or on blocking their action with other chemicals that occupy sites the biochemicals normally occupy to carry out their functions.

The first chemotherapeutic approach to controlling high blood pressure came with what are known as ganglionic blocking agents. Warner-Chilcott introduced the first oral preparation in 1952. It and others were considered at the time to be breakthroughs in treating the ailment. They influence the action of acetylcholine in ganglionic cells, probably becoming bound to the acetylcholine receptors and thus preventing acetylcholine from acting. They effectively lowered blood pressure, but undesirable side effects such as impairment of cerebral circulation and severe constipation limited their general acceptance by the medical community.

Several other chemicals for treating high blood pressure followed the ganglionic blocking agents in the early 1950's, but their successes were soon overshadowed by that of a newly found agent, reserpine. It was first offered for sale late in 1953, and physicians still use it, either alone or combined with diuretics or other agents, to lower blood pressure.

Reserpine is an alkaloid produced in the Rauwolfia plant. Medicinal chemists embarked on a systematic study of the plant in the 1930's, as it had long been used in India to treat a variety of diseases. Two Indian chemists isolated the first alkaloids from Rauwolfia roots in 1931, and their achievement made possible more detailed pharmacological studies. Workers at the Swiss laboratories of CIBA first isolated reserpine in 1952, and chemical studies there and at such locations as CIBA's U.S. facilities, Lilly, Squibb, University of Wisconsin, and Iowa State College led to elucidation of its structure.

Scientists now know that reserpine acts by causing norepinephrine to be released from its binding sites and making it available for destruction by monoamine oxidase. Reserpine also inhibits the synthesis of norepinephrine by blocking the conversion of its precursor, dopamine, to norepinephrine. The depletion of norepinephrine lessens the contraction of arterioles that norepinephrine normally causes. With less constricted arterioles, blood pressure falls.

Scientists at CIBA in the U.S. contributed a second important chemical to treat high blood pressure when they synthesized guanethidine in 1958 and found it lowered pressure. It acts by impeding release of norepinephrine. Introduced in the U.S. in 1960, it proved to be as potent as the ganglionic blocking agents developed earlier, but it lacked their disadvantages.

Another approach to controlling release of norepinephrine came from Merck. In 1954, scientists at Merck first described α-methyldopa as an inhibitor of dopa decarboxylase. This enzyme acts to decarboxylate dopa to dopamine, a precursor of norepinephrine. They believed that a nontoxic inhibitor of the biosynthesis of dopamine and related catecholamines (a group of compounds that include norepinephrine) would be an effective agent for lowering high blood pressure. In 1960, the results of human trials at the National Heart Institute established the antihypertensive effectiveness of the drug, and Merck introduced it in 1963. Although the basic mechanism of action does not depend on decarboxylase inhibition, α-methyldopa has turned out to be an excellent hypotensive agent. In fact, it is probably the most widely used agent today for controlling high blood pressure.

Vasoconstrictor Agents

Raising blood pressure is necessary in other conditions, such as in hemorrhagic or cardiogenic shock and during anesthesia, surgery, or hemodialysis. A substance that has a prompt and potent pressor action in such emergencies is the vasoconstrictor angiotensin II amide. It acts by increasing the resistance of the arteries in the periphery to the flow of blood. One of its advantages is that its action lasts only a short time, because the enzyme angiotensinase chemically destroys it in the organism.

Angiotensin II amide was first marketed in 1962, but its developmental history dates from 1939 when it was discovered in the U.S. and independently the following year in Argentina. It was not until 1954, however, that pure angiotensin was obtained by workers at the Cleveland Clinic. They incubated a pig renin with a renin substrate found in horse plasma. Later, two forms of angiotensin were discovered: angiotensin I, which is a decapeptide formed by the action of renin on renin substrate; and angiotensin II, which results from the removal of histidylleucine from angiotensin I by a peptidase in plasma.

In 1957, two groups working independently, the Cleveland Clinic in the U.S. and CIBA in Switzerland, reported the synthesis of angiotensin II peptides. The Swiss group later synthesized angiotensin II amide, an octapeptide that differs structurally from the naturally occurring angiotensin II by a minor variant. It is the amide that is made commercially for medical use.

Antiarrhythmic Agents

Drugs that control arrhythmias, or variations in the normal heartbeat, are very important in controlling cardiac fibrillations. A number of substances have antiarrhythmic activity, but they probably do so by acting on different cardiac parameters. This difference makes it hard for medicinal chemists to draw conclusions about relationships between structure and activity. The result is that despite numerous investigations, advances in antiarrhythmic agents have been relatively slow.

One of the first and still very important antiarrhythmic agents is quinidine. It is the result of research that began in Germany in 1914 with the finding that quinine alkaloids, derived from cinchona bark, control heart rate in a limited number of cases of fibrillation. By comparing several cinchona alkaloids, the Germans four years later also found that a quinine relative called quinidine was most effective. Quinidine is obtained from cinchona bark along with quinine, but today a considerable quantity of quinidine is prepared by isomerization of quinine in a chemical process developed in the U.S. at Columbia University.

No other rival of quinidine existed until procaine amide became available in 1951. Its chronicle began in Germany in 1926 when it was shown that procaine amide compared favorably with quinidine in controlling fibrillation of the heart. Ten years later, German scientists also observed that procaine hydrochloride depressed the excitability of heart muscle during surgery. The next link came in 1948 when scientists at the National Institutes of Health found that diethylaminoethanol, a metabolite of procaine, exerted an antiarrhythmic action on the heart. Following that discovery and after extensive research, scientists at Squibb suggested and synthesized procaine amide.

Procaine amide spurred a continuing search for more effective but less toxic drugs. Of the many compounds tested, some have been of use. Among them are an anesthetic, lidocaine; an antihistamine, diphenhydramine; *beta* adrenergic blocking agents propranolol and others; and an antiepileptic, diphenylhydantoin. While quinidine and procaine amide remain the two most frequently used antiarrhythmic agents, lidocaine is probably used mostly in hospitals. More recently, propranolol is beginning to gain wide acceptance for the control of arrhythmias.

Diuretics

Edema, the swelling that occurs when the body retains too much water and salt, figures importantly in congestive heart failure, hypertension, and kidney and liver diseases among others. Diuretic drugs correct edema, mainly through their effect on the kidneys. Such natural diuretics as digitalis, theophylline, and caffeine have long been known, but the modern era for diuretics began in the 1920's with the discovery of organomercurial compounds and their diuretic properties.

By studying structural features responsible for both activity and toxicity of the organomercurials, U.S. chemists contributed to the development of several improved organomercurial diuretics. The first of these, mercaptomerin sodium, was synthesized at Campbell Pharmaceutical Co. and introduced in 1949. While this innovation resulted in an agent that provided less cardiac toxicity and local irritation, it and other mercurial diuretics are seldom used now, because the less toxic and orally active nonmercurial diuretics are available.

The first sulfonamide diuretic was made available in 1953 by American Cyanamid Co., where a search was under way for compounds that inhibit the enzyme carbonic anhydrase. The work eventually yielded acetazolamide, whose effectiveness and safety offered some improvements over the organomercurials.

Early diuretic drugs showed low potency or loss of activity after chronic use. Thus, one of the 1950's major developments occurred with the discovery of the first orally effective nonmercurial diuretic that could be used continuously. Synthesized at Merck Sharp & Dohme, it was called chlorothiazide.

Chlorothiazide was closely followed in 1958 by hydrochlorothiazide from CIBA. In the next several years, many more diuretics emerged from U.S. pharmaceutical laboratories as chemists expanded the thiazide discovery.

The thiazides and related compounds facilitate elimination of sodium and chloride ions in nearly equimolar amounts along with an osmotic equivalent of water. Thus, they give physicians control over what had for some time been recognized as an important factor in hypertension: the amount of electrolytes in body fluids. In addition, long-term studies have revealed that thiazides lower blood pressure more than their salt-depleting activity accounts for.

The thiazides had one side effect that needed to be controlled, however. Along with the excretion of sodium ions, considerable potassium ions are also excreted. Excessive potassium imbalance is undesirable. Therefore, a search for diuretics without this liability continued. Diuretics that spared excretion of potassium arrived with the introduction of spironolactone by Searle in 1959 and of triamterene by Smith, Kline & French in 1964. These two diuretics appear to be useful combined with hydrochlorothiazide in causing good saluresis with minimum potassium loss.

Still more recent are two compounds that added a new dimension to the usefulness of diuretics: the treatment of pulmonary edema. The compounds are furosemide (introduced by a West German company, Hoechst, in 1966) and ethacrynic acid (introduced by Merck Sharp & Dohme in 1967). The latter resulted from a program to design a compound that would mimic the organomercurials. Chemists at Merck reasoned that the effectiveness of an organomercurial comes from its chemical affinity for compounds found in the kidney with a molecular structure containing the sulfhydryl grouping of atoms. The structural class from which ethacrynic acid evolved resulted from the search for compounds that would react with sulfhydryl groups under physiological conditions in a manner similar to the organomercurials.

Steroidal Drugs

Steroid drugs are used as oral contraceptives, anti-inflammatory agents, sex hormones, adrenocortical hormones, and anabolic agents. In the form of male and female sex hormones, they are also used to treat such life-threatening conditions as cancer and in the form of adrenocortical steroids to treat shock and Addison's disease. Female sex hormones are also used for replacement therapy, the estrogens particularly during menopause and the progestogens in pregnancy and menstrual disorders. Male sex hormones are useful for replacement therapy and as anabolic agents in the debilitated. In terms of quantities used, contraceptives and anti-inflammatory agents rank highest.

Early work (1926–27) on steroids involved demonstration of the vital nature of secretions of the adrenal cortex and isolation of the active hormone, cortisone. American endocrinologists and chemists made some major contributions during this phase.

The second stage of adrenal cortex hormone research involved making large enough quantities to study in relation to disease. The steroid of the adrenal cortex that turned out to be most important is cortisone. Its synthesis was first achieved by a combination of efforts at the Mayo Clinic and Merck. Merck undertook the task of making large quantities of cortisone available for clinical studies, and it was this outstanding developmental research that resulted in what is now recognized as one of the historic discoveries in medicine.

At the time, cortisone was foreseen as having an important application in conditions of adrenal insufficiency such as Addison's disease. But the major clinical application of cortisone followed Philip S. Hench and E. C. Kendall's dramatic discovery of its antirheumatic properties in 1949 at the Mayo Clinic.

The Hench-Kendall discovery set off an intense effort among chemists throughout the world to modify the structure of cortisone to improve potency and reduce side effects. In 1953, Squibb announced the first of these successful efforts with 9-α-fluorohydrocortisone, the first synthetic steroid with greater potency than the natural one. A year later, a group at Schering found that prednisone and prednisolone, two derivatives of cortisone and hydrocortisone, are both more potent in antirheumatic activity and produce fewer undesirable side effects than the natural product. Since then, modifications of the steroid structure by U.S. chemists have led to many major drugs in this field.

The discovery of cortisone's effect against rheumatism and arthritis set off major efforts among another group of U.S. chemists to find a plentiful, low-cost source of raw material from which to make cortisone. Bile from cattle was Merck's starting material for making the cortisone used by Hench and Kendall. While Merck developed this process into a high yielding and practical one, many believed that supplies of animal glands weren't sufficient to meet expected demands. A new starting point in the search for other raw materials, therefore, was a process developed in the early 1940's at Pennsylvania State College. Scientists there had found an efficient way to degrade the side chain of compounds called sapogenins to change them to steroids that could then be converted to steroid sex hormones.

Sapogenins usable in the process come from the roots of many plants. The search for the most economical source of the sapogenin eventually narrowed to Dioscorea, a yam that grows in Mexico. Today, it is the source of diosgenin, the starting material for most steroid drugs, and some 700 tons of diosgenin are processed annually to make them.

Elaborate chemical processes are needed to convert diosgenin to the adrenocortical hormones and anti-inflammatory agents, since these compounds have an oxygen atom at a position in the complex steroid nucleus that is particularly awkward for synthetic organic chemists to work on. Nevertheless, some adroit methods have been developed. After the synthesis of cortisone in 1946 at Merck, research turned to a process for making the anti-inflammatory corticosteroids from bile acids, and the process eventually was put to use to make cortisone and related steroids. Upjohn scientists, meantime, used a mold to attach an oxygen atom at the proper place in progesterone and also developed an efficient alternative path to corticosteroids from diosgenin. Pfizer and Squibb also used fermentation techniques.

Most of this research preceded development of oral contraceptives, the other major use of steroids. Long before oral contraceptives were marketed, scientists had known that a combination of two natural hormones, progesterone and estradiol, blocked ovulation in normal women. Unfortunately, these compounds are essentially inactive orally, and chemists turned to seeking inexpensive progestins and estrogens that would control ovulation effectively when taken orally.

The problem of producing an orally effective estrogen was solved in Germany in 1938 with the finding that ethinylestradiol and its methyl ester are active when administered orally. In the early 1950's, a group at Searle prepared norethynodrel and a group at Syntex prepared norethindrone for the progestin component. Combinations of these progestins with the estrogens developed in Germany were highly effective contraceptive agents. Searle's combination reached the market first (in 1960), and it was followed by products from Ortho Pharmaceutical (in 1962) and Syntex (in 1964) that contained the Syntex progestin.

Since that time, many other progestational agents have been prepared and marketed throughout the world. Perhaps the most potent progestin available today is norgestrel, which Wyeth Laboratories introduced in 1968. In contrast to the other progestational agents that have been prepared from natural sources, norgestrel is a totally synthetic material.

Development of oral contraceptives represents one of the most discussed drug achievements of the 1960's. Despite questions that have been raised about their safety, they are the most commonly used form of contraception in the U.S., and an estimated 9 million women in the U.S. rely on them for fertility control.

Nonsteroidal Anti-Inflammatory Agents

Arthritis is a degenerative disease that is characterized by inflammation of the joints, and finding safe and effective anti-inflammatory agents has occupied many medicinal chemists in recent years. Although the goal is far from reached because of the complex nature of arthritis, there have been some successes with steroidal drugs, as has just been discussed. There have also been some successes with nonsteroidal drugs, and indomethacin and the phenylbutazones are among those that U.S. chemists have discovered or developed.

Indomethacin originated at Merck, where scientists first became interested in indole derivatives as potential anti-inflammatory agents because of the possible role of an indole, 5-hydroxytryptamine, as a mediator in the onset of inflammation. Another clue was an observation at Edinburgh that some metabolites of tryptophan (another indole relative) are excreted at higher levels by rheumatic patients than by people without rheumatic symptoms. A number of indole derivatives were tested as anti-inflammatory agents, and some activity was noted in a few cases. More than 350 analogs were then prepared and tested before the program culminated in the synthesis of indomethacin in 1963. Merck established the compound's anti-inflammatory activity in the same year.

Another important nonsteroidal anti-inflammatory agent, phenylbutazone, originated outside the U.S. (at Geigy in Switzerland). Shortly after its introduction in the U.S. in 1952, a U.S. group at the National Institutes of Health proved that the principal metabolite of phenylbutazone is oxyphenylbutazone. The Geigy group then began clinical trials of the oxy compound, which proved to be a potent anti-inflammatory agent and appeared to have fewer side effects than the parent compound. Phenylbutazone, its oxy derivative, and indomethacin are major adjuncts to the treatment of rheumatoid arthritis today.

Psychotropic Agents

Mental illnesses rank among the top five ailments that strike Americans (cardiovascular diseases, cancer, arthritis, and allergies are the other four). For much of history, the medical profession rarely succeeded with treatments, and confinement proved the ultimate solution all too often. Conditions typical of old-time asylums very early gave the English language the word for wild disorder and confusion (bedlam). Even today, public dismay at conditions in mental institutions still is noteworthy. It is not the evil intent of administrators and medical personnel that brings about the deplorable conditions, however, but rather the hopelessness of maintaining large numbers of mentally disturbed people in hospital communes with any semblance of human dignity. It is not unexpected, therefore, that chemicals to treat mental illnesses were hailed as they were or that such chemicals have become the largest selling drugs in any therapeutic category.

Sufferers of mental disorders may exhibit many symptoms, but medical personnel specializing in mental problems have classified the illnesses into various categories. The most prevalent illness is schizophrenia, with perhaps half of the persons confined to mental institutions being schizophrenic. They tend to live in an inner world, have hallucinations and delusions, and often act incoherently and in a bizarre fashion. Manic-depression is another major category. Manic-depressives suffer severe deviations in mood. They may be either highly excited or depressed, or they may alternate from one extreme to another. A number of other categories have been defined as well, among them such less serious conditions as psychoneuroses and character disorders.

Persons deeply affected by such serious disorders as schizophrenia or manic-depression often require hospitalization and extensive treatment. Those suffering from the less serious disorders rarely require hospitalization, and they often respond well to drug treatment.

The Rauwolfia alkaloids and the phenothiazines became the first drugs used successfully to treat psychiatric disorders. The alkaloids are isolated from extracts of the plant *Rauwolfia serpentina*. The Rauwolfia plant had been used in India for centuries for treating a variety of diseases, as mentioned earlier, but it was not until 1953 that medical scientists recognized the therapeutic value of the Rauwolfia alkaloids. Introduction of a pure Rauwolfia alkaloid, reserpine, triggered the recognition.

At first, reserpine was used only to treat high blood pressure, but research, mainly at CIBA, soon followed to establish reserpine firmly as a psychotropic agent. Important at the time, it is no longer the drug of choice for schizophrenia in the U.S., however.

Almost simultaneously with reserpine's development the first of the synthetic tranquilizers became available. It is a phenothiazine derivative, chlorpromazine. Chlorpromazine was first synthesized in France and used there as a component of an anesthetic in surgery. The French soon saw it could be used to treat severe mental disorders, but U.S. scientists first instituted broad clinical trials and introduced the compound as a psychotherapeutic agent.

Chlorpromazine set off a flourish of activity in pharmaceutical company laboratories everywhere as chemists synthesized and biologists evaluated an eventual total of 2,000 to 3,000 compounds related to phenothiazine in the hope of improving on chlorpromazine. Their efforts uncovered many derivatives with valuable and differing clinical attributes. The variety gave physicians a wide range of choices for treating mental and nervous disorders.

Meanwhile, scientists at Wallace Laboratories in the U.S. followed a completely different route to drugs for treating mental conditions. They were searching for muscle relaxant drugs. They studied 1,200 compounds in all, from which they selected a propanediol carbamate derivative (meprobamate) that they made in 1951. It is used particularly to help relieve anxiety, distinguishing it from the antipsychotic drugs.

Still another class of tranquilizers arrived in 1960 with the introduction of the 1,4-benzodiazepines, which are now used extensively to treat nervous disorders such as anxiety and neuroses. This group of compounds originated in the laboratories of Hoffmann-La Roche during exploratory chemical studies aimed at finding new classes of drugs that act on the central nervous system. Three of the compounds have been introduced. They are chlordiazepoxide hydrochloride (1960), diazepam (1963), and a hypnotic agent, fluorazepam (1970). Hoffmann-La Roche chemists prepared more than 2,000 benzodiazepine derivatives during their exploration. More patients throughout the world are probably now being treated with diazepam (Valium) than any other drug.

Antidepressants

The 1950's saw a concerted chemotherapeutic attack against another mental disorder—depression or melancholy. Toward the end of the decade, success came in the form of antidepressants in several classes of chemical compounds.

The first chemical compound to show pronounced antidepressant properties was iproniazid, which was synthesized at Hoffmann-La Roche. During the course of the compound's evaluation as an antitubercular drug, clinicians noticed that it exerted a strong mood-elevating effect. Clinical studies were then started on its psychotropic properties, and iproniazid turned out to be the first specific antidepressant.

Iproniazid is a potent inhibitor of the enzyme monoamine oxidase, and this property became the test by which other compounds were screened for antidepressant activity. Unfortunately, iproniazid and other monoamine oxidase inhibitors that were placed on the market by other pharmaceutical companies were found to cause complications in many patients needing antidepressants. These drawbacks have greatly curtailed their use, but some psychiatrists still favor them highly for certain patients.

Antidepressants of another type, structurally related to the phenothiazine tranquilizers, replaced the monoamine oxidase inhibitors. They differ from the phenothiazine tranquilizers in that the six-membered thiazine ring is replaced by a seven-membered ring system. The first drug to be really applicable in the treatment of endogenous depression was imipramine, introduced by Geigy in 1959. Imipramine, amitriptyline (Merck Sharp & Dohme, 1961), and others are structurally related to the phenothiazine tranquilizers. The molecular modification is small but profound, and these drugs are now often used as a substitute for electroshock in order to prevent suicide by manic-depressors.

Anticonvulsants

The oldest effective agents to treat epilepsy are the barbiturates, and pharmaceutical workers have continued to search for new and more efficacious agents within this chemical area. Notable examples are Abbott's metharbital (1952), Ayerst's primidone (1954), and Parke, Davis's succinimides (1953–62).

A significant advance in the treatment of seizure disorders came when the anticonvulsant activity of an old chemical, diphenylhydantoin, was recognized at Harvard Medical School. Parke, Davis made it available in the U.S. in 1938. Since then, a number of other hydantoins have also been developed as anticonvulsants. Eli Lilly introduced the hydantoin relative, phethenylate, as an anticonvulsant in 1950, following its synthesis at the University of Texas, while Abbott marketed ethotoin in 1957. However, diphenylhydantoin remains the most widely used.

Research has also been directed toward completely freeing epileptics from seizures. Parke, Davis chemists synthesized phensuximide, methsuximide, and ethosuximide, and these succinimides were produced as anticonvulsants between 1953 and 1960. Earlier, Abbott chemists prepared the oxazolinediones, two of which (trimethadione and paramethadione) are still in general use. Phenylacetylurea also came from Abbott.

Other substances useful for their anticonvulsant activity although not principally intended for this purpose are two tranquilizers, chlordiazepoxide hydrochloride and diazepam, and a diuretic, acetazolamide.

Muscle Relaxants

Muscle relaxants, or drugs acting on certain central neuronal pathways, are used to reduce muscle spasm associated with injury, tetanus, and orthopedic procedures. None of the presently available drugs is totally successful, but they do often produce measurable relief.

Muscle relaxant research goes back to 1910, when 1-phenoxy-2,3-propanediol was observed to cause flaccid paralysis in animals. Interest in this property of the glyceryl ethers revived in the late 1940's with the study of mephenesin. This compound is not in much use now, however, because it is rapidly metabolized and its activity is only of short duration. In 1956, the synthesis of methocarbomol decreased this shortcoming of mephenesin. Now sold by A. H. Robins Co., it is one of the most frequently used muscle relaxants.

Another widely used muscle relaxant is chlorzoxazone, synthesized at

McNeil Laboratories in 1956. Wallace Pharmaceuticals has still another approach to muscle relaxants in the drug isopropylmeprobamate. Developed in 1960 in a continuation of the work on the tranquilizer meprobamate, it is now also one of the most widely used muscle relaxants. The tranquilizer diazepam is also finding increased use as a muscle relaxant, although its use is still far below that of the others.

Anti-Parkinson Agents

Another group of agents affecting skeletal muscles are the drugs used in treating Parkinsonism. These drugs reduce the rigidity and tremor characteristic of the disease and often allow an otherwise incapacitated person to lead a relatively normal life.

Natural spasmolytics of the belladonna alkaloids and certain antihistamines have long been known to have some muscle relaxing effect. A synthetic spasmolytic, trihexyphenidyl, was developed simultaneously in several pharmaceutical companies in the late 1940's. Among the U.S. companies involved were Sterling-Winthrop and American Cyanamid. One antihistamine with muscle relaxant properties is orphenadrine, which came from Riker Laboratories.

In a basic study, chemists discovered in 1961 that brain tissues of persons suffering from Parkinson's disease showed a reduction of biogenic amines, especially dopamine. The discovery has led to one of the most exciting of the pharmaceutical research projects currently in progress. A logical extension of the discovery was to try to replace dopamine, but dopamine does not readily cross from the blood stream to the brain cells. Next, its biological precursor, L-dopa, was tried. This approach also failed at first because the amounts of L-dopa required caused excessive nausea and vomiting.

In 1967, scientists in the U.S. discovered that tolerance to L-dopa could be built up by starting with small doses and gradually increasing them over several weeks. Eventually, the dosage was raised to the point that the dopamine content of brain tissues was actually increased, and the patients showed marked improvement.

L-Dopa is now available to patients in the U.S. and most other countries to treat Parkinson's disease. Because of the large dose required (5 to 10 grams per day), the supply situation has been tight and the cost to patients very high. Good production processes have been developed in the U.S., however, and production of large quantities of L-dopa constitutes a fine achievement.

Clinical work is now under way using an inhibitor of L-dopa decarboxylation before the L-dopa reaches the brain. The scientists involved hope that such inhibitors will reduce the quantity of L-dopa needed by the patient and hence lower the cost per day for treatment.

Analgesics

The search for a drug to control severe pain as effectively as morphine, codeine, and their relatives but without the dependence-producing hazard of the morphine family has long been an active project in many of the world's pharmaceutical laboratories. The first significant advance, however, did not come until 1953, when scientists at Eli Lilly synthesized d-propoxyphene. d-Propoxyphene proved to be about as effective as codeine when taken orally. In cases of mild to moderate pain, its addiction potential is low enough to keep it off the narcotic list, and it is used extensively.

d-Propoxyphene cannot substitute for morphine for cases of severe pain, however, and that part of the problem remains. There was a brief glimmer of hope in 1943 when Merck prepared the first authentic sample of N-allylnormorphine (nalorphine). Nalorphine is an antagonist of the pharmacological effects of morphine and related strong analgesics, and its analgesic powers equal morphine's. But side effects, particularly hallucinations, at effective doses preclude clinical acceptance.

Even the successful synthesis of morphine at the University of Rochester in 1952 has not so far pointed the way to an acceptable morphine substitute. Scientists at Sterling-Winthrop Research Institute decided it would be more practical to investigate the benzomorphan nucleus, for which a simple synthesis had been worked out at NIH, than to use the brilliant but complicated synthesis of morphine as a starting point. They prepared and evaluated a series of narcotic antagonists of the benzomorphan type. Out of their research came pentazocine, which controls moderate to severe pain, is clinically acceptable from the standpoint of side effects, and is not tolerant with the opiates. There is an addiction potential with pentazocine, unfortunately, even though it is much lower than the addiction potential of morphine.

A pentazocine analog, cyclazocine, is being investigated as a way to treat heroin addicts. Naloxone, a morphine derivative synthesized at Endo Laboratories, is also being investigated for the same purpose. When either drug is given to a detoxified heroin addict, it blocks the action of heroin and, in effect, prevents reestablishment of the addictive state. In contrast to methadone, these two compounds do not act as opiate substitutes, and abrupt withdrawal by chronic users produces only mild symptoms and no manifestation of drug-seeking behavior. Cyclazocine is 40 times more powerful than morphine as a pain killer, and it must be administered carefully or it produces hallucinations. Naloxone does not produce hallucinations nor does it have analgesic properties.

Antihistamines

Research on antihistamic drugs started in 1933 in France. The aim was to find a way to treat a variety of allergic disorders that presumably result from the release of histamine, such as hay fever, dermatoses, and drug reactions. In the U.S., CIBA and Parke, Davis did the research that resulted in the first antihistamines being put on the market shortly after the war. Their success prompted most of the nation's drug companies to start projects searching for other antihistamines, and a great number of products resulted.

The greatest success was achieved at Schering Corp., where chemists synthesized pheniramine and its chloro and bromo derivatives. Chlorpheniramine, by far the most widely used antihistamine, is more potent than earlier products, and even that potency has been doubled by using the compound's dextro form.

For some time, antihistamines were thought to abort the common cold, but this view has since been disproved. However, they do seem to relieve some cold symptoms, and so they are a major ingredient of many cold remedies being sold throughout the U.S. today. A principal side effect of antihistamines is their sedative action, which is used in proprietary non-barbiturate sedatives.

Antidiabetic Agents

Insulin's discovery in the early 1920's provided diabetics with an effective form of treatment, but a diabetic treatment that could be administered orally was sought for decades. Eventually, two chemical classes of compounds, the sulfonylureas and the biguanides, proved to be highly effective.

The sulfonylureas date from 1942, when French clinicians noted while studying sulfa drugs that a sulfathiadiazole caused severe hypoglycemia, or a deficiency of glucose in the blood. This finding stimulated scientists to investigate sulfa compounds for the control of diabetes and led to the sulfonylureas.

Tolbutamide was one of these, first reported in 1956 in Germany to possess hypoglycemic activity. Upjohn introduced the drug in the U.S. in 1957. It has proved to be useful for a large number of diabetics and has shown relatively few side effects. It is much more widely used than insulin, which is now restricted largely to forms of diabetes that do not respond well to tolbutamide.

Another successful sulfonylurea is chlorpropamide, made at Pfizer. Introduced in 1958, it has been found to be a longer-acting agent.

The biguanides, meantime, have a longer history, dating back to 1918 with the observation at Yale that guanidine sulfate reduces blood sugar levels in animals. Guanidine sulfate proved to be too toxic to be used clinically, and an attempt to use the hypoglycemic properties of guanidine was the short-lived use of decamethylenediguanidine in 1926. It also proved too toxic for use in man.

Then in 1958, U.S. Vitamin Co. reported on the oral hypoglycemic activity of phenethylbiguanide in animals. This compound was much less toxic than the previously studied biguanides. It was found that phenethylbiguanide effectively manages diabetes in some patients. It was introduced in the U.S. in 1959, and it has proved to be a valuable adjunct in the care of diabetics.

Tropical Diseases

U.S. pharmaceutical companies over the years have mounted a major research effort to cure tropical diseases even though the vast majority of sufferers live outside the U.S. A great many tropical diseases are caused by helminths, or worms, that infect the intestines. Frequently, patients suffer from multiple infections, and a broad-spectrum antihelminthic drug is therefore desirable. A compound that comes close to satisfying the need is thiabendazole, discovered at Merck Sharp & Dohme Laboratories. It has been used successfully to treat ascariasis, enterobiasis, strongyloidiasis, trichuriasis, and hookworm infections.

Thiabendazole, however, has not been successful against schistosomiasis, which infects perhaps 100 million people in the tropics and semitropics. For many years, physicians mostly used tartar emetic and other trivalent antimonials. These drugs had to be injected for several days to effect a cure. Then in 1939, lucanthone, a derivative of thioxanthenone, was found at I. G. Farbenindustrie to be effective in animals. Clinical trials proved it was also effective in man and could be given orally. Further research on lucanthone and how it acts suggested that the body's cells metabolize or absorb lucanthone and change it to an active form before it attacks snail fever.

Some years later (in 1963), scientists at Sterling-Winthrop Research Institute succeeded in preparing hycanthone, a highly active metabolite of lucanthone. Hycanthone proved to be an effective agent given orally or parenterally. Clinical evidence gathered so far indicates that a one-dose injection is all that may be needed to treat infections caused by snail fever organisms.

Antimetabolite research has also led workers to a preventive agent for malaria, one of the most feared tropical diseases. The discovery was pyrimethamine, an inhibitor of dihydrofolate reductase. Dihydrofolate reductase is an enzyme of the malaria parasite. Pyrimethamine is 2,000 times more powerful against this enzyme than the one produced naturally by mammals, and it is now in wide use in Africa as a malarial prophylactic.

Veterinary Medicine

Chemistry's most important contributions to veterinary medicine are agents for control of coccidia and helminths, parasites of major economic significance in poultry and livestock production. A variety of drugs is now available, but five coccidiostats and one antihelminthic stand out.

The antihelminthic is thiabendazole. This was the first broad-spectrum agent effective for elimination of gastrointestinal nematodes in sheep, cattle, horses, swine, goats, dogs, and poultry. It enormously stimulated further developments. Antihelminthics that followed include tetramisole, pyrantel, and dimethyl dichlorvinyl phosphate incorporated in polyvinyl chloride. The last ingeniously lets an organophosphate be used safely.

Five major developments have occurred in coccidiostats. They date from the 1939 observation that sulfanilamide was effective against coccidia in poultry. Wide testing of sulfa drugs followed. Sulfaquinoxaline, developed at Merck, is one that emerged. It is still used in severe outbreaks.

Sulfaquinoxaline has limitations for prophylactic use, the day-in, day-out protection regarded as ideal. Merck scientists found an answer in nicarbazin, discovered in 1953. The even performance of nicarbazin against various species of coccidia, the absence of severe resistance problems, and the excellent immunity development in treated birds have impressed parasitologists greatly. It is a molecular complex of two compounds, a carbanilide and a hydroxypyrimidine. The carbanilide component provides the coccidiostat activity, and the hydroxypyrimidine enhances absorption of the extremely insoluble carbanilide.

Although many coccidiostats have been discovered since, nicarbazin remains of significant value. It has two disadvantages, however. It adversely affects egg production, limiting its use for layers, and it sensitizes birds to severe heat stress. An agent that can be used safely in laying hens emerged from Dow Chemical several years after nicarbazin's introduction. This was 3,5-dinitro-o-toluamide (Zoalene). At about the same time, Salsbury's Laboratories, working independently, recognized the effectiveness of 3,5-dinitrobenzamide, a lower homologue of Zoalene.

Then in 1960, Merck discovered amprolium, a broad-spectrum agent that has prophylactic and curative powers. Furthermore, amprolium can be used in turkeys, cattle, sheep, and goats, as well as chickens. Veterinary scientists were interested in amprolium for its mode of action as well as for its effectiveness. Amprolium works by inhibiting coccidial utilization of a vitamin, thiamine. The antivitamin activity is apparently selective, for the drug is well tolerated by both broilers and layers. It is now marketed as Amprol-Plus, in which it is combined with methyl 4-acetamido-2-ethoxybenzoate or ethopabate. This additive provides improved control of the so-called intestinal species.

The most recent coccidiostat is metachlorpindol, developed at Dow. Strategies for broilers now are alternate use of Zoalene and metachlorpindol or Amprol-Plus and nicarbazin.

FUTURE DEVELOPMENTS

Worldwide drug research faces a more favorable future today than at any time in the past. Scientists have never had a greater supply of information nor has their information ever appeared more soundly based. Forecasts in both the U.S. and U.K. predict continued advances against classical diseases for which good drugs already exist. They also dare to suggest that drug therapy will advance in treating virus diseases and cancer in the next 25 to 30 years. While their prophecies are admittedly speculative, the forecasters also confidently expect far-reaching achievements in controlling heart and related diseases, which today are major killers. Even in the face of great difficulties in finding new drugs to alter mental states, an English study predicts many people will be taking new drugs by 1990 to correct mental disorders.

Not all authorities agree with all aspects of this promising picture for drug research in the future, however. Some are particularly concerned with the fund of basic knowledge, which is always a prerequisite for applied drug research. As a countermeasure to the belief that fundamental advances are being made less rapidly, at least one pharmaceutical company is investing heavily in basic research institutes both in the U.S. and Europe. These institutes will seek only new basic concepts of disease causes to provide scientists in applied drug research with information for further advances, while the feedback to basic concepts from applied research will continue to provide information, as it has in the past.

Speculation about the future and the possible dearth of basic concepts aside, it is clear that dramatic advances in medicinal chemistry and biology in the past decade have laid a sound base for proceeding. We now know that the giant deoxyribonucleic acid (DNA) molecule of the cell chromosome is nature's universal genetic code, and the discovery is becoming one of the great determinants of mankind's future. DNA transmits all the heritable information accumulated since the genesis of life. The broad outlines are now clearly visible of the mechanism whereby the information from DNA is read and used ultimately to form a human being from a single fertilized cell. Charting the infinitely complex pathways will occupy scientists beyond the foreseeable future, but each advance they make will provide an opportunity for new drugs. Small molecules have already been shown to interact with DNA, ribonucleic acid (RNA), and other components of the genetic system. Understanding the nature of these interactions will undoubtedly lead to drugs with greater specificity and with less toxicity for treating many diseases.

A young and vigorous science called molecular biology has emerged in recent years in which chemistry, physics, and mathematics interface with classical biology and genetics. Much research in this field will focus on viruses and cancer. Viral infections generate many cancers in animals, and it will be strange indeed if scientists don't learn that viruses also generate many cancers in people. A virus's core is similar to the genetic system (DNA) of normal cells. It affixes itself to that system and changes normal cells into cancerous ones that multiply without control. There are many steps in the transformation from passive, inactive virus to formation of a tumor where drugs could interfere and thus prevent the flare-up of virus infection or the occurrence of cancer.

The great hurdle is to find drugs that work selectively and that don't interfere with the host's normal cells. Scientists already know a great deal about the structure of viruses and the way they propagate as parasites of normal cells. They have made some progress in finding drugs that interrupt the propagation of some viruses and maintain them in a quiescent state until the body's immune response mechanism can finally destroy them.

Manipulating the body's immune response itself may also prove to be another approach to viral and cancer control. Scientists have made much progress in recent years in understanding this defensive system. They know the structure of human immune globulin at the molecular level just short of having a precise blueprint for the tiny regions of specificity that recognize foreign cells and unnatural chemicals. By inhibiting the immune response with drugs, physicians can now transplant some body organs (which in effect are invading agents to the recipient). They cannot as yet use drugs to suppress the immune response to prevent rejection of organ transplants, however, without risking lowered resistance to infections and toxicity to normal cells. The goal is more specific immune-suppressant drugs that allow the immune response to regard cells from an organ donor as acceptable while the immune response of a cancer patient regards cancerous cells from his own normal cells as invaders.

The immune response also looms as a factor in dealing with other diseases. The body from infancy learns to discriminate between foreign invaders and its own cells, but then it appears to forget. The change can occur at any stage in a person's life, but the changes increase with age. Many degenerative diseases often associated with aging thus likely involve impairment of the immune response, particularly a decrease in its normally high specificity. In the future, progress in immunology (the study of immune response) will very probably provide much needed drug therapy for degenerative diseases, notably arthritis. These advances will also yield drugs that will effectively treat allergies, in which the immune response is too sensitive and is badly timed.

In contrast to organ transplants and degenerative diseases, where the immune response must be inhibited, treatment of cancer requires a stimulated and sensitized response to produce antibodies to cancer cells. Ultimately, the cure of cancer must involve the selective destruction and scavenging of cancerous cells regardless of the cause of their formation.

Research in hormones will also have an impact on progress in cancer therapy. There are strong indications that hormones influence genetic processes to accomplish their mission of regulating growth of the human body and of protecting the animal body from an inimical environment. Advances in endocrinology (the study of hormones) will provide drugs that will control many serious diseases, probably including arteriosclerosis. Above all, such study will give better control of fertility and propagation.

The genetic system in cells directs and controls the production of enzymes, and scientists are gathering precise information on the structure of many enzymes that control body functions. Many drugs either slow down or speed up enzyme activities, and they owe their effect to their affinity for certain sites of these complex chemicals. Research on the inhibition and harnessing of key enzymes involved in transmitting nerve impulses will certainly lead to important drug discoveries.

Advances continue to be made on the role of vitamins, hormones, and antibiotics even though new discoveries of such substances are now few and far between. During the past decade, for example, a whole new class of hormones has been found that many tissues produce locally. Originally thought to be made only by the prostate, they are known as the prostaglandins. Actually, the testes make them in relatively large amounts, and many other organs make them in smaller amounts.

The prostaglandins provide a new research approach in many important body functions, including reproductive, cardiovascular, and degenerative. Most recently, scientists have shown that a variety of drugs, most notably aspirin, influences their rate of synthesis. Among other benefits, this finding may open the way to determining the mechanism that, for the first time since its discovery in 1899, explains the manifold benefits of aspirin. Understanding aspirin's mechanism undoubtedly will lead to discovery of more potent drugs with greater specificity.

The greatest cause of death in the U.S. and many other countries is coronary artery disease. Research to date has broadened the understanding of factors that predispose persons to it, and attention as a result has been focused on atherosclerosis. This disease is now associated with high lipid and cholesterol levels in the blood, and many laboratories have research projects under way to find drugs that lower the levels of both. The most promising drug to appear so far is one named clofibrate, but whether prolonged administration of it actually reduces deaths has not been completely determined. Future investigations with it and other drugs should show whether such forms of therapy are beneficial.

While research with clofibrate and similar drugs deals with prevention, other research deals with what should be done after a thrombosis occurs. At present, the approach is to use enzymes that dissolve clots. One of them is streptokinase, a bacterial enzyme that has shown some promise. Unfortunately, it causes antibodies to form, since it is not of human origin. Another thrombolytic enzyme, urokinase, has been isolated from human male urine. It doesn't cause antibodies to form, and it has shown some promise in treating pulmonary embolism. Problems associated with suitable collection of urine, however, limit the availability of purified enzyme.

Whether either of these clot-dissolving enzymes will have a place in medical practice is yet to be determined. Meantime, the events leading to clot formation are becoming more clearly defined. Drugs that affect the adhesiveness of blood platelets may well be developed to prevent the formation of thromboses in the cardiovascular system.

Fertility control is another topic currently generating high medical interest. Scientific findings on the physiology of reproduction over many years led to the introduction of the oral contraceptives in the 1960's. As scientists continue to improve their knowledge in this very important field, medical applications are likely to occur more rapidly. For example, recent research has shown that a hormone secreted by the hypothalamus triggers release of the pituitary hormone that controls ovulation. Research is under way to develop a new type of contraceptive to exploit this discovery.

Another area of interest is the development of a male contraceptive that is orally effective. Several years ago, a relatively nontoxic group of compounds became available that inhibited spermatogenesis. Unfortunately, they induced an intolerance to alcohol, and interest dwindled. The work, however, did serve to point out that an oral male contraceptive is a possible approach to population control.

Beginning in the mid-1950's, scientists launched a more modest but very promising new approach to developing drugs with maximum efficacy and safety for the largest range of patients. Exploiting this new discipline known as drug metabolism has really just begun. Research workers increasingly take advantage of the extremely sensitive analytical methods that have become available for determining in detail the fate of drugs from administration to elimination. Now, study of heritable traits among individuals shows that people differ genetically in their abilities to eliminate foreign substances. Persons studying drug fates must take this new fact into account when they study metabolism of new agents. Fortunately, the outlook for success in such research is excellent.

Drug Addiction

The foregoing at best only summarizes some of the more important directions for developments in drug therapy in the future. It deals with long-standing problems, and medical science will doubtless make much

progress in the coming years in solving many of the major—and minor—
ailments that people still suffer from. Meantime, these long-standing ills
have in the past few years been joined by another that is getting high
priority for national social reform. That problem is drug addiction.

No one knows what drug addiction now costs the U.S. With considerable effort, it might be possible to add up what every federal, state,
and local agency spends directly for narcotics control and rehabilitation.
But no one could calculate the sum that would include these direct costs
and also account for lost productivity from absenteeism and inefficiency,
for crime, for welfare associated with addiction, and for the drain on the
energy and health of families and others that addicts come in contact with.
In dollars and in suffering the cost is staggering.

Scientists have fortunately recorded some successes in developing drugs
that separate relief of pain from the tendency to induce addiction and
that help treat and rehabilitate the addicted. They have also developed
methods to detect morphine and heroin in body fluids and urine, important
in discovering addicts in the first place and then in monitoring their progress
during treatment.

Relieving pain is by far the most legitimate use of some drugs subject
to abuse. Scientists have developed drugs that relieve mild to moderate
pain but that are far less addicting than morphine and its derivatives.
These advances are substantial. More importantly, when buttressed with
the new and highly refined analytical methods, they strongly indicate
that drug research might finally solve the physiological aspects of drug
addiction. By understanding fully the mechanisms of addiction and physical
dependence, scientists might well succeed in designing drugs that would
reverse the physical dependence on and craving for morphine and heroin.
Meantime, for real prevention of addiction, the medical profession still
needs a deep analgesic that will not induce uncontrolled craving and a
need for higher doses.

While the search continues for less addictive drugs, other advances
have occurred in treating and rehabilitating the addicted. The synthetic—
and somewhat controversial—analgesic methadone is receiving serious
attention for treating hard-core addicts. Many see it as substituting a
more acceptable dependency on it for dependency on heroin.

Another approach lies in such chemicals as cyclazocine and naloxone,
which are related to morphine. They antagonize the action of morphine
and save addicts from death because of respiratory failure from over-
doses. By eliminating the pleasurable effect of morphine, they also serve
as steps in treating addicts.

Unfortunately, methadone itself is addictive and thus subject to abuse.
Naloxone would be more valuable if it acted longer, while cyclazocine
when administered continuously causes hallucinations and other unpleasant side effects. Even the most enthusiastic proponents of methadone and
the antagonists accordingly agree much room still exists for improving the
treatment and curing of addiction.

Medical approaches alone, of course, will not solve drug addiction, since
it is a complex problem with major social and economic components.
Nonetheless, progress in solving the problem would be speeded if drug
research could develop less addicting drugs for medical needs and more
effective drugs to cure and rehabilitate the addicted.

Of the more than $600 million the U.S. pharmaceutical industry spent
for research in 1970, about $30 million probably went for analgesic research.
Government agencies, universities, and medical schools probably spent an

equal sum for similar work. Society might well insist that drug research organizations multiply their efforts to provide nonaddicting drugs that would displace all narcotics from medical practice. It would have to help improve the climate for drug research, however, for such a campaign to be successful.

Government Regulation

When viewed with an international perspective, the future for developments in drugs can only be optimistic. When viewed from the slightly narrower perspective of prospective developments in the U.S., however, some in the pharmaceutical industry find themselves less optimistic than they would like to be. They see themselves as spending ever-growing sums for research, but they also see their work as being less obviously effective than it was earlier. In 1939, the U.S. pharmaceutical industry spent an estimated $6 million for research and development. Now it is spending more than $600 million a year. However, new regulations put in force following the passage of the Kefauver-Harris bill in 1962 have profoundly affected the way those funds are spent. One change (and one probably not anticipated) has been that many smaller, independent pharmaceutical companies have abandoned research, and larger companies in several instances have absorbed some of the smaller companies. Perhaps more significant, however, has been the law's effect on the effort devoted to preclinical toxicological and pharmacological studies. The new regulations have made it more difficult to develop drugs, especially those that must be administered for long periods. Among drugs approved for marketing in the U.S. in the past few years, there is a preponderance of antibiotics and a dearth of agents for treating such ills as hypertension, angina pectoris, and bronchial asthma. Yet it is the chronic diseases that stand out as the major unsolved therapeutic problems.

Many in the pharmaceutical industry are concerned that the trend of fewer drugs being introduced in the U.S. will not be reversed. On the contrary, they see the sharp public debates of the 1960's on drug safety, drug effectiveness, and drug prices as increasingly causing the Food and Drug Administration (FDA) to delay acting, and even not to act, on applications for clinical trials and for marketing new products. With prospects apparently dimming for eventual permission to introduce new drugs in the U.S. after years of research, the more pessimistic in the pharmaceutical industry are less frequently recommending research by their companies on a number of diseases, including some of the more promising and important fields covered in this chapter.

The drop in new product introduction in the U.S. during the past decade may not account for the currently poor standing of the U.S. in infant mortality and life span when compared to those rates in a few other developed countries. Such statistics do suggest, however, that the country's national health policies haven't been the successes they might have been. Be that as it may, if a decline in effective drug research and in introduction of new medicinals does eventually affect national health adversely, the reversal of the trend will be exceedingly slow and costly (as the nation found so quickly in the space program).

Drug research is a highly complex and difficult endeavor, too important not to be a matter of concern in a progressive society but too intricate to flourish in what many view as an atmosphere of repression and harassment. On the other hand, the practices of the pharmaceutical industry are too important to the health of the nation to be free of governmental control. Approval for marketing a new drug involves responsibility of monumental proportions. It must be based on the most carefully planned and executed clinical research. One of the major tasks the country therefore faces is to improve the quality and reliability of clinical research. Various parts of the National Institutes of Health support research in clinical pharmacology in a number of universities and medical schools throughout the country. Since 1965, the Pharmaceutical Manufacturers Association has awarded almost $2 million for the same purpose. The sums are far from sufficient, however.

In addition to improving clinical pharmacology, the nation has also undertaken to review the effectiveness and safety of all existing drugs as well as review these properties of new drugs before they are released for sale. It also is monitoring promotional material more strongly than ever before. These goals enjoy the endorsement of the overwhelming majority of scientists in industry, government, and universities. Only the more thoughtful, however, are acutely aware of the enormity of the task that FDA has undertaken under the guidance of congressional committees and of the potential for harm in an unbalanced approach to the problem.

Drugs have helped increase longevity. They have reduced the need for hospitalization. By improving health, they have raised the standard of living in the U.S. The decline in discovery and marketing of new drugs in the U.S. should concern those in drug research, in the medical profession, in government agencies, and in the public at large (the actual and potential patients). It is therefore most timely that the basic reasons for the decline in finding and marketing new drugs in the U.S. be sought and that new national policies be formulated, if Americans are to have an even higher level of health care in the future.

8 / **Soaps and Detergents**

SOCIAL AND ECONOMIC IMPACTS

Production of soaps and detergents in the U.S. in 1970 amounted to 6.6 billion pounds. While production of that much of any product is a matter of direct economic significance, the social and economic impacts of soaps and detergents are measured by more than the tangible statistics of commerce. New products developed by chemists in the past half century or so have established new standards of cleanliness for clothing, furnishings, and other articles. In addition, some industries such as the food and beverage industry depend completely on high levels of sanitation made possible by today's soap and detergent products. Many other industries, meantime, find the use of soaps and detergents essential in various degrees to their manufacturing processes.

Of equal economic significance have been the changes that have occurred within the soap and detergent industry itself as a result of chemical innovations. The industry still sells products for cleaning, and it still makes soap, as it has for many years. It buys most of its raw materials today from the chemical and petrochemical industries, however, rather than exclusively from agriculturally-based industries, as it did in the past.

Soaps and detergents[1] are products that exert their effects in dilute aqueous solutions. Some of them, such as soap and the more common synthetic surface active agents, function reasonably effectively when used alone. To give their products optimum cleaning power, however, soap and detergent manufacturers formulate nearly all of them with a number of other ingredients that serve definite purposes. They add what are known as builders, for example, to soften water and make the surface active agent perform more effectively. They also add relatively smaller amounts of what are known as auxiliaries to provide specific properties, such as to prevent redeposition of soil, brighten fabrics, remove special kinds of stains, and reduce corrosion.

The soap and detergent industry is considered part of the chemical and allied products industry. It uses soaps and synthetic surface active agents, builders, and auxiliaries to make products for many different cleaning purposes. Most of the companies in the industry make soap. Some of the larger ones also make synthetic surface active agents, but they and all of the smaller ones buy most of such products from other companies in the chemical and petroleum refining industries that make intermediate organic and inorganic chemicals. Few soap and detergent producers make builders and auxiliaries but buy them instead from chemical and other companies.

Of all the chemical innovations that have occurred in the soap and detergent industry, the development of synthetic surface active agents that outperformed the traditional soaps based on fatty acids clearly ranks as the most significant. Fatty acid soaps and other formulated cleansers containing them may have performed with reasonable effectiveness and may have reached a fairly good level of development up through roughly the late 1920's. They suffered from decided drawbacks, however, among which relatively poor performance in hard waters was probably the most unfor-

---

1. In this article we shall use the term soap to refer to a product that is based on naturally occurring chemicals, such as fatty acids, that are recovered from vegetable or animal sources. We shall use the term detergent to refer to products based on surface active agents that have been made synthetically from other chemicals.

tunate. Then chemists in the soap and detergent, chemical, and petroleum refining industries made and patented thousands of new surface active agents from the 1930's on. In fact, during roughly 10 years from the mid-1930's to the mid-1940's, patent activity in surface active agents probably exceeded that in any other single field of chemical technology.

Among companies with research and development groups that contributed in the U.S. were Allied Chemical, American Cyanamid, Armour, Atlantic Refining, Colgate-Palmolive, du Pont, Emulsol, General Aniline and Film, Monsanto, National Oil Products, Ninol Laboratories, Procter & Gamble, Rohm & Haas, Sharples Solvents, Standard Oil of California, and Union Carbide. The large chemical, specialty chemical, and petroleum companies were surprisingly more active in the field than the already-established soap companies, incidentally.

In time, chemists in these and other companies showed the strong interdependence between chemical composition of a detergent and the surface or substrate it will clean. Alkalies alone clean some hard surfaces, such as tile and glass, and no surface active agent is needed. On the other hand, organic fibrous materials, such as the common textiles, require a cleanser with a high proportion of surface active agent. When the requirements of low cost and convenient physical form are combined with the chemical composition that will clean different substrates, it is possible that a unique and optimum formulation could exist for every cleaning job. No industry could go that far in custom tailoring its products, of course. The soap and detergent industry, however, has increased the number of products at least 10-fold since the time some 40 years ago when soaps were the only surface active agents available, not many builders were produced, and the auxiliaries were limited to colors and perfumes.

CHEMICAL ACCOMPLISHMENTS

The first synthetic surface active agents were sulfated fatty alcohols, fatty acyl taurides, and isethionates. They were all based on fats, as was traditional soap. Then in the mid-1930's, chemists began making surface active agents from chemicals that could be recovered from petroleum. The first commercial successful surface active agent of this type was alkylbenzene sulfonate (ABS). It was made from benzene and a refined paraffin distilled from petroleum. The refined paraffin was in time replaced by trimers and tetramers of propene and butene, and more recently they in turn have been replaced by straight-chain hydrocarbons. Regardless of the details of the molecular composition, however, alkylbenzene sulfonate has eventually become the most important single synthetic surface active agent to be made.

Successful formulation of ABS and other chemicals into cleaning products of all types has profoundly changed the economics of the soap and detergent industry and of industries selling raw materials to it. For example, the market for inedible fats, particularly inedible tallow, has been changed completely. Bought mainly by soap manufacturers before development of ABS and other synthetic surface active agents, it is now largely used in animal feeds and as a base for special chemicals and lubricants. In addition, the soap and detergent industry itself changed its manufacturing practices. It has been transformed from what was largely a chemical converting industry to one that buys much of its ingredients from chemical and petroleum refining companies and formulates and blends them into products sold to consumers and industrial users. Finally, new products have been designed for novel noncleaning uses by industries that had not previously used soap products. These new uses in such industries as rubber, petroleum, mining, metals, and agriculture and food processing were discovered and developed almost exclusively by chemists in companies that make surface active agents.

The products that consumers and industrial users buy for cleaning, sanitizing, and other applications are rarely composed of simply one chemical. Instead, as mentioned earlier, they are formulated from surface active agents, builders, and auxiliaries. Of the three general categories of ingredients, surface active agents and builders are used by far in the largest volumes. Of these two, surface active agents are probably the most important chemically.

Surface active agents, or surfactants for short, are chemicals that concentrate at surfaces and solubilize materials with little natural affinity for one another. The phenomena involved are physical rather than chemical reaction, and they include adsorption, surface tension, and micellization. To carry out their functions, surfactants combine in their molecules a highly polar (or hydrophilic) part that is soluble in water and a highly nonpolar (or hydrophobic) part that is soluble in oily materials. The chemical groups that form each part and the relationship between their hydrophobic and hydrophilic strengths determine where and how individual surfactants function.

The industry classifies its products into four categories based on the electrical charges present on surfactant molecules in solution. *Anionics* have a negative ion as their polar part, while *cationics* have a positive ion for theirs. *Nonionics* have no charge, and *ampholytics* have either a negative or positive charge depending on the pH of the solution. Each has as its nonpolar constituent a hydrocarbon part that differs in molecular size and complexity.

Soap, the surfactant known since antiquity, is anionic. Its polar part is a carboxyl ion ($-COO^-$), while the nonpolar part is a hydrocarbon of 12 to 18 carbon atoms. Synthetic anionic counterparts of soap most often have a polar part consisting of a sulfonate ($-SO_3^-$) or a sulfate ($-OSO_3^-$) ion. Dodecylbenzene is the most common nonpolar hydrocarbon used with sulfonates, while a variety of alkyl groups are used with sulfates.

The other three classes of surfactants are used far less than the anionics. Briefly, salts of long-chain amines and quaternary ammonium compounds are the most prevalent cationics. Hydroxyl, amide, ether, and ester groups provide the hydrophilic character of nonionics, which depend on structure for activity and not on ionization in solution. Ampholytics contain both anionic and cationic groups.

In the early 19th Century a French chemist elucidated the nature of glyceride fats and the saponification process for making soaps from them, and he laid the basis for modern scientific soapmaking. From then until after World War I, however, the soap industry changed very little technically, as mentioned earlier. It bought inedible fats and oils (by-products of agriculture), rosin and coconut oil (also agriculturally related), and alkalies used to saponify fats and to build soap. The industry marketed primarily household and commercial laundering and personal care and cosmetic products, with maintenance and industrial cleaners clear seconds.

Textile manufacturing at the time was the only industrial activity that consumed any appreciable quantities of soap, and it was the demands of the textile industry that stimulated the development of synthetic detergents. Converting raw fiber into finished fabric involves much wetting and cleaning, and many problems occur. In some processes for bleaching cotton, for example, air bubbles remaining in the goods can spot them and also mechanically weaken the fibers. When woven cotton and synthetic fibers are scoured before they are dyed and finished, residues of sizing must be removed completely, and the goods must be free of hydrophobic material. If they aren't, they likely will dye unacceptably spotty or uneven.

With soaps as the only available surfactants, wetting and scouring were tedious operations. Even a moderate amount of hardness or transient lack of alkalinity in the water could result in hydrophobic lime or acid soap spots, making reprocessing necessary. Textile finishers, therefore, had a long-standing and explicit desire for a surfactant that hard or acidic water would not precipitate or render ineffective.

The First Synthetics

The answer to textile finisher needs and the start of modern synthetic surfactant technology came during a relatively short time in the late 1920's with the development of the first commercially feasible sulfonated surfactants. Soapmakers, however, did not invent these compounds that resisted lime and acid conditions, but chemical and dye manufacturers supplying the textile industry did. Groups at Boehme Fettchemie and I. G. Farbenindustrie in Germany did much of the pioneering work, but chemical manufacturers in the U.S., U.K., France, and Switzerland immediately took up the ideas and expanded the work enormously.

Chemists at I. G. Farbenindustrie in Germany and Union Carbide in the U.S. first reported on the ethoxylation of alcohols in 1926, and chemists at I. G. made the first surface active ethylene oxide condensates in 1929–30. While much of the early research on these products occurred in Germany, chemists in the U.S. gave nonionic surfactant developments their greatest impetus in the 1930's. Union Carbide began producing ethylene oxide commercially in the U.S. at that time. In 1937, workers at Rohm & Haas disclosed the ethoxylates of the alkylphenols. In the same year, Union Carbide began the first commercial production of ethylene oxide by direct oxidation of ethylene at South Charleston, W.Va., and the way was open to low-cost ethylene oxide. This reactive chemical has become the basic ingredient in a great many nonionic surfactants.

Another important class of nonionic surfactants are the fatty amides, developed at Ninol Laboratories in 1935. These materials, derived from fatty acids and alkanolamines, are now widely used as surfactants and as foam modifiers in surfactant products.

Arrival of the early synthetic surfactants brought a vision of freedom from well-recognized disadvantages of soap in all laundering, but the hope soon died. Cotton at that time made up the largest part of the average household wash, and the synthetics unfortunately performed poorly in such heavy-duty laundering. Fortunately, the synthetics performed as well as soap in light-duty laundering of items such as woolens, lingerie, and other fabrics that suffer from too much heat, alkalinity, and mechanical action. Furthermore, their instant solubility, neutral pH, high foam, and clarity in hard water suited them very well for washing dishes by hand.

In the middle 1930's, the soap and detergent industry introduced the first light-duty synthetic surfactants to the household market. They were flaked or beaded solids of synthetic surfactant and sodium sulfate, a builder for washing wool and also a useful standardizing agent and filler. These light-duty products competed with the well-established flakes and powders of pure soap, and they gained popularity steadily. Their composition has changed little since then, although light-duty liquids, introduced about a decade later, are more popular today than the solids. They are concentrated aqueous solutions of a mixture of surfactants and are compounded to provide maximum cleaning effect and suds (a desirable indicator of activity in hand dishwashing) at minimum concentrations.

Detergent Builders

If invention of synthetic surfactants first revolutionized the soap and detergent industry, discovery that condensed phosphates build these surfactants certainly revolutionized it again. Until the end of World War II, a powdered or beaded mixture of soap and soda ash (sodium carbonate, or washing soda) served as the main household product for heavy-duty laundry. Although an excellent builder for soap, soda ash did not serve very well in making the sulfated and sulfonated synthetics clean cotton well. The alkaline silicates, borates, and orthophosphates commonly used with soap were also only moderately effective. The synthetics thus could not compete with soap in heavy-duty service despite their other advantages. Then in the mid-1940's, chemists found that pyrophosphate and tripolyphosphate strongly promote the cleaning action of synthetic surfactants and also soften hard water without precipitating the problem ions. Procter & Gamble became the first American company to exploit the finding commercially (although some evidence exists that groups at Lever Brothers and Colgate-Palmolive also recognized the effect almost simultaneously).

Discovery of condensed phosphate building opened the heavy-duty laundry market to synthetic surfactants, especially alkylbenzene sulfonate. Production of alkylbenzene sulfonate increased roughly 20-fold in the two decades of the 1950's and 1960's, and ABS's performance as a heavy-duty cleaner has been unmatched by any other product, at least at its price.

Discovery of phosphate building also led to a large increase in the demand for phosphates, and sodium tripolyphosphate became a tonnage chemical. As the 1960's ended, however, the possible role of phosphates in the degradation of streams and lakes became a matter of sharp controversy. Chemical companies, soap and detergent manufacturers, and government agencies all have been actively trying to determine just what the role of detergent phosphate is in possibly causing ecological problems and to find solutions. Meantime, partly under the stimulant of government regulations and partly voluntarily, the soap and detergent companies have reduced the amounts of phosphates in their products. In addition, a few companies have introduced detergents that are built with other compounds and contain little or no phosphates.

Alkylbenzene sulfonate consists of a benzene ring to which a paraffin of about 12 carbon atoms and a sulfonic acid group are attached. It is prepared by first reacting benzene with the hydrocarbon to make alkylbenzene. Reacting the alkylbenzene with concentrated sulfuric acid and neutralizing the resulting sulfonic acid with sodium hydroxide yields sodium alkylbenzene sulfonate.

The first hydrocarbon used for ABS was a kerosine fraction recovered from distilling petroleum, in a development pioneered by Allied Chemical Corp. This distillate consisted largely of straight-chain hydrocarbons. It was soon replaced with (for the most part) propylene tetramer, a highly-branched hydrocarbon. Then in the mid-1960's, it became clear that detergents having improved biodegradability were required. The original approach to making ABS returned when the soap and detergent industry and its suppliers spent several hundred million dollars to switch from propylene tetramer to a linear paraffin for the hydrophobic side chain.

At that time, it had become possible to duplicate natural fatty hydrophobes at relatively low and stable cost with chemicals derived from petroleum. The two most important techniques used today are polymerization of ethylene and a separation of straight-chained hydrocarbons with molecular sieves. The linear paraffins are then generally converted to olefins or alcohols and reacted with benzene to make today's alkylbenzene base material, LAB (for linear alkylbenzene).

Households consume by far the largest tonnages and dollar volumes of detergents, and laundering accounts for the largest use in homes. The development of phosphate-built ABS, the availability of more convenient washing machines, and growing affluence all combined from the late 1940's on to change drastically the laundering practices in the U.S. Automatic washer sales were 700,000 in 1948, but within six years had risen to 2.3 million. By 1970, annual sales exceeded 4 million, and the total number of automatic washers in use approached 40 million.

This growth in home laundering has posed a number of challenges for soap and detergent manufacturers. Short washing cycles of 10 to 15 minutes and the limited capacity for hot water normally available in mass market homes required efficient detergents to cope with the usual soils, for example. Products for automatic washers had to dissolve quickly, and manufacturers had to develop special spray-drying methods for synthetic detergents that would give quick-dissolving powders. They also had to develop other ingredients that would be compatible with the relatively severe and high-speed processing conditions.

As the revolution in home laundering progressed, the nature of fabrics in clothing began changing from the familiar cotton to a predominance of polyester and resin-treated cotton. The change helped to free housewives from laborious ironing, but it placed new requirements on detergents, since polyester and resin-treated cotton tend to hold soil tenaciously.

Heavy-duty synthetic detergents have been improved in many ways in the 25 years of their existence as a result of these and other pressures. One of the early improvements, made at about the time of the condensed phosphate building discovery, called for the addition of a very small amount of carboxymethyl cellulose (CMC) to detergent formulations. The CMC virtually eliminated the gradual graying of cotton caused by redeposition of soils during repeated soiling-washing cycles. More recently, inclusion of fluorescent whitening agents has enhanced the wash's brightness. Most formulations now contain anticorrosion agents to protect washing machine parts. Enzymes were found to remove certain problem stains. Finally, changes in surfactant composition have improved detergency, controlled foaming as desired, given good performance in cool water, and—most important of all—avoided unwanted effects on the environment.

Use in Other Products

Synthetic surfactants have also entered other products besides heavy-duty laundry formulations. Scouring powder, for example, was a well-established product a generation or two ago. Synthetic surfactants with and without condensed phosphates have replaced soap in these products. General-purpose liquid cleaners, designed primarily to clean painted, plastic, and ceramic surfaces, are relatively new products based on synthetic surfactants and pyrophosphates. Most leading brands of toothpaste and shampoos have been based on synthetic surfactants for many years.

The familiar bar of toilet soap is one of the few cleaners in which soap hasn't given way to the synthetics. Many difficulties have been encountered in developing synthetic bars. Although there are now a few successful ones, soap still commands the major part of the market.

Possibly the most important development in bar soap has been the introduction and growth of deodorant antibacterial soaps. In contrast to the older soaps containing cresol, these soaps may be perfumed and colored conventionally. The first to be marketed contained hexachlorophene, but several other antibacterial agents have also been used. While the use of hexachlorophene has very recently been questioned by some medical experts, its use up to the present has been significant.

Households and the textile industry may consume most of the country's soap and detergents, but other uses are also important. Cleaning buildings, public places, buses, aircraft, and railroad cars as well as a variety of industrial equipment such as dairy and food processing machinery consumes tons of surfactants. It is difficult to realize the full extent to which standards of cleanliness have risen and cleaning operations have been facilitated since the introduction of nonsoap surfactants and condensed phosphate builders. A whole new branch of the soap and detergent industry, now grown to healthy maturity, formulates specialty cleaning compounds. These individually tailored products of selected surfactants, builders, and auxiliaries give the desired effect at optimum cost efficiency. Chemists in soap and detergent industry laboratories develop them, working closely with consumers to define and solve different cleaning problems.

One very important advance in maintenance cleaning has been the development of sanitizing and disinfecting cleaners of high germicidal potency. Before synthetic surfactants became available, the only such cleaners were alkaline hypochlorite combinations and soap–phenol or soap–pine oil mixtures. Hypochlorites cleaned poorly, and the phenol/pine oils killed few bacteria. Then in 1935, chemists found that quaternary ammonium surfactants were highly effective against many of the common infectious bacteria. Their use as germicides has grown steadily since that time. Alone they are not suitable in detergents, but they have been combined with other surfactants, notably the nonionics, to make widely used disinfectant cleaners.

Another type of disinfectant cleaner made possible by availability of synthetic surfactants combines a surfactant with a phenolic or halogenated aromatic germicide. Not all surfactant-germicide combinations are compatible, and working out the most effective ones was a chemical development of great practical importance. It was done largely by chemists working for makers of surfactants and specialty cleaning compounds.

In many industries, cleaning operations occur as a necessary part of the manufacturing process. Textile manufacturing has been mentioned as one such industry. Another in which thorough cleaning is of utmost importance is metal finishing. Metal articles that are to be plated, enameled, or painted must be freed of all surface dirt. The development of special cleaning compounds that reduce both the time required and also the incidence of rejects was made possible by the advent of surfactants and sequestering builders. Another process that modern surfactants have speeded up and made more efficient has been cleaning fruits, vegetables, eggs, and other foods before packaging.

Noncleaning Uses

In general, surfactants are useful wherever interfaces between different phases must be modified. They are therefore economically important in many noncleaning uses for wetting or water repelling, emulsifying or demulsifying, and flocculating or deflocculating. Since surfactant molecules concentrate naturally at the interfaces requiring their action, small quantities do the work and the cost remains low relative to the economic benefit. In road aggregates, for example, they make bituminous blacktop adherent, and in fertilizers they prevent caking. They emulsify insecticides, fungicides, and herbicides for easy application in dusts, wettable powders, and sprays. Shortenings, cake mixes, baked goods, salad dressings, and various other emulsified preparations benefit from their presence in food processing.

There are, in fact, few heterogeneous systems in modern chemical technology where surfactants of one type or another have not at some stage proved useful. Emulsion polymerization, for example, requires an emulsifying agent. Soaps were the first surfactants used in this process to make large tonnages of vinyl polymers, and they are still used on a large scale, especially to make synthetic rubber. Monomers respond quite specifically to surfactants, however, and for each monomer a surfactant or mixture of surfactants gives optimum results. As more surfactants became available, chemists at polymer and plastics companies formulated such mixtures and replaced older, more cumbersome polymerization processes with the emulsion process.

The petroleum industry also uses surfactants extensively. Surfactants adjust drilling muds to the desired viscosity and serve as anticorrosive agents in acid fracturing in oil fields and in handling crude oils high in sulfur. They aid secondary oil recovery when injected into slow-flowing formations. As emulsion breakers, they help separate water that frequently is emulsified in crude petroleum.

Surfactants of various types are also used as conditioners and collectors in ore flotation in the mining industry. The variety of surfactants available, particularly with regard to ionic type, makes precise adjustment of conditions possible. Mineral values or gangue may be floated or suppressed. Surfactants are also used in sprays for allaying dust in coal and other mines. High-sudsing light-duty liquids originally developed for household dishwashing generate foam in mine tunnels and shafts to help fight underground fires. After metals are won, surfactants are extensively used as anticorrosion agents, affording great economic saving at low cost.

FUTURE DEVELOPMENTS

The soap and detergent industry is highly competitive and markets many of its products directly to individual consumers. It is therefore very sensitive to changing consumer needs, desires, and attitudes. Its major current trends toward products of greater effectiveness per unit cost and greater convenience will continue. Combination products that not only clean but also have special stain removers, soil repellents, softeners, sanitizers, and other adjuncts can be anticipated. Introduction of new fibers and fabrics and use of new finishes to retard flames, repel soils, and provide other properties will continue to stimulate development of detergent formulations to clean and maintain them. Biochemical agents are also apt to be used more widely in cleaning.

Such innovations may well have a huge economic impact on the industry. Consider, for example, what the recent introduction of enzymes in formulations has meant. After several years of use in Europe, proteolytic enzymes began to be used in the U.S. in the late 1960's. Within a few years, the industry was spending millions of dollars annually for them even though they are generally used at levels below 1% in formulations.

The specialty surfactant business, like other specialty chemical businesses, thrives by being not only sensitive to its customers' needs but anticipatory of them. We can therefore expect continued diversification of surfactant types and increased effectiveness and diversity in their application. New surfactants from the newly available α-olefins are on the horizon. Olefin sulfonates and paraffin sulfonates in particular are of increasing interest and near commercial reality.

Surfactants and the Environment. Like many another industry, the soap and detergent industry will place more emphasis on developing products having minimum impact on the environment and maximum safety in use. Detergents, like other substances that are not consumed in use, constitute a potential waste disposal problem. Being soluble, they simply run down the drain and are diluted indefinitely with other waste water. If natural processes in streams and ground or artificially accelerated processes in sewage treatment plants convert them to innocuous end products, no waste disposal problem exists. Such is the case with soap, which bacteria readily biodegrade to carbon dioxide and water.

Bacteria similarly biodegrade many synthetic surfactants, particularly those having linear hydrocarbon nonpolar groups. Unfortunately, the first ABS household surfactants were not of this type. Resisting biodegradation, they persisted and foamed in sewage disposal plants and even in streams wherever they were used in sufficient quantities. By the late 1950's, the problem had become serious in many localities. The industry and its suppliers converted at considerable cost to biodegradable surfactants as a result. The conversion involved the uninterrupted replacement of one large-volume chemical with another, and it occurred at no inconvenience and at minimum cost to consumers.

More recently, some evidence indicates phosphates may under certain circumstances contribute to the eutrophication of lakes. Phosphates in waste waters can be removed by special treatment in municipal sewage plants. In addition, in view of phosphates' proved lack of toxicity and their unequaled cost efficiency and versatility as builders, their replacement may not be the most desirable solution. Nevertheless, the soap and detergent industry and its suppliers are presently engaged in developing a satisfactory substitute for phosphates that will not stimulate growth of algae. Since the industry now uses about $150 million worth of phosphates a year, the economic impact of their replacement will not be minor.

As water continues to become a scarcer resource, elimination of detergent residues from waste water will become increasingly important. Interlocking detergent formulations with waste disposal processes either on a municipal or individual home scale may be necessary. Using recycle washing systems employing water or some other solvent may also become economically feasible. Under such circumstances, application of ultrasonic energy could prove useful in cleaning processes. Use of disposables to replace washable fabrics is also likely to increase (although such a substitution would increase solid waste disposal problems).

In any event, the soap and detergent industry and its suppliers will face an enormous task of testing new materials for all possible effects on man and environment. To handle the problem of such complexity and magnitude, the industrial, governmental, and scientific communities will obviously have to cooperate in carrying out the extensive research that will be needed.

SOCIAL AND ECONOMIC IMPACTS

Buying and using makeup preparations, personal soaps, deodorants, perfumes, hairdressings, and a host of other cosmetics, toiletries, and beauty and hair care implements have long marked the ways of affluent members of any society. In the U.S. today, however, persons at nearly all income levels are buying and using such products at an unparalleled rate. Everyone uses at least one or two personal care products every day, and many use perhaps up to a dozen or more, if not daily at least during the course of a week. Personal grooming as a result has become a major manufacturing and service industry in the U.S., and Americans now spend nearly $10 billion a year for personal care products sold at retail and for periodic visits to barber shops and beauty salons.

The personal grooming industry consists of three main parts:
- Making, distributing, and selling at retail a wide variety of personal care products that people apply to their hair, skin, and nails to clean them, soften them, help control their shape or contour, change their color, protect them, and mask or prevent unpleasant odors.
- Making, distributing, and selling implements at retail for use in homes for applying or otherwise using personal care products effectively.
- Providing professional services for personal grooming in barber shops and beauty salons.

According to *Drug Topics,* a publication serving drugstore personnel, sales of personal care products in all retail outlets in 1970 in the U.S. exceeded $4.7 billion. *Drug Topics* also estimated 1970 retail sales of personal care accessories, such as brushes of many types, shaving implements, hair dryers, and cosmetic applicators, amounted to nearly another $1.3 billion. Meantime, the most recent Census Bureau survey of business (1967) put American spending in beauty salons and barber shops at $3.4 billion for professional hair, face, and hand care.

In the three main sectors of the industry providing grooming products and services, chemistry is most intimately involved in the largest—the formulation and production of personal care products. Chemistry has helped improve products and services in the other two sectors, of course, but not to the extent it has for personal care products themselves. We shall accordingly concentrate on chemistry's role in that sector.

People have long sought to improve their appearance with oils, ointments, powders, dyes, and even muds. Application of cosmetics has been common, for the rich at least, from the earliest of times, and records from Egypt, the Near East, and the Bible often describe the use of natural oils, pigments, and aromatic compounds. The Greeks and later the Romans indulged in elaborate baths followed by massages with perfumed oils.

Persons returning from the Crusades introduced many cosmetics to western Europe, and in succeeding years the use varied from country to country. Within individual countries, popularity of cosmetics also rose and fell with the styles of fashion set by the royal and ruling families. Later in the American colonies and then in the U.S., their use also varied by area. Puritan New England generally frowned on them, for example, while French New Orleans tended to be more liberal.

The barriers to more liberal use of personal care products, especially cosmetics, began dropping in the U.S. in the 1920's, but the changed attitudes of those years raised consumption to nowhere near the levels it has reached today. Even immediately before World War II, retail sales of personal care products were only one tenth today's level. Growing affluence that opened markets among those with moderate incomes, more effective products, especially after World War II, and strong promotion have resulted in the $4.7 billion personal care products industry of 1970. Such high demand for what in some respects are occasionally luxury products consistently places the personal care products industry among the most profitable in the U.S. (calculated as return on investment). Only beverages and drugs consistently rate as high.

In contrast to the social and economic impacts of many of the other industries dealt with in this study, the impact of the personal care products industry is in some respects less easily measured. The impact of some types of its products, for example, may rest largely on psychological and subjective values. Nevertheless, personal care products do have an economic impact in the U.S., for making and selling them is big business. Of the half dozen major categories into which they may be grouped, those for the hair lead in sales by a slight margin. Retail sales (excluding sales to and by beauty salons and barber shops) in 1970 amounted to just over $1.2 billion for shampoos, conditioners, tints and dyes, fixatives, and waving and other preparations for use in the home. Personal care products for hands and faces ranked second, with retail sales of just over $1.1 billion. They were followed by personal cleanliness items (deodorants, soaps) with sales of $730 million; oral hygiene products (toothpastes, mouth washes), nearly $710 million; fragrance preparations (perfumes, colognes), about $375 million; and shaving products (creams, lotions), nearly $340 million. Products that border on having medicinal properties, such as suntan lotions, lip protectors, foot salves and ointments, acne preparations, and eyewashes, ranked last at about $185 million. Among individual products, personal external deodorants ranked first, with 1970 retail sales of nearly $400 million. Dentifrices at about $360 million came second, followed by shampoos at about $350 million, bath soaps and detergents at about $330 million, spray hair fixatives at $295 million, mouth washes and fresheners at $280 million, and hair coloring products at $245 million.

CHEMICAL ACCOMPLISHMENTS

People use personal care products to perform a variety of cleaning, softening, concealing, tinting, controlling, and freshening functions. Face and hand creams, for instance, reduce skin dryness, and some contain detergents that clean the skin when they are removed but without leaving it dry as simple soaps might. Face and hand creams may also help control and conceal blemishes and serve as powder bases. Properly formulated shampoos clean the hair and remove dandruff without leaving the hair unduly dry or with a sheen-dulling film. Other hair products help control contour by permitting it to be waved, and hair fixative sprays hold the style in place once it is devised. Hair tonics and rinses make hair more manageable, and dyes permit the dissatisfied to enhance existing color or try different colors. Powders absorb secretions from sebaceous and sweat glands and make the face velvety smooth and not shiny. Deodorants control body odor with bacteriostats or with astringents, while antiperspirants retard perspiration itself. Men's shaving preparations either soften the beard for those who shave with blades or they make the beard stand up for those who use electric shavers. After-shave lotions give a feeling of freshness and act as mild antiseptics, as do mouth washes and gargles.

While all personal care products serve useful—if occasionally psychological—functions, some more directly and immediately affect personal health. Regular use of toothpastes, for example, deters formation of caries and development of other tooth and gum problems. Suntan lotions filter the radiation that burns, and some themselves react chemically with the skin to produce a color very similar to that of tanning. Foot powders and lotions help control athlete's foot, while acne aid topicals assist in some cases in reducing the lesions that are a cause of lasting psychic stress in adolescents.

To achieve their effects, chemists in the personal care products industry may choose among a wide array of ingredients. At one time, their raw materials consisted mainly of items gathered from natural sources. Over the years, they have subjected such products to increasingly stringent refinement. In more recent times, chemists in the industry have also done much to convert to compounds they and others have synthesized to perform specific functions. Included are polymers used in hair sprays, dyes used in hair coloring products, and specialty detergents, emulsifiers, and viscosity controlling agents used to extend the range of effects that personal care products can achieve.

The chemistry of personal care products has also become much more exact than it once was. The physical chemistry used today, for example, rates among the most sophisticated used anywhere. Successfully formulating personal care products rests heavily on applying the principles of detergency, emulsification, and chelation. Thixotropy, viscosity, absorbency, and hardness must be adjusted appropriately, while the drying time, adherence, and reflectance achieved in preparations may mean the difference between success and failure. Since most products are applied to proteinaceous substrates, knowledge of protein chemistry and surface chemistry is of considerable importance. Successful packaging of products also requires expert knowledge of corrosion chemistry and plastics technology. Improvements in product performance previously judged on purely subjective grounds are now frequently measured with complex analytical, physical, and biochemical techniques. These and other advances not only have resulted in the industry's having a fuller understanding of what personal care products accomplish but have given its personnel much information on the mechanisms by which its products produce their effects.

Permanent Waving

Of all their features that people spend time and money on, the hair ranks No. 1. Its appearance is probably more important to pleasing grooming than any other single attribute, at least at first glance. They wash it, wave it, color it, cut it, and comb it into a variety of styles. The modern personal care products industry, in fact, may be considered as dating from the invention of the Marcel wave (by a French hairdresser, Marcel Grateau, in the 1870's). Some three decades later (in 1906) a German living in London, Charles Nessler, introduced the first supposedly permanent wave. These and other approaches to waving the hair took a lot of time, heat, and money, however, and they hardly thrived in a large market. Until the early 1940's, women who wanted permanent waves or straightened hair sat through cumbersome, time-consuming, and sometimes dangerous processes. They either used heat or potentially unsafe chemicals such as ammonium bisulfide in cold wave processes. Then in 1941, thioglycolate cold waving lotions were introduced to the U.S. market. Since then, a completely new industry has been created in the U.S. and elsewhere to make and distribute cold wave lotions for home and beauty salon use. While this market fluctuates widely with the demands of fashion, retail sales of home and professional waving and straightening units probably exceed $100 million a year in the U.S.

Success with cold permanent waving dates from the 1930's when scientists at Rockefeller Institute showed that either sulfides or mercaptans at or near room temperature and at alkaline pH's would split the disulfide bonds in proteins and especially in the hair protein, keratin. Based on this knowledge of disulfide chemistry, the way was open to the personal care products industry to develop chemicals that people could use to wave or straighten hair without heat. Ammonium bisulfide was introduced in 1939 as a cold wave lotion, and it was followed in the early 1940's by the faster acting (in the cold) ammonium and sodium thioglycolates.

The cold wave market didn't grow much until the end of the war, however, at which time a good quality thioglycolic acid became commercially available in the U.S. The existing professional market for permanent waving preparations used this mercaptan widely, and it created a new market, that of the home permanent wave.

Introduction of thioglycolate home permanent waving in the U.S. influenced not only the beauty industry but also the chemical industry. American chemists continued to improve the earlier products to make them more convenient to use. The original liquid neutralizers (or oxidants) based on solutions of potassium bromate or hydrogen peroxide were succeeded by solid neutralizing agents such as sodium bromate or sodium perborate, and introduction of a self-neutralizing wave product made waving even easier. This latter development stemmed from an observation in 1936 that air readily oxidizes thioglycolates and mercaptans when certain heavy metals are present.

Hair straighteners and depilatories generally use the same technology as permanent waving products, but both are relatively small markets. The depilatory market, however, has grown in a limited way in recent years. Such products weaken the hair shaft, so that water can readily wash it away. Mercaptan depilatories apparently were first used in the early 1930's to dehair hides. At one time, people used strontium and other sulfides to remove hair from arms and legs, but their rotten egg odor was a drawback. Calcium thioglycolate has proved to be more satisfactory, and depilatories are an $11 million retail market today.

Hair Sprays

Throughout the ages, women have sought ways to create and maintain their coiffures. In ancient days they resorted to mud or mechanical devices to build and hold intricate styles. Later, they replaced mud with gums and other natural extracts. Still later they turned to hot irons for curling hair. Then in modern times the first commercial chemical cold waving product was introduced in the U.S. in 1939, followed by the thioglycolates in the early 1940's, and a variety of styles was available to nearly everyone, as already noted.

It was not until the arrival of hair sprays in 1950, however, that women could hold intricate hair styles in place simply by spraying them with tiny droplets of resin. Such spraying placed droplets at hair juncture points and bound hairs together to make the style structurally strong.

Today, hair fixative sprays represent a retail market of more than $295 million a year, not including sales to beauty salons. They are based on three important technologies: propellant synthesis, synthetic resin synthesis, and packaging and valve developments.

The major propellants used today are fluorinated hydrocarbons. They are compatible with alcoholic solutions of many resins and have the proper vapor pressures to produce ideal sprays. Before propellants were synthesized in the late 1920's, however, people used a variety of other materials for aerosol applications (of products other than hair sprays, of course, which had yet to arrive on the market). At the turn of the century, methyl and ethyl chloride were used to spray solutions of gums, resins, and collodion using heat generated by the hand. In 1903, carbon dioxide was used to propel perfumes from atomizers. Dimethyl ether, isobutane, vinyl chloride, and low-boiling petroleum distillates such as rhigolone and cymogene were in use in the 1930's.

Then in 1933, Thomas Midgley, Jr., and co-workers under the auspices of Frigidaire patented fluorinated hydrocarbons (specifically dichlorodi-fluoromethane) as propellants. Their discovery resulted from research aimed at developing new and improved refrigerants to replace sulfur dioxide and ammonia.

Shellac, a natural resin, was the first resin to be used in a commercial hair spray and polyvinylpyrrolidone (PVP) the first synthetic resin to be so used. Chemists at I. G. Farbenindustrie developed PVP during the 1930's, and it was used as a blood plasma expander during World War II. Its commercial development in the U.S. began in 1950.

Following these first products, other resins have been introduced into hair spray compositions to improve their holding power, water sensitivity, and touchability. Such resins include copolymers of polyvinylpyrrolidone and vinyl acetate, dimethylhydantoinformaldehyde, copolymers of esters of polyvinyl methyl ether and maleic anhydride, copolymers of polyvinyl acetate and crotonic acid, and terpolymers of vinyl monomers containing ester, heterocyclic, and carboxylic groups.

Of all the developments affecting the personal care products industry, few have had the impact of aerosol dispensing. Hand atomizers might have done the job in hair sprays, for example, but they would have sadly lacked the convenience of pushing a button. Moreover, after World War II, aerosol spraying became the key to many another personal care product's growth, and few escaped its influence. Aerosol-dispensed shaving creams sent the shaving brush into history. Deodorants and antiperspirants and toilet waters and colognes are all more conveniently applied (if perhaps somewhat wastefully occasionally) from the ubiquitous aerosol can. Few stores have enough shelf space to display the range of products and number of brands that the personal care products industry probably wishes.

Development of aerosol hardware is primarily an accomplishment of mechanical engineering. Since so many products developed by chemists are dispensed as aerosols, however, brief coverage of hardware development appears worthwhile here.

A patent issued in 1862 disclosed the basic concept of aerosol hardware, that of a valve and dip tube. Between that time and the beginning of World War II, such hardware benefited from a number of improvements. Engineers produced finer sprays by introducing an expansion chamber between the actuator valve and the container's main body, with a small capillary tube separating expansion chamber and main body. They soon further improved this design by eliminating the capillary tube and regulating the spray pattern by controlling the relative sizes of valve and nozzle orifices. Somewhat later they got an even better spray by imparting a spiral rotation to the jet stream just before it discharged through the nozzle orifice.

The real impetus to aerosols came during World War II, however, when U.S. Department of Agriculture personnel developed a portable aerosol insecticide bomb for use by U.S. troops. They used dichlorodifluorometh-ane as the propellant to obtain a fine spray and a heavy metal container to hold the room temperature pressure of 70 p.s.i.g.

Immediately after the war, these heavy dispensers appeared on the market, and consumers quickly accepted them despite their high cost. Such success prompted can manufacturers to develop cheaper, less bulky containers similar to those then used as beer cans. By the time the first hair spray based on shellac was introduced in 1950, the major problems of propellant and hardware technology had been solved. Development of synthetic resins, meantime, had paralleled those inventions, and all components converged for use in the sophisticated products we know today as hair fixative sprays.

Shampoos

Care and cleaning of hair has long held a prominent place in the cosmetic arts, but only in this century have modern technical and marketing methods greatly affected the preparation and distribution of shampoos. During the early part of the 1900's, most people washed their hair with ordinary bar soap or with liquid shampoo soaps. Such soaps cleaned the hair, of course, but they also reacted with calcium and magnesium ions in hard water to leave a dull scum behind. It was not until 1933 that the availability of synthetic detergents changed shampoo cleaning ingredients and shampooed hair began to emerge shiny as well as clean.

Synthetics replaced soaps in the 1930's and 1940's, and the shampoo market expanded rapidly. New detergents with different chemical structures and performances have since caused a steady evolution in shampoo forms. Cream or paste, lotion, and gel shampoos have joined the traditional liquid shampoos to give consumers a wide choice based on performance and convenience.

A shampoo's main job is to remove dirt and also residues of hair and scalp preparations, but it should also promote effective styling and coiffuring. Formulators accordingly have added such compounds as lanolin, proteinaceous components, glycerin, fatty esters, and silicones to help condition the hair and also make the products aesthetically pleasing.

Until fairly recently, traditional medicinals of limited efficacy, such as coal tar and sulfur, were used in medicated shampoos to control common dandruff, a grooming problem for many people. With the development of the broad-spectrum antimicrobials such as selenium sulfide and zinc pyrithione, formulators were able to prepare shampoos that clinical tests have demonstrated much improved control of common dandruff scaling.

Hair Coloring

Sales figures of recent years show that hair coloring has been one of the fastest growing sections of the personal care products industry, with annual sales having grown from $65 million in 1960 to $245 million in 1970. The products in use today are generally classified into three categories based on the durability of the coloring achieved. These are appropriately termed permanent, semipermanent, and temporary.

For permanent coloring, intermediates such as p-aminophenol and p-phenylenediamine that have low molecular weights and are usually colorless are used with hydrogen peroxide. Because they are small molecules, they readily penetrate the hair, where they are slowly oxidized by hydrogen peroxide to much larger, colored molecular forms. The permanent colors have good fastness to shampooing, and they completely cover gray hair.

For semipermanent colors, formulators use many types of dyes. These dyes are generally used in their colored form, and most have been developed for textiles. Because large dye molecules are not formed within the hair, as they are in permanent coloring, the dyes used for semipermanent colors are not as firmly bound within the hair. They do resist shampooing moderately well, but they do not cover gray hair as well as do the permanent colors.

For temporary colors, formulators must use products that one shampooing removes completely. Penetration of the hair is therefore not desirable. The colors instead must be substantive to hair, must cover gray, and must be fast to bleeding with water and free from crocking (rubbing off). Preformed dyes are used in temporary hair coloring products as a result.

Dyed hair today more nearly matches the appearance of naturally colored hair than ever before. Among significant innovations by U.S. chemists that have led to such results are development of shampoo-in colors that eliminate the need to section the hair during application and development of stabilized dye solutions that do not deteriorate on contact with air (before the addition of peroxide). Products that thicken on addition of hydrogen peroxide keep the solution from dripping off, and new formulations give colors that do not crock and do not change hue on aging or shampooing. Hair conditioners decrease hair damage. New surface active materials, solvents, and carriers or boosters improve dye addition in semipermanent colors, and they make such colors faster but stain scalps less.

Nail Lacquers

Sales of nail polishes and enamels and nail enamel removers exceed $90 million annually, and about 50% of the women in the U.S. probably use such preparations either regularly or for special occasions. Most users, moreover, buy more than one shade to complement their dress colors.

Before World War I, nail polish was just that—a mildly abrasive composition based on powdered tin oxide and wax that users rubbed across the nails with a chamois buffer to polish off ridges and give nails a smooth, glossy finish. Liquid nail polishes, or more correctly nail lacquers, were introduced soon after 1910 as an improvement over the slow and tedious process of buffing and waxing nails with a powder, paste, or cake polish. These first nail lacquers were natural resins dissolved in alcohol, and they were either their natural color or tinted pink or red. Although readily accepted, they lasted only a short time, because they resisted abrasion poorly and soapy water soon washed them off.

A significant improvement in water and abrasion resistance came in 1918 with introduction of fast-drying lacquers of nitrocellulose, plasticizer, and solvent. To improve adhesion, wear resistance, and gloss, manufacturers added resins to the lacquers. They used natural resins such as dewaxed damar gum at first, but in the 1920's and 1930's turned to the synthetic polymer p-toluenesulfonamide formaldehyde as a superior modifying resin. At about this time, opaque or cream nail lacquers were introduced in which highly colored but insoluble pigments or lake colors were used in place of soluble dye.

Besides improving lacquers with modifying resins, manufacturers have introduced a number of other innovations. The first pearl nail polishes were sold in 1923. They were an adaptation of the pearly lacquers containing fish-scale guanine crystals that had been used for years to make simulated pearl beads. Two decades later, chemists also patented synthetic pearl crystals made by coating platy crystals of substances such as mica with thin layers of materials with high refractive indexes such as titanium dioxide. Such lacquers were subsequently improved by others who controlled coating thickness to produce interference colors. Such pearly crystals are now available in gold, red, blue, and green interference colors, and they are widely used in nail lacquers.

The industry has also recently developed nonsettling lacquers based on thixotropic systems formed by thickeners such as organic derivatives of bentonite. Such lacquers eliminate customer dissatisfaction with earlier pearly lacquers in which the pigment and crystals had densities considerably higher than the vehicle and accordingly settled into cohesive layers that were hard to redisperse by shaking.

Cuticle removers, nail polish removers, and nail menders round out the nail product line. Cuticle softening compositions were developed as early as 1911. These products, based on dilute alkaline solutions, are still used today. Nail polish removers became necessary with the advent of the nitrocellulose lacquers. Early products were composed mainly of solvents such as acetone and ethyl acetate. Improvements in later years included addition of an oily ingredient to limit removal of oil from nails and skin and addition of a small amount of water to provide a no-smear feature. Finally, nitrocellulose lacquers containing short sections of plastic fibers are used as builders to mend or lengthen broken nails. Acrylic resins have also been used as the film former for such menders.

Suntanning Preparations

Suntan lotions are becoming increasingly important to a population that both spends more time in the sun and at the same time consists of relatively more older people. All persons with fair skin know that they burn before they tan, and the lotions give them a measure of control over burning while they await the tanning reactions that result from exposure to ultraviolet rays. Moreover, sun ages the skin prematurely, and there is a high correlation between exposure to sun and skin cancer. With properly formulated and applied suntan lotions, however, people can enjoy the benefits of more exposure without the high attendant risks that such exposure leads to. Some in the medical profession, as a matter of fact, are recommending increased applications of suntan lotions almost as one would wear clothes to prevent premature aging and to reduce skin cancer incidences.

Before World War I, fashion dictated long-sleeved clothes and hats, and people were largely protected from the effects of the sun's rays. As fashions changed after the war, various oils came into use supposedly to prevent sunburn, but they were all ineffective. Then the first effective sunscreen cosmetic reached the U.S. market in 1928 (based on a European patent). During the next 30-odd years chemists gradually improved such products, and today a leisure society in the U.S. daubs itself with suntan lotions that retail for nearly $60 million a year.

At first, it appeared that suntan lotions should reflect the sun's rays to be effective. As chemists studied sunlight's spectrum, however, they learned that only wavelengths ranging from 297 nm. to 330 nm. burned and tanned the skin. The two effects of burning and tanning nearly parallel each other until wavelengths of about 315 nm. are reached, after which the longer wavelengths tan more than burn. What is needed is not total reflection but absorption of wavelengths from 297 to 315 nm.

The first effective sunscreen contained benzyl cinnamate and salicylate. It was followed in the 1930's with products that successively contained menthyl salicylate, menthyl anthranilate, and menthyl p-aminobenzoate. All these products functioned reasonably well, but they suffered a decided drawback by staining bathing suits, beach towels, and other such items.

Following World War II, the industry made many attempts to replace the salicylates as sunscreens. Among candidates claimed to stain less were a variety of other p-aminobenzoate esters and derivatives. Another very successful candidate was 2-ethoxy ethyl p-methoxy cinnamate, which was introduced in 1958. It stained fabrics much less than earlier products yet screened harmful radiation without preventing tanning.

The idea of providing an instant tan from a bottle was introduced in the 1920's in the form of highly tinted skin stains in water-alcohol solutions. Then in the late 1950's, chemists in the U.S. found that dihydroxyacetone in a water-alcohol solution acted on amino acids in the outer horny layer of the skin to produce a chemical reaction leading to a tan not unlike suntan. Glyceryl aldehyde is the most recently patented chemical that performs the same function as the original dihydroxyacetone.

Deodorants and Antiperspirants

As judged by frequent historical references, personal malodor has been a human social problem throughout the ages. The use of perfumes and colognes probably developed, at least partly, to help make close human contact more pleasant. More recently, the approach has been to stop odors from forming rather than to mask them, to retard perspiration, or both.

The first reference to a specific deodorant in the U.S. is apparently a cosmetic cream marketed in 1899. Zinc oxide, a mild antiseptic, presumably contributed to the product's efficacy, but it was not until 1947 that the bacterial origin of underarm odor was demonstrated. With that knowledge, the way was open to development of bacteriostats for odor control. Until recently, the most widely used deodorant ingredient was hexachlorophene, a highly effective antibacterial agent. It was synthesized in 1941, and its first important commercial application came in 1949 when it was marketed nationally in a deodorant soap. Other manufacturers introduced additional products containing hexachlorophene, but recent action by the Food and Drug Administration has limited its continued widespread use in deodorants. As a result, it has now largely been replaced with other phenolic and nonphenolic germicides.

While underarm products dominate the deodorant market, they are aimed at a social problem. Products for the feet, however, are concerned with physical as well as odor problems. The first well-known foot product was marketed in the U.S. in 1909. A salicylic acid–collodion corn remedy appeared in The National Formulary as early as 1916, but use of this combination probably preceded that date. Salicylic acid and benzoic acid was used for tinea pedis, or athlete's foot, at least as early as 1929. Then in the 1940's, chemists turned to undecylenic acid for this condition, an agent still in use today. Finally, aerosol foot sprays were introduced in about 1955, their advantages being cooling relief and convenience.

Stopping perspiration is another control approach to preventing odors. The power of astringent aluminum salts to control perspiration was first recognized and patented in 1921, and the first generally recognized use of this discovery was incorporation of aluminum chloride in an antiperspirant lotion sold in the 1920's.

The aluminum compounds used originally were salts of strong mineral acids. Although highly effective, they required sophisticated formulation to overcome their high acidity. Then in 1947, a new compound, aluminum chlorhydrate, was introduced that had the advantage of good antiperspirancy and low acidity. This favorable combination of properties has led to its becoming by far the most popular antiperspirant ingredient.

Since then, major developments in deodorants and antiperspirants have been in new and more convenient methods of packaging, such as roll-ons and aerosols.

Dentifrices

Americans spend more than $3.5 billion a year for dental health care, more than $360 million of which goes for dentifrices. Highly significant changes have occurred in dental health care during the past two decades. In the 1960's particularly, the emphasis shifted from reparative care to preventive care, and changes in dentifrice formulations made their contributions.

Dentifrices enhance appearance by maintaining cleaner teeth, reduce the incidence of tooth decay, help maintain healthy gums, and reduce the intensity of mouth odors. Moreover, good dental health increases the possibility of good general health, an important secondary result of cleaning teeth.

In the early part of this century, formulators used chalk as the abrasive to clean and polish the teeth. Then in the mid-1930's, dentifrice research workers developed a stabilized dicalcium phosphate and also an insoluble sodium metaphosphate. These ingredients were followed by other polishing agents, the principal ones being alumina, calcium sulfate, anhydrous dicalcium phosphate, and calcium pyrophosphate.

In the 1930's, formulators made dentifrices more acceptable by including foaming agents, mainly soap. Soap wasn't entirely satisfactory, however, and the industry turned to the synthetic detergents when suitable ones appeared for dentifrices. Today, the most prominent ones are sodium lauryl sulfate, sodium N-lauroyl sarcosinate, and sodium coconut monoglyceride sulfonate.

Substantial efforts were made during the 1950's and 1960's to develop dentifrices that would be more effective in reducing caries and maintaining gum health than the then used cleansing dentifrices. Special ingredients studied for this purpose included urea, dibasic ammonium phosphate, water-soluble copper chlorophyllins, penicillin, and fluorine compounds. Of these, stannous fluoride and sodium monofluorophosphate have been accepted (by the American Dental Association's Council on Dental Therapeutics) as useful agents to increase the anticaries effectiveness of cleansing dentifrices, and at the present time half of the dentifrices bought in the U.S. contain a fluorine compound.

FUTURE DEVELOPMENTS

The last 30 years have seen a rapid increase in the application of physical, organic, analytical, and surface chemistry to the field of personal care products. Formulation, once principally an art, has been largely revolutionized by the science of cosmetic chemistry. While it is difficult to foresee the areas of major developments in the future, it is likely that the physical sciences will continue to play an important role in defining and controlling the physical aspects of personal care product functionality.

The development of biologically active cosmetics is only recently emerging. Personal care products containing active ingredients that control dandruff, acne, oiliness, pigmentation, sweating, hair growth, and aging of the skin are likely to be among the major new improvements of the future. These may require expansion of the industry into the drug field. However, it represents a potentially large opportunity for new growth in which cosmetic scientists will still face largely unsolved problems.

10 / **Fertilizers**

SOCIAL AND ECONOMIC IMPACTS

The chemical fertilizer industry is an indispensable part of American agriculture. A century ago, farmers in the U.S. used about 320,000 tons of fertilizers and a half century ago about 7.2 million tons. Today, American farmers apply more than 40 million tons of fertilizers worth more than $2 billion. Population may have increased fivefold since 1870 and twofold since 1920, but fertilizer tonnages have grown 125-fold since 1870 and sixfold since 1920. Coupled with many advances in plant genetics, pesticides, and mechanization, the widespread production and use of chemical fertilizers have made the country's crop lands among the most productive anywhere. Chemical fertilizers, as a matter of fact, may account for more than a third of the annual U.S. harvest. A point could probably also be made that Americans enjoy better health because of better quality food that contains more minerals and proteins as a result of fertilizer use.[1]

Plants need nearly 20 chemical elements for growth, according to present knowledge. All occur naturally in the air and earth. Those the air supplies are unlimited, but some soils have either a limited supply that must be raised or a supply that must be replenished regularly to restore what growing crops remove.

Plants need nitrogen, phosphorus, and potassium in the greatest amount, and they are accordingly classified as major nutrients. Plants also need smaller amounts of calcium, magnesium, and sulfur, the so-called secondary nutrients, as well as very small amounts of boron, copper, iron, manganese, molybdenum, zinc, and possibly chlorine, silicon, and sodium, the so-called micronutrients. (Carbon, hydrogen, and oxygen round out the list of known chemicals needed by plants for growth, but they are not considered nutrients in the fertilizer sense.)

Some chemicals become part of the plant structure while others are essential for growth but do not become part of the structure. A few apparently are needed to make various plant enzyme systems function properly. Deficiencies in any of the major, secondary, or micronutrients may limit plant growth, and a fertilizer's job is to remove that limit and make soils fertile. Farmers put nitrogen, phosphorus, and potassium on their fields in amounts and at times determined by such factors as soil conditions, crops grown, season, and weather. Secondary nutrients are often part of major nutrient fertilizers and usually require no special allowances for application. Meantime, micronutrient deficiencies occur occasionally. When they do, farmers correct them by applying very small amounts either as solids to the soil or as liquids to the crop foliage. Since secondary and micronutrients rank far below the major nutrients, both economically and as potential problems, we shall confine this study to innovations in nitrogen, phosphorus, and potassium fertilizer technologies.

---

1. For a fuller appreciation of how fertilizer production and use have impacted on the economy, readers should also see the chapter in this volume on food processing. While that chapter deals largely with food processing as such, it is important to realize in reading it that the availability of many fertilizer formulations and of knowledge of how to use them have helped the U.S. grow much more food per acre with relatively fewer workers than in former times. Accordingly, improved fertilizer technology ranks quite high in importance among innovations affecting the country's food supply and the changes that have occurred in how that supply is processed and made available for consumption.

At the dawn of the agricultural age, farmers moved their crops from field to field as the soil's fertility declined. But farmers have long known that applying some substances to their fields would make their harvests bigger. For several thousand years they limed acid soils and applied manures, bone chips, wood ashes, and organic wastes. They also grew legumes and rotated crops. No one at the time understood why such practices worked, and only some two to three centuries ago did a more scientific approach begin evolving to determine where and how plants get nutrients. The birth of a technically-oriented fertilizer industry soon followed.

CHEMICAL ACCOMPLISHMENTS

Among his many other chemical accomplishments, the German chemist Justus von Liebig showed in about 1840 that nutrients in bones became more available to plants if the bones were first treated with acid. Then in 1842, a John B. Lawes in England patented a process using sulfuric acid to do so.

Modern fertilizer technology dates roughly from Liebig's and Lawes' days. Natural forms of fertilizers served adequately at the time. Bones provided phosphorus, for example. Manures and other commonly available organic materials supplied most of the nitrogen, although rainwater added some that had been fixed in the atmosphere by lightning. Legumes and nitrogen-fixing bacteria also served as an important source of nitrogen. In the early 1800's, the vast deposits of nitrate in Chile became a nitrogen and potassium source of worldwide importance. Wood ashes and sugar beet wastes also supplied potassium.

As population grew, however, it became apparent that the world could not meet its need for food without a much expanded plant nutrient supply. Organic sources and even deposits such as those in Chile would hardly suffice, and scientists and engineers accordingly turned to investigating inorganic sources for plant foods. In time, insoluble calcium phosphate from mined ores yielded useful phosphorus fertilizers when treated with sulfuric or phosphoric acids. Direct synthesis processes fixed nitrogen from the air into usable compounds. New discoveries and technological advancement helped upgrade to workable concentrations the low-grade forms of potassium minerals. In a (relatively) short period of about 100 years, chemical innovations made each of the major plant nutrients available to whatever extent they were needed and could be used economically.

Phosphate Fertilizers

Today, phosphate rock deposits provide all fertilizer phosphorus. Major deposits occur in a number of countries, and the U.S. has known reserves of about 13.5 billion metric tons, about 30% of the world reserves. They are located mostly in Florida, Tennessee, Montana, Wyoming, and North Carolina.

The principally used mineral form in the deposits is fluoroapatite, which has a nominal composition of $Ca_5(PO_4)_3F$. Phosphorus in apatite is virtually unavailable to growing plants, but treating rock with sulfuric acid converts it to normal superphosphate, which is a form that is available. The process, incidentally, is analogous to Liebig's and Lawes' pioneering ones with bones.

Normal superphosphate was apparently made for the first time in the U.S. in the early 1850's, a decade after the pioneering work in Europe. William Davison, emigrating from Ireland, founded Davison Chemical

Co. (now part of W. R. Grace & Co.) in Baltimore, Md., and acidulated old bones and later phosphate rock during this period.

The technology—or more accurately, the mechanics—of producing normal superphosphate has improved greatly since the early days. Briefly, the process involves mixing phosphate rock with sulfuric acid, holding the slurry in a container until it solidifies, and then storing the material for some weeks to allow the chemical reactions to go to completion.

The three steps are known as mixing, denning, and curing. The practice of hand mixing had been largely supplanted by mixing in power-driven equipment by the turn of the century. Mechanical means for excavating normal superphosphate from dens replaced manual labor also about then. In time, the practice of mixing acid and rock in batches, discharging the semisolid into a den, and then conveying the solidified product to storage has largely been superseded by continuous mixing and denning, although batch equipment is still very much in evidence and competes effectively with continuous equipment.

One early innovation in phosphate fertilizer technology was production of triple superphosphate. In this process, manufacturers first solubilize rock with sulfuric acid, as in normal superphosphate production. They filter off the gypsum (or calcium sulfate) and then use the remaining phosphoric acid (known as wet-process acid) to solubilize another batch of rock. The resulting triple superphosphate analyzes about 46 to 48% phosphorus pentoxide (P_2O_5) compared to 15 to 20% for normal superphosphate.

Companies at first commonly used wet-process acid containing 40 to 45% P_2O_5 to make triple super, and they dried the cured product artificially. In recent years, however, they have used more concentrated acid and have eliminated artificial drying. Tennessee Valley Authority accounted for the first large-scale manufacture of self-drying triple superphosphate (in 1934) using phosphoric acid made from elemental phosphorus (furnace acid) containing about 54 to 57% P_2O_5. Subsequently, other producers used wet-process acid raised to about this P_2O_5 concentration.

Wet-process acid is the base material of the phosphate industry. Improvements in its technology have occurred during the past half century, and they have accelerated in the past decade or two. In early practice, manufacturers treated phosphate rock with very dilute sulfuric acid and filtered the calcium sulfate from the dilute phosphoric acid. They understood the crystallization mechanism poorly, and their batch filter presses were quite primitive. Moreover, the filtered acid was very weak and had to be concentrated to the required strength by heating in open, lead-lined pans.

Then in 1915, Dorr Co. developed a continuous process in which sulfuric acid and rock reacted in a series of tanks and the calcium sulfate separated first in settling tanks and then in a continuous vacuum filter. The new system lowered costs and also produced acid containing 22 to 23% P_2O_5, considerably more concentrated than the 10 to 15% P_2O_5 content of acid made by the batch process.

The industry has made steady progress in understanding crystallization chemistry and in increasing the concentration of filter acid, and it has made outstanding improvements in filters and evaporators. Horizontal filters have proved most satisfactory for removal of calcium sulfate from the phosphoric acid.

Filter acid in today's modern plants runs about 30% P_2O_5, and it is now concentrated in several types of equipment. A forced-circulation vacuum process with steam heating is the most popular.

Phosphorus may also be produced for conversion into fertilizers by high-temperature smelting of phosphate rock (with coke and silica) in an electric arc furnace. Credit for first use of furnace acid in fertilizer manufacture in the U.S. apparently belongs to Piedmont Electro-Chemical Co., which made triple superphosphate and ammonium phosphate with it at Mt. Holly, N.C., in the 1914–16 period.

When TVA produced superphosphoric acid (76% P_2O_5) in 1956 and demonstrated its utility as a fertilizer, it stimulated interest in wet-process superphosphoric acid. The phosphate fertilizer industry today makes the wet-process super acid product by concentrating ortho acid to 69 to 72% P_2O_5. About half or more of the P_2O_5 exists as polyphosphates or condensed phosphates formed by removal of water from orthophosphate. Not only is the superphosphoric acid more concentrated than merchant grade acid, which runs about 54% P_2O_5, but the ammoniated acid (ammonium polyphosphate) is much more soluble. This characteristic has made possible the production of a new generation of high-analysis liquid fertilizers.

American Cyanamid, Consolidated Mining and Smelting (Cominco), Olin Mathieson, and Missouri Farmers Association have pioneered the use of ammonium phosphates for fertilizer use. Dorr designed and installed most of the early plants, the first one in 1933 for Cominco at Trail, B.C.

In the mid-1950's, Colorado Fuel and Iron, TVA, and others began production of crystal diammonium phosphate from furnace acid. Promotional efforts of these organizations led to a sharp upturn in demand and expanded efforts to develop new processes and greater markets.

A particularly desirable goal was to find ways of making diammonium phosphate from wet-process acid, a cheaper and more widely available material than furnace acid. In 1959, U.S. Phosphoric Products Division of Tennessee Corp. first produced such diammonium phosphate in its large Florida plant. About a year later, TVA introduced the process that is used in most new ammonium phosphate installations.

In the 1960's, production of ammonium phosphate fertilizers increased rapidly. Their natural advantages of high analysis, solubility, and good physical characteristics have helped them overtake both normal and triple superphosphate and reach top spot.

Nitrogen Fertilizers

Natural organic materials such as meat and garbage waste products, sewage, dried animal manures and guano, and agricultural commodity by-products served as the first sources of nitrogen for fertilization. The fertilizer industry clearly needed synthetic sources, and American chemists and engineers joined in the efforts to find alternatives.

Early approaches to fixing atmospheric nitrogen included a reaction between nitrogen and oxygen under an electric arc to make nitrogen oxides and a reaction of calcium carbide with nitrogen to make calcium cyanamide. Neither proved important for making fertilizers in the U.S., however.

In 1916, about a year before beginning to build a cyanamide plant, the U.S. Government took the first significant step in nitrogen fixation in the U.S. when it built a plant at Muscle Shoals to make ammonia by the then new Haber-Bosch process.

Fritz Haber of the Kaiser Wilhelm Institute and Karl Bosch of Badische Anilin- & Soda-Fabrik and their co-workers had developed the process in Germany before World War I. In it, nitrogen reacts with hydrogen at high temperature (up to 500° C.) and pressure (up to 1,000 atmospheres) over a catalyst to yield ammonia. The ammonia may then be reacted with sulfuric acid to make ammonium sulfate, with nitric acid to make ammonium nitrate, or with normal superphosphate, triple superphosphate, or phosphoric acid to make ammonium phosphate, all of which are excellent fertilizers.

The U.S. Government's 1916 effort failed, because the designers and operators knew too little of the chemistry, particularly of the catalysts, and of the high-pressure equipment required, but not much time was to pass before the Haber-Bosch process would make an impact on American chemical technology as few other developments have ever done.

After the war, the U.S. Fixed Nitrogen Laboratory and several large industrial organizations conducted research, notably on catalysts, and gathered the information needed to design and build ammonia synthesis plants. In short order as a result, several plants were built during the 1920's, the two largest being the plant at Atmospheric Nitrogen Corp. (later Allied Chemical) at Hopewell, Va., and the plant of E. I. du Pont de Nemours & Co. at Belle, W.Va.

In the intervening years, the process has been improved in many ways. Among the most important has been perfection of processes for reforming natural gas to supply hydrogen for the synthesis, as hydrogen made from natural gas has proved to be much less costly than hydrogen made from coal. Another significant step, and a recent one, has been the development of centrifugal compressors to reach the pressures required in the synthesis. Centrifugal compressors not only made much greater scales of operation practical, but they also required them. With these and other improvements, the industry has turned to building huge plants making 1,500 tons and more per day of low-cost ammonia that now provides most of the nation's fertilizer nitrogen.

Ammonia from the plants built up to the early 1940's generally cost too much to compete in fertilizers with by-product ammonium sulfate from the steel industry, with mined sodium nitrate, and with various organic wastes. Beginning shortly before World War II and continuing to the present, however, more output and improved processes have lowered the cost of ammonia until it has now become the cheapest and most popular source of fertilizer nitrogen.

Availability of large amounts of synthesis ammonia at competitive prices has helped spur the long-term trend to higher analyses in nitrogen fertilizers. At first, the industry made the then standard forms of calcium nitrate, sodium nitrate, and ammonium sulfate, but they all analyzed fairly low (15 to 21% nitrogen, or N). Then ammonium nitrate, urea, and ammonia itself began replacing them.

Tennessee Valley Authority (TVA), Spencer Chemical, and others heavily promoted ammonium nitrate (33.5% N) as a fertilizer during and after World War II. Produced mainly for munitions during the war, ammonium nitrate today has become the leading form of solid nitrogen fertilizer.

Urea has been a second route to higher analysis nitrogen fertilizers. Produced by reacting ammonia with carbon dioxide under pressure and over a catalyst, urea analyzes about 45% N. It accounts for much of the new and planned nitrogen fertilizer capacity, and it may soon overtake ammonium nitrate as a supplier of nitrogen.

Ammonia itself is the ultimate form of nitrogen fertilizer, analyzing 82% N. Shell Chemical Co. pioneered direct injection of ammonia into the soil in California in the 1930's, and the Mississippi Agricultural Experiment Station in cooperation with TVA advanced the technology during the 1940's. Today, more than half of the fertilizer nitrogen in the U.S. is applied as anhydrous ammonia, aqua ammonia, or nitrogen solution. Ease of shipment, distribution, and application accounts for the popularity of ammonia and nitrogen solutions as manufacturers, distributors, and farmers have sought to reduce labor costs in an industry that seems squeezed more than most between rising costs and falling prices.

Potassium Fertilizers

Potassium, the third of the major plant nutrients, is recovered mostly as potassium chloride by fractional crystallization and other processes from such U.S. resources as deposits in New Mexico and Utah and brines at Midland, Mich., Bonneville, Utah, and Searles Lake, Calif.

U.S. fertilizer producers relied heavily on potash from Europe until World War I. When that event cut off these sources, two plants produced much of the nation's potash from the brine of Searles Lake in Southern California. Their operations were very inefficient, and costs were too high to compete with the cheap potash imported after the war, however. One of the companies closed its plant, but the other began research that resulted in a bench-mark achievement by American chemical technology.

Chemists at American Potash and Chemical Corp. (now part of Kerr-McGee Corp.) began studying the extremely complex brines of Searles Lake in 1919. They discovered new salts and crystallization systems over a period of several years and improved equipment and operating methods as a result. Major advances were scored in developing and applying phase rule chemistry, and after several starts and stops, Ampot put the plant into continuous operation in 1922. It has functioned since then to make potassium and boron compounds for fertilizers and for industrial uses.

Further advances in potash technology have been primarily in methods of mining and beneficiation following the discovery and development of deposits in the Permian Basin of New Mexico, in Utah, and in southwestern Canada. Recovery of underground deposits of Permian Basin potash represents fairly straightforward mining technology from deposits some 800 to 900 feet underground, and these deposits have served as the major U.S. potash source since their discovery in the 1930's. Meantime, potassium deposits were found in 1947 at depths from 4,000 to 12,000 feet in Saskatchewan. Some are recovered by solution mining.

Solid Mixed Fertilizers

Combining several materials to make multinutrient fertilizers has dominated U.S. fertilizer practice from the beginning. The industry formulated the first mixed goods from such materials as normal superphosphate, guano, sodium nitrate, potash salts, packinghouse wastes, fish meal, cottonseed meal, and peanut hulls. It often added ground limestone as a filler and soil conditioner. The formulator usually made his own superphosphate and bought the other ingredients from outside suppliers. He mixed the formulations with simple equipment and allowed the mixture to cure a few weeks in storage piles, during which time some chemical reactions occurred. After curing, the mixtures were often more or less caked, and they had to be broken up and pulverized before being bagged.

In the 1930's, the industry began ammoniating normal superphosphate to make a fertilizer containing both nitrogen and phosphorus. By reacting concentrated solutions of ammonia and ammonium nitrate with superphosphate, companies made improved products at considerable savings over mixes in which ammonium sulfate supplied most of the nitrogen. The practice spread rapidly. The use of various waste organic materials diminished, often because a better use was found for them. Cottonseed meal, for example, became more valuable for stock feed. The analysis of products increased as fillers were eliminated.

With the higher analysis came increased caking problems. A growing use of mechanized application equipment by farmers called for dependably free-flowing fertilizers. These demands led to general adoption of granulation in the 1950's. What was then Davison Chemical Corp. was an early innovator (late 1930's) in granulation techniques. The TVA ammoniation-granulation process became the most popular as it combined mixing, ammoniation, granulation, and partial drying in a single unit.

Analysis of mixed fertilizer rose steadily as triple superphosphate replaced ordinary superphosphate and as ammonia and ammonium nitrate (supplied as ammoniating solutions) replaced ammonium sulfate or sodium nitrate. Later, still higher analyses were attained by liberal use of phosphoric acid and ammoniating solutions in the formulations. Mixed fertilizer preparation was no longer a simple mechanical job but a complex chemical engineering operation. Small local mixers either expanded, became bulk blenders or dealers, or went out of business.

Bulk blends became popular as a means by which a local mixer could prepare in simple equipment at low cost whatever mix the farmed needed. Granular materials such as diammonium phosphate, potassium chloride, and ammonium nitrate were produced in large modern plants. These plants were located where raw material costs were favorable, and they were operated by large companies that had the facilities and capabilities for low-cost production. The products were shipped in bulk to blenders who offered prescription mixing, custom application, and many other services to the farmers. The success of this system depends to a large extent on dependable, low-cost transportation. Covered, hopper-bottom rail cars protect the material from the weather, minimize spillage, and make unloading easy. Barge transport is also suitable between points on waterways.

Liquid mixed fertilizers became significant during the late 1950's, after the introduction of superphoric acid, as mentioned earlier. Analysis of the liquid fertilizers has risen steadily over the years, and this form of fertilizer now offers many of the advantages of bulk blends. Liquids are useful because of their flexibility of formulation and because they offer the farmer advantages of convenience and trouble-free handling.

A disadvantage of liquid mixed fertilizers is the comparatively low analysis of the clear liquids, or solutions, particularly those grades containing potash. Suspension fertilizers are a means of overcoming the disadvantages of low analysis. They are liquids containing solids held in suspension by inclusion of gelling-type clay.

Suspensions are advantageous in several ways. The analysis can be on a level comparable with solid fertilizers and about twice that of the solutions. In addition, formulation is flexible, as the materials need not be soluble. There are some disadvantages, however, and suspensions are not quite as simple, convenient, or trouble-free as clear liquids are.

A number of large phosphoric acid producers are heavily involved in research and process development on improved fluid fertilizers. Production of liquids is increasing rapidly. Several large primary plants and more than 2,700 small field units are producing an estimated 4 million tons of liquid mixed fertilizers annually.

FUTURE DEVELOPMENTS

Mature though it may be, the U.S. fertilizer industry has nonetheless not stopped growing and changing. However, any new technology will not likely depart as radically from earlier practices as did some developed in the past three decades. The industry may not have stopped looking for totally new processes, products, and techniques, but it isn't looking nearly as hard as it once did. Instead, it is concentrating more on finding cheaper ways to make and distribute its present products. Saving a few cents a ton on today's large-volume products will almost surely bring more profit than introducing a new product to a market already well supplied with ones that perform more than adequately.

The Environment

As the population of America grows and the U.S. becomes more crowded, we generate pollutants in ever greater quantities to foul air, water, and land. The fertilizer industry has not had to recoil before this problem quite as much as have a few other industries. It has offered a good target of opportunity, however, for the pollution it does cause and also for being "chemical" in an era approaching hysteria over "chemicals in the environment."

The public has complained about fumes, dusts, and other effluents from fertilizer plants—and often rightly so. The industry has in large part responded and corrected many intolerable situations. Little doubt exists, however, that opportunities remain for further improvements.

Of greater importance to the environment is the effect, for good or bad, of fertilizers applied to the soil. The record here is mostly good. Studies by chemists throughout the country are showing that phosphate from fertilizer is a minor factor in eutrophication and that nitrogen and phosphate fertilizers when properly used pose practically no threat from runoff and leaching.

Conversely, the benefits of chemical fertilizers are many and obvious. The pressure for food and greater agricultural efficiency has led to some misuse of arable land. Erosion, sedimentation of rivers, floods, and a decline in fertility are a few results. Fertilizer can reduce soil erosion by stimulating cover crops that protect the soil and prevent washing. Applying plant nutrients also slows the decline of fertility. Remarkable advances have been made in both areas, but the job is far from finished. Continuing effort and ever greater quantities of chemical fertilizer will be required to hold and then increase the gains.

Fertilization contributes other, more obscure benefits. Lush, well-nourished crops absorb greater quantities of carbon dioxide from the atmosphere than sparse growth and release more oxygen for animal life. Their growth helps correct the balance between carbon dioxide and oxygen in the air, an environmental problem of growing concern. Also, most crops (with the help of rainfall) remove and profitably use sulfur oxides generated by burning fossil fuels.

Despite past advances and demonstrated benefits from chemical fertilizers, some extremists call for a return to natural materials. Many years ago, when fewer people lived and fertile lands were plentiful, natural farming with organic fertilizers was efficient enough. It would be impossible today.

Manures, organic wastes, sewage, and other nutrient-containing materials don't exist in the amounts needed by American agriculture in the first place. The task of gathering them, processing them, and shipping them to consumers looms as a monumental and prohibitively expensive undertaking in the second. Organic fertilizers contain perhaps an average of only one tenth the major nutrients that chemical fertilizers do. Handling and transporting fertilizers with 10 times the weight and more than that in volume in an already overtaxed transportation and storage system clearly is out of the question. Moreover, a recycling system for municipal wastes would expose people to the danger of toxic accumulations of heavy metals, such as lead, mercury, and cadmium. Finding a technical solution to avoid accumulations would be a difficult feat at best.

Fertilizers for Developing Nations

World population may well double by 2000 A.D. and double again within another 20 to 30 years. Today's roughly 1 acre of arable land per person could become $1/2$ acre by 2000 A.D. and $1/4$ acre by 2030 A.D. Food output per acre must obviously be greatly increased or population growth sharply controlled—or a combination of both—if widespread famine well beyond what already occurs is to be avoided.

In the late 1950's, some in the U.S. fertilizer industry became enamored of preventing world famine through greatly expanded production and use of fertilizers. Among solutions discussed were growing much more food in the U.S. (by using much more fertilizer, among other things) and shipping the surplus to food-poor countries. Another solution was to be large increases in exports of fertilizers made in the U.S. Finally, American companies were to participate extensively in building and operating fertilizer plants where shortages existed abroad.

The marriage between the U.S. industry's technical strength and growing food shortages elsewhere hasn't worked quite as visualized. Almost nonexistent transportation systems in have-not nations often keep imported food near the ports. Shipping fertilizers raises costs, and when they get there the lack of transportation keeps them near ports, too. Even when fertilizers reach farms, local agricultural practices delay widespread use. Moreover, with understandable nationalism, developing countries prefer to build and operate their own plants rather than to depend on imports or on plants owned and operated by foreigners.

Not everyone in the U.S. fertilizer industry viewed the opportunities abroad as optimistically, of course. Now most agree that the future business of the U.S. industry is in the U.S. and that solving economic, political, and cultural problems elsewhere lies clearly beyond the means—and the immediate interests—of this industry. This conclusion does not mean that the industry sees no place for itself in helping increase food production elsewhere. Its contributions will come as advice and information rather than as investment and production, however.

American organizations involved include the Department of Agriculture, Agency for International Development, and Tennessee Valley Authority. Many U.S. chemists and chemical engineers are participating, often in teams with agricultural specialists. Typical activities include studies of comparative economics of production or importation and of alternative sources of raw materials. They advise developing countries on problems of shipping, handling, storing, and using fertilizers.

Many of these American chemical and agricultural specialists travel to countries around the world to advise on modern fertilizer technology. Some of these groups spend months in other countries to assist in setting up a local fertilizer industry or in making an existing one better. In addition, delegations from various countries frequently visit the U.S. for classes on fertilizer manufacturing, marketing, and application.

Energy and Materials Resources

The U.S. applies about 40 million tons of fertilizer to its crop lands annually and the rest of the world more than three times that amount. World consumption has grown threefold in the past 15 years, and it will probably double in the next 12 to 15 years. To provide such great tonnages, the world will obviously need huge sources of nitrogen, phosphorus, and potassium as well as hydrogen and sulfur (used in nitrogen and phosphorus fertilizer manufacture) and essentially inexhaustible sources of energy.

The world fortunately has natural deposits of phosphate rock great enough to last for centuries at the present rate of consumption. Morocco has nearly 50% of world reserves, the U.S. about 30%, and the U.S.S.R. about 15%. Smaller deposits occur elsewhere in North and South Africa, the Middle East, South America, and the islands of the Pacific and Indian Oceans. Unfortunately, some of the heavily populated regions, such as Asia, lack this important fertilizer raw material. One chemical challenge will be to provide have-not regions with adequate levels of P_2O_5.

Producers need sulfuric acid to make most phosphate and some other fertilizers, such as ammonium sulfate. Elemental sulfur, pyrites, and sour gas (natural gas containing hydrogen sulfide) are the raw materials. Elemental sulfur mined by the Frasch process from underground deposits along the Gulf Coasts of the U.S. and Mexico costs the least. Regrettably, increasing demands for sulfur, probably industrial chemistry's most basic single chemical, are speedily exhausting the deposits. Pure and inexpensive Frasch sulfur quite possibly will be gone before the end of the century.

Sour gas sulfur is becoming a factor in the supply picture and, if anything, is a slightly better raw material for making acid than is Frasch sulfur. Reserves in pyrites as alternative sources are more plentiful, but sulfuric acid made from them costs more.

When the lower-cost forms of sulfur become limited or are no longer available, manufacturers will turn to by-product sulfur oxides from burning fossil fuels and to the enormous deposits of gypsum (calcium sulfate). Research has quickened on both routes to find methods with reasonable costs. Most of the activity is concentrated on processes for removing sulfur oxides from stack gases, mainly from power plants that burn fossil fuels. The work is urgent, not so much for making fertilizers now as for reducing air pollution.

No country will ever lack for nitrogen, but it must be combined with hydrogen to get it into forms plants can use. Hydrogen, though abundant, is not readily available. Hydrocarbons such as natural gas and naphtha usually supply it now, and the U.S. and other countries are consuming petroleum and natural gas reserves at a high rate. Such reserves appear to be ample for several more decades, and greater reserves of oil shale and tar sands back up the readily available gas and liquid hydrocarbons. They, too, are exhaustible and irreplaceable sources, however.

The world fortunately has abundant reserves of potash. Since the discovery of the very large deposits in Germany and France more than a century ago and in the U.S. and Canada more recently, ample supplies are assured for centuries. Opportunities for improvements lie in finding better mining and refining methods and in devising better systems for transporting potash.

Lack of secondary and micronutrients can limit plant growth. However, major nutrient fertilizers and low-cost materials such as limestone and dolomite provide secondary nutrients, while micronutrients are not likely to be exhausted or to become limiting factors in the foreseeable future.

The world thus has a plentiful supply of raw materials to make the quantities of fertilizers it will need except for impending shortages of fossil fuels (for hydrogen) and sulfur (for sulfuric acid). Fortunately, a solution is in sight in the form of nuclear power.

Nuclear power can replace both sulfuric acid in making phosphate fertilizers and fossil fuels in making ammonia. When sulfuric acid becomes too expensive, for example, producers could turn to electric furnace smelting of phosphate rock to make elemental phosphorus. As fossil fuels also become too expensive or even disappear, producers could obtain hydrogen for ammonia synthesis by electrolysis of water. The phosphorus industry already smelts rock in furnaces on a large commercial scale. Electrolysis of water is not as well developed commercially but doubtless will be when the need becomes critical.

Both changes offer advantages beyond alleviating raw materials problems. Among them is removing restrictions on plant site locations. Elemental phosphorus, the equivalent of 229% P_2O_5, could be made near rock deposits and then shipped long distances to be converted to fertilizer at the destinations. Ammonia, on the other hand, would no longer have to be made near natural gas and petroleum refining centers but could be made virtually anywhere that water is available.

Energy generated with uranium fuel will be limited by the extent of uranium reserves, and these are exhaustible. Fortunately, a promising advance—the breeder reactor—gives reasonable assurance of making energy at the expense of very little fuel. The breeder reactor, which makes about as much fuel as it consumes, is far along in development and is likely to be ready for large-scale operation in another 10 to 20 years. When fully developed and widely adopted, breeder plants will assure power generation indefinitely without fuel shortages.

In one of the more notable responses to the challenge of the future, plans have been made for huge nuclear complexes for multiple purposes. A very large nuclear power plant would be built at an appropriate seacoast site. Waste heat would be used to desalt seawater to provide clean water for industrial and agricultural uses, which are growing needs in many areas. Some of the great amount of electricity generated would be used in heavy-power-consuming industries to manufacture such products as aluminum, chlorine, caustic, alloys, and the fertilizer basics of phosphorus and ammonia. Construction of complexes of this type is likely during the next decade or two. They could be important contributors to agricultural and industrial needs of developing countries.

11 / Pesticides

SOCIAL AND ECONOMIC IMPACTS

The greatest benefit of pesticides is their ability to help provide a varied and nutritious diet for an ever-increasing number of people. Pesticides in addition protect many nonfood crops and products against a variety of pests. In certain parts of the world they have transformed the statistics on human illness and death by controlling the pests that carry diseases such as malaria and yellow fever. Pesticides also can produce a variety of unintended environmental effects, some good, some bad. For this reason it is vital today that their usefulness be assessed on the basis of overall social benefit versus overall risk, instead of on the basis of purely economic factors related directly to the intended use.

In 1970, U.S. companies produced 1.03 billion pounds of synthetic organic pesticides, valued at $1.07 billion at the manufacturers' level. Data for inorganic pesticides are limited, but production is relatively small, and the organics constitute an estimated 90% of all pesticides output. Of the organic materials produced, insecticides, fumigants (including nematocides), and rodenticides together accounted for about 47%. The vast majority of this fraction was insecticides. Of the remainder produced, 37% was herbicides, and 16% was fungicides. In 1970, the U.S. imported some 13.3 million pounds of synthetic organic pesticides, valued at $10.6 million. Pesticides exports were 404 million pounds, valued at $220 million.

Much of the world's population remains ill nourished, but chemicals have been a prime factor in agriculture's ability to keep pace with the mushrooming population of the past quarter century. The great Irish potato famine of the 1840's was caused by a then-uncontrollable outbreak of a potato fungus. The famine has not recurred in the past century, primarily because of the availability of fungicides used almost universally to control this plant disease. Similar instances are common. In the U.S. corn belt today, for example, a farmer cannot raise a crop economically without treating for corn rootworm.

Phenomenal gains in unit productivity in agriculture have been achieved in the past 30 years by the integrated use of chemicals, genetically improved crop varieties, better crop management and nutrition (fertilizers), farm mechanization, and irrigation. Wheat yields have almost doubled in Great Britain. Rice yields in the U.S. are up 100%, potatoes are up 60%, corn yields have tripled. Between 1952 and 1962, Japan reduced the labor required to produce an acre of rice by more than 25%.

Such progress has sharply reduced the amount of labor required to produce an adequate diet. In the U.S. in the period 1945–64, about 7% of the population worked in agriculture, and the nation spent about 18% of its income on food. In India in the same period, about 73% of the population worked in agriculture, and the nation spent about 66% of its income on food. Although chemicals are only one of a number of factors that contribute to such a contrast, it is perhaps significant that India uses roughly 10% of the weight of pesticides used per acre in the U.S.

The variety of foods provided by modern intensive agriculture is important to well-balanced nutrition. Agricultural chemicals play a vital role in making many popular types of food practical to produce on a scale that makes them available to all.

Much of the earth's land surface is ill suited to intensive crop production but if well managed will produce, dependably, sufficient grass and forage herbs to support ruminant animals. The productivity of such pasture and rangeland often is limited by growths of woody and herbaceous plants that are unpalatable, nonnutritious, or poisonous. Herbicides are proven tools for controlling such growths and thus creating better grassland to produce more meat and milk. Both can be important in overcoming the protein deficiencies that threaten the health of so many people in less developed countries (although such deficiencies often are at least partly sociopolitical problems and not truly agricultural problems). Meat and milk, moreover, are vital to maintaining the high level of nutrition now enjoyed in the more fortunate parts of the world.

Chemical tools also improve the efficiency of production of nonfood products of farm and forest. In spite of the tremendous growth of synthetic fibers, cotton and wool are still widely used. Without the chemicals that prevent excessive loss from insects and disease, neither cotton in the U.S. and Egypt nor sheep in Australia could be raised at a profit to the grower or economically enough to permit widespread use of cotton and wool. Agricultural chemicals are used regularly against forest insects in the U.S., and the use of herbicides in forestry is developing steadily. Foresters, like farmers, find herbicides useful for controlling selectively the growth of unwanted plants, including trees and shrubs that can greatly retard growth of species needed for future timber and pulpwood.

Stored grain insects have been known for centuries. Supplies sealed in tombs in ancient Egypt were found on opening, centuries later, to have been infested by some of the same insect species we know today. Early American colonists suffered great losses from grain insects, and the colonial army of the Revolutionary War is reported at times to have had nothing to eat but wormy flour. One of the most troublesome species infesting cereal grains, the Angoumois grain moth, became widespread during the 16th Century as a result of global exploration and the international commerce that followed. So serious was this pest during the 18th Century that it was the object of some of the first recorded research on insect control. Chemical fumigants protect stored grain against such pests, as do certain nonfumigant insecticides of very low toxicity to mammals.

Rats and other rodents are notorious destroyers of stored food. Holding their populations in check is one of man's perennial battles. Experience has shown that losses to rodents can be kept at a reasonable level by sustained use of chemical rodenticides integrated with good sanitation to minimize the animals' food supply. In view of the excessive losses of stored food to rodents that still prevail in some parts of the world, rodenticides must be considered among our most beneficial pest control chemicals.

Most utility poles, railroad ties, and dock pilings are now treated to prevent destruction by rot and termites. Fence posts and structural lumber increasingly are being treated before use, particularly where termites are present. About 600 million cubic feet of wood is treated annually in the U.S.

Control of termites in structures already in place is another important aspect of wood protection. Perimeter treatment with a residual insecticide for the subterranean termite and fumigation for the drywood termite prevent destruction of many homes.

Pesticides, particularly insecticides, have helped remarkably to control such insect-borne diseases as malaria, yellow fever, encephalitis, and typhus. The highly-debilitating malaria, which has contributed so much to the backwardness of entire nations, has been greatly reduced almost everywhere, primarily by the use of DDT against mosquitoes.

Flies transmit several diseases, and cockroaches transmit *Salmonella* and possibly hepatitis. These pests are controlled by several chemical products used widely by pest control operators, particularly in restaurants and in food processing, storage, and distribution facilities. Rodents, besides destroying food, carry the insects that transmit a number of diseases, including typhus and bubonic plague. Without rodent control programs, including chemical controls, such diseases would be far more prevalent.

Microbe-controlling agents are used routinely in dairies and food processing plants as well as for water treatment. Pesticides control nuisance insects such as mosquitoes and fleas in homes, gardens, and recreational areas, as well as the many pests jeopardizing pets and livestock.

Besides their use in agriculture, herbicides have been used to eliminate poison ivy and poison oak from many recreational areas. They've been used also to control allergenic plants such as ragweed. Railroad safety has long benefited from the use of herbicides on the gravel in track ballast. Control of woody brush along utility lines, railroads, and highways improves visibility, eases inspection, and protects wires and drainage ditches. Herbicides are used also to reduce the growth of combustible plant material in firebreaks, along forest roads, and around industrial sites.

Benefits and Risks

The past decade has seen rapid realization of the need to weigh carefully the benefits of technology on the one hand and the risks in its application on the other. No event has driven this lesson home so forcefully, perhaps, as the belated discovery that some of the insecticides that degrade slowly tend to be concentrated biologically by organisms other than their target organisms. These "persistent" insecticides generally present no hazard immediately upon application. They can build to hazardous levels during their residual life, however, in host organisms that bioconcentrate them over long periods or that stand at the end of a food chain that bioconcentrates them stepwise.

Research and regulation have done much to limit the use of persistent pesticides and replace them with ecologically safer materials. Much research is still under way. With encouragement from industry, government, and the public, research can produce safer, more effective pest control based on chemicals integrated with other methods. Pesticides have been evaluated for safety for many years, but until roughly the past decade the emphasis was primarily on protecting the pesticide user and the consumer of the food treated. Today, expanded research programs are aimed at defining the safety and the potential hazard of specific agricultural chemicals in each projected use.

In the current drive to identify the risks of pest control chemicals, it is possible to lose sight of the broad benefits that man has derived from the use of chemicals to control destructive insects, weeds, and other organisms. If wise decisions are to be made on the net value of a given agricultural chemical or on the benefits and risks of a specific use, the benefit side of the balance must be kept clearly in focus.

Agricultural chemicals by definition are a threat to specific parts of the environment, because they are toxic to pests, which are part of the environment, and sometimes are harmful to nontarget organisms. The seriousness of the threat, however, is not nearly so clear-cut nor so well understood as might seem to be the case. This is particularly true of the threat of long-term exposure of living systems to minute amounts of certain pesticide residues. In some species of birds and fish, accumulated pesticide residues have been shown to cause harm.

Many pesticides have the potential to cause toxicological problems for wildlife. In this country, however, the registrations of these products, which must be approved by the Environmental Protection Agency, are designed to minimize the hazard by the use of proper registrations and precautions in the application of the pesticides.

Agricultural chemicals can produce ecological benefits, too, and these, like the risks, are not always well understood. The crop yield revolution of the past 30 years is enabling the technically advanced countries to produce their crops on half the acreage that would otherwise be required. Vast areas are thus saved for ecologically desirable, noncrop uses such as permanent pastures, forests, wildlife preserves, and recreational areas.

The yield revolution in the U.S. stems largely from a system of monoculture, in which crops are grown in the optimum geographical areas and in the optimum acreages. This form of agriculture has proved so efficient that roughly 5% of the work force can meet the nation's need for food and fiber. At the same time, monoculture tends to tilt the ecological balance in favor of crop pests. Without agricultural chemicals, therefore, improved crop varieties, intensive mechanization, fertilizers, and the like would not contribute their full increment to the crop yields that can be achieved by intensive monoculture.

Without these yield increases, some fraction of the additional acreage that would be in crops today would be on erodible land, ill adapted to tillage. The result would be increased silting of streams and greater destruction of valuable soil. Today's high yields involve close plant spacing and vigorous growth. More unharvested parts of crop plants, including roots, are left in and on the soil when yields are high than when they are low. These heavier crop remains improve the physical condition of the soil and rainfall penetration and reduce erosion. Thus, yield increases have contributed to conservation of the land that still is farmed as well as of the land that they have freed for noncrop uses.

Soil runoff is related closely to tillage practices. It can be greatly reduced if a continuous cover or mulch is kept on the land. Herbicides provide a new horizon in soil conservation by making it possible to grow crops without tillage. The primary purpose of plowing and cultivating is to control vegetation at planting time and the weeds that come up later. By killing existing vegetation with a herbicide, planting corn in the dead sod, and following with a postemergence selective herbicide, many farmers in hilly country have been able to retire their tillage equipment. At the same time they practically eliminate soil erosion and silting of streams.

Pest control chemicals contribute to a better environment in numerous other ways. Selective control of water-stealing, deep-rooted woody plants in semiarid areas not only improves grazing land but increases stream and spring flow to the benefit of wildlife as well as of man and his livestock. Herbicides have been used to improve the habitat of specific species of wildlife by selectively inhibiting vegetation that competes with their food plants or is otherwise detrimental. An example is the opening up of waterfowl landing strips in cattail thickets. Insecticides also have been used effectively to help control the vast swarms of locusts and caterpillars that erupt periodically and defoliate much of the vegetation in their paths.

CHEMICAL ACCOMPLISHMENTS

Insect Control

Insects and mites are invertebrate animals that predate man and today are found all around the world. Many insects are beneficial. Some are predators or parasites of insect pests. Others produce valuable products such as honey, wax, shellac, silk, and dyes. Insects are also essential for pollinating crops and other plants.

But a few of the thousands of insect species damage man and his environment extensively. They destroy crops and stored food, and they transmit or harbor the organisms that cause dread diseases. The book of Joel describes the ravages of locust swarms:

> That which the palmerworm hath left hath the locust eaten; and
> that which the locust hath left hath the cankerworm eaten; and
> that which the cankerworm hath left hath the caterpillar eaten.

Insects in the U.S. caused average estimated losses of $3.8 billion per year in 1951–60. They cost the state of California alone an estimated $159 million in crops and livestock in 1969, despite control measures that cost an additional $125 million. These losses are at least paralleled overseas, although economic factors there have caused the use of insecticides (and other pesticides) to grow more slowly than in the U.S. In some foreign areas, where hand labor is very cheap, insects are still picked from crops by hand.

Not surprisingly, man through the ages has sought methods of insect control, including chemical methods. Homer (about 1000 B.C.) spoke of "pest averting sulfur fumigation." Succeeding centuries saw the use of arsenic compounds, nicotine, pyrethrum, rotenone, and a variety of other chemicals, both natural and synthetic.

Despite these developments, insect control was still in its infancy before World War II. It was dominated by inorganic compounds, spray oils, phenolics, natural products, and the fly swatter and flypaper. Spray oils and pyrethrums were reasonably effective and safe for a few special uses and are still used today. Most of the chemicals, however, were neither very effective nor safe and could not be used for most of the serious insect problems.

Then, in the late 1930's, came two developments that were to revolutionize insect control. The first was the discovery of the insecticidal powers of DDT. The compound was the first member of the first major class of synthetic organic insecticides, the organochlorines or chlorinated hydrocarbons. The second development was the discovery of the organophosphorus compounds, the second major class of organic insecticides. Development of the third major class, the carbamates, did not begin until the late 1940's.

The synthetic organic materials were far more effective than the inorganics then used. The inorganic calcium arsenate, for example, was applied at roughly 2.5 pounds per acre 10 to 20 times a year to control cotton insects. The organics are applied, typically, at about 10% of that amount. One result of the massive use of inorganics was that their residues in soil built up to levels that are still high enough in some areas to affect the growth of lawns and shrubs in subdivisions that once were farmland. The organic materials avoid this problem.

Organochlorine Insecticides. DDT, the first of the organochlorine insecticides, was synthesized initially by Zeidler in Germany in 1874. Research on mothproofing agents in Germany in the 1920's led ultimately to the discovery of the compound's insecticidal properties. The discovery was made by chemist Paul Müller in 1939 at J. R. Geigy, S.A., in Switzerland. The first samples of DDT reached the U.S. in 1942, and U.S. industry quickly made it the cheapest insecticide available.

DDT has had a fantastic impact on human health (Müller received the Nobel Prize in Medicine and Physiology in 1948). The compound was first used in public health in the early 1940's in Italy and North Africa to control typhus-transmitting lice. The Americans and British used DDT widely during World War II to control the *Anopheles* mosquito, the vector of malaria, in North Africa, the Middle East, and Southeast Asia. The compound also was used to control the mosquito that transmits yellow fever, as well as houseflies, which transmit gastrointestinal diseases. The World Health Organization has estimated that DDT, in only its first eight years of use, prevented about 100 million illnesses and 5 million deaths.

DDT reduced the number of malaria cases in India from an estimated 75 million in 1952 to 100,000 in 1964. In the Soviet Union the number of cases fell from 35 million in 1946 to 13,000 in 1956. In Ceylon, malaria was virtually wiped out by the early 1960's, but then the DDT spray program was discontinued for budgetary reasons. Before the program could be re-established in late 1969, more than 1 million cases of malaria had reappeared in a population of 8 million.

DDT was first used in commercial agriculture in 1940 in Europe to control Colorado potato beetle and then to control insects on cole crops. In the U.S. it was first used commercially in 1945 on grape insects in California and later on vegetable crops.

Benzene hexachloride (BHC), another important organochlorine insecticide, was first synthesized in 1825 by the British chemist Michael Faraday. The compound's insecticidal activity was discovered independently in France and at Imperial Chemical Industries in England about 1942. U.S. industry contributed to the development of BHC by developing improved methods of producing lindane, by far the most active of the six structural forms in which BHC can exist.

Also important among the organochlorine insecticides are the chlorinated cyclodienes, which include chlordane, heptachlor, aldrin, dieldrin, endrin, and toxaphene. The first of these compounds was discovered about 1944 in the U.S. by Julius Hyman at Velsicol Chemical Corp. Other members of the group were developed by Hyman's own company, Julius Hyman & Co., and by Shell Chemical Co., Hercules, Inc., and B. F. Goodrich Co.

The chlorinated cyclodienes were a valuable addition to DDT and BHC. Aldrin and heptachlor, for example, were soon found to control a group of about 20 insects that comprise the soil insect complex in corn. Only 1 to 2 pounds of insecticide per acre were required, whereas DDT required 5 to 10 pounds per acre and BHC did not perform consistently.

The residues of certain organochlorine insecticides are relatively stable to biological and chemical attack, and this stability is the cause of much of their environmental difficulty. DDT, which is very safe to use, persists in the environment for months to years, depending on where it's deposited. It may then be concentrated biologically in the food chain. DDT in a lake, for example, can be taken up by plankton, which are eaten by small fish, which are eaten by lake trout, with the concentration of the DDT residues building up at each step. Humans in the U.S. and elsewhere, as a result of such phenomena, carry in their body fat some 10 to 20 parts per million of residues of DDT and some other organochlorine insecticides.

The use of these compounds has declined in the U.S. for three main reasons: concern over the buildup of residues; the increasing tendency of some insects to develop resistance to the materials; and the advent of insecticides that can replace the organochlorines. U.S. consumption (domestic disappearance) of the aldrin-toxaphene group fell from 76 million pounds in 1960 to 39 million in 1968. Consumption rose sharply in 1969, to almost 90 million pounds, but dropped again in 1970 to 62 million pounds, 18% below the 1960 level. Domestic consumption of DDT fell from 70 million pounds in 1960 to 25.5 million in 1970 (although world-wide demand probably exceeds that for any other single insecticide). U.S. sales of organophosphorus insecticides, on the other hand, rose from 55 million pounds to 73 million in 1965–68, including exports.

The shift away from organochlorines seems certain to increase the cost of insect control. The compounds that are replacing the organochlorines generally cost more, have less residual activity, and often control fewer insect species. Larger dose rates or several applications or both are sometimes needed to do the job. Grasshoppers, for example, could be controlled with ⅛ pound of dieldrin per acre, while 2 pounds per acre are required for the organophosphorus replacement. Costs also are increased by the tendency of the replacement compounds to require more sophisticated handling and application techniques because of higher acute toxicity.

Organophosphorus Insecticides. The organophosphorus compounds, the second of the three major classes of insecticides, grew out of research that began in the early 1930's in Germany at I. G. Farbenindustrie. Close to 50 organophosphorus compounds are registered in the U.S. today as pesticides. They encompass a range of characteristics: toxic and safe, moderately persistent and nonpersistent, specific and broad-spectrum.

Gerhard Schrader, who was long involved in the German work, synthesized the organophosphorus compound TEPP in 1938. It was the first synthetic insecticide to be as active as nicotine. Work on organophosphorus compounds continued during the early 1940's in Germany, the United Kingdom, and the U.S. The goal shifted to toxic war gases (nerve gases), but much was learned that later was turned to advantage in insect control.

After World War II, Allied scientists learned of the German development, parathion, useless as a war gas but a promising broad-spectrum insecticide. The knowledge was exploited in several countries, and by 1948 U.S. companies were making parathion commercially. They began also about then to make methyl parathion. It costs more than parathion but is somewhat safer and proved ultimately to be just as useful an insecticide.

Both parathion and methyl parathion are relatively toxic to mammals. Therefore, the discovery of the much safer malathion by American Cyanamid Co. in the early 1950's was important. This organophosphorus compound has broad insecticidal activity but very low toxicity to mammals.

The organophosphorus compounds also provided, in the systemic members of the class, an important new route of application. The parathions, malathion, and other nonsystemic compounds are applied to surfaces or spaces and kill the insects that touch or ingest them. Systemic compounds are absorbed by the plant through its roots, stem, or leaves and move throughout the plant in its vascular systems. The plant thus becomes temporarily lethal to insects that ingest its juices in the process of feeding on it. The first systemic insecticide was OMPA, discovered by Schrader and his colleagues in Germany. American companies which have taken part in developing systemics include American Cyanamid and Chemagro (now part of Farbenfabriken Bayer).

The organophosphorus insecticides generally degrade rapidly in the environment compared to "hard" insecticides such as the organochlorines. Many organophosphorus compounds, in fact, degrade more rapidly than is desirable for fully effective insect control. This limits the usefulness of such compounds, but they can be used to advantage to control insects that threaten crops just before harvest, while avoiding toxic residues in the harvested crop. A few organophosphorus compounds, on the other hand, are stable enough in soil to provide many weeks of protection against such soil insects as corn rootworm.

Organophosphorus compounds are used widely not only on crops but to control animal pests such as cattle grubs, horn flies, fleas, and ticks. Ronnel, discovered by Dow Chemical Co., was the first systemic animal insecticide. Such compounds enter the animal's circulatory system and are effective against pests at a concentration of about 2 parts per million in the blood. DDVP, discovered by Shell Chemical Co., is used in resin formulations that are hung in rooms for insect control or used as flea collars on pets.

The organophosphorus insecticides are used widely also to control disease vectors, particularly mosquitoes. The compounds used must be safe for humans and include malathion and naled (Chevron Chemical). These two were used in mid-1971 in a vector control program to help combat the spread of Venezuelan equine encephalomyelitis, which had entered this country from Mexico. During a 35-day period the chemicals were applied to 13.5 million acres along the Gulf Coast of Texas and adjacent areas to control mosquitoes and other insect vectors of the disease. Research in the U.S. and abroad lately has produced additional organophosphorus compounds that combine outstanding insect control with remarkably low mammalian toxicity.

Data are scarce on domestic consumption of organophosphorus compounds, but they account currently for an estimated 25% of the poundage of insecticides used, and their share of the total is rising as they replace the organochlorines. As mentioned earlier, total U.S. sales of the organophosphorus materials, including exports, rose 33% in 1965–68, to 73 million pounds. U.S. production in 1970 exceeded 132 million pounds.

Carbamate Insecticides. The first carbamate insecticides, the third of the three major classes, were developed in the late 1940's at J. R. Geigy, S.A., in Switzerland. Geigy scientists made a series of insecticides based on the carbamate structure of the alkaloid *physostigmine,* which occurs naturally in the poisonous Calabar bean. The chemical structure of the alkaloid had been known since 1925.

In the early 1950's, Kolbezen and Metcalf at the University of California, Riverside, established the insecticidal activity of another series of carbamates. Drawing partly on this work, Union Carbide Corp. developed carbaryl (Sevin), the first major commercial carbamate. A variety of carbamates have since been developed by U.S. companies.

The carbamates show a wide range of toxicities to mammals. Furadan, for example, is about 50 times more toxic to rats than is Sevin.

Most of the carbamate insecticides act topically, but a new class introduced recently shows a high degree of systemic activity. Two such compounds are Lannate (du Pont) and Temik (Union Carbide). This class of carbamates appears to be headed for fairly specific uses on crops.

The economic importance of the carbamates is clear in the widespread use of Sevin, whose sales are still growing 5 to 10% per year after 15 years on the market. The uses of the compounds vary from broad-spectrum utility for Sevin on cotton, vegetables, fruits, poultry, and pets to narrow, but still important, soil applications such as for Buxton on corn. Other uses will develop for existing and new carbamates, but they are likely to be aimed at one complex of insects rather than to broad-spectrum activity.

Chemists can build subtle variations in structure into the carbamates to make them more specific than the parent generations of insecticides. The compounds thus offer economically sound means of protecting certain crops in an ecologically safe manner. This factor already is a reality in Japan, where several carbamates help control rice insects but do not harm fish.

Miscellaneous Insecticides. Insecticides other than the organochlorines, organophosphorus compounds, and carbamates are limited in both numbers and amounts used. Such products nevertheless fill important chinks in the wall against pests.

Two classes of natural products used for years as insecticides are the rotenoids (from the derris plant) and the pyrethroids (from a species of chrysanthemum). Two synthetic pyrethroids, cyclethrin and allethrin, have been made commercially, and S. B. Penick & Co. has lately started to make a third one licensed by Britain's National Research and Development Co.

Another naturally-derived insecticide is *Bacillus thuringiensis,* a biological control. It's applied as spores on dusts or in liquid suspension and produces a toxin that kills the insect. Commercially, *B. thuringiensis* has been used mainly against the cabbage looper. Another bacterial insecticide, which causes milky spore disease, has been used commercially against the Japanese beetle.

Certain other insecticides that belong to none of the classes mentioned thus far have been used for many years in specific applications. Among them are materials used against mites, such as Kelthane and Karathane (Rohm & Haas), Morocide and Tedion (FMC Corp.), and Aramite (Uniroyal). Sulfur is still used as a miticide in some citrus and vegetable applications. Both the organophosphorus and carbamate insecticides include compounds that are also active against mites. Mites tend to develop resistance to organophosphorus and organochlorine compounds after prolonged use, but they can then generally be controlled with another type of miticide. Resistance is not yet a problem with the carbamate miticides.

Weed Control

Rapid acceptance of chemical weed control did not begin until the discovery of the herbicides 2,4-D and 2,4,5-T during World War II and their commercial introduction shortly thereafter. The need for chemical weed control was delayed no doubt by the partial control achieved by cultivation, hand hoeing, and other methods. Nor was it widely recognized how seriously weeds can reduce the yield and quality of crops, that they are havens for harmful insects and plant diseases, that certain of them are poisonous to livestock and even to man, and that pollen from weeds causes widespread suffering among allergic humans.

Scientists knew by about 1900 that chemicals such as sodium arsenite kill weeds. Such compounds, however, tended to kill all vegetation, not just the offending species. The new synthetic organic herbicides, on the other hand, were selective. 2,4-D, for example, controls many broadleaf weeds in cereal grains, which can't be cultivated, whether for weed control or otherwise. Even in crops such as corn and cotton, which can be cultivated, chemicals can control weeds at lower cost. Their use, moreover, does not disturb the soil and thus release valuable moisture.

Herbicides may be applied to crops before they emerge from the ground (preemergence application), or after they do so (postemergence application). Some can be used either way. The use of preemergence herbicides accelerated rapidly in the 1950's and 1960's for controlling annual weeds in cotton, corn, soybeans, sorghum, and peanuts, the major agronomic row crops. Much of the U.S. acreage of these row crops is now treated with one selective herbicide or combinations of two or more. Such herbicides are available also for sugarcane, sugar beets, rice, flax, and most vegetable crops. Overall, herbicides are applied to an estimated 122 million acres in the U.S., more than for all other classes of pesticides combined.

The target species for herbicides in the past have been annual weeds rather than perennials. Annuals are the more widespread in cultivated crops, and they are easier to control. Annuals start with the germinating seed, which is quite susceptible to preemergence treatment. Perennials can develop from seeds, too, but once they're established, their underground root systems simply regenerate whatever top growth is destroyed by contact herbicides or other methods. Some herbicides will control certain perennial weeds, but the future is likely to see more emphasis on the development of chemical controls for perennials.

Modern herbicides generally control only grassy weeds or only broadleaf weeds, and they permit certain weeds to escape. For this reason the weed ecology is changing. Some species, such as nutgrass, that once were minor problems in some areas have increased to the point of economic importance. 2,4-D, for example, is credited partly for the buildup of annual grasses in corn, where the herbicide has long been used to control the broadleaf weeds that normally would compete with the grasses.

Partly to ease such problems, scientists in the late 1960's developed combinations of herbicides that utilize the strengths of each. Some such products are now on the market, and more can be expected.

The development of weed control chemicals has required the parallel development of suitable application equipment. Most herbicides are available as either emulsifiable concentrates or wettable powders that can be dispersed in water and sprayed from ground rigs or aircraft. Invert emulsions and formulation thickeners have been developed to control spray drift, especially from aircraft. The more recent granular formulation eliminates the need to handle large volumes of water, although it does not lend itself to all herbicides. Granular herbicides are largely of the preemergence type and are applied directly to the soil. Equipment has been developed to apply granules in bands on row crops, much the preferred method, and they can also be applied broadcast fashion by ground or air equipment.

Herbicides as a class are not without ecological problems, but they are rarely of the type experienced by the organochlorine insecticides. Of the 120-odd herbicides on the market in the U.S., only a few persist longer than one growing season, and they are seldom detected in pesticide monitoring programs. A considerable outcry has arisen over the ecological effects of the military use of herbicides in Vietnam. Quite apart from the merits of using herbicides as a weapon, however, the purpose and methods differ markedly from those characterizing use in domestic agriculture.

Herbicide production in the U.S. grew at an average annual rate of 17% in 1965–69. In 1970, herbicide sales in the U.S. came to 308 million pounds, valued at $498 million. Herbicides in that year accounted for almost 57% of the total sales value for pesticides.

Phenoxy Herbicides. The herbicidal activity of the phenoxy compound 2,4-D, the most widely used of all herbicides, was discovered at Imperial Chemical Industries in England in the early 1940's. Important phenoxy herbicides developed since include 2,4,5-T, silvex, and MCPA.

The phenoxy herbicides have been used on millions of acres of land for a variety of purposes. 2,4-D and MCPA selectively control broadleaf weeds in grasslands and rangelands and in major crops such as wheat, barley, oats, flax, rice, corn, sorghum, and sugarcane. 2,4-D, 2,4,5-T, and silvex have played a leading role in increasing the livestock-carrying capacity of rangelands infested with noxious brush and poisonous plants. The cost of treatment by phenoxy herbicides is often less than $1.00 per acre for the herbicide, and the return to the farmer is very favorable.

Extensive use of 2,4-D, 2,4,5-T, and silvex to control noxious brush has lowered maintenance costs on thousands of miles of rights-of-way for public utilities, highways, and railroads. Herbicides are now an integral part of many right-of-way maintenance programs, where they reduce manual labor and overall costs significantly.

The economy of using 2,4-D arises partly from the fact that it is made from low-cost commodity chemicals: phenol, acetic acid, and chlorine. The same is generally true of other phenoxy herbicides. The influence of raw materials is clear in the popularity of MCPA over 2,4-D in Europe. While both compounds require acetic acid and chlorine, the third raw material for MCPA is cresol from coal distillation, which traditionally has been a more popular chemical building block in Europe than in the U.S. The third raw material for 2,4-D is phenol, synthesized from petroleum hydrocarbons, which leads to the herbicide's greater use in this country.

Widespread use of the phenoxy herbicides seems assured for many years. They show high ratios of benefits to costs and low toxicity to mammals and are degraded relatively rapidly by microorganisms in the soil.

Carbamate Herbicides. The carbamate herbicides, like 2,4-D, began at Imperial Chemical Industries, where the herbicidal properties of IPC were discovered in about 1945. IPC was the first compound to control grassy weeds selectively, and its use spread quickly in the U.S. and abroad. In the U.S., Columbia Southern Chemical Co. (now part of PPG Industries) received a patent in 1954 on CIPC. This chemical relative of IPC was more effective than the latter and was important for many years until pushed aside by better herbicides.

The many carbamate herbicides that have been developed since the first ones include a very selective sort of compound, carbyne, which can distinguish between two species of grasses, wheat and wild oats. Carbyne was discovered at Spencer Chemical Co. It was important because the use of the combine to harvest wheat had intensified the wild oat problem by distributing the seeds in the fields instead of removing them to the thrashing areas. Carbyne is used mainly in the North Central States and Canada.

In 1958, Stauffer Chemical Co. introduced the thiocarbamate herbicides, and Monsanto Co. has since entered the field. These compounds are distinguished by the manner in which their properties can be varied by making small structural changes in the molecule. The highly selective, special-purpose compounds that thus can be made are doing much to broaden the use of herbicides.

The thiocarbamates control many important weeds, both grassy and broadleaf, in corn, sugar beets, rice, potatoes, and other important food crops. They are useful also against nutgrass (nutsedge), an important weed through much of the tropics and in parts of the temperate zones. No chemical control for the sedge had previously been available, and the thiocarbamates themselves have shortcomings in this use. Because of the worldwide problem associated with nutgrass, a need exists for more specific agents to control the weed under diverse agricultural, geographic, and climatic conditions with safety to many crops.

Urea Herbicides. Another milestone in herbicide technology was the introduction of the substituted ureas in the early 1960's by du Pont. A key member of the group, diuron, is used extensively to replace hand labor for weed control in cotton. Diuron thus removed the last obstacle to complete mechanization of cotton growing, which was a boon to the hard-pressed U.S. cotton industry.

New types of substituted urea herbicides continue to be developed, mainly for preemergence weed control in cotton. The compounds are used also, however, to control weeds on industrial sites and in a number of crops. The urea herbicides will remain important for some years, and their use will increase moderately in the next decade. Increasingly they will be integrated with other measures to achieve optimum weed control.

Amide Herbicides. In the mid-1950's, Monsanto Co. developed the first of the amide herbicides. These compounds selectively kill annual grasses and some broadleaf weeds in both grassy and broadleaf crops. They filled an important need in controlling giant foxtail, barnyard grass, and crabgrass, which were serious problems in corn and soybeans in the U.S.

The future is good for the amide herbicides, which are used today on a number of crops in addition to corn and soybeans. They produce up to a four-to-one return on investment and are highly selective, biodegradable in a single growing cycle, and have relatively low toxicity to mammals.

Triazine Herbicides. The triazines are an important group of herbicides discovered by J. R. Geigy, S.A., in Switzerland. The first triazine herbicide was introduced in the U.S. in 1956. The outstanding feature of these compounds is that some of them couple high herbicidal activity with excellent crop selectivity. This property was demonstrated first by Simazine, which gives outstanding weed control in corn without damaging the corn. Biochemists in the U.S. showed later that a natural constituent of corn destroys the herbicide before it can exert its toxic action. Most weeds lack this mechanism for destroying Simazine.

Because of its unusual persistence, Simazine has been largely replaced by a close chemical relative, Atrazine, which has similar herbicidal properties but disappears from the soil by the end of the season in which it is applied. Atrazine today outranks all other herbicides in the U.S. in dollar value of sales and probably is exceeded only by the phenoxies in the amount used.

Dinitroaniline Herbicides. Three dinitroaniline herbicides, trifluralin, nitralin, and benefin, are used widely in agriculture. Trifluralin ranks third in the U.S. in total usage, after the phenoxies and atrazine. Trifluralin, benefin, and related compounds were discovered at Eli Lilly & Co., which introduced trifluralin in 1961 and benefin in 1965. Shell Chemical developed nitralin and introduced it in 1966. The most important features of these herbicides are their high potency, selectivity, and dependability when put in the soil. They are essentially nontoxic to man or animals.

Wide use of the dinitroanilines has virtually eliminated hand hoeing of cotton and hand weeding of a number of vegetable crops. Control of weeds, particularly grass types, in cotton, soybeans, peanuts, safflower, and other crops has increased yields and lowered costs. Additional uses are being developed in orchards, citrus groves, vineyards, and elsewhere.

Other Herbicides. A number of companies have developed herbicides in addition to those already described. None of these compounds enjoys the widespread use of materials such as those previously discussed, but each has characteristics that make it useful in some aspects of weed control. Chloramben (Rorer-Amchem, Inc.) has been used on as many as 9 million acres of soybeans in a single year to control broadleaf weeds and grasses. Dinoseb (Dow Chemical) is used on about 3 million acres annually in the U.S. to control weeds in a number of crops and in orchards. Phytar, MSMA, and DSMA (Ansul Chemical) control Johnsongrass, which resists many other herbicides.

Plant Disease Control

Plant diseases—blights, rusts, mildews—have affected man's food supply for centuries. In Ireland, a potato fungus was a root cause of the great famine of 1845–49. In the 25 years starting in about 1867, coffee rust essentially ended coffee growing in East Asia and made a nation of tea drinkers of Britain, which previously had consumed about as much coffee as tea.

The U.S. today loses an estimated $3.2 billion annually to plant diseases, although the nation spends well over $100 million per year to control them. Better farming practice and various biological approaches may partly reduce the loss, but chemicals certainly will remain essential weapons against plant disease.

Demand for agricultural fungicides in this country is estimated at one quarter to one third of a billion pounds annually. Consumption of synthetic organic materials is roughly 150 million pounds, valued at about $65 million. Farmers continue to use inorganic materials, however. In 1964, for example, they used 137 million pounds of sulfur and 7 million of copper compounds, primarily as fungicides. The inorganics are cheap and well established in some uses, but they will likely continue to decline as research produces better organic materials.

Dithiocarbamate Fungicides. The dithiocarbamate compounds were discovered about the middle of the 19th Century. Subsequent knowledge of their chemistry developed particularly in the rubber industry, where they were useful vulcanizing accelerators and oxidation inhibitors. Tisdale and Williams at du Pont discovered the fungicidal properties of the dithiocarbamates and received a patent in 1934. Use of the compounds grew steadily after about 1940, following the necessary field research.

Among the dithiocarbamate fungicides used today are ferbam and ziram, developed in the 1930's by du Pont. They combine the thiocarbamate molecule with iron (ferbam) or zinc (ziram), an approach tried initially because of the usefulness of certain inorganic fungicides based on metals.

Other thiocarbamate fungicides are based on sodium (nabam), zinc (zineb), and manganese (maneb). The development of these compounds started at Rohm & Haas.

Other Agents. Another important class of fungicides is exemplified by Captan, first developed by Stauffer Chemical Co. and Chevron Chemical Co. These compounds both protect against and eradicate various plant diseases and are used widely to control scabs, rots, mildews, and the like in fruits, vegetables, berries, and ornamentals. They're used at a rate of about 10 million pounds per year in the U.S.

The variety of other organic fungicides used today includes three that act systemically. These compounds, Plantvax, Vitavax, and Benlate, are not well characterized yet in terms of use, but they are able to eradicate plant diseases as opposed to the purely protective action of most other fungicides. The farmer who has only protective fungicides available is willing often to risk a certain level of loss from plant diseases. The advent of systemic eradicants, however, promises capabilities that could change this attitude, with consequent changes in fungicide use patterns.

Antibiotics also are used in plant disease control. Streptomycin, for example, is the agent of choice for controlling fire blight, a serious bacterial disease of pear and apple trees. Griseofulvin is active against certain diseases of vegetables and ornamentals, and cycloheximide is used as a mildewcide in ornamentals.

Nematode Control

Nematodes are a large group of thread-like invertebrate animals that probably surpass insects in numbers of individuals. The relatively few species that attack plants are problems almost throughout agriculture. They attack the roots, reducing yield and quality and, in extreme cases, killing the plant. Nematodes grow more damaging, moreover, as agriculture grows more intensive, and their effects often facilitate attack by other organisms.

U.S. farmers today spend an estimated $60 million annually to treat some 1.7 million acres with nematocidal fumigants. All of the Hawaiian pineapple acreage is treated, for example, as are substantial acreages of southern tobacco, cotton, sugar beets, peanuts, and potatoes. Smaller acreages of a number of other crops are treated also. The dollar return to the farmer varies widely, but appears to average about $100 per acre, or about four times the average investment.

Until the development of modern chemical nematocides in the past 30 years, the farmer fought nematodes with relatively inefficient methods. He could flood or fallow his land, use cover crops, properly rotate crops, and carefully time planting. Carbon disulfide was tried on nematodes in Europe in 1871 but did not work well. Other chemicals were tried in succeeding years, but all were too costly.

Hawaiian growers fumigated pineapple fields with chloropicrin for several years in the mid-1930's. Chloropicrin is effective, but it is costly, highly irritating, and must be applied to the soil under an impermeable paper cover. For these reasons its use today is restricted to a limited number of high-value markets.

The modern nematocide industry began in 1943, when Shell Chemical Co. introduced D-D Soil Fumigant, a mixture of dichloropropene and dichloropropane. D-D, an excellent nematocide, is a by-product of the manufacture of allyl chloride. The latter is used to make synthetic glycerol, and creation of a market for D-D thus improved the overall economics of both markets.

In 1945–55, Dow Chemical introduced EDB, and Shell and Dow introduced DBCP. These compounds and D-D remain the three major soil fumigants used today. A fourth material, methyl bromide, is highly effective in special uses, such as seedbeds, and also has been used for years to fumigate stored and imported commodities for insect control.

The three major fumigants are volatile liquids. They must be applied 8 to 12 inches deep, so that their vapors will remain in the soil long enough to kill the nematodes. Their use thus requires a fairly high level of equipment and technical sophistication, which tends to limit their markets.

Despite the progress made in nematode control, U.S. losses to the organisms total an estimated $1.65 billion per year. The existing products are effective but far from perfect. Considerable incentive exists, therefore, to develop nematocides that combine broad-spectrum activity, season-long effectiveness, relative ease of application, and moderate cost. A new generation of nematocides is appearing that includes organophosphorus and carbamate materials. These compounds are largely by-products of insecticide research. They do not perform outstandingly, but they do indicate the promise of research on nematodes.

Rodent Control

Rodents destroy annually an estimated 3.5% of the world's production of stored grain and cereals, or enough to feed 100 million people. Rats in addition harbor fleas and lice, the vectors of bubonic plague and typhus fever, diseases that still are serious wherever they occur.

The most effective means of eliminating rats and mice from a given area is by poisoning with anticoagulant rodenticides. The leading product is warfarin, developed by the Wisconsin Alumni Research Foundation in the early 1950's. The anticoagulants impair the clotting of blood and cause death by internal bleeding. Their main advantage over acute poisons such as arsenic is that they act slowly, so that rodents are not frightened away from the bait by a few quick deaths.

Birds are little affected by the anticoagulants, and single accidental feedings offer little hazard to man and domestic animals. Several doses are required to kill, and Vitamin K_1 is a good antidote.

The anticoagulants date back to the work of K. P. Link at the University of Wisconsin in 1941. Link and his students isolated dicoumarol from spoiled hay and showed that it causes hemorrhagic disease in cattle. The compound's rodenticidal properties were discovered in the U.S. Department of Agriculture, and the chemically-related but more effective warfarin was developed later. Warfarin and related compounds have been used successfully for 20 years to control rodents. They're more effective on rats than on mice, but populations of both that resist warfarin have turned up in some parts of Europe. The problem is not serious yet but indicates a continuing need for new, effective rodenticides.

A more recently-developed rodenticide is norbormide, developed by McNeil Laboratories and introduced in 1964. This acute poison is unique in being highly toxic to rats but practically harmless to all other animals, including mice. Norbormide causes bait shyness, however, which has limited its utility to a degree.

Plant Growth Regulators

Plant growth regulators are organic compounds that, in small amounts, modify physiological processes in plants. Plants themselves produce substances that regulate their physiological activities. One is auxin or indole acetic acid, discovered in the 1920's. Others that have been chemically isolated include gibberellins, cytokinins, ethylene, and abscissic acid.

Chemists also have synthesized a number of plant growth regulators. Many studies have disclosed few that have commercial utility, and those that do are used relatively little or remain in the laboratory. One problem is that the compounds produce different and unpredictable effects on different plants and thus are difficult to find. Considerable research is under way nevertheless on plant growth regulators because of their great potential value. They might be used to improve a plant's resistance to drought and cold, to control blossoming and sugar content, and to perform other functions that would improve agricultural efficiency. The estimated potential market for such compounds at least equals the market for all existing agricultural chemicals combined.

The indole acids, whose effects on plant growth were recognized in the 1930's, are among the handful of commercial growth regulators. Mainly they promote the rooting of asexually propagated plants. Commercial propagators and many home gardeners use root-inducing substances, but the total sales volume is small.

Another class of commercial growth regulators is the naphthalene acids. Apple trees, for example, are sprayed with naphthaleneacetic acid a few weeks before harvest to prevent apples from dropping prematurely from the trees. In some years, an estimated 1 million pounds of the compound have been sold for this purpose. The herbicide 2,4-D is a plant growth regulator. It has been used to prevent preharvest drop of certain varieties of apples that do not respond to naphthaleneacetic acid.

Maleic hydrazide is a plant growth regulator that's used on more than 80% of U.S. tobacco to control shoots from the roots or lower parts of the plant. TIBA, a benzoic acid, is used to increase yields in soybeans. An aminosuccinamic acid is being developed to bring young fruit trees, particularly apple, to bearing at an earlier age. Chloroethyl phosphonic acid shows promise for stimulating latex production in rubber trees and inducing uniform flowering in pineapples.

FUTURE DEVELOPMENTS

The pesticides industry stands at a critical juncture as it enters the 1970's. It faces changes in patterns of farming, widespread environmental concern, rising costs for research and development, and uncertain administration of increasingly stringent regulations.

To discover and develop a new pesticide today requires $4 million to $10 million and six to nine years or even more. (These figures include the time and money expended on products that fail commercially.) The costs of pesticides research and development have more than tripled in the past 15 years, according to one authority, partly because of the more extensive testing required and partly because of inflation. In the same period, the cost and time required to do toxicological studies on pesticide compounds have climbed from $10,000 and 30 to 90 days to as much as $700,000 and four years or more.

Profits are restricted, moreover, by the nature of patent law combined with the long gestation period for new pesticides. The world market for pesticides is still expanding profitably but is growing steadily more competitive. Because of these factors, many experts feel that research and development on new pesticides already is subject to diminishing returns. A study made in 1971 showed that 33 companies increased their R&D spending on pesticides by 33% in 1967–70, to a total of $69.9 million. The estimated figure for 1971 was $71.6 million, but a 5% allowance for inflation reduces the number to about $68 million, a slight decline from 1970. The companies surveyed accounted for $693 million in pesticides sales in 1969 or 81% of the industry total.

Insecticides. Despite the current problems, the need for effective insect control will always exist. The future of the industry lies in developing cheap insecticides that are highly specific to insects, as opposed to other forms of life, that selectively control undesirable species, and that degrade to ecologically acceptable products within a reasonable time. Among the urgent aims of pesticide manufacturers, for example, is to develop more insect control measures that do not nullify the role of beneficial insects in regulating the populations of many destructive or potentially destructive insects and that do not unduly affect the pollinating insects essential to man's welfare.

The need to develop safer, more selective insecticides conflicts with the increasing difficulty of earning a reasonable return on the required research and development. This situation may dictate a reexamination of existing insecticides. The goal would be to increase the safety and broaden the spectrum of compounds already well studied by developing new combinations and formulations, improving application methods, and by other means.

The immediate future will see strong emphasis on the so-called substitute insecticides that will displace the persistent organochlorine materials in some uses. Research and development probably will expand on the carbamate and organophosphorus materials and on other compounds that act by inhibiting the action of an enzyme, cholinesterase. The persistent compounds, meanwhile, will continue to play a key role in certain uses where long-term control is essential to public health and economical production of food and where short-term emergency control is required.

Eventually the pesticides industry will find and develop more pesticides that can be integrated with biological controls, such as sterilants, sex attractants, pathogens, and insect-resistant plants. Biological controls have limited capability for insect control in most cases at present, and they are not sufficiently well developed to indicate clearly how effective they may become. Consequently, agriculture will continue to depend on pesticides, integrating them with all useful biological, chemical, and cultural methods to a greater extent than at present.

A particularly challenging aspect of chemical insect control lies in isolating and characterizing those substances in biological systems that endow them with natural resistance to insect attack. Similarly, insects themselves secrete growth-controlling compounds such as juvenile hormone that might be mimicked by man and used against them. Research and development of this kind can be exceedingly difficult, and the compounds involved are likely to have bioidiosyncrasies of their own that create toxicological and ecological problems. Such work may open nevertheless a new and exciting phase of insecticide chemistry whose products would move steadily toward the goal of great effectiveness combined with minimum environmental impact.

A considerable amount of research already is under way that indicates more specifically the directions that insect control might take. In April 1972, Zoecon Corp. reported successful control of mosquitoes in large-scale field tests of an experimental juvenile hormone-type substance with a very good margin of safety for man, fish, and nontarget organisms. In 1971, a cotton belt–wide field trial showed good promise for control of the bollworm by a product based on *Heliothis* virus and developed by International Minerals and Chemical Corp. Improved production procedures have yielded new and more active strains of *Bacillus thuringiensis*, a bacterial pathogen that is toxic to many insects. The new strains already are in commercial use. The U.S. Department of Agriculture, meanwhile, is undertaking a number of activities designed to find more integrated methods of pest control.

Herbicides. Herbicides have not experienced quite the degree of environmental difficulty as insecticides. They are subject to the same stringent government requirements on safety and efficacy, however, and must go through the same lengthy and expensive development period imposed on other pesticides. But of all classes of pesticides, herbicides appear to have the most encouraging future.

The highly selective herbicides developed over the past decade have shown remarkable ability to control broadleaf weeds in grasses and to control grasses in other crops, including corn, itself a member of the grass family. This high selectivity, however, has suppressed the natural plant competitors of certain weeds, which thus have become more troublesome than in the past. The development of specific herbicides to control such weeds, which include Johnsongrass and often nutsedge, probably will occupy herbicide chemists for some time to come.

Increasing attention will be directed toward determining the stage of crop growth at which weed control is most essential and the level of weed infestation that demands control. A good deal is known in this area already, but significant further progress is possible. The possibility of inhibiting the growth and reproduction of weeds without necessarily killing them is another approach to be considered in the future.

A great deal remains to be learned at the molecular level about how herbicides kill weeds and undesirable vegetation. Once the mode of action of herbicidally-active compounds is more clearly understood, the synthesis of molecules with more specific and desirable activity will become possible. This is the case, for example, with the organophosphorus insecticides that inhibit the activity of cholinesterase. Random screening of compounds for herbicidal activity, however, will very likely continue. It is not the most significant cost in developing new products, and almost 20 years of research, moreover, have identified very few chemical reactions as the key reaction in herbicidal activity.

At relatively high concentrations, many herbicides inhibit certain classical chemical reactions. Unfortunately, these reactions are not significantly unique nor sensitive enough to explain the selective toxicity that the same herbicides display in minute concentrations toward the species of weeds they affect. Most research on new compounds consequently will be directed toward inhibiting those processes that are unique to green plants, such as photosynthesis. One advantage of this approach is that herbicides that inhibit processes common to most plants generally, though not always, exhibit low toxicity to mammals.

Some work is in progress on integrated control of weeds. In Florida, for example, two insect enemies of the alligatorweed have been established. In West Virginia, a weevil native to parts of Germany and France has been introduced to reduce stands of the musk thistle.

Plant Disease Control. Success to date in combating diseases of plants has been achieved almost exclusively by protective measures—by attacking the parasite before it enters the plant. The future development of protectant fungicides will depend on solving specific plant-disease problems.

The current concern over mercury in the environment may well bring an end to the use of organic mercury fungicides in this country, although agricultural fungicides account for only about 3.4% of the total mercury used in the U.S. The amount used in seed dressings is probably less than 1%.

The future of fungicides as a whole will depend on how successfully science can exploit certain concepts conceived around 1940. Generally these embrace the idea that systemic materials can reach a fungus that is not influenced by the surface protectants and might be useful against subsurface pathogens. "Systemic" fungicides include compounds that are not necessarily toxic to fungi but that, when taken up by the plant, protect it from fungal attack in a way that indicates generally that they increase the plant's resistance. Such a substance might, for example, inactivate the fungus's weapons of attack.

One such weapon is the pectolytic enzymes that fungi use in most cases, if not all, to destroy the pectin in the middle lamella of the plant cell. The fungus must destroy the pectin to enter the parenchymatic tissue, where plant growth takes place. This suggests that greater knowledge of specific inhibitors of pectolytic enzymes might be highly significant to the development of systemic fungicides.

Before this concept of plant disease control can be developed much further, however, it will be essential to understand the complex biochemical interactions that occur between plant and parasite. It is difficult today to design compounds that inhibit specific metabolic processes in fungi but that are not toxic to plants. One possibility lies in the fact that the invading fungi grow much faster than the invaded plant. It might thus be feasible to attack the fungus when it is in a particularly sensitive stage of growth and the plant is not.

SOCIAL AND ECONOMIC IMPACTS

At one time, American families for the most part processed their own food. Today, an entirely separate industry known as the food and kindred products industry does it for them. Food preparation and consumption differ markedly as a result, and the food supply is much more healthful and varied than at any time in the past. Much of the work formerly done in every kitchen has been moved to central factories, and many different foods are available year-round. Families spend much smaller parts of their incomes for food than they once did, and homemakers spend many fewer hours preparing meals.

The U.S. food industry sells about $105 billion worth of products a year. Processing these items for 205 million Americans and nearly 30 million people in other countries provides employment for about 1.8 million people. The food industry operates about 33,000 plants, and about 150 million tons of its products are sold annually. It now invests more than $2.5 billion in new plants and equipment and nearly $200 million in research and development each year.

Of all manufacturing industries, the food industry ranks among the most complex and involved. Its raw materials vary widely both as to type and within each type, and they frequently grow many miles from where they will be used. Many of these materials have very desirable but often very fragile properties that the industry must preserve, sometimes for long periods, before consumers use them. Its products must keep people healthy when eaten in proper balance, must be uncontaminated when used, and must be consumable safely day after day by people whose constitutions may differ. The less work its products require once bought the better, and the more they provide needed nutrients and simple eating pleasure the greater the possibility for economic success.

As is true in many industries, scientists and engineers trained in many disciplines have helped find the facts and discover the principles the food industry uses today. Chemists and chemical engineers must be credited with much of the industry's progress, however, as their accomplishments underlie much of today's food technology. They have studied foods at the molecular level, for example, and we now know a great deal about proteins, carbohydrates, fats, vitamins, and minerals and how they relate to nutrition and health. They have also determined the chemical and physical properties of many food components, and we can now build some foods almost to order, interchanging raw materials, depending less on single sources of supply, and using ingredients in ways current market economics dictate.

Work based on chemistry has led to the use of many special chemicals that preserve, emulsify, flavor, color, leaven, acidulate, and the like. Traditional food ingredients from harvested crops and slaughtered animals seldom limit options for improvements. Amino acids, vitamins, and other chemicals made from nonfood raw materials have for some years supplemented nutrients of plant and animal origin. Prospects now exist that complete proteins, carbohydrates, and fats may eventually be made on relatively large scales by nonagricultural chemical processes. If successful, such processes may produce foods that will complement in an important way the food now produced by the biosphere.

Of all the innovations the food industry has recorded to improve its products and processes, those dealing with food preservation doubtlessly have had the greatest impact on society. Food spoils quickly, and people have long tried to preserve it and depend less on freshly harvested crops and newly killed animals. Seeking to tide themselves over from days of plenty to days of scarcity, they salted or smoked meat, parched cereals, dried vegetables and fruits, and fermented milk. In more recent times, the opportunity to feed growing numbers of people leaving farms and settling in cities plus the demands of national security and exploration led to efforts to improve on the old ways. Large industry sectors based on canning, refrigeration, and dehydration are the result. Today, illnesses and deaths from contaminated or spoiled foods almost never occur even when we eat foods months after harvest. In addition, chemicals that retard growth of microorganisms or rates of degradative reactions reduce spoilage sharply. Without them, there would be a high rate of consumer dissatisfaction because of rancidity and other forms of chemical spoilage. Also, many foods would lose some nutritive value, since oxidation of fats and oils may destroy some vitamins. Moreover, changes in color and texture would make many products unattractive.

Lacking solutions to such problems, food manufacturers would have to take remedial steps, such as removing some food items from the market, developing new packaging methods, and using raw materials less subject to oxidation or bacterial attack. These extra efforts might often lead to a less nutritious food supply and to higher prices for many foods. Consumers might even face a more limited selection of higher priced items.

While preservation levels out agricultural cycles and makes more foods available to more people everywhere, convenience lessens the labor required to prepare them. Homemakers a century ago started with simple raw materials and cooked them or used them to compound their own finished foods. To feed their families, they worked many long hours, but leisure time for homemakers has become so accepted today that no one finds it remarkable.

Cereals, for example, provide much of our protein, but no one can eat and enjoy them as harvested. Today, new processes provide cereals in a wide variety of forms that either require no cooking in dry, ready-to-eat-cold forms or a minimum of cooking in instant forms. Machines in factories do the work, and people have a more varied diet as well.

Baking bread was once a weekly requirement for homemakers who gave their families more than the simplest of diets. Today, aided by many processing chemicals, central plants bake bread and distribute it fresh daily to thousands of outlets. Meanwhile, at one time only the truly expert cook could bake a cake successfully. Even casual cooks can now do so by using premixed formulations that require at most the addition of an egg and milk or water, some brief mixing, and a hot oven.

The list of conveniences goes on and on. Beverages of many types and flavors have replaced homemade root beer and sassafras tea. Cheeses come in many natural and processed forms. Gelatins, puddings, and other desserts are prepared instantly or eaten directly from plastic or metal containers. Spreads, seasonings, and salad dressings exist for any palate. Instant coffees and teas eliminate brewing and prepared toppings eliminate whipping. Milk with controlled fat content and cottage cheese and other cultured milk products aid the diet-conscious.

The food industry has long known that the most successful foods are those that retain most of the appearance, aroma, flavor, and texture of freshly prepared forms. Preserved foods might taste good, they might be easy to prepare, and they might keep people healthy. If they don't have much farm-fresh quality, however, they are often judged inferior and threatened with oblivion.

Canners accordingly strove for years to match the qualities of freshly cooked foods. They eventually found that the original long cooks destroyed flavors and changed textures, two factors that greatly affect acceptance. They turned to shorter cooks at higher temperatures to get the same microbial kill yet keep as much of the character of properly prepared foods as possible.

If heat needed to preserve foods drives off aromas, changes flavors, and ruins textures, one might naturally wonder about using low temperatures instead. As it happens, cold (36° F.) and frozen (0° F.) storage does keep food from spoiling for long periods. Moreover—and more important—such foods retain much of their farm-fresh character when warmed and cooked. As a result, a new industry has grown around the concept.

First, refrigeration held foods in cold storage to help the industry move products from farms to distant cities and have them arrive fresh. Next, the industry turned to frozen storage to lengthen storage times, level out processing peaks, and increase shipping distances. But the real innovation occurred when perishable foods were frozen in individual packages ready for retail sale. Following that step, every household became potentially a miniature depot for frozen foods. Each could have vegetables, meats, baked goods, and fruit juices throughout the year and with a minimum of effort. From a very small beginning in northeastern states in 1930, the retail frozen foods industry as a result has grown in 40 years to sales of nearly $5.8 billion.

As noted earlier, chemists and other scientists have determined the chemical and physical properties of many food components. New processes and chemical modifiers now often permit them to exchange one raw material for another. Nowhere among broad categories of foods and food components are their advances more striking or the results more significant to the economy than in edible fats.

Fat is one of the three major components of the diet (proteins and carbohydrates are the others), and Americans ate about 11.5 billion pounds of fats in 1970. At one time, most edible fats came from animal sources, but today most come from vegetable sources. Chemical innovations have for the most part made the shift possible, and the impacts on American agriculture and food processing have been marked. Farmers have gained new sources of income, for example, processors have grown and diversified, and consumers have varied their diets even more.

All of these advances in preservation, convenience, and quality occurred as people were leaving farms and settling in cities. The food industry didn't make the U.S. urbanize because of its developments, to be sure, but without doubt the country could not have done so without those developments.

In colonial days, nine out of 10 Americans who worked did so on farms. They therefore fed and clothed themselves and only one other person. By the time of the Civil War, each American farmer produced enough to feed and clothe himself and three others. By 1900, the ratio had risen to one farmer and five others.

While relatively fewer farmers in 1900 fed relatively more city dwellers, no one could claim to have an extensive menu to choose from. Food available for a typical turn-of-the-century midwinter meal consisted of pork or beef, potatoes, turnips, coleslaw, pasta, fried apples, and stewed tomatoes. For dessert, families could choose among Indian pudding (corn-meal, milk, and molasses), rice, cake, and pie. To supply his family with this type of diet, the average wage earner had to spend about 40% of his income.

The situation differs today, to say the least. One farmer now produces enough food and fiber to supply himself and 42 others, 37 here and five in other countries. The food industry now often converts raw agricultural commodities into foods the same day they are harvested and preserves them for use many months later and often many miles away. Yet with all the variety and convenience, the average wage earner now spends less than 20% of his aftertax income on food.

If chemists and other scientists hadn't changed how we produce, process, and distribute food, our first concern in the U.S. undoubtedly would be where do we get our next meal, as it is in so many other parts of the world. With inadequate food supplies, many problems of nutritional deficiencies, disease, and death would still plague us, as they did for so many centuries and as they still plague millions in many parts of the world. Moreover, people no longer needed as farmers have served as the labor force for expanding manufacturing and service industries; for education; for medical, public health, and legal services; and for the arts and sciences. Progress in these activities, some of which we record in other chapters of this volume, otherwise would have been meager indeed without that source of labor.

CHEMICAL ACCOMPLISHMENTS

Canned Foods

We now know that heat kills or inactivates the microorganisms and enzymes that cause illnesses and spoil food. The canning industry existed for half a century, however, before Louis Pasteur proved the point. Today, we may at times joke about housewives who cook with can openers, but the fact remains that ubiquitous cans (and more recently ubiquitous glass bottles) have probably carried more food and drink from processor to consumer than any other containers. In recent years, sales of foods and beverages in cans, jars, and bottles have been growing nearly 4.5% a year, and they amounted to $16 billion in 1970.

The canning industry dates from the work of Nicolas Appert, a Parisian confectioner, who showed in 1810 that food would keep if heated in corked bottles. Another early pioneer was Peter Durand, who filed a patent in England (also in 1810) citing a vessel made of tin or other suitable material for canning foods. Manufacturers soon began making tin cans commercially, and canned foods were shortly in use, especially by explorers and military personnel. Commercial canning began in the U.S. in 1820, and the first patent on a tin container was issued in the U.S. in 1825.

While canning became commercially established in Europe and America in the early 1800's, it lacked a scientific basis until 1873 when the French chemist Louis Pasteur found that bacteria caused spoilage. His fundamental finding probably contributed more to the growth of the canning industry in later years than any other single one.

Many others followed Pasteur to relate cause to effect and establish the relation between times and processing temperatures that make foods safe to eat. Once it had established basic processing principles, the canning industry became concerned largely with improving processes to yield products that retained the nutrition and flavor typical of freshly cooked foods and with improving containers to provide greater convenience for consumers. Work along these lines is still under way today.

In commercial canning as originally developed, the processor filled the can, sealed it, and then cooked the food in the can. Processors at first boiled the cans in water, and heat processing took four to five hours. In 1861, calcium chloride added to the water raised the boiling point from 212° to 240° F. and shortened processing times to 25 to 40 minutes. Production in a typical cannery rose from 2,500 cans per day to 20,000 cans per day. In 1874, the steam pressure retort was invented, raising processing temperatures and reducing times once again. A still later step to improve flavor and nutrition by reducing overcooking was to shorten autoclaving or retort cooking cycles by agitating the can.

An improvement more basic than any of these, however, has been to package food aseptically. In this method, the processor fills a pre-sterilized container with cooked and sterile food and then seals the container in a sterile atmosphere. The food thus needs no heating after canning to sterilize it or to inactivate its enzymes. Furthermore, since the food is cooked for shorter times (but at higher temperatures) than in conventional canning, its flavor, color, texture, and nutrition are improved compared to foods processed at conventional canning temperatures.

The process is actually a commercial adaptation of the old-fashioned home canning process, and it was invented at James Dole Engineering Co. in 1948. Many food companies have since adopted the system, but most food is still canned in the conventional way. Chemists, engineers, and food formulators have solved many of the technical problems in aseptic canning related to food quality, but they have yet to develop the high-volume and low-cost process that will compete fully with conventional canning. Some new aseptic systems that make containers in-line offer cost savings along with product advantages, and it is in this direction that the potential of aseptic canning may be fully realized.

Dehydrated Foods

Of all the ways to preserve foods, canning and refrigeration rank first in volume and in value. Dehydration, however, is by far the oldest. It is also growing fast. More than 100 companies are major dehydrators. They plus many smaller companies process and package some 30 billion pounds of dehydrated foods in the U.S. each year, and supermarket shelves hold a wide variety of dehydrated foods that have helped change marketing and eating habits. Coffee, tea, milk, eggs, fruits, meat, vegetables, and fruit juices are among foods and beverages that are dried. What were once largely materials handling industries often depend heavily on chemical engineering today.

Microorganisms and enzymes that spoil food thrive in moisture, but they may die or at least become much less active when there is little or no moisture. Other reactions that harm food also proceed at negligible rates or not at all when moisture is low or absent.

People from the earliest times dried meats, fruits, vegetables, and cereals and empirically developed much of the technology of drying. No one necessarily dehydrates a food successfully simply by removing water, however, since drying may harm a food's chemical and physical properties. Moreover, some chemical reactions that deteriorate a food, such as fat oxidation and nonenzymatic browning, can occur slowly even with reduced water contents. Progress in dehydrating foods, therefore, especially recent progress, has come largely through the efforts of chemists and other scientists to overcome problems that dehydration may introduce. Only in the past 200 years have scientists and engineers studied artificial drying very much, and only in this century have they developed methods that improved on primitive air and smoke drying in more than small steps. Likewise, only in this century have they studied in a concerted way the chemistry, physics, and microbiology of the drying process.

A British patent issued in 1780 for treating vegetables in hot water and then holding them under drying conditions was one of the first moves toward today's modern industry. Wartime supply problems dominated most of the work from that time until after World War II, and many of the advances in the field lie hidden as anonymous contributions from government research agencies.

A notable exception to the military motivation for drying research was the development in 1890 of the "Oregon tunnel" for prune manufacturers of Oregon and Washington. With this dryer, prune processors depended less on Pacific Northwest weather, and when California processors adopted the dryer it became the ancestor of present-day tunnel dryers.

At about that time, other novel drying methods were first being investigated as solutions to specific problems. Most of them, however, did not come into extensive use until World War II or after. Since then, several have attained wide use within the food industry. Although dehydrators combine processes and products in many different ways, typical combinations include vacuum-drying fruit and coffee and spray-drying milk, eggs, and coffee. They also freeze-dry meat, fruit, coffee, and vegetables and fluid bed– and drum-dry mashed potatoes. Fruit juices are often foam mat–dried.

Dehydrated foods often turn an undesirable brown because of enzymatic and nonenzymatic reactions. Both reactions have been the subject of much research, and both are controllable today.

Blanching inactivates enzymes. It was first mentioned in 1780, but processors apparently did not use it widely until after 1917, when research in British and American government laboratories showed its utility.

As for nonenzymatic browning, its chemistry is very complex. One of the more important routes to making brown pigments is a reaction of aldoses and amino acids that produces aldosylamines. These aldosylamines rearrange to ketoseamines that then decompose to deoxyosones, reductones, and other reactive chemicals. By a variety of routes these products then react further to produce the brown pigments so objectionable in dehydrated foods.

One way to interrupt this reaction chain is by adding sulfite. Sulfiting has been known from antiquity, but it was not until after World War II that its action began to be understood, even in part. Even now the action of sulfite is incompletely delineated, but it apparently interferes with ketoseamine decomposition. A lag phase is induced, after which browning resumes when sufficient sulfite has been consumed.

Frozen Foods

The idea of preserving foods for retail marketing by freezing them and then distributing them frozen probably stems from a trip Clarence Birdseye made to Labrador in 1915. He noted there how people used the area's climate to their advantage to preserve foods. He made no immediate use of the observation, but seven years later while working in New York he recalled his Labrador experience and began developing a belt froster to process foods. He subsequently lay no claim to having invented the quick-freezing process, but he did refine it and make it workable as a marketing technique. Today, all types of vegetables, many meats, some baked goods and juices, and even complete meals are sold frozen. They give flavor and freshness at prices that don't fluctuate with the seasons and in convenient forms that require much less labor in the home.

Much of the success of the frozen foods industry lies in developments in processing and distribution equipment, first to freeze the crop at harvest and then to transport, warehouse, and otherwise distribute it to the point of sale, all the time holding the food at 0° F. or below. Thus, food scientists and engineers developed portable freezers that could be moved from plant to plant to follow the harvest, multiplate freezers that increased production and put the industry into year-round production, and super-insulated railroad cars and trucks to transport foods from plants to warehouses and stores.

The industry also has depended on other developments besides those in processing and distributing equipment. The idea of quick-freezing foods is relatively simple in concept, but many scientists had to gather many data on times, temperatures, freezing rates, and heat transfer methods before fully satisfactory results were obtained. In addition, they had to answer some critical questions about microbial contaminants and spoilage before the industry could consider itself on a sound footing.

Vegetables, for example, were initially frozen fresh (that is, as picked). It soon became obvious that enzymes quickly spoiled the vegetables even when they were held at 0° F. Chemists found that a short, controlled heat treatment inactivated these enzymes and led to the now commonly used blanching procedure. Chemists also developed catalase and peroxidase enzyme tests to determine negative enzyme levels and storability of frozen foods.

During the early days, no one had information on the behavior in frozen foods of *Clostridium botulinum*, which is a recognized hazard in improperly canned foods. Food chemists were thus much concerned about the action of bacteria in frozen foods, especially if the foods were to be inadvertently or unknowingly exposed to temperatures other than the ideal of 0° F. during distribution and storage. As a result, frozen foods were very carefully tested to be sure this organism would not be a problem. As it developed, frozen foods never have been sterile, and no adverse results have occurred.

The industry has found that besides the expected characteristics of convenience, broad appeal, and quality equal to or better than that of similar products marketed in other forms, frozen foods also seem to be subject to a market readiness factor. In the early days, for example, meats were included in company product lines, but they were largely abandoned in the mid-1930's. Today they are making a comeback. Also, frozen baked goods of good quality were produced in the early periods, but they did not become popular until the 1960's. Properly packaged baked goods have a shelf life of several months compared to only a few days for unfrozen products, and frozen baked goods of excellent quality are now retailed widely as a result.

Meats

Americans today eat more than 35 billion pounds of meat a year. They make their buying decisions expecting tenderness, juiciness, and full flavor for the species involved. Of these variables, they generally rate tenderness most important. As a result, chemists have devoted much time in recent years to studying how to measure it, how meat composition and commercial handling affect it, and how to increase it.

Consumers also base buying decisions on a meat cut's color, since they associate freshness with the bright red of newly slaughtered meat. Studying the chemistry of meat color and how to retain the freshly-killed appearance at the point of sale, therefore, has also absorbed much research effort in the industry.

Part of the research on meat tenderness has involved delineating the chemistry of rigor mortis. Meat is very tender before rigor mortis sets in and very tough during it. Then the toughness gradually declines as the meat ages, but it rarely reaches the prerigor state of tenderness.

As a result of their findings, chemists in the meat industry now believe that the same complex biochemical reactions by which muscles in living animals contract and relax also probably lead to rigor mortis following death. In living animals, the presence of oxygen facilitates the reactions that reverse muscle contractions. In dead animals, however, with no oxygen available, these reactions are limited. Muscles contract, and only after a period of time do they relax.

So far, the meat industry hasn't succeeded in applying these findings commercially, and the research continues. When (or if) industry scientists find reagents that control muscle reactions themselves or their rates and that are also safe for use in foods, they may be able to produce consistently tender meats. Such meats will also lose less fluid when frozen, thawed, cooked, or canned.

Meantime, the industry has made some progress in restoring some of the tenderness meats lose following slaughter. Enzymes that are present naturally act to tenderize meat proteins and muscle fibers as they age. Recognizing this fact, scientists at Swift & Co. patented a tenderizing process in 1959 that complemented these natural enzymes with papain. This enzyme derived from the papaya plant had been approved and used in homes and institutions for some years as a tenderizer. In Swift's Pre-Tendering process, slaughterhouse workers inject a carefully standardized enzyme solution directly into the live animal's vascular system a few minutes before slaughter in an amount determined by the animal's weight, age, grade, and sex. In the short interval before slaughter, the blood carries the enzyme to all meat tissues. Because distribution occurs uniformly and completely, a very small amount of enzyme produces the desired tenderness.

Swift and Armour have also developed other methods to treat meat with tenderizing enzymes. In these processes, workers inject an enzyme solution through multiple needles into carcass meat or meat cuts. Solution pressure forces the enzyme to penetrate the tissues. Alternatively, nitrogen gas injected under pressure opens the meat structure and the solution is then injected.

Such treatments eliminate the special handling called for by various aging processes. Because pretendered beef doesn't require aging to become tender, it remains bright red and appears fresh. Furthermore, such beef has its full, natural flavor, and its keeping qualities aren't affected in either market display cases or refrigerators and freezers. Housewives may use their regular cooking methods (although some cuts cook in less time than comparable cuts of nontendered beef).

Since a good red color represents to a meat retailer the difference between a salable and unsalable package, meat industry chemists have for some years worked to retain the bright red color of freshly killed meat. As with rigor mortis, they have found that complex biochemical reactions are involved. These reactions are associated with the blood's transport of oxygen. A meat changes from bright red to dull gray-brown when the partial pressure of oxygen at the meat-air interface is less than that of the atmosphere but more than pressures near zero. The meat industry now takes advantage of this finding by avoiding intermediate partial pressures of oxygen in handling meat from slaughterhouse to retail counter. It either wraps meat with films that have high diffusion rates for oxygen, or it packages meat under vacuum with films that allow virtually no oxygen to pass. Satisfied retailers and consumers are the result, as is less waste from apparent spoilage.

Electrostatic deposition of smoke on hams, used to a limited extent since the mid-1950's, eliminates the need to heat smokehouses, since the hams are cooked after canning. In the early 1960's, a "liquid smoke" became available to add a smoked flavor to meats not usually smoked. It now may be used on any meat. The result is a wider range of meats with the desirable smoked flavor, better quality control, and reduced processing costs. Sausagemakers can produce sausages with better texture now that they have learned more about the capacity of meat to bind water. Knowledge about water binding also has given the meat industry better control of juice loss in packages and cans, drip release on freezing or thawing, and moisture loss during cooking.

Ready-to-Eat Cereals

In few sectors of the food industry has the trend to convenience and diversity made more of an impact than in breakfast foods. Today, supermarkets stock some five to six dozen breakfast cereals that grew out of pioneering work in the U.S. around the turn of the century, and Americans in 1970 spent more than $970 million for nearly 2 billion pounds of hot, cold, and instant cereals and cornmeal and hominy grits.

Cereal grains have traditionally provided much of man's protein. They supply nearly 40% of the protein to Americans and about 70% worldwide. People don't find raw cereal grains palatable, however, and few eat them as harvested. At one time, only laborious grinding and extensive cooking prepared cereals for eating, such as in breads, gruel-like mushes, and fermented beverages. Frontier and rural Americans accordingly emphasized meats and starches in their diets. They boiled or baked rice, wheat, oats, and corn into porridges or breads, and they fried much of their food. Technology for making ready-to-eat cereals had not yet been developed, and primitive kitchens discouraged people from cooking their own. Reflecting the trend from rural to urban life styles, Americans by the end of the 19th Century were more than ready to shift to a lighter and more easily digested diet. There is little wonder that a new industry succeeded as few had succeeded before.

The first of the breakfast cereals so familiar today, shredded wheat, was invented in 1893, and it was quickly followed by corn flakes (in 1894) and Grape Nuts (in 1897). Each was invented by a devotee who preached health through diet. Corn flakes and Grape Nuts especially were used first as cures and later as preventives in health spas in Battle Creek, Mich. It turned out, however, that the time for ready-to-eat cereals had come, and they were soon being marketed as a new class of foods.

Following the original round of breakfast cereals at the turn of the century, the industry introduced a number of variations in the next 50 years based on different grains and using a variety of cooking, forming, toasting, and puffing processes. In the early 1950's, the presweetened cereal came into being as a new concept. Shortly after that, the industry returned to the concept of its inception by promoting cereals as health foods. Cereals high in protein for low calorie diets arrived in the mid-

1950's, followed in the early 1960's by cereals high in vitamins to meet minimum daily requirements for these essential nutrients.

Presweetened and nutritionally fortified cereals are the two growth components of the cereal market, which apparently has plateaued out. Thousands of permutations are still possible, however, depending on the processing steps and the variations within each step selected. Moreover, the technology of ready-to-eat cereals may lead to new formulated foods, not only for Americans but for people in other parts of the world whose diets aren't so complete or so varied.

There's more to processing cereals than providing convenience, of course. Once a grain's outer coating is torn or cut or the embryo in it damaged by crushing, for example, the grain deteriorates rapidly. Insect eggs that lie dormant on the bran coating hatch as the food becomes available to the larvae. Fat-splitting enzymes are released and hydrolyze the grain fats. Oxidation turns the fats rancid.

Thus, ground whole grain has a short shelf life and must be converted into end uses immediately to avoid deterioration. Grain processors have accumulated considerable information on how to alter grain kernels so they won't deteriorate, and the importance of their knowledge can hardly be overemphasized. Underdeveloped countries where such knowledge isn't put to use may lose up to 25% of their grain crops to spoilage and another 25% to insects and rodents. The loss in the U.S., by contrast, is about 12%.

Some beneficial vitamins and minerals are lost as parts of the bran and germ components are removed to make flour more stable. Food processors in the U.S. today, therefore, generally fortify ready-to-eat cereals and some other cereal products with vitamins and minerals to raise their nutritional quality back to that of whole grain. Doing so, however, may also cause problems. Some vitamins and minerals are difficult to add to processed foods without altering the palatability or texture of foods. Other vitamins are reactive chemicals and are difficult to stabilize in processed foods to prevent their being lost during storage. Food scientists thus have had to devise formulations and processes quite carefully to minimize such problems.

Edible Oils

Fats provide essential nutrition and also simple eating pleasure. We get them from plants, animals, and fish. While some may be eaten directly, most require processing to remove undesirable colors, flavors, and odors and also to change their chemical and physical properties to adapt them to a wide range of uses. Included are such processes as refining, bleaching, deodorizing, winterizing, fractionating, hydrogenating, and interesterifying.

Before about 1900, the edible fats and oils industry tended to treat its products with fairly simple processes that were not too sophisticated by today's standards. To refine oils, for example, mills treated them with caustic. Free fatty acids were removed as soaps, and other impurities settled and were drawn off. Bleaching was often an integral part of refining. Processors in the early days exposed oils to sunlight, heated them with steam coils, or blew air or steam through them. Operations were carried out in batches, always a slow process, and the oxidizing treatments harmed oil flavor considerably.

Continuous centrifugal refining was one of the first major technological advances in refining. It has proved to be of major economic importance and was developed in the U.S. in the early 1930's. It made oil refining automatic rather than dependent on skill and judgment, and U.S. refineries have adopted it almost universally. More than likely the U.S. could not refine the volume of oil it now does with batch methods and current labor costs. By using continuous refining, companies have held prices in line in a way that would not have been possible otherwise.

Before an oil can be processed, of course, it must be extracted. Here, too, chemical technology has had an impact. At first, extractors mechanically pressed batches of oil-bearing materials that they usually had pretreated (cooked) to render the oil more readily available. These treatments, including the high temperature developed by mechanical pressure, degraded both oil and meal (an important protein product). Also, mechanical pressing left some oil behind, economically important when dealing with an oil source that had a relatively low oil content to begin with.

A major advance in extraction in the U.S. occurred in the mid-1930's when Archer-Daniels-Midland, Glidden (now part of S.C.M. Corp.), and Central Soya adapted to American conditions the solvent extraction processes and equipment originally developed in Germany. Solvent extraction is used profitably now on many oilseed crops, but it is uniquely responsible for the tremendous expansion of the U.S. soybean industry. By increasing soybean oil yield and quality at economical costs, it opened a market not only for a new cash crop but for a new cash crop during the Depression. Farmers reacted swiftly. In 1930, they planted only 1.1 million acres of soybeans. By 1935, soybean acreage had risen to 2.9 million, and it has increased every year since, reaching some 42.4 million acres by 1970.

It is difficult to assess all the impacts that the growth of the soybean industry has had on the economy. Nearly 14 million of the 16 million tons of oilseed meal fed to livestock in 1969 came from soybeans, however. In addition, the cash value of soybeans at the farm level in 1970 amounted to about $3.2 billion, and soybeans now rank second (to corn at $5.5 billion) as a crop in the U.S. They would never have reached that position without effective and economical processing methods worked out by numerous chemists and chemical engineers, at least not with the prices that prevail today.

Extraction, refining, and related processes may have helped change the edible oils industry, but none has changed it nearly as much as has hydrogenation. Hydrogenation is responsible for growth of the shortening and margarine industries. It is also responsible for increasing the value of cottonseed oil and for development of the huge soybean industry. Moreover, hydrogenation has had a most decisive economic impact by making oil and fat resources substantially interchangeable. As a result, plants have largely supplanted animals as sources of fats in the U.S.

Being highly unsaturated compared to animal fats, most vegetable oils exist as liquids at normal temperatures. The unsaturation makes their flavor less stable, since the fatty acids in unsaturated oils oxidize more rapidly than their saturated counterparts do. Hydrogenation lessens the amount of unsaturation and also hardens the oils. The resulting oils look better, taste better, and have more stable flavors because they react less with oxygen. By controlling how far hydrogenation proceeds, the edible oils industry can produce oils with different properties and adapt them to many different uses.

Liquid phase hydrogenation over a nickel catalyst was developed in England in 1903. Procter & Gamble bought rights to the English patent in 1909, and improvements by P&G plus subsequent invalidation of the patent in 1915 opened the way to rapid and extensive development of hydrogenation throughout the entire U.S. edible oils industry.

Chemists and others have also improved processing of edible oils in other ways. Adsorptive clays at first and then vacuum processing later improved bleaching efficiency, and oils were deteriorated much less than by early oxidizing methods. Steam sparging reduced oxidative decomposition during deodorizing, and vacuum systems later controlled decomposition even more, permitting higher temperatures to be used that deodorized oils more efficiently. Sequestering agents and antioxidants retard oxidation caused by trace metals in oils, making the oils last longer on the shelf and in processed foods. Interesterification and acidolysis rearrange fatty acids on fat triglycerides. The resulting triglycerides have different melting points and thus serve in different uses.

All of these and other improvements in the edible fats and oils industry have changed the use of table spreads and shortenings markedly. A half century ago, per capita consumption of butter and lard was just under 27 pounds a year, but it had dropped to just under 10 pounds by 1970. Meanwhile, per capita consumption of margarine and shortenings other than lard had risen from just a few pounds in 1920 to nearly 30 pounds by 1970.

Consumers not only changed their sources of supply of fat and oil products. They received price benefits as well. Butter, for example, sold for about 70 cents a pound in 1920 and for about 87 cents in 1970. Margarine, however, declined from about 43 cents in 1920 to about 30 cents in 1970.

Margarine. Of individual edible oil products, margarine in some ways occupies a unique position. Its volume is large—1.8 billion pounds in 1970—but more importantly it was one of the first of the formulated foods. As such, it challenged an established food, and the competition has had several important effects.

For one thing, reputable margarine manufacturers had to undertake the chemical research that would prove their product nutritionally equal to butter and would remove legal restrictions to its use. Today's margarines doubtless have qualities they would not have had without that pressure for research. For another, margarine's development helped change an important part of American agriculture. In the early days, coconut and palm oils served as the raw materials, but they were imported. Margarine foes decried such imports as detrimental to American agriculture. Margarine producers therefore set out to develop domestic sources, and they eventually succeeded with cottonseed and soybean oils. Farmers and processors found these oils to be so economically attractive that they joined in the battle to remove legislation that restricted margarine's use. The result was repeal of all federal and most state taxes and license fees and other restrictions in 1950 and the years immediately thereafter.

Today, a century after margarine's invention, Americans use nearly 1.8 billion pounds, well above their use of butter (1.0 billion pounds). The difference in price during most of this time helped spur that growth for margarine. In more recent times, so has concern over the role of fats in the diet. One of margarine's advantages is the ease with which its composition may be changed as we learn more about nutrition and its relationships to health. In the past two decades, some nutritionists and heart specialists have become convinced of the desirability of substituting polyunsaturated fats for saturated fats in American diets. Producers have promoted margarines heavily, based on the concept that with higher levels of polyunsaturation, margarines are more acceptable from a health standpoint than is butter (although admittedly the link between heart disease and saturated and unsaturated fats in the diet is controversial and still not settled).

Just over a century ago, in 1869, Hippolyte Mège-Mouriès, a chemist in France, started what has grown into a big industry when he patented an emulsion of meat fats as an inexpensive but crude substitute for butter. Margarine arrived in the U.S. five years later, and today it is supplied here by more than 30 companies.

The technical history of margarine's development involving chemists and chemical engineers consists of many small steps and several big ones. Among the latter were development of hydrogenation, which eventually led to vegetable oils' replacing animal fats, and determination of flavor constituents, which eventually led to synthetic flavors.

Margarine began life as an emulsion of mostly beef fat (known as "oleo," hence "oleomargarine"). Manufacturers later supplemented beef fat with pork and other animal fats. Then in the early 1900's, hydrogenation of liquid vegetable fats produced fats with higher melting points. By 1915, the all-animal-fat margarine had been replaced in the U.S. with blends of animal fat and coconut oil. The new vegetable-fat margarines held their flavor better and had melting points that gave them the firm and stable shape that consumers associated with butter.

The trend toward all-vegetable-oil margarines continued in the 1930's and 1940's. Best Foods introduced hydrogenated cottonseed oil in the mid-1930's, and domestic cottonseed oil replaced imported coconut oil. Then in the late 1940's, Girdler and Refining Unincorporated improved the methods for deodorizing soybean oil. Their new methods allowed producers to use less expensive soybean oil, and the trend became complete with the introduction then of margarines based only on blends of vegetable oils.

To make margarine, oils are refined, hydrogenated to the proper consistency, deodorized, and then churned with milk or an aqueous phase. As when making butter, producers often use starter cultures to get the desired flavors. They used cultured milk as the flavor ingredient at first in the U.S., but they later shifted to the European technique of using strains of bacteria that produce lactic acid to get uniform acidity and flavor.

In 1919, scientists at Iowa State College characterized the microorganisms responsible for the flavor and aroma present in starter cultures. During the next several decades, scientists in the U.S. and Europe found biacetyl and acetylmethyl carbinol (acetoin) to be the principal flavor and aroma ingredients in butter, carried out basic work on aliphatic lactones as flavorings for margarine, and determined the volatile carbonyl compounds found in butterfat. All of these advances, as well as a number of minor advances by others, led to greater understanding of the makeup of butter and margarine. The result was products, especially margarines, with enhanced flavors.

Until 1966, producers could use only biacetyl and the flavors found in water distillates of starter cultures to flavor margarine. Since then, they have been able to use chemical flavor ingredients, and producers have taken advantage of flavor work such as that described to produce today's margarines.

Establishing the nutritional qualities of margarine has also been important to its acceptance. A series of studies in the period 1915–25 demonstrated the nutritional equivalence of fats. The need for linoleic acid as one of the fatty acids essential for maintenance of growth was shown in 1929–30 and the presence of this essential fatty acid in margarine demonstrated in 1938. The nutritional equivalency of margarine with butter, which had been questioned for years, was conclusively proved in 1940, and the Food and Drug Administration issued standards of identity for margarine the following year.

Food Additives

Food is not the most long-lived commodity after harvesting, and it is not the most perfect as grown, as noted earlier. In recent decades, therefore, chemists and other scientists in the food processing industry and in the industries supplying it have developed numerous organic and inorganic additives to preserve foods and to improve or widen the range of food properties whenever possible.

They have obtained some of these chemicals from sources in nature, while they have made others from nonliving raw materials. Some of the chemicals are high polymers, such as modified starch, that stabilize and thicken. Others are small molecules, such as sodium benzoate, that preserve. Still others add flavor or color or change texture. Many have made large-scale processing easier and economically feasible.

Modified Starch. Food processors today use many products to modify food properties. Among the more important is starch, a product of photosynthesis and a reserve carbohydrate found mainly in plant seeds, tubers, and roots. The paper, textile, adhesive, cosmetic, explosives, and mining industries, among others, use millions of tons of starch annually. Its most important use, however, is as man's major source of food energy.

People eat most starch directly in such foods as potatoes, cereals, and cereal products. Since starch doesn't cost much and has unique properties, especially when processed and modified chemically, the food industry also adds it to many foods to supply more than simple (if important) food value. Extracted from raw materials, purified, and often changed chemically, starch thickens soups, sauces, and pie fillings. It also stabilizes salad dressing emulsions, holds moisture in toppings and icings, and forms gels in confections. It aids in binding breadings to fish and chicken before frying, coats and glazes candy and nuts, and dry-dusts bakery products.

Chemically, starch is a macromolecule ranging from several thousand to more than 1 million dextrose units. In the natural state it exists as granules with diameters from about 2 to about 100 microns. The granules don't dissolve in water, but cooking swells them into small balloons that eventually burst. On cooling, the cooked pastes set to an opaque gel. As opaque gels aren't desirable in many foods, chemists have studied the chemical and physical properties of starch. By changing them, they have expanded the ways the food industry can use it in foods.

Starch has been known for several thousand years, probably first having been prepared from wheat flour or from other cereals. Processes for separating pure starch from grain date at least from the 2nd Century B.C., but it is only since the Industrial Revolution that starch industries as such have developed to any extensive degree.

The modern era for starch probably began in the early 1800's when a German scientist attached to The Imperial Academy in what was then St. Petersburg in Russia found that acid hydrolysis produces dextrose and related products. Processors built many starch sugar plants in Germany following that work, but some 60 years passed before large-scale development started in the U.S.

At first, the U.S. used mainly potatoes and wheat as raw materials. Then at the turn of the 20th Century, a chemical development led to the first major change in the U.S. At that time, Corn Products (now CPC International) introduced continuous corn steeping. In time, this process virtually eliminated potatoes and wheat as major sources of starch. Corn millers now grind more than 150 million bushels of corn (about 8.4 billion pounds) to make nearly 7 billion pounds of numerous starch products. Potatoes and wheat furnish only about 300 million pounds of starch (and about 300 million pounds of imported tapioca starch round out U.S. supplies). The U.S. food industry uses some 4.4 billion pounds of starch and ranks first among the starch consuming industries (followed by paper and allied products at 1.5 billion pounds and textiles at 500 million pounds).

The 19th Century acid process for converting starch to corn syrups and dextrose has since been joined by enzymatic hydrolysis, and the wet corn milling industry in the U.S. now converts about 4 billion pounds of cornstarch each year. The corn syrups and dextrose are cheap, sweet, and desirable as partial replacements for sucrose (man's main sugar from sugarcane and sugar beets). Food processors use them in jams, jellies, preserves, and ice cream, where they inhibit sucrose crystallization, and in brewing and baking, where they ferment readily and react with protein to give an attractive brown to beverages and crusts.

To use starch other than to sweeten, the food industry has had to learn much more about how structure relates to function. Based largely on work at the University of Geneva and Corn Products in the early 1940's, food chemists now know that one starch fraction, amylose, has a long straight molecule of dextrose units connected end to end. The other fraction, amylopectin, also has molecules of dextrose units connected end to end but with many branches of dextrose polymers along the chain. Amylose consists of several thousand dextrose units, whereas amylopectin is estimated to contain more than 1 million.

Solutions of the two molecules exhibit quite different properties when heated and cooled. Amylose solution cools to a rigid gel, while amylopectin solution cools to a clear, nongelling paste whose viscosity often drops when it is cooked and sheared. Most natural starches contain 70 to 80% amylopectin and 20 to 30% amylose, and cooked pastes always reflect the presence of the two components. Such starches are thus limited to uses in which the natural properties happen to fit. If a processor wants a clear gel, then amylose's opaque gel won't do. Similarly, if the processor's procedure calls for much cooking and pumping, then amylopectin's viscosity drop may be a decided deficiency in the final product.

Once chemists knew about starch's basic structure, the way was open to them and to other scientists to overcome drawbacks. During such work, they achieved what are now judged to be the two most important advances in commercial starch chemistry in recent years: development of new genetic varieties of crops with different ratios of amylose to amylopectin and chemical modification of starch granules.

Waxy starches are genetic variations that consist of essentially all amylopectin. Although known for centuries in the Orient, waxy cornstarch remained a curiosity in the U.S. until World War II cut off the supply of tapioca from the East Indies. Searching for a tapioca replacement, geneticists at Iowa State Agricultural Experiment Station developed high yielding waxy corn hybrids. American Maize-Products and National Starch and Chemical produced them commercially in 1942 and Corn Products followed a decade later with waxy sorghum starches.

Although waxy starches work better than regular starches in some food uses, the unmodified cooked products are too stringy, cohesive, and unstable for many uses as food thickeners. But chemists found that if they very slightly crosslinked the starch granules with phosphorus oxychloride, they toughened the granules sufficiently, so that they swelled rapidly but did not disintegrate during cooking. Such cooked pastes are less stringy and cohesive and, of great importance, their viscosity doesn't break down when they are cooked, pumped, and stored. Following this discovery, chemists soon perfected other crosslinking agents such as sodium trimetaphosphate, epichlorohydrin, and acrolein to accomplish the same results.

Another significant aspect of this discovery was the extremely small amounts of crosslinking agent required to obtain the desired effect and the ease of purification of the crosslinked starch. Only one crosslink for each 1,000 to 2,000 dextrose units in the amylopectin is required. The treated starches can be easily purified by washing with water, since the treatment is applied to intact starch granules that are not soluble in water.

While some geneticists and chemists were developing tailor-made food starches from waxy or amylopectin starches, others were developing high amylose starches and their modifications. A major step came in 1952 with the discovery of an amylose gene extender. This discovery provided the basis for developing corn hybrids with up to 80% amylose and only 20% amylopectin.

High amylose starch was first commercialized in 1958, and starches are now being produced with 50 and 70% amylose. Because of their high amylose content, these starches form strong gels and films. They are potentially valuable for gelled desserts, gum confections, and edible films, and markets are now being developed for these uses.

While some were developing high amylose starches, others were developing commercial processes to separate amylose from amylopectin. In the U.S., chemists at Mississippi State University in the early 1960's developed a steam injection cooker that instantly gelatinizes the starch, following which amylose crystallizes at controlled temperatures and may be recovered.

Amylose, like high amylose starch, forms good films, resists grease well, and is impermeable to oxygen. These properties make it potentially useful as an edible food packaging material and protective coating for oxygen-sensitive materials, such as those that contain unsaturated fatty acids, vitamins, and flavors. As with high amylose starch, the full potential has not been realized as yet, however.

Emulsifiers. Successful cooks for years have known that to bake, fry, cream, whip, or carry out any of a number of other food preparation steps they need to use eggs or shortening or both. Many foods are finely dispersed emulsions of water, air, proteins, carbohydrates, and fats and oils. Complex chemicals in shortenings and eggs supply the necessary colloidal forces to keep the nonmiscible phases dispersed in each other.

Emulsions are essential in biological processes of both plants and animals, and it is only to be expected that the first emulsifiers came from natural food materials. As food chemists learned more about the organic, physical, and colloidal chemistry of such emulsifiers, they began to improve them, first with crude concentrates of the original natural emulsifiers and then with synthetic chemicals.

Shortenings based on natural emulsifiers worked only within narrow tolerances, and their seasonal and local availability limited industrywide use. Chemists at Procter & Gamble scored the first important success when they transesterified natural saturated fats and glycerol to make monoglycerides, whose food emulsification value they recognized around

1930. One part of monoglyceride replaced three parts of the older type of semisolid shortening mixtures. Among other things, shortenings with monoglycerides also emulsified more reliably, giving impetus to mechanization of the baking industry.

Organic chemists also made diglycerides and then followed them in the 1950's with a number of new edible emulsifiers to contribute more to mechanization of the baking industry and to the nationwide mass production of high quality baked goods. Some of these emulsifiers, however, were synthetic, and some synthetic ingredients in foods may lead to metabolism problems. Chemists in the late 1950's therefore turned to making emulsifiers from fully metabolizable molecules to minimize toxicity problems.

Antioxidants. Many types of foods deteriorate because their fatty portions oxidize when in contact with air. Through laboratory experiments and empirical observations, chemists and others have found that some compounds, generally with phenolic structures, inhibit oxidation effectively. The first step in oxidative degradation of a fat molecule may involve loss of a proton from the molecule at a point of chemical unsaturation. A reactive free radical forms that atmospheric oxygen readily attacks. Since phenolics can donate protons, they suppress the initial loss of protons from fat molecules and delay the oxidation that ultimately leads to rancidity.

People have used several ways in the past to extend the keeping quality of foods before oxidative rancidity makes them unpalatable. They dried meat over open fires in early times, for example. Somewhat later (in the 1800's), they apparently used substances such as oat flour and powdered bark from slippery elms. Smoking probably deposited phenolics that acted as antioxidants. Other substances also worked because of high concentrations of naturally occurring antioxidants.

In the four decades before 1940, food scientists worked extensively to trace oxidative reactions and mechanisms of substances that inhibit them. n-Propyl and longer-chain esters of gallic acid had previously been recognized as effective antioxidants for fats and oils and had been used to a very limited extent (primarily in Europe). The development of butylated hydroxyanisol (BHA) in the late 1940's, however, marked the real beginning of present-day commercial use of chemical antioxidants.

n-Propyl gallate was frequently used in combination with BHA at the time. Then in 1954, butylated hydroxytoluene (BHT) was established as a food antioxidant. Today, these three phenolic chemicals provide the basic line of antioxidants for use in edible fats and fatty foods. However, the Food and Drug Administration has recently expressed some reservations on the use of BHA and BHT, and they are now being reevaluated for safety.

Preservatives. Many foods are preserved by canning, freezing, or dehydrating and therefore do not require chemical additives. Some products cannot be treated by these procedures, and hence chemical preservatives must be used. Compounds for this application must inhibit or inactivate spoilage agents likely to occur and also must be harmless at concentrations used.

Approved chemical preservatives are limited, and uses of each are generally restricted to a narrow spectrum of foods. Benzoic acid and its sodium salt are most active at pH values of 2.5 to 4.0 and are more inhibitory to yeasts and bacteria than to molds. They are most suitable for use in highly acid foods such as fruit juices and drinks, soft drinks, salad dressings, pickles, relishes, olives, and sauerkraut. Alkyl esters of p-hydroxybenzoic acid (parabens) possess antimicrobial activity. They are active at higher pH values than benzoic acid, and their use is limited to baked goods, beverages, beer (n-heptyl ester only), fruit products, olives, pickles, and syrups.

Certain unsaturated fatty acids serve as antifungal agents. Generally used as the potassium or sodium salts, these materials are most active at pH values below 6.5 and are used in fruit drinks and other fruit products, soft drinks, baked goods, some meat products, cheese and cheese products, fresh and dried fruits and vegetables, salad dressings, pickles, olives, relishes, and sauerkraut.

The antimicrobial action of propionic acid has been known for some time, one of the earliest reports having come in 1913. Sodium and calcium salts are used most commonly to inhibit mold growth in baked goods and dairy products.

Sulfur dioxide has also been used in foods for many centuries. The ancient Egyptians and Romans apparently used fumes from burning sulfur as a sanitizing agent in making wine. Sulfurous acid, derived from sulfur dioxide or sulfite salts, inhibits yeasts, molds, and bacteria, although some resistant yeasts are encountered. A low pH enhances antimicrobial activity, and sulfiting is used with dried fruits and vegetables, fruit products, and wine.

Other preservatives include acetic acid (vinegar) and acetates, which inhibit bacteria and yeasts. They are used in some cheese products, baked goods, salad dressings, ketchup, pickles, and relishes. Nitrites are included in curing mixtures for meats to develop and fix color. They are always used with sodium chloride, and the combination has antimicrobial activity. Diethyl pyrocarbonate (DEPC) was isolated and identified in 1938. It is used for the cold pasteurization of wine, since it inactivates yeasts. In aqueous solutions, it hydrolyzes rapidly to ethanol and carbon dioxide, leaving only trace residues. Recently, urethanes have also been found as decomposition products of DEPC. They are potential carcinogens, and the safety of DEPC is therefore likely to be reevaluated.

Flavors. Americans now flavor their foods with more than 1,100 different ingredients, making flavors the largest single category of food additives. Such widespread use reflects interest in varied and flavorful as well as convenience foods. It also reflects the considerable contributions of flavor chemists, since some two thirds (or about 750) of today's flavor ingredients are synthetic.

In primitive times, people flavored their foods with natural spices and herbs. As time passed, they developed more sophisticated forms of natural flavors, such as essential oils and plant extracts. Then in the past half century as the food industry evolved into its modern form, the demand for flavors made processors acutely aware of the importance of a more varied supply of quality flavors. It is not surprising, therefore, that flavor manufacturers have responded in the past two decades with increasing efforts to replace or augment many natural flavors with synthetic equivalents.

Synthetic flavors have certain advantages over natural ones. For one, they are uniform and not subject to variations in character and intensity of flavor and color caused by differences in soil, climate, plant age, and harvest time. They also have stable prices and can be produced in adequate amounts. Moreover, they are free of insect infestation. Finally, they are adaptable and stable, since chemists can tailor them to withstand processing and storage conditions.

Success came slowly at first, but recently, and particularly during the past decade, advances in synthetic organic chemistry and in analytical chemistry and instrumentation have resulted in considerable improvements in both the number and the quality of synthetic flavors. Chemists now routinely separate complex natural flavors into individual components with gas chromatography and then identify them chemically using ultraviolet and infrared spectroscopy, mass spectrometry, and nuclear magnetic resonance in addition to the classical analytical methods.

Despite these advances, however, no one easily formulates a flavor that successfully duplicates a natural one. The volatile fraction of a natural flavor, for example, may contain 100 to 200 compounds. Some of these occur in very minute but well-balanced amounts. Moreover, some ingredients present in extremely small quantities contribute importantly to the total impact of some flavors, while others present at relatively high levels contribute little. The flavor chemist's task has been to learn how the sensory importance of individual components relates to the overall flavor sensation and then to devise the formulation that achieves the desired result.

Colors. People have colored some of their foods for many years, spices and condiments having been tinted as long as five centuries ago, for example. Originally, they used colors from natural vegetable, animal, and mineral sources. Then in 1856, the British chemist William Henry Perkin produced mauve, the first synthetic dye, and his work stimulated chemical research in many countries that has led to the synthesis of thousands of dyes.

Most dyes have been used in textiles, paints, and similar products. A number, however, have been found suitable for use in foods. They are vastly superior to natural colors, especially those of vegetable origin that usually vary widely, as do natural flavors, because of differences in soil, climate, plant age, and harvest time.

The use of synthetic dyes has grown considerably in the past 30 years, and in 1969 the Food and Drug Administration certified more than 3.6 million pounds for use in foods in the U.S. They are used widely in such products as candies, confections, and desserts; cereals and bakery goods; ice creams and sherberts; dairy products; sausage surfaces; snack foods; and beverages.

Breeding Grains and Oilseeds

The Austrian monk Gregor Mendel first set down the principles governing hybridization slightly more than 100 years ago, but his work went largely unnoticed until 1900, when scientists in Germany, Holland, and Austria independently rediscovered his basic tenets. Since then, geneticists, plant pathologists, entomologists, chemists, and other scientists have worked extensively on breeding improved characteristics into food plants (and into animals and other plants as well, of course). Today more than ever before, their work is showing results not only in better yields per acre, which was their early concern, but also in better nutritive value in the food part of the crops involved.

Breeding new hybrids may be primarily the function of geneticists, but chemists contribute to the successes achieved. They answer such questions about plants under genetic investigation as:

• How does the new variety respond to fertilizers?
• What is the composition of the new variety and more particularly what is the nutrient content?
• Are toxic materials present, and if so in what amounts and how might they be eliminated economically in food processing?

- How effective will be chemical control of rust, ergot, leaf blight, and other diseases where resistance cannot be bred into the plant itself?
- What mutations might chemicals or radiation induce?
- What chemical gametocides effectively control plant sterility in producing hybrids?

Corn was the first major crop to be hybridized, and hybrid corn became a commercial crop in the 1930's. Yields rose 25% immediately, and plant scientists have since increased yields even more. Moreover, the hybrids have other desirable characteristics. They tolerate drought well, resist diseases and other pests, and respond favorably to different climates and to fertilizers. Today, virtually every U.S. corn field is planted with hybrids.

In addition to yield, the amino acid balance is also important, especially in a protein food such as corn. Corn is normally deficient in lysine, and lysine is one of the amino acids that the body cannot synthesize. It must either be present in the food as harvested, or the food must be fortified with it during processing.

Scientists at Purdue University in 1965 observed that two particular corn mutants have more lysine than other lines. Since then, these mutants have been bred into many other lines. Yields, however, are still somewhat less than those from regular hybrids, and they are unsuitable for making some products traditionally made from corn. As a result, the variety's potential for feeding people as well as animals has yet to be realized fully, but the original finding's importance is assured.

Sorghum is a grain most Americans are probably unfamiliar with, since it is not one they eat directly. Yet it is the grain produced in the fourth largest volume (at 700 million bushels in 1970, following corn at 4.1 million bushels, wheat at 1.4 million bushels, and oats at 900 million bushels). It resists drought and hot climates, and farmers grow it extensively in the Great Plains and the Southwest where such conditions prevail.

An event of more than passing significance occurred in 1954 when scientists at the U.S. Department of Agriculture discovered cytoplasmic control of pollen production, a major finding in practical cross-pollination of normally self-pollinating crops. Grain yields rose by 25%, and sorghum quickly moved into new areas to become a major commercial crop. About half the crop is used for forage and silage and half for grain to feed cattle and poultry. It is thus an important supplement to the country's corn supply.

Wheat has also benefited from genetic work. In the U.S., scientists at agricultural experiment stations in Kansas, Nebraska, and Washington have made significant contributions, but private seed companies have also made major efforts to develop hybrid wheats. Norman Borlaug and his associates at the Rockefeller Foundation in Mexico extended the work by breeding varieties based on germ plasms collected from around the world, and use of these varieties in underdeveloped countries has been partly responsible for the recently hailed "green revolution." For his work, Borlaug received the 1970 Nobel Peace Prize, the first agricultural scientist to receive the honor.

Another factor in the green revolution has been a short strawed variety of rice that a research team at the International Rice Research Institute in Manila developed. This variety responds to fertilizers without lodging and gives yields from 50 to 300% greater than other varieties. It also matures faster and allows farmers to grow two crops per year.

Scientists working to improve oilseed crops want not only to increase yields per acre but also to breed seeds with more oil and with a better balance of nutrients in the oil. Soviet scientists have been particularly successful in breeding sunflower hybrids with high yields of seeds and oil. Scientists in the U.S. have made progress with safflower along similar lines. They have also bred a variety with a high oleic acid content. Such an oil should be more stable against development of rancidity. Canadians have pioneered in methods of breeding for different fatty acid composition in oilseeds such as rapeseed, an important oil crop in northern climates. They announced a rapeseed with zero content of erucic acid in 1967 and followed it with varieties having different ratios of oleic and linoleic acids. The low erucic acid variety is expected to become an important oil crop. Meantime, further research is being carried out in the U.S. on soybeans to produce beans having more protein and a more complete balance among the amino acids.

Nutrition

In 1810 when the Parisian confectioner Nicolas Appert revealed how to preserve foods by canning, he took what historians later judged to be one of those major steps taken in any field only a few times in a century. The French chemist Pasteur contributed the 19th Century's second major step for the food industry when he showed how microorganisms cause spoilage. Appert's work marked the beginning of the food processing industry, and Pasteur's work put it on a scientific basis.

Man wasn't interested only in eating the foods that canning and other preserving methods provided, of course. He was also interested in nutritious food, and around the turn of the century and in the first few decades of the 1900's chemists and other scientists took the third major step in the food industry. They found the relationships between specific nutrients and several diseases.

Today we know that man needs at least some 50 nutrients to live and enjoy adequate health. All occur naturally (or are synthesized in the body from naturally occurring foods) and provide adequate nutrition when eaten in the proper amounts. Throughout the world, however, many specific disease states can be identified readily with dietary deficiencies in one or more of these 50 essential nutrients, ranging all the way from a lack of adequate macronutrients (proteins, carbohydrates, and fats) to a lack of a single vitamin or mineral.

As chemists and others found the causes of these diseases, they also worked out ways to prevent or cure them. Iodized salt, for example, prevents goiter in people who do not get adequate amounts of iodine otherwise. Diets fortified with proteins prevent kwashiorkor. Iron supplements reduce anemia. Knowing what these nutrients do and having an industry able to produce foods fortified with them, no country need have citizens who suffer from problems associated with them.

Diseases associated with deficiencies in the vitamins have probably had the most far-reaching effects and have probably received the most research attention. Many scientists worked for years before the relationships between diseases and vitamins were established, the essential chemicals isolated and identified, and food fortification methods worked out. Their approaches and their results with knowledge of vitamins typify progress in the field.

Vitamin C. Scurvy (or scorbutus) is a classic dietary deficiency disease. A lack of Vitamin C (ascorbic acid) causes it. It is probably the first disease to be associated with a dietary deficiency. Sailors were the traditional sufferers of fact and fiction, but people on land also often suffered during long winter months when fresh fruits and vegetables weren't available. Then in 1747, a Scottish naval surgeon, James Lund, brought about dramatic cures among sailors by feeding them fresh oranges and lemons. The successful battle against scurvy had begun, but it was not for another 150 years that scientists in several countries began to show that diseases such as scurvy were caused by diets lacking specific nutritional factors.

By the early 1930's, Vitamin C was isolated, identified, and a practical synthesis developed. Today, synthetic ascorbic acid is widely available at low cost. It is used extensively to fortify foods and beverages, and scurvy is no longer the peril it once was.

Vitamin D. The bones of children suffering from rickets don't develop properly, and adults suffering from the disease find their bones soften. Such skeletal deformations had been noted in people for centuries, and millions were afflicted with the disease throughout the world. No explanation for its cause existed until the 1920's when scientists in England and the U.S. established that a factor found in fats effectively prevented its occurrence. The factor eventually became known as Vitamin D.

Now we know that the vitamin is essential in the biochemical pathways by which the body absorbs and uses calcium and phosphorus. Such foods as eggs and fish oils provide it, and it is also formed in the skin by ultraviolet rays.

Before discovery of Vitamin D, rickets was a major cause of bone deformations in infants, even among those getting enough milk. Then in the 1930's the food industry began to fortify milk, butter, and margarine with the vitamin, and by about 1940 rickets had been virtually eliminated in the U.S. Unfortunately, in the late 1960's when malnutrition in the U.S. became a matter of public concern, welfare teams found cases of rickets in distressed areas of the country where people are too poor or spend their money unwisely and do not buy vitamin-fortified milk. Technology may have succeeded in solving a problem, but other factors have prevented effective use of that technology.

Other Vitamins. Similar stories of discovery can be told for all the other vitamins. Scientists have shown Vitamin A to be essential for sight, growth, and health of the skin and mucus membranes, for example. Niacin is now known to be essential in preventing pellagra and thiamine essential for normal growth and for a healthy nervous system. Riboflavin has been shown to be essential for energy conversion and growth, while pantothenic acid is known to be involved in energy transformations from fats and carbohydrates.

Total annual production of vitamins in the U.S. today exceeds 17 million pounds, nearly six times the production of 20 years ago and nearly two times the production of 10 years ago. Multivitamin preparations take part of the output, while food (and feed) fortification takes the balance. Vitamin A is used most extensively in fortification, and dairy products and margarine are its most important outlets. Milk and milk products are fortified with Vitamin D, while the vitamins thiamine, riboflavin, and niacin are used extensively in grain products and Vitamin C in beverages.

Other Accomplishments

It has been impossible in the space available to cover many other interesting and important developments in the food industry based on chemical innovations. Instant and freeze-dried coffees, for example, are products the food industry has developed during the 20th Century that have given consumers an important item of their diet in very convenient form. They have allowed raw materials to be used efficiently, the beverage to be prepared easily, and the waste products to be disposed of economically. Furthermore, development of the processes has contributed important knowledge about making water-soluble powders, about analyzing and conserving fugitive flavors and aromas, and about designing equipment to percolate, spray-dry, freeze-dry, and package in high-speed lines.

As a result of these advances, consumers can make a cup of instant coffee for considerably less today than the cost of brewing roasted and ground coffee, and instant and freeze-dried coffee now account for about 30% of all coffee consumed in the U.S. Moreover, the popularity of the product has influenced coffee-growing countries to establish their own manufacturing plants, and many such plants in these countries contribute significantly to their economies as well as help them obtain much needed manufacturing capabilities.

Not too many years ago the family that enjoyed a cake for dessert did so only after many hours of work by the housewife. She measured out the basic ingredients and then ground them, sifted them, whipped them, and otherwise melded them to make the chemical reactions occur at the right time and to the right degree.

Today, the situation is quite different. Supermarkets stock dozens of prepared cake mixes that require only mixing in water or milk, perhaps an egg, and sometimes butter or shortening before being put into the oven to bake. Cake mixes are now big business, amounting to some $200 million at retail in the U.S. in 1970.

The forerunner of today's mixes was General Mills' Bisquick, a biscuit mix first marketed in 1931. Products more closely resembling today's mixes were Dromedary's gingerbread mix and a devil's food cake mix, both on the market in 1942. General Mills followed with a gingerbread mix in 1947, and it and several other companies, including Pillsbury and General Foods, introduced a variety of other cake flavors in the next few years. These products for the first time gave the homemaker a relatively convenient means of preparing high quality, freshly baked cakes.

Cake mixes didn't emerge from some striking scientific finding such as pasteurization or the relationship of vitamins and health. Rather they emerged from many findings by many food scientists and engineers who investigated the parameters involved and who gradually worked out the combinations of ingredients and preparation methods used today.

Americans eat about 46 million loaves of bread a day. A widely dispersed baking and distributing system supplies them fresh daily to thousands of grocery stores six and even seven mornings every week. As baking moved from tens of thousands of rural kitchens to hundreds of urban center bakeries, food chemists gradually changed traditional formulations. Emulsifiers now aid mixing and provide more uniform loaves. Dough conditioning agents speed up and improve processing. Softening agents and preservatives lengthen shelf life. Finally, about half the bread baked in the U.S. comes from continuous mix operations in contrast to the former entirely batch production.

In earlier days, fresh fluid milk not only had a short shelf life but more importantly frequently carried germs that contributed significantly to the high incidence of infant mortality then prevalent. Today, it lasts up to two weeks under refrigeration, and no one questions its suitability. Of all the improvements that have led to modern milk processing in centralized plants, the discovery of the relation of bacteria and other organisms to disease and spoilage has had the most far-reaching effect. Original pasteurization processes required many improvements, however, before pasteurized milk became universally accepted.

No one questions homogenization today, either, but they did in the early 1930's when the process was introduced in the U.S. Practically all milk is now homogenized in the U.S., and it has advantages most consumers are unaware of. For one, it provides a uniform, soft curd that people digest more readily than curds from untreated milk. For another, the industry introduced homogenization just as it began fortifying milk with Vitamin D. Homogenized milk's uniformly distributed fat was just what was needed as a carrier for the oil-soluble vitamin.

Other process improvements besides pasteurization and homogenization include development of equipment free of crevices and dead spots where milk can collect and contaminate later batches. Adoption of stainless steel and development of sanitizing chemicals also improved plant cleanliness. High-temperature, short-time pasteurization processes replaced cumbersome vat methods. These processes produce milk with a shelf life of up to 14 days, and dairy scientists and engineers are now working on processes that will sterilize and package milk aseptically and produce milk with a shelf life measured in months instead of days.

In addition to improving processing equipment and methods, chemists and chemical engineers also improved packaging. First, they worked with paper chemists to make caps free of flavor for the then-standard glass bottle. Later, petroleum chemists supplied improved coating wax to allow single service paper containers to be formed, waxed, filled, and sealed automatically. These paper bottles, which replaced glass bottles that had to be returned for washing and reuse, were in turn improved by plastics and paper chemists with a plastic-coated paper container that didn't wick and didn't crack as did waxed containers. Using plastics also eliminated controversy over possible health hazards from the wax coatings (although it may only have been replaced with similar controversy over hazards from plastic coatings). Now the industry is about to switch to using plastic bottles that are blown, filled, and sealed in one short operation entirely protected against bacterial contamination.

At one time whipped cream was the sole product of its type, and even today with competitors in the market it would probably be preferred were taste the sole basis for judgment. However, food chemists in the past three decades have developed substitutes to provide lower cost, uniformity of whipping, convenience in buying and preparation, and storage stability both before and after preparation.

Appearance of substitute products has introduced heavy competition in the traditional topping market. The aerosol products arrived first (in the 1940's), followed by dry mixes based on vegetable fats (in the late 1950's) and frozen ready-to-use toppings containing vegetable fats (in the mid-1960's). Toppings based on vegetable fats now hold 80% of the market, partly because of their convenience but probably largely because they sell for 50 to 60% of the price of the dairy-based product.

Making cheese was long regarded as an art, but as in many other areas, scientists found the bacteriological and chemical principles involved. Cream cheese and Swiss cheese have been particular beneficiaries, but the development of process cheese undoubtedly ranks as the most significant American innovation in the cheese industry. It not only provided cheese in a new form, but it also changed the marketing of cheese from a local enterprise to a major national industry. Mixing an emulsifying salt with natural cheese and heating it produces a cheese with a smooth, plastic consistency. Such a cheese can be packaged and marketed in convenient sizes without rind or waste. Furthermore, the process makes a cheese that keeps for many months, an advantage that increased consumption, especially in warmer regions of the country where grocers did not normally sell cheese in the summer months.

Finally, improvements in animal feeds must not be overlooked. In developed countries such as the U.S., animals make up an important part of the diet. Producing more animals on less land with fewer workers has contributed significantly to the country's food supply and to that supply's year-round variety.

The major changes have occurred in the past 60 years or so. Instead of feeding whole grains and forages to animals and poultry, growers now feed them scientifically blended rations containing the right nutrients for growth and the proper drugs to protect them against diseases they are all too susceptible to.

For example, chickens eating whole grain and roaming the barnyard some years ago once required 10 weeks to reach the 2-pound state called spring fryers. No one can calculate what such birds cost, since they got a fair part of food free from nature. Today, their offspring some generations removed are raised in a broiler house with up to 30,000 birds in one pen. They reach 3 pounds in only seven weeks. The feed conversion ratio is slightly more than 2 pounds of ration to 1 pound of live bird, and the cost to the poultryman is less than 13 cents a pound. In the process the poultryman no longer risks waking up one morning to find his flock decimated by coccidiosis or any of a host of other diseases.

This same change in efficiency and speed of production, coupled with greatly improved quality in the end product, applies to feeds for the rest of the animal industry (although to a somewhat lesser degree). Feed chemists are now seeking to give commercial feeds even greater efficiency. They are constantly using new ingredients such as urea and amino acids that will produce animal protein faster and at less cost.

FUTURE DEVELOPMENTS

New Foods and Nutrition

Americans may still be eating proteins, carbohydrates, and fats as they always have, but they are doing so in many new forms and with much less labor in the home. Where the neighborhood grocery of yesterday carried an inventory of a few hundred items, many of them in bulk, the supermarket of today stocks several thousand different food items, each individually packaged in convenient sizes for protection of quality and wholesomeness. Developed largely by chemical science and engineering, these new foods and their packages have given the consumer the widest of choices of high quality, nutritious food. They have also reduced the hours of work daily to minutes of ease. Yet food's share of the cost of living is lower than ever before.

Past successes notwithstanding, chemists and chemical engineers in the food industry still face many challenges. In the coming years, most of the food industry's investment for research and development, for labor, and for plants and equipment will be to accelerate trends that have dominated its recent past—more built-in convenience, increased emphasis on nutrition in new forms of foods, more farm-freshness at the table for more foods. As in the past, chemical science and engineering will contribute in a critical way to developing these new products and the new processes needed to make them.

The food industry now develops about 5,000 new products each year. Admittedly, only about one in 10 survives for as long as a year under the stern test of the market. Five hundred successful new products per year, however, is no mean accomplishment, and we have not seen the end of such accomplishments. Included among typical products now being considered, some of which may find supermarket shelf space in the future, are:

- Prepared foods sold hot, from individual entrées so prosaic as hamburger to more exotic foreign (to American palates) ones and even to complete meals.
- New foods for high-speed cooking in electronic microwave ovens.
- Foods low in calories for waist-conscious dieters, yet otherwise nutritious to maintain their general health.
- Packages that control humidity and atmosphere to get foods from grower to consumers at peak freshness.

While chemists and chemical engineers in the food and other industries will help provide new convenience and new combinations of food ingredients, they will not forget that their ultimate goal is nutrition. Man evolved over the centuries eating what he has come to consider "natural" foods. As it happened, many of the so-called natural foods found to be most palatable have proved for the most part to be the most nutritious and appropriate for a balanced diet. People in the U.S. will continue to eat more processed foods, however, as food chemists select and blend food ingredients into new forms that are palatable (and therefore, among other things, marketable), that provide more convenience, or that lower (or at least don't increase) costs.

A public that is more information conscious will want more meaningful labels that list nutrient contents. Nutritionists, food chemists, and other scientists in industry and government will have to determine what information is important, what can be put on a label practically, and what consumers want and will use.

Even with the best of informative labels, however, many people will continue to eat improperly. Some won't know what to eat, as they don't today. Others will lack the motivation to eat properly, as they lack it today. Both industry and government must provide the proper environment, so that everyone is encouraged to eat adequately. Education may be the most obvious way to do so, but many other pressures in daily activities readily turn off those who need education the most. Experience, therefore, has proved education to be one of the most difficult ways to change eating habits.

New and untried educational approaches still to be developed may help overcome distractions to which people now succumb. The job is so massive, however, that only the most strenuous of efforts by industrial, educational, and governmental organizations will lead to success. Chemical experts will contribute in part to such efforts, of course, but their greatest help will come in producing foods that provide a nutritionally adequate diet. In fact, cooperation between industry and government to supplement basic foods soundly and judiciously may be the most practical way almost to assure proper nutrition without any overt effort on anyone's part and possibly without their even knowing about it.

Fortifying cereals, dairy products, and beverages with vitamins and amino acids has become established practice. The foregoing suggestions, however, could take us well beyond present practice. In fact, some proponents suggest that even "fun foods" might be supplemented with nutrients, since calories from these foods do replace calories from foods that provide more essential nutrients other than calories.

No one, of course, should expect industry and government to correct the most bizarre and eccentric diets by supplementation. Moreover—and more importantly—food supplementation of the magnitude suggested would require the soundest scientific base to eliminate reasonable possibilities for harm to people, to the environment, and to the economy.

Chemists and others can also contribute to other aspects of nutrition besides helping provide basically adequate diets. No one knows the optimum diet for good health, longevity, and a more useful life, for example. Overweight probably leads among nutritional problems in the U.S., and controlling weight and appetite still challenge chemical science. So does the relation between nutrition and degenerative processes, such as heart disease, arthritis, cancer, and even aging itself. On a more specific level, medical and nutritional experts would also benefit from knowing how dietary protein affects kidney function and health.

Meantime, no one has written the final report on the full nutrient content of foods, particularly on trace elements but not omitting the more familiar vitamins and minerals. People can also use more information on nutrient requirements at all ages, and we still may not have developed the best schedule for eating.

Product Safety

Chemists and chemical engineers in the food industry may work most on developing new foods, and they may still have much to accomplish in the science of nutrition. However, they face two other major challenges that will call for an increasing share of their time. Providing more food for more people and coping with malnutrition is one, and we shall deal with that subject in the final pages of this section. Product safety is the other, and we shall deal with it here.

Proteins are chemicals. So are carbohydrates, fats, minerals, and water, the other components that make up the bulk of foods. Plants, meat, and fish also contain smaller amounts of other chemicals, such as vitamins, antioxidants, antimycotics, buffers, thickeners, emulsifiers, chelating agents, colors, and flavors. Because he has long used unmodified foods with apparent safety, man has come to consider them natural and therefore automatically good. Natural foods, however, are not always the ultimate in nutrition, variety, taste, aroma, and texture. The food industry has accordingly improved them, so they will keep longer, provide more nutrition, and have properties that make them more palatable (and therefore more acceptable). The industry has also made them more economical to produce and distribute in the volumes needed by today's urban society.

At first, food chemists used as additives the products they separated from plants, animals, and fish. As they learned more about how to use these chemicals from nature effectively in foods, they turned first to making many of them in laboratories and factories and then to making others that have no exact counterparts in nature.

Man tampers with his food supply at his peril, of course, and when one tags a product for use in foods with pejoratives such as *chemical* or *additive*, storms quickly rise. Few storms have buffeted any industry as "chemicals in our foods" has buffeted the food industry in recent years. No responsible person in the food industry or in government would settle for less than the greatest of care when introducing products into food, to be sure. Part of the controversy rests on emotion—"nature is natural and therefore good; synthetic is unnatural and therefore bad." Part of the controversy, however, also rests on honest differences of opinion over how to achieve optimum safety in foods—and even over what optimum safety is.

The food industry has long evaluated the safety of both intentional and unintentional food additives. Not only must it prove that intentional additives function usefully but that people (or animals that are food sources) may eat them safely at recommended and even highly exaggerated levels. Unintentional additives, such as substances that migrate from packaging or that may remain as residues from food processing or use of pesticide, fertilizer, or animal husbandry chemicals, must cause no harm either, also at allowed as well as highly exaggerated levels.

In today's climate of awareness of and concern over possible hazards from exposure to chemicals, industry and government chemists and other scientists face a monumental task. As new information and methods become available, they must continuously review additives that extensive testing already has allowed to be used. Recently, additives "generally recognized as safe" (the so-called GRAS list of chemicals) have also been placed under review. Additives considered for use for the first time require more meticulous pretesting than ever. Metabolic biochemists, enzymologists, and biochemical geneticists still have much to learn about how deleterious agents act. While new analytical methods and instruments have lowered detection limits sharply in recent years, food scientists still ask for more sensitivity.

Meantime, chemists will also help determine how safe is safe. Chemicals as possible causes of cancer have received much public attention in the past two decades or so. Other chemicals that possibly lead to birth defects or cause mutations have recently joined them. The Food Additives Amendment enacted in 1958 requires FDA to ban any substance that produces cancer in people or animals regardless of the amount of additive tested or the test procedure followed. Some have suggested that this clause (known as the Delaney clause, after its Congressional sponsor) be revised to eliminate zero tolerance. They reason that absolute safety doesn't exist and believe that safety should be decided on the basis of least risk and "reasonable" interpretations of toxicity data. Others have proposed expanding the clause to ban chemicals that cause birth defects or produce mutagenic effects, also regardless of amount or test method. Chemists and other scientists will obviously have to consider these proposals very carefully to provide a workable solution that assures a safe food supply.

The law—statutes, court opinions, and government regulations—reflects public policy and in effect codifies the will of society. The law has significantly affected the food industry in the past, but it will do so to a much greater extent in the future. When the U.S. had largely an agrarian society, little concern existed over food safety. The law ignored problems arising from food processing. As food processing became a business, the law said let the buyer beware. In more recent times, however, and especially in the past decade or so, the law has increasingly held the food industry absolutely liable for its products.

This development is occurring in the common law, which is being codified into product liability statutes, and it is not regulatory law developed by governments. Regardless of type, however, it binds just as strongly. Looking ahead, the food industry sees itself losing as a defense the exercise of reasonable care and becoming an insurer to its customers. This increasing liability signals the need for better evaluation procedures.

While the common law is being codified into liability statutes, regulatory law is increasing in scope and substance. The trend lies clearly toward complete biodegradability of food by-products at the manufacturing level and such associated products as packaging at the consumer level. Plants face the requirement of zero contamination of the environment.

The tightening legal climate will place substantial and new demands on chemical personnel in industry and government. The type of training people need and the employment opportunities open to them will change. Formation of consumer associations, for example, is leading to more policing of the food industry, involving both food quality and environment contamination. Added policing, growing regulation, and countermeasures against them will call for chemical and other personnel to meet the legal and moral requirements while their organizations still operate in a capitalistic economy.

As research and development continue to cost more for these and other reasons and as individual companies become unable to support work by specialists in all fields, the food industry expects joint research will grow. Joint R&D companies may be formed, new industry groups organized, and more consultants and independent laboratories utilized. Chemical experts will be increasingly asked to define carefully the scope of such undertakings to protect the rights of those concerned. They may also perform defensively in light of continuing study and some criticism by government of joint development companies and group-sponsored research. Meantime, the rising cost of R&D will lead entrepreneurs to seek stronger patent protection, and food industry scientists and engineers will be called on to help secure it.

Expanded Sources of Food

Providing more food for more people in the U.S. seems feasible, at least for the foreseeable future, based on fertile land, favorable climate, and a strong technological base. As for malnutrition, it will place different demands on the U.S. food industry depending on whether the concern lies with Americans or with persons in underdeveloped areas elsewhere. In the U.S., malnutrition is almost entirely a matter of economics. People don't go hungry because there isn't enough food. They go hungry because they don't earn enough to buy the food available. By contrast, people in Africa, Asia, and Latin America go hungry because there simply isn't enough food. (In the main, correct, but somewhat oversimplified, to be sure. Some Americans, for example, are poorly nourished not only because they lack money but also because they are poorly educated and don't know how to get the most nutrition for the dollars they do have. Others simply spend their limited funds unwisely. As for the poorly nourished in Africa, Asia, and Latin America, their incomes and education often beggar description.)

In the U.S., one solution will be to help people earn more, so they can afford foods that provide an adequate and pleasing diet. Another will be to educate them to know what to buy and to have better priorities for buying. Such solutions are long-range and not ones the food processing industry itself can solve directly. The food processing industry, however, can develop inexpensive foods that are adequate nutritionally, that are pleasing to eat, and above all that don't have the stigma of being poor people's food.

In the underdeveloped countries, meantime, the problems are vastly more complex. Populations are large, jobs are scarce, productivity is low, and education is backward. Food supplies are scant, governments are weak, and industry is embryonic. While the outlook is indeed dismal, some progress is being made, however.

Growing More Protein. Diets may be short of calories, they may be short of certain essential components, or they may be short of both. Where simply one or more vitamins or a few essential trace elements are short, fortifying foods with the missing ingredients solves the problem. Since vitamins and trace minerals are readily available and since only very small quantities are involved, food chemists can correct such deficiencies fairly easily with technology presently available (technically easily, that is, but shortages of funds, bureaucratic or industrial inertia, diet habit, and social custom may make the solution politically hard or impossible).

In the more common form of malnutrition, however, people either don't have enough total protein or enough of certain essential amino acids. Much of the effort to overcome malnutrition, therefore, consists of increasing the total protein supply, upgrading the amino acid content of the present protein supply, and of developing new foods whose protein will supplement that available from present sources.

To increase food supplies, countries normally try to grow more grain, and to improve protein quality they normally try to raise more animals. We shall consider the future for animal production later. Meanwhile, much has been done in various parts of the world in recent years to grow more grain (the most important single source of protein), first by using more fertilizers, pesticides, and other modern agronomic techniques and more recently by breeding strains that produce more grain per acre and are more resistant to diseases and insects. Possibilities for improvement following these two approaches are far from exhausted, however, and chemists and other scientists have much to do in both areas.

Upgrading the amino acid content of present protein foods enjoys a major advantage in that people don't need to change their diet habits, a far from insignificant hurdle. It suffers from two disadvantages, however:
• Where breeding is the approach, the process is tedious and the outlook for success is uncertain.
• Where fortification is the approach, the hard realities of rising food costs and shortages of capital (to say nothing of bureaucratic weaknesses and similar political problems) can prove to be deterrents.

Problems in breeding may be formidable, but scientists recorded some notable successes as the 1960's drew to a close. Massive importation of a dwarf variety of wheat developed in Mexico raised India's harvests of food grains to new highs. A new variety of rice made the Philippines self-sufficient in that grain, essentially for the first time since 1903. Iran became a net exporter of wheat, and Turkey expects to be self-sufficient by 1975. Kenya has a surplus of corn. These and similar stories elsewhere have come to be hailed as the green revolution, and much research is still under way to provide even greater improvements.

Fortifying Grains. The main cereals are wheat, rice, corn, oats, barley, and grain sorghum. Most of them, however, are not entirely satisfactory as protein sources, since they are deficient in one or more essential amino acids. One solution has been to fortify them with amino acids made synthetically, principally with lysine for wheat, lysine alone or with threonine for rice, and lysine and tryptophan for corn and grain sorghum.

Processors fortify cereal foods widely with lysine and methionine now that large-scale production has dropped the selling price to about $1.00 a pound. They might also use more of two other amino acids, tryptophan and threonine, to fortify cereals, but prices of $20 a pound for tryptophan and $7.50 a pound for threonine rule out widespread use (except for fortifying corn with both lysine and tryptophan, since little tryptophan is needed).

Chemists have worked out a chemical synthesis for tryptophan, and methods exist for producing threonine by fermentation. However, cost is still a hurdle, and research continues for methods with target prices of

$5.00 a pound for tryptophan and $3.00 for threonine. Food industry experts estimate the potential world market will be about 10 million pounds a year should prices for all four amino acids reach the $1.00-per-pound level. Anticipating this demand, the Japanese have been particularly active in research on the production of amino acids by fermentation.

Finally, the discovery that certain corn genes exist that give grain with a higher than normal level of lysine and tryptophan will lead to a revolution in corn eating areas, principally Latin America. Geneticists are now working hard to introduce these high lysine genes into corn that will have satisfactory growth and yield characteristics for different growing conditions. Their success could bring good nutrition, or at least a balanced protein source, directly to the corn grower and to the city slum dweller, who eats practically nothing but corn.

Oilseed Protein. Another approach to improving world protein supply and quality besides fortifying grains with amino acids is to develop other sources of protein. Included is protein from soybeans, cottonseed, peanuts, sunflower, sesame, safflower, fish protein concentrate, whey, green plants, and single-cell microorganisms such as algae, fungi (yeast), and bacteria. Some of these, such as soybean and cottonseed, are already major protein sources. For them the effort is on preparing new forms of food products. Others, such as fish protein concentrate and single-cell microorganisms, are just about to go commercial on a major scale. The balance, meantime, await further development.

Oilseeds represent the cheapest possible source of protein now, with both oil and meal either edible or processable to edible products. Soybeans, cottonseed, and peanuts are the products used most, while safflower, sesame, and sunflower have yet to become economically important.

For nearly a century, food chemists have sought to substitute raw materials from vegetables for milk and milk products. Margarine, in which vegetable oils replaced butterfat, was one of their early major successes. More recently they have used vegetable fats to produce many palatable and functional cream substitutes. The substitution of vegetable fat for butterfat in fluid milk and ice cream is now technically feasible and economically attractive, but it is limited mostly by legislative standards controlling ingredients in products labeled "milk" and "ice cream."

Food chemists face a more complex problem, however, in replacing both fat and protein or even protein alone in milk products. They have considered a number of approaches but have worked most on using whole soybeans to make a milk based entirely on vegetable sources. In doing so, they are following the lead of the Far East, where the Chinese and Japanese use foods based on soya milk as common, everyday staples to supply their protein and to some degree their fat needs.

Soybean products are used in many ways in formulated foods in the developed countries. They contribute not only protein nutrients but also desirable texture, water absorption capacity, and binding, whipping, and gelling qualities. By taking advantage of these properties, food chemists in the developed countries will continue to replace ingredients in present products with them and to formulate entirely new products from them.

New beverages based on soybeans offer great promise in the fight against malnutrition in Latin America, Africa, and Asia. The U.S. food industry hasn't made large inroads on the problem yet, and people haven't always accepted the new beverages with unbounded enthusiasm. However, the goal of producing such nutritionally balanced beverages has been pursued more diligently in recent years, and several companies are now test marketing and selling low-cost beverages based on soy protein.

Considering the magnitude of the problem, efforts are still modest indeed. However, the food industry expects progress to pick up in the future. It is carrying out research on supplementing soy beverages with methionine or with methionine-rich sources such as fish protein concentrate or sesame. In addition, another food industry goal is to develop formulated concentrates that companies can make at a central point and distribute to local bottlers to reduce costs even more.

Growing more cotton and processing its oil into more food products without upsetting agronomic practices and native dietary habits too much may improve protein supplies in areas it is native to but that are deficient in protein. Cottonseed, however, is not the most easily processed oilseed. It contains a toxic pigment, gossypol, that must be reduced to a low level before people or animals can eat its products. Gossypol can be removed or detoxified during processing, but that raises the price of oil and meal.

To increase the use of cottonseed, geneticists have recently bred a glandless variety containing very small quantities of gossypol, but it has yet to be planted in significant amounts commercially. Moreover, gossypol is either a pesticide or a repellent, and plants lacking it are susceptible to insects and diseases. Cotton yields are also low. Geneticists thus are continuing their work, aided by the analytical skills of chemists, to develop gossypol-free varieties that also have the plant's traditional insect and disease resistance and that at the same time preserve yield and quality of the cotton lint (which is, after all, the major economic component).

Meantime, what is now the U.S. Department of Agriculture's Southern Marketing and Nutrition Research Division at New Orleans has developed two processes for making edible cottonseed concentrates and isolates from cottonseed flour. These processes separate intact pigment glands, which contain the gossypol, from the flour. Air classifies a finely divided flour in one process, and a liquid cyclone fractionates a hexane slurry of finely divided flour in the other. Both produce two fractions, one analyzing about 70% protein and sufficiently low in gossypol for people to eat and a second fraction somewhat lower in protein but higher in gossypol that animals may eat. These processes, while not used commercially yet, promise to provide high quality protein at relatively low cost. Dorr-Oliver, manufacturer of the liquid cyclone, is actively pursuing a development program in India with them.

Protein from the Ocean. Planktonic algae in the oceans may carry out as much as 90% of the photosynthesis on all the earth. Man gets less than 10% of his protein from the sea, however. Faced on the one hand with starving millions and seeing on the other tremendous potential primary food production, food experts hailed the sea a few years ago as man's salvation. Although the world fish catch has tripled in the past two decades, the protein potential of the sea is far from being realized. Meantime, the possibility does exist of using more efficiently the marine protein we do harvest by processing some of the catch into fish protein concentrate (FPC). FPC hasn't become fully commercial yet, but much of the development work is behind us. Food scientists have shown they can make acceptable and nutritious products. Their problem now is to develop foods that economically incorporate FPC in them. Once the economic problems are solved, FPC should find its way into a number of products, including fortified cereals, beverages, sauces, and possibly textured products, in some instances in a major way by the mid-1970's.

Meat Protein. Animals provide man with about 25% of his protein worldwide and thus are the second (to cereals) most important protein source. Animals rank so high because of several advantages but largely because most people judge animal protein most desirable (i.e., most palatable). Since eating is a pleasurable experience as well as a simple necessity, delectability cannot be overlooked when evaluating the food chain and considering changes to it.

Some experts often criticize animals as inefficient converters of raw materials to food. These critics see a better salvation in changing grazing lands to cereal growing lands (where climate and other factors permit) rather than in raising more animals.

In some places such conversion from grazing to farming is doubtless the right short-term solution, and animal production can be encouraged some years later when local incomes make eating meat economically feasible. Meanwhile, animals as a source of more protein should not be dismissed out of hand. Livestock can eat food that man can't, such as forages and by-products from the cereal, oilseed, canning, and fermentation industries. Furthermore, ruminants can use urea as a source of nitrogen, and urea requires no land (other than plant sites) for its production. Urea costs less than any other form of organically bound nitrogen, and animal nutritionists estimate that in the long run the nutritionally desirable, high quality milk proteins can be produced for 40 cents a pound by using feeding regimens of urea, carbohydrate, and cellulose. Finally, animals produce other desirable products such as wool and hides whose value must not be overlooked.

As animals themselves aren't a complete solution because of high prices for meat, some food scientists have turned to making meat substitutes. When they do so, they almost automatically turn to the oilseeds for raw materials, since oilseeds are the least expensive protein sources now available. Oilseed proteins are floury materials. They yield gruel-like foods, although in some cases they can be put in suspension and consumed as beverages. In either form, however, they lack the one key feature that can make the difference between being merely nutritious (no mean feat in itself, to be sure) and being delectable as well. That feature is texture.

It is texture that gives meat or bread its appeal, and it is texture that makes rice accepted as a steady diet by more people than any other food. Introducing texture into new foods using low-cost raw materials can thus be counted as a major advance for chemistry and chemical engineering.

Producing meat analogs (the industry shies from the terminology meat substitutes) long defied the efforts of food chemists, however. Then in the mid-1950's, a process developed first at Ford Motor Co. and then at Drackett Co. was patented that spun protein fibers using techniques similar to those used by the man-made fibers industry. The spun fibers are precipitated or coagulated with acid and then impregnated with flavors, colors, and other ingredients to give edible structures similar to meat products. Swift, General Foods, National Biscuit, Worthington Foods, Ralston Purina, and General Mills developed the technology further, and today several products are made in commercial quantities, mainly for institutional and school lunch programs. The manufacturers predict increasing use of the products in prepared convenience foods.

To date, most of the technology of meat analogs has been based on soybeans. There is no reason, however, that protein from other sources might not be equally suitable. Fibers have been spun from corn, cottonseed, and peanut protein. When combined with soy protein fibers, they may give meat analogs with improved texture, flavor, and nutrition. Other possible protein sources include processed cereals, fish protein concentrate, and protein from single-celled organisms grown on carbohydrate or hydrocarbon (petroleum) media.

The economic advantage of bypassing animals is marked. Animals are only about 10% efficient in converting vegetable protein to meat. Meat analog processes are 85% efficient. When adequately developed, they will provide a technology that can feed people pleasantly and nutritiously using readily grown crops.

Furthermore, meat analogs aren't the only possibilities for future development. Pastas are textured wheat, and new technology permits the production of macaroni from mixtures of corn and soybean protein. Such innovations, if exploited, could open the way for greater use of corn protein in human foods.

Food from Nonfarm Processes. Practically all nutrients come from the biosphere and depend directly on traditional agriculture for their production. Agriculture will continue for the foreseeable future to be by far the cheapest way to provide most food. Scientists and engineers, however, may some day succeed in making significant amounts of nutrients efficiently and economically in nonagricultural processes. Their first successes have come in recent years with production of amino acids and vitamins. Now they are considering more ambitious possibilities for supplementing nutrient supplies in the future and are approaching commercial production with various degrees of success. Among the ways are harvesting single-celled organisms grown on waste carbohydrates or petroleum fractions and synthesizing carbohydrates and fats from simpler chemicals.

Much information has been developed in the past few decades on growing yeasts on waste carbohydrates such as paper mill solubles, whey, and molasses. It is clear that the growth rate of yeasts and protein recovery from them is many times faster than the production of food protein by any other plant or animal system. Furthermore, the process doesn't depend on sun and soil, it doesn't require as much labor as traditional agriculture, and raw materials are generally available and inexpensive. The protein concentrates have a fair amino acid pattern and are rich in some water-soluble vitamins.

Disadvantages, however, have more than offset these advantages. No one has achieved unquestioned palatability, for example. Moreover, a highly developed technology is required to build and operate complete manufacturing plants, and such plants also require large capital investments. Finally, yeasts produced by such methods contain unusually high levels of nucleic acids, and people don't metabolize nucleic acids fully. The reaction stops at uric acid, which may accumulate in the body when excessive amounts of yeast are eaten. The excess uric acid may cause physiological disturbances, such as gout and renal stones. Chemistry's challenge for making yeast into a suitable source of protein for people, then, lies in developing palatable products free of or low in nucleic acids and that may be produced in economically competitive plants.

Another approach to producing protein in factories rather than on farms lies in fermenting petroleum hydrocarbons. Production of what some people popularly call petroproteins was first conceived at a British Petroleum (BP) facility in France in 1960, and a number of major petroleum, food, and chemical companies soon joined BP in research and development work on the subject. What the French government considered satisfactory margins of safety in feeding experiments with swine and poultry led it to clear petroproteins for animal feed use recently, and BP has built petroprotein plants in Scotland and France to supply that market. Meantime, toxicological studies for use of petroproteins by people have been satisfactory on experimental animals using purified products, but long-range investigations are still needed before decisions can be made on their use by people. Also, many chemical and chemical engineering problems must be solved before such proteins are widely made and used.

Some technology already exists for making carbohydrates. Offering the most immediate promise is the so-called formose reaction. In this reaction, formaldehyde condenses in the presence of weak bases to form a complex mixture of sugars. The reaction unfortunately is only partially suitable as a route to edible products, since it produces many toxic intermediates. Fractionation and recently developed chromatographic isomer resolution techniques offer some hope for making such six-carbon sugars as glucose, galactose, and fructose. However, this or any similar approach will need some strong stimuli and a much changed economy before it becomes commercially feasible, especially in the face of world sugar prices of 2 to 3 cents a pound.

As with proteins and carbohydrates, chemists can also make fats, either to tailor-make them to meet some needs that now exist for special fats or possibly to increase the supply for growing numbers of people. Included are special confectionary fats, frying fats with optimum stability, stable and highly polyunsaturated fats, fat-containing convenience foods, and acceptable substitutes for dairy fats.

Long-range needs to be met include developing fats or fat substitutes that have few calories but at the same time satisfy the appetite and fats having unusual properties for persons having special dietary needs.

To make fats, one first makes the two major parts, glycerol and fatty acids. Esterification of glycerol with the fatty acids then leads to synthetic fats. Not any fatty acid is suitable in food, of course, and one must produce the essential ones. These include linolenic, linoleic, and arachidonic. Of these, linoleic is the most important. It is the principal fatty acid in many edible oils, but its synthesis except by tedious chemical means eludes the grasp of industrial chemistry. A future task for chemists and chemical engineers, then, is the practical production of linoleic acid, preferably from saturated fats such as waste fats, fats based on petroleum, by-product fats from animal slaughter, or hydrogenation fractionation systems.

Glycerol may be made by synthesis from propylene via allyl chloride or acrolein or by directing the formose reaction to it. Glycerol itself is a promising energy source that is metabolizable. It is well tolerated in human feeding, but it is more nutritious when it is esterified with acetic acid to triacetin (glyceryl triacetate).

The Fischer-Tropsch synthesis developed in 1923 makes aliphatic hydrocarbons with and without oxygen from carbon monoxide and hydrogen. It opened the possibility for making edible fatty acid–like compounds, but the reaction has never proved to be very suitable for edible fats. The most significant nutritional limitation is the presence of branched chains, which interrupt normal oxidative chain degradation. Hydroxy and keto acids also heavily contaminate Fischer-Tropsch products.

A more practical approach may lie in using the general olefin polymerization reaction that Karl Ziegler discovered in Germany in 1954. The significance of his discovery to the food industry is that the reaction produces long-chain hydrocarbons of controllable length without branching.

A Ziegler-based process for food production would start with carbon monoxide and hydrogen (chemicals readily available as a result of normal petroleum refining) and condense them to ethylene. This reaction is exothermic and maintains itself without further energy input. Ethylene would be recovered by liquefaction and then polymerized in a Ziegler reaction (at low pressure over a metal alkyl catalyst). The product would be a mixture of C_{12} to C_{18} olefins. Ozonation of these olefins would lead to saturated fatty acids similar to those in animal and vegetable fats.

Technology has thus advanced to the point of making saturated fatty acids from petroleum possible. However, competition from natural fats, the onus of eating products from "unnatural" sources, and concern over the role of saturated fats in the diet have until now kept such products off the market. The first two objections will become increasingly academic as per capita fat supplies dwindle. The third objection, concern over the relative roles of saturated and unsaturated fats in atherosclerosis, is less easily overcome and will depend on results from numerous research projects now under way and still to be undertaken.

The ultimate solution for best nutritional effect may be a balance of saturated and polyunsaturated fatty acids properly distributed in glycerides. Meantime, a longer-range problem is concerned with the increased intake of fat calories in the western world. If clinical nutritionists are correct, we should, as a people, reduce our daily intake of fats. However, our physiological makeup and our eating habits apparently are such that we continue to eat more and more fat. Ideally for the long term, then, we should design compounds that are efficacious and physiologically harmless for use as near-zero-calorie substitutes in our total fat diet.

Food chemists have already explored some approaches to this type of compound. Some low-calorie fat derivatives now available are sugar esters, diglyceride esters, hindered alcohol esters, and polyoxyethylene esters. They provide few calories because the body either assimilates or metabolizes them poorly. Thus, they can serve as fat-like fillers in our foods.

Meantime, fat products other than low-calorie ones have been proposed for meeting special needs. Medium-chain triglycerides are available for people who don't absorb fats properly. These products have roughly 10% fewer calories than normal triglycerides, and this approach points to some of the possibilities for the future for chemists, clinicians, and nutritionists to design fat-like products that are aimed at specific dietary needs.

Developing synthetic fats provides a unique challenge to chemists, chemical engineers, biochemists, food technologists, and others. Their development will also pose significant problems to economists and sociologists, since new products introduced on enormous scale must of necessity have an impact that will reverberate throughout the economy and throughout society.

SOCIAL AND ECONOMIC IMPACTS

Petroleum refining is the heart of the petroleum industry, whose total assets are exceeded in the U.S. only by those of agriculture and the combined public utilities (electrical, gas, communications). Petroleum, including natural gas and natural gas liquids, fills about 75% of the nation's energy needs today. Petroleum products fuel automobiles, aircraft, and trains; they provide heat for homes and buildings; they fire steam-electric generating plants; they provide almost all of the lubricants needed to keep a highly industrialized society running smoothly. Petroleum in addition is the source of close to 90% of the organic raw materials that the petrochemical industry converts to synthetic fibers, plastics, detergents, and a multitude of other products.

The degree of the nation's reliance on oil is evident in the consumption of petroleum products, which has more than doubled in the past two decades. In 1971, consumption reached 5.5 billion barrels—nearly a quarter of a trillion gallons or more than 1,000 gallons per person. This huge amount of material moves from source to point of use through the four main segments of the petroleum industry: production, transportation, refining and petrochemical processing, and marketing. Chemical technology is widespread in these operations. Examples include the formulation of drilling muds used in sinking oil wells, the use of acidizing to increase production from oil-bearing formations, application of the chemical engineering principles of fluid flow in pipeline work, and the development of means of extracting oil from new sources such as tar sands and oil shale. The focal point of chemistry in the industry, however, is the petroleum refinery, the classic example of continuous flow operation, and the work of industrial chemists and chemical engineers accounts very largely for the modern petroleum refiner's ability to convert crude oil into a host of products at an average total refining cost of less than 1 cent per pound of product.

The U.S. in 1971 had roughly 270 petroleum refineries whose processing capacity ranged from 300 to more than 400,000 barrels per day. The industry's total refining capacity exceeded 13 million barrels daily and accounted for more than 25% of Free World capacity.

Of the refinery products consumed in the U.S. in 1971, roughly 40% by volume was motor gasoline, 33% was fuel oil, and 7% was jet fuel. The remaining 20% encompassed a broad range of products, including kerosine, aviation gasoline, liquefied petroleum gas, lubricating oils and greases, and asphalt. Overall, some 3,000 products are made wholly or in part from petroleum as well as roughly 3,000 petrochemicals.

The raw materials that enter the petroleum refinery are complex mixtures of chemical compounds that number in the thousands. The refinery uses a variety of physical and chemical processes to separate and modify these hydrocarbons into usable form. Most products of the refinery are still mixtures, but the physical and chemical properties of the compounds they contain lie within a precisely defined range. Most of the hydrocarbons in gasoline, for example, boil at from 100° to 400° F., contain five to 12 carbon atoms per molecule, and have been chemically modified to improve the performance of the finished gasoline in the automobile. Some refinery products, such as benzene, are single compounds, processed where required to purities that approach 100%. Refinery products also may contain additives, such as the dispersants in diesel fuel and the oxidation inhibitors in home heating oil.

Most of the petroleum refining processes used around the world were developed by chemists and chemical engineers in the U.S., which today holds a preeminent position in petroleum refining technology. This technology is rooted in the many years of refining research and development that began to gain impetus in this country around 1910. The petroleum industry spends on the order of $500 million annually on research and development in a variety of subject areas. Broadly they include the composition and properties of petroleum, its occurrence and recovery, refining equipment and processes, petroleum products and their uses, and related problems such as those involving the environment. While nearly all aspects of such work involve chemistry in greater or lesser degree, this study confines itself to the contributions of the science in petroleum refining processes and their products.

CHEMICAL CONTRIBUTIONS

The modern petroleum refinery is designed to produce usable products, with minimum waste, from the thousands of different molecules in hydrocarbon raw materials or feedstocks. The methods used to achieve this goal fall into two broad classes: separation processes, which do not change the chemical nature of the molecules involved; and conversion processes, which do change the chemical nature of the molecules. The two are complementary in that a separation step normally is used to segregate the molecules to be chemically converted and often is used to segregate the new molecules created by the conversion.

Chemists and chemical engineers have upgraded these processes steadily over the years in response to changing economic conditions and market demand. The results can be seen in a general way in the yields of various products per barrel of crude oil. In 1920, the average yield of gasoline, the industry's premium product, per 42-gallon barrel of crude was 11 gallons or 26%. By 1971, the yield was approaching 50%, and the gasoline produced was of much higher quality than it was in 1920. In the same half century, of course, the yields of other petroleum products have changed as well. Refiners adapted their processes, for example, to meet the sharply growing demand for aviation gasoline in the 1930's and 1940's and for jet fuel in the 1950's and later.

The conversion processes used to raise the yield of gasoline and other products from crude oil function in part by breaking big molecules into smaller ones. One result has been a steady decline in the yields of certain heavier fractions of the raw crude. The yield of residual fuel oil, for example, was about 7% in 1971, less than half the yield in 1920. Today, in consequence, imports fill about two thirds of domestic demand for residual fuel oil.

Separation Processes

All separation processes aim at separating a mixture of molecules into two or more different streams or "cuts." A number of characteristics of the molecules are used to differentiate one stream from another. Among them are:
- Size: smaller molecules may be separated from larger ones.
- Linearity: molecules with a simple linear skeleton of carbon atoms may be separated from those with a more complicated, branched structure.

• Aromaticity: aromatic molecules, which contain the very stable benzene ring structure, may be separated from paraffinic molecules, which do not. Such characteristics are important in various products. Small molecules tend to boil or to vaporize at a lower temperature than do large molecules. The branched-chain paraffinic molecules in gasoline tend to have a higher octane rating than do the linear paraffins. In fuel oil, the branched molecules tend to crystallize less readily as filter-plugging waxes. Aromatic molecules also tend to impart higher octane ratings to gasoline.

Distillation. The fundamental process in refining, and the one most widely used by far, is distillation. In the past half century or so, chemical engineers have transformed that traditional art of the whiskeymaker into a highly efficient science. To do so they have developed extensive knowledge of the theory of fractional condensation and heat transfer and of the application of thermodynamics to fractional distillation. Progress has depended also on the development of materials of construction that withstand the heat and corrosion that accompany modern distillation processes.

Noteworthy among the many contributors to the field were chemical engineer W. K. Lewis and his co-workers at Massachusetts Institute of Technology. In the 1920's and 1930's, they advanced the theory of fractional distillation to the point at which equipment could be designed that would split multicomponent hydrocarbon streams predictably and consistently into the desired fractions.

Practically all crude oil that enters a refinery today goes to distillation units, where most of it is vaporized by heating to as high as 700° or 800° F. The refinery streams obtained by this fractional distillation process are called "straight run" products. Almost always they undergo further separation and/or conversion processes. The liquid remaining is drawn off to be processed into asphalt, heavy fuel oils, and the like.

The vapors ascend through a fractionating column, cooling as they rise and condensing when they have cooled to approximately their boiling points. The resulting liquids are withdrawn at heights on the column predetermined to yield fractions of the entering hydrocarbons that boil (or condense) in particular ranges of temperature. Since molecular size correlates roughly with boiling point, the molecules in each fraction or "cut" contain numbers of carbon atoms that fall within the corresponding range. The fraction withdrawn lowest on the column normally boils at 550° to 800° F. and contains 15 to 28 carbon atoms per molecule. Lubricating oils can be prepared from this cut. The cuts withdrawn progressively higher on the column contain smaller and smaller molecules. Highest on the column comes gasoline, and uncondensed hydrocarbon gases are taken from the top of the unit.

Many refineries also use vacuum stills, which operate at less than normal atmospheric pressure. The lower pressure allows more of the hydrocarbon charged to the still to vaporize at a given temperature. As a result, the still yields more of the economically desirable lighter hydrocarbons and less of the unvaporized products or residuum. Vacuum distillation was used as early as 1923, but only since World War II have refiners used it on a large scale. The main incentive has been to convert more of the incoming crude oil to distillate gas oil that can be subjected to catalytic cracking, one of the major conversion processes.

Solvent Extraction. Another important separation process used in petroleum refineries is solvent extraction. This process uses one liquid to dissolve specific molecules out of another, multicomponent liquid stream. The two liquids can then be separated into layers by gravity and the solvent layer drawn off and separated from the desired hydrocarbon by distillation.

In 1907, L. Edeleanu described the basic principles of a solvent extraction process using liquid sulfur dioxide as the solvent. In 1924, Associated Oil Co. of California adapted a European sulfur dioxide extraction process to refine kerosine. The company applied this first continuous solvent extraction process to lubricating oils in 1927.

The idea caught on rapidly, and in 1933 several large plants were built that used various selective solvents. E. Eichwald patented the use of furfural for extractive refining of lubricating oils in 1925. The compound is particularly effective, for example, in removing color bodies and molecules that contain sulfur or oxygen. Furfural has a decided affinity for such polarizable molecules and is used in a process pioneered by Texas Co. and Shell Oil in 1933. A number of other extractants have since been developed for refining lubricating oils.

Solvent extraction permits lubricants to be made from many different types of crude oil. Previously it had been possible to make them only from light-colored or paraffinic crude oils that did not contain unwanted contaminants. In 1970, the petroleum industry used various solvent extraction processes to make some 3 billion gallons of lubricating oils.

Solvent extraction also is used to separate the aromatic or benzene-toluene-xylene (BTX) fraction from hydrocarbon streams. This fraction is a major raw material for petrochemicals. For some years, the BTX fraction was separated from other hydrocarbons by distillation, but in the early 1950's distillation began to give way to the Udex solvent extraction process. This process, licensed by Universal Oil Products, uses diethylene glycol to extract the aromatics selectively. The extractant is then separated from the aromatics by distillation and recycled to the process. The Udex process produces high-purity aromatics at lower cost than straight distillation and has been used widely since it was introduced.

The growing demand for petrochemicals (and high-octane gasoline) has compelled refiners steadily to increase their extraction capacity for aromatics. They have tended to do so by developing more effective solvents, as opposed to building new and larger units. A trend has developed, for example, toward using tetraethylene glycol and other glycols instead of diethylene glycol in the Udex process, and, in the mid-1960's, Shell Oil commercialized a process based on another extractant, sulfolane.

Adsorption. A third separation process used in petroleum refining is adsorption. The adsorbent may be one of several different types of solid materials. Placed in a stream of liquid or gaseous hydrocarbons, it adsorbs and holds certain molecules on its surface and rejects the others. The adsorbed compounds can be recovered, if desired, by driving them from the adsorbent with heat and by other methods. Adsorption had significant though relatively limited utility in refining for some years, but in the past decade industrial scientists have increased its importance markedly by developing what are known as molecular sieves.

Molecular sieves are synthetic zeolites (aluminosilicate materials) whose highly-porous structure can be controlled closely in the manufacturing process. Pore diameter can be varied but lies generally in the range of a few angstroms (10^{-8} centimeter). This is in the same range as the cross-sectional diameters of various hydrocarbon molecules, a fact that is the basis of the molecular sieves' utility as a method of separation.

An early and important use of molecular sieves, developed in the early 1960's, was the separation of linear paraffins from branched and cyclic hydrocarbons. The linear compounds were needed to make the biodegradable detergents that came into production in the U.S. in 1965, and the sieves provided the means of obtaining them economically on a large scale. In essence a stream of hydrocarbon vapors passes through a bed of molecular sieves whose pore diameters are slightly larger than the cross-sectional diameters of the straight-chain, detergent-range paraffins (10 to 12 carbon atoms per molecule). The sieves adsorb these paraffins but reject the larger branched and cyclic molecules in the stream. The detergent paraffins are then recovered from the sieves for further processing.

Molecular sieve processes have been developed in this country by Union Carbide, Universal Oil Products, and Standard Oil Co. (New Jersey). The UOP process differs from the other two in using a liquid rather than a gaseous stream.

Crystallization. A fourth separation process used by refiners is fractional crystallization, in which the molecules to be separated are induced to form solid crystals that can be filtered from the hydrocarbon stream. Dewaxing, a form of crystallization pioneered by Texas Co. in 1927, is often aided by using antisolvents such as propane or various ketones. In a similar process, deasphalting, propane can be used to help the large, asphalt-like molecules clump together and settle from a refinery stream in separable form.

Crystallization processes also have been developed for obtaining p-xylene, the starting point for polyester synthetic fibers. The boiling points of p-xylene and m-xylene, which occur together in refinery streams, differ by less than 2° F. Thus, the two cannot be separated economically by distillation, the method that normally would be used. p-Xylene's melting point (55.9° F.), however, is about 110° F. higher than that of m-xylene. The difference is more than enough to have made fractional crystallization a likely separation process when a demand for pure p-xylene from petroleum became apparent. Standard Oil Co. (Calif.) put the first p-xylene crystallization unit on-stream in 1950, and other refiners have since developed versions of the process. The method consists basically of cooling a stream of xylenes and related compounds to about 0° F. The p-xylene crystallizes (freezes) and is filtered or centrifuged from the rest of the stream, which remains liquid. Such processes can produce p-xylene in 99.5% purity.

Conversion Processes

Early in the 20th Century, petroleum refiners saw that separation processes, no matter how sophisticated, would be unable to satisfy the radically changing market for petroleum products. The primary problem was the automobile, which showed every sign of creating an enormous demand for gasoline. Distillation could produce only the gasoline that crude oil contains naturally and would meet the demand only if refiners processed an uneconomically large volume of crude. Thus, industrial chemists began to seek means of converting other hydrocarbons in the crude to molecules in the gasoline range.

The first major result of this work, thermal cracking, appeared in 1913. Refiners since then have greatly expanded their ability to manipulate the hydrocarbon molecules in petroleum. Modern conversion processes not only yield products of much higher quality than could be obtained by straight-run distillation; they yield products that could not be obtained at all by that means, and they allow the refinery's product mix to be adjusted readily to accommodate changes in market demand. These processes, whose mainstay is cracking technology, account largely for the flexibility and economic efficiency of the petroleum industry today.

Conversion processes operate on hydrocarbon molecules made up of a skeleton of carbon atoms to which hydrogen atoms are attached in various numbers. The molecules in petroleum, however, may also contain atoms of other elements, such as nickel, vanadium, sulfur, oxygen, and nitrogen. These extraneous atoms generally are undesirable in petroleum products. To change the naturally-occurring molecules of petroleum into more useful ones, conversion processes aim at several different types of chemical change:

- A change in the number of carbon atoms in the molecule.
- A change in the ratio of hydrogen atoms to carbon atoms.
- A change in the arrangement of the carbon atoms, that is, in the structure of the molecule.
- Elimination of atoms other than carbon and hydrogen.

Thermal Cracking. The first commercial process for reducing the number of carbon atoms in petroleum molecules was the Burton thermal cracking process. It was invented by chemist William Burton and his associates at Standard Oil Co. (Ind.) and introduced in 1913. The primary purpose of the process (and of most other cracking processes) was to make gasoline molecules, containing primarily five to 12 carbon atoms each, from larger hydrocarbon molecules in crude oil.

Burton and his colleagues began their research in 1909. They found initially that the gas oil fraction of crude, when heated under pressure, yielded much more than the normal proportion of gasoline and much less coke. Further work defined the critical relationships between charge stock, boiling range, and the yields of gasoline and coke. Commercialization of the invention required also the development of adequate pressure control and mechanical handling systems. The first commercial Burton unit consisted of a battery of 8,250-gallon stills operating at 75 pounds per square inch pressure and yielding 23% gasoline. By 1919, U.S. refiners were producing more than 600 million gallons of gasoline annually by the Burton process.

Competitive thermal processes materialized quickly. Work by C. P. Dubbs at Universal Oil Products resulted in an improved process in which uncracked gas oil was removed from product gasoline continuously and recycled to the cracking still. Licensing began early in 1919. A number of other thermal cracking processes were introduced at about the same time. In 1920, Texas Co. introduced the Holmes-Manley process, a truly continuous one. Over the next 10 years, U.S. production of gasoline by thermal cracking increased to 6 billion gallons annually.

Thermal cracking reached its zenith in the 1930's, when the use of catalysts to facilitate and control the cracking process came on the scene. In 1937, the year that catalytic cracking was introduced, U.S. refiners produced 11 billion gallons of gasoline by thermal cracking. The process by now has been displaced entirely by catalytic methods, but its development represented nevertheless a significant contribution to the U.S. economy.

Catalytic Cracking. Although thermal cracking sharply improved the yield of gasoline from crude oil, it did not significantly improve the quality (octane rating) of the gasoline produced. Both goals were achieved by the catalytic cracking process. Catalytic cracking, like thermal cracking, produces gasoline-type molecules from larger hydrocarbons in petroleum. The catalyst, however, facilitates different types of chemical reactions than thermal cracking. These reactions yield both more gasoline and more higher-octane aromatic and branched-chain paraffinic molecules.

Catalytic cracking was developed by a French engineer, Eugene Houdry, who did much of his work in this country with financial support from Sun Oil and Mobil Oil. Mobil built the first semicommercial plant in 1937 followed by Sun Oil's first commercial Houdry unit. By 1971, U.S. catalytic cracking capacity had reached 120 million gallons of feedstock per day. Without the octane improvement obtained by the process, further development of the internal combustion engine would have been hampered severely.

In the Houdry process, hydrocarbons were cracked as they flowed over a stationary bed of catalyst at roughly 900° F. and up to 30 pounds per square inch pressure. The catalysts were both synthetic and natural silica-aluminas, such as clays. Tars and coke produced in the cracking process deposited on the Houdry catalysts and rapidly deactivated them. It was thus necessary to reactivate the catalyst by burning off the deposits, which was achieved by frequently switching the streams flowing through the catalyst bed from the hydrocarbon feedstock to the air required for burning and back to the hydrocarbon. Engineering development soon improved this alternating cracking-regeneration process. An example is Mobil's TCC process, in which the catalyst pellets were moved from a reaction zone to a regeneration zone and back again.

The next major step was the development of fluidized bed cracking. The step was taken partly under the impetus of the then-approaching U.S. involvement in World War II with its expected huge demand for high-octane aviation gasoline. The fluid bed was an outgrowth of the pioneering work of an independent inventor, W. W. Odell, in 1929. It used a powdered catalyst with particles only 0.0002 to 0.008 inch in diameter. The catalyst was maintained in a quasifluid state by the flow of air or hydrocarbon vapor and would flow readily from a reaction vessel to a regenerator and back again. The motion of the "fluid" catalyst within each vessel, moreover, helped to maintain the desired uniform temperatures. W. K. Lewis and his co-workers at MIT were significantly involved in the development of the process.

Standard Oil Co. (New Jersey) put the first fluid bed cracker on-stream in 1942, and the process today accounts for some 90% of U.S. catalytic cracking capacity. The principles of fluidization have also been applied to many other types of chemical processing.

Petroleum cracking catalysts basically have remained silica-alumina material but of a much more sophisticated type than was used originally. Intensive research on catalysis has led to great improvement of the physical and chemical performance of catalysts used in cracking and other catalytic refining processes. A major development in catalytic cracking has been the synthetic zeolites commercialized by Mobil Oil in the early 1960's. These catalysts are of the same class as the molecular sieves used in some hydrocarbon separation processes. They have allowed petroleum refiners to increase still further the yield of gasoline from a given amount of crude oil.

Hydrocracking. The development of specific families of catalysts has accounted in part for the success of hydrocracking, probably the most versatile of modern petroleum conversion processes. Hydrocracking is a form of catalytic cracking in which middle- or high-boiling hydrocarbons are heated to 500° to 800° F. at seven to 140 times atmospheric pressure in the presence of hydrogen and catalysts. The products include components for blending into gasoline, jet fuel, and diesel oil as well as charge stocks for further refinery processing. Hydrocracking was introduced in 1959. By 1970, U.S. capacity on-stream or under construction was approaching 1 million barrels of feedstock daily.

Hydrocracking can be used with feedstocks that range from the relatively light naphthas to the very heavy residual materials from crude distillation. Among the useful characteristics of the products are high chemical stability and, in the low-boiling gasoline fraction, high octane rating. The hydrogen used in the process acts in part to inhibit the undesirable reactions that lead to the formation of coke and gases. The hydrogen also reacts with the generally objectionable sulfur, nitrogen, and oxygen in the feed to form hydrogen sulfide, ammonia, and water, all of which can be readily removed.

Polymerization and Alkylation.   The thermal and catalytic cracking processes that came into wide use in the 1930's produced substantial amounts of light hydrocarbons such as ethylene, propylene, butylenes, and isobutane. These gases had to be used mainly as fuel, a relatively low-value market. At the same time demand for gasoline was growing steadily, and petroleum chemists thus began to seek means of converting these small molecules (two, three, or four carbon atoms each) into larger molecules in the gasoline range. The processes that resulted were polymerization and alkylation.

Polymerization produces gasoline-range molecules by joining two smaller olefins, hydrocarbons that contain fewer than their full complement of hydrogen atoms and thus are termed unsaturated. The product, polymer gasoline, was first made commercially by Pure Oil in 1931 by a thermal process. In 1934, Shell Oil introduced a process for polymerizing butylenes using sulfuric acid as a catalyst. The following year Shell commercialized a catalytic polymerization process using phosphoric acid as the catalyst. It was based on the work of V. N. Ipatieff at Universal Oil Products.

By 1939, nearly every U.S. refinery with cracking facilities was producing polymer gasoline by catalytic polymerization. The process provided a use for the growing amounts of gases from crackers as well as a great improvement in the octane rating of gasoline. During World War II, polymer gasoline was a most significant factor in meeting the demand for aviation gasoline. As recently as 1951, more than 150 UOP catalytic polymerization units were operating in the U.S. As the octane requirements of premium gasoline continued to increase, however, refiners found it more advantageous to make alkylate rather than polymer gasoline from the olefinic gases produced by crackers. Polymerization as a result is no longer a significant refinery operation in this country.

Alkylation resembles polymerization in that it joins two small molecules to make a larger molecule in the gasoline range. The difference is that alkylation joins not two olefins but an olefin and a paraffin, a hydrocarbon that contains its full complement of hydrogen atoms and is termed saturated. Pioneering work by V. N. Ipatieff and Herman Pines at UOP first produced an alkylate in 1932. It was the addition product of an olefin, butylene, and a paraffin, isobutane, and was made using aluminum chloride/hydrochloric acid as the catalyst. This research led to the development of the two major alkylation processes, one using sulfuric acid as the catalyst, the other, hydrofluoric acid.

Independent work by several petroleum companies, which then combined their knowledge, led to the first commercial alkylation unit. It was put on-stream in 1938 by Humble Oil, domestic affiliate of Standard Oil Co. (New Jersey). It used sulfuric acid as the catalyst. Ipatieff and Pines developed the hydrogen fluoride process, which was first used commercially in 1942. It operated at higher temperature than the sulfuric acid process, thus reducing the need for cooling equipment. Also, it eliminated the problem of acid sludge disposal inherent in the sulfuric acid process.

The sulfuric acid method has advantages of its own, however, and has remained the dominant process.

Alkylation, like polymerization, was developed just as the wartime demand for aviation gasoline began to materialize. By the end of World War II, U.S. refiners were producing more than 3 million gallons of alkylate daily. It remains today a prime ingredient in high-octane motor gasoline and is made at virtually every refinery of any size in the country. Current U.S. production exceeds 30 million gallons per day. More than two thirds of it is made by the sulfuric acid process and the remainder by the hydrogen fluoride process.

Catalytic Reforming. Catalytic reforming supplements catalytic cracking by converting relatively low-octane naphthas into high-octane gasoline components. In part the process changes the structure of hydrocarbon molecules, but the main reactions involve changes in the ratio of hydrogen to carbon in such molecules. These changes are made by removing hydrogen atoms from the molecules (dehydrogenation). Catalytic reforming, for example, converts hexane, a straight-chain, six-carbon compound, into cyclohexane, a cyclic six-carbon compound. Then it removes six hydrogen atoms from the cyclohexane to yield benzene, a stable, high-octane aromatic compound.

Since the process increases octane rating mainly by converting other hydrocarbons to aromatics, it is also an important source of aromatic raw materials for the petrochemical industry. A reformer can be operated, for example, to produce relatively high amounts of toluene and xylenes for petrochemical use. Because of the dehydrogenation that takes place in catalytic reforming, the process also is an important source of hydrogen (in about 80% purity). Refiners use this off-gas in hydrocrackers and other refinery processes.

Standard Oil Co. (Ind.) developed the first catalytic reforming process, which came into use in 1940. It used a silica-molybdena catalyst in the presence of hydrogen. The process was a major step forward, but the catalyst lost its activity rapidly and had to be regenerated at frequent intervals. Other refiners developed reforming processes and catalysts, but the next major step was the introduction of the platinum catalyst by Universal Oil Products in 1949.

The UOP process, Platforming, was a radical departure from traditional practice. It was the first process to use a precious metal catalyst. Like earlier processes using molybdena catalysts, it used hydrogen in a system designed in part to remove that element partially from hydrocarbons.

Platforming was extremely effective and in improved form remains the dominant catalytic reforming process today. Virtually all such processes developed since then employ platinum-containing catalysts of one type or another. The petroleum industry now uses catalytic reforming to treat more than 100 million gallons per day of feedstock, or close to 25% of the crude oil that enters refineries.

Dehydrogenation. Of the several conversion processes that occur together in a catalytic reformer, dehydrogenation is one that, in more intensive form, is used widely on its own. Important applications include the conversion of butane to butadiene and of ethylbenzene to styrene. Both butadiene and styrene are used in large volume to make plastics and synthetic rubbers. Dehydrogenation, like catalytic reforming, also produces significant amounts of hydrogen in about 80% purity for recycling to other refinery processes.

Thermal dehydrogenation has been used with gaseous feeds, such as ethane and isobutane, but catalytic methods are generally more selective. Early workers on thermal dehydrogenation included Hague and Wheeler. They showed in 1929 that thermal pyrolysis of the paraffinic compound, ethane, drives two hydrogen atoms from the molecule to form the olefinic compound, ethylene. (Both paraffins and olefins are straight- or branched-chain hydrocarbons, but paraffins are saturated with hydrogen and olefins are not.) In the 1930's, the chemical equilibrium between saturated hydrocarbons and unsaturated hydrocarbons plus hydrogen was studied extensively by Frey and Huppke. Their work was instrumental in the development of selective catalytic dehydrogenation and gave impetus to the commercial production of olefins from paraffins.

Among the first catalytic dehydrogenation processes were those developed by Houdry Process Corp., by Shell Oil Co., and by Standard Oil Co. (New Jersey). They were first used commercially during World War II to convert butane to butadiene for use in styrene-butadiene rubber.

A related group are the hydrogenation processes in which hydrogen is added to olefins to produce the more stable saturated paraffins. A good deal of the isooctane needed to fuel the Spitfires in the Battle of Britain was obtained by hydrogenation.

Isomerization. Branched-chain hydrocarbons have higher octane ratings than do the corresponding straight-chain compounds, and petroleum chemists have thus developed methods for rearranging the structures of gasoline-range molecules accordingly. The process is called isomerization and normally involves a catalyst. Isomerization was first used commercially to convert straight-chain or n-butane to branched-chain or isobutane. The isobutane was then added to aviation gasoline directly or reacted with olefins to make alkylate for use in aviation gasoline. Many isomerization processes have been developed since. In general they are used to isomerize straight-chain hydrocarbons containing four, five, or six carbon atoms per molecule.

Isomerization is used also for purposes other than improving the octane rating of gasoline components. One example involves the eight-carbon aromatics, ethylbenzene and the three isomers of xylene, which normally occur together in refinery streams. One of the isomers, p-xylene, is a raw material for polyester fibers, whose sharp growth in the 1950's (and since) began to outrun the supply of p-xylene that could be obtained from the aromatics stream by the normal separation methods. One solution to the problem was the Octafining process. It was developed by Atlantic Richfield Co. and Engelhard Minerals and Chemicals Corp. and was first used commercially in 1960. The process isomerizes part of the ethylbenzene, o-xylene, and m-xylene in the aromatics stream to p-xylene, which then can be obtained in greater-than-normal yield by the usual separation methods.

Coking and Hydrotreating. Crude petroleum often contains relatively small amounts of extraneous elements such as nickel, vanadium, sulfur, and nitrogen. These elements usually are undesirable in petroleum products and processes, and methods have been developed to remove them.

Metal atoms usually are associated with the very largest molecules in crude oil. Most metals are thus removed from the main petroleum streams in the residual material from vacuum distillation or in a deasphalting process. Another means of demetallizing a refinery stream is coking. In this process the heavy feedstock is heated in the absence of air until most of the hydrocarbons have cracked off. The material that remains consists of very heavy tar or coke deposits along with most of the metals. The petroleum coke from this process can be burned as fuel or calcined for use as carbon or graphite.

The U.S. produces some 8 million tons of petroleum coke per year, more than 80% of the world's supply. About half of it is used to make carbon anodes for electrolytic production of aluminum and about one tenth to make graphite. Nearly all the rest is used as fuel.

Of this country's output of petroleum coke, roughly 88% is made by delayed coking and 12% by a fluid bed process. Delayed coking was developed by Standard Oil Co. (Ind.) in 1931. It yields several different forms of coke, depending on the feedstock. Fluid coking was developed by Standard Oil Co. (New Jersey) in 1954. It produces less gasoline than does delayed coking, but the gasoline is more easily reformed to higher octane numbers.

Sulfur and nitrogen, like the metals in crude oil, are present primarily in the large hydrocarbon molecules. They occur also, however, in smaller molecules and, if not removed, would find their way into the premium petroleum products. Sulfur and nitrogen usually are removed by treating refinery streams with hydrogen at high temperatures and pressures in the presence of a catalyst. This process, generally called hydrotreating, has been known since the early 1930's. Its use did not grow rapidly, however, until the 1950's, when large supplies of relatively cheap hydrogen became available as a by-product of catalytic reforming.

In hydrotreating, hydrogen atoms combine with sulfur, nitrogen, and oxygen to form hydrogen sulfide, ammonia, and steam, which can be easily removed. The hydrogen also tends to saturate the petroleum molecules, leading to more stable products. Many hydrotreating processes are available to refiners. The current emphasis on improving the quality of the atmosphere has required lower and lower sulfur content in fuel oils, including the heavy residual fuel oils, and hydrotreating processes for these heavy feedstocks have become available. Desulfurizing installations now handle more than 30 million gallons a day.

Refiners use a variety of chemical treating processes other than hydrotreating. Their goal in general is to upgrade refinery streams and products by eliminating undesirable characteristics: color, odor, acidity, and others.

Petroleum Products

Petroleum products traditionally are divided into four categories:
* Fuels: gasoline, diesel fuel, jet fuel, heating oil, residual fuel oil.
* Lubricants: engine oils, industrial lubricants and greases.
* Asphalts: paving and roofing materials, other watertight sealants.
* Specialties: solvents, agricultural spray oils, hydraulic oils, waxes.

The formulation of the 3,000-odd products made wholly or in part from petroleum depends heavily on thorough knowledge of the chemistry of the application. Thus, the chemistry of liquid-solid interfaces is vital in formulating not only lubricants but many other products: gasolines that keep carburetors clean; asphalts that adhere properly to crushed rock; jet fuels that do not pick up electrostatic charges from filters; diesel fuels that do not precipitate large, filter-clogging wax crystals. Among the variety of other chemical problems in formulating petroleum products is the development of additives: antioxidants, rust preventives, pour-point depressants, and the like.

Motor Gasoline. Motor gasoline is the primary product of U.S. petroleum refiners, who now make more than 250 million gallons of it daily. The quality of gasoline, as measured by octane rating, has improved steadily over the years in parallel with the increasing efficiency of the internal combustion engine. The improvement has resulted mainly from two developments: the use of tetraethyl lead (and related compounds), whose antiknock properties were discovered in the 1930's by chemist Thomas Midgley at General Motors; and refiners' growing ability to use chemical processes to convert low-octane hydrocarbons into high-octane blending stocks for gasoline.

Increasingly stringent air pollution regulations are causing continued reduction in the amount of lead antiknocks that can be used in gasoline. These compounds reduce the efficiency of prospective exhaust emission control systems, and considerable controversy exists over the effects of lead itself as an air pollutant. Chemical conversion processes, as a result, will be used ever more widely to supply the alkylate and aromatics required to achieve optimum octane ratings in gasoline.

In addition to knock-free performance, automotive gasoline should provide easy starts and rapid warm-up. These characteristics are built into the fuel by including suitable amounts of volatile hydrocarbons that provide combustible fuel-air mixtures to the engine even before it is fully warmed up. Volatility must be controlled carefully, however, because too much can cause vapor lock or carburetor icing.

Gasoline additives are used not only to improve octane rating but other properties as well. In the late 1920's and early 1930's, Standard Oil Co. (Ind.), among others, developed several antioxidants and metal deactivators to stabilize gasoline against oxidation during storage. Mobil and Sinclair pioneered with corrosion inhibitors for gasoline to protect metal surfaces in pipelines and storage tanks against rust. The first carburetor detergent additive, developed by Standard Oil Co. (Calif.), was added to gasoline in 1953. The use of detergents has expanded to the majority of gasolines today. These materials minimize deposits in the venturi and throttle body section of carburetors, thus helping to maintain uniform adjustment. Proper carburetor adjustment improves gasoline mileage, especially in stop-and-go town driving, and also reduces exhaust pollutants.

Many of the carburetor detergents function also as anti-icing additives. Carburetor icing occurs primarily in cool, damp weather. The evaporating gasoline causes atmospheric moisture to form ice, which deposits around the throttle valve and starves the engine for air, particularly during idle. The first anti-icing gasoline was developed in 1951 by Standard Oil Co. (New Jersey). Du Pont introduced an additive that modifies the surface of the ice crystals, keeps them small, and prevents them from adhering to the metallic surfaces of the throttle.

The first combustion-control additive was tricresyl phosphate (TCP), introduced by Shell in 1953. TCP and related phosphorus compounds aim to modify combustion chamber deposits and reduce their tendency to initiate combustion before the spark occurs. Mobil used the first engine detergent/dispersant in 1968. Among recent motor gasoline additives is hydrogenated tallow amine (HTA) introduced in 1971 by Humble, the domestic affiliate of Standard Oil Co. (New Jersey). HTA is designed to minimize the amount of gasoline that clings to the walls of the intake manifold and thus provide more even distribution of the fuel-air mixture to the cylinders to prevent starving the engine for fuel during acceleration.

Diesel Fuels. Diesel fuels account for roughly 40 million gallons per day of U.S. refinery output. Of this amount, about 50% is consumed by trucks and buses, about 30% by railroad locomotives, and about 20% by industrial, marine, and military power plants. Major performance criteria of diesel fuels are sulfur content, cetane number (a measure of ignition quality), flow at low temperature, and ease of volatilization.

Sulfur contributes to engine deposits and wear, and its reduction by refinery treatment of diesel fuels has been a significant improvement.

Cetane improvers, largely alkyl nitrates, were first introduced by Ethyl Corp. They provide easier starting, quicker warm-up, and decreased roughness.

A variety of additives is used to improve the storage stability of diesel fuels, which permits the use of otherwise unstable petroleum stocks. Amines, sulfonates, and other types of additives have been used as detergents and dispersants. The detergents maintain the cleanliness of the fuel pump, injector, and nozzle and can markedly increase time between nozzle overhauls. The dispersants have achieved a similar improvement for fuel filters. Many diesel fuels also contain rust preventives.

Recent air pollution legislation has increased the interest in using additives to reduce diesel smoke. The most effective of these additives are barium compounds. Effective concentrations of barium additives are quite costly, however, and the ash that forms in the combustion process could itself be a problem in diesel engines.

Aviation Fuels. U.S. consumption of aviation fuels currently is some 55 million gallons daily, of which about 90% is jet fuel. The demand for jet fuel first exceeded that for aviation gasoline in 1958, and the gap has widened steadily since then.

Commercial kerosine was the fuel used in early development work on jet aircraft in the U.S. It was chosen over gasoline because its low volatility minimized the danger of vapor lock in certain flight conditions and because it was available as a commercial product of uniform characteristics. The first military jet fuel, JP-1, was highly refined kerosine having a very low freezing point (−76° F.). Current commercial jet fuels are mainly kerosine, while most military jet fuels approximate a blend of one third kerosine and two thirds gasoline components.

Jet aircraft are designed so that the fuel, in addition to providing energy, serves as a coolant for lubricating oil and other components of the aircraft. The fuel must, therefore, be sufficiently stable, so that the heat involved does not cause it to deposit lacquer and gums. Such deposits would reduce the efficiency of heat exchangers and clog the filters and valves in aircraft fuel-handling systems. Current jet fuels have satisfactory thermal stability for aircraft operating at speeds of up to about Mach 2 or twice the speed of sound. Jet aircraft of the future, operating at up to about Mach 3, may require more stable fuel. The development of turbojets that operate at Mach 3 or 4 and of ramjets that operate at Mach 6 or more will impose even more stringent requirements on jet fuels.

U.S. demand for aviation gasoline peaked in 1958 and has been declining since then. The technology for producing aviation gasoline is essentially the same as that for automotive gasoline. That is, the same chemical conversion processes are used to produce gasoline molecules with the necessary characteristics.

Home Heating Oil. More than 25 billion gallons per year of No. 2 heating oil are burned in the U.S. to heat homes and small commercial buildings. This heating oil must satisfy a number of requirements. It should form no sediment in storage and leave no measurable quantity of ash or other deposit when burned. It should remain fluid at the temperatures encountered in winter storage. Its composition must be controlled to help reduce smoke emission. The fuel should be relatively low in sulfur to help control both corrosion and the emission of sulfur oxides upon burning. Domestic heating oils in addition should have a light color, an attractive appearance, and an acceptable odor.

These requirements for heating oil are met partly by refinery processing and partly by additives. Refining techniques have reduced both the carbon residue and sulfur content of domestic heating oils by more than 30% in the past 15 years. Ever lower sulfur requirements are being imposed now by air pollution regulations, and refiners are meeting these, too, although at some economic penalty.

The additives of prime importance for No. 2 heating oil are those that reduce sludge formation and prevent filter plugging. The first of these was introduced by Mobil about 1930. Most additives of this kind are oxidation inhibitors and dispersants that inhibit the formation of organic sediment. A secondary effect of such compounds is to keep suspended, in finely dispersed form, the small amount of sediment that persists in heating oil. Metal-based materials were used at first, but most inhibitors today are nonmetallic amines. Low-temperature pumpability is essential in No. 2 heating oil, whose low-temperature flow properties are improved by pour-point depressants. Other additives in common use include rust inhibitors, color stabilizers, and combustion improvers.

The improvements thus effected in No. 2 heating oil have resulted in more efficient, less costly heating systems. The fuel is cleaner burning and causes less line plugging and other problems associated with outdoor storage in consumers' tanks. The improvement in the odor of heating oil benefits those who store it indoors. The oil is more stable in storage, and the time between tank cleanings and oil filter changes has been extended significantly.

Residual Fuels. Residual fuels contain some fraction of the residuum from crude oils—the heavy cut that does not vaporize even at the high temperatures used in refinery distillation equipment. Burning such fuels requires more sophisticated equipment and control procedures than are available to the home owner and other small users. Thus, residual fuels are used typically to provide steam and heat for industry and large buildings, to power ships, and to generate electricity. U.S. consumption of residual fuels averaged more than 90 million gallons daily in 1970. U.S. refiners usually find it attractive to convert as much of their crude as possible to more valuable products, so that about two thirds of the residual fuel oil burned in the U.S. is imported.

Residual fuels contain fractions that are difficult to burn quickly under "cold" conditions. Accordingly, such fuels are burned as a rule in equipment that permits relatively steady operation in an environment where firebox temperatures can be high. Extraneous constituents of the residuum, such as sodium, vanadium, sulfur, and nitrogen, cause problems in burning residual fuel by contributing to equipment corrosion and air pollution. Ways used to minimize such problems include equipment modifications, fuel additives, fuel desulfurization and demetallization, and flue gas cleaning.

Engine Oil. Engine oils consist of refined oil-base stocks plus additives. They must meet a multitude of service requirements that vary with the application: automotive engines, the several types of diesel engine, outboard and other two-cycle engines, reciprocating and gas turbine aircraft engines, and others. Among the requirements to be met are wear protection, lubrication under heavy loads, stability against oxidation and heat, control of corrosive acids, resistance to foaming, detergency for removing deposits, and performance over wide ranges of temperatures. Moreover, lube oil additives function in very complex ways, detergents alone achieving their effect by at least three different mechanisms.

Roughly two dozen basic types of additives are used in engine oils. Several varieties of many of them have been developed to maximize desired performance characteristics. At least six zinc dialkyldithiophosphates, for example, are used as antioxidation and antiwear additives.

An important development of the 1950's was the perfection of the multi-viscosity-grade oils, which offered the motorist a lubricant that performed well at both winter and summer temperatures. Very severe processing in the refinery, plus large amounts of viscosity-improving additives, produced engine oils having the operating characteristics of both SAE 10W and SAE 30W grades. These products combine the quick flow and easy engine starting of 10W oils with the lubricating-oil mileage and engine protection of 30W oils.

Improvements in automotive engine oils have extended the oil drain period from an average of 2,000 miles in 1957 to more than 4,000 miles in 1971. The problem was complicated by the heavier load imposed on auto engine oils by the positive crankcase ventilation (PCV) valves, which were introduced to minimize hydrocarbon emissions from the crankcase. The new ashless dispersant additives, however, have made possible extended-drain oils that keep the PCV valves in good operating condition. For the most part the service has been satisfactory.

Chemists and other technologists continue to solve problems of the same general nature for nonautomotive engines. Among other products, they have developed ashless detergents of the amino-amide or succinimide type for two-cycle engines; specially formulated oils that do not corrode the silver wristpin bearings in railroad diesel engines; oils with reserve alkalinity to give cleaner operation and less corrosion in marine diesel engines; synthetic oils, usually polyesters, for aircraft gas turbines.

The importance of chemical additives in engine oils is evident in the growth in the number of companies that sell them to the oil industry. During World War II, three companies at the most had significant business in this field. Today there are at least 10 such companies that, besides selling individual additives, often formulate packages or combinations of additives for their customers.

Automatic Transmission Fluids. When the first automatic transmission, Oldsmobile's Hydramatic, was introduced in 1938, the only transmission fluid available was the one developed for the purpose by Mobil and Oldsmobile. The Hydramatic fluid performed satisfactorily, but much more effective automatic transmission fluids have since been developed.

Such a fluid nowadays must be a mild, extreme-pressure lubricant having a variety of other properties. It must be nonfoaming, resistant to oxidation at up to 300° F., nonrusting and noncorrosive, free-flowing at −20° F. or below, capable of allowing smooth engagement of the transmission clutches and bands without excessive slipping or grabbing, and able to transfer up to 1,500 British thermal units (B.t.u.) of heat per minute to the oil cooler and transmission case for dissipation to the atmosphere.

Lubricating Greases. A lubricating grease is a lubricating oil or other lubricating fluid that has been thickened to control flow. The greases containing oil are the most important commercially, and essentially all of the grease made by the petroleum industry is of this type. Grease improvements to a large degree have followed the pattern of lubricating oil. Base oils and additives are selected to give maximum resistance to heat and oxidation and to provide antirust and extreme-pressure lubricating properties. Thickening agents are selected to give the proper balance of cost and resistance to heat, water, and mechanical degradation. And the grease manufacturing process itself has undergone extensive development.

Since about 1946, improvements have been made in all three components of a grease: thickeners, fluid, and additives. New thickeners that have been introduced include modified clays, polyureas, and complex soaps of calcium and aluminum. All of these are multipurpose in that they are both water resistant and high melting. Lithium soap greases were developed during World War II and today constitute more than 50% of the volume of grease sold in the U.S.

Opposing trends have caused grease production to remain fairly constant in recent years. The general rise in industrial activity has been counterbalanced by the extended relubrication intervals made possible by improvements in greases. "Lifetime" greases are in particular demand for automobiles and electric motors. The manufacturing and mobile equipment industries are asking for greases that allow them to reduce still further the frequency of relubrication without endangering their operations or equipment. Greases with improved resistance to oxidation, heat, water, and mechanical breakdown are finding an expanding market, as are greases that protect against rusting and lubricating-film failure.

Asphalts. The heaviest fraction of a great many crude oils includes natural bitumens or asphaltenes and generally is called asphalt. This material, the oldest product of petroleum, has been used throughout recorded history. The U.S. in 1970 consumed more than 32 million tons of asphalt products. About 80% of the total went into paving products, and the rest was split about equally between roofing products and a variety of other uses. About 90% of the existing paved highways, roads, and streets in the U.S. are surfaced with asphalt.

Crude asphalt, like nearly all other petroleum products, is upgraded in utility by various chemical processes and additives. Asphalt products for roofing and other waterproofing service are generally modified by airblowing. Blowing hot asphalt with air increases the temperature at which the material becomes too soft and tacky, thus increasing its suitability for many services. Asphalts also are emulsified with water or modified with certain chemicals to upgrade their quality as cold coating and paving materials. These materials can be used with wet aggregate and thus increase the time during which roads can be paved in areas of normal or high rainfall and increase the likelihood of obtaining a good pavement.

Asphaltic cements, liquid asphalts, and asphalt emulsions meet a variety of requirements for paving materials. The ancient application of asphalt in hydraulic structures has received new and expanded emphasis in recent years in the lining of reservoirs, canals, swimming pools, and sewage lagoons for seepage control. Other new uses include the facing of earthen dams and river banks for erosion control and stabilization of breakwaters, seawalls, and the like.

Further uses for asphalt are developing as an extender in polymer-asphalt systems. Thermosetting and thermoplastic compositions are used as special coatings, cements, and adhesives. Promising results also have been obtained with asphalt in soil stabilization, as a mulch, and as an underground moisture barrier in sandy soil.

Specialty Products. Products of the petroleum industry that are produced in relatively low volume compared to the fuels, lubricants, and asphalts are classified generally as "specialty products." They include solvents, agricultural spray oils, hydraulic oils, electrical oils, rubber extenders, waxes, and the like. Often they are custom-made to fit a particular customer's unique specifications.

The most widely used of the specialty products are the solvents. Their uses range throughout industry, from perfume extraction to rubber cements, from dry cleaning to degreasing, from paints and varnishes to printing inks. The critical specifications of solvents depend largely on end-product use. Most, however, have comparatively narrow boiling ranges. The absence of low-boiling compounds reduces fire hazards, and freedom from high-boiling compounds reduces the amount of residue left after evaporation. Air pollution regulations in recent years have imposed new requirements on solvents. In some areas, for example, solvents for certain applications must not undergo oxidant-forming photochemical reactions in the air.

Waxes constitute another widely used group of specialty products. They are used mainly for coating and impregnating packaging papers, where they are often combined with modifying resins. The waxes find uses also in candles and polishes, in waterproofing electrical equipment, in cosmetics, and in rubber compounding.

FUTURE DEVELOPMENTS

Many technical trends can be singled out in petroleum refining, but the four most significant are:
• Decreasing use of lead antiknock compounds in gasoline.
• Increasing conversion of various petroleum fractions into gaseous fuels.
• Increasing reliance on "artificial" crudes from shale, tar sands, or coal.
• Increasing emphasis on larger, more automated, cleaner refineries.

For petroleum refiners today, the central fact is that stringent regulation of automotive exhaust emissions may well require the elimination of lead antiknock compounds from gasoline, as lead compounds tend to limit the effectiveness of advanced auto emission control devices.

Without the lead compounds, today's overall average gasoline pool would drop from the current 96 octane rating to about 88, a level well below that needed for current or expected cars. As lead is reduced, refiners will have to increase the octane rating of their gasolines. They will do this by more intensive processing to give them more of the high-octane blending stocks—primarily more aromatics obtained from advanced reforming processes. Isomerization and alkylation will also see increased use.

The future will see decreasing availability of cheap, high B.t.u. (high heating value) natural gas. The gasification of petroleum fractions, particularly the lighter ones, to high B.t.u. gas is now used or planned in a number of U.S. locations to meet peak natural gas demands. Longer range, these fuels will generally be too costly for manufacturing high B.t.u. gas, but the gasification of some residual oils, particularly high sulfur- and high metals-containing materials, to low B.t.u. gas will occur. This low B.t.u. gas will generally be a new refinery product to be sold to nearby power stations for use as a clean boiler or turbine fuel.

Chemists and chemical engineers will help petroleum refiners in the U.S. learn to handle an ever increasing amount of "artificial" crudes. These crudes, obtained from shale, tar sands, or coal, are made up of considerably different mixtures of molecules than the "natural" petroleum crudes now generally being refined. Shale oil, for instance, has a high fraction of nitrogen-containing molecules, and oils made from coal contain many molecules where the carbon atoms form multiple rings.

Shale oil is obtained by retorting, that is, distilling oil from crushed shale rock heated in the absence of air. Technology is available for producing shale oil at somewhat above competitive costs in the U.S. if satisfactory shale resources are made available and if environmental requirements are reasonable. Operation of the first few plants would provide answers to remaining technological and economic questions. An attractive possibility for significantly reducing the cost of shale oil would be the development of in situ retorting, that is, the recovery of shale oil from heating unmined shale rock.

The technology for the conversion of coal to liquid products still requires a major step forward before it can generally be applied economically without special incentives.

Tar sands technology is further developed than that of the other synthetic liquids base materials, and commercial ventures are now under way. One plant operated by Great Canadian Oil Sands is producing nearly 2 million gallons a day. Improvements in tar sands recovery will be made by improved mining and overburden removal in a hostile climatic environment.

Another continuing trend in petroleum refining is the one to larger refineries. The investment-related costs for refining are escalating rapidly. In addition to the pressure of inflation, extra costs will occur because of the need for more stringent environmental control and the need to reduce utilities consumption as the costs of energy and water increase. The economies of scale will dictate ever larger plants. Ecological pressures to prevent acquisition of new refinery sites will reinforce this incentive to build larger plants or to expand existing plants where possible.

These larger plants, however, will be carefully designed to avoid excessive costs and potentially huge losses because of poor service factors. Improvements in equipment sizing and construction techniques will lead to refineries of increasing capacity—approaching 40 million gallons of crude capacity per day in the near future. Developments in computer control technology will help make these large refineries possible. The economic driving forces toward the "push-button" refinery will be improved yields of more valuable products, higher throughputs, improved stability, and lower manning per unit of capacity.

Advanced chemical engineering technology will permit these larger refineries of the future to be essentially pollution free. Particulate emissions will be controlled by advanced electrostatic precipitator designs, scrubbers, or other techniques. Sulfur oxide and nitrogen oxide emissions from process furnaces will be controlled by new furnace burner designs and fuel cleanup processes. Atmospheric pollution from hydrocarbons will be held to a minimum in strictly regulated areas by utilizing totally enclosed systems. For tankage, tanker, and tank truck/car loading operations, vapor collection systems will be commercially available to control hydrocarbons from these sources almost completely. New rotating equipment and valve sealing devices and equipment erection techniques will be available to minimize leakage from flanges and seals.

Biological units for waste water treatment will be replaced to a large extent by more reliable and flexible processing schemes. Development of suitable water treatment processes using activated carbon will be accomplished by the mid-1970's. These processes will be augmented by the application of sophisticated treatment processes using ion exchange, liquid membranes, reverse osmosis, ozonation, and electrochemical treating.

Waste refinery solids, comprised largely of catalysts and desiccants of different types, will be shipped to catalyst manufacturers, metals recovery businesses, or to other industries that can use the solid as a raw material in the production of a usable product.

Noise pollution will receive considerable attention in the future. The ability will exist to predict and attenuate plant noise for much larger systems than exist today. Improved hardware will be available to abate noise from traditionally troublesome facilities such as rotating equipment, pressure reducing valves, burners, and air fins eventually to a level not in excess of 45 dBA in residential areas adjacent to plants. This level is approximately equivalent to the rustling of leaves from a light wind.

SOCIAL AND ECONOMIC IMPACTS

The nuclear industry is based on naturally-occurring fissionable and potentially-fissionable materials that originally were characterized and now are obtained and processed by methods that rely heavily on chemistry and chemical engineering. The industry was born in 1938. In that year, several decades of research on the structure of the atom was climaxed by the discovery of nuclear fission by the German chemists Hahn and Strassman. The subsequent massive U.S. effort on the atomic bomb laid the scientific and technological base from which the nuclear industry has grown.

In 1970 in the U.S., shipments of nuclear products from privately-owned establishments were valued at just over $900 million. Roughly 85% of that amount involved nuclear power plants and related equipment. The other 15% involved products used in a broad range of applications. Such products include radioactive isotopes, radiation sources, isotopic power devices, radiation detection and monitoring devices, and control and measuring devices that contain radioactive isotopes.

Nuclear power currently accounts for just under 3% of U.S. electric generating capacity. The Atomic Energy Commission and other authorities expect that by 1990 nuclear power plants using basically today's technology will be supplying about half the nation's demand for electricity or about 25% of the total demand for primary energy. The fast breeder reactor, the next major technological step in nuclear power, is expected to come into commercial use after 1985. Controlled fusion seems unlikely to be a significant source of nuclear electric power until well into the 21st Century. The currently projected growth of nuclear power, at any rate, will involve a variety of chemical problems in areas such as fuel preparation and reprocessing, waste treatment and disposal, corrosion, and heat transfer media.

These forecasts for nuclear power are conditioned by various environmental difficulties, a problem which is not, of course, unique to that source of electricity. Nonnuclear generation of electricity is expected by the year 2000 to be consuming more than 800,000 tons of coal annually, 60% more than is used for the purpose today. Mining and burning coal at that rate presents a considerable environmental problem in itself. These and other aspects of the nation's energy-environment dilemma are not properly within the scope of this study, but it is certain that chemical science and technology will play a large part in resolving that dilemma.

The 15% of the nuclear industry that does not involve nuclear power plants is based mainly on radioisotopes, the radioactive species of the chemical elements. The radioactive isotopes have a variety of uses in areas that include agriculture, crime detection, industrial measurement and control, medicine, and space and terrestrial power sources.

The commercially-useful radioisotopes have four general sources. Some, such as radium-226, are found in nature. Some, such as cobalt-60 and carbon-14, are made intentionally in nuclear reactors. Some, such as strontium-90 and cesium-137, are made unavoidably in nuclear reactors through the fission process. Some, such as sodium-22 and iodine-123, are made in accelerators (cyclotrons). Regardless of source, however, chemical methods must be used as a rule to separate the isotopes in usable form.

Among the uses of chemistry is the synthesis of compounds in which an atom of the normal carbon-12, for example, is replaced by an atom of radioactive carbon-14. Detection and measurement of the radioactivity from these labeled compounds has become a powerful tool in research on the course of complex chemical and biochemical reactions. The isotopes used most commonly in labeled compounds are carbon-14, hydrogen-3 (tritium), phosphorus-32, and iodine-131. The current retail market for labeled compounds is estimated at $14 million annually.

This market does not include labeled pharmaceuticals, whose sales currently are estimated at $40 to $45 million annually in the U.S. As with other labeled compounds, detection and measurement of the radiation from radiopharmaceuticals has sharply improved medical scientists' ability to study the behavior of pharmaceuticals in the body, whether for research or diagnosis. Labeled pharmaceuticals are one of several aspects of nuclear medicine, which has grown very rapidly and now is a recognized medical specialty. An estimated 25% of all patients admitted to U.S. hospitals today receive a radioactive tracer as part of their diagnosis, and U.S. investment in medical nuclear equipment has grown to about $50 million annually.

Isotopes are used also to make radioactive sources for nuclear gauges used to control various industrial process parameters, such as flow rate, thickness, level, and density. Annual sales of such gauges are estimated currently at $53 million. Other uses for radioactive sources include activation analysis systems, instrument calibration, and radiation processing equipment.

Radioisotope power generators convert to electricity the heat released by the decay of isotopes such as plutonium-238. The main market for such power sources has been in space vehicles and other out-of-the-way devices. They show promise, however, for use in cardiac pacemakers and artificial hearts.

Chemical Impact

The U.S. effort on nuclear weapons in World War II generated many chemical problems that had never previously been faced. In the process of solving these problems, chemists and chemical engineers built a body of new science and technology that gave chemistry in the broad sense a tremendous forward thrust. The results can be seen today not only in the nuclear industry but in areas far removed from it. A new era in inorganic chemistry was started, for example, by the chemists who unraveled the complexities of the chemistry of uranium, plutonium, other transuranium elements, and the fission products and evolved methods for isolating and purifying plutonium. Two other notable examples are analytical chemistry and fluorine chemistry.

Analytical Chemistry. No classification of chemistry was more influenced by the atomic energy project than was analytical chemistry. The problems of determining trace amounts of impurities in the materials used to produce the first nuclear reactors were so crucial to the success of the project that chemists were forced to concentrate on them first.

Mass spectrometry came of age as an analytical tool because of its peculiar applicability to the isotopic analysis of uranium. Emission spectroscopy had been essentially a qualitative technique. It was extended to a reliable quantitative technique that became the standard method for determining many of the elements of high neutron capture cross section, such as the rare earths. Analytical methods for these elements previously had been unknown. The vital question of boron in reactor-grade graphite was solved by emission spectroscopy.

The pressures of the atomic energy program transformed venerable electrochemistry into modern electroanalytical chemistry. N. Howell Furman and his students at Princeton University investigated potentiometry as a substitute for the less precise colorimetric indicators as a means of determining the endpoint in oxidation-reduction reactions, particularly those of uranium. The various electroanalytical techniques—polarography, amperometry, controlled potential and controlled current coulometry—were applied with great success in the atomic bomb project and later in nuclear materials analysis. With the continued use of these techniques came the development of instrumentation to today's refined state. The electroanalytical instrumentation industry whose elegant instruments find so wide a use now is clearly an outgrowth of the atomic energy program.

The concept of analytical separations that existed before the mid-1940's was predicated largely on precipitation of quantitatively insoluble sulfides. Many basically simple yet efficient separation techniques grew from efforts to replace the precipitation technique. Ion exchange resins were used widely for uranium. Solvent extraction, using various organic extractants, was adapted as a standard separation method. Paper chromatography in the form of cellulose strips and pulp columns also gained much acceptance.

The overall consequences of the nuclear energy program can be recognized in the highly sophisticated instrumentation and methodology of modern analytical chemistry. The difficult, seemingly almost impossible analytical problems of the project had to be solved. The chemists who solved them started a renaissance in analytical chemistry.

Fluorine Chemistry. One of the significant chemical milestones of the past 30 years has been the development of the technology for producing fluorine and fluorine-containing compounds. The work began in the early 1940's, stimulated primarily by the need for uranium hexafluoride. It was the only volatile compound of uranium suitable for use in separating uranium-238 and the fissionable uranium-235 by gaseous diffusion.

To obtain the large amounts of materials required for the gaseous diffusion process, the main separation method, completely new technology had to be developed for producing fluorine and uranium fluorides. These materials are highly reactive, and it was thus necessary also to develop the inert fluorocarbon polymers for use as gaskets and valve seats. Also required was the development of the inert fluorocarbon coolants used to remove the heat of compression resulting from pumping uranium hexafluoride through the porous barriers of the gaseous diffusion plant.

Before 1940, the reaction of fluorine with other materials was considered virtually impossible to control, and production facilities for the element did not exist. Under government contract, however, extensive development was undertaken by Harshaw Chemical, Hooker Electrochemical, du Pont, and, later, Union Carbide. As a result, fluorine plants with multiton daily capacity were installed at the three gaseous diffusion plants. These plants use electrolytic cells and a fused salt electrolyte of potassium fluoride/ hydrogen fluoride. Because of the growing demand for uranium hexafluoride for producing nuclear reactor fuel, fluorine production facilities have been installed in the past few years by Allied Chemical, Gulf General Atomic, and Kerr-McGee in the U.S. and by others in Canada, the United Kingdom, and France. The Atomic Energy Commission estimates that the nuclear industry will require more than 150,000 tons per year of uranium hexafluoride by 1980. The concomitant demand for fluorine, an element that was not in commercial production in 1940, will exceed 50 tons daily.

Early in the development of the gaseous diffusion process, chemists determined that the completely fluorinated hydrocarbons or fluorocarbons were sufficiently stable to be used safely in contact with the highly reactive uranium hexafluoride. At the time, the organic fluorochemicals industry was only 10 years old and devoted entirely to the manufacture of the simple one- and two-carbon chlorofluorocarbons used as refrigerants. Development was thus started on fluorocarbon liquids and solids for use as lubricants, coolants, plastics, and elastomers.

Before World War II, du Pont had begun development of the first successful perfluoroplastic, a polymer of tetrafluoroethylene now known as Teflon TFE. The demands of the gaseous diffusion process greatly speeded the progress of this material to the commercial production stage. Research in many industrial and university laboratories, meanwhile, produced a variety of fluorocarbons and processes for making them on an industrial scale. Some of these products have never found a market, but many are used widely today as plastics, oils, and greases. The extensive study devoted to all of these compounds under the pressures of the nuclear energy program has had a continuing influence on the growth of the flourishing fluorochemicals industry of today.

CHEMICAL ACCOMPLISHMENTS

The discovery of radioactivity in 1896 stimulated a substantial effort in those aspects of uranium chemistry which had to do with removing the metal from its ores. The goal was to isolate the more radioactive daughter products, the uranium itself being a waste product of this processing. The limited studies devoted to uranium metal produced few applications beyond some uses in glaze and glass and as a reagent for the quantitative determination of sodium in analytical chemistry. Even the melting point of the element was not known within hundreds of degrees.

By 1940, the U.S. atomic bomb (Manhattan) project was under way, and the Office of Scientific Research and Development was organizing a small chemical effort involving only a few tens of thousands of dollars. Both uranium processing and analytical technology were under study at the National Bureau of Standards (NBS) and Columbia University. It was realized early that very high purity materials, especially uranium and certain nonfuel materials, would be needed for the first nuclear reactor. By late 1941, Metal Hydrides Co. had made small amounts of uranium by powder metallurgy techniques, but the costs were too high and the purity too low. Westinghouse Electric, meanwhile, had made a few grams of satisfactorily pure metal by electrolysis of the fused fluorides of uranium and potassium.

As the project expanded through 1941, increasing supplies of about 95% pure U_3O_8, the black oxide of uranium, were coming from Canadian Radium and Uranium Co. By 1942, NBS had perfected a process for purifying the black oxide by diethyl ether extraction of uranyl nitrate from concentrated nitrate salt solutions. Mallinckrodt Chemical Works was using the process to supply 1 ton per day of superlatively pure uranium dioxide. Harshaw Chemical Co. and du Pont converted the dioxide to uranium tetrafluoride, the feed for electrolytic production of uranium metal by high-temperature hydrofluorination. The electrolytic process proved difficult and expensive, however, and was replaced by the bomb reduction of tetrafluoride by magnesium, a process developed independently at Iowa State College and NBS.

As the uranium program proceeded, related work was under way at the Metallurgical Laboratory at the University of Chicago under the general direction of A. H. Compton. Scientists there initiated the first nuclear chain reaction in December 1942. They developed the methods used to separate plutonium from other products of the chain reaction and the data required to effect an explosive chain reaction with either uranium-235 or plutonium. They worked out in addition the plans for the Hanford (Washington) plutonium production plant, which started up in September 1944.

The uranium metal production problem, at any rate, was solved by the end of 1942, and work was under way on uranium salvage and recycle problems. The processes involved included solvent extraction, precipitation of the metal (cation) from acid solutions with hydrogen peroxide, and the solution of uranyl compounds in alkaline carbonate solutions.

The establishment of uranium tetrafluoride capacity for making metal also fitted into the needs of the gaseous diffusion plant used to separate uranium-235 from uranium-238. The feed to the plant was uranium hexafluoride, made by reacting the tetrafluoride with fluorine gas.

Much chemical research was done at the University of California, Berkeley, in support of the electromagnetic method of separating uranium isotopes (which in the event was not used on a large scale). The feed selected was uranium tetrachloride, and by 1943 Eastman Kodak had perfected a production method for the compound. It involved reaction of uranium dioxide with carbon tetrachloride to produce uranium pentachloride, which then was decomposed to tetrachloride. The method required large amounts of uranium dioxide, which came from the metal production cycle. Much of the chemistry of the uranium dioxide process had to be modified, however, to meet the needs of the electromagnetic separation plant.

The uranium chemistry necessary to produce weapons hardware was further perfected at the Los Alamos laboratory. Chemists there made a major contribution by reducing to gram-size systems the scale on which the bomb reduction of uranium tetrafluoride to very pure uranium could be performed in essentially quantitative yields. The procedure used calcium metal reduction, boosted by iodine. The gram-size systems were required to avoid the dangers of approaching the "critical mass" of uranium-235, the point at which the fission process becomes self-sustaining and starts the chain reaction that is the basis of the atomic bomb. Quantitative yields were required of the reduction process because uranium-235 was an extremely valuable material.

Thus, between 1940 and late 1944, all of the basic chemical knowledge needed to operate one of the most complex and demanding large-scale chemical operations ever attempted had been collected and reduced to practice for the element uranium. The effort had started with information that could well have been summarized in a few pages of a standard textbook.

Uranium Feed Processing

Very early in the Manhattan project it was recognized that uranium hexafluoride was the only uranium compound suitable for use in separating the natural isotopes, uranium-238 and the fissionable uranium-235, by the prospective separation processes: gaseous diffusion, gas centrifugation, and thermal diffusion. The hexafluoride is highly corrosive, but its unique physical properties made its use mandatory. Gram quantities of uranium hexafluoride had been made as early as 1931 by reacting uranium metal with fluorine. This method, however, was not feasible for the large amounts of the compound that would be required.

A more practical process resulted from the work of Philip Abelson at the National Bureau of Standards. He found it easier and safer to make uranium hexafluoride by reacting fluorine with the tetrafluoride rather than with uranium metal. During 1941, Abelson's small plant produced all of the hexafluoride for the isotope separation experiments at Columbia University, the University of Virginia, and the Naval Research Laboratory. By the spring of 1942, Harshaw Chemical was producing 10 pounds of uranium hexafluoride daily in a pilot plant. By 1944, the company had completed the production plant that supplied all of the hexafluoride feed for the Oak Ridge gaseous diffusion plant. (It had been necessary also to develop a means of producing fluorine gas on a large scale, and an electrolytic process was perfected in 1943.)

Mallinckrodt Chemical Works made the uranium tetrafluoride that was fluorinated by Harshaw. The process involved purification of uranyl nitrate by solvent extraction, conversion of the uranyl nitrate to uranium trioxide by calcination, reduction to the dioxide with hydrogen, and conversion of uranium dioxide to uranium tetrafluoride with hydrogen fluoride.

These processes remained in use until the late 1940's, when an extensive development program was started at the Oak Ridge gaseous diffusion plant to improve their efficiency and economy. As a result, the fluidized bed technique for gas-solid reactions was adapted to the reduction of uranium trioxide to the dioxide with hydrogen and the conversion of the dioxide to uranium tetrafluoride with hydrogen fluoride.

The key to large-scale preparation of uranium hexafluoride had been the development of a continuous flame fluorinator in which the tetrafluoride is reacted with fluorine. Uranium hexafluoride from the gaseous diffusion plant, enriched in uranium-235, was then reduced to the tetrafluoride. A batch process was used to avoid criticality (the area of the critical mass), which at the time was not thoroughly understood. The development process of the late 1940's, however, made it possible to replace this process with a continuous method. Fluorine mixed with uranium hexafluoride passes through the center pipe of a concentric nozzle, and hydrogen passes through the annular pipe. The self-igniting reaction of hydrogen with fluorine supplies the heat required to sustain the reduction of uranium hexafluoride to tetrafluoride with hydrogen.

In 1971, the cost of converting uranium concentrates to uranium hexafluoride in the U.S. came to about $25 million. The figure is expected to reach $150 million by 1980. Both values are based on the fuel requirements of nuclear power reactors.

Uranium Ore Processing

With the advent of the large demand for uranium after World War II, the existing processes for treating uranium ores were cumbersome and gave poor recoveries and low-grade products. In the late 1940's and early 1950's, the development of ion exchange and solvent extraction revolutionized the situation. The new methods accounted largely for a reduction of 50% in the cost of recovering uranium from its ores.

The limited percentage of uranium ores that are high in lime must be leached with sodium carbonate and thus are not amenable to ion exchange or solvent extraction. Most uranium ores are leached with sulfuric acid, however, and essentially all such plants worldwide now use ion exchange or solvent extraction. The eventual, in fact inevitable, processing of very low-grade ores will rely almost entirely on the two processes.

Ion Exchange. Synthetic ion exchange resins were becoming available for use in water treatment in the late 1940's, and chemists at Iowa State

University made a major advance by using cation exchange resins to separate rare earths. Studies at Battelle Memorial Institute, under AEC contract, indicated that anion exchange resins might well be useful in recovering uranium from sulfuric acid leach liquors. The development effort was joined immediately by others, including the AEC Winchester Raw Materials Laboratory, Rohm & Haas, Dow Chemical, and Oak Ridge National Laboratory. The combined efforts of process chemists and industrial chemists concerned with tailoring resins for hydrometallurgical use produced ion exchange processes that soon were accepted widely by uranium producers in the U.S., Africa, Canada, and many other countries. The information and advanced resins resulting from the uranium program hastened the extension of ion exchange technology to other hydrometallurgical applications. The uranium work contributed also to improved utility in other nuclear uses, notably the recovery and purification of plutonium from spent reactor fuels and the treatment of low-level radioactive wastes.

Solvent Extraction. The first concerted effort to use solvent extraction to recover uranium from ores was started about 1950 by Dow Chemical under AEC contract. Dow's principal interest was in the uranium-bearing Florida phosphates. Shortly thereafter, Oak Ridge National Laboratory launched an extensive program aimed at using solvent extraction on the uranium ores that were being found in increasing abundance. This program uncovered several extraction reagents that proved to be well adapted to processing sulfuric acid leach liquors from uranium ores. The principal reagents were di(2-ethylhexyl) phosphoric acid (Dapex process) and long-chain alkyl amines (Amex process).

The first uranium millers to install solvent extraction were Kerr-McGee Oil Co. and Climax Uranium Co. (Dapex process) and Texas-Zinc Minerals Co. (Amex process). Solvent extraction currently is used in at least 50% of U.S. uranium milling capacity, and numerous plants have been built in other countries. In many mills, the process is used not only in mainline processing but also to recover uranium from solutions used to remove it from ion exchange resins.

Since its development in uranium ore processing, the Dapex process has been applied in treating ores of thorium, vanadium, beryllium, scandium, and nickel-cobalt; in recovering and separating rare earths; in separating zinc from copper ore concentrates; and in recovering fission product strontium and rare earths from spent nuclear reactor fuels. The Amex process has been developed for treating ores of thorium, vanadium, beryllium, rhenium, and tungsten; for purifying beryllium concentrates; and for recovering and purifying plutonium and fission-product palladium from spent nuclear reactor fuels. Both the Dapex and Amex processes, together with resinous ion exchange, have been developed for recovering and separating the man-made transplutonium elements.

Nuclear Fuel Processing

The growth of nuclear power is creating in this country a mature fuel services industry in which an important step is chemical separation of fuels, both in refining and spent fuel reprocessing. The fuels involved for the near future will be mainly those for pressurized-water (PWR) and boiling-water (BWR) reactors, with a small amount for high temperature gas-cooled (HTGR) reactors. The PWR's and BWR's use fuel elements containing uranium or uranium–plutonium oxide pellets clad with zirconium alloy. The HTGR's use a graphite block with fuel holes filled with spherical particles of thorium and uranium carbide coated with pyrolytic carbon and silicon carbide and held together with a carbonaceous binder.

The internationally dominant separation methods in nuclear fuel cycles today are various solvent extraction processes in which the organic extractant and complexing agent is tri-n-butyl phosphate (TBP). Such TBP processes are used both in feed materials plants for refining natural uranium and thorium and in spent fuel reprocessing plants for recovering, separating, and decontaminating neutron-irradiated uranium, plutonium, and thorium.

TBP is best known for its use in the Purex Process, an extractive separation method for uranium/plutonium-239. This process, appropriately modified, also seems destined for use in fuel recovery for advanced breeder reactors such as the Liquid Metal Fast Breeder Reactor. The feasibility of such reactors is expected to depend in part on the ability to minimize fuel inventory charges by reprocessing highly-burned fuels after only short cooling periods. Thus, fuel recovery would have to be done at extremely high radiation levels, to which the Purex Process is relatively well adapted.

Somewhat similar to the Purex Process is the Thorex Process, an extractive separation method for thorium/uranium-233. A patent on the Purex Process issued in 1961 to T. C. Runion, C. V. Ellison, et al. Patents on the Thorex Process were applied for by A. T. Gresky, et al., in 1956 and by R. H. Rainey and J. G. Moore in 1960. The Rainey-Moore patent issued in 1962. (All such patents were assigned to the Federal Government.)

Until mid-1948, the dominant solvent extraction processes in AEC's studies of spent fuel recovery used basic or neutral aluminum nitrate as the aqueous salting agent and hexone as the plutonium and/or uranium extractant. Nitric acid offered certain advantages as a salting agent, but it reacted to an objectionable degree with the hexone. Thus, in the summer of 1948, a search began for an extractant that would be stable in the presence of nitric acid.

AEC had an immediate need for such an extractant in the postwar recovery of uranium from the diuranate slurries that had accumulated in the radioactive aqueous waste tanks at Oak Ridge National Laboratory (ORNL). A metal recovery process using TBP was soon developed and became a welcome alternative to the existing processes. Soon afterward, the uranium refinery at Fernald, Ohio, also converted to TBP-based processes to minimize solids handling and improve product purification.

The use of TBP allowed nitric acid to be used as either the principal or only aqueous salting agent. Continuing studies at ORNL with natural uranium reactor fuels soon demonstrated that a TBP-based process offered considerable advantages in waste handling and in the overall economics of nuclear fuel reprocessing. In the late 1940's, therefore, a number of chemists at Oak Ridge actively pursued various studies of plutonium reduction-oxidation and extraction in aqueous nitric acid–TBP systems. Both the Purex Process and a TBP-25 Process evolved rather quickly from this work. They provided valuable alternatives to the other processes available in 1949–52.

By the early 1950's, studies at ORNL by E. C. Bohlman and D. C. Overholt had indicated the general feasibility of TBP processes for extracting thorium as well as their potential for separating thorium from uranium-233 and fission products. Then, the wide versatility of TBP processes was demonstrated during a crash pilot plant program at ORNL in the summer of 1952. This program produced the Interim-23 Process, which combined solvent extraction and ion exchange. This process provided for recovery of fissionable uranium-233 from neutron-irradiated thorium and for decontaminating the uranium from highly radioactive protactinium-233 and fission products. Laboratory and pilot plant work on the Interim-23 Process were especially helpful in the development, by late 1953, of the Thorex Process and two modified TBP-25 processes. The TBP-25 processes were for use with fuel elements of uranium-235 clad with zirconium or stainless steel.

Radioactive Waste Management

Chemists and chemical engineers have been heavily involved in developing methods for treating and storing radioactive wastes, which can be defined broadly as waste products that contain radionuclides. The radiation from the radionuclides is harmful to man and other living species, and the wastes must thus be isolated from the biological environment. Very long isolation periods are required, moreover, because the half-lives of the radionuclides range up to thousands of years.

The principal source of radioactive wastes, now and in the foreseeable future, is the nuclear fuel cycle: mining and milling of ores, isotopic enrichment, fuel fabrication, reactor operations, and spent fuel processing. By the year 2000, the fuel cycle will have generated in the U.S. alone an estimated 80 million gallons of concentrated, high-level, liquid wastes, billions of gallons of low-level wastes, and hundreds of millions of cubic feet of solid wastes. Another important source of radioactive wastes is the production and use of radioisotopes in medicine and industry.

In the past 20 years, the U.S. Atomic Energy Commission and private industry have spent millions of dollars developing suitable methods for managing these complex, ill-defined wastes. The goal is to treat and dispose of the wastes in such a manner that they do not increase the background radiation of the environment significantly above the natural level derived from the sun, cosmic rays, and natural radionuclides in the earth.

High-level liquid wastes at present are concentrated by evaporation and stored in underground tanks. Advanced methods being studied for managing such wastes include solidifying them to ceramic form and storing them underground in steel containers in natural salt formations. Areas of chemical research associated typically with these methods include the relative volatilities of waste constituents, corrosion of containers and equipment, solubilities and thermal properties of the ceramic solids, and thermodynamic analysis. In May 1972, the Atomic Energy Commission announced a new, long-range program for storing solidified high-level radioactive wastes from the nuclear power industry.

Low-level liquid wastes are treated by concentrating the radionuclides by scavenging and coprecipitation, ion exchange, or evaporation. The radionuclides are stored, and decontaminated fluid is discharged into the environment. Areas of chemical research include solubilities, adsorption of nuclides on precipitates, thermodynamics of ion exchange with natural and synthetic zeolites and resins, and solid-liquid separations.

The storage of solid radioactive residues also requires chemical knowledge in several areas. They include the diffusion rates of nuclides in salt or other minerals, the mechanisms of spontaneous combustion of some solids, and the leach rates of radionuclides into natural waters.

Nonfuel Materials

From the start of the atomic era it was clear that the known chemistry of the nonfuel materials that might be used in nuclear devices was not a sufficient guide to their utility. The behavior of such materials had to be reassessed in terms of the then-unfamiliar parameter, radioactivity.

The resulting research, which in fact is still in progress, produced a variety of materials requirements of unprecedented stringency as well as means of meeting them. Among the important materials involved have been graphite, boron, zirconium, water, and heavy water.

Graphite. The development of nuclear power required extremely pure graphite both to make fundamental measurements and to design feasible reactor systems. Chemists had learned after years of research to make high-purity graphite electrodes for spectroscopic use, generally by exposure to chlorine. This process was carried out in small lots and with some difficulty, however, and it did not effectively remove boron, a particularly troublesome contaminant in graphite for nuclear use. The process thus had to be scaled up to produce the large amounts of high-purity graphite that would be required and, in addition, the boron-removal step had to be improved.

The inexpensive procedure that was developed involved the use of both chlorine and Freon 12 (dichlorodifluoromethane). Reaction with the latter compound removed the boron as boron trifluoride. This process produced high-grade nuclear graphite simply and efficiently. L. Brooks of United Carbon Products Co. is identified with much of the development of the process, but credit must go also to scientists at National Carbon Co. and to numerous workers in thermodynamics and the measurement of fundamental physical constants.

Boron-10. The element boron occurs naturally as a mixture of two isotopes, about 80% boron-11 and 20% boron-10. The thermal neutron absorption cross section of boron-11 is less than 0.05 barn, and that for boron-10 is 4,000 barns. That is to say, boron-10 absorbs thermal neutrons about 80,000 times more efficiently than does boron-11. It is thus advantageous to separate boron-10 for use in nuclear applications such as boron trifluoride neutron detector tubes and the like. A particularly important application of boron-10 is in the control rods that provide the essential means of controlling the chain reaction in nuclear reactors.

Research on separation of the two boron isotopes was first undertaken at Columbia University in 1943 under the direction of chemist H. C. Urey. From this research came the basic conditions for separating the isotopes by equilibrium distillation of a complex of dimethyl ether and boron trifluoride as well as a process for reducing the complex to elemental boron. As a result of the work at Columbia and experience with a related pilot plant operated by Standard Oil Co. (Ind.), AEC authorized construction of a plant in 1953. It was to produce 500 kilograms per year of elemental boron enriched to 90 to 95% boron-10. The plant was built at Model City (Niagara Falls), N.Y., and operated by Hooker Electrochemical Co.

The plant's monthly production rate varied with the isotopic purity of the product. The rate was about 140 pounds of elemental boron at 90% boron-10 and about 45 pounds at 95% boron-10. Total cost of the final product in elemental form averaged about $3.00 per gram. Operation of the Model City plant has been suspended intermittently since 1958 because of inventory buildup, and the plant is now closed permanently for the same reason.

Hafnium-Zirconium Separation. U.S. scientists saw very early that the metal zirconium might be useful in water-cooled nuclear reactors because of its physical properties and, especially, its resistance to corrosion by high-temperature water. In those days, the metal's only drawback seemed to be its moderately high rate of absorption of thermal neutrons. In the late 1940's it was recognized that this property was due to contamination of zirconium by the element hafnium. It was then clear that an economical separation of hafnium would make zirconium a valuable nuclear material.

All minerals in which zirconium occurs contain 1 to 3 parts of hafnium per 100 parts of zirconium. Partial separations of the two had been demonstrated following the discovery of hafnium, in 1924, but the processes available in the mid-1940's were laborious and almost obviously hopeless for large-scale, economical use.

Research on improved methods was greatly accelerated by persons responsible for developing water-cooled reactors for submarine propulsion. By early 1949, many industrial, university, and government organizations were at work on the problem.

Chemist Werner Fisher, in Germany, had had some success in preparing enriched hafnium by extraction from aqueous solution with a solution of thiocyanic acid in diethyl ether. A small group under W. R. Grimes at Oak Ridge National Laboratory began to examine this process in May 1949. They found quickly that the method was promising, especially after substituting methylisobutyl ketone for ether as the extraction solvent, and defined the effects of numerous process variables. By about six months after the laboratory work began, a pilot plant was producing 10 pounds per day of zirconium containing less than 200 parts per million of hafnium. By March 1950 the process had been scaled up to produce about 450 pounds per day of zirconium containing less than 100 parts per million of hafnium.

It became apparent, meanwhile, that the hafnium-free product must be freed of other common impurities. The research group under Grimes demonstrated that precipitation of zirconium salicylate at controlled acidity gave a product that met all specifications. This method was used in 1950 to prepare the first large (25,000-pound) lot of reactor-grade zirconium oxide. (The salicylate procedure was supplanted later by precipitation with ammonium phthalate with recycle of the organic precipitant and a more economical product.)

The Oak Ridge plant produced more than 200,000 pounds of hafnium-free material in the succeeding two years. Then the process was turned over to private contractors, who since have produced many millions of pounds of reactor-grade zirconium for the nuclear industry.

Many metallurgical developments in metal reduction, alloying, and fabrication were necessary before the use of zirconium in nuclear reactors became widespread. Plainly, however, the chemist's contribution to hafnium removal and zirconium purification was a vital link in the development of the submarine reactor from which has grown the power industry based on water-cooled nuclear reactors.

Water Coolant Chemistry

The use of water as a coolant is basic to most nuclear reactors in the U.S., whether for research, production of weapons materials, naval propulsion, or electric power. The development of water coolant technology, including its chemical components, has thus been a keystone in the atomic energy program in this country.

Among the significant chemical contributions is the development of an understanding of corrosion mechanisms and of the ability to control them by controlling the chemistry of the cooling water. Controls have been worked out for a variety of materials and a wide range of conditions (as from ambient temperature to 300° C.). Such control assures the integrity both of the containment of highly radioactive isotopes from fissioned fuel and of the coolant circulating system itself. Suitable controls in addition promote efficiency of operation by preventing fouling of heat transfer surfaces by corrosion products.

These goals have been achieved in spite of the special problems and limitations imposed by the very high radiation fluxes in reactor cores and the need to restrict the chemicals used to those not made radioactive by neutron bombardment. The resulting minimization of corrosion and of radioactive species in the cooling water contributes substantially to safety and economy of maintenance.

Chemists have developed additives for reactor cooling water which optimize the utilization of the fuel and thereby appreciably reduce the cost of the electric power generated. Similar chemical systems have been developed for emergency shutdown and cooling in case of accident. A third chemical contribution has been the development of purification processes which keep the coolant free of radioactive and other impurities while maintaining its composition at the optimum.

Heavy Water

Nuclear power reactors fueled with natural uranium, all of which require heavy water (deuterium oxide) as a neutron moderator, are now economically competitive with fossil-fueled power generating plants in most areas of the world. Power plants of this type having some 5,600 electrical megawatts of capacity are scheduled to come on-stream during the 1970's (all of them outside the U.S.). They will require, typically, an inventory of about 1 ton of 99.8% heavy water per rated megawatt.

The natural content of heavy water in normal water is about 0.015% (150 parts per million by weight). A number of different methods may be used, in principle, to concentrate heavy water to the required 99.8%. Almost all of the world's present supply, however, has been concentrated by a chemical process developed at Columbia University in the early 1940's by H. C. Urey, who discovered deuterium.

The process is based on the exchange of hydrogen isotopes between counterflowing streams of gaseous hydrogen sulfide and liquid water. To produce reactor-grade heavy water by this reaction normally would require costly equipment and large energy input. The process can be operated, however, at temperatures and pressures at which deuterium transfers from the aqueous to the gaseous phase, thus eliminating the need to convert water chemically to hydrogen sulfide. This thermal transfer of deuterium between the two phases is largely responsible for the economy of the process. (The price of heavy water produced at AEC's Savannah River Plant, near Aiken, S.C., was pegged recently at $39 per pound.)

At one time, the AEC plants at Savannah River or Terre Haute, Ind. (Dana Plant) produced essentially all of the Free World's supply of heavy water. U.S. production currently is about 180 tons per year, down from a peak of 900 tons per year in the mid-1950's. The anticipated demand for heavy water in the next decade has prompted the construction in Canada of three plants having a combined production capacity of 1,600 tons per year. Each is based on the chemical exchange method described in the foregoing.

Sol-Gel Processes

Sol-gel processes have been developed in roughly the past decade to prepare ceramic nuclear fuels for reactors and in a variety of other special forms. These processes produce dense ceramic microspheres of the fuel, which are then fabricated into fuel elements. The outstanding asset of the sol-gel processes is their versatility. By adding carbon to oxide sols, for example, and by modifying the sintering and conversion conditions, carbides and carbonitrides of a wide variety of metals can be prepared.

Ceramic nuclear fuels are important to the economics of power reactor fuel cycles. Sol-gel processes for preparing the ceramics have the advantages of using liquid process streams that can be easily monitored and shielded and of being carried out at ordinary temperatures and pressures in conventional equipment.

Sol-gel processes involve, first, the preparation of the sol, an aqueous dispersion of electrically-charged oxide particles ranging from about 0.001 to 0.1 micron. The sol is then dispersed uniformly in the gel, a semirigid or rigid structure of a noncrystalline, colloidal nature. When the gel is heated, the water is driven off, and the oxide particles sinter together to form a dense ceramic body in the configuration of the original gel body.

The production of dense thorium oxide by a sol-gel process was first discussed in late 1959 by D. E. Ferguson, O. C. Dean, and K. H. McCorkle of the chemical technology division at Oak Ridge National Laboratory. Their work culminated in 1962 with a patent on the process. Sol-gel processes currently are being used and evaluated by Gulf Energy & Environmental Systems, Atlantic Richfield, Babcock & Wilcox, and W. R. Grace.

The development of sol-gel processes has required chemists to work out methods for preparing sols, to solve the problems of sol stability, to evaluate process variables, and to establish optimum process flowsheets. The uranium and plutonium in fuels for the Liquid Metal Fast Breeder Reactor are unstable in each other's presence when each is in a valency (reactive) state of IV. Chemists thus had to devise means of fixing the valence states, so that only negligible reaction occurs between the two elements during sol-gel processing.

Sol-gel processes have been used in a variety of nonnuclear applications. Spheres of rare earths have been made this way for use in fundamental research on diffusion in ceramics and on the sintering of oxides. Ceramic microspheres with controlled porosity can be made with relative ease by sol-gel processes. Such spheres have been studied as selective adsorbents, especially of biochemical materials which may find use in chromatographic separation of certain compounds of interest to research biochemists and others.

Sol-gel processes have potential uses in preparing free-flowing powders in which high density, high strength, homogeneity, low impurity levels, and minimum dusting are important considerations. Such applications include plasma torch spraying, the preparation of porous substrates for catalysts, and making concentrated sols, or grogs, for use in extruding rods, tubes, and fibers.

New Nuclides for Technology and Medicine

Soon after the discovery of uranium fission in 1939, scientists found that it was the isotope uranium-235 that underwent fission when struck by a thermal neutron, releasing two or more neutrons in the process. (Natural uranium contains about 0.005% uranium-234, 0.71% uranium-235, and 99.28% uranium-238.) This suggested the possibility of a neutron chain reaction, and studies of nuclear systematics suggested that other heavy nuclides which could be made by nuclear reactions might also fission with thermal neutrons.

In studies along this line, chemist G. T. Seaborg and his co-workers at the University of California, Berkeley, irradiated normal uranium and thorium with neutrons from the 60-inch cyclotron in search of the products of two sequences of reactions that were expected to yield plutonium-239 and uranium-233. The chemical procedures used were very elaborate, since only micrograms of the products were formed in kilogram quantities of target material, and since a very small amount of normal uranium impurity would make it impossible to measure the fissionability of the products. Plutonium, moreover, was a new element of unknown chemistry.

Seaborg, et al., succeeded in discovering plutonium-239 early in 1941 and uranium-233 early in 1942. They showed that both of these nuclides behaved as did uranium-235 in the presence of thermal neutrons. Thus, three nuclides were known which, at least in principle, could be used as large-scale sources of energy, whether for bombs or peaceful purposes.

At the end of 1942, scientists in the Metallurgical Laboratory of the University of Chicago showed that natural uranium could be made to undergo a self-sustaining chain reaction. In the process, uranium-235 fissioned and emitted enough neutrons to sustain the chain and to produce plutonium-239. The chemists under Seaborg then worked out a procedure for isolating the plutonium from the much larger amount of uranium and from the highly radioactive fission products. (The plutonium work culminated in the full-scale plutonium production plant that started up at Hanford, Wash., in September 1944.)

At the same time, chemists largely under the direction of C. D. Coryell identified various fission products as to element, mass number, and decay characteristics. This latter work opened the door to large-scale use of radioactive nuclides (at least of fission product elements) in research, in industry, and in medicine. Examples include strontium-90 in small, portable out-of-the-way energy sources, such as satellites and buoys, and iodine-131 in diagnosing thyroid ailments and in therapy for irradiating tumors, in which the isotope concentrates.

In 1944, Seaborg's group discovered isotopes of the elements americium and curium. Since that time, isotopes of all elements through 105 have been reported. Of these, curium-244 has been used as a portable energy source, while californium-252 (which fissions spontaneously at a high rate) and mixtures of americium-curium-beryllium have been used as portable neutron sources. Plutonium-238 (discovered before plutonium-239) has been used as a portable energy source, as a portable neutron source, and as an energy source in a cardiac pacemaker. In all, more than 100 radioisotopes are used today in research, in industry, and in medicine.

Breeder Reactors

Since World War II, considerable attention has been given to breeder reactors that involve one or the other of two reaction sequences. In one of these, neutrons from the fission of uranium-233, the fuel, convert thorium-232, the fertile material, to more uranium-233, using up the thorium in the process. In the other sequence, plutonium-239 is the fuel, and uranium-238 is the fertile material. The breeder reactor produces at least as much fissionable material as it consumes, and thus, in principle, makes all of the natural thorium and uranium available as fuels. The breeder has been termed "our best hope today for meeting the nation's growing demand for economical clean energy."

The development of the breeder reactor involves a number of technical problems, however. The fuel and fertile materials must be chemically repurified many times to remove fission products, which act as neutron poisons, and for other reasons. If the reactor is to breed at all, moreover, the chemical losses of fuel must be kept at a bare minimum. This problem is so enormous that it has occupied many chemists and chemical engineers for almost three decades and will continue to do so.

The U.S., Great Britain, France, Germany, and the Soviet Union are actively developing fast breeder nuclear reactors. Principally these are liquid metal fast breeder reactors (LMFBR's), cooled by molten sodium. The U.S. is committed to demonstrate a commercial-size LMFBR by 1980, and several are planned or under construction abroad. The first such reactor to begin operating started up in the U.S.S.R. in 1972, and the second is scheduled to start up in the United Kingdom in 1973.

The use of molten sodium to cool the LMFBR permits operation at high temperature, with the consequent high thermal efficiency. In addition it conserves neutrons for breeding, since sodium has a relatively low capture cross section for fast neutrons. The fast breeder reactor has three flow systems: the primary sodium system for removing heat from the reactor core, the secondary sodium system, and the water or steam-generating system. Neutron activation makes sodium in the primary system intensely radioactive, and the secondary sodium system acts as a radiation barrier between the primary and steam-generating systems. Also, should water leak from the steam system, the secondary system prevents it from reaching the primary sodium with the consequent danger of the explosive sodium-water reaction in the reactor core.

Knowledge of the chemistry of sodium and its impurities is important in three major areas: corrosion mechanisms, especially the effects of impurities on corrosion or degradation of structural metals; the transport and deposition of neutron-activated corrosion products or fission products that escape to the sodium from leaking fuel elements; and the solubility relationships of gases used in the cover-gas systems of the reactor and of inert fission-product gases.

Over the past several years, the major U.S. effort in sodium chemistry has been in the development of sampling and analytical methods. Approved chemical methods for the analysis of sodium and cover gas have been compiled and published. Methods and devices have been developed that permit continuous, on-line monitoring of major nonmetallic impurities—oxygen, hydrogen, and carbon—in sodium in reactor systems.

Attention now is shifting to elucidation of the chemical behavior of impurity elements in sodium: the compounds formed, their interactions, and their importance in corrosion phenomena in sodium-steel systems. Much work remains to be done on corrosion and fission product chemistry. Considerable experimental and theoretical work has been done on the solubility of helium, argon, krypton, xenon, and nitrogen in liquid sodium. This work has been completed and has contributed importantly to reactor design, especially the design of systems for detecting fuel-element failures.

Molten Salt Breeder Reactors.   The molten salt breeder reactor, while not as far advanced as the liquid metal fast breeder reactor, has been the subject of considerable research. Alvin Weinberg of Oak Ridge National Laboratory recognized around 1950 that a promising fuel for a breeder reactor would be a solution of uranium in molten salts. At his instigation, a group of chemists under W. R. Grimes began to look into the possibility of developing such a fuel.

This effort produced a fuel that was stable and noncorrosive and had almost no vapor pressure. To develop it, however, the Oak Ridge group had to learn much about the chemistry of molten salts. The body of known chemistry at the time was based largely on what happened in water, so that much of the same ground had to be covered for molten salts.

The work resulted in a demonstration, the Molten Salt Reactor Experiment (MSRE) at Oak Ridge. MSRE operated virtually flawlessly for four years. By December 1969 it had achieved its major objectives and was closed down. The next step, the Molten Salt Breeder Technology Program, is concentrating on development of fuel, materials, and fuel processing and components and will include design studies of large molten salt reactors.

FUTURE DEVELOPMENTS

The dominant aspect of the nuclear industry in the foreseeable future will remain the production of electric power. The uses of radioisotopes will continue to multiply, in science, in industry, and in medicine, but that part of the industry is unlikely to approach nuclear electric power in terms of dollars or of the scientific and engineering manpower directly involved.

Complex problems remain to be solved in bringing the breeder reactor to commercial status. The compulsion to do so is great, since by making all natural uranium and thorium available as fuel the breeder would provide a very long-range source of power in a world of diminishing resources. A Liquid Metal Fast Breeder Reactor Program Plan has been developed by the Atomic Energy Commission with the cooperation of industry and electric utilities. Its goal is to establish "a safe, reliable and economic breeder reactor system on a broad technical and engineering basis with sufficiently comprehensive industry involvement to assure a strong and competitive industrial capability in the early 1980's." The program includes operation of a commercial-size LMFBR by 1980.

Further in the future is the controlled nuclear fusion reactor. Fueled by deuterium from the oceans, it promises an almost unlimited source of electric power. The Federal Government has supported research on controlled fusion for nearly two decades and currently is spending about $30 million annually on the problem. Worldwide, an estimated $150 million annually is being spent on research on controlled fusion. No one yet has demonstrated the scientific feasibility of controlling a sustained fusion reaction that releases more energy than it consumes. Evidence is accumulating that it can be done, however, and some scientists believe that a fusion power plant could be operating as early as the year 2000.

Radioactive Waste Management.   By 1979 or 1980, meanwhile, the nuclear power industry expects to begin delivering solidified, high-level radioactive wastes to the Atomic Energy Commission for storage. AEC announced in May 1972 that it proposes to design and build engineered surface facilities for the purpose. The Commission has used this method of storage safely for 10 years at some of its facilities. The surface facilities would be under constant surveillance and control and could be used to isolate the wastes for centuries if need be.

At the same time, AEC is reorienting its research and development program on storing wastes in salt beds, several hundred feet underground, or in other geologic formations. The plans include a pilot plant to verify the results of some 15 years of laboratory and field work on this method of storage. By the mid-1980's, the Commission hopes to be in a position to decide whether storage on the surface or in geological formations is the method that should be expanded for use in later years.

Work is proceeding also on other methods of radioactive waste management. These include storage in very deep holes in the earth and conversion of the transuranic radioisotopes into shorter-lived materials.

15 / Pulp and Paper

SOCIAL AND ECONOMIC IMPACTS

The pulp, paper, and paperboard industry in the U.S. converts a renewable resource, cellulosic fibers from trees and other sources, into more than 5,000 types of products in six broad categories: newsprint; printing papers; fine papers; specialized industrial papers; tissue and sanitary papers; and linerboard, boxboard, foodboard, and construction board. Industry sales in 1971 were about $22 billion; production was 54.4 million tons, about 91% of total production capacity. Companies in the industry paid some $5 billion in wages and salaries to almost 700,000 employees in more than 5,000 cities, towns, and suburban communities.

Pulp and paper companies use annually as raw materials about one third of the nation's timber consumption. In 1971, they produced some 44 million tons of wood pulp, the industry's primary fiber. They recycled about 12.9 million tons of wastepaper and paperboard; 12.2 million tons of the recycled material went back into industry products, accounting for 22% of the fiber consumed by U.S. mills. The industry also makes a small amount of paper from nonwood cellulosic fibers, such as bagasse, the waste from sugarcane processing.

Paper companies use both hardwoods and softwoods drawn from more than 500 million acres of commercial timberlands in all parts of the country. The companies themselves own about 13% of the acreage and the state and federal governments about 28%. The remaining 59% is owned by farmers and small landowners who sell timber to the companies. The companies manage almost all of their own timberland scientifically to achieve the maximum yield of pulpwood per acre. Many help small timber owners to do the same, providing free seedlings and advice on planting, cultivation, and harvesting. Companies operate most of their own timberlands on a multiple-use basis, allowing the public to use them for hunting, fishing, hiking, and the like.

Paper is a very old product. It was made first in China—in 123 B.C., according to some historians, and in 105 A.D. according to others—and reached Europe around 1050 A.D., when the Moors introduced it in Spain. Until roughly the mid-1800's, however, the high cost and scarcity of paper limited its use to writing and printing. The product was made until then almost entirely of linen and cotton rags. The processes used were slow and costly, and the supply of rags was limited, partly by inability to use colored rags to make writing and printing papers.

The first significant chemical development in the paper industry, in fact, was the bleaching process and its use on colored rags. The process was based on the discovery of chlorine in 1774 by the Swedish chemist, Scheele, and the recognition of the element's bleaching powers by the French chemist, Berthollet, in 1785. The use of potassium, sodium, or calcium hypochlorite to bleach colored rags began in 1790, but even so the raw materials for paper remained in short supply.

The solution to the problem was to obtain cellulosic fiber from wood. Mechanical pulping was tried experimentally in this country in 1840, two groundwood mills were operating in Germany in 1852, and production began in Canada in 1866. The groundwood content of paper was limited, however. Too much groundwood weakened the paper and imparted a serious tendency to yellow. The remedy was chemical woodpulping. In

1854, in England, Burgess and Watt patented the soda process for reducing wood to pulp, thus opening the door to low-cost, volume production of paper from wood. In 1867, Benjamin Tilghman of Philadelphia patented the sulfite pulping process, which assured the use of wood as the major raw material for paper.

Developments in producing pulp were paralleled by those in the mechanical devices that convert the pulp to paper. The paper machine was invented in 1799 in France by Robert. The method was improved by the Fourdrinier brothers, who installed their first commercial machine in England in 1803. By 1920, high-speed machines were attaining production rates of 1,000 feet of paper per minute.

Continuous chemical and mechanical development by now has transformed paper into a high-volume, low-cost product. Mills using the kraft or sulfate process, the dominant pulping method in the U.S. today, had an average production capacity of roughly 750 tons per day in 1971. Paper machines, mainly of the Fourdrinier type, can turn out continuous sheets on the order of 300 inches wide at 1,500 to 3,000 feet per minute. Paper and paper products in addition are subjected to a variety of chemical and other treatments to control printability, whiteness, wet strength, and other properties.

Specific contributions of chemistry to the utility of paper products are evident not only in pulping processes but in a variety of end uses. The classic example is photographic paper, which is coated with compounds of silver that react to light. Paper used as insulation in electrical transformers may be treated with acrylonitrile to upgrade its resistance to heat. Chemical treatment flameproofs paper convenience items such as clothing and drapes and lends fluid repellency to paper hospital fabrics. Physical and chemical treatments can produce paper multiwall shipping sacks that withstand submergence in water and can be stored outdoors for more than a year.

Conservation of materials and energy is crucial to the economics of the pulp and paper industry. Bark removed from trees before pulping is commonly burned to generate steam for use in the mill. Fly ash recovered from bark-burning boilers is used to help purify public water supplies. Pulping chemicals are recovered and recycled extensively. Companies have done a good deal of research on industry by-products, which include tall oil, turpentine, rosin, fatty acids, and others. Lignin compounds freed by pulping are used to make surface active processing aids for many industries. Lignin compounds are used also to make special products such as vanillin and dimethylsulfoxide. Carbonaceous substances recovered from pulping liquors are converted to active carbons, which are used widely to control tastes and odors in drinking water.

Pulp and paper is a capital-intensive industry, requiring as much as $1.50 of capital investment per $1.00 of annual sales. In 1947–69, U.S. papermakers spent more than $17 billion on new plant and equipment. In 1970, they spent $1.65 billion, a new single-year record. Capital spending declined sharply in 1971, to an estimated $1.23 billion.

Increasingly the industry has been putting capital into pollution abatement facilities to meet the steadily more stringent requirements of federal, state, and local laws and regulations. In 1971, such facilities accounted for almost 20% of the industry's total capital spending, compared to less than 15% in 1970. Mills built in recent years have been designed to include pollution abatement equipment, but installing such equipment in older plants can be costly. Such costs have been a factor, although not necessarily the only one, in forcing some older, smaller mills to close.

The pulp and paper industry is, of course, a heavy consumer of the products and services of other industries. Of the 20 major industry groups in the U.S., paper and allied products stands sixth in consumption of electric power. The group uses more than 7.4% of the dollar value of all fuel and energy consumed by U.S. manufacturers. The industry's annual rail transportation bill exceeds $1 billion, almost 10% of the gross freight revenues of U.S. railroads.

The pulp and paper industry also is the sixth largest buyer of chemicals in the U.S. For pulping and bleaching alone, the industry consumes about 70% of all the sodium sulfate (salt cake) produced in this country, about 17% of the chlorine, 14% of the sodium hydroxide (caustic soda), and 9% of the sodium carbonate (soda ash). In addition, something like 10% of the cost of making paper and paperboard is represented by chemicals other than those used in pulping and bleaching. These chemicals—starches, pigments, and many others—have so varied and improved the properties of paper and paperboard that, without them, consumption today might readily be only a tenth of the actual amount.

The pulp and paper industry spends about 0.5% of sales on research and development. The figure is relatively low, compared to the average of about 3% for all industry, but is typical of a very old, commodity industry. And applied to total sales, 0.5% provides some $100 million per year for research and development industrywide.

The industry draws heavily on chemistry and on a variety of other sciences and technologies. Multidisciplinary research has shed new light on the properties of the complex components of the tree, has steadily upgraded papermaking technology, and has created a range of new products with diverse characteristics and uses. Industry problems today lie often at the frontiers of the scientific and engineering disciplines. Probing the structure of lignin, the primary binding material in wood, fully extends the capabilities of chemical analysis and synthesis. The unknowns in photoconductive papers are at the limits of present knowledge in solid state physics. Hydrodynamic instabilities in the papermaking machine tax the full resources of chemical engineering science and technology. Other difficult problems lie in colloid chemistry and physics, physical chemistry, materials science, the engineering sciences, and the theory and practice of computer control.

CHEMICAL ACCOMPLISHMENTS

The pulp and paper industry today comprises four broad processes:
* Growing and harvesting timber.
* Pulping the timber to separate cellulose fibers in usable form from the lignin and other components of wood.
* Papermaking to form the cellulose fibers into sheets.
* Converting the raw paper into marketable paper products by coating, impregnation, fabrication into bags and boxes, and other methods.

Chemists and chemical engineers have contributed to each of these broad areas. They have worked to greatest effect, however, in pulping (and bleaching), in certain aspects of papermaking, and in conversion processes.

Pulping Processes

Wood is converted to cellulose fiber or pulp by several different processes: sulfate (kraft), sulfite, soda, mechanical or groundwood, and semichemical, which combines chemical and mechanical pulping. The resulting pulp is bleached, if necessary, to remove coloring materials and produce a pulp of the desired whiteness and physical and chemical properties. Common bleaching agents include chlorine and its compounds, hydrogen peroxide, and sodium peroxide.

Of the pulping processes used in this country, by far the most important is the kraft or sulfate process. It accounts currently for some 70% of the wood pulp produced. The process was first developed in Germany in the last quarter of the 19th Century, when the sulfite process also was being developed. The dominant factor in the success of the kraft process is the superior strength of paper made from kraft pulp. ("Kraft," in fact, is German for "strong"; "sulfate" comes from sodium sulfate, the makeup chemical used in the process.) Among other advantages of the kraft process is its ability to pulp different species of wood. The disadvantages include problems with air pollution, heavy capital investment in equipment for recovering and recycling pulping chemicals, and the dark color of the unbleached pulp. Continuing chemical and equipment development have brought the kraft process to such a high level of efficiency that it has been the basis of practically all new pulp mills built in this country for close to two decades.

Kraft or sulfate pulp mills may differ among themselves in certain process details, and they differ substantially from other types of pulp mills. The process will serve well enough, however, to illustrate the essentials of producing wood pulp.

Like most pulping operations, the kraft process starts with the tree, which is harvested and cut into short lengths. At the mill, the pulpwood is debarked hydraulically, washed, and chipped into pieces not much larger than a quarter. The chips go to pressure cookers or digesters, where they are cooked in "white liquor," a water solution of sodium hydroxide and sodium sulfide. The cooking process dissolves the lignin (and other materials) that binds the wood together, thus freeing the cellulose fibers for conversion to paper. The fibers are separated, washed, screened, and bleached, if necessary, to the required visual whiteness or "brightness." After further refining, the fibers may be combined with various substances, such as pigments, dyes, sizings, and resins, depending on the type of paper to be made. The resulting product is diluted with water to less than 1% fiber and other solids, and this is the fiber "furnish" that is fed to the paper machine.

Spent pulping liquor or "black liquor" from the digesters, which contains a large fraction of the original pulping chemicals, is oxidized with air to convert at least part of its content of sulfur compounds to less volatile forms. (Certain of the sulfur compounds are extremely malodorous, and the human nose can detect them in air at concentrations of only a few parts per billion.) The oxidized black liquor is concentrated in evaporators, and sodium sulfate or salt cake is added to replace the pulping chemicals lost up to that point in the process.

The concentrated liquor is fed to a recovery furnace where its organic constituents are burned. The heat of combustion generates steam for use elsewhere in the mill and converts the dissolved pulping chemicals to a molten slag. The recovered slag is dissolved in water to produce "green liquor," which is treated with slaked lime to produce white liquor for recycling to the digesters. The treatment of green liquor with slaked lime yields a lime sludge which is converted back to slaked lime for recycling to the green liquor.

The modern kraft pulp mill integrates to an unusual degree the characteristic unit operations of chemical engineering: evaporation, fluid flow, heat transfer, and the like. The economics of the process rely heavily on conservation of materials and energy, and years of chemical and mechanical research and development have raised the kraft process to a high level of efficiency. Roughly 80% of the pulping chemicals in the white liquor, for example, are recovered and recycled. Heat generated in the recovery furnace, moreover, is the source of essentially all of the process steam and electricity required in the mill.

Bleaching. Close to 40% of the wood pulp made in the U.S. in 1971 was bleached to improve its brightness. Bleaching chemicals usually are oxidizing agents, although reducing agents are used to some extent. Their primary function is to remove residual lignin compounds and destroy coloring matter. Chlorine and its compounds are the most common bleaching agents, their main advantage being low cost. Other oxidizing-type bleaches are hydrogen peroxide and sodium peroxide, which are used most often in the last stage of a multistage, chlorine-based bleaching process. Reducing-type bleaches include sulfur dioxide and sodium bisulfite and hydrosulfite. The bleaching agents used in a given mill are determined by such factors as cost, the nature of the pulp, the degree of brightness to be attained, and the properties desired in the finished paper.

For a century after bleaching was first commercialized it remained a one-stage, hypochlorite process. Around 1920, pulpmakers began to use two-stage bleaching, which reduced the consumption of chemicals. Up to six stages are used commonly today. Gaseous chlorine was first used in 1915 (in a process in which sodium hydroxide was used in a second stage to dissolve and remove chlorinated lignin compounds). This concept required the use of corrosion-resistant steel, developed around 1910, to handle the highly-active chlorine. Chlorine bleaching was a boon to the kraft industry. Until it was developed, strong, fully-bleached pulps could be made only from sulfite pulps because of the severity of the hypochlorite treatment required to bleach kraft pulps.

Subsequent years have seen steady progress in bleaching processes. A major factor in this development has been the chemical industry's ability to produce elemental chlorine in bulk at low cost. Gaseous chlorine may be injected into the water suspension of cellulose pulp fibers and also is used to make hypochlorite bleaching compounds from calcium or sodium hydroxide solutions.

Although the reactions that take place in the bleaching process are not understood thoroughly, chemists have developed considerable know-how in the operation of multistage bleaching processes. A modern installation consists of two or more stages, each followed as a rule by a pulp-washing step. A fully bleached kraft pulp, for example, might be produced in five stages: chlorination, hypochlorite bleach, chlorine dioxide bleach, peroxide bleach, and a second chlorine dioxide bleach.

Papermaking Processes

The machines used to convert cellulose pulp to paper are of two general types: Fourdrinier and cylinder. Both types form a paper on woven wire mesh. The basic difference between them lies in the methods used to support the wire mesh and to control the drainage of water from the sheet as it is formed. The huge and complex Fourdrinier machine is the dominant type. Such machines can be designed to make papers that range from light tissue to heavy foodboard.

Pulp fibers ready for forming into paper enter the feed equipment of the Fourdrinier machine as a suspension of less than 1% cellulosic fibers in water. The fibers flow onto the flat, moving wire screen of the machine and mat together into a web or continuous sheet of paper. Much of the water and 40 to 60% of the fiber are drawn through the screen, separated, and recycled. Heavy rollers press more water from the moving sheet of paper, which then moves over steam-heated cylinders that evaporate the remaining moisture. The paper or paperboard leaves the dry end of the Fourdrinier in huge rolls that normally are slit and rewound into smaller rolls before further processing.

Paper from the mill rewinder almost always must go through a series of finishing and converting operations to prepare it for the intended use. Finishing includes operations such as supercalendering to improve the surface of the paper, rewinding into rolls of the proper size, and cutting into sheets. Converting includes coating, impregnating, laminating the paper, and using various chemical additives. It also includes conversion of the paper into bags, boxes, and the like.

Chemical Additives in Paper

Most papers contain not only cellulose fibers but various additives whose purpose is to improve a property of the paper or paperboard that otherwise would be deficient. These materials are added while the pulp is being prepared for papermaking or after the sheet is formed. In many instances, huge quantities of chemical additives are used. The paper industry is the largest consumer of white clays (kaolin), for example, and is about equal to the food industry in consumption of starches.

The size of the paper industry's output—some 55 million tons annually—attracts many new products from the chemical industry each year. The use of even small percentages of an additive based on such a large tonnage presents an attractive market. And in some products, such as coated and impregnated papers and paperboards, the additive may become a major component of the final structure.

Paper additives are used for a variety of purposes. Starches, gums, and some synthetic polymers are used to supplement the mechanical beating and refining of the pulp and to improve fiber-to-fiber bonding and dry-strength properties. Urea-formaldehyde, melamine-formaldehyde, poly-amides, and other synthetic polymers impart wet strength to paper or board. Cationic or positively charged polymers are coming into wide use to improve paper's retention of fiber fines, pigments, and bonding adhesives. Alternatively, such polymers may be used in systems that collect and recover suspended solids which then are returned to the papermaking system rather than ending up in the mill effluent. Internal sizing agents make papers that resist penetration of water, rain, blood, ink, oils, and greases. Clays and other white pigments improve brightness, opacity, and printing properties. Surface active agents improved the absorbency of paper tissues, towels, and sanitary products.

The paper industry's use of additives, currently about 5.5 million tons annually, is growing at about 6% per year. Chemical additives account for an estimated 10% of the cost of making paper and board or just over $2 billion annually. The true utility of the additives, however, is hardly evident in such figures. Without them, production of paper would be confined to uncolored tissues with little wet strength, a few absorbent papers, some glassine, unsized writing papers, and perhaps unsized newsprint. Total annual production might be 5 million tons instead of the current 55 million tons of paper and board in a multitude of grades. Thus, by incorporating about 5 million tons of various chemicals in its products, the paper industry has so increased their versatility as to command a market 10 times as large as would otherwise be the case.

Utility of Paper Products

The specific contributions of chemistry to the utility of paper products are evident in a number of end-use classifications: communication, industry, convenience products, health and hygiene, packaging, and shelter. Considerable overlap exists in these classifications, in that communication papers are used in industry, convenience products in health and hygiene, and so on. The groupings nevertheless will serve to highlight the many diverse services provided by paper.

Communication. The purpose of paper historically has been communication. Today, transformed by chemistry, paper provides a remarkably flexible means of recording and transmitting information across time and distance—quickly, accurately, inexpensively. The chemist contributes first his basic knowledge of cellulose fibers and of the pigments and adhesives often coated on paper to improve its brightness and the clarity of print on it. He has developed as well paper-treating materials that react to heat, electric current, light, pressure, and chemicals. Paper so treated can be used to produce single or multiple copies of words or pictures stored on paper, on film, or even in a computer memory.

Photographic paper, a classic chemically-treated paper, is coated with compounds of silver that react to light. To this system chemists have added photosensitive dyes to produce images in full color.

Blueprints, a communication medium in construction, are made by photoreproducing drawings on paper coated with iron compounds and dyes. Drawings stored on microfilm can be used to produce working blueprints by transferring the image to paper treated with photosensitive chemicals.

The pressure of a stylus on a sheet of carbonless carbon paper ruptures small capsules of chemicals on the down-face of the paper. These chemicals then react with other chemicals on the top-face of the sheet below to form the duplicate image.

Thermography relies on dyes that respond to heat to produce copies rapidly and conveniently. The more recent xerography reproduces original copy on paper coated with photoconductive zinc oxide pigments and insulating binders. Chemistry has brought lithography within the reach of everyday office copying needs. The chemistry of the sheet can be adjusted to let the user choose the most effective materials for a few or many thousand copies. The lithographic printing plate may incorporate one of several chemical imaging systems that respond to light, heat, pressure, or sequential application of chemicals.

Industry. Industry uses paper components widely in its products. Paper so used may require not only chemical treatment but special processing controls as far back as the pulping stage. Dissolving pulps, for example, must be free of materials that may cause haze or other flaws in the rayon fiber or cellulose acetate film made from them. Insulating papers used in electrical equipment must be free of metallic particles that would degrade their electrical resistivity.

Antistick papers are used as the backing for pressure-sensitive tapes, as the surface for casting plastic films, and in other ways. The basic structure of such papers must be tight to prevent penetration by the product and to avoid loose fibers that might become embedded in the product. The surface of antistick papers often is treated with silicones or other materials to prevent adhesion of wax, rubber, asphalt, and the like.

Many papers are impregnated with heat-setting resins to produce hard, water-resistant materials suitable for use as electrical panels or counter tops. The conditions of treatment can be varied to produce stiff, porous, water-resistant materials for auto air filters or plywood overlays.

Paper is impregnated with latices to produce tough, abrasive papers, tapes, or gasketing. It is treated with resins or other binders and densified under pressure to make automobile board and binder board. Papers used as insulation in electrical transformers may be treated with acrylonitrile to reduce their tendency to degrade in the heat generated in the transformer. The many other industrial uses of paper include specialized filter papers, cellulosic sound deadening materials, battery separators, molded cellulosic parts, and paper shells for commercial explosives.

Convenience Products. "Convenience" has long been characteristic of paper products ranging from cuffs and collars to towels and napkins. The concept lately has become a powerful influence in the U.S. marketplace, however, and the versatility of paper, enhanced by chemistry, has created a variety of new convenience products, from diapers to dresses, from bath mats to drapes. In 1968, the U.S. convenience market consumed an estimated 700 million square yards of yard goods, valued at $150 million. The estimate for 1973 is 2.2 billion square yards, valued at $515 million.

Nearly all the convenience products share four characteristics: They are designed for a limited number of uses; they are priced within the reach of most consumers; they are composed of nonwoven sheets of natural fibers (such as cellulose), synthetic fibers, or both; and they replace the traditional woven fabrics. Paper itself is a nonwoven sheet of fibers, and papermaking technology is thus heavily involved in the manufacture of nonwoven convenience items. For this reason, the nonwoven products have been developed largely by paper companies in cooperation with fiber and chemical suppliers and machinery designers and builders.

Nonwoven materials find their way into many different end products, from soft, light, facial tissues to dense, stiff, rather boardy lining materials. This wide variation in physical properties is attained through particular manufacturing processes, such as air laid, dry laid, or spun bonded techniques, and the use of chemical additives such as adhesives, thermoplastic and thermosetting binders, opacifiers, water repellents, and flame retardants.

Some estimates show that U.S. production of these nonwoven materials will climb from 250,000 tons in 1970 to 1 million tons in 1980, thus creating substantial markets for the fibers and chemicals used to make them. Current special challenges in nonwovens include chemicals that impart improved flame resistance, a water repellent that functions in the presence of binders for use in rainwear and similar materials, binders that adhere better to the smooth surfaces of synthetic fibers, synthetic fibers that disperse readily in water, and fast-curing binders and special finishes.

Health and Hygiene. The impact of paper on health and hygiene is exemplified by toilet tissue, which was becoming a well-established household sanitary product by the end of the 19th Century. In the ensuing years, paper has been put to a multitude of sanitary uses, ranging from paper towels and wipers to food containers and protective wraps. In recent years, paper products have begun to achieve very large usage in medicine and health care.

Paper is inherently a good sanitary material. Cellulose is readily purified and nonallergenic, and pulping and papermaking take place at temperatures high enough to ensure a product free of disease-producing organisms. This plus paper's other physical properties and low cost have made it useful for some years in such items as medicine cups, pill envelopes, dental bibs, and covers for examining tables. The major recent development in health-care paper products has been the advent of nonwoven fabrics that combine wood pulp and fibers such as rayon. The resulting material is soft, strong, and inexpensive and is used widely today in hospital sheets, pillowcases, towels, examination and surgical gowns, and lab coats. Such products can be discarded after a single use, which both eliminates the need for laundering and helps to prevent the spread of infectious diseases.

One problem with hospital wear is that the natural interfiber bonding that normally would provide the necessary mechanical strength must often be minimized to obtain the necessary softness and flexibility. The problem can be solved by using natural or synthetic adhesives to improve interfiber bonding strength at lower sheet densities. Another use of chemistry is to produce hospital fabrics that repel water and other fluids. The main purpose of making the fabric fluid repellent is to prevent or at least minimize cross infections.

Packaging. Almost all packaging in the U.S. contains paper or paperboard in one of their many types and grades, and packaging consumes roughly half the production of these materials. In 1970, the nation used 51 million tons of packaging valued at $19 billion. Paper and paperboard accounted for about half of both tonnage and dollar volume. Also in 1970, 20% of all packaging paper and paperboard was reused, and 18% of the fiber used in packaging was recycled fiber.

Paper packaging materials fall into five broad groups: unbleached kraft papers, bleached papers (mainly kraft), unbleached paperboard, bleached paperboard, and combination boxboard. These basic materials are modified physically and chemically and combined with other materials, such as metal foil and plastics, to meet the requirements imposed on packaging.

Unbleached kraft paper, whose largest use is the common grocers' bag, can be modified physically to provide a range of porosities and to increase its extensibility and impact resistance. It can be treated chemically to enhance its dry strength, wet strength, and water repellency. It can be coated, impregnated, or laminated to improve function and looks.

Multiwall sacks, the second largest use of unbleached kraft paper, are used to ship and store more than 2,000 products: cement and lime, agricultural chemicals and supplies, and many food products and industrial chemicals. The requirements for multiwall sacks are numerous and often severe. The physical and chemical treatments available, however, can produce multiwall sacks that withstand submergence in water and can be stored outdoors for more than a year. Multiwall sacks can be made that resist gas, heat, flame, cold acids, alkali, oil, grease, sunlight, insects, and mildew and mold.

Bleached papers are used primarily to package and merchandise foods. They are growing more important, largely because of the development of the chlorine dioxide process for bleaching kraft pulp. The process gives kraft paper the bright appearance and whiteness required in many functional and promotional uses with minimum degradation of kraft's inherent strength. The properties of bleached papers can be upgraded by many of the chemical and physical methods used on unbleached kraft papers. Bleached papers also can be made pliable with chemical and physical treatments and combined with film and/or foil to form pouchlike packages on automatic machines that form and fill the packages.

Unbleached paperboard is used in the largest tonnage of all packaging materials. Most of it is made into corrugated board, sheets of linerboard bonded by adhesives to a central fluted medium. More than 90% of all packaged freight in the U.S. is shipped in corrugated containers. The corrugated structure has changed little for decades, but chemical and other treatments have made corrugated containers increasingly versatile.

Unbleached paperboard also is used in fiber drums and the fast-growing composite cans. The latter, which generally have metal ends, are spirally-wound combinations of board with specialty papers, films, and foils. The familiar beverage six-pack is made from kraft linerboard that has been clay-cooked to create a good printing surface.

Bleached paperboard is used in sanitary food containers, such as cartons for milk, ice cream, and frozen foods. It's used also in a growing variety of folding cartons, which are precision-made, low-cost packages supplied in knocked-down form. Almost all folding cartons are made on special order for customers who erect and fill them on custom-made machines. Special coatings and decorative treatments can be applied before, during, or after printing, cutting, and creasing.

Combination boxboard, multiple plies of recycled fiber and bleached or semibleached virgin pulps, is used mainly in folding cartons and setup boxes. The outer plies are made of virgin pulps or high-grade recycled fibers to provide the required appearance, scuff resistance, and bending qualities. The inner plies are made of lower-grade recycled fibers and serve primarily as fillers. Coatings and laminations can be used to achieve special surface properties and appearance.

The particular requirement of folding cartons is that the boxboard not crack when scored, bent, and folded. Setup boxes are manufactured in their final form, so the boxboard must be rigid and resist abuse. It can be decorated with coverings or wrappings applied with adhesives. Setup boxes are strong and durable and are used to package items that aren't consumed at once, such as candy, stationery, dental supplies, and machine parts. They're used as prestige packages and are well suited to attractive displays. Most setup boxes are custom designed, and the low cost of tooling makes them economical in small quantities.

Shelter. Paper and paperboard shelter components have become important products of the pulp and paper industry in the 20th Century. Such materials include hardboard, structural fiberboard (insulation board), composition roofing, and gypsum core wallboard.

Production of most of these products involves treatment with water or steam, whether for pulping wood or other cellulosic materials or reclaiming fiber. The methods range from low-temperature treatment for reclaiming wastepaper to treatment with high-pressure steam to make hardboard fiber. The pulping processes originally were worked out empirically, but chemists now are learning a good deal about the complex reactions of water with wood and other fibrous materials.

Resins and water-resistant compounds are used in hardboard to improve dry and wet strength. Hardboard also is given prime and finish coats for exterior exposure and special resin or laminate finishes for interior use. Structural fiberboard receives all the treatments used on hardboard and others as well. An emulsion or ground suspension of asphalt is used commonly in structural fiberboard sheathing. Interior grades take a range of coatings, including intumescent finishes that impart fire resistance to special products such as acoustical tile.

A relatively new grade of structural fiberboard is made expressly for exterior siding. It contains an unusually high proportion of water-resistant resin, is pressed to high density, and is finished with durable prime coats.

Composition roofing is made by saturating and coating coarse, highly porous paper with special grades of asphalt and embedding natural or synthetic mineral granules in the resulting surface. The chemistry involved is complex. The mineral granules, for example, must be opaque to ultraviolet light to protect the asphalt from photochemical degradation.

Environmental Control

The biggest single challenge facing the pulp and paper industry today is the need to reduce its contamination of the environment to a level that represents the optimum balance of economic and ecological factors. Where that optimum point might lie is the subject of considerable debate, but the overall problem is evident in the industry's outlays on capital equipment and research and development. Pollution control facilities absorbed almost 20% of total capital spending in 1971, compared to less than 15% in the previous year. Research and development on environmental matters currently absorbs $10 to $15 million annually, industrywide, or 10 to 15% of the total R&D outlays.

Each of the paper industry's four basic operations—harvesting, pulping, papermaking, and converting—presents its own potential pollution problems. Those that involve chemistry and chemical engineering, however, are most prevalent in pulping and papermaking.

The primary pulping method, the kraft process, poses mainly an air pollution problem. Its potential emissions include large amounts of particulate matter and smaller amounts of malodorous organic sulfur compounds (mercaptans). The latter are not highly toxic, but the human nose can detect them at concentrations as low as five to 10 parts per billion. The second most important pulping method, the sulfite process, poses primarily a water pollution problem, since it is a potential emitter of waterborne wastes with very high biochemical oxygen demand. Papermaking, too, is primarily a water pollution problem. More than 90% of the water that carries the cellulose fibers onto the machine is recycled, but the remainder, containing suspended fiber and sometimes clay and other solids, constitutes a significant effluent problem.

To clean the water they discharge to the environment, companies use a variety of methods, such as clarification, screening, flotation, and vacuum filtration. Among areas that require development are improvement of effluent clarity by use of chemical additives, optimization of fiber recovery in the mill against the overall cost of sludge disposal, and finding new ways to remove the water from certain types of sludges that are particularly difficult to handle. To remove nonsettlable solids from effluent water, companies use standard biochemical methods ranging from nonaerated, long-term storage basins through high-rate activated sludge treatment. Development needs here include special methods to remove more than 95% of the biochemical oxygen demand, better means of disposing of excess activated sludge, and better control of the activated sludge process.

The paper industry's leading air pollution problem, the kraft process, has an emission potential of well over 200 pounds of particulate matter and about 50 pounds of sulfur compounds per ton of pulp produced. Control measures in common use can reduce particulate emissions to under 5 pounds per ton, and 6 to 8 pounds is commonly encountered. Significant needs include particulate monitoring devices to determine that equipment is operating satisfactorily and more study of other control methods to be certain that optimal technology is being applied.

The kraft process emits malodorous sulfur compounds at two stages: the digesters and succeeding multiple effect evaporators and the recovery furnace system. Methods are available that can be used in combination to eliminate such emissions from the digesters and evaporators. The use of these methods is growing and already is the norm for new mills in some areas. Emissions from the recovery furnace system, a more difficult problem, can be reduced by several means. Each is a combination of techniques, and early in 1972 the industry was reviewing the alternatives intensively. The objective was a control technology capable of reducing emission of sulfur compounds by the furnace system from a previous norm stated to be nearly 50 pounds per ton of pulp to less than 0.1 pound.

The pulp and paper industry overall has had a continuing need for better means of controlling pollution abatement processes and for new and more selective processes. To this end, a group of Wisconsin paper mills formed the Sulfite Pulp Manufacturers' Research League in 1939. (In 1969, the league merged with the Institute of Paper Chemistry, Appleton, Wis., a degree-granting graduate school that is subsidized by the industry.) Additional environmental research has been coordinated since 1940 by the industry-supported National Council for Air and Stream Improvement.

Paper companies nowadays regard the cost of pollution abatement as a normal cost of doing business, but the industry is concerned nevertheless that the benefit should justify the cost. This is a complex and controversial concept, and in varying degree it confronts most manufacturing industries as well as the legislators and regulatory authorities who deal with environmental questions. A discussion of the benefits to be derived from limiting particular pollutants to particular levels of emission is beyond the scope of this study. For conventional treatment systems in the pulp and paper industry, however, the relationship between cost and emission level can be described roughly in terms of two principles of chemical process engineering.

The first of these principles is that doubling the size of a facility increases its cost to about the six-tenths power. Per unit of capacity, therefore, bigger units are more economical. The second, and counteracting, principle is that the less concentrated the waste, the more difficult it is to remove. Doubling the size of the equipment, as a result, will fall significantly short of doubling the amount of pollutant captured. The second control unit in a series, in fact, will capture in general only one half to one tenth as much pollutant as will the first unit of equal size.

Thus, per pound of pollutant removed from the waste stream, it costs three times as much to go from 70 to 90% removal as it costs to remove 70%. To remove 1 additional pound of pollutant at the 95% level costs 30 times as much as removing 1 pound at the 70% level. Costs will vary for specific installations, but as the percent of pollutant removed increases, the cost per pound tends quickly to overshadow whatever economies result from using larger abatement facilities. The problem, then, is to balance the benefits of removing the last few percent of pollutant against the relatively high cost of doing so.

Wastepaper Reclamation

The paper industry in the U.S. has used wastepaper extensively for many years as a source of papermaking fiber, but until recently it has done so primarily for economic reasons. The current emphasis on the environment and conservation of resources, however, has cast the reclaiming process in a new role. Increasingly, for example, requirements are emerging that various papers contain a specified minimum content of recycled wastepaper.

In 1969, the U.S. consumed about 59 million tons of paper and paperboard. The paper industry recycled some 10.5 million tons of this amount into new paper and paperboard, while nonpaper industry reused something less than 1 million tons. Some 9 million tons of the 1969 consumption were papers that do not lend themselves readily to salvage: glassine, parchment, greaseproof papers, roofing paper, wallboard, and the like. Thus, unrecycled paper in 1969, less the "untouchables," came to 39 million tons. This equals 31 million tons of virgin fiber or the annual pulpwood from about 117 million acres of commercial forest land.

Wastepaper to be recycled, usually called "paper stock," falls into four general categories: containers; news, mainly newspapers; pulp substitutes, generated mainly by paper and paperboard converters; and mixed papers. Except for mixed papers, these categories represent paper stock segregated by grade. Mixed papers, in contrast, contain a variety of papers of widely varying fiber quality as well as nonpaper contaminants. They find only limited use in recycling and command the bottom price—$2.00 to $10 per ton—in the paper stock market.

Most unrecycled wastepaper is generally considered to be mixed papers. Thus, if more of it is to be used as a source of papermaking fiber, new markets for mixed papers must be developed or technology must be developed for removing contaminants and separating conglomerate wastepaper into grades that have wider application and command a higher price than mixed papers.

Wastepaper is recycled in the paper industry in four ways, depending on its quality:

• Direct pulp substitution: The paper is of such a quality that it can be substituted directly for virgin pulp. No special equipment or processing is required.

• De-inking stock: The paper is of such a quality that it can be processed economically in a de-inking plant. De-inking, in a sense, is a special kind of pulping process. It removes the ink and brightens the resulting stock by bleaching. De-inked stock is used in place of virgin pulp to make a variety of products, such as book papers, newsprint, tissue paper, and paperboard.

• Polyethylene stripping: A significant tonnage of paper milk bottles is recycled as a substitute for virgin pulp. The bottles are made of relatively heavy, bleached kraft paperboard, extrusion-coated on each side with 4 to 6 mils of polyethylene. The polyethylene is stripped mechanically or by solvent (trichloroethylene). The resulting fibers are of excellent quality and can be substituted for virgin pulp in making new paper and paperboard.

• Direct repulping: About three quarters of all recycled wastepaper is used to make paperboard. The paper is repulped in equipment that can remove wire, string, wood, and other contaminants. The repulped stock is then cleaned in various types of equipment, depending on the end use, and used in the normal way to make new paperboard.

For many grades of paperboard, the primary end-use requirement is strength. The strength of paper depends on the intrinsic strength of the individual fibers and on the number and strength of the molecular bonds that hold the fibers together in the sheet. Each time a cellulose fiber is recycled it undergoes a small but irreversible reduction in bonding potential and intrinsic strength relative to virgin fiber. Thus, although very strong paper can be made from recycled fiber, finite limits exist on the number of times a fiber can be recycled while maintaining the required strength in the resulting paper.

Recycling involves quality considerations in addition to strength, but it should be possible nevertheless to expand the amount of paper that is recycled. One approach would be merely to expand the present practice of segregating and collecting wastepaper in the hope that the mills could use such wastes. This approach would necessitate working inferior sources, which would tend to make the operation less profitable, and it is not really clear how much more wastepaper would be used if it were available at the current levels of quality. Wastepaper generated by paper converters, for example, already is recycled almost entirely.

A second approach requires the development of new technology in such areas as collection, mechanical or other types of sorting (bleached from unbleached, etc.), cleaning, strength development, and fiber classification. The goal would be means of processing conglomerate wastepaper so as essentially to restore the original quality of each type of fiber. Reclaimed fibers would thus be usable in the same general grades in which they were used as virgin fibers and would command a ready market. No ready market exists for conglomerate fibers as such.

The paper industry and others are doing a limited amount of research on new means of processing wastepaper so that more of it can be recycled. Studies also are being made of the use of wastepaper as a source of cattle feed, protein, and alcohol and as a fuel.

FUTURE DEVELOPMENTS

Among the trends in the pulp and paper industry that will involve chemistry and chemical engineering is the continuing effort to develop new or previously unused sources of cellulosic fiber. Serious local shortages of wood, high wood costs, and forest damage caused by wood residues are problems that can be eased by "whole tree logging" and similar measures. The yield of pulp from hardwood trees, for example, can be increased 30% by pulping unbarked limbs that are chipped in the forest. In recent mill trials, specks of bark in the pulp have been eliminated by new bleaching and cleaning techniques. Chemical debarking may be applicable to hardwoods with tough barks. Well established is the separation of bark from pine slabs in flotation equipment. The burning of sawdust, which can pollute the air, has been replaced by pulping the sawdust with special processes. Wood residues, overall, supplied 20% of the pulpwood used in this country in 1970.

A possible new source of papermaking fiber is kenaf, the East Indian hibiscus whose fiber is used widely for cordage. Chemical research has overcome the main disadvantage of kenaf, the fact that its use reduces the speed at which the paper machine can operate. Kenaf can be harvested every nine months when grown year-round in a suitable climate, such as the southeastern U.S. and California. An area of 9,800 acres can supply the green kenaf for 570 tons of pulp per day, and up to 70% kenaf fiber can be used in many printing papers.

A limited amount of paper has been made for many years from bagasse, the residue from sugarcane. The paper originally was of relatively low quality, but its properties have been improved by biological pretreatment of the bagasse and modern chemical pulping techniques. Increased production of bagasse pulp is feasible if an economical fuel can be found to replace the bagasse, which is used commonly now to generate steam in cane sugar refineries.

High Yield Pulping. The paper industry has worked steadily on pulping processes to increase their yield, the fraction of the cellulose in the wood that is converted to usable papermaking fiber. The highest yields—93 to 98% without bark—are achieved with groundwood pulp, which is made by defibering logs mechanically on rotating stones. Such pulp can be used in newsprint but not in stronger papers. Tough papers, such as for shipping bags, are made of pinewood pulped by the kraft process with a yield of about 42%. The net effect, besides differences in quality, is that a ton of kraft pulp requires twice as many trees as a ton of groundwood.

By far the largest tonnage of packaging paper is used in corrugated boxes with a top and bottom layer of pine kraft linerboard separated by a corrugating medium made from neutral sulfite semichemical (NSSC) hardwood pulp. The NSSC process was invented at the U.S. Forest Products Laboratory, Madison, Wis., in the 1920's. It combines mild pulping with fiberization of the softened wood chips by precision disc refiners. NSSC hardwood is made at an average yield of 75% and is used mainly in corrugating medium. The recovery of sodium sulfite for recycle in the process is complex, and several mills have switched to ammonium sulfite. A few mills have bleached the ammonium sulfite pulp, but the effluent from the bleaching process caused water pollution problems. New bleaching chemicals may solve this problem.

Several years of mill trials have gradually increased the yield of pine kraft pulp for the linerboard used in corrugated boxes. The yield was increased 35% recently by using a milder kraft digestion and breaking up the resulting fiber bundles in a large disc refiner of new design. Further increase in the yield of kraft pulp reduces the toughness of the linerboard, but the loss can be restored by adding synthetic resins to the board. In addition, the yield can be increased by several percent by using chemical additives to retain waste cooking liquor solids in the paper sheet.

Higher yields of kraft pulp at acceptable strengths can be obtained by modifying the process with hydrogen sulfide or sodium polysulfide pretreatment. Many laboratories are investigating means of recovering these chemicals for recycle without emission of malodorous air pollutants. Paper as tough as kraft has been produced, at higher pulp yield, by increasing the alkalinity in sodium sulfite pulping of softwoods. Pilot plant work on the process was planned as of early 1972.

Great progress has been made in high-yield pulping of the chemical pulp fraction of newsprint, which also contains mechanical (groundwood) pulp. Yields of 65 to 70% are being obtained, and some 20 mills have achieved good recovery of the magnesium bisulfite waste liquor from the digesters. The high-yield newsprint processes use mainly single-stage pulping, but even higher yields, at higher strength, are being obtained in a few mills with two-stage pulping. It has been proposed that the pulp from the first stage be disc refined under pressure, thus increasing the efficiency of the second pulping stage and raising the overall yield. Precedent exists in building board pulp, which is produced in large tonnage by mild, single-stage pulping to 78 to 80% yield, followed by pressure disc refining.

Several processes have been developed to improve groundwood. Disc refiners produce a considerably stronger groundwood pulp than do the older stone grinders. The overall yield of newsprint is thus increased, because more disc-refined groundwood at maximum yield and less chemical pulp at lower yield can be used to make a paper strong enough to sustain the required high operating speeds of paper machines and printing presses. The strength of groundwood can be increased with a minor drop in yield by several different chemical pretreatments of the chips before disc refining. These techniques permit the production of several types of groundwood.

The industry will continue to press for higher yields from its pulping processes, consistent with the properties required in the resulting paper and paperboard. The competition between chemical and chemimechanical processes for highest yield, in fact, is gradually being resolved by better understanding of the real marketing requirements of the industry's many end products.

New Papermaking Materials. The paper chemist historically has dealt with fibers from living plants. As plastics and other synthetics came on the scene, however, he learned to use them in paper. Blends of synthetic and cellulosic fibers were used in some cases; in others, the entire sheet was noncellulosic.

The paper chemist today uses fibers made of rayons, acrylics, polyamides, polyesters, acetates, polyethylene, and polypropylene. In addition, he may alter cellulose chemically to make it essentially a plastic fiber. These synthetic fibers may be used as such, either separately or in combination, or they may be blended with plant fibers. By such means the strength of paper can be improved both when wet or dry. Its life expectancy can be increased or, if desired, sheets can be constructed that dissolve completely in water. The electrical properties of paper can be enhanced enormously, and its ability to withstand extreme temperature can be increased. Sheets can be made to resist attack by acids and alkalies. These improved properties fill widespread requirements in personal and commercial life.

The paper chemist has learned also to make paper-like structures from metal fibers, asbestos, glass wool, and carbon and ceramic fibers. "Paper" webs have been made experimentally of man-made, superstrong crystals (whiskers). Metal fiber structures may be used where cellulose itself will not suffice, where a need exists to extend the conductivity of heat or electricity to very high or very low extremes, to impart magnetic properties to webs, or to make them capable of absorbing or reflecting electromagnetic radiations. Metal fibers can contribute substantially to the durability, modulus of elasticity, and impact strength of paper products, thus expanding their uses in sound absorption, vibration dampening, friction control, wave guide channels, heat exchangers, and elsewhere.

The past five years have seen yet another chemical development, a "paper" made not from fibers but from a solid sheet of plastic, such as polyethylene. The impetus came from Japan, where industry/government evaluation determined that a plastic substitute for cellulosic paper should be developed to conserve timber and, at the same time, to capitalize on worldwide pricing and availability of petroleum, the raw material for most plastics. The paper and chemical industries of the U.S. have responded by developing their own plastic papers, by cross-licensing, and by licensing foreign inventions. In all of this, the chemist has played a vital role in paper technology, in polymer chemistry, and in printing ink formulation.

Environmental Control. Although it is rarely safe to predict what research might achieve in a given area, there would seem in the paper industry to be little that is not fairly predictable in the treatment of air by purification, filtration, precipitation, or scrubbing. Thus, the opportunity seems limited for breakthroughs in air pollution control technology. Careful monitoring and control of process units and abatement devices will remain necessary and important. For the most part, however, the future of air pollution control is most likely to entail better design of present methods and equipment and better maintenance and control of such equipment. There exists in addition the possibility of developing process modifications and new processes having significantly lower air pollution potential than those in use today.

The future of water pollution control presents quite a different picture. Liquid effluents cover a much broader range of chemical structures and physical states than do airborne emissions. Treatment technology is much more varied and complex, and the results depend much more heavily on the physical and chemical nature of the specific waste stream. An indication of the need for fundamental research in water pollution control can be seen in the technique of removing color from kraft effluents by treatment with an excess of lime, which is then recovered and reused. One of the earliest publications on the method is more than 20 years old, but only recently have chemists studied the fundamental chemical and physical composition of the color in kraft effluent, the effect of lime treatment on these characteristics, and the practical barriers that must be surmounted to use the process on an industrial scale.

Because of water pollution problems, oxygen has begun to replace chlorine for bleaching chemical pulps. One mill in South Africa and one in Sweden are now using oxygen, and a new bleach plant in Virginia planned to start using it in late 1972 as a partial substitute for chlorine. The use of oxygen reduces the biochemical oxygen demand, color, and chloride load in the bleach effluent.

SOCIAL AND ECONOMIC IMPACTS

The strength of an industrial nation rests heavily on its capacity for making iron and steel. Nonferrous metals, concrete, wood, and plastics may compete with ferrous metals in many uses, but none is so fundamental to a country's production of goods and services. Steel combines strength, hardness, and formability to a degree that gives it strong advantages in most uses and the only advantages in many uses. Add to its properties a price that ranges down to 6.5 cents a pound for some common grades, and economical competitors are often hard to find.

The world's steel producing countries now make some 650 million tons of raw steel a year. In 1970, the U.S. continued to hold the top producing spot (with 131.5 million tons, or about one fifth of world production), as it has for many years. However, the U.S.S.R. was only 4 million tons behind. Considering recent growth rates for the U.S. and U.S.S.R., 1971 probably proved to be the first year the U.S. slipped into second position.

With gross revenues of $19.1 billion in 1970, the steel industry employed about 530,000 people directly in iron- and steelmaking operations. Perhaps as many as another 200,000 also earned their livings in related activities, such as mining, preparing raw materials, and in transporting raw materials to steel mills and steel mill products to customers.

Steel sheet and strip, structural shapes and plates, bars and tool steel, pipe and tubing, and tin mill products are the industry's primary products. What the industry calls steel service centers and distributors are its largest single class of customers (17.7% of shipments in 1970), followed by the automotive industry (15.9%), construction and contractor products (14.7%), machinery and equipment (9.6%), containers and packaging (8.6%), forgings, bolts, nuts, etc. (6.1%), household and commercial equipment (4.3%), rail transportation and shipbuilding (4.3%), and gas and oil industry products (3.9%).

Making ferrous metals requires a large amount of both heat and mechanical energy. Casting, rolling, forging, drawing, and machining fabricate raw steel into many finished products. Chemistry is little used in these processes, except in such auxiliary roles as in providing coolants, lubricants, and inert atmospheres. Making iron and steel themselves differs, however. The reactions are all chemical, and they depend on processes that often resemble unit operations of chemical engineering.

The processes that reduce ore to pig iron and then convert pig iron to steel have remained basically unchanged for many years. In making pig iron, oxygen in a blast of air blown into a furnace containing ore, coke, and limestone burns carbon in the coke to carbon monoxide. This burning liberates the large quantities of heat needed for the reactions and for keeping the reduced iron molten. Meantime, moisture at the temperatures reached in the air blast reacts with some of the coke to produce hydrogen and more carbon monoxide. The carbon monoxide from both reactions plus the hydrogen then reduce the iron oxides in the ore to iron.

The primary product, pig iron, contains small amounts of impurities,

---

Some of the technical material in this article has been taken from "The Making, Shaping and Treating of Steel," published by United States Steel Corp., which the author found to be an invaluable source of detailed information.

mainly manganese, phosphorus, and silicon from the original ore plus carbon and sulfur mostly from the coke. To convert pig iron into more useful steel, the impurities must be lowered to carefully controlled levels. In the first historical processes to make steel, producers refined pig iron in furnaces of several types to convert it to wrought iron that consisted of nearly pure iron. The wrought iron was never molten, but producers obtained it as a sponge that contained a good deal of entrapped slag. They could then convert bars of wrought iron into steel in relatively small quantities at high cost by either what were known as the cementation process or the crucible process. In the cementation process, they packed wrought iron in air-tight boxes with carbon and heated it for several days until sufficient carbon had diffused into the bars to convert them to steel. In the crucible process, they melted wrought iron to free it of slag and added the necessary carbon to the molten metal.

In the mid-1800's, Henry Bessemer in England showed that oxygen in a blast of air could be used to convert molten pig iron directly into molten steel. The Bessemer converter then served for some years as the principal route to steel. In the late 1800's, however, the Siemens regenerative open hearth furnace, also developed in England, began to replace the converter, and today little Bessemer capacity remains. Now, the open hearth furnace is being superseded by what is known as the basic oxygen process (BOP), which uses what is essentially a converter blown from the top.

Regardless of the details of the process—Bessemer, open hearth, or BOP—the chemistry is the same. Oxygen combines with unwanted elements, except for sulfur, in molten pig iron, and the oxidized products leave as gases or enter the slag. Sulfur dissolves in a slag that is maintained at a basic pH.

In the Bessemer process, a blast of air provides the oxygen. In the original open hearth processes, oxygen was supplied by iron ore, oxidized scrap, or excess air in the combustion space above the charge, while jets of oxygen often enhance the oxidizing potential in open hearths today. In the BOP method, 99.5% oxygen is blown directly on the molten pig.

CHEMICAL ACCOMPLISHMENTS

The chemistry of reducing ore first to iron and then refining the iron to steel is fundamentally the same as it has always been. Details of process engineering have changed, however, often greatly. Recognizing that a blast furnace, for example, is basically a packed column with a countercurrent flow of rising gases and descending solids and liquids has led to more effective controls and greater productivity, while studies of the physical chemistry and thermochemistry of making steel have laid a basis for applying thermodynamic principles to metallurgy. In addition, studies of equilibriums and rates of reactions in metals have provided information from which numerous alloys with new properties have been prepared, while phase studies of refractory oxide systems have been of great value in clarifying the behavior of slags and refractories in the processes. Making iron and steel today, as a result, is in many ways a chemical science and engineering process.

Making Iron and Steel

Much of the more scientific basis with which the ferrous metals industry operates is best illustrated by developments of the past 40 years or so, and the discussion that follows will cover mostly innovations in this period. Important developments occurred before 1930, of course, but that year represents a good inflection point in the history of the industry. It was at about that time, for example, that the industry began to realize more widely than before that metallurgy and metallurgical engineering are essentially special branches of chemistry and chemical engineering. Also, the desirability of converting from batch to continuous processes was becoming clear. Finally, significant improvements in mechanical equipment, notably in rolling mills, were opening the way for many advances.

Up to about 1930, blast furnace operators did not understand too well many of the chemical and physical features of the blast furnace process. The temperature of the hot blast rarely exceeded 1,200° F., and operators seldom tried to control the blast's moisture content. Under these conditions, small changes in humidity affected furnace operations adversely because of the endothermic reaction between steam and coke. Variations were particularly noticeable in the seasonal changes between summer and winter. Moreover, in later years as the industry developed better hot blast stoves and began using higher blast temperatures, furnaces operated irregularly because the hearth temperature was higher than necessary.

Operators then realized that adding steam to the blast to make its moisture content uniform also helped control flame temperature and thus improved furnace operation. They obtained another benefit in that the reaction between steam and coke produced more reducing gas per unit of carbon, yielding one unit of carbon monoxide and one unit of hydrogen, instead of just one unit of carbon monoxide in the carbon-oxygen reaction. The industry now generally keeps the moisture content of the blast at a fixed level, and higher blast temperatures and moisture control have contributed significantly to increased blast furnace productivity.

Another innovation of the period involves injecting a supplementary hydrocarbon fuel into the hearth of the blast furnace to replace coke. Coke was becoming increasingly expensive at the time, and engineers sought not only to reduce costs directly but also to obtain a reducing gas higher in hydrogen. Early trials didn't succeed, and examination of the reaction's thermodynamics soon showed why. Whereas the reaction between coke and air is highly exothermic, the one between hydrocarbons and coke is highly endothermic. Furnace temperatures accordingly dropped.

The solution lay in providing more heat by raising the temperature of the blast. Natural gas, coke oven gas, fuel oil, pulverized coal, and slurries of oil and fine coal have all been used successfully. Selection of the best supplemental fuel for a given plant today depends chiefly on the cost at that plant of the auxiliary fuel relative to that of coke and the cost of the equipment for injecting it.

Blast furnace productivity has also been increased by raising the pressure. During World War II, the industry tried a suggestion that increasing pressure within the furnace's shaft would allow a greater amount of air to be blown and increase a furnace's throughput. Although the first attempt (in 1944 by Republic Steel Corp.) had to be discontinued because of mechanical difficulties, successful operation was achieved in 1946. High top pressure is now in general use throughout the industry.

The U.S. steel industry 40 years ago made nearly nine tenths of its product in open hearth furnaces and slightly more than one tenth in Bessemer converters (with electric furnaces accounting for the very small balance). The open hearth furnace of the time had a large heat size and long heat time, could use up to 50% scrap, and had a roof of silica refractories that melted at a relatively low temperature. Process temperature and metal and slag compositions were not easily controlled, and control of residual elements in the metal was especially troublesome at times.

Moreover, the industry had relatively little scientific background for the process, although its empirical practice was remarkably good. It knew nothing in those days about thermodynamic activities in metallic solutions or in slags. Designers and operators used the mass action law exclusively to represent the equilibrium constants of metallurgical reactions. To compensate for the nonideal behavior of slags, they represented the composition in terms of "free" and "combined" oxides.

Beginning in the late 1920's and continuing through World War II, the industry applied the concepts of thermodynamics more rigorously to its processes. Outstanding contributions during the 1940's marked the beginning of a new era of better understanding of the physical chemistry of steelmaking. Compilations of data on the thermochemistry of materials provided basic information for applying thermodynamics to metallurgy.

Also of great value were phase diagrams of refractory oxide systems determined during these years. Chief among these were those for the systems $CaO-Al_2O_3-SiO_2$, $CaO-MgO-SiO_2$, $CaO-MgO-Al_2O_3$, and $Al_2O_3-MgO-SiO_2$. These and the diagrams for other systems determined in later years have been of great value in clarifying the behavior of slags and refractories. Work on the iron-oxygen system was another notable contribution.

A few years after World War II, relatively cheap oxygen became available in tonnage lots. It served as a powerful tool in steelmaking. At about the same time here and abroad, an interest developed in improving basic refractories, especially those that might be used in open hearth roofs. Such refractories permitted the use of higher temperatures in the furnaces and combined well with the introduction of oxygen through a lance in the roof to give a much greater rate of production in open hearths.

At about this time (in the early 1950's), a successful top-blown oxygen converter for making steel was developed in Europe. It was at first geared to the scrap situation in Europe, and it employed small vessels. Adaptation to American conditions has been rapid, however, and what is now known as the basic oxygen process (BOP) in the U.S. is replacing open hearths here. BOP furnaces for the first time outproduced open hearths in the U.S. in August 1969, and they have continued to widen their margin.

Casting, Pickling, and Coating

Once open hearth, BOP, or electric furnaces make molten steel, it is cast into ingots or into slabs or billets by continuous casting and is then processed into a variety of plates, bars, pipes, wires, and the like. Much of the work is mechanical, as noted earlier, but chemistry is involved at some points, sometimes in a critical way.

In 1930, interest in improving casting and structure of ingots was just beginning, for example. The problems then recognized ranged from the occurrence of blow holes to flaking (the presence of small internal ruptures in large ingots that lead to imperfections in products made from them). Flaking is generally ascribed to high internal stress caused by hydrogen in the steel. The first efforts to prevent it involved very slow cooling of the ingot to allow the hydrogen to diffuse out. Later, the industry realized that an alternative method would be to cast the ingot in a vacuum and remove dissolved hydrogen at the casting stage. Investigations simultaneously but independently here and abroad led to the development of a method known as stream degassing. In this process, molten steel flows from a ladle through a special seal into an ingot mold contained in a large evacuated chamber. Later, variations on this technique, including ladle degassing and recirculation degassing methods, were developed by many different organizations.

Vacuum degassing also removes some oxygen from the steel and helps to free it of undesirable oxide inclusions. A third use of vacuum treatment is vacuum carbon deoxidation. If an undeoxidized molten steel is exposed to a vacuum, a reaction between carbon and oxygen produces carbon monoxide, which readily escapes. In this way, it is possible to produce steels containing less than 0.03% carbon, which are difficult to make by conventional operations.

By 1930, the modern, wide, continuous hot-strip mill and the continuous cold-reduction mill began to come into general use. They made it necessary to design and develop suitable equipment to remove oxides from hot-rolled strip and to prepare it for cold reduction in coil form. Such continuous mills called for continuous pickling lines to remove the oxides, yet those built at first were not truly continuous. A number of tanks in series each contained a different concentration of the pickling agent (sulfuric acid), and the strip passed from one tank to another, starting with that containing the weakest solution. Operators saw that such an arrangement could be made continuous, however, by having the pickling solution flow from tank to tank countercurrently to the movement of the strip. This innovation increased line speed, and hence output, considerably, and it also gave better control of the operation.

In the mid-1960's, the industry began switching to hydrochloric acid for pickling. The move was accelerated by an increase in the price of sulfuric acid and a reduction in the price of HCl, a by-product in making vinyl chloride and polyvinyl chloride by the chlorination of ethylene. Besides a cost benefit, the industry also achieved faster and cleaner pickling, lower acid consumption, and less attack on the base metal. Today, producers use HCl to pickle about two thirds of the steel used in sheet and tin mill products.

Unfortunately, both sulfuric and hydrochloric acids present disposal problems. There still is no economically satisfactory answer for sulfuric acid. Hydrochloric acid, however, can be regenerated and reused and some plants do so, although it requires extensive and costly facilities.

In 1930, all tinplate produced commercially was made by hot dipping. Although electrodeposition of tin on steel had been known for many years, it could not compete economically with hot dipping and accordingly was not used. However, with the introduction in the early 1930's of continuous cold-reduction mills, the prospects for continuous high-speed electrotinning became much better. American Sheet and Tin Plate Co., a subsidiary of U.S. Steel, by 1935 had designed and built an experimental unit. In 1938, the first commercial unit went into production at USS's Gary (Indiana) Works.

This process uses an alkaline sodium stannate bath. It deposits tin more uniformly than hot dipping and permits much lighter coatings. When Far Eastern supplies of tin were cut off in 1942, the U.S. stockpile of tin was so small that military needs would have consumed nearly all, leaving virtually none for civilian use. However, rapid installation of electrolytic lines resulted in a significant saving in tin, and by the end of the war about one third of total production of tinplate was electrolytic.

Today, virtually all tinplate is made electrolytically. A number of different baths are now in use, including phenol sulfonic acid, a fluoride-chloride mixture, fluoroborates, and sulfate, as well as the alkaline sodium stannate bath.

In recent years as tin has become more expensive and a market has grown for containers that do not require the usual type of tin coating because they are lacquered before use, several producers developed methods for making a new kind of steel. This material, usually called tin-free steel, is more correctly termed electrolytic chromium plate. It is coated with a very thin layer (0.3 microinch) of chromium that has a film of chromium oxide on it. Lacquers and other organic coatings adhere extremely well to it, and it resists undercutting of the coating to a high degree. It is used largely for beer and soft drink cans.

Significant improvements have also been made in both methods of production and quality of galvanized (zinc-coated) sheet. For example, the metallic coating is now more uniformly thick, and paints and other coatings adhere better to it. It is also treated to retard formation of white corrosion products that sometimes mar its appearance. Various treatments have been developed, but the most common are washes with dilute aqueous solutions of sodium silicate (water glass) and sodium dichromate, chromic acid, and mixtures of phosphates and chromic acid. If paint adherence is of primary importance, proprietary phosphate treatments are available.

Because of the rapid changes in organic coatings, use of such coatings on steel is not only extensive and complex but is in a constant state of flux. However, among the salient developments in this field have been the introduction of alkyd resins, phenolic resins, vinyl resins, usually copolymers, and epoxy resins cured with amine or polyamine hardeners. Epoxy resins are now being mixed with coal tar to give a coating of good chemical resistance at moderate cost. Acrylic resins are also coming into wider use. Bituminous coatings—coal tars, asphalts, or asphaltums—should also be mentioned, since such coatings perform well underground and in contact with water.

Metallurgy as a Chemical Science

Ferrous physical metallurgy, like ferrous process metallurgy, has profited greatly from the work of physical chemists, especially in equilibriums and rates of reaction in metals and alloys. The foundation for the study of equilibrium dates from the work of J. Willard Gibbs (1839–1903) at Yale University, but it was not until his work was made much more accessible in the form of the phase rule by H. W. B. Roozeboom (1854–1907) at the University of Amsterdam and others that progress began to be made. In fact, it was soon evident that this relationship could be applied quite easily and profitably to condensed systems, and particularly to metals and alloys. By 1930, the basic features of most binary and some ternary ferrous systems had been worked out, though not always with the accuracy and fine detail desired. They were, however, unquestionably useful and were, in general, being steadily refined. For example, by 1930 the details of the iron-rich end of the iron-chromium diagram had been worked out, and the fact that chromium had a limited solubility in austenite had been recognized. Likewise, the existence of an intermetallic compound in iron-nickel-chromium alloys had been demonstrated.

Studies of the rate of reaction in solid metals, on the other hand, just beginning. Yet it was not long before chemists and metallurgists showed that many such rates could be understood on the basis of the nucleation and growth of particles of a precipitate. Later, they demonstrated that a microconstituent, called martensite, formed in iron-carbon alloys by a shearing mechanism in the crystalline lattice.

As all this fundamental information became available, it was clear that the hardness, ultimate strength, and yield strength and ductility of a steel or ferrous alloy were determined by its microstructure, including the grain size. It then became possible to express the influence of an alloying element in steel by its effect on the microstructure. This, in turn, led to a more general understanding of the behavior of steels that has been of great value in developing steels with high ultimate strength and high yield strength and alloys for use at high temperature. In fact, it now appears possible that computer programs can be devised in the not-too-distant future to write a prescription for the composition of a steel or alloy with certain specified properties.

Physical chemistry has also made a significant contribution in studying and preventing corrosion in certain steels. An interesting and somewhat typical example is the intergranular attack of the 18-chromium-8-nickel alloy in which corrosion takes place along the grain boundaries. In the worst cases, the grains simply fall apart. The industry found by careful study that this effect was due to the precipitation at certain high temperatures of a chromium-rich carbide that depleted the grain boundaries of chromium and so made them susceptible to attack by the corrosive environments. One successful solution was to add an alloying element that formed a carbide more stable than chromium carbide, thus avoiding the depletion of chromium. A second successful solution was to keep the carbon in the alloy at a very low level.

In recent years, there has been growing use of a special class of high-strength, low-alloy steels that develop a strongly adherent protective oxide coating when properly exposed to the atmosphere. Such steels, sometimes referred to as weathering steels, have been known for many years but have recently attracted the attention of architects who like the unique color and texture of the oxide coatings and the lack of maintenance such as periodic paintings. The color of the oxide film may take several years to develop, the exact rate depending largely on the environment, but there is little additional corrosion of the steel after the initial formation of the coating.

Blast Furnace Raw Materials

Iron ore, coke, and limestone are the ferrous metals industry's primary raw materials. While many of the chemical advances of the past 40 years have occurred in the iron and steel processes themselves, preparation of raw materials for charging to furnaces has also benefited from new chemical process technology.

Burning coke furnishes most of the carbon monoxide for reducing ores to metallic iron and most of the heat needed by the reactions, for example. It is made from coal. In 1930, coke plant operators usually worried primarily about producing coke of sufficient strength to withstand the weight of the burden in the blast furnace. They gave relatively little attention to the by-products. They removed ammonia from the coke oven gas, since it tended to form ammonium chloride and clog the gas mains. Passing the gas through sulfuric acid led to the production of large quantities of ammonium sulfate. Fortunately, there was a ready, and in 1930 profitable, outlet for this material as a fertilizer. Other by-products recovered included benzene, toluene, xylene, naphtha, and creosote. Small amounts of phenol, pyridine, thiocyanates, and naphthalene were also recovered in some cases.

External market developments after World War II brought many changes to the production of coal chemicals by the steel industry. The growing need for intermediates to make plastics and resins, for example, placed new demands on the industry for refined benzene, toluene, and xylene. Some of the sulfa drugs opened a market for pyridine. The availability of synthesis ammonia made the production of ammonium sulfate unprofitable, and it became essentially a waste product.

As a result, a modern coke plant little resembles its predecessors, and it does much more than merely make coke (although coke, of course, is still its primary product). Instead of using sulfuric acid to remove ammonia from the off-gases, for example, some plants spray monoammonium phosphate in solution into the coke oven gas. The resulting solution moves to a steam stripper where the mono salt is regenerated for recirculation. The steam and water are condensed to a solution of ammonia and water that is fractionated to specification grade ammonia and ammonia-free water. The process has a recovery efficiency equal to, or better than, that of sulfate systems.

After removal of ammonia, hydrogen, among other chemicals, can be separated from coke oven gas by cryogenic processes. A tonnage liquid oxygen plant that supplies oxygen to nearby steel mills also produces nitrogen. Nitrogen from the oxygen plant and hydrogen from coke oven gas are mixed to form a synthesis gas that is reacted to yield anhydrous ammonia. Part of the coke oven gas can be desulfurized to make elemental sulfur and a sulfur-free fuel gas in a process that significantly reduces air pollution.

Recovery of naphthalene in such modern plants has been increased to meet the demand for conversion to phthalic anhydride. Benzene, toluene, and xylene are now much more highly refined. Part of the pitch recovered from coke oven off-gases is also upgraded to make it suitable for use as a binder in electrodes for the production of aluminum.

Iron ores have also benefited from chemical innovations. Ores used in 1930 were still largely direct-shipping ores. They were mined on the Mesabi Range and a few other places, and they received little treatment beyond a limited amount of washing and then blending to obtain a more uniform composition.

Today, these direct-shipping ores are virtually exhausted. Furnace feeds instead must be prepared with great care, including sizing to obtain the uniform distribution that the packed column prefers. Meantime, the Mesabi Range still has large quantities of taconite that can be ground to a very fine powder to liberate silica and then concentrated to a high-grade iron ore. One method of concentration is flotation, which is made more effective by additives such as fatty acids, resin acids, soaps, and alkyl sulfates and sulfonates developed by both the steel and chemical industries.

Chemical Analyses

In the early 1930's, the methods of chemical analysis used in process control or in the determination of product quality were mainly those employed for many years. These included the gravimetric determination of carbon on combustion and a variety of wet methods for iron, manganese, silicon, phosphorus, sulfur, and the alloying elements. Oxygen content of the steelmaking bath for some time had been determined chiefly by adding aluminum to a spoon sample of liquid and then extracting the alumina thus formed by preferential dissolution in acid. The vacuum fusion process for oxygen was coming into use and by the late 1950's had become more or less routine, though it was neither as precise nor as rapid as desired. This hampered not only process control but also much of the research being done on deoxidation of steel. Methods for analyzing coal chemicals were relatively crude or, in some instances, nonexistent.

The situation today is very different. There has been a virtual revolution in that many instrumental techniques and sophisticated methods are now available that provide more rapid and more accurate results. As one example, neutron activation analysis for oxygen in steel, known but not developed in the 1930's, now enables one to make a rapid analysis with a reproducibility of a few parts per million. Great strides have also been made in developing an immersion-type, disposable oxygen probe for measuring the oxygen content of liquid steel. These oxygen-galvanic cells, which are rugged enough for industrial use, are becoming increasingly important as oxygen steelmaking increases.

A number of new and improved spectrographic methods permit more rapid and more accurate analyses for many materials. For example, in slag control, the lime-silica ratio can now be determined in about five minutes. Phosphorus can be determined by a direct-reading vacuum spectrograph. Alloying elements and residual elements can be readily determined spectrographically. There are rapid chemical methods for iron, ferrous oxide, ferric oxide, and manganese. In the combustion method for carbon, the use of a thermal conductivity cell to analyze the resulting gas yields accurate results in two minutes or less. In making alloy steels, X-ray fluorescence is being used to determine the major alloying constituents of the bath.

Nowhere are the advances more evident than in the analysis of coal chemicals. The use of infrared spectrophotometry and gas chromatography has made it possible to identify constituents and impurities to a degree that was not possible only a few years ago. In some cases, especially in the development of processes for the production or treatment of coal chemicals, the mass spectrometer is proving very useful.

Of special significance in the last few years is the development of very sophisticated automatic methods for the control of basic oxygen steelmaking. In one such system, the carbon content of the charge to the furnace is determined and this information is fed to a computer. The gases flowing out of the converter are then analyzed for oxides of carbon, and the cumulative loss of carbon is calculated by the computer from which the carbon remaining in the bath at any moment can be derived.

FUTURE DEVELOPMENTS

The outlook for the American steel industry depends to such an extent on economic, political, and technical factors that it is difficult to foresee with any clarity what lies ahead in most areas. Nevertheless, certain broad trends are evident that are not likely to change quickly, and there are obviously some serious challenges to be faced.

The first clear trend is that the domestic demand for steel has been increasing about 2.5% a year since 1955, whereas elsewhere the demand has been increasing about 6.5% a year. Moreover, prospects for a marked acceleration of our growth rate are slight, even taking a fairly long view.

In contrast to demand, U.S. production has not increased significantly since 1965 (although production has fluctuated fairly widely above and below the average during the intervening years). This lack of real growth in domestic production stems in large part from increased imports, which have grown faster than exports. Furthermore, the pattern of products imported has been shifting toward those of higher value that are generally more profitable.

Another trend of growing concern is that productivity in man-hours per ton has not increased during the past five or six years. As a result, the U.S. has virtually lost its advantage over major foreign competitors. This lag in productivity has come in spite of some $18 billion spent between 1960 and 1970 on improving and modernizing facilities. Relatively little has been spent on new capacity because of the slow rate of growth in domestic demand.

Reasons for this disappointing return on such large investments are many and varied. Steel mill equipment is usually large and heavy and requires not only an extended period of construction, but because each new plant is virtually unique it also requires a prolonged period of debugging before it attains design capacity. A considerable lag therefore occurs between the outlay of money and a significant improvement in performance. In addition, a very substantial part of this capital spending has gone to meet the growing demand by customers for higher quality and narrower tolerances. Such spending has had little, if any, effect in improving productivity and has in some cases reduced it by increasing finishing and inspection costs and by slowing production rates to achieve high quality. (From an overall U.S. point of view, however, it should be noted that these improvements usually mean higher productivity in customer plants.) Another factor has been the increasing proportion of capital spending for pollution abatement, which raises costs per man-hour but doesn't raise output. Finally, the trend to lighter products directly influences man-hours per ton.

In the face of these conditions, steel companies are looking hard at the best places to put their money and are evaluating the profitability of every plant and each product line. Operations that cannot contribute adequately to profits are being discontinued. Old-line steel executives used to argue that a company must be prepared to provide whatever products its major customers need on the grounds that if the customer has to look elsewhere for small or unprofitable items he might not order the big, profitable ones. This viewpoint is changing—and rapidly. The number of producers of railroad products, for example, is now very small. Some companies have left the tinplate business. Others have phased out older bar, pipe, and plate mills.

These trends have also resulted in a great move to diversify. Steel companies are now becoming suppliers of an assortment of materials, such as coal, limestone, cement, and ores of many kinds. A few are entering the plastics business in a sizable way. Many have gone into real estate, including residential and commercial land development involving condominium and hotel projects in resort areas and prefabrication of homes, schools, and hospitals. Such activities are usually not yet very significant in relation to the total dollar volume of the companies' activities, but they are growing.

Diversification directly influences both number of employees and employee mix. The influence varies greatly with each situation, and it is difficult to generalize. However, many developments in pollution abatement and further involvement in plastics, fertilizers, and the like will provide openings for those with backgrounds in chemistry and chemical engineering.

In the face of little incentive to expand capacity, the steel industry is paying more attention to its continuing search for new and cheaper methods of making steel, and doing so may lead to processes that will employ more chemically trained people. For example, the industry still needs to improve methods of preparing blast furnace feed to make its size and composition more uniform and to give it a higher content of iron. One way to achieve the latter is to prereduce the ore, which seems likely to play a more important role in the future.

Closely related to prereduction are the so-called direct reduction processes. In these processes, ore is reduced with a solid or a gas without making a liquid product. Hundreds of processes have been tried, and much chemical ingenuity has been used. In 1930, the industry was giving direct reduction a great deal of attention, and the subject is still active today. Of particular interest are those processes that use a gaseous reductant in a fluidized bed of iron ore, because they have been developed largely by chemical engineers. Among the most prominent are the H-Iron process of Hydrocarbon Research, the Nu-Iron process of U.S. Steel, and the ADL process developed by Esso Research and Engineering and Arthur D. Little. These processes differ chiefly in the range of temperature and pressure at which the reduction is carried out. A modification of the ADL process, known as the FIOR process, is being used at a 300 ton-per-day plant in Nova Scotia to obtain necessary data for the design, construction, and operation of a much larger plant. A modification of the Nu-Iron process is being put into operation in Venezuela by U.S. Steel.

So far, no proposal for large-scale production of steel by such a process has been successful, but this situation could easily change. For example, a reduction in the cost of hydrogen comparable to the reduction in the cost of oxygen in recent years would have wide ramifications and might well cause some radical changes.

In steelmaking, there is still much to be learned about the operation and control of the BOP method. In addition, continuous steelmaking is being investigated by a number of people.

The industry will also give much attention to its needs for energy. One suggestion has been that the high temperature heat available from a nuclear reactor might be useful as process heat in metallurgical operations, but so far no satisfactory scheme for using it has been developed. One problem has been to develop a refractory container to move a heat transfer fluid at high temperature from the reactor to the process that is to use it.

On the other hand, as one looks at what can be done with raw materials such as coal, the prospects seem brighter. The companies that have turned from making intermediates from coal chemicals for the plastics industry to entering the plastics business directly are finding that their coke plants are no longer able to supply some materials in the quantities needed. The supply of naphthalene from which to make phthalic anhydride, for example, is no longer sufficient in some cases, and other sources are being sought.

As coke becomes more expensive, the steel industry will increasingly try to reduce the amount needed. Here again, prereduction may offer some advantage. Another problem that will receive more attention is the increasing sulfur in coking coals as the low-sulfur coals become depleted or exhausted. Washing removes much of the inorganic sulfur, present chiefly as iron sulfide, but no satisfactory way now exists for removing organic sulfur. A process that is technically and economically feasible for doing so will certainly be sought and if found will have far-reaching effects. Finally, there is increasing interest in coal gasification, perhaps even underground. The product might be useful in a number of chemical and metallurgical processes.

The current expansion and diversification in coal chemicals will continue. More sophisticated processes and controls for them and new products will emerge. An illustration of the trend is that even with the improvements in the processing of coal chemicals the supply of naphthalene from coal may not be able to meet the demand posed by phthalic anhydride. Attention is therefore turning to other methods, such as the oxidation of phenol.

There is also a growing realization that coal may soon become more valuable as a source of carbon and its compounds than as a source of energy. Just what this may lead to is difficult to say, but whatever it may be it will certainly mean more jobs for chemists and chemical engineers.

In the abatement of air and water pollution, there is a wide opportunity for new or improved methods. An obvious one is the removal of sulfur dioxide from stack gases. The steel industry will continue its efforts to solve pollution problems, and many of the answers will require intensive investigations by chemists and chemical engineers. One special class of polluting agents is waste by-products, such as ferrous sulfate and slags. Again, an economically successful process for disposing of them will undoubtedly require considerable work by chemists and chemical engineers.

Clearly, the recent advances in analytical methods for process and product control will continue. The result will doubtless also have a usefulness in efforts toward abatement of pollution. For example, local governments in setting limits on the amount of certain compounds that can be discharged as an effluent, such as phenol in waste water, sometimes set a maximum value that is beyond the capability of existing methods of analysis.

SOCIAL AND ECONOMIC IMPACTS

The U.S. glass industry today makes and ships more than 25 billion pounds of product every year worth more than $3 billion. Containers comprise the largest single market (nearly half of the dollar value), followed by consumer, technical, and scientific markets (about one quarter of the dollar value). Glass fibers and flat glass share the balance of the total market about equally.

Hard, transparent, inert, and strong in tension and compression, glass ranks as one of the most useful materials of construction and decoration. Houses, buildings, and vehicles take advantage of its transparency and strength for use in windows. Its inertness and imperviousness to liquids and gases make it an ideal material for containers of all types, but especially containers where such fragile properties as taste, odor, and product stability and safety are involved. Industrial piping and process equipment made from it or lined with it make many processes practical that otherwise could not be operated or operated only at prohibitive expense. Cords made from it have helped tire manufacturers within the past few years produce some of the strongest and longest wearing tires ever mass-marketed in the U.S., while textiles woven from glass fibers have added to the range of home decorating textiles. Plastics reinforced with mats of glass fibers are fabricated into many consumer and industrial products, and many homes, offices, and industrial buildings are insulated against heat and cold with thick blankets of glass fibers.

The glasses from which these and many other products are made are in effect liquids that have cooled without crystallization to the point that for practical purposes they are considered solids. Many elements in the periodic table may combine to form glasses, but the oxides of sodium, calcium, and silicon are by far the most commonly used. Such soda-lime-silica glasses account for some 95% by weight of all glass made in the U.S.

Soda ash (sodium carbonate), limestone (calcium carbonate), and sand (silicon dioxide) are the raw materials. The common soda-lime-silica glasses may also contain small amounts of other chemicals that are added for several reasons. Some chemicals, for example, aid in removing bubbles before products are formed from the melt. Others prevent undesirable crystallization during cooling. To make decolorized glass, manufacturers add some chemicals to neutralize the green caused by iron impurities in the sand, while they add other chemicals to impart desirable colors.

Major departures from the basic formulation of oxides of sodium, calcium, and silicon lead to a wide variety of specialty glasses that account for the balance of 5% by weight of all glass made in the U.S. Use of rare earth elements, for example, has led to much improved optical glasses. Manufacturers formulate such glasses to adjust refractive index and control dispersion (the difference among refractive indexes for the different wavelengths of transmitted light). Lens systems of excellent quality are the result. Boric oxide and aluminum oxide in formulations produce glasses that don't break when subjected to sudden thermal shock. Every laboratory, most kitchens, and some industrial processes rely heavily on such ware. Lead monoxide produces glasses with high sparkle and brilliance, important in more expensive tableware and decorative crystal.

The economical commercial production of glasses may pose difficult technical challenges, but production of molten glass is a continuous process that may be described rather quickly and simply. Operators mix ingredients mechanically, charge them to a furnace, and heat them to their melting point. The carbonates, nitrates, and sulfates decompose, and carbon dioxide, nitrogen oxides, and sulfur trioxide evolve. Operators raise the melt's temperature to help drive out all bubbles, and after partial cooling, the molten glass is ready for forming.

The glass industry forms by hand only custom products or products calling for a high degree of artistic skill. Most glass products are formed mechanically in equipment that is entirely automatic. In such equipment, the molten glass may be pressed, blown, cast, floated, or rolled. Once formed, it may receive a variety of annealing, cutting, and fire or mechanical polishing operations. Mechanically-formed products are either finished articles, such as bottles, bulbs, and inexpensive household or other ware, or they are semifinished articles, such as sheets, fibers, and tubing that glass manufacturers and others later fabricate into many end products.

CHEMICAL ACCOMPLISHMENTS

Window panes, food and beverage containers, kitchen- and dinnerware, laboratory and optical goods, insulation, household and industrial textiles, and other products make glass a very common article of commerce. While much of the work in the glass industry that produces such articles is mechanical, chemists and chemical engineers make their contributions. They study the relationships that exist among compositions and properties, devise glass formulations, and develop commercial-scale processes. Chemistry, therefore, is critical to the successful production of glasses of all types to meet the requirements of the many uses to which glass is put.

Today's common soda-lime-silica glasses, for example, are technically superior to their counterparts of 50 and more years ago. Their production at today's prices would hardly be possible without the innovations in formulations and in production methods provided by chemists and chemical engineers. Each innovation has resulted in better products, of course, but we shall single out only three for the soda-lime-silica glasses to serve as representative of the work that has led to today's products. These improvements have dealt with inertness, durability, and production speed, and each has contributed important savings to consumers. Each involves production of containers, but that is perhaps not surprising. Containers provide such a large market for glass that manufacturers can be expected to emphasize such research and development.

Ordinary soda-lime-silica glass is so inert that users rarely think about its ability to withstand attack by products stored in it. Yet some liquids in long contact with it will leach out very small amounts of its ingredients, and chemists have spent much time working to increase durability under such conditions. While many users of glass containers have benefited from the results, those using health-related products are particular beneficiaries.

As it happens, water stored in ordinary soda-lime-silica glass bottles will gradually extract chemicals from the glass and within a year will be contaminated to the extent of perhaps 30 parts per million (p.p.m.). Similar storage in a more inert borosilicate glass will also result in some contamination, although less than 1 p.p.m. While 30 p.p.m. represents only slight contamination, it was considered sufficient in the 1930's to require that medicines and other liquids that might be affected be stored in borosilicate glass, even though such glass was (and still is) comparatively costly.

In seeking to overcome the problem and improve soda glass's durability chemist Oscar G. Burch at Owens-Illinois, Inc., at the time recognized that treating soda glass with sulfur gases at an appropriate temperature greatly improved the glass's resistance to water and to many other liquids. His studies disclosed that the sulfur combined with alkali from the glass's surface layers to leave the surface highly resistant to attack by neutral and acidic solutions. Considerable development was required before a successful process was developed, but eventually scientists at Owens-Illinois perfected a controlled process in which soda-lime-silica glass containers are treated with sulfur during the annealing step.

Containers produced by this process have been marketed for some 35 years. Their use permits medicinal and parenteral solutions to be packaged at a fraction of the cost of the borosilicate containers required previously. The U.S. today uses about 150 million containers annually for packaging intravenous and irrigating solutions. If manufacturers had to use borosilicate glass containers instead of sulfur-treated soda-lime-silica ones, their cost for containers would be four times what it is. Instead, Americans save more than $85 million annually on their medical bills.

The sulfur treatment does not improve soda-lime-silica glass's resistance to alkaline solutions, and Owens-Illinois chemists continued to work on the problem, most notably under the direction of Frank R. Bacon. They found no dramatic answers to make soda-lime-silica glasses completely resistant, but their delineation of the interaction of glass with many chemicals eventually made it possible for users to package particular products completely assured as to their safety and to avoid packaging in glass where the effects are undesirable.

The strength of glass in a pristine condition exceeds that of many metals, but abrasion reduces that strength materially. Prevention of the loss of strength caused by surface abrasion as glass containers rub against one another or other hard surfaces has therefore been an important research objective of most glass manufacturers. The surface treatments they have devised are now used to protect some 80% of the glass containers made in the U.S. For consumers the result has been a significant cost reduction, since glass containers no longer need be made as thick and heavy.

Among the important parameters glass production personnel must deal with is the change in viscosity as glass cools. A low viscosity at melting temperatures assists in production, while a higher viscosity at forming temperatures helps the glass stiffen more quickly and allows the products to be removed sooner from forming molds. Burch and his associates at Owens-Illinois found that substituting (within limits) alkaline earth oxides for the alkali metal oxides reduced the viscosity at melting temperatures but increased it through the temperatures at which glass is formed into ware. Since the holding time in molds was a limiting factor in forming-machine speed, the finding increased the speed at which containers could be made. Other container manufacturers eventually adopted the required composition modifications, and the change contributed considerably to lowering the cost of glass containers.

Shock-Resistant Glasses

Many fundamental chemical developments have resulted in the production of a wide variety of specialty glasses. While individually their tonnages are often extremely small, many vital glass products would not exist without the innovations recorded by chemists. Included are glasses that resist thermal shock, that control or manipulate a wide range of light and other radiant energy, and that reduce risks to personal injury.

Of all the specialty glasses the average consumer comes in contact with, the heat-resistant glasses used in many kitchens are probably the most familiar. Their development dates from the early 1900's, when the glass industry sought to solve a problem for the railroads—the breakage of hot lantern globes by rain and snow. Eugene C. Sullivan and William C. Taylor at Corning Glass Works eventually found that glasses containing borax resisted thermal shock far better than ordinary glasses. By 1912, they had perfected a low-expansion, chemically durable borosilicate glass that cut lantern globe breakage by 60%. Moreover, its resistance to acids more than doubled the life of acid-battery jars. Then others at Corning suggested that the new glass might serve in cooking, and the wife of a Corning scientist soon baked a cake in a sawed-off bottom of a battery jar. The nonmetallic cookingware market today has grown to probably well in excess of $100 million a year from that beginning.

Borosilicate glasses with low coefficients of expansion, excellent chemical resistance, and high service temperatures constitute a major discovery of the glass industry. The early successes with such glasses encouraged Corning's staff to look for other applications as well. After some modifications to the original railroad glass, Corning scientists developed a series of glass compositions for laboratory ware whose trademark was to become a laboratory (and eventually a household) byword—PYREX.

In its PYREX ware, Corning not only had a new product but one timed for immediate commercial success. At the time, Germany had supplied nearly all chemical ware, and World War I cut off that country as a source. Corning's new glass not only filled the void but was superior. PYREX ware expanded less than that made from the German glass, which meant it broke less often under heat shock and hence could be made more rugged. Finally, its simpler composition and superior durability meant less danger of contamination, often highly important in chemical laboratories.

This type of borosilicate glass is widely used throughout the world today, and many companies make it. Its resistance to chemical attack has put it into many chemical and food-handling processes. Glasses containing boric oxide also have excellent electrical insulation properties, and they have been widely used as insulators in power and telephone lines.

The original borosilicate development led to a number of subsequent innovations. The first glasses, for example, lost resistance to acid after abnormal and prolonged heat treatment. The study of this problem led to 96% silica glass. Discovered by Harrison P. Hood and Martin E. Nordberg at Corning in the 1920's, it was the first new way of making glass since the days of the Egyptians. It has made available many technical products that otherwise would not exist, including laboratory ware, heat-resistant windows, and many other articles requiring low expansion. In the 1930's, silicones were developed from work that had originally sought to combine hard, heat-resistant silica with malleable plastics. The result has been the multimillion-dollar field of silastics. As an outgrowth of this work, scientists also found that silicon tetrachloride sprayed through a hot flame produces a fused glass. It was used at first in radar delay lines and is now used in wind tunnel windows, crucibles, and laboratory instruments, where its optical and chemical purity are important. Finally, among other innovations, the development of a titanium dioxide–silicon dioxide system has given ultralow-expansion glasses for telescope mirror blanks.

Architectural Glass

Glass has long been an important material of construction for buildings of all types. In the past two decades or so, however, it has become more important than ever, especially in large office buildings where it may account for a very significant proportion of the exterior paneling's area. Since the heating and cooling system for such buildings may represent a big part of the building's final cost, the job that large areas of glass do in controlling heat transfer can be a very important factor for builders and for owners.

The advent of widespread use of tinted transparent glasses in the early 1950's, therefore, represents a development of some note in the glass industry. Offered in green, gray, and bronze, such glasses provided not only thermal comfort and economy but visual comfort as well.

These tinted glasses are made by adding small amounts of metallic oxides to the basic soda-lime-silica glass formula. The oxides reduce the light that the glasses transmit, reducing glare and also absorbing some of the solar heat. In addition, as the glasses become warm, they reradiate the heat, with only part of that heat being radiated indoors.

The tinted glasses of the 1950's gave architects and owners some control over transmission of light energy, but the development of reflective glasses in the 1960's provided even better control. At that time, chemist Richard G. Miller and his associates at PPG Industries concluded that the most effective way to cope with solar heat and brightness would be to reflect the sun's rays. They accordingly developed a thin, transparent metallic coating for glass. The resulting muted mirror-like coating subdues solar light extremely well and reduces significantly the loads on air conditioning and heating systems. These glasses also give better visual comfort indoors and at the same time provide soft reflections of a building's surroundings outdoors.

The first double-glazed reflective glass product was introduced by PPG Industries in 1965. Since then, a number of other reflective glass products have been developed that use various metallic or metal oxide coatings. Some of the coatings are durable enough for use in single glazing rather than having to be protected in the air space of a double-glazed unit.

High-Strength Glasses

Concern over strength often deters designers from using glass in many applications, and the glass industry has long looked for ways to make glass stronger. Producing compression in the surface was the usual approach, and thermal tempering for many years was the only practical means available. Then in the late 1950's, chemists at Corning found a chemical way to produce compression in the surface and extend considerably the range of strengthening that manufacturers could accomplish. As a result, glasses (and glass-ceramics, of which more in a moment) are being used much more widely than ever before.

Some glass articles, it turns out, can be made stronger by a chemical exchange at glass surfaces that contain sodium or lithium. These smaller ions can be exchanged for larger alkali ions, causing compressive stresses in the surface layer that make the articles stronger. Ion exchange processes now commonly strengthen products such as aircraft and automobile windows, laboratory ware and bench tops, and building cladding.

Auto Windshields

The automobile industry has long used much glass. Hard and clear, glass protects drivers and riders from the weather while letting them see. Unfortunately, glass becomes extremely hazardous when hit, and automobiles unfortunately are prone to being hit or to hitting something. To overcome the hazard, the auto and glass industries developed safety glasses in the mid-1920's, and by the late 1930's new autos had safety glasses for windshields, side windows, and backlights. They were either laminated glasses, with two plates of glass bonded by plastic sheets that held glass fragments on impact, or tempered glasses, which on impact shattered with nonjagged edges.

Despite the obvious advantage of safety glass over ordinary glass and even after a number of improvements following its introduction, safety glass used in windshields continued to be hazardous until the mid-1960's. When riders hit the old laminated windshields and their heads penetrated them partly, they received severe cuts. If they penetrated the windshield more completely, they often found themselves with a windshield collar and badly cut necks. On the other hand, if manufacturers made the windshield so strong and unyielding that heads wouldn't penetrate it, people would receive serious head, neck, and brain injuries. Clearly, the auto industry needed a resilient windshield that would absorb impact, hold sharp glass fragments in place, and yet not act like a stone wall.

In examining the problem in the early 1960's, Paul T. Mattimoe at Libbey-Owens-Ford found that the plastic interlayer did not stretch appreciably before tearing because of the tight bond between glass and plastic. He and his associates accordingly developed a new plastic interlayer to overcome the deficiency. It was thicker than the old plastic, but more importantly it adhered less strongly to the glass. Even when sandwiched tightly under pressure, the plastic slid along the glass surfaces. As a result, the plastic bulged rather than tore to absorb impact energy. The glass shattered, of course, but it still adhered to the bulged interlayer.

An important feature of the L-O-F windshield design is its high resistance to penetration. Findings at Cornell Aeronautical Laboratory show that more than 50% of the old windshields were penetrated at least partially, with people suffering deep, extensive, and disfiguring cuts on their heads. On the other hand, the new windshields are often cracked but rarely penetrated, and injuries most often are confined to minor cuts and bruises.

The L-O-F windshield design has been standard on U.S. and Canadian autos since the 1966 model year. In 1967, L-O-F began making aftermarket windshields in the new design, and motorists can now replace windshields on all U.S. and Canadian autos back through 1960 models.

Scientists at PPG Industries have also contributed to making safer glass for autos. Their work resulted in a thinner, lightweight windshield that the auto industry introduced on 1971 models. This windshield consists of inner and outer panels of float glass separated by a thicker polyvinyl butyral interlayer that is highly resistant to penetration. PPG tests have shown that cuts (other than those from penetration) in most cases are reduced compared to those caused by the 1966 windshield when tested with impacts in the 20- to 30-m.p.h. range. The thinner windshield also has shown some superiority over the 1966 windshield in resisting stone breakage. Since the new construction weighs about 0.7 pound per square foot less than the former windshield, auto manufacturers save about 10 pounds (or about 25%) on some of their popular models.

Glass-Ceramics

As mentioned at the outset, glass is a liquid that has cooled without crystallization, and preventing crystallization often concerns glass manufacturers in some of their processing. In recent years, however, manufacturers have found that controlled crystallization can lead to desirable new materials and products. Two of these materials are photosensitive glasses and glass-ceramics.

Manufacturers and users have long known that bottles and windows turn a deep purple after years of exposure to sunlight. By 1938, a chemist at Corning Glass Works had verified the belief that ultraviolet radiation produced the effect by oxidizing manganese from the 3^+ state to the 4^+ state. A co-worker, S. D. Stookey, then soon found that a change in the oxidation state of copper in heat-treated glass produced ruby glass. By changing the formulation, he sensitized a copper ruby glass with ultraviolet radiation on selected areas that then turned red when heat treated. By 1943, Corning had made a glass photographic medium. Further composition study resulted in a photosensitive glass in which sodium fluoride crystals made a white image.

As in the case of low expansion glasses, work on radiation-induced changes in glass led to a number of other useful products. One was a chemically machinable glass, trademarked Fotoform by Corning. In such glasses, an image is formed that is more soluble in acid than is the undeveloped clear glass. The composition lies in a lithium aluminosilicate system, and the image phase is composed of lithium metasilicate crystals that are highly soluble in hydrofluoric acid.

In the course of developing the chemically machinable glass, Stookey accidentally subjected a plate to extreme temperature in a runaway furnace. Instead of the expected blob on the hearth floor, however, he found a completely crystallized and intact plate that was much stronger than the original glass. Heat alone had triggered crystal growth and bypassed the photosensitive process. Trademarked Fotoceram by Corning, the product is used in the new technology of fluidic devices and in making masters for flexographic printing plates.

This discovery also led to a new family of crystalline materials that now bear the generic name of glass-ceramics, and research on such materials is now worldwide. Trademarked PYROCERAM by Corning, they are called Neoceram in Japan, Sitall in the U.S.S.R., and Vitrokeram in Germany. They are formed as true glasses. Heat treatment then causes heterogeneous nucleation and crystallization to transform them into crystalline materials having properties quite different from those of the original glass. Glass-ceramics are used in such diverse applications as radomes, dielectrics, thermal-setting solder glass, regenerative heat exchangers for gas turbine engines, telescope mirrors, building cladding, laboratory bench tops, dinner- and cookware, and electric range tops with integral heating elements. They rank as the most important glass discovery since the borosilicate discovery of the early 1900's.

Radiation Control

Among the specialty uses of glass, those involving control of radiation depend most on variations in chemical composition. In such uses, the sciences of optics and chemistry combine to devise items that control radiant energy waves by absorbing them, transmitting them, and changing their directions. Applications range from simple eyeglasses to complex photographic and astronomical lens systems and from simple shielding in home television sets to massive radiation shields in atomic energy installations.

With the discovery of X rays by the German physicist Wilhelm Roentgen in 1895, intensive study of their nature and uses quickly followed. Scientists soon recognized the need to protect personnel, and Corning made the first high-lead glass for this purpose in 1911. This glass was first used commercially in 1916, the same year that Corning chemists developed glasses to absorb ultraviolet radiation.

Since then, chemists at Corning and elsewhere have developed an extraordinary variety of glasses to control radiation. Their use is indispensable in scientific and therapeutic X-ray equipment, color filters, light-transmitting but heat-absorbing glass, infrared-transmitting glasses, massive windows for cells containing highly radioactive materials, and the like.

Radiation-controlling glasses have also become important in more consumer-oriented devices. Early television tubes, for example, contained lead. Lead oxide, however, was not only expensive but somewhat scarce. In the late 1940's, therefore, Corning developed a composition containing barium that had all the properties of the lead glass. As a result, the glass industry was the largest user of barium carbonate for many years. Then with the need for additional absorption of X rays from color TV's, Corning substituted strontium for barium. Overnight, a completely new market developed for strontium carbonate, which had had little commercial use up to that time.

The most recent glass to control radiation also comes from Corning. It is a photochromic glass that darkens under ultraviolet radiation and then clears when the radiation stops. This type of glass is now used worldwide for ophthalmic lenses and sunglasses. Eventually, building windows, auto glazing, and erasable memory parts in electronic devices may result from the development.

Railroad Signal Lamps

Well before the glass industry reached its present level of formulation competence in controlling radiant energy, of course, it recorded other accomplishments besides the foregoing. Among them was a problem it solved for the railroads around the turn of the century. In the 1890's, railroads used 29 shades of green because of a lack of standardization and of the inability of manufacturers to control colors well. At the time, green meant *caution,* white (or clear) meant *go ahead,* and red meant *stop.* Unfortunately, when rain or snow hit a hot lamp and the signal lamp broke, *stop* or *caution* immediately became *go ahead.* The results were often disastrous.

In 1899, railroads introduced yellow for *caution* and changed green to *go ahead.* Engineers still became confused, however, because of poor delineation between reds and yellows and because of greens that appeared yellow. In the same year, Corning began research that combined studies of optics and color perception with glass compositions and color control. Within a few years, Corning workers had selected ideal red, yellow, and green for signals and had developed the glass compositions that would reproduce the hues precisely. In 1908, the first Railway Signal Association specification based on this work was approved as a standard for lenses, roundels, and glass slides. The latest revision was approved in 1938, and official limit glasses and certificates may be obtained only from the National Bureau of Standards today.

Optical Glass

Men have used lenses for centuries. Eyeglasses date at least from the 13th Century and possibly earlier, while microscopes and telescopes date from the turn of the 17th Century. These and other optical goods have undergone tremendous improvement in the intervening years, but it is only in relatively recent times that American scientists and engineers have participated to any important degree. As a matter of fact, all optical and ophthalmic glass used in the U.S. up to the beginning of the 1900's was imported. Bausch & Lomb initiated experiments on such glasses in 1903 and soon accounted for the first production in the U.S. By 1915, suitable crown, dense flint, and light flint optical glasses were being made, and the following year B&L was producing astigmatic lenses from these glasses.

In 1917, B&L became the center for research and development activities on optical glasses in the U.S., and the work involved not only B&L personnel but also scientists from the University of Chicago, the Geophysical Laboratory, and the National Bureau of Standards. In the following several years, this group solved many problems associated with providing satisfactory raw materials of sufficient purity for making glasses of high transmission, determining melting and annealing schedules to provide glasses with reproducible properties, and evolving stirring and casting techniques to make sufficiently homogeneous glass.

These and other developments in the 1920's resulted in remarkably improved lenses, but production of quality lenses still entailed certain difficulties. Part of these difficulties stems from the fact that light waves passing through a lens are not all easily focused at one point. Moreover, a lens bends light waves of different lengths (and hence different colors) different amounts. Lensmakers correct focus either by grinding or by combining two or more lenses or both. They correct unbalanced refraction of different wavelengths (known as dispersion) by combining lenses with different refractive indexes. Best results are obtained only at high cost, however, largely because of the labor involved.

Then in the early 1940's, a project pioneered by Charles W. Frederick of Eastman Kodak and George W. Morey of Carnegie Geophysical Laboratory led to a Kodak announcement of lenses with higher refractive indexes than formerly possible but without increased dispersion. The most obvious advantage to be gained from such high-index, low-dispersion glasses is to obtain better field correction and to reduce aberrations. In retrospect the finding is now viewed as a milestone in improving glass composition, and it proved to be of great benefit to all applications of geometrical optics.

The new glasses had been made by using certain rare earth elements, notably lanthanum. With them, designers significantly improved lens formulations, and production occurred just in time for extensive use in World War II. Lenses consisting of rare earth glasses advanced both resolution and aperture, and military observers and aerial photographers both benefited. Improved sighting instruments also provided greater fields of view and brighter, sharper images.

Recently, combining computer calculations with the use of rare earth glasses has simplified lens production by reducing the number of lens elements while maintaining or improving performance. The combination has also made it feasible to design highly complex lenses quickly that perform at levels once thought impossible. In another recent development, Bausch & Lomb scientists have introduced a new method for providing a controlled refractive index gradient in glass. It allows index differences up to 0.1 over a distance of about 10 mm. and appears to lend itself especially to the preparation of aspherical lenses.

Lasers

When light shines on certain atoms that serve as the active ingredient in a host material, it raises these atoms to new energy levels. When they return to their lower energy state, they all do so in phase and give off radiation of the same frequency. This light then emerges as a laser beam, a very intense beam that hardly diverges over very long distances.

Scientists first produced laser action in 1960 in ruby (the host) containing chromium (the active ingredient). At American Optical, Elias Snitzer, who had been following the development, realized that glass is amenable to production in many shapes and in large sizes, and he accordingly began studies to develop glass as a laser host material. He chose rare earths as potential active ingredients, since rare earth ions in glass fluoresce with narrower lines than do other fluorescent ions. By late 1961, he had produced the first working laser with neodymium glass.

In the past decade, some tens of millions of dollars of laser glass and glass-laser systems have been sold by a number of companies sharing the market. Notable among uses for the neodymium glass laser is its application in research for producing controlled thermonuclear reactions for power generation.

Glass in Electricity and Electronics

Glass and electricity became partners as early as 1879 in the Edison light bulb, but few paid much attention to the electrical properties of glass until World War I. Since then, however, its applications have widened considerably in these industries.

One of the first innovations was a new glass with high electrical resistivity that could be melted in continuous tanks to make light bulbs. In the 1920's, phone and power companies began using glass insulators on which a conducting coat of tin oxide eliminated static interference. In the 1940's, conducting films on glass were used in industrial radiant heaters, followed by uses in drying panels, frost-free windshields, backlights, and rearview mirrors. Glass resistors, which incorporate conducting films, achieved volume production in the late 1940's. At about the same time, glass ribbons half the thickness of newspaper were successfully produced for use in capacitors, where they replaced mica.

The advent of color television called for an entirely new system of sealing the viewing panel to the funnel of the picture tube. Because of the size and complexity of the assembly that has to be inserted in a color tube, panels and funnels are shipped unsealed. During assembly, the tube-maker seals the unit at a low enough temperature to avoid the slightest deformation of the panel or damage to the assembly. Once he seals it, he must then bake the tube at about the same temperature under vacuum. Chemists at Corning Glass Works met these apparently contradictory requirements by developing thermal-setting solder glasses. As a result of this development, devitrifying solder glasses are available from several sources for a number of applications in the electronics industry. Meantime, no other means of sealing panel and funnel has ever proved to be satisfactory.

FUTURE DEVELOPMENTS

Chemists in the glass industry may have examined many elements of the periodic table, but they haven't found all glass-forming systems yet. Stronger glasses are among their many goals for future products, for example. Glass's intrinsic strength appears to be about 3 million pounds per square inch, far above that of any other known material of construction. The industry has reached only about 10% of this potential in massive glass, however, and then only as a laboratory curiosity. Much research and development still remains to be done to make the process a commercial reality. These stronger glasses will be especially suitable at first as replacements for metals because of glass's superior resistance to corrosion.

Brittleness has also limited the use of glass in structural applications and in places where it must be machine-finished. Chemists in the glass industry, therefore, hope to develop nonmetallic materials that are malleable and ductile and that yield under excessive load rather than fracture. The nearest they have come are machinable glass-ceramics that Corning announced in 1970. All standard metal-working tools will shape these glass-ceramics, but they are still brittle.

To control electromagnetic radiation, the first steps are being taken now to transmit information over optical wave guides. The radio-frequency spectrum is becoming increasingly jammed as the communications industry grows. By using optical frequencies, it could expand manyfold the number of communication channels. Glass fibers in telecommunications should become a reality throughout the world within the next decade.

In other areas, as chemists increase glass's thermal resistance, the industry will find new uses in high-temperature applications. Findings in surface chemistry have given scientists the ability to stabilize enzymes for use as catalysts in biochemical reactions and in food processing. Work in this area is still very new, but it might lead to very important applications for glass in immunology, chemical analysis, and processing. Chemists and electronic engineers are continuously expanding glass uses in instrumentation. They have produced many new sensors for specific ions in the past, and new sensors in the future will be especially important in studying body chemistry. Surface treatments may permit glass-ceramics having great strength to substitute for bones in artificial implants.

Meantime, while it may search for such more exotic uses for its products, the glass industry faces a more prosaic problem: What to do with glass containers, which have become a monumental disposal problem. Among potential solutions are plans in California to make building brick from crushed bottles and manure. Tests have also been made on the use of crushed glass as an aggregate in highway pavement. Glass might also be used as a loam builder to improve the quality of poor land.

SOCIAL AND ECONOMIC IMPACTS

The millions of Americans who include photography among their pursuits took some 4.75 billion still pictures and consumed countless feet of movie film as amateur cinematographers in 1970. For film and processing, they spent about $1.2 billion. When purchases of camera, flash, and other equipment are added in, the American amateur photography market in that year exceeded $1.6 billion. [1]

The family Christmas photo or the family vacation photo may be the most widely recognized use of photography in the U.S., but they rank in value—and probably in total impact—somewhat behind the nonamateur market for photographic industry products. Since photography by definition can include any process in which exposure to radiant energy induces chemical change, the nonamateur photographic market includes not only the professional still and motion picture market directly comparable to the amateur market. It also includes many copying, duplicating, and even printing processes as well as a wide range of scientific applications of the photographic process. As a result, the nonamateur market for photographic equipment and supplies exceeds the amateur market by nearly 40%, having amounted to somewhat more than $2.2 billion in 1970.

Successful commercial photographic processes date from the 1839 announcement in France of the daguerreotype and the 1841 announcement in England of the Calotype process. In the intervening 130 years, the original methods based on the chemistry of silver have been refined and improved drastically. They have also been joined by a number of processes based on the chemistry of other compounds that respond to radiant energy to form images in continuous-tone or line form in either black and white or full color. Chemists and chemical engineers, as well as other scientists and other engineers, made many improvements during the first 100 years. The innovations dating from roughly the mid-1930's, however, have been the ones that have largely led to the expanded photographic products industry of today. Photographic equipment and supplies now rate as one of the most rapidly growing industry sectors in the U.S., as a matter of fact, having grown nearly 650% since 1950 while the Gross National Product grew somewhat less than 250%.

The sciences of chemistry, physics, and optics have been the most important in the development of successful photographic processes, while chemical engineering techniques rate high in making radiation-sensitive films and papers and in developing the images formed on them. Mechanical engineering, meantime, has been critically important for designing and producing cameras, copiers, and other equipment. Personnel expert in electronics have also contributed in recent years, especially with the proliferation of computer applications for information processing.

Personnel trained in many disciplines have therefore contributed to the photographic products industry. It probably is not unreasonable to consider chemical contributions to be pivotal, however, since the entire image-forming process from the production of raw materials to appearance

---

1. All figures on sales, except for those for individual companies and for a few individual commodities, are derived from "The U.S. Photographic Industry, 1970–75," published by Arthur D. Little Co., Cambridge, Mass., October 1971.

of the final image is chemical. In the most widely used photographic processes, those based on silver, making the film or paper support relies on cellulose chemistry, for example. Making the sensitive layer calls for knowledge of the chemistry of gelatin and silver halide salts plus that of a wide range of chemical adducts. Exposure itself involves photochemistry, while processing the exposed emulsion is chemical from development of the image to drying the film or print. Even the design of lenses depends on understanding the chemistry of glass.

The sensitivity of silver halide to light energy may underlie much of the photographic products industry, but other chemical systems have become important, too. Thousands of engineering drawings are produced every year, for example. Some are made by the diazo process, which is based on the reaction under light of diazonium salts with phenolic or enolic compounds. Others are made by the blueprint process, which is based on the reduction by light of ferric iron to ferrous iron and the subsequent formation of a blue compound. The chemical change caused by heat (which in turn has been generated by light energy) was used in equipment that started the revolution in office copying. Making use of selenium's different electrical conductivity under exposure to light then made office copying the very large activity it is today. Silver-iron salts, bichromated colloids, zinc oxide, titanium dioxide, and various polymeric systems are among the other chemical compounds being used today in commercially successful photographic processes.

All of these processes have had an economic and social impact in the U.S. Silver halide processes stand out, of course, largely because still and motion pictures are so important to recreation and entertainment as well as to publishing, advertising, and even art. Silver halide and non–silver halide processes, however, have also had an impact in many other applications. The tracks left by atomic and subatomic particles in photographic film are an important tool in atomic energy studies, while astronomers have long found photography essential in their work. Photography helps scientists and engineers analyze processes operating much more rapidly (or occasionally much more slowly) than the eye can detect, while photomicrography and radiography are used to improve metals and detect flaws. X rays have long been an important diagnostic tool for the medical and dental professions, while photography using ultraviolet, visible, and infrared radiation are increasingly used in criminology for identification, security, and detection. Aerial and space photography provide more accurate maps, aid weather forecasting, and allow studies of land use (and abuse) to degrees never before possible.

In the past decade, these applications of what might be considered traditional photography have been joined by a number of processes in which chemical sensitivity to radiation has been used to produce documents to meet several different needs. In a society that depends heavily on recording and exchanging facts and ideas, individuals and organizations have needs ranging from producing one or two copies (copying) to several hundred copies (duplicating) to several thousand and more copies (printing). In addition, once a document has been produced in anywhere from one to thousands of copies, occasions arise for making one or more copies from the original.

The office typewriter and carbon paper at one time largely served for communications among a few individuals, while Mimeograph, Ditto, and similar machines provided for the duplicating-sized audience. Only hot metal typesetting generally served for audiences numbering more than several hundred (or where relatively high quality products were wanted), and only fairly expensive photostating served for direct copying of documents already in existence. Then in the 1940's, several companies made commercial attempts to simplify the production of documents from graphic originals, such as letters in office and interoffice communications. In the early 1950's, their efforts started a major shift in copying from the previously accepted tray processing or centralized photostat-type operations. In quick succession, 3M introduced a thermographic process (in 1952), American Photocopy Equipment Co. (Apeco) a diffusion transfer process (also in 1952), and Kodak a silver halide process (in 1953). Meantime, Xerox had introduced electrostatic copying during the 1950's also, and then followed with its 914 copier in 1960. As a result of these and other developments, copying of all types of documents has become extremely convenient and widespread throughout all organizations.

The printing end of the communications spectrum has also felt the impact of photographic innovations. The high cost of setting type in lead and of letterpress operations generally, for example, has helped stimulate improvements in other printing methods, most notably in lithography, phototypesetting, and photopolymer printing plates. Each depends directly on photography. The success the printing industry has achieved in producing high quality printing with lithography, especially in color, has led to a major shift within that industry from letterpress to offset. The shift to phototypesetting from hot metal typesetting has not progressed quite as far, but it is nonetheless well launched. Photopolymer printing plates, meantime, continue to take more of the market previously monopolized by metal type. In addition, the development of inexpensive offset film and plates and of easy-to-operate equipment has helped offset invade the duplicating market once held by Mimeograph, Ditto, and similar methods for jobs where the number of copies is not very large and the demand for quality is not too great.

Photographic processes have also had a role in yet another aspect of information recording and exchanging. The increasing need to store information and to have it readily accessible and the increasing use of computers to do so have led to a rapidly growing market for microfilm. The chemical contributions have been the production of films with extremely high resolving power to achieve the size reduction called for plus the production of films with the speed to keep pace with the rate at which computers can generate characters for projection on film. Marriage of microfilm readers with equipment that produces hard copy for reading under normal conditions has helped round out the system.

Prior to World War II, consumers of photographic products were more or less dependent upon the services provided by others. A typical exception was the professional motion picture business in which the producers exposed, processed, and printed their films. Since World War II, there has been a strong trend toward devising systems in which the sensitized materials, the processing chemistry, and the equipment were designed together to produce compact, convenient, automated, and lower-cost operations with increased productivity, if possible. This trend has had a strong economic impact relative to the number of companies participating and to the increased uses of the photographic process.

CHEMICAL ACCOMPLISHMENTS

The term photography embraces many processes using light-sensitive layers that are exposed to some kind of radiation and then developed to form a visible image. Making and using photographic materials calls for applying all of the chemical disciplines, as practically all photographic process steps rest entirely on chemical reactions and even the few exceptions rest partly on them. Manufacturing light-sensitive chemicals and placing them on supports that are themselves chemical materials, for example, require the use of both chemical and chemical engineering principles. Exposure relies on photochemical reactions, while developing and printing exposed film involves chemistry and also chemical engineering. Photographic materials, the equipment in which they are used, and production and processing machinery, however, must each be adapted to the other's possibilities and limitations. As in other technologies, therefore, photography makes use of techniques and principles from many other disciplines, including optics, physics, electronics, and mechanical engineering.

Photographic processes fall logically into two broad groups. One group is based on silver halide chemistry, such as common black and white and color photographs. The other group is based on non–silver halide chemistry, such as the more specialized copying methods of diazo, blueprint, Van-Dyke, thermographic, electrostatic, vesicular, polymeric photoresist, and polymeric etching. Silver halide chemistry dominates the amateur and professional still and motion picture film and related markets, while the others function in business and industrial copying and duplicating and spill over into printing applications.

Silver Halide Photography

Chemically, silver halide photography works because light striking silver halide crystals in colloidal suspension changes their crystal structure in such a way that an appropriate developing agent will reduce the crystals to metallic silver. The developing agent, however, will not reduce unexposed crystals, and they are removed to leave a negative image of developed metallic silver on the film.

The process dates from experiments performed in Europe in the early 1700's showing that exposure of certain salts of silver to light causes them to darken. About 100 years elapsed, however, before a French physicist, Joseph Niepce, used the observations in the early 1800's to make the first negative with sensitized paper in a camera. Niepce was soon joined by Louis Daguerre, a French painter and physicist who had independently begun similar work. The result of their collaboration was the daguerreotype that Daguerre announced in 1839 following Niepce's death and ceded to the French Academy of Science.

Daguerreotypes were one of the forerunners of today's photographic prints. John W. Draper and Samuel F. B. Morse are generally credited with introducing them to the U.S. not long after the public announcement in France. Draper, a philosopher, historian, holder of an M.D. degree and professor of chemistry at the University of the City of New York, among other accomplishments improved the Daguerre process and made the first satisfactory photographic portrait with it in 1840. Although Morse, an American inventor and painter, also promoted daguerreotypes here, he achieved more fame for his work with telegraphy.

The daguerreotype was one of the first practical techniques for taking pictures, and its name is probably the early one most familiar to the public. However, it is not generally considered the basis for modern photography. Daguerreotype chemistry involved fuming iodine (or iodine and bromine) onto a copper sheet that had been plated with silver. The resulting light-sensitive surface was then ready for exposure in a camera. Work that led more directly to today's silver halide photography occurred elsewhere at about the same time that Niepce and Daguerre were developing their process. Notable among such efforts was the finding in 1837 by J. B. Reade that sodium thiosulfate removed unexposed but still light-sensitive chemicals in negatives and prints. Reade's work helped eliminate a major early problem—gradual obscuring of the image by the originally unexposed silver halide crystals. Then in 1841, William Henry Fox Talbot announced the Calotype process, in which the image was developed mostly chemically following a short exposure. This process gave pictures that lasted for extended periods and involved photographic development essentially as known today.

Many scientists, engineers, and inventors in Europe and the U.S. gradually improved on Fox Talbot's process, and photography in the middle and later 1800's gradually evolved toward today's products and methods. In those years, however, all methods required elaborate equipment. It was not until George Eastman in 1888 in the U.S. introduced a roll film of light-sensitive chemicals in gelatin on a paper backing that photography as we see it today began to emerge.

Other major chemical innovations in succeeding years include:
• Gaining a clearer understanding of the properties of light-sensitive materials from work in Europe beginning in the 1890's.
• Introduction in the U.S. of positive color film in 1935 and of color negatives in 1942.
• Direct production in 1948, also in the U.S., of positive prints in cameras in only seconds after exposure.

Concurrent innovations in lenses, cameras, bulb and electronic flash, and related equipment and materials have led to the photographic products industry today that has sales in the U.S. net of foreign trade of nearly $4 billion.

Silver halide photography begins with the manufacture of light-sensitive emulsions and their supports. The emulsion consists of a suspension of one or more of the silver halides (silver chloride, silver bromide, and silver iodide) in a protective colloid such as gelatin. The halides are formed by precipitation in a gelatin solution from a reaction between silver nitrate and the potassium or ammonium halide. Workers in England first demonstrated this relatively simple system in about 1880 when the dry gelatin photographic plate was introduced. Today's emulsions are hardly simple, however, and manufacturers add a number of chemicals besides the silver halides to perform different functions and to control the photochemical reactions properly for different applications. The result of their research and development in chemistry and chemical engineering is the availability of a wide variety of silver halide film and paper products.

The highly specialized physical chemistry of making emulsions involves the sequence of emulsification and physical ripening, washing or coagulation, and digestion or chemical sensitizing to obtain film of the desired sensitivity. Chemicals added during emulsion-making produce the desired chemical, physical, and photographic properties of the end product, but much of such procedures is proprietary information that manufacturers guard carefully. However, two very significant chemical innovations among others that can be discussed involved the role of gelatin in a film's sensitivity and the use of dyes for what the industry calls spectral sensitization.

Some years ago, manufacturers found that some gelatins affect an emulsion's sensitivity to light and others do not. Then in 1925, a report was published that described the isolation of a sulfur-bearing sensitizer from gelatin. This compound resembled mustard oil, or allylisothiocyanate, which contains sulfur. This observation had very important theoretical and practical consequences. Theoretically it provided the basis for an explanation of the relationship between exposure and chemical development. Practically it provided one means of control of sensitivity in the manufacture of photographic emulsions that is known today as chemical sensitization.

In addition to employing chemical sensitization with sulfur compounds such as allylthiocarbamide, manufacturers also sensitize films in other ways, including the use of what they term spectral sensitizing. In this process, they add dyes to make emulsions sensitive to longer wavelengths of the spectrum than only the ultraviolet, violet, and blue wavelengths normally absorbed by the silver halides.

Early experiments and observations in Europe along these lines include the discovery in 1873 of spectral sensitizing itself, development in 1883 of erythrosin dye for the green region of the spectrum, combining quinoline and cyanine in 1884 for sensitizing across the spectrum to orange, inclusion of orange sensitization in 1901 with the use of isocyanines, and finally the addition of carbocyanine as a red sensitizer in 1905.

World War I shut the U.S. off from such developments in Germany, and work accordingly began here on the subject. In succeeding years that work has resulted in the development of many new and useful dyes as well as an understanding of the spectral sensitizing mechanism. Today, film manufacturers can add spectral sensitizing dyes to make photographic emulsions that are sensitive to selected regions of the spectrum with wavelengths from 100 millimicrons to 1,300 millimicrons. (A millimicron equals 1 millionth of a millimeter. Visible light wavelengths range roughly from 400 millimicrons at the blue end to 700 millimicrons at the red end.) Supersensitizing is also possible by combining dyes, and the synthesis of dyes that do not migrate has been very useful in making three-color film.

Once the emulsion is prepared, it must be supported on a base for use. Cellulose nitrate modified with plasticizers and other chemicals served among the earliest flexible films to support emulsions. Unfortunately, cellulose nitrate is not only chemically unstable but also a serious fire hazard. It has not been used in film since 1951, finally disappearing under competition from various cellulose acetate, acetatebutyrate, and triacetate films introduced some 50 years ago and polyester films introduced in the mid-1950's.

Cellulose acetate and related films were primarily developed by Kodak. They solved the fire hazard, and they soon became the standard film base. The polyester film that du Pont introduced is also a safety base, but it provides additional benefits. It is at least twice as strong as conventional film bases. This strength permits its use in printing applications as well as in conventional photographic ones. Lithographic film of polyester is some 30% thinner than regular lithographic film. In addition, polyester film is so little affected by moisture and heat that it can be used where the most accurate reproductions are desired. In many cases it can replace glass plates for making separation negatives used to make color plates for printing. Moreover, polyester films can be thin enough for special printing techniques yet strong enough to be used for other purposes as well, providing valuable savings in both stock and storage space among other advantages. Finally, polyester movie film can be used without failing at temperatures from $-60°$ to $180°F.$, a point of interest in military photo reconnaissance and photogrammetric applications.

Another major accomplishment in the chemistry of supports for photographic emulsions was the manufacture of photographic paper supports from wood pulp rather than from bleached cotton rags. This was mainly a Kodak contribution and a quite different problem from producing high-grade rag stock paper. Wood pulp having cellulose purity equal to that of new-grown cotton proved to be a more uniform raw material. When combined with appropriate sizing chemicals, the wood pulp paper and the emulsion coated on it had keeping properties equal to the same emulsion coated on glass.

Once the film or paper has been exposed, it must be developed and fixed. Regardless of the developing agents used in various processes and the complicated series of reactions involved, the end result must be the reduction of silver ion to metallic silver. Hydroquinone is the main reducing agent used today, but developer solutions now contain many other ingredients. Among them are chemicals such as potassium bromide, which restrains reduction of unexposed silver halide and prevents fogging of negatives, and sodium sulfite, which reacts with the quinone formed during reduction to prevent staining and other undesirable side reactions.

Once the exposed film has been developed, it must be fixed. The most commonly used fixing agents are sodium and ammonium thiosulfate. In an intricate sequence of reactions they form complexes with the unexposed silver halides and permit their removal from the developed film by washing.

Tanning Development. Some photographic developers produce a hardened gelatin image when the silver image is developed. Both the hardened gelatin and silver images coincide, and the thickness of the gelatin hardened in the process is proportional to the amount of silver developed. Warm water readily removes the unhardened nonimage gelatin by melting.

The formation of a gelatin relief image is known as tanning development. Many photographic processes have been based on its chemistry, but two of particular interest are a dye transfer process for high quality color prints from Kodak and motion picture color film from Technicolor.

In Kodak's dye transfer process, advantage is taken of the fact that the gelatin relief image formed by tanning development can absorb dye in proportion to the gelatin's thickness. This absorbed dye can then be transferred to special receiving materials by bringing the dyed relief into close contact with the receiving surface. Separation negatives are made on a black and white film using red, green, and blue filters to record these colors of the object. The gelatin relief images are dyed with the corresponding complementary cyan, magenta, and yellow dyes and the dyes then transferred in order and superimposed in register for a color print.

Technicolor employs the same principles for color motion picture film, except that the absorbed dyes are transferred by contact with a film having a gelatin coating that contains a mordant. This imbibition process has been used extensively for professional motion pictures in the U.S. and abroad since its introduction in 1932.

Reversal Processing. In reversal processing, exposure and development of silver halide are normal, but instead of removing the unexposed silver halide, the developed silver image is removed by treatment with a bath containing an oxidizing agent such as potassium chromate. The remaining unexposed silver halide is then developed to form a positive image either by exposing it to light and using conventional development or by use of a fogging developer.

Although the chemistry of reversal processing had been demonstrated as early as 1862 and the idea of amateurs using such film for home movies considered, it was not until Eastman Kodak introduced a process for amateur cinematography in 1923 that the idea became commercial. Kodak followed the original 16 mm. equipment with an 8 mm. process in 1932, and the growth of amateur cinematography has been greatly accelerated since those days with the introduction of suitable color films.

Color Photographs. Eastman Kodak revolutionized color photography when it introduced 16 mm. Kodachrome color film for amateur cinematography in 1935 and 35 mm. film for color slide transparencies in 1936. Although various schemes had been suggested and demonstrated earlier for making color pictures, only the integral tripack system Kodak introduced has been successful commercially. General Aniline and Film Corp. (now GAF Corp.) joined Kodak as the second producer of color film in the U.S. The two companies were able to bring about the change by succeeding in finding ways to coat thin emulsion layers in contact with one another, to make sensitizing dyes that would not migrate, and to make nontoxic developing agents and couplers.

In color processing, the oxidation by-products of the developing agent are available to react with dye formers called couplers. Couplers that are chemically different react with the by-products of developing agent oxidation to form dyes of the various colors needed. During development of color film, a negative image forms in each layer that consists of silver and dye produced in proportion to the original exposure. As in reversal processing, the unexposed silver halide in each film layer is then flash-exposed to light and developed to produce a positive image. Bleaching removes the silver and produces a positive dye image.

In processes such as Kodachrome, each layer is processed in the appropriate dye coupler developer to produce the yellow, magenta, and cyan colors. In processes such as Ektachrome, the dye couplers are incorporated in the emulsion layers. This approach leads to relatively simplified processing. It uses only a single developer solution for all three color layers instead of using three separate color developers as are used in the Kodachrome process.

The color processes just described are direct positive or reversal processes. There are also negative-positive color processes based on the same chemical principles. After exposure, processing produces color negatives composed of the complementary colors yellow, magenta, and cyan. A positive film or paper coated with essentially the same kind of emulsion as the negative film is processed to produce a color picture looking like the original.

Diffusion Transfer Processing. As George Eastman's roll film had revolutionized photography in the 1880's and Kodak's color film had revolutionized it again in the 1930's, so Edwin Land's Polaroid camera and film of the late 1940's revolutionized the photographic products industry once more. Amateur photographers provided the first large market for Polaroid photography, but the system's virtually instant reproduction has led to its widespread adoption by scientific laboratories and professional photographers. Today, about 90% of Polaroid's sales of more than $500 million comes from applications of diffusion transfer photography.

One-step photography uses diffusion transfer processes for the first time to provide full-scale photographic images without the use of darkrooms and the multiple processing baths required by conventional photography. In the process, the amount of exposure governs the diffusion of substances from the photosensitive layer to an adjacent image-receiving layer. In processes producing black and white prints, the image-forming substances are soluble silver complexes. In those producing color prints, the substances that form the images are either compounds that contain dyes or that generate dyes. In the Polacolor process, which is the only commercial one-step color process today, the substances that form the images contain both chromophoric and developing groups within the same molecule.

Much of the early work on diffusion transfer photography occurred at the time of World War II as independent studies by André Rott of Gevaert

Photo-Producten, N.V., in Belgium, Edith Weyde in the Agfa laboratories of I. G. Farbenindustrie in Germany, and Edwin Land of Polaroid Corp. in the U.S.

Rott and Weyde directed their work toward the photocopy market, but Land sought a system that would let a photographer view continuous-tone pictures within a minute after taking them without having to perform elaborate processing operations. The goal called for a compact, light, and simple camera and a high-speed film that would contain all the resources to produce a picture comparable in quality to that obtained from conventional methods.

Land's experiments showed that a thin film of processing solution containing a developing agent, a silver complexing agent, and alkali could develop a negative image rapidly while simultaneously generating a positive transfer image of soluble silver complexes from the unexposed silver halide grains. If these complexes were then allowed to diffuse to an image-receiving layer adjacent to the negative, nuclei in the receiving layer would catalyze the reduction of the complexed silver ions to form a positive silver image. Adding a polymeric thickening agent to the processing solution solved the problem of spreading the thin but uniform layer of reagent between the negative and the receiving sheets.

The first Polaroid camera resembled a conventional bellows camera, but it contained two steel rollers for breaking a pod to spread the processing solution and a small, light-tight chamber where the film unit remained for a brief processing interval. Opening the chamber and stripping from the negative revealed an essentially dry, finished print.

Polaroid announced the system early in 1947 and had the first cameras and film on the market late in 1948. The first sepia prints were followed by black and white prints in 1950, high-speed (3000 ASA) film in 1959, and 10-second processing in 1961.

In 1963, Eastman Kodak introduced the Kodak Bimat Process for providing in-flight processing of black and white aerial roll films. The chemicals are preimbibed into a role of transfer film which is laminated to the exposed negative film during processing. Upon separation, a positive image is obtained in the transfer film and a negative image in the exposed film. The process was used in the Lunar Orbiter flights in 1966 and 1967. No use was made of the positive image, but instead the negative image was scanned electronically for transmission to earth.

Also in 1963, Polaroid introduced a film unit that produced color prints within one minute. Likewise based on diffusion transfer chemistry, it uses a subtractive color process that required some 10 years to develop. For it, Polaroid scientists made molecules with chromophore groups connected to developer groups. These dye developers are paired with photographic emulsions of complementary spectral sensitivity by coating each type of dye developer in a separate layer beneath its emulsion layer.

After exposure, advancing the film spreads an alkaline processing solution between the negative and receiving sheets. The alkali penetrates the negative to solubilize and activate the dye developer molecules by ionizing the hydroquinone groups. As the dye developers diffuse toward the receiving sheet, they pass through their paired emulsions, reducing the silver halide grains and being oxidized to an immobile form.

The receiving sheet consists of the mordant and a timing layer with polymeric acid layer combinations, which optimize picture quality by controlled reduction of alkalinity. The dye developer molecules that do not encounter exposed grains diffuse to the mordant layer of the receiving sheet and form the positive image. The amount of each subtractive dye that transfers is therefore determined by the exposure its associated emulsion receives. Thus, the three separate subtractive images form and transfer essentially independently of each other to form the color print.

Non-Silver Halide Reprography

Many methods exist today for reproducing graphic originals besides the photographic processes based on silver halide chemistry. Some use different chemical reactions sensitive to light to form images, while others use processes in which reagents are sensitive to heat or to electrostatic charges. Business offices and industrial plants use the reprographic methods widely to provide anywhere from one copy to hundreds of copies in a spectrum of applications that ranges from copying to duplicating and that impinges on printing.

Of all the changes that have occurred in office procedures in the past century or so, the advent of the office copying machines must surely rank at the top. The practical commercial typewriters first developed in the latter half of the 1800's and their subsequently considerably refined successors speeded written communications enormously, of course. Office copying machines not only increased the number of copies readily reproduced, however. They opened the field to copying from any original of reasonable size. They also in effect eventually put what are nearly small printing plants in many an office and broadened considerably the access to information by larger numbers of people at much lower costs per copy than previously possible.

Thermography. The first office copying devices drove many clerks and secretaries to despair, no doubt, and elimination of what was admittedly a messy chemical process became an early goal of companies in the field. Their first success came from applying thermography to the problem. Thermography provided the first feasible method for rapidly copying graphic originals with a dry process, and as such it laid the basis for the modern office copying business. Although supplanted today in many ways as an office copying method, thermography still possesses unique advantages, particularly in cost and convenience for certain copying situations. Moreover, it remains preeminent for making transparencies for overhead projectors, and its utility has also been extended to multiple copying (duplicating) applications.

Thermography involves recording images that exist as heat patterns. Historically, it was originally used to record continuously with a heated stylus. Heated styli are still employed and are even enjoying some return to popularity in some applications. Use of heat-sensitive papers and exposure to heat, however, are the more normal thermographic reproductive methods used now.

The heat-sensitive layer of a thermographic copy sheet contains a heavy metal soap and another component that reacts with the metal when heated to form a visible reduction product. One such system consists of ferric stearate and gallic acid, which produces a colored image of the original. Another is silver behenate and an organic reducing agent, which produces a black image of metallic silver on a white background.

In thermographic copying machines, the copy sheet is juxtaposed with the original and transported at a controlled rate past an intense infrared light source. The printed areas of the original absorb heat, and that heat is absorbed by the adjacent copy sheet to form the reproduced image.

In addition to using copy sheets based on ferric stearate and gallic acid or silver behenate and an organic reducing agent, manufacturers have also devised copy sheets that act by having a fusible material melt in image areas and by inducing dyes or dye precursors to transfer under heat from an intermediate sheet to a receptor sheet.

In another thermographic variation developed and introduced by 3M Co. (formerly Minnesota Mining & Manufacturing Co.), heat development follows exposure by reflex light. It provides high quality black on white copies of originals in any colors. The light-sensitive intermediate sheets contain suitable photosensitizing dyes, a photodesensitizable co-reactant for the reaction that produces color, and generally a film-forming polymer. The receptor sheets contain the other co-reactant, auxiliary reducing agents, and suitable film-forming polymers.

While 3M reached the market first (in 1952) with the "no messy chemicals" concept in its Thermofax office copying machine, its position immediately came under competition from Kodak's Verifax process the next year. The Verifax process actually used silver halide chemistry,[2] but it did so without the use of bothersome plumbing. It used paper coated with a silver halide emulsion that contained a special tanning developing agent and a special colorless dye former. In the machine, the exposed emulsion was processed in an alkaline activator solution and the exposed silver halide reduced to form a silver image in tanned or hardened gelatin. A dark blue dye formed simultaneously along with some silver in the unhardened, unexposed areas and was transferred by short contact with a suitable receiving sheet.

Kodak first described the Verifax process in March 1952 at a time when a simple, economical copying method that had no plumbing was greatly desired by secretaries for use under normal room illumination. Kodak had machines on the market in 1953, and they made a real impact on office copying for seven to eight years until other copying processes gained more favor.

Electron Beam Recording. In this method, a focused electron beam in a vacuum impacts on a material sensitive to electrons and yields a visible image after processing. Scanning the modulated electron beam produces the image, much as an image is formed in a cathode-ray tube.

Chemistry has implemented electron beam recording by providing unique materials sensitive to electrons. Of special importance in making electron beam recording useful commercially has been 3M Dry Silver film. It has succeeded in removing what some workers have felt to be an output limitation of electronic data processing machines. With the film, up to 100,000 alphanumeric characters can be recorded per second with electron beams, which is as fast as modern computers can generate them.

Electron beam recorders that 3M introduced in 1967 are designed to generate computer output in microfilm and take advantage of the film's high resolving power. Microfilm in turn is the ideal format for reference information, which accounts for more than 75% of the computer output in many large installations.

Another important use for electron beam recording occurs in television. Recording on 3M Dry Silver film takes advantage of its higher capacity to store information compared to magnetic tape, which has been the traditional medium for TV.

Dry Silver technology also serves as the basis for a unique series of photographic-sensitized materials that are processed by heat alone. Materials of this type provide high resolution and provide near real-time image access (that is, available nearly as formed). They consist of a metal salt,

---

2. Although both Verifax and 3M Dry Silver film (which is discussed next) are based on silver halide photography, we have included them in this section because of the similarity of application to non–silver halide photographic processes.

such as radiation-sensitive silver chloride, juxtaposed with a radiation-insensitive image-forming compound, such as an organic silver salt and a reducing agent. The latent image formed in the silver chloride by radiation serves as the catalyst for the image formed thermally by the second component.

Electroconductive Electrophotography. Certain photoconductors can maintain latent images as states of increased electrical conductivity following exposure, and they exist long enough for them to be developed and amplified by electrodeposition of metal, dyes, or other image-forming compounds. 3M Co. commercialized the process in 1957 in a series of microfilm reader-printers, and they have gained preeminence in the field of making readable copies from microfilm.

In the process, a photoconductive layer of insulating zinc oxide dispersed in a polymeric binder and coated on a conductive support is exposed to the image to be copied. The exposed sheet then serves as a cathode for the differential electrodeposition of a metal image, such as of silver, from a plating solution. The result is a negative silver image that is widely used to make copies from records, periodicals, and other documents stored on microfilm.

Electrostatic Copying. Of all the office copying machines used in the U.S. today, those manufactured by Xerox Corp. are clearly the most ubiquitous. Few offices lack one (or one of its subsequent competitors), and larger organizations may count them in tens and even hundreds. Moreover, the availability of Xerox and similar machines has given birth to countless small businesses in all large cities and even in many smaller towns that offer copying services to other small businesses and to individual consumers. They do so on a scale far beyond and at costs well below those once provided by photostat and similar copying methods.

Chester Carlson, a physicist and patent lawyer, invented electrophotography in 1938 in a crude laboratory in Astoria, Long Island, N.Y. His basic patent issued in 1942. It described a process in which an electrostatic image is formed on an insulating surface that is photoconductive, the image developed with finely divided pigmented particles, the particles transferred to a permanent support such as paper, and the powder image fixed into a permanent form.

Battelle Memorial Institute began development work on Carlson's invention in 1944, and Haloid Co. (Xerox Corp.'s parent company) obtained a license in 1947 to develop the process commercially. Those involved coined the term xerography (dry writing) for the process, and it was demonstrated publicly in October 1948. Xerox marketed the first manually operated copier in 1950 and followed it with the first automatic machine that could make copies on ordinary paper. Introduction of the Xerox 914 Copier followed in 1960, and the revolution in the office copying industry accelerated sharply.

In 1954, Radio Corp. of America announced a binder-photoconductor variation of electrophotography it called Electrofax. The process, later ruled to be covered by the Xerox patents, was based on a paper coated with a photoconductive zinc oxide–resin binder layer. In this process, the paper surface is electrostatically charged in the dark and exposed to an optical image. The resulting latent image is developed by spreading powder on the surface with a magnetic brush or by using liquid electrophoretic methods. The developed image is finally fixed in place on the zinc oxide paper surface.

Since that time, many companies in the U.S. and elsewhere have developed and marketed a large variety of copiers based on electrostatic technology. Together, these machines consume thousands of tons of paper. They also use significant volumes of chemicals. Consumption of zinc oxide (ZnO) by the copying industry more than doubled in the short period between 1966 and 1968, for example, and today copying accounts for more than 10% of U.S. ZnO use. Copying also uses more than 10% of the annual U.S. demand for selenium (the element used for the photoconductive surface in xerography). Probably more significant to the country's economy, however, has been the business generated by manufacturing, leasing, selling, and servicing such equipment. Xerox Corp. itself serves as the prime example of economic success. With revenues of $21 million in 1955 and $66 million in 1961, Xerox counted revenues of $1.48 billion in 1970.

Diazotypes. In any society based on massive production of machinery and buildings and other structures, thousands of man-hours are invested annually in producing the necessary engineering drawings. Each such original drawing represents a not inconsiderable investment that must be protected against damage. Machine shops and construction sites may require hundreds of copies for use by engineers, machinists, and other shop and construction personnel, and various reproductive methods have long been relied on to provide such copies. Since most drawings are still made with pencil on vellum, they present difficult reproduction problems.

Before the turn of the century, and even well into the 1900's, the blueprint process ranked as the most widely used for reproducing engineering drawings. It is based on the reduction of ammonium ferric citrate to the ferrous salt on exposure to light and the subsequent reaction of that ferrous salt with potassium ferricyanide to form a blue compound. In making copies, the exposed (nonimage) areas of paper treated with ammonium ferric citrate and potassium ferricyanide accordingly turn blue, and the unexposed (image) areas appear white when the unreacted ammonium ferric citrate is washed away with water. The result is white lines on a blue background.

In 1890, A. G. Green, C. F. Cross, and E. J. Bevan in England described a form of the diazotype process that, with refinements, would supplant the blueprint process some three to four decades later. (Chemists Cross and Bevan, incidentally, are the same Cross and Bevan whose viscose process for making rayon fiber two years later helped begin a major change in the world's fiber and textile industries.) Successful marketing of diazotype papers began in Germany in 1923 and in Holland in 1927, and the process rapidly gained in popularity abroad and replaced blueprinting there.

In the U.S., the mechanization of blueprinting has kept it alive, but today it is far less popular for copying engineering drawings than is diazotype. In 1944, for example, manufacturers coated some 31,000 tons of blueprint paper, but by 1971 their total had dropped to only 2,800 tons. By way of contrast, consumption of diazotype paper in 1971 amounted to more than 72,000 tons, and sales of diazotype film and paper products amounted to more than $100 million.

The diazo process relies on the coupling reaction that diazonium salts undergo in an alkaline medium with colorless phenolic or enolic compounds to form colored azo dyes and on the decomposition of diazonium salts to colorless, noncoupling reaction products by exposure to light at the blue end of the visible spectrum. These reactions allow a differentiation between exposed (nonimage) and unexposed (image) areas and lead to positive copies.

World War II cut off many U.S. manufacturers of diazotype products from the technology that had come mainly from Europe, and American companies accordingly embarked on considerable efforts to develop new and improved products. Among some of the more important during the first postwar decade were precoats of colloidal silica (H. P. Andrews Paper Co., 1947), colloidal silica incorporated in light-sensitive solutions (GAF Corp., 1951), use of noncolloidal silica and a binder as a precoat (GAF Corp., 1953), and the use of rice starch as a pigment to improve diazotype paper appearance (Charles Bruning Co., 1957).

During the same period, many industrial laboratories worked to develop new diazonium salts, couplers, and stabilizers to improve development rates and both sepia imaged and black imaged papers. The use of resorcinol as a coupler in diazotype papers, for example, generally leads to brown or sepia dyes, but the images are black when it is combined with couplers yielding blue dyes. Resorcinol unfortunately migrates into the paper after coating, and as the coated paper ages the images become more blue than desired. The problem was corrected in the mid-1950's by workers at GAF Corp. with a manufacturing process that produced nondiffusing brown couplers.

The use of diazotype coatings on film (instead of paper) bases today is a segment of the copying industry that is enjoying great growth. The process was described abroad in 1938, but it remained for American companies such as GAF Corp. and Tecnifax Corp. to develop products and processing equipment that popularized the use of such films in microfilm duplicating and for making colored slides for overhead projector and color proofing applications.

These developments occurred in the late 1940's and 1950's. The growing importance of diazotype products on film and the use of formulations containing solvents to coat film bases have in turn led to considerable chemical research on diazonium salts and couplers that are soluble in solvents. In addition, the use of polyester as a film support has required the investigation of methods of anchoring diazotype coatings to this relatively inert substrate.

Vesicular Products

In April 1956, Kalvar Corp. was formed in New Orleans to develop, produce, and market "new" vesicular materials based on research at Tulane University and at Arthur D. Little, Inc. These materials can be developed by heat alone and are in the diazo speed range.

A typical vesicular film consists of a fine thin layer 0.005 inch thick of a highly crystalline polymer in which amorphous noncrystalline regions also exist coated on a polyester support 3 to 5 mils thick. A diazonium salt that will form gaseous decomposition products upon irradiation is uniformly dispersed or dissolved in the polymer. Upon exposure to ultraviolet light, the diazonium compound decomposes and liberates nitrogen, carbon dioxide, and water. Heat development at 240° F. causes softening of the polymer and, in the image areas, a pattern of tiny bubbles or vesicles ranging in size from 0.5 to 2 microns in diameter. The vesicles act as light-scattering centers, and the image areas therefore transmit light differently than the unexposed areas.

Printing

The printing industry has long used photographic processes in its operations. They are based on the fact that light will polymerize certain monomers and light-sensitive polymers into polymers and that further processing can result in images on substrates suitable for printing. For many years, one of the most familiar applications has been photoengraving, in which photographs and drawings are transferred to metal printing plates. In more recent times, photopolymerization has been adapted to processes that make complete plates for printing either by letterpress or by lithography. Both applications have grown quite rapidly in the past dozen years or so, and both promise to be major growth areas in printing in the coming years.

In letterpress printing, paper is pressed against a raised surface that has been coated with ink. The raised surfaces have traditionally been prepared by deep-etching of images into metals such as copper, lead, or magnesium or by casting type in lead. In the 1940's, chemists at du Pont began searching for ways to use light to form images by processes other than those used in photography. Within a short time, the goal had been narrowed to making printing plates. About a decade later (in 1958), du Pont introduced the first photopolymer printing plate with a relief (or raised) surface for use in letterpress printing.

Sold under the trade name of Dycril, the plates are composed of a layer of light-sensitive photopolymer plastic that is bonded to flexible steel, rigid aluminum, or flexible polyester plastic. To prepare the plates for printing, a negative of the original is first made by standard photography and then placed against an unexposed plate. During exposure of the plate, ultraviolet radiation penetrates the clear areas of the negative and causes monomers in those areas to polymerize. A spray of sodium hydroxide solution washes away the unexposed (and soluble) monomers and leaves exposed (and insoluble) polymer behind as a relief replica of the original.

When first introduced, photopolymer printing plates wore and had to be replaced more often than metal plates. They have since been improved to give much longer wear, however. While they may still not wear quite as well as metal in a few applications, they otherwise provide all the high quality associated with letterpress printing. In doing so, they offer significant cost savings to an industry that is highly labor-intensive. Since photopolymer plates can be made from any graphic original, their use can eliminate dependence on the relatively slow process of casting type in lead, assembling the type in page format, and then preparing an electrotype from a plastic impression of the original type. Instead, photopolymer plates can be used in a system that takes advantage of the high speeds and sharp images provided by photocomposition, which is growing in importance in the printing industry. In addition, the use of photopolymer printing plates saves much of the time required to make heavy metal printing plates and then to prepare presses for printing, both of which are high labor cost operations.

Other companies in addition to du Pont have also developed plastic printing plates. Eastman Kodak, for example, introduced its Relief Printing Plate in 1966, and W. R. Grace & Co. announced its Letterflex photosensitive plate system in 1968. Grace, incidentally, designed the latter at first mainly for use by newspapers but has since modified it for use by commercial printers. Two other relief printing plates are Nyloprint and Dynaflex.

Other Photopolymerization Systems

A number of other applications of photopolymerization have also recently become of considerable commercial importance. Lithography, for example, now accounts for a remarkably large amount of printing. Its growth is largely due to the ease with which lithographic plates can now be made. Originating in Austria in 1796, lithography for years required laborious hand work in stone. Now, however, a sensitized plate coated with a light-sensitive polymer is exposed with the negative of the image to be printed. Development produces a plate on which the image and non-image areas are differentiated by their relative hydrophobic/hydrophilic properties. Ink adheres to the hydrophobic image areas, and it is transferred from these areas to an inert medium that in turn places the images on paper.

The chemistry involved is primarily the photocrosslinking of polymers by exposure to light. Such crosslinking usually occurs at conjugated double bonds and is aided by spectral sensitizers. The exposed areas become insolubilized to provide ink differentiation in printing. There are many chemical approaches in this field as evidenced by the patent literature, including "backbone" polymers, functional groups, spectral sensitizers, etc.

There are two lithographic plate systems, one from du Pont (Lydel) and one from Kodak (Polymatic LN). The significant difference between the two systems is the use of a monomeric substrate in Lydel compared with the use of a light-sensitive polymer substrate in Kodak Polymatic LN plate.

The photocrosslinking of polymers was also the basis for the first of what are known as resists used in photofabrication (microcircuitry, printed circuits, and chemical milling). The polymeric coatings are hardened by exposure to light, permitting the formation of an etching mask by photographic means on any metal surface to which the resist is applied. After chemical development, the coatings are ready for etching.

Investigations begun at Kodak in 1927 resulted in a series of resists for special applications. Then Kodak Photo Resist (KPR) was introduced in 1952, Kodak Metal Etch Resist (KMER) for chemical milling in 1960, and Kodak Thin Film Resist (KTFR) in 1964 for thin film technology applications. Improved products followed in 1966, 1967, and 1969, and Micro-Neg Resist in 1971. Du Pont introduced Riston in 1968, a dry film lamination photoresist process which provided some practical advantages.

FUTURE DEVELOPMENTS

The photographic products industry will continue to grow in the years immediately ahead, although industry observers expect that growth in the first half of the 1970's will most likely be about 8% compared to the 15% growth rate the industry recorded during the mid-1960's. In the amateur market, simplified picture taking and processing will continue to be a major trend affecting growth, while the increasing demand for economical storage and retrieval of information and for less costly printing will dominate the nonamateur market. In both markets, continued chemical innovation will be required for success.

The amateur market has from the first responded directly to simplification. George Eastman pioneered with a simplified roll film in 1888, and since that time chemists and chemical engineers as well as other scientists and engineers have continuously lowered the level of skill required and increased the convenience by improving both films and cameras. The change began with black and white film, but it achieved true importance with the advent of color photography in the 1930's for movies, 35 mm. slides, and prints. Concurrent improvements in both films and cameras in subsequent years have broadened the amateur market to the point that Americans bought nearly 9 million still cameras and nearly 850,000 movie cameras in 1970. The use to which they put these and those bought in earlier years, moreover, has been particularly affected by the availability of simple and inexpensive flash equipment, especially of the flashcube ignited by battery that Sylvania pioneered with (in 1965) and the Magicube ignited mechanically that Sylvania also pioneered with (in 1970). As a result, amateurs probably take about 40% of their pictures with flash equipment today.

Of the chemical innovations that have simplified photography and broadened the market among amateurs, the in-camera processing that Polaroid introduced a quarter century ago doubtless ranks among the most important. The in-camera process now not only accounts for half of the dollar value of amateur film sales by manufacturers, but expansion of the technique offers probably the most intriguing chemical challenge for the industry's near-term future. Among possibilities are significant improvements in picture quality, lower cost, and reduction in the skill needed to take good pictures. More intriguing, however, is the successful adaptation of in-camera photography to slides and movies. Polaroid's success rests strongly on the appeal of the instant print, and the amateur market will doubtless change drastically should the concept be extended to include slides and movies. Chemists at Polaroid, Kodak, and elsewhere are actively at work on the idea. Polaroid scientists have already patented a movie film cassette and a slide film camera for in-camera photography, while Kodak scientists also have a patent on a camera that exposes and processes slides.

The spread of in-camera processing in the amateur market (and spin-off into traditional professional photography and into recognized industrial and scientific applications) appears well assured for the next several years. Somewhat more long-range possibilities for photographic innovation lie in the field of video recording and playback. Involved are such concepts as home playback of movies through color TV sets and recording programs for later playback, with both film and magnetic tape being considered as the recording medium. The possibilities have not advanced much beyond embryonic stages as yet, however, and much development and refinement, market evaluation, and—most notably—cost reduction will be required before clear patterns emerge.

In the nonamateur market, the opportunities for combining computers and microfilm appear to offer the most immediate potential for market growth, closely followed by expansion of lithography and phototypesetting. Among recent innovations typical of those generally being worked on are increases in film speed for microfilm (now up to 120,000 characters per second), the use of phototypesetters for making microfilm directly without going through full-sized intermediate copies, and making offset plates for producing hard copies directly from microfilm. In more conventional printing, meantime, lithography and phototypesetting will continue to carve out a larger market at the expense of letterpress and hot metal.

SOCIAL AND ECONOMIC IMPACTS

"I have been in the habit of taking the first of January 1896 as marking the end of the old period in the development of physics and the beginning of the new," Robert Millikan was to write in his autobiography in 1950. "*Electronics*," he continued, "is the single word which best describes the new period . . ."

Millikan ranks among the top American physicists, to be sure. He measured the charge on the electron (in 1912), and for that and for his work on the photoelectric effect he received the Nobel Prize in Physics (in 1923). Others, however, suggest the invention of the two-electrode vacuum tube in 1904 or the three-electrode vacuum tube in 1906 as better birthdates for electronics. Regardless of the merits of any specific date, however, several events clearly occurred in the last few years of the 19th Century and the first few of the 20th that led to a totally new and important technology. Within the space of a very few years:

• Heinrich Hertz for the first time showed that electromagnetic waves exist (1886–89).
• Guglielmo Marconi sent the first long-wave radio signals more than a modest distance (1895).
• Joseph J. Thomson proved that electrons exist (1897).
• John Fleming's two-electrode tube detected radio waves electronically (1904).
• Lee De Forest's three-electrode tube detected and amplified radio waves (1906).

In its first 75 or so years, what we now call the electronics industry that grew from these pioneering inventions and from the many others that followed them has obviously become a very important part of the economy. In fact, some observers view society during this period and especially more recently as experiencing a second industrial revolution, with electronics as its cornerstone.

The science of electronics and the growing technology based on it have been almost from the outset highly interdisciplinary. The very first pioneers and many of their successors, at least those generally most widely recognized, tended to be physicists and electrical engineers. Chemists, metallurgists, ceramists, and chemical engineers have made major contributions as well, however. During approximately the first half century of the electronics industry's history, this latter group probably had more of a supporting role, although a growing one toward the end of the period as electronic equipment became more sophisticated. Then in the late 1940's with the development of solid state semiconductor technology, chemistry and its closely allied disciplines began assuming a role comparable in importance with physics and electrical engineering. The result has been the economical production of very complex electronic circuits that perform efficiently, reliably, and to standards impossible to achieve earlier.

Social Impacts. The places where electronics impinge on the nation's economy are too varied and extensive to examine in detail here, and we shall mention only some of the most broadly based applications. Telephone (and telegraph), radio, and television are probably the most easily seen and widely used products based on electronics. As of 1972, Americans had about 125 million phones in use, and they made about

470 million calls a day with them. First radio and then television revolutionized both the entertainment and the news industries, and both became essential tools in selling consumer products. In addition, industrial and defense uses of radio and television may not directly involve the same number of people as does commercial broadcasting, but economically they are probably equally important.

Electronic products have affected society in many other ways, of course. Phonographs and tape recorders, for example, make high fidelity music available outside the concert hall and opera house. Electronic computers record millions of financial transactions daily as we head toward a largely cashless society, and they handle data of many other types. Computers also record travel and hotel reservations; edit manuscripts and operate photocomposition equipment for publishing books, periodicals, and telephone directories; and provide scientists and engineers with computational capacities well beyond that of many man-years of effort by persons working with simpler equipment. The combination of electronic sensors, telemetering, and computers enables control of processes as varied as metal working, assembly lines, chemical production, and petroleum refining. The result—the automated factory—is to a large degree dependent on electronics.

Satellites launched by electronically guided rockets and performing with electronic circuits make weather forecasting faster and much more accurate. Such fast and accurate forecasts are critically important to industry and to agriculture. They are also important for more than mere personal comfort, as many fewer people are killed and property damage is often much less than formerly when severe weather hits. In 1900, for example, the hurricane and subsequent flood at Galveston, Tex., killed about 5,000 people. Yet in 1961, equally severe Hurricane Carla in the same region killed fewer than 50 people. In more general terms, hurricanes killed about 160 people for every $10 million in property damage some 50 years ago. By the early 1960's, deaths were averaging fewer than two per $10 million in damage.

Radar, another product of electronics, serves in both military and civilian uses. It detects ships, aircraft, and missiles; aids navigation; and helps control aircraft and surface vehicle traffic. Space exploration depends on electronic systems for guidance, communication, and navigation. Spotting metal flaws, matching color tints, and examining minute areas with electron microscopes rely on electronic circuits. Electronic instruments of many types have been developed to provide data to persons working in the physical and life sciences. Current rectified electronically is important in such industrial-scale processes as electroplating and the production of aluminum.

With the advent of inexpensive miniature electronic circuits, hand calculators have become available to consumers at realistic retail prices. Watches powered by batteries and using electronic circuits record time highly accurately and make gears and springs obsolete. Electronic light meters using solid state light-sensitive chemicals allow even casual amateurs to take photographs with a fair degree of ease.

In the field of health, applications include monitoring patient conditions, analyzing specimens, and providing programmed diagnoses and treatment schedules. Even artificial limbs can now be operated by electronic equipment that detects and then amplifies nerve impulses in the stumps of amputees.

Economic Impacts. As in the other industries this study deals with, the products of the electronics industry show up in so many uses and directly influence society and the economy in so many ways that showing the industry's impact with numbers is difficult. Be that as it may, the largest single component of the electronics industry is communications (telephone and telegraph systems and radio and television broadcasting). The value of this component's shipments amounts to about $30 billion a year. Among individual products, nearly 10 million television sets were sold in the U.S. in 1970. So were about 8 million radios, 8 million tape recorders, and 4 million phonographs. Their combined retail value approached $4 billion. In the same year, more than $4 billion worth of electronic computers were shipped. Among component parts, electronic tubes and semiconductor devices each recorded sales that exceeded $1.3 billion in value in 1970, while switchgear and switchboard apparatus exceeded $1.7 billion and wiring devices and supplies exceeded $1.5 billion.

The electronics industry also consumes large tonnages of many materials. Each of the 10 million telephones made yearly contains some three dozen different metals and alloys, a dozen types of plastics, another dozen different adhesives, and nearly two dozen semiconductor devices. In 1970, the Bell Telephone System alone (which accounts for about 80% of the U.S. telephone network) used 520 million pounds of copper, 190 million pounds of steel, and 280 million pounds of plastics. It also used some 350,000 troy ounces of palladium and 365,000 troy ounces of gold. Other U.S. telephone companies consumed proportionate amounts, while producers of radios, television sets, computers, and hundreds of other electronic industry products also used large quantities of materials. Electronics industry companies also use large amounts of specialty materials. The Bell System, for example, prepares more than 30,000 pounds per year of quartz single crystals.

Improved technology has allowed the electronics industry to increase productivity constantly. Today, what are now its standard products and services generally cost less than they did in earlier years, if not always in current dollars at least almost always in constant dollars. Radios, television sets, and high fidelity phonograph and tape equipment all perform at higher standards than formerly but grade for grade sell for less. Telephone subscribers can now reach many more phones, but they pay less in constant dollars for local service than they did formerly. Moreover, long distance calls also cost less—and in current dollars as well. A daytime call station to station from New York to San Francisco, for example, cost $2.50 for three minutes in 1950. By 1970, the same call could be made for $1.35.

Chapter Scope. In the following pages we shall discuss some of the innovations that have spawned these commercially successful products and services. In doing so, we shall concentrate on the work of chemists and chemical engineers. Readers should realize, however, that persons trained in many disciplines have contributed to the advances. In addition to chemists and chemical engineers, people working as metallurgists, ceramists, physicists, electrical engineers, and mathematicians have all shared in making important inputs. The melding of these different disciplines increasingly has come to be known as *materials science* in recent years, and electronics may have well become the classic interdisciplinary science. Interestingly enough, in this new interdisciplinary field, people not formally trained as chemists or chemical engineers, or not considering themselves as working as such, often use the techniques of chemistry much as all scientists and engineers use mathematics.

In this chapter we shall discuss mostly advances in communications and often specifically those in telephone communications. Many of the original chemical impacts, for example, took place in telephony. Also, telephony in many ways is representative of the whole of electronics. In addition, the telephone system is the largest single part of the electronics industry in sales, plant investment, and employment.

While radio, TV, and computers (a specialized form of communication) all use the same electronic principles and share many circuit concepts, telephone electronics is in some respects different. Two unique features of the telephone network, for example, are the long lives required of components (typically 20 to 30 years and possibly more) and the compatibility each new component added to the system must have with all the then existing components. Moreover, the complexity and extent of the telephone network pose special problems not faced elsewhere. Indeed, some of the most fundamental advances in electronics involving chemistry have been made in telephone company laboratories.[1]

Many contributions have been made by chemists and chemical engineers throughout the electronics industry, of course, as well as in government laboratories and universities. Although we shall not mention many of them explicitly, they are omitted because of the limitations posed by space rather than any desire to gloss over work done elsewhere that has scored impacts equal to the developments we shall cover.

CHEMICAL ACCOMPLISHMENTS

Michael Faraday (1791–1867), the English chemist and physicist whose name appears so often in these chapters on chemical accomplishments, undertook some of the first experiments in a long series that would eventually culminate in modern-day electronics. Other early pioneers include such persons well known to students of science as the French physicists Charles de Coulomb (1736–1806) and André Marie Ampère (1775–1836), the Danish physicist and chemist Hans Christian Oersted (1777–1851), and the German physicist Georg Simon Ohm (1787–1854).

They and many others in Europe provided experimental details, but it remained for a Scottish physicist, James Clerk Maxwell, to provide the mathematical basis for electrical concepts. His publication in the early 1870's, "A Treatise on Electricity and Magnetism," in time permitted others to extend the utility of electricity far beyond the then envisaged use for power and light and into a new field, that of electronics.

Maxwell's equations gave a much better understanding of electromagnetic phenomena. He also concluded that electrical and magnetic energy travel in waves and that light is composed of similar waves. In the late 1880's, the German physicist Heinrich Hertz showed that invisible electromagnetic waves indeed exist and that they can be generated, transmitted at the speed of light through space without wires, and detected by receiving equipment. Shortly after Hertz's work, the Dutch physicist Hendrik Lorentz postulated the existence of the electron, and in 1897 the English physicist Joseph J. Thomson established its existence experimentally.

While these and other scientists were developing the hypotheses, theories, and laws of electronics, persons with inventive minds were applying the findings to commercial equipment. Among them was an Italian

---

1. Some of the information on accomplishments at Bell Laboratories and elsewhere is adapted from "Impact," which describes innovations in science and engineering by the Bell System that have helped create new industries and new products. It was prepared by the staffs of Bell Laboratories and Western Electric Co., with M. D. Fagen serving as editor. We appreciate permission to use this material.

electrical engineer, Guglielmo Marconi. He invented a practical antenna, sent electromagnetic waves more than a mile in 1895, and applied for a British patent on wireless communication in June 1896. Five years later (on Dec. 12, 1901), Marconi sent the first trans-Atlantic wireless message (from England to Newfoundland). Instant, worldwide communication to what were eventually to become unlimited audiences was at last possible.

As of the early 1900's, only two more major developments remained to be recorded before the electronic age became truly practical. In 1904, an English electrical engineer, John Fleming, invented the two-electrode vacuum tube that detected radio waves electronically. Two years later, an American inventor, Lee De Forest, inserted a third electrode into Fleming's tube and made it possible to generate stronger electromagnetic waves and also to detect and then amplify the resulting weak currents with much greater facility than previously possible.

Technically termed a triode amplifier oscillator and more familiarly known to Americans simply as a tube, De Forest's invention permitted both wire (telephone) and wireless (radio) communication systems to function over much greater distances. In addition, it and its subsequent refinements in such roles as oscillator, detector, amplifier, rectifier, switch, and voltage and current regulator opened the way for many other uses in electronics. Communications, however, continued to be the single most important application. Today, a multiplicity of underground and overhead wires, submarine cables, microwave relays, radio and TV antennas, and communication satellites puts nearly everyone within reach of telephone, radio, and television.

Electronic Principles

Electronic communication usually involves changing a sound or visual symbol into an electrical signal, transmitting that signal as electromagnetic radiation, receiving the radiation and converting it back into an electrical signal, and then reproducing the symbol in its original form. The techniques now appear familiar and simple, but many problems have had to be overcome. Signals, for example, generally are weak, and they become weaker with distance. Increasing distance requires either higher power or greater sensitivity or both. Power, however, must be increased exponentially to achieve even a linear increase in distance, and both materials of construction and the electronic characteristics of generating equipment eventually impose limits. As for increasing sensitivity, extremely good design is required to minimize losses and to control unwanted signals that might interfere with the desired signal. Again limits are eventually reached.

Meantime, the need to amplify signals poses another set of problems. In telephony, for example, signals have to be amplified periodically to achieve any realistic distance, either by wire or microwave relay. In radio and television, they are amplified on receipt to drive speakers and light picture tubes. Unfortunately, not all parts of the complex waves amplify equally, and the regenerated signal may be distorted, sometimes to uselessness. To eliminate (or at least reduce) distortion, circuit performance must be modified with devices such as tubes (or transistors), resistors, and capacitors.

In addition to these problems, the need for more communication channels eventually reaches the limit imposed by the capacity of the frequencies used. The solution has lain in using higher and higher frequencies, since the higher the frequency the greater the information capacity.

Finally, any communication system works only when the desired audience is actually reached. In radio and television, the problems have been to provide power and sensitivity for distance and enough frequencies to prevent signals from interfering with one another. Telephone shares these problems as the system spans greater and greater distances and as it includes more users. The telephone system has an added problem, however, in that it operates person to person. As the number of phones grows, so does the switching load. At the ultimate, the number of possible switching connections equals about half the number of stations multiplied by itself. Obviously, such complexity requires extensive circuitry for selection and concentration of the paths that make up the network.

To oversimplify somewhat, then, the driving forces in electronic communication systems have included a demand for producing waves with higher frequencies that have greater information capacities, a need for higher power inputs to provide both higher frequencies and stronger signals, and a requirement for more switching capacity. In such an environment, physicists and electrical engineers worked to design circuits and circuit components, while chemists, chemical engineers, metallurgists, and ceramists helped provide the improved materials that would perform properly in circuits and components.

By the end of World War II their cooperative efforts had extended the spectrum of useful frequencies from the original AM radio band of kilometer waves at low frequencies to microwaves at frequencies up to perhaps 10,000 megahertz.[2] Commercial (entertainment) radio programs were broadcast everywhere, and television was a technical reality. The telephone system extended into many regions, and it was becoming much more automatic. Other electronic equipment such as radar had become an essential military and civilian tool, while the first large electrical digital computers had been built during the war years in the U.S. and analog computers had been conceived and developed during the same time.

For all their success, however, scientists and engineers in electronics had developed their products and services about as far as possible with the then available devices and concepts, and a totally new approach was needed. That approach came in 1947 when scientists at Bell Laboratories demonstrated what was called "the transistor effect." Within a very short time, invention of the junction transistor followed, and the electronics industry entered a largely new era. Particularly important insofar as this study is concerned, the electronics industry began relying on chemistry and metallurgy to a totally new degree. Before discussing the contributions of chemists (and also metallurgists and ceramists) that are associated with transistors and other solid state innovations, however, we shall cover some of their earlier work.

Early Chemical Contributions

In making their inputs to electronics, chemists and metallurgists initially worked on such properties as the dielectric constant and loss

---

2. The electromagnetic spectrum extends from very short waves having lengths of perhaps 0.0000001 millimeter and shorter and frequencies of 10^{18} and more cycles per second to very long waves spanning a kilometer or so and having frequencies of a few tens to a few hundreds of kilocycles per second. The very short waves are gamma waves, and successive parts of the spectrum are grouped under such familiar names as X rays, ultraviolet rays, light rays, infrared rays, millimeter waves, and so forth. Light occupies the part of the spectrum in the micron (0.001 millimeter) region with frequencies of about 10^{14} per second. UHF and VHF television operate roughly in the centimeter-meter region at cycles of a few billion to a few million per second, while AM radio operates with waves several hundred meters long at frequencies of several hundred kilohertz.

in natural and synthetic organic materials, the role of impurities in the conductivity of metals, and the magnetic properties of alloys. They also dealt with the electron emission properties of metals and inorganic compounds and the thermal, corrosion resistance, and mechanical strength of structural materials and circuit elements. In the early years (and to some extent still today), materials they worked with included glass, porcelain, rubber, wood, cotton, silk, linen, mica, paper, and metals and alloys for electrical, magnetic, and structural uses. As they became available, celluloid and phenolic and alkyd resins were among man-made materials that chemists adapted to electronics industry uses.

Among the accomplishments of chemists and others in the pre– and post–World War I years were better insulators for telephone and telegraph service. Silicon carbide varistors (or resistors that change resistance with voltage) helped prevent damage by surges in lines caused by lightning. New plastics eventually provided better dielectric properties than other materials and were more readily fabricated into junction boxes, terminals, and the like or served as more stable potting compounds for components.

Chemically related developments also led to improved circuit components themselves in those years. Tubes, for example, benefited from better glasses and metals as well as from better seals between glasses and metals. In the 1930's, a porous aluminum oxide insulator for cathode heaters in tubes improved performance characteristics. A need in those years for low dielectric loss in supports in tubes that operated at high power and temperature led to the development of magnesium metasilicate crystals bonded with a glass based on the oxides of barium, magnesium, aluminum, and silicon. Also in the 1930's, zinc die castings were found to become brittle with age and fail. Metallurgists solved the problem by devising alloys containing aluminum and magnesium. Good solders and electrical contact alloys were also among important contributions of the pre–World War II years. So was developing a lead-antimony alloy to replace lead used as the sheath in communication cable, while development of impregnants and stabilizers for paper helped it retain its position as the most widely used capacitive component.

In the quarter century since the end of World War II, many new materials have become available to the electronics industry from the plastics, synthetic fiber, synthetic rubber, and nonferrous metals industries. As in the earlier years, many of these materials have been adapted to uses in electronics. Some of the new materials fitted readily into the industry, such as the plastics used to mold radio and TV cabinets. Others, however, had to be modified by persons in both the electronics and chemical industries to meet demands peculiar to electronics uses.

Modifications to polyethylene provide a prime case in point. Invented in England in the late 1930's, polyethylene became available for civilian use in the U.S. immediately after the war. In the electronics industry, an early use was to extrude it onto the individual wires making up the larger telephone cables strung overhead or placed underground to connect subscribers and central station exchanges. Having good electrical properties when extruded pinhole-free and readily color coded to aid installers, polyethylene began replacing pulp paper previously used in such cables.

Another early use for polyethylene was as a replacement for and an improvement on lead alloy cable sheaths. Successful use in this application, however, required considerable research and development to eliminate both the stress cracking and radiation and oxidation degradation that originally caused polyethylene to fail. Polymerization to the right molecular weight and molecular weight distribution stopped the stress cracking caused by mechanical deformation and also by service in environments containing lubricants and wetting agents. In addition, adding 1 to 2% of a specially developed carbon black stopped radiation degradation, while addition of small amounts of organic sulfur compounds stopped thermal oxidation degradation. As a result of these developments, virtually all new telephone cable is jacketed with polyethylene now, and accelerated weathering tests indicate such cables will have service lives of at least 40 years.

Use of polyethylene has also helped make coaxial cables perform better and last longer, both in aerial and underwater installations. In fact, polyethylene's use in aerial coaxial cable was its first application in the telephone system following World War II. Up to that time the inside conductor in such cable had been insulated from the outside conductor by cylindrical ceramic beads strung on the inside conductor. As for undersea coaxial cables, gutta-percha provided insulation for the first voice cables installed after the war, following the practice used for many years for telegraph cables. Then in the 1950's after polyethylene had proved its suitability in land installations, it began being used to insulate subocean coaxial telephone cables. Since then, all subocean coaxial cables have been insulated with polyethylene. Their design calls for the use of high voltages and many repeaters, which in turn calls for many seals along the cables. Chemical developments helping permit the use of polyethylene in this application include finding ways to produce good seals between polyethylene and metals and between polyethylene and cast resin insulation.

Successful adaptation of polyethylene and other plastics to wire and cable insulation also called for the development of special antioxidants. Antioxidants are generally polar compounds. If their molecular configurations increase the dielectric loss of an insulating material, they may cause some loss of signal. At low frequencies, such losses normally cause no problems, but at higher frequencies, such as those used in transocean cables, the losses may be significant. Appropriate antioxidants had to be developed for these applications.

While polyethylene has become the most important single insulator for telephone cable, it has by no means replaced paper as insulation for the individual wires in cable. In fact, paper-insulated wire is being made today at the same and possibly at a faster rate than ever. A thinner layer of paper than of polyethylene produces the necessary insulation between wires, so using paper allows more wires to be enclosed per unit cable diameter. The difference can be critical when cable must be placed in conduit beneath city streets, since the cost for increasing conduit diameter in any major city would be prohibitive. Thus, paper is still used in high-density applications. Its properties, meantime, have been improved by chemists, who have found that "purer" papers perform better. Trace impurities in paper lead to signal losses, and the removal of ionizing salts has permitted paper to retain a market as insulation in telephone cable.

Plastics in Other Uses

Plastics are also important for telephone sets themselves. In this use, plastics have primarily a material of construction function rather than an electronic function. Successfully adapting plastics to telephone sets, however, has provided a significant market for plastics. In addition, the problems that have had to be overcome are typical of many that chemists have been called on to deal with.

The first telephones were made of cast metal and painted, but cast metal was eventually replaced, first by phenolic resins and then by cellulose acetate butyrate. Cellulose acetate butyrate served for a number of years, as it resisted impact well. It resisted scratching poorly, however, and also readily became stained by lipstick, nicotine, and other personal use items. It has now been replaced by a styrene-acrylonitrile plastic in which a small amount of butadiene has been incorporated.

This acrylonitrile-butadiene-styrene plastic has a number of advantages. It molds readily, has high impact resistance, and costs less than its potential competitors. It also has disadvantages, however, and they are numbered among current problems that telephone companies are seeking solutions to.

Butadiene may provide the rubbery characteristic essential to good impact resistance, but its use leaves some unsaturation in the polymer molecules. Unsaturation, unfortunately, enhances degradation by light. Light degradation would concern no one if all phones were black, as black would not change color. However, the popularity of colored phones for decorative effects makes any color changes undesirable.

When photooxidized, butadiene changes from colorless to yellow. A blue phone, say, then begins turning green. Such a change in itself would be undesirable. It becomes especially undesirable, however, because the wire connecting the handset to the base is covered with polyvinyl chloride, which doesn't change color under the same conditions. The end result is a blue cord on a green phone. While the problem hasn't been solved as of this writing, chemists are actively working on it.

Semiconductors and Transistors

As mentioned earlier, the comunications industry had exploited available devices and concepts about to the extent possible by the end of World War II. In fact, by the 1930's it had become clear to communication scientists and engineers that among limitations that would soon hamper further advances were those inherent in tubes.[3] While tubes perform admirably in many uses, they do have disadvantages. Tubes, for example, function only at high temperatures and consume much energy. The tube current to be modulated might be as much as 1 million times greater than the incoming signal. Not only is much of this energy lost as heat, but in large-scale applications the heat becomes a costly disposal problem.

Made of glass, containing fragile metallic filaments, and operating in a vacuum, tubes require careful handling in manufacturing, shipment, and use. In addition, even when handled carefully, they have only limited lives. A typical tube might operate for perhaps 10,000 hours. That is certainly adequate for noncritical products such as home radio and television sets, where the use per year might amount to 1,000 hours or so and the total number of components is small. That life is often very inadequate for industrial and military products, however. The required reliability level per component increases as the system size increases. Since vacuum tubes have poor reliability, their use limits total system size. In such equipment, the hazards of its not operating and the cost for tubes and for the labor to find and replace burned out ones would be completely unacceptable.

Tubes are also bulky, making electronic equipment bulky as well. While scientists and engineers had reduced some tubes to thumb size and even somewhat smaller by the late 1930's and early 1940's, they could go no farther and still have filaments, grids, and plates spaced far enough apart

---

3. For a discussion of how tubes (and also transistors) function, see the box on pages 380–81.

to keep electrons from flowing when not desired. For such products as radio and television sets, a larger size is not necessarily a great disadvantage. In other products such as computers and telephone exchanges, however, size and power needs become costly—and often limiting—items.

Finally, as communication scientists in the 1930's began using higher and higher frequencies, especially for the then developing field of radar, tubes simply could not provide all the properties needed in some instances. With all these handicaps in mind, therefore, industry scientists began searching for replacements for tubes. In time the search centered in the field of solid state physics and on materials now familiarly known as semiconductors.

Semiconductors are materials that have electrical properties between those of conductors and insulators. At the time the investigations began in earnest, the concept of semiconduction was already being put to use in the industry. Microphone carbon, for example, is a semiconductor. So are the materials found in copper oxide and selenium rectifiers, variable resistors made of silicon carbide, and resistors made of uranium oxide and silver sulfide, all of which were in use before World War II.

None of these semiconductor applications compares in importance with the transistor, however. The transistor effect was first demonstrated at Bell Laboratories late in 1947 by John Bardeen, Walter H. Brattain, and William Shockley. Then in 1949, Shockley published a theoretical discussion of the junction transistor, and Morgan Sparks demonstrated it experimentally in 1950.

Bardeen, Brattain, and Shockley shared the Nobel Prize in Physics in 1956 for their work. Today, there are field effect transistors and unijunction transistors in addition to the original bipolar junction transistors. In all three types, one flow of electrons controls another, and they perform the same functions as do vacuum tubes. They do so, however, with few of the disadvantages of tubes.

How Tubes and Transistors Work

The common tube consists of a filament as the cathode, an intervening electrode called the grid, and a plate as the anode, all enclosed in glass and under a vacuum. Applying a current to the filament causes it to become hot and emit electrons, which flow through the vacuum to the anode. That flow, however, is controlled by a (different) voltage fed to the intervening grid. If a very weak voltage, such as that generated by sound waves in a microphone or by radio waves in an antenna, is fed to the grid, its fluctuations modify the (relatively) large current flowing from filament to plate. The current emerging from the plate as the tube's output then reproduces the signal imposed on the grid but is much stronger.

Transistors function similarly, but they do so in the solid state. They consist of crystals of semiconductor materials. Germanium and silicon from Group IV of the Periodic Table are the most common elements used, although others may be used, especially certain pairs of elements from Groups III and V.

Using germanium as an example, if it exists as a pure crystal, with no other element present, then almost no current will flow through it. However, if a very small amount of another element, such as arsenic,

Only tiny fractions the size of vacuum tubes, transistors permit equipment sizes to be reduced markedly. In addition, transistors work at room temperature. They need no warm-up period and waste no energy from hot filments. Nor are there delicate filaments to break, hazards from broken glass, or manufacturing problems associated with mechanically positioning conductors, providing vacuums, or sealing glass to metal.

Of all the advantages, however, the greatly reduced need to dispose of heat and the possibility for building long-lived equipment are perhaps the most important. For example, given the lifetime of tubes, a large-capacity tube computer would be impractical. Not only would its size and the attendant heat dissipation problems be formidable. In addition, the probabilities of failures among the many tubes required would be so high that the computer would be down for repairs most of the time.

(While transistors thus at once have long lives, sidestep the tube heat problem and use energy very efficiently, resist shock, and occupy much less space than tubes, they are not entirely free of *all* problems. As it happens, they individually produce little heat, but it turns out that dissipation of the heat they do produce is one of the most severe constraints designers face in devising the very congested integrated microcircuits being dealt with today.)

With these advantages there is little wonder that transistors rapidly changed the electronics industry. While scientists and engineers still dealt with such problems as size, reliability, and cost of multiple components, the use of transistors (and associated solid state technology) permitted a quantum jump forward in dealing with them. Moreover, the economics of production became totally different in the industry. Prior to the use of transistors, an electronic circuit typically consisted of tubes plugged into sockets, carbon compounds as resistors, and metal plates or foils separated by insulation as capacitors, all connected by wires soldered to connection points. Transistors, however, are small single crystals

---

is introduced, the resulting crystal has a surplus of free electrons that will move when a voltage is applied. If on the other hand the different element is, say, gallium, then a deficiency of electrons exists, and the "holes" thus created move when a potential is applied.

A semiconductor with an excess of electrons is known as an n (donor) conductor, while one with a deficiency is a p (acceptor) conductor. If an n crystal is interfaced with a p crystal and both are appropriately connected to a potential, then electrons and holes move toward each other from n and p regions respectively to the n-p junction. If the potential is applied in the opposite direction, however, almost no current flows, since it is "uphill" (on an energy basis) for both electrons and holes. Because of this property of permitting current to flow only in one direction, n-p semiconductors function as rectifiers to change alternating current to direct current.

If a three-part device is made with n, p, and n regions in sequence, the result is the junction transistor that operates as does a vacuum tube. The relatively large current comparable to the tube's filament-to-plate current is connected to flow from n to n region. The relatively weak current, as from a microphone or antenna, is fed to the intervening p region, which then functions as does a tube's grid. Fluctuations in the p signal current are imposed on the much larger n-to-n current, and the transistor's output is a much amplified signal.

that may measure no more than a square millimeter or so. Simply replacing tubes with them brought the first reduction in circuit size. Then in the dozen or so years of the 1950's and early 1960's, the industry moved successively from simply replacing tubes with transistors to using integrated circuits in which components were connected with deposited films instead of bulky wires and finally to the techniques used today. For the past five or six years, the electronics industry has been making entire circuits commercially in which thin layers of materials deposited on ceramic substrates less than a square centimeter or so may contain hundreds of diodes, transistors, resistors, capacitors, and their thin film interconnects. In what may so far be the ultimate in microminiaturization, as many as 660,000 semiconductor diodes are now fabricated in an area slightly more than 1 centimeter in diameter for use in American Telephone & Telegraph's Picturephone camera tube.

As shall be discussed in more detail in a moment, the techniques of chemistry have become among the primary production tools for electronic circuitry. Raw materials for preparing silicon and germanium diodes and transistors are purified chemically, and single crystals of silicon and germanium are grown and then further purified by successive recrystallizations. The chemical processes of deposition, lithography, and etching are used to place chemicals on supporting ceramics for later transformation into resistors, capacitors, and connectors. In making semiconductor devices, manufacturers use lithography and diffusion to deposit the impurities in the right places and in the right amounts to transform single crystal silicon or germanium to appropriate devices. Manufacturers also anodize thin film materials such as tantalum and tantalum nitride to produce resistors and capacitors having the electronic values called for by the circuit's intended function.

While the contributions of chemical techniques to production have thus undoubtedly aided solid state electronics significantly, the contributions of chemical concepts, viewpoints, and tools are probably more important. Purity and perfection of materials and good chemical analyses, for example, are absolutely essential to electronics technology. Among other important examples, a concept developed by scientists at Bell Laboratories states that holes and electrons obey the law of mass action; that is, they can be treated in the same manner in equilibriums as ordinary chemical species, such as hydrogen or hydroxyl ions. Since holes and electrons determine conductivity and conductivity type in semiconductors and since their control is all important in making useful semiconductor devices, the concept lies at the root of the industry's ability to fabricate devices with controlled conductivity.

Within the period of little more than a decade, these and other innovations in solid state technology removed many of the barriers formerly imposed by tube technology. Among them was the size barrier. Generally speaking, as devices become smaller their uses increase. Television sets have shrunk to truly portable sizes, increasing their markets significantly, while radios have become pocket sized—and in the process have added a new word to the language, transistor, as the synonym for small radio. Hearing aids are now no larger than ear canals, while heart pacemakers are small enough to be implanted at no discomfort to the users. Small appliances, electric ranges, washers, and dryers all have become more versatile through the use of microelectronic circuits in their controls. Similar examples of bigger markets and new markets exist for many other products.

Solid state technology has had even more important impacts on industrial products. In no application is its commercial use probably more significant than in computers. Computer circuit parts once the size of small suitcases now occupy a few cubic inches. Not only are very large computers much less bulky than formerly and not only is heat disposal not a problem. They also have much larger memories because of concurrent advances in magnetics and as a result carry out many more' operations in their even smaller sizes. Moreover, they do so at speeds formerly impossible. The time required for signals to travel around relatively long circuits had become one limiting factor in computer speed even after semiconductors had begun replacing tubes and other large-scale circuit components. Microcircuits, however, reduced signal distances markedly, and computer speeds accordingly rose. Probably most important of all, however, microelectronics in many instances has provided much higher reliability at lower costs.

In another example, solid state electronic technology helped make not only space probes and moon landings possible but also has provided immediate practical results with communication, weather, and reconnaissance satellites. Reduced weight for equipment that carries out highly complex computations and operations allows limited payloads to be devoted to other equipment. Resistance to shock permits launchings without failures in fragile equipment. Once vehicles are in space, the relatively little fuel remaining leaves scant power available to correct navigation errors. Moreover, since little electric power is available to operate transmission and receiving equipment, spacecraft must be oriented very accurately if signals are to be received or sent with the highly directional antennas that must be used for such relatively weak signals. Only microelectronic computers and other navigation aids are small enough yet sophisticated enough to provide the degree of accuracy necessary in both trajectory and orientation with the fuel available. In addition, only microelectronic circuits can gather, process, and transmit information handled either by space probes or the more immediately practical satellites in earth orbit.

Elsewhere in the economy, alternators and auto ignition and fuel injection systems depend on semiconductor microcircuits. Light dimmers in small applications and speed controls for large applications such as locomotives and paper mills all use continuously variable control elements made with solid state circuits. Microwave ovens for commercial and home cooking add a new dimension to food preparation.

In telephony, two of the problems mentioned earlier are a need for more switching capacity to handle the number of calls and a need for more channels to handle the growing volume of communication. Solid state devices offer the possibility of electronic switching with equipment that occupies little space but that is fast and has a large capacity. As for the availability of more communication channels, invention of the laser in 1960 opened an entirely new region of the electromagnetic spectrum for communication use. While many problems remain to be overcome in adapting lasers to communications, these products of solid state electronic technology do appear likely eventually to alleviate crowding among communication channels.

Single Crystal Germanium and Silicon

As mentioned in the foregoing paragraphs, the electronics industry today fabricates much of its circuitry with chemical methods. Development of these production methods evolved from laboratory research to commercial production over a period of years beginning in the early 1950's. For example, zone refining to purify single crystals of germanium and silicon was invented at Bell Laboratories in the early 1950's, and by 1952 transistors were introduced commercially into Bell System equipment. Texas Instruments in 1954 produced the first commercially available silicon transistors using the method. Impurity diffusion controlled by a technique known as oxide masking came in the late 1950's, as did the development of tantalum thin film technology for making capacitors and resistors. Meanwhile, resolution limits using photolithography have been continuously reduced, and now in the early 1970's a new method employing electron beam lithography promises to reduce circuit dimensions still farther into the submicron region.

Today, sales of integrated circuits produced by these and other means total about $560 million, up from about $80 million half a dozen years ago. They have reached this level from a very small beginning, however, and they are derived from research that started shortly before World War II on semiconductor devices and that culminated in the operation of the first junction transistor in 1950.

Scientists and engineers working in electronics have long sought to use shorter and shorter electromagnetic waves, as noted earlier. In the 1930's, few research projects seeking to do so were more critical than those devoted to the then fledgling field of microwave radio and its application to radar. The impending war in Europe and the fact that aircraft had become too fast for detection and range estimation by either sight or sound made the work particularly urgent.

In the course of this research, scientists found that vacuum tubes were somewhat inadequate at the frequencies being considered, and they began reexamining crystal detectors such as those used in the very early crystal radios. Silicon and germanium were among the materials studied as likely to yield better crystal detectors, and it was as a result of work on them before World War II that scientists at Bell Laboratories developed a silicon point-contact diode that could detect centimeter waves. The great importance of this device for microwave radar systems stimulated intensive development of the diodes and of the semiconductor materials used in them. Among subsequent developments was the invention, also at Bell Labs, of the silicon junction rectifier.

This early program disclosed the role of impurities in silicon in producing p or n conduction. The program also made possible the provision of samples of silicon and germanium with controlled purity to the work which resulted in the discovery of the transistor and the birth of the modern semiconductor industry.

As work on semiconductor materials progressed, it became apparent that crystals of silicon and germanium would be needed having unprecedented purity and perfection. Detecting, removing, and controlling impurities; determining solubilities of impurities in single crystals; and elucidating the kinetics of crystal growth were among chemical research projects undertaken.

Some chemists worked on means of purifying raw materials from which single crystals could be made. Silicon tetrachloride, for example, is the raw material from which pure silicon is produced (by hydrogen reduction). One method developed to purify it involved adding a chemical complexing agent to reduce the vapor pressure of the unwanted impurity. Distillation then purified the tetrachloride more effectively.

More important than purifying raw materials, however, was purifying the single crystals themselves. In fact, devising methods for growing large single crystals ranks among the most important developments in solid state electronics following invention of the transistor itself. The solution lay in what is known as zone melting or zone refining. In this method, a previously grown crystal is melted in a zone that is moved along the crystal's length. With an appropriate distribution co-efficient, atoms of the matrix crystal preferentially deposit in the lattice as the molten phase solidifies. The impurities thus tend to remain in the molten phase. Successive passes of the molten zone along the crystal segregate the impurities in the molten zone and eventually leave a purer single crystal behind.

W. G. Pfann invented zone melting at Bell Laboratories in the early 1950's. First applied to germanium, it successfully removed unwanted impurities and provided an extremely uniform crystal for controlled addition of the right amounts of impurities. Molten silicon, however, is highly reactive and attacks all known crucible materials. Therefore, zone melting was adapted to silicon by a method called float zone refining in which the silicon crystal itself serves as the crucible.

Once a sufficiently pure crystal is obtained, it must be "doped" with elements that cause semiconductors and transistors to function as they do. In silicon's case, for example, it is doped with an impurity having either one more or one fewer electron (that is, with an element either to its right or left in the Periodic Table). Those having one more electron provide n-type conductivity, while those with one fewer (a "hole") provide p-type conductivity.

The importance of chemical diffusion processes as a way of forming p-n junctions was recognized in the early 1950's. In time, large-area diffused p-n junctions were used to make efficient solar batteries, and improved transistors were also made by forming the base and emitter regions by diffusion.

In today's technology of silicon integrated circuit devices, diffusion methods are widely used to form p-n junctions and to control conductivity. Among the several methods used, the technique known as oxide masking has become of widespread importance in integrated circuit technology, particularly where very small geometrics are involved. It was first demonstrated in 1957 when chemists showed that thin layers of silicon dioxide grown on a single wafer of silicon crystal would inhibit the diffusion of impurities. In oxide masking as now practiced commercially, photolithographic processing of the silicon dioxide allows holes to be etched where diffusion is wanted. Repeated deposition and etching make it possible to build arrays of integrated circuits in which many different electronic functions are carried out by numerous semiconductor devices.

Thin Film Materials in Integrated Circuits

Thin film materials are used extensively in what are known as monolithic and hybrid integrated circuit technologies. Monolithic devices are those in which such components as transistors, diodes, and low-precision resistors perform the circuit functions. These devices are manufactured on single silicon chips by photolithographic processes. Thin films of highly conductive metals such as gold or aluminum are used to interconnect the separate components and also to act as connections to components off the chip.

In many analog circuits, however, there is a need for resistors and capacitors that have very precise values. Since such devices cannot be fabricated readily on silicon chips, scientists and engineers have developed a technology in which such precision resistors and capacitors (which also have carefully controlled temperature coefficients) are fabricated on ceramic substrates. Such components are also interconnected using thin films of a conductive metal such as gold. When monolithic silicon devices are also interconnected with these precision thin film resistors and capacitors, the assembly is referred to as a hybrid integrated circuit.

In producing monolithic circuits, manufacturers first deposit a thin film of the required conductive metal, usually by vacuum evaporation. Aluminum adheres well to oxides in this use. When gold is used, however, a film of titanium is usually deposited first to act as an adhesion promoter for the gold. Following deposition, the required array of film interconnects is then formed by photolithography.

In photolithography, manufacturers apply a film of a photosensitive polymer over the metal conductor film and then expose it to ultraviolet radiation through a mask having a negative image of the desired interconnect pattern. In regions exposed to radiation, the polymer film crosslinks. On subsequent treatment with a developing solution, only the unexposed regions of the polymer film dissolve away. The polymer film remaining then acts as a protective mask for the underlying metal film when the part is exposed to a metal etchant solution.

Final removal of the crosslinked polymer film after etching has been completed leaves the desired network of conductor lines on the silicon chip. Typically these are 10 microns (0.01 millimeter) wide and 1 micron (0.001 millimeter) thick, dimensions which place great demands on the photolithographic art.

In preparing hybrid circuits, the thin film resistors are often made from tantalum nitride. The film is photolithographically defined as just described into what are termed meander path lengths. These depositions have resistance values somewhat lower than those required by the circuit. In a step known as trimming, manufacturers then anodize the surface of the resistor film to increase its resistance until precisely the required value is reached. The process, incidentally, has the added advantage of giving the film resistor a stable and protective oxide coating.

As for capacitors for hybrid circuits, they are formed by anodizing the patterned tantalum film to build up the required thickness of oxide. When the appropriate thickness has been reached, a counter electrode film of conductive metal is then deposited on top of the oxide.

Thin Film Reliability. Since the reliability of integrated circuits is of paramount importance, the study of the properties of materials in film form is an active area of research. Film materials involve extremely small amounts of matter, and effects that may be trivial for materials in bulk can frequently be of great importance for the same materials when they are in film form.

Such, for example, was the finding in connection with corrosion. In some circuits, gold is used to conduct the signal from one component to another. It has excellent conductivity, and it also bonds readily to other materials in the circuit. Unfortunately, gold does not bond well to oxide substrates, and this difficulty was overcome in early production methods by first depositing titanium, which does bond well to oxide substrates, and then overlaying it with gold.

Producers soon found that such circuits failed, with pieces of the gold conductor flaking off. In examining the problem, teams of chemists and metallurgists recognized that the exposed edges of gold and titanium formed a potent electrochemical cell. In the presence of atmospheres containing just water vapor and trace amounts of contaminants typical of those in polluted urban environments, bimetallic corrosion caused the circuit to fail in short periods of time. In bulk materials under the same conditions, the effect was negligibly small.

Research soon established that placing a small amount of palladium between the gold and titanium stopped the corrosion. This passivation technique has not only resulted in protecting millions of dollars worth of circuits but has also provided an added bonus in the form of higher production yields. The work has since been extended to show that such electrochemical degradation affects many pairs of film materials. Such effects, however, can generally be obviated by interposing a very thin film of platinum or palladium between the other layers.

Lithographic Improvements

Of all the steps required to make integrated circuits, those involving photolithography are among the most critical. Since manufacturers are under economic pressure to increase component packing densities, they continuously seek to improve all aspects of lithography to produce patterns that are more exact and have the smallest dimensions possible.

Lithographic processes used today are based on the exposure of photosensitive polymers to ultraviolet radiation through a mask bearing the required pattern. The basic requirements of a mask material are that it be capable of high resolution, durable, and contain few defects (since defects on the mask are replicated in the circuit). Masks of chromium on glass or silver emulsion suffer from defect and durability problems, and recent work by chemists and physicists at Bell Laboratories has led to a novel use for films of iron oxide as a mask material that has many advantages over the older materials.

Bell scientists have worked out a process involving chemical vapor deposition of iron pentacarbonyl followed by reduction to iron oxide on a heated glass substrate. Such iron oxide masks are now in use by Western Electric, the manufacturing arm of American Telephone & Telegraph Co., and many other manufacturers are now shifting from chromium to them. In addition, several companies now supply iron oxide masks commercially.

While development of iron oxide masks has improved optical processes for microcircuits, there are some applications where resolutions below the approximately 1-micron diffraction limit of optical lithography are desirable. Based on recent research, it appears that electron beam processing will soon help solve the problem.

In electron beam processing, a computer will be programmed to focus and direct an electron beam with magnetic lenses to "write" the pattern desired. Such a process has required the development of materials highly sensitive to electron beams that are analogous to photoresists sensitive to light. Moreover, these "electronresists" must be highly sensitive to electrons, for obviously the faster the pattern can be traced the more economical the process.

The project has called for close cooperation among the physicists and electrical engineers who have been developing the electron beam equipment and the chemists who have been developing materials of the necessary sensitivity for the equipment. As a result, polymer chemists have now produced two new families of electronresists for electron beam lithography, one in which electron beam processing causes polymerization and another which causes depolymerization. Both function at exposure times measured in minutes, and both produce patterns measured in the submicron region. The process was still developmental as of the end of 1972, but commercial use is anticipated in 1973. Early applications will be for circuits where resolution is really a premium. Whenever cost vs. performance warrants, the process should then begin taking over more of the market now served by optical lithography.

Magnetic Materials

The electrical and electronics industry uses large amounts of magnetic materials, and the annual production value now exceeds $600 million. Magnets are classified as permanent (or hard) or nonpermanent (or soft). Permanent magnets serve as compact sources of magnetic fields in such devices as electron tubes, telephone receivers, and loudspeakers. They also serve as magnetic memory units and in small direct current motors. Soft magnets, meantime, are used extensively as cores in solenoids, relays, electrical motors, and transformers. Both types are highly essential in electronics technology.

Although magnetic phenomena have been known for many years, it has been only in relatively modern times that a more scientific understanding of them has been developed. The French scientist Pierre Curie established the first important law of magnetism in the 1890's during the course of his work on the effects of temperature on magnetism. Others contributed in the next three decades, and then applying quantum mechanics in the early 1930's led to the emergence of modern magnetic science.

During these and more recent years an important contribution of chemists to magnetics has been the development of the discipline of magnetochemistry. Magnetochemists endeavor to understand magnetic properties of materials, predict properties, and design new and useful materials by blending chemical bonding and structure ideas with the physical theories of magnetism.

Permanent Magnets. Strong fields are required to magnetize permanent magnets completely, but such magnets retain their properties once the fields are removed. Among characteristics that scientists in the industry try to improve are resistance to demagnetization (which is measured as coercive force) and the amount of magnetic energy that can be stored per unit volume (which is measured as energy product). The higher the coercive force the shorter the magnet, and the higher the energy product the smaller the magnet for any given performance.

In earlier years, carbon steels at first and then low alloy tool steels and later high alloy steels were used to make permanent magnets. Carbon was a required ingredient for magnetic hardness, and producers used typical steel industry practices to harden such steels. Then in the early 1930's, scientists in the U.S., Germany, and Japan found that alloys other than those of carbon steels could be used to make permanent magnets, and modern permanent magnets date from about that time.

The evolution of these modern magnets followed the discovery that alloys of iron with beryllium, titanium, molybdenum, and tungsten could be processed by age hardening to give carbon-free permanent magnets. Following this discovery, many more complex carbon-free alloys were studied, leading ultimately to discovery of the outstanding permanent magnet properties of alloys of iron, nickel, and aluminum. These alloys later were drastically improved by incorporation of additional alloying elements, notably cobalt, and by the use of special heat treatments in a magnetic field. These alloys were known as the Alnicos, and they dominated the field of permanent magnets for some years. More recently, permanent magnets based on the structure of the mineral magnetoplumbite have also been widely used (as will be discussed in what follows under nonmetallic magnetic materials).

Research and development in permanent magnets continue active today. The newest developments are based on magnetic materials that contain rare earths. These materials became commercially available in the late 1950's, and they have since been improved to have properties well above those of what had been the standard commercial products. Alnico magnets, for example, typically provide intrinsic coercive forces of 600 to 800 oersteds, while the best of them have energy products of perhaps 10 million gauss-oersteds. Among recent developments in rare earth magnets are formulations reported from such companies as General Electric, Raytheon, Philips, and Bell Laboratories with intrinsic coercive forces of 28,000 oersteds and energy products of 20 million gauss-oersteds. Incidentally, these developments are based on chemical procedures for extracting and purifying rare earth elements that led to their commercial availability.

Soft Magnetic Materials. Materials that become magnetized readily in external magnetic fields and that become demagnetized quickly when the fields are removed are known as soft magnets. Since electronics involves weak electrical currents and weak magnetic forces, materials that become magnetized and demagnetized easily and quickly are important in numerous applications. Many unusual uses in electronics, and especially in communications, therefore, stemmed from the discovery in 1917 of the very good soft magnetic properties of nickel-iron alloys.

These alloys, known as Permalloys, are composed of 45 to 80% nickel in iron, with small amounts of third elements, usually molybdenum, added to increase resistivity and to reduce electrical losses. A special heat treatment developed a few years after their invention increased their permeability (or response to very weak fields) even more, while an improved manufacturing process for making Permalloy powder on a large scale in the early 1920's further contributed to their utility. When used instead of iron powder for loading coils, Permalloys reduced sizes drastically and simultaneously greatly improved performance. Transformers made of Permalloy and operating at carrier frequencies provided entirely new levels of performance, while the use of Permalloy in the continuous loading of submarine cable increased such cable's message capacity sevenfold. These and other advances have resulted in the alloy's being considered indispensable in today's telephone and radio systems.

Research on these magnetic materials also showed the ill effect on magnetic properties of impurities, particularly carbon, nitrogen, oxygen, and sulfur. Straight nickel-iron alloys, for example, have initial permeabilities of about 8,000, while those containing molybdenum have initial permeabilities of about 20,000. When purified chemically to remove the harmful elements, however, they have initial permeabilities above 100,000.

Chemists and metallurgists have also found that these same impurities harm the magnetic properties of other materials such as iron, which is used extensively for cores in electromagnetic relays. Production procedures worked out to purify cold rolled steel eventually permitted it to be substituted for the more expensive magnetic iron in telephone relays. In just this one application, the finding has resulted in savings of more than $800,000 per year.

Another soft magnetic alloy discovered in the mid-1920's had at the time the highest known magnetic saturation at room temperature. Consisting of 50% cobalt and 50% iron, the alloy unfortunately could not be fabricated into useful strip form because of its brittleness. Other compositions consisting of ternary alloys of iron, cobalt, and vanadium were found later, however, that could be rolled into strip when heat treated properly. More recently (in the late 1950's), the alloy formulation was further modified to improve even more the coercive force, hysteresis loss, permeability, and other magnetic characteristics.

Applications for these alloys include telephone receiver diaphragms, transformers, and magnetic amplifiers, where they provide such advantages as smaller sizes and greater power outputs. Production is now more than 5 million pounds yearly. While that might not at first be considered a large amount, it must be realized that the alloys appear in a large number of applications as thin gauge wire and foil and in fine particle form in very many small parts.

Nonmetallic Magnetic Materials. Both hard and soft ceramic magnets are widely used throughout the electronics industry. In fact, their value exceeds $300 million annually, and they account for about half the value of all magnetic materials. Recent sales figures for individual applications include about $180 million for magnetic recording tape, $55 million for computer memories, $20 million for television receivers, and $20 million for communication and radio components, recording heads, and transducers.

Although lodestone, a natural mineral with permanent magnet properties, was known to the ancients, it was not until the early 1940's that crystal chemists began preparing analogs of lodestone based on its crystal structure (that of the mineral spinel). Just before World War II, workers in Japan began investigating such materials. During the war, investigations carried out in the Netherlands led to the rapid development of the first main group of oxide magnets, the ferrites, which were based on the spinel structure. Their development was particularly important in that they provided properties useful to the communications industry as carrier frequency transformers and inductors of very high quality. They also contributed to the extension of the usable spectrum of electromagnetic radiation to the higher frequencies associated with FM radio and television. Means were also found to adjust their magnetic properties for useful application in the microwave region of the spectrum, where they made possible a number of devices (such as isolators, circulators, and the like) necessary to modify signals in appropriate ways.

Later, analogs of the mineral magnetoplumbite led to a class of permanent magnet materials that also has been widely used. These materials are frequently called hard ferrites or barium-strontium ferrites (although erroneously so, since their crystal chemistry is based on a different mineral, as mentioned earlier). Still later, magnetic analogs of the mineral garnet were prepared (as in yttrium-iron garnet, for example). These, too, found initial application in microwave devices, and from current research and development interests they will no doubt find other applications as well.

Much of the early work was concerned with the crystal structure and ionic nature of ferrites, and it led to an understanding of the mechanism of electric conduction in oxides of this type. These findings in turn led to the addition of compensating ions that resulted in making low-loss ferrites that were needed for microwave transmission.

Some of the early work on oxide systems also dealt with chemical equilibrium. In particular, the work on phase boundaries and dissociation pressure was of prime importance in making not only ferrites but also in making garnets and magnetoplumbites, the other two main groups of ceramic magnetic materials.

The properties of inductor ferrites depend strongly on microstructure.

Here again the role of chemistry in finding suitable additives was important. Included were additives for inhibiting grain growth and for enhancing densification, additives that segregate to the grain boundary to improve loss factors markedly, and additives that extend the useful frequency range of ferrites. In addition, additives that refined the microstructure and others that controlled the distribution of the second phase improve the switching speed of memory magnetic materials. Similarly, microstructure and in particular the control of crystalline orientation are of marked importance in the magnetoplumbite materials.

Ceramics

Ceramics are important insulation materials used by the electronics industry. Among their applications are supports for wires on poles, lead-out bushings for oil-filled transformer housings, supports for electrodes in vacuum tubes, bases and junction points for wired circuits of all types, and many dozens of other similar uses. In more recent years, ceramics have also become very important parts of microelectronic circuits, since they serve as the principal substrate on which all functioning circuit devices are deposited by today's thin film technology.

Ceramists, chemists, and others working with ceramics for communication and other electronic applications are concerned with providing such properties as low dielectric loss, high mechanical strength, and high thermal conductivity. In the communications industry, where increases in transmission wave frequencies called for ceramics with better insulating properties, ceramists and chemists over the years developed such improved products as electrical porcelain, then steatite, and now within the past decade or so high alumina ceramics.

Microcircuit Ceramics. While ceramics remain important materials of construction for use throughout the electronics industry, their more interesting use today is probably as the base material for microcircuits. In this use they must have not only extremely good dielectric properties, but they also must be chemically inert to a higher degree than for other uses and extremely stable dimensionally in their intended uses. They must also be quite smooth and be capable of being made within very small dimensional tolerances.

To achieve these properties, the ceramic's microstructure must be controlled very closely, and ceramics also must be made in highly dense compositions very near to theoretical density (that is, to the density of a single crystal). Among many improvements recorded in recent times along these lines was the use of inhibitors to control grain growth. In the late 1950's and early 1960's, scientists at General Electric found that a chemical such as magnesium oxide added to the extent of 0.1 to 0.25% to high alumina ceramics led to dense, compact structures with uniform grain sizes. Such an inhibitor segregates at the grain boundaries and prevents the aluminum oxide crystals from growing too rapidly during sintering. With slower crystal growth, all gases in the porous shape can evolve, and the result is a more uniform grain size throughout.

Another important step to attaining small and uniform grains well knit together without imperfections and voids at grain boundaries has been to prepare the oxides chemically. In this approach, producers start with water solutions that contain the desired cations in the ratios wanted in the final composition rather than using the traditional methods of grinding coarse, rock-like oxide minerals to fine particle sizes and then mixing them mechanically. Coprecipitation from the water solutions produces the oxides in the form of compounds such as oxalates in very uniform compositions. These compounds can then be calcined to form oxides and fabricated into the necessary shapes by conventional ceramic processing methods.

Freeze Drying. Another representative approach for making ceramics chemically with the requisite microstructure involves freeze drying precursors for the oxides rather than using spray or other heat drying methods. In freeze drying, the water solution of salts containing the desired cations is sprayed as a fine mist into a hydrocarbon such as hexane or a fluorocarbon such as Freon that is held at very low temperature. The mist freezes instantly in the cold liquid. It is then collected and dried by sublimation while frozen. Once dried, the beads are calcined in a furnace to convert the salts to the appropriate oxides.

During sublimation, the salts do not migrate within the very small spheres as they tend to do during spray or other heat drying methods. In addition, some formulations calcine at temperatures several hundred degrees (C.) below the temperatures needed for conventional mixtures of the same components. Most importantly, however, the dried and calcined products have better uniformity, finer particle size, greater chemical reactivity, and a more exactly controlled chemical composition than they do when dried by heat. Ceramics produced from them as a result have better properties, and the properties are more reproducible from run to run.

Although the technique of freeze drying has been proved feasible experimentally, it is not yet being used commercially. At the moment it is still a relatively expensive way of producing homogeneous compositions for electronic ceramics. Addition of freezing as an extra manufacturing step imposes costs for labor and equipment that are not as yet balanced out by the gains, except in possibly the most exotic applications for ceramics with highly exact performances. However, scientists involved in the development expect that in time freeze drying may be made much more inherently economical than may now appear feasible.

FUTURE DEVELOPMENTS

Clearly the electronics industry is assured of continuing to grow and possibly even to advance in its relative ranking among manufacturing industries. Much of the change in the next several years will come from greater electronic product proliferation. Trends started by the availability of miniature circuits operating on very little power and now so evident in transistor radios, television sets, cassette recorders, and battery-powered calculators will continue and be extended to other products. Computers will grow in capacity and versatility, and they will continue to become smaller. Lasers will become less costly and will find many new applications. Utilization of energy produced from solar cells, a product of solid state electronics, will become much more common.

In telephone communications, improvements in size and cost will lead to increased applications of telephone handsets, call directors, mobile telephones, and paging sets. Electronic switching will continue to be extended throughout the U.S. Based on routing calls in part with transistors, diodes, and other solid state devices, the method has now been installed in some 400 switching centers in the U.S. and serves perhaps 6 million subscribers. Its continued expansion will probably be one of the main tasks for telephone companies in the next decade and a half.

Electronic switching will make the phone system more automatic and will handle the larger switching load imposed by simple growth in telephones installed, of course. However, it will also be more flexible and will handle new operations such as transferring calls from one dialed number to another, holding calls while other calls are made, and dialing frequently called numbers using only three or four digits. Moreover, it is the only method that will provide the capacity necessary for the anticipated transmission and processing of data as more operations become automated for both business and consumers, such as buying, selling, billing, and paying.

Carrier wave availability for such greatly expanded communications will be provided by extending the usable part of the electromagnetic spectrum to infrared and visible light frequencies. The shift to microwave radio and coaxial cable transmission systems some years ago increased carrier capacity from about 2,000 voice channels to upwards of 90,000 channels. Now, millimeter wave guides being considered for field evaluation in New Jersey will have nearly 250,000 and possibly half a million voice channels. When laser and optical wave guide technology is sufficiently developed and the infrared-visible region of the spectrum is used, communication scientists see up to 100 million channels being available.

In communication, meantime, use of these higher frequencies will provide an added advantage in their not being disturbed by solar-induced phenomena in the upper atmosphere as are conventional high frequency radio transmissions. In addition, high frequency radio waves are limited to frequencies reflected by the ionosphere, whereas microwave and laser transmissions can provide high-capacity long-distance communications with satellite relays.

Persons expert in chemistry and closely related disciplines such as metallurgy and ceramics, as well as in physics, electricity, and electronics, will work on the research, development, and eventual production of these and numerous other electronic products. Electronics depends largely on the control of solid state phenomena. Improvements today increasingly require the production and use of new materials that are both complex and difficult to prepare. They also are requiring the production and use of older materials at higher levels of purity and perfection. Thus, electronics generally appears to be materials limited on a number of vital fronts. Synthesis, purity, and perfection have traditionally been the role of chemistry. As the industry moves beyond what might be viewed as the comparatively simple elements such as silicon to much more complex inorganic and organic compounds, research and development programs will increasingly require interdisciplinary efforts of chemists, metallurgists, and ceramists interfacing with physicists and electrical engineers.

That chemistry is a prime science for the electronics industry is amply confirmed by both the industry's recent history and the trends already evident in newly developing materials. The solid state revolution introduced by the transistor, for example, dealt at first with germanium as a major electronic material and then with silicon. Both are products of chemical technology, as are the capacitors, resistors, and current carriers in today's microcircuits. The first maser action was produced by a stream of ammonia molecules entering an evacuated chamber resonating with microwaves. The concept was extended to laser action by using single crystals of ruby (aluminum oxide containing low levels of uniformly distributed chromium). In the decade since the first experiments, laser action has been demonstrated with a number of materials, including samarium in calcium fluoride, uranium in barium fluoride, neodymium in calcium tungstate, neodymium in glass, and benzophenone and naphthalene in organic glass. Again, all have been made in the necessary purities and with the necessary crystal structures with the techniques of chemistry.

Among other electronic materials produced with chemical techniques are gallium arsenide and gallium phosphide for use as light-emitting diodes and electromagnetic wave generators. Still newer single crystal materials such as yttrium aluminum garnet doped with neodymium, lithium niobate, and lithium tantalate are of interest for solid state lasers and for such devices as modulators and oscillators in optical communication systems.

As chemists learn more about how chemical bonding and structural relationships influence electronic properties, they will be able more and more to engineer materials at the molecular level and thus come to depend less on trial and error among the nearly infinite combinations they might examine for electronic utility. In fact, the prediction of properties by applying bonding and structural concepts to predesign materials will be the main route to progress in the next several decades.

As they work with new materials, however, either molecularly engineered or empirically found, chemists will often find their tasks more and more difficult. These difficulties will occur because the newer materials are tending to be harder and harder to process and because ever higher specifications must be met. Germanium, for example, has a melting point of nearly 940° C., but silicon has one of about 1,420° C. Yttrium aluminum garnet that is now of interest for laser applications has a melting point of about 1,980° C. These higher melting points as well as higher vapor pressures, a need for better stoichiometric control, and a requirement to reach purities of fractional parts per billion and to fabricate at dimensions well into the submicron region will provide major challenges.

Light-Emitting Diodes

Gallium phosphide, gallium arsenide, and gallium aluminum arsenide have been found to have several unusual properties, including the ability to emit light. Pioneering work on this important property of electroluminescence began in France in the 1930's using zinc sulfide and other chemicals. By the 1950's, scientists in Germany had progressed to examining junction electroluminescence in Group III-V elements. Bell Laboratories and IBM in this country, Services Electronics Research Laboratory and the Royal Radar Establishment in Great Britain, and Philips in the Netherlands have been among organizations performing important research on the subject.

Scientists now know that when a semiconductor such as gallium phosphide is fabricated with p-n junctions and a current is applied, it will emit red light if it contains zinc and oxygen as impurities and green light if it contains nitrogen as an impurity. The devices are particularly intriguing for electronic applications, because they produce light with only a few milliamps of current at only 2 volts, a condition that is very compatible with most semiconductor circuitry. The Bell System is beginning to replace the millions of small hot-filament lamps used in switchboards and telephone sets with them. Such "cold light" sources should be more economical than hot-filament lamps. They consume less current, and they have the advantages of long lives and resistance to shock typical of solid state circuitry. In addition, they operate from current available in telephone lines themselves instead of having to depend on standard 110-volt utility current, as do lamps used up to now.

While replacing hot-filament lamps in the telephone system is one major potential use, there are many others for light-emitting diodes. In telephone switching, for example, it might be possible to deflect the light electronically, such as by altering the transmitting medium's refractive index. The deflected light then would be detected with light-sensitive materials such as cadmium telluride and thus electronically establish circuit paths.

Proper deflections can also permit the light to be displayed as digits for readouts from all types of instruments. Instrument displays today are either dials and pointers or mechanically operated digits. Digital displays are almost always read more accurately than dial displays, but instruments with mechanically changed numbers are necessarily bulky and often slow. Digital displays using light-emitting diodes, however, function instantaneously. In addition they require little current and their circuits occupy little space.

Much longer range applications for light-emitting diodes include flat display panels for television and possibly even general room illumination. More likely for the more near-term future, however, are as light sources for optical communications. This application requires more complex light-emitting diodes to be made than ones used simply as replacements for hot-filament bulbs. They consist of a layered structure in which gallium arsenide is sandwiched between two layers of gallium aluminum arsenide, with p-n junctions being formed between the two different materials. Such a structure puts a channel of high refractive index between material with a lower refractive index. The result is both an electron guide and a light guide. When a current is applied to such a device having the right geometry and properly polished and mirrored, it will lase and emit coherent light as a beam from one end. Such laser light would then be available for modulation in an optical communication system.

Chemists have been performing research on these concepts for about 15 years now and have so far succeeded in developing the purification methods needed to make the various crystals containing gallium and in improving the junction forming methods. They also have raised efficiencies in converting electrical energy applied to light energy emitted from early levels of 0.0001% to 1 to 2% today (and with red light as high as 15% experimentally).

Light-emitting diodes are now used commercially in digital displays in hand calculators, and digital display will doubtless be extended to many other uses in the next few years. Meantime, widespread replacement of hot-filament incandescent bulbs is probably not far off, while laser applications in communications are probably still some time in the future. Among work to be done is gaining a better understanding of the fundamental chemistry and physics of the materials as well as improving the methods for growing crystals, forming junctions, and otherwise fabricating devices and incorporating them into circuits. Chemists are also seeking to improve efficiencies in generating light, to increase the service life of devices, and to generate light with other wave lengths.

Optical Communications

What is known as the optical electronics industry now grosses more than $200 million annually from its products, and it is growing at more than 15% per year. It makes a number of commercial products that span a spectrum of uses from radar ranging, highly accurate surveying, and precision alignment in machine tool work to micromachining, retinal eye surgery, self-cauterizing surgical knives, and tools to inspect inaccessible places in industrial as well as medical applications. Ultrahigh-speed photography, photocopying systems, and dashboard displays in autos to show whether lights are functioning properly are among other applications.

Among future developments in optical electronics, the one generating the most interest is probably optical communications. In an optical communications system, messages would be impressed on carrier waves of optical frequencies that are generated by lasers or by light-emitting diodes. These signals would then be transmitted in glass fibers to receiving equipment, where they would be detected and amplified electronically and the messages reproduced in their original forms.

By making use of the much higher frequencies in the infrared and visible parts of the spectrum, optical communication systems would provide an almost limitless number of new channels, as pointed out earlier. Only 0.1% of the optical frequencies, for example, have the same capacity as all of the frequencies from AM radio through microwaves. In addition, optical communications also offer other important advantages. Among them:

- The very small diameters of glass fiber optical wave guides would make many more channels available in highly congested conduits in metropolitan areas.
- Expensive and increasingly scarce copper for cables would be replaced by plentiful (and eventually economical) glass.
- Cables of glass fibers possibly would not need to be pressurized as do copper cables, and problems associated with temperature changes and water intrusion into cables would be reduced.
- Since glass fibers are noninductive, protection against lightning would not be needed.
- Circuits would not have to be equalized, since some glass fibers disperse signals only very slightly.

When research began on optical communications about a decade ago, it appeared that long distance transmission would be the first application. It now appears, however, that first applications may come at much shorter distances, such as between central offices in a city or within office buildings.

Chemists, physicists, and others have performed much research on components that would comprise an optical communication system, but much still remains to be accomplished before such a system becomes a commercial reality. Among the many needs now undergoing research and development are:

- Processes for making glasses with the necessary purity and optical quality in large quantities.
- Manufacturing methods for fabricating the clad and coated glass fibers, also with the necessary purity and freedom from imperfections in the very small diameters contemplated.
- Suitable techniques for joining fibers to generators, repeaters, and detectors and also techniques for installing circuits in field-scale applications, including field splicing of individual fibers at the small tolerances required.

As has been true in the electronics industry generally for the past quarter century, chemistry will also be critically important in both research and development for optical communication systems. The operating parts of lasers and light-emitting diodes rest directly on chemistry for their discovery and for their preparation in the proper crystal configurations and with the right content of elements. Moreover, successfully controlling refractive indexes of glasses and producing glasses having the necessary purities also rely on chemistry, as does fabrication of the circuits involved and joining one circuit to another.

Optical Wave Guides. As just noted, one of the main needs for communicating with optical frequencies is a means of conducting the infrared or visible radiation that will carry the messages. Since atmospheric conditions would often prevent direct transmission above ground, some type of enclosed conductor is necessary. Two possibilities are pipes in below-ground trenches in which beams would be directed by a series of either fused silica lenses or gas lenses. Either would be expensive to install and complicated to operate, and neither is being seriously considered.

A third possibility, and the one under very active consideration, is glass fiber. Such a wave guide might consist of a very small core of glass surrounded by a cladding of glass having a lower refractive index, with the entire assembly then jacketed by a relatively thin but opaque material such as plastic. The cores would be drawn with diameters ranging from a few to perhaps 100 microns and the claddings in thicknesses from perhaps 15 to 50 microns.

Such optical wave guides operate on the principle that when a medium transparent to light is surrounded by another medium having a lower refractive index, the light shone along the inner medium's axis is totally reflected at the boundary with the surrounding medium. This principle was first demonstrated about a century ago in an experiment in which water flowing from a hole in a container guided out part of the light shining into the container. The possibility of using a glass fiber coated with another glass having a lower refractive index was not considered until the 1950's, however. Since that time, use of glass fibers to conduct light has drawn an increasing amount of attention, and the results have included the practical devices mentioned at the beginning of this section.

From roughly the mid-1950's to the mid-1960's, the principle was exploited in devices where transmission paths were relatively short (a few meters at most). Transmission losses, therefore, were not critical, and losses typical of commercial optical glasses of 1,000 decibels per kilometer (a measure of loss with distance) were tolerable. Such losses in communication systems where distances may be measured in many kilometers become intolerable, however. In fact, the chances for using glass fibers in an optical communication system were believed to be remote until research in England in the mid-1960's showed that the inherent loss in silica glass amounts to less than a few decibels per kilometer. Now, chemists and other scientists have shown that the losses in glass fibers are due mostly to absorption losses induced by the presence of primarily transition ion impurities and somewhat to scattering losses caused by scratches, bubbles, and nonhomogeneous compositions along the fiber that produce different refractive indexes.

Of these losses, those associated with nonhomogeneous compositions and mechanical imperfections are the ones most readily prevented by appropriate manufacturing methods. Losses associated with the presence of impurities, however, are less readily handled, but chemists now have come close to succeeding in doing so, both in producing the initial raw glass and then in fabricating the clad and jacketed fibers.

Glass fiber optical wave guides produced on laboratory scales at Corning Glass Works have losses of about 4 decibels per kilometer, well below the 20 decibels considered to be the upper practical limit. Considerable work still remains to be done before production becomes commercial, however. In addition, much research and development is still needed to perfect field fabrication methods involving cable installation, splicing, and joining cables to generators, modulators, repeaters, and detectors.

Signal Generation and Modulation. Two other general needs involving chemical research and development on optical communication systems are to generate carrier waves in many different frequencies in the optical region of the spectrum and also to find a means of impressing on the carrier frequencies the signals to be transmitted. Although the solid state lasers known so far generate only a limited number of frequencies, crystals do exist that can function as harmonic generators and double the frequency of a given laser light. Meantime, crystals also exist in which imposition of a very small voltage (as from a signal) changes the index of refraction and thus produces a large change in the speed of light passing through the crystal. Fortunately, the same crystals can produce both effects. The search has not been easy, however, since requirements in such matters as crystal structure, refractive index, and transmission of polarized and unpolarized light differ for the two effects.

Be that as it may, research during the past half dozen years has resulted in the discovery of several crystals having the necessary properties. They include lithium niobate, lithium tantalate, barium sodium niobate, and barium strontium niobate, which were mentioned earlier. Lithium tantalate and barium strontium niobate undergo sufficient changes in refractive index to make them practically useful as modulators in optical communication systems, while barium sodium niobate and lithium niobate can be used to produce new coherent frequencies by harmonic generation from today's solid state lasers. As with glass fibers, however, much work remains to be done before a completely practical commercial system can be installed.

Resources and the Environment

While many chemists in the electronics industry are dealing with research and development associated with the exotic materials just discussed, others are dealing with what might be considered the more prosaic materials long used by the electronics industry. Included in this category are copper, aluminum, lead, zinc, and steel; gold, silver, platinum, and palladium; and plastics and other synthetics.

As in the past, the substitution of one for another of these will continue, such as the substitution of plastics for metals in general or the substitution of aluminum for copper more specifically. These and other substitutions will be influenced in part by improvements made to individual materials, of course, but they may often be more influenced by the simple economics of cost vs. performance.

This trend in substitution is hardly new, nor is it peculiar to the electronics industry, for that matter. What has become new within the past few years, however, and what promises to be increasingly important for these and other mass materials of construction are their forecast disappearances and the impacts their use and disuse will have on the environment.

Prognosticators have spoken for years of an ominous outlook for natural resources, but it does indeed now appear that the outlook is becoming grim. Only aluminum, calcium, chromium, iron, magnesium, potassium, silicon, and sodium are present in the earth's crust with some semblance of abundance, and all other metallic elements must be considered scarce. Projections for world reserves of minerals especially important to electronics suggest exhaustion for lead by 1995, copper by 2020, and aluminum (from bauxite, presently the only practical source) by 2100. Gold, silver, mercury, and tin are all among metals projected for exhaustion before this century ends, if current recovery and consumption trends continue. Alloying elements such as chromium, cobalt, manganese, nickel, and tungsten are not expected to last much beyond the next century, if that long.

As accessible and rich deposits disappear, recovery will become more difficult and expensive. The rising costs, increased power consumptions, possibly greater damages to the environment, and eventual exhaustion of specific minerals are of concern to all of industry, of course, not just the electronics industry. Problems faced because of rising costs, the need to substitute one material for another, and the development of other materials will differ from industry to industry, however.

The kind of work chemists and chemical engineers have been called on for the replacement of copper with aluminum is typical of what they will face in the future in electronics. Aluminum has now replaced copper perhaps to the extent of 5% in all conductor cable. In making the shift, the electronics industry had to consider not only the direct cost of copper vs. aluminum but also the introduction of unforeseen problems. When the first attempts to shift were made in the early 1950's, for example, neither polyethylene nor polyvinyl chloride was available as insulation. Pulp paper was used instead. With a direct current potential present, electrolytic cells developed when water entered the aluminum cable, as it almost inevitably did. The result was corrosion at a rate never experienced with copper. In time, better insulations kept water out, and that particular problem no longer exists. Meantime, substituting aluminum for copper also called for changes in soldering and joining technologies, but these problems, too, have been solved.

Other substitutions will doubtless bring similar problems for chemists to solve. Meantime, some substitutions may not occur as readily as aluminum for copper. For example, there is no known substitute for mercury in switch gear contacts wetted with the metal, for cadmium in fuse alloys, for beryllium in submarine cable hardware, and for beryllium oxide in certain triodes. Gold, platinum, palladium, and silver are all highly efficient current carriers and perform this important function in microcircuits. All are in short supply, and the outlook for supplies in the future is very dim. Moreover, platinum and palladium in particular may have a very demanding market in catalytic mufflers for reducing air pollution by cars, trucks, and buses.

Chemists and chemical engineers will have to develop substitutes for these and other materials that are at once economical, adequate in performance, and compatible with materials now used and with circuits as now designed and operating. Diffusions that may occur from material to material, the potential for galvanic corrosion, and matched thermal characteristics are among the parameters to be dealt with. Meantime, chemists and chemical engineers will have to be concerned with the impact on the environment of the production of standard or replacement materials as well as the disposal of worn-out products. The Clean Air Act of 1970 calls essentially for zero concentrations in effluents of such chemicals important to electronics as beryllium, cadmium, mercury, and asbestos. Effluents containing arsenic, boron, barium, chromium, copper, manganese, nickel, selenium, vanadium, and zinc are also to be strictly controlled. Emission of sulfur oxides from smelters that process such sulfide ores as copper, lead, and zinc also must be considered.

Recovering usable materials from worn-out products and devising systems for disposal of solid wastes are also problems that chemists and chemical engineers in the electronics industry will deal with. Large recycle activities are already under way for such metals as copper, lead, and zinc, but more, clearly, will have to be done. For the electronics industry, these recycle processes are often complicated by the presence of many types of plastic insulations. Since plastics generally cannot be recycled, they usually must be stripped and disposed of otherwise. More attention during initial design periods will have to be spent on subsequent disposal methods than has been true in the past, as will attention to recycle methods and the costs and potential impacts on the environment.

SOCIAL AND ECONOMIC IMPACTS

Chemistry is an essential discipline in the electrical equipment industry but like electricity itself is visible mainly through its effects. Even the modern insulating materials to which chemistry has contributed so heavily are normally hidden from view in electrical equipment except on cables and leads. Few housewives would see "chemists" or "chemistry" in their self-cleaning ovens, but the devices are based on two chemical processes, pyrolysis and catalysis. One might even note, though speciously, that the first electric generator was built by the British physicist-chemist Michael Faraday in 1831.

More to the point, electric power was first supplied to the public in London in January 1882, 15 years after Faraday died. New York City followed in September of that year. Early power plants had relatively little capacity and range and usually supplied direct current at low voltage. Their output was used mainly for electric lights. Electric power grew very rapidly, however, and starting around 1910, the booming electrical equipment industry provided a major market for the first synthetic insulating materials, the phenol-formaldehyde polymers developed by Baekeland. By now, sales of electric motors and generators alone in this country exceed $2.5 billion annually. Switchgear sales are roughly $2 billion annually, transformers $1.5 billion, and electric lamps (bulbs) $900 million.

The range of chemical problems has grown with the industry as producers seek to improve reliability and performance and to reduce cost, size, and weight. Research in surface chemistry upgrades the performance of carbon brushes and other types of electrical contacts. Analytical chemists have had much to do with determining the effects of additives and impurities on the characteristics of magnetic material. The preparation of lamp phosphors normally involves solid state chemical reactions at high temperature. These and other chemical aspects of electrical equipment are treated more fully in the material that follows.

CHEMICAL ACCOMPLISHMENTS

Insulation

Through about 1930, the predominant insulants or dielectric materials used in electrical equipment were natural products such as air, mica, asbestos, and cellulose paper. Demand was growing by then, however, for insulants with improved mechanical properties that could withstand higher operating temperatures and stronger electric fields. The consequent development and application of synthetic polymers as insulants have permitted significant improvements to be made in transformers, switchgear, capacitors, motors, cable, and other products used to transmit and distribute electricity. The sales of such products using synthetic insulants are very large, a rough estimate being $4 billion annually.

Excepting cable and leads, insulating materials generally are not visible in electrical equipment, and their volume and cost are not large compared to those of other materials of construction. Yet they are tremendously important in both design and operation. Partly because of advances in synthetic, heat-resistant insulants, electric hand tools weigh much less than in the past. Modern insulants have been particularly important in electrical equipment for aircraft and space vehicles, where weight and space are at a premium. In some cases, improved insulants have helped to reduce the size of equipment by half or even more, with significant savings in cost. Even a few cents' savings is important in electric motors, which are manufactured by the millions. Also apparent is the potential saving in a turbine generator, which may sell for $50 million or more.

Synthetic insulants first came on the scene in 1910–20 with the discovery and development of the phenol-formaldehyde polymers (Bakelite) by the Belgian-born chemist Leo Baekeland (president of the American Chemical Society in 1924). The systematic development of synthetic polymers did not begin, however, until the mid-1930's with the thermoplastics: polyethylene, polyvinyl chloride, polystyrene, and polymethyl methacrylate.

The outstanding dielectric properties of polyethylene and its low absorption, chemical inertness, and extrudability made it the outstanding material for shielding high-frequency (radar) cable during World War II. It is still the leading material for that use. Polyvinyl chloride (PVC) has poor electrical properties compared to polyethylene, but it is tough and abrasion resistant. These properties have brought PVC wide usage as insulation in low voltage applications such as lamp cords. Polystyrene and polymethyl methacrylate have proved valuable as dielectric films and solid insulation and can be used for moldings and structural purposes when the operating temperature is low.

Heat Resistance. Electrical equipment generates heat, and the importance of heat-resistant insulants and associated products was recognized as early as 1913 by Steinmetz of General Electric and Laume of Westinghouse in their sponsorship of proposals for classifying insulants:

Organic materials	Class A: 100° C. (now 105° C.)
Organic combined with inorganic materials	Class B: 150° C. (now 130° C.)
Inorganic materials	Class C: above 150° C.

These simple, arbitrary definitions actually were being made obsolete at about that time by the development of phenolic resins. The crosslinked (thermoset) phenolic polymer had intrinsically greater temperature resistance than the asphalts and other natural resins then in common use.

The alkyd resins, developed by General Electric in the 1920's, were the start of a host of developments in the use of synthetic resins in electrical equipment. They included the unsaturated polyesters (about 1940), the silicones (early 1940's), the epoxies (about 1950), and polytetrafluoroethylene (early 1950's). The pressure of these events led finally, in 1954, to the first basic changes in U.S. and international standards for temperature classification of insulating materials. The new standards specified test conditions for different classes of insulation that were defined much more rigorously in terms of oven temperature, temperature control, time, relative humidity, and the like, which permitted more reliable comparisons between tests performed at different times or in different laboratories. Thus, the stage was set for the use of a great many new high-temperature insulating materials in electrical equipment.

The number of synthetic insulants has grown explosively, until today the insulation engineer has hundreds of materials at his disposal. These insulants are used in various forms, including sheet, board, enamels, rods, and "papers." Papers have become important in recent years because of the effort to build the inherently porous structure of cellulose paper into the synthetics, with the concomitant improvement in electrical properties. The pores in such a structure provide a wicking action for liquid impregnants, making impregnated papers an excellent insulant for many high-voltage applications. Used in this manner, "paper" occupies today, as it did 60 years ago, a prominent place in the insulation of electrical equipment.

Filled and Reinforced Plastics. When Baekeland first developed phenolic resins he found quickly that their physical properties could be improved by fillers of cellulose fiber (wood flour). A little later came the concept of applying partly cured resin from a solvent to paper or fabric reinforcement and laminating the combination in a hot press. Both filled and laminated phenolic products found ready application in the young and rapidly expanding electrical industry. Many companies manufacture such products, whose trade names by now have become household words: Bakelite (Union Carbide), Formica (American Cyanamid), Micarta (Westinghouse Electric), and Textolite (General Electric).

Laminates of phenolics with paper and cotton fabric almost completely replaced the heavy and expensive marble and slate used previously as panel boards for switchgear and other electrical equipment. The laminates were used also in rotating electrical machinery as collars for field coils, as wedges to hold coils in the slots, and for many other purposes. Filled, molded phenolics were used in switchgear and myriad other applications.

The special needs of electrical equipment led to many developments in filled and reinforced phenolic laminates. Where heat resistance was needed, asbestos fiber and fabric replaced cotton as the reinforcing material. Mineral and mica powder, used as fillers, created phenolic products with improved electrical properties and decreased sensitivity to moisture. The greatest expansion in the use of filled and reinforced plastics in electrical equipment, however, began in the late 1930's. The precipitating factor was the advent of glass fiber reinforcement from Owens Corning Fiberglas, coupled with the development of unsaturated polyester resins and, somewhat later, the epoxy resins. Polyester resins were developed by chemists at a number of companies, including American Cyanamid, Allied Chemical, General Electric, Koppers, and Reichhold. Ciba and Shell Chemical led the development of epoxy resins.

These new synthetic insulants possessed great mechanical strength and excellent electrical properties. They have since been tailored, chemically and physically, to a very large variety of specific applications. They have helped significantly to reduce the size and improve the reliability of equipment ranging from the smallest, portable "double insulated" electric tools to huge 1,000-megawatt turbine generators.

Since about 1955, glass-fiber reinforced polyester rings, typically 8 to 9 feet in diameter, have replaced the stainless steel rings used to hold in place the stator end-windings of large turbine generators made by General Electric. The steel rings were laboriously insulated with a handwrapping of mica tape to withstand a voltage stress of 25 kilovolts. The polyester rings are insulators in themselves and markedly decreased the cost of the end-winding. The new boron and graphite fibers are just beginning to be used to reinforce plastic epoxy compressor blades in military jet aircraft. They will also undoubtedly extend the use of reinforced plastics to blades in gas turbines for electrical applications.

The adhesion of resin to reinforcing material in reinforced plastics has posed difficult physical-chemical problems. Chemists have developed many kinds of coupling agents to improve the adhesion of a variety of organic resins to glass fibers. Silicone compounds (Dow Corning, General Electric, Union Carbide) have had important uses in such coupling agents. Silicone resins themselves have not been used extensively with glass fiber reinforcement, mainly because they bond poorly to the fiber but also because of their relatively high cost and low strength. Very recent developments indicate that these limitations can be overcome and that reinforced silicone products will find greater use in electrical equipment.

Fillers can contribute special and desirable properties to plastics for electrical use. Tracking and erosion degrade the surface of most polymers and limit their use in high-voltage applications out-of-doors. The resin industry developed the use of aluminum trihydrate in butyl or ethylenepropylene elastomers and epoxy resins to improve their resistance, and very recently barium sulfate filler has been shown to impart properties similar to those achieved with aluminum trihydrate. As a result, strong reinforced plastics can sometimes replace the traditional porcelain in bushings and line insulators.

The enormous variety of filled and reinforced plastics available today poses serious problems to the design engineer who must select the materials to be used in particular applications. These products nevertheless have contributed significantly to the variety, quality, reliability, and economy of electrical equipment.

Wire Enameling. Wire for electric motors was insulated in the 1920's by dip coating in oleoresinous enamels which then were cured by heating. The enameled wire was generally wrapped with silk or cotton filaments. By the early 1930's, this kind of wire had become unsatisfactory in terms both of the space it occupied and of its properties. Chemists at General Electric Co. then developed a new type of enamel, a mixture of polyvinyl formal and phenol- or cresol-formaldehyde resins in organic solvents. The trade name for the insulated wire was Formex.

This new enamel could not be applied to bare copper wire by conventional dip coating. GE developed a new method, using a die that floated on the enamel. The die centered itself automatically to control the concentricity of the enamel film that adhered to the wire drawn upward through the die. The die method made it possible to coat magnet wires for electric motors with high-viscosity, high-molecular-weight polymer solutions. This technology became standard in the industry and is used still with the newer enamels that have evolved in recent years.

After the Formex-type enamels came the polyesters, such as General Electric's Alkanex, which provided superior heat resistance. Later, the polyesters were supplemented by overcoatings of polyamide (nylon) to give superior mechanical properties. In the past decade have come the polyimide wire coatings (du Pont's ML), which have outstanding resistance to heat and thermal cut-through (a measure of the softening characteristic of the resin relative to electrical short-circuiting). To improve processing and reduce costs, chemists are copolymerizing these materials to make polyesterimides and polyamideimides. The copolymers cost less than the polyimides but at a moderate sacrifice in high-temperature capability.

These developments in wire coating have made it possible to increase operating temperature and markedly decrease the size of a large part of all electrical equipment. Among other advantages is the toughness of the newer wire coatings, which has made it possible, for example, to apply enameled wire automatically to fractional horsepower motors. Formex wire was introduced in 1938 and rapidly replaced both older enameled wires, which were thicker and too fragile to be coiled on automatic equipment, and silk- and cotton-covered wires, which required bulky equipment for automatic winding and were much thicker. Formex and other enamels are usually applied at a thickness of 0.002 inch, compared to fiber-insulated wires in which the layer thickness may be 0.008 inch.

Extruded Polymer Insulation. More than 1 billion pounds of extruded synthetic polymers are used annually in the U.S. to insulate electrical wire and cable. For this purpose the polymers have two primary advantages: unique electrical and physical properties and ability to be extruded readily onto conductors. Synthetic polymers have combined with extrusion processes to improve sharply the speed of production, performance, reliability, and safety of electric wire and cable.

Wire and cable were insulated originally with natural polymers such as cotton, jute, and paper, impregnated with asphalt, varnish, or oil. These materials were applied to the conductor by wrapping or braiding, and the cable often had to be enclosed in a lead sheath.

A major change came in the mid-1930's with the development of flexible polyvinyl chloride compositions and thermoplastic extrusion technology. PVC's toughness, flame resistance, and oil resistance have made it one of the major insulants, more than 400 million pounds being used annually on wire and cable. Plasticized PVC in addition can be extruded at speeds up to several thousand feet per minute.

Polyethylene, too, has become a major insulant for wire and cable at more than 450 million pounds per year. Wire can be extrusion-coated with several mils of polyethylene at speeds of several thousand feet per minute.

Another growth area in extruded insulation has combined new synthetic elastomer compositions with a continuous vulcanization (crosslinking) process. Uncured insulation is extruded onto the conductor, which then is passed through a high-pressure, high-temperature steam pipe where the vulcanization reaction takes place. This approach has been used with a variety of synthetic elastomers: styrene-butadiene rubber, butyl rubber, silicone rubber, neoprene, and, more recently, the ethylene-propylene rubbers. This range of materials offers the wire and cable user a broad spectrum of properties which can be tailored to specific end uses.

Within the past decade, the boundary between thermoplastic insulations, which melt, and the vulcanized elastomers has been erased by the invention of crosslinked polyethylene. Crosslinking adds to the excellent properties of polyethylene by tying the polymer chains to each other, so that the insulation cannot melt. A further advantage is that special properties such as abrasion resistance and flame resistance can be designed into the insulation by adding fillers, creating an entirely new family of insulants.

Considerable attention is being paid now to the extrusion of multiple layers of polymer on wire and cable in one pass. This technique provides a void-free and more uniform interface between the dielectric and the shields. Higher-voltage cables with such extruded insulations are under development to help eliminate overhead power transmission lines.

Powdered Resin Technology. The use of the chemical engineering technique, the fluidized bed, to apply thermoplastic or thermosetting resins as electrical insulation was conceived some 15 years ago. The resin, in powdered form, is held in a fluidized state by bubbling air or nitrogen through a diffuser at the bottom of the holding tank. The part to be coated is heated above the fusion point of the resin and dipped into the bed, much as if it were being dip-coated in a liquid.

Many millions of small motors, bus bars, stamped metal pieces, and the like have been electrically insulated in this fashion. The fluidized bed not only avoids the use of solvents; it applies thicker coatings and covers sharp edges better than other methods. The process has been improved substantially by inducing an electrostatic charge onto the part to be coated before dipping it into the bed. Alternatively, the part can be sprayed with powder charged by passing it through a high-voltage field. Either method minimizes loss of powdered resin and avoids problems caused by instability in the fluidized bed. The electrostatic charge also enhances the "throwing power" or ability of the method to coat partially hidden surfaces.

The use of powders to apply electrical insulation is useful particularly where relatively heavy layers of insulation are required. Considerable progress is being made with tougher, quick-curing, thermosetting plastics, which should lead to increased usage of the technique in the future.

Solventless Resins. One of the problems with modern thermosetting polymers used as insulants has been that they had to be shipped, stored, and used as solutions. All low-boiling solvents are hazardous, and solvent retained in the resin after application reduces its insulating efficiency. The insulating process thus required that the solvent be removed by heat and the fumes either vented or processed in costly removal equipment.

To solve those problems, chemists have developed solventless resins, liquid monomers that can be applied and polymerized to a solid film. Besides eliminating solvent fumes, they increase greatly the amount of resin that can be applied per application and eliminate the voids that can form during solvent removal. Solventless resins are now available commercially in increasing numbers and types: epoxies, unsaturated polyesters, polyurethanes, silicones, and imides. The use of these materials is growing rapidly. As their economics become more favorable and as solvent emission requirements become more stringent, the solventless polymers should replace all solvent-containing resins.

Use of Synthetic Polymers with Mica. Mica in itself is one of the most valuable of electrical insulating materials, but it is virtually useless without help from polymers in some form. Polymers serve to put the mica where it is needed and, sometimes, to hold it in position. Until about 1950, the only useful form of mica was the delaminated flake. In this period, the synthetic polymers merely replaced the natural resins, mainly shellac and asphalt, used with mica.

But then mica became available in forms not found in nature, forms that overcame many of the limitations of conventional mica flake. In particular a paper was developed made wholly from muscovite (potassium aluminum silicate) ground into platelets and reconstituted into a uniform, flexible, continuous sheet (General Electric's Micamat, Isola's Samica, and 3M's Isomica). Like the older, interlayered flakes, this shingle-like structure required help from other materials to put it into place, and here the special properties of the synthetic polymers have come into play. Some serve directly as binders; others are used as adhesives to combine the mica sheet with backing materials; still others are the bonding agents in laminates.

Depending on the requirements, these functions are filled by a variety of resins, such as unsaturated polyesters, epoxies, silicones, vinyl polymers (especially acrylics), acetates, and many proprietary compositions. The backing for the mica sheet may also be a synthetic polymer, in either continuous film form or as a woven or nonwoven fabric. Glass cloth also is an important backing, because it is relatively inextensible and thus will not permit the mica sheet to be stretched to the point of separation.

Some families of resins find little use with mica. The phenolics tend to grow brittle with age, and their inherent charring characteristics tend to produce conductive carbon tracks as a result of electrical arcing or corona effects. The polystyrenes are too thermoplastic. Nylon is sensitive to moisture, which causes it to undergo dimensional changes.

Radiation Crosslinking of Polymers. It was known as early as 1927 that high-energy (ionizing) radiation would crosslink or vulcanize polymers, but extensive interest did not awaken until World War II. The main interest then was in selecting the polymers (for use in nuclear piles) that best resisted ionizing radiation. Serious commercial interest in the radiation crosslinking of polyethylene began in the early 1950's with the pioneering work of A. Charlesby at the United Kingdom Atomic Energy Authority.

The first irradiated insulation to be commercially available was Irrathene, an irradiated polyethylene tape introduced in the mid-1950's by General Electric. Its primary use has been as a wrapping for jacketed cables. The earliest radiation-crosslinked extruded wire insulations were flame-retarded polyethylenes introduced by Sequoia Corp. in the mid-1950's and, subsequently, in an improved form by Raychem Corp. in the late 1950's. These wires were designed for use in the range of −100° to 300° F. They found their initial uses in the military aerospace field.

Lightweight miniature coaxial cables using foamed irradiated polyethylene were introduced in 1958. These cables were considerably lighter and smaller than other RG-type miniature and subminiature coaxial cables.

The need for a strong, lightweight wire rated at 135° C. became crucial in the aircraft field in 1964. To fill the need, Raychem Corp. developed a sandwich construction using radiation-crosslinked polyalkene and polyvinylidene fluoride. The rating of this wire was upgraded to 150° C. in 1967. The wire has since gained general acceptance in the military field and also in the commercial aircraft market. At about the same time, International Telephone & Telegraph introduced a radiation-crosslinked polyvinyl chloride wire insulation to the electronics market. Similar products can be obtained today from any of the major manufacturers of irradiated wire and cable insulants.

Irradiated, flame-retarded, heat-shrinkable polyolefin tubing was introduced in 1957 and found its first uses in the aircraft, military, and aerospace fields. The number of irradiated polymers used in heat-shrinkable products has broadened steadily and now includes perhaps a dozen different types. Among those available are polyvinyl chloride, neoprene, and silicones.

Most of these products were used at first primarily as mechanical protection or strain relief, as on wiring harnesses, and their electrical insulating properties, though useful, were regarded mainly as secondary or standby protection. In recent years, however, these materials have begun to be used to provide both primary insulation and strain relief at junction points. One example is the terminations in appliances, the points at which the wire lead is attached to the appliance and to the plug by a continuous molding. Another involves the recently enacted federal mine safety laws, which set severe limits on the useful life of mine power cable as then used and repaired. The introduction of a mine cable splice kit using radiation-crosslinked heat-shrinkable tubing for primary insulation and mechanical protection has sharply improved reliability in an area critical to mine safety. The kit makes it much easier and quicker to make a safe, reliable splice.

Radiation-crosslinked specialty products include Raychem Corp.'s Solder Sleeves, heat-shrinkable plastic tubes containing solder and flux. When heated, these sleeves shrink, solder, seal, insulate, and provide strain relief at terminations and branching junction points. Coaxial terminations which can be applied automatically are another example of this new technology.

Thermally-Upgraded Cellulose. Cellulose is a keystone of many electrical insulation systems, particularly in transformers, because of its unique combination of purity, insulation strength, versatility in use, and low cost. Cellulose is not without drawbacks, however. One is that it absorbs water readily, with adverse effects on its insulating qualities. Another is its relatively low thermal stability. By the 1950's, thermal stability was becoming a problem because of the increasing power demands on transformers. Insulation was required to withstand relatively high temperatures for long periods, often decades, without losing efficiency.

As a result, companies that make electrical equipment developed chemical treatments to improve the thermal stability of cellulose insulation for transformers. One such treatment is the General Electric Permalix system, in which the cellulose molecules are modified by reacting them with acrylonitrile. A second type of treatment is the Westinghouse Insuldur or McGraw-Edison Thermecel processes. In these, finished cellulose paper is saturated with a mixture of nitrogenous organic compounds, such as urea and dicyandiamide. These chemical treatments have enhanced the advantages of cellulose, leading to smaller, less costly, and more reliable distribution transformers, the type seen on power poles in residential areas.

Synthetic and Modified Liquid Insulants. For more than 75 years, the electrical equipment industry has used insulating liquids to fill and impregnate high-voltage equipment such as transformers, capacitors, cables, switches, and terminals. These materials often act also as coolants. The most commonly used fluid has always been petroleum-base oil. Synthetic liquid insulants and chemically-improved natural liquids, however, have allowed electrical equipment to be improved in a number of ways: size and weight reduction; higher, more stable voltage capabilities; extended operating temperature range; nonflammability; longer, more reliable life.

The electrical industry today uses more than 50 million gallons per year of petroleum-base oils as insulants. Chemical research has revealed that specific properties of such liquids can be enhanced by choice of base crudes (naphthenic or paraffinic), special refining techniques, and the addition of small amounts of certain chemicals.

Transformer oils are based on naphthenic crudes, which provide relatively low viscosities and pour points and relatively good sludging characteristics. Addition of a synthetic butyl cresol to transformer oil improves its operating life by inhibiting oxidation, the main mechanism of oil degradation. These inhibited oils are used in distribution-type transformers, which are likely to breathe air during periods of heating and cooling. Uninhibited oils, on the other hand, are used in power transformers, which normally are sealed or have a nitrogen cushion to protect the oil from contact with air.

Cable and capacitor oils are made from paraffinic crudes. They are refined carefully, however, to control their content of aromatic and olefinic hydrocarbons, which determine the oils' electrical stability.

Chemical developments in hydrocarbon polymers produced a family of synthetic, oil-like fluids, the polybutenes. Because of their low electrical losses, these materials are used in capacitors and cables. The electrical equipment industry consumes some 6 million gallons per year of polybutenes, which are made by Amoco Chemical, Cosden Oil and Chemical, and Chevron Chemical.

Another chemical development was the askarels, synthetic insulating liquids that evolve only nonflammable gases when decomposed by an electric arc. The dielectric constant of these materials is more than twice that of oil. For this reason they could be substituted for oil in capacitors as a means of reducing size and weight by some 40%. In transformers, for the same reason, they made possible a more equalized distribution of voltage stress between paper and liquid in composite insulants. General Electric introduced the askarels in 1930 under the trade name Pyranol (a series of chlorinated biphenyls or blends of these compounds with chlorinated benzenes). Askarels now are produced only by Monsanto Co.

Synthetic liquid insulants also include fluorocarbons made by 3M Co. These are low viscosity fluids with a nonflammability characteristic similar to that of the askarels. The fluorocarbon liquids volatilize readily, and the dielectric strength of the vapor approaches that of the liquid. These properties make the fluorocarbon liquids useful in a special spray-cooling technique for transformers.

The silicone fluids (Dow Corning, General Electric, Union Carbide) are semi-inorganic polymers that show excellent thermal stability over a wide temperature range. Their pour points are exceptionally low, less than $-50°$ C., and they are stable to $150°$ C. in air or $200°$ C. under an inert gas. The viscosity of the silicone fluids changes relatively little with temperature, and they resist oxidation well. These properties make them suitable for use in small transformers and certain capacitors that can take advantage of their unusual physical constants.

The class of compounds called organic esters has been used as liquid insulants in capacitors for many years. The class includes a natural product, castor oil, but also certain synthetics, such as dibutyl sebacate and ethylene glycol. Dibutyl sebacate is used in parallel-plate capacitors for operation at high frequencies, where its electrical losses are low. Ethylene glycol is used in electrolytic capacitors.

Synthetic Gaseous Insulants. Air at ambient pressure is the oldest and still the most common gaseous insulant used by the electrical industry. For requirements that air cannot meet, gases such as nitrogen, hydrogen, argon, and helium have been used, often at up to several hundred pounds per square inch. Neither air nor these simple elemental gases, however, can fill certain needs imposed by the industrial demand for ever more compact electrical apparatus operating at higher temperatures and voltages. Hence the use of synthetic gaseous insulants.

The main synthetic gaseous insulants are sulfur hexafluoride and the halogenated hydrocarbons. In general, the synthetics offer relatively high insulating ability and chemical stability, good heat transfer properties, low flammability, and acceptable toxicity. They have found considerable use in transformers and cables.

Sulfur hexafluoride was first synthesized in 1900 in Europe. The gas was first used in electrical equipment in 1937, when scientists at General Electric discovered the advantages of its exceptionally high dielectric strength in high-voltage transformers. By 1950, sulfur hexafluoride was being used routinely as an insulant in lightweight X-ray transformers for the commercial market.

The first of the halogenated hydrocarbons were the Freons, commercialized by du Pont in the 1930's. The initial and still a major use of the compounds is as refrigerants. They do, however, offer a wide range of dielectric strengths, all higher than that of air, and were introduced by du Pont for electrical applications in 1961. The halogenated hydrocarbons are made also by Allied Chemical (Genetrons) and Union Carbide (Ucons).

Electrical Contacts

The electrical properties of materials used as electric contacts depend on surface chemistry in the contact zone, including the compositions of films in the zone. Thus, all electrical machinery and devices that use electrical contacts are affected by chemical influences in the contact zone during service and by the chemical composition of the contact elements.

Carbon Brushes. The unique properties of carbon for electric contacts were recognized in 1885 by Van dePoele in the U.S. and Forbes in England. They discovered that the massive wear of motor commutators by metallic brushes could be avoided by substituting carbon brushes, which could conduct electricity to a sliding surface without abrading it. There followed the development of a wide range of carbons adapted to the specific requirements of current density and polarity of electric motors of all types and sizes. The carbons, through chemical processing, could be designed to absorb residual inductive electromagnetic energy, which otherwise could destroy a motor commutator through arc erosion. Addition of specific metals to the carbon brush led, in 1900–20, to successful operation of high current capacity machines such as those required for electroplating and for synchronous converters and inverters in power plants.

Despite the great improvement of carbon brushes over metal, excessive wear of the carbon brush plagued the electrical industry for years. Basic understanding did not begin until about 1940, when surface chemistry studies showed that atmospheric moisture was essential to the smooth, low-wear operation of the brush. It was found also that oxygen inhibited wear but at considerably higher pressure than required by moisture.

Until the early 1940's, carbon brushes in airborne electrical equipment were unreliable above 25,000 feet, where the vapor pressures of both oxygen and water were too low to inhibit catastrophic wear. This problem was solved by adding organic chemical impregnants and inorganic adjuvants to the brush. Brushes for use at high temperature were not fully practical until about 1950, when molybdenum disulfide was combined with carbon. Brushes of this kind are required in all space vehicles for rotating electrical equipment and sliding contacts.

Despite these advances, however, the high wear of carbon brushes at low humidity, as in desert and arctic regions and at high altitude, remains a problem in electrical equipment. The use of additives, such as molybdenum disulfide in aircraft brushes, has limited utility as yet.

Other Electric Contacts. The contact microphone grew out of the discovery, in 1861–75, that the electrical resistance of assemblages of carbon particles varied with the pressure on the mass. The change in resistance was found later to be associated directly with changes in contact area caused by fluctuations in pressure. The resulting contact or carbon-grain microphone proved to be the basis of the telephone.

The sensitivity of telephonic response was perfected later to a major extent by chemically modifying the structure of the carbon particles. This development extended over several decades. Basically it involved controlling the architecture of the particles by modifying the carbonization processes used to generate them. Even today, however, scientists do not really understand the principles involved because of the difficulty of investigating the structures of amorphous and semicrystalline materials.

A related development in contact devices originated with the coherer action, in which the contact resistance between particles is modified enormously by voltage gradients upon them. The coherer effect has been known since 1835. It was not exploited, however, until 1890, when it was used to detect Hertz waves, a key development in the evolution of wireless communication. The chemistry of the surface film on the particles was basic to the further development of the detection of Hertz waves in this way, and the effectiveness of a coherer device depended on controlling the nature of the film by direct chemical action, such as gold plating.

Contact rectification, another useful effect, depended on surface films similar to those responsible for the coherer effect. The ability of semiconducting oxide films to rectify alternating current—to convert it to direct current—was first exploited commercially in 1928 in the copper oxide rectifier. The efficiency of this contact rectifier depended on controlling the growth of the boundary layer at the copper–copper oxide interface.

The development of the transistor can be traced to results of investigations of coherer and rectification phenomena, involving thin films, chemically developed and controlled.

Direct studies of contact reliability in switches began in the mid-19th Century. It was learned early that bad contacts on copper could be improved by wetting with mercury, while good contacts could be protected by a thin film of petroleum grease in the contact zone, which prevented oxidation. Later investigations led to the use of semiprecious and noble metals for higher reliability in switches. Noble metals also were electroplated on the contact surface where economy so required.

Chemical additives played a crucial role in developing contact materials for high-current, high-efficiency devices. Doping of metals with cadmium oxide reduced the adhesion of contacts. Tungsten made contacts more refractory, thus reducing erosion. Tungsten carbide improved the hardness and sometimes the wear resistance of the contacts.

The surface chemical properties of contacts provide the key to reliability. These properties reach a peak of technical importance in computer circuitry where, over the life of the equipment, contacts may be required to cycle through 1 billion operations without malfunction.

Contacts must be operated sometimes in chemically-controlled atmospheres to shield them from contamination. An example is reed switches in telephone circuitry, which operate in a nitrogen atmosphere.

The high production silver switch, expending silver at a catastrophic rate, awaits a substitute solution through chemistry.

Lubrication

Lubrication may seem to have little connection with electricity, but it is essential in many electrical devices ranging from clocks to steam turbines. Furthermore, the problems of carbon brushes mentioned in the preceding section have also necessarily involved studies of friction, lubrication, and wear. Two unusual lubricating materials have been developed largely as by-products of industrial research undertaken for other purposes: Silicone lubricants were developed by the electrical industry, and polytetrafluoroethylene was developed by a branch of the chemical industry built up to serve the needs of a working fluid (Freon) for the electrical refrigerator.

The silicones grew out of chemists' attempts to generate a system of chemistry based on silicon, much as organic chemistry is based on carbon. This work led to early efforts to prepare compounds with silicon-carbon bonds, which do not occur in nature. Much of the classical work on organic derivatives of silicon was done by F. S. Kipping in the early 1900's. Kipping, however, regarded the polymers that formed frequently in his reactions as useless by-products. It remained for E. G. Rochow and W. J. Patnode at General Electric and J. F. Hyde at Corning Glass Works to begin to appreciate the possibilities of the organosilicon polymer, in the late 1930's. Rochow's discovery of a direct synthesis of organochlorosilanes provided the impetus for large-scale development of the polymers we know today as silicones.

Silicone oils and greases are not considered general-purpose lubricants but a special class reserved normally for severe environments. They are stable at high or low temperature, resist weather, ozone, or electrical discharges, and possess constant viscosity over wide ranges of temperature and pressure. Polymethylphenylsilicone oils, for example, may have pour points as low as $-50°$ C. but can be used for short periods at up to $400°$ C.

Silicone lubricants often can be tailor-made. The lubricating properties of the simpler methylsilicone oils are improved, as a rule, by incorporating phenyl groups in the polymer. Further improvement, particularly in metal systems under high contact pressure, can often be obtained by halogenating the phenyl groups, as with chlorine. Silicone greases are used to best advantage in rolling contacts (bearings) which must operate at either very high or very low temperatures. The most common of these greases consist of silicone oils thickened by adding metal soaps such as lithium stearate.

Polytetrafluoroethylene (PTFE), discovered in the early 1940's, has a wide range of unusual properties, and its uses appear to be increasing exponentially. PTFE is characterized by temperature stability up to about $250°$ C., good toughness, high dielectric strength, excellent resistance to water, corrosive agents, and solvents, and by the lowest coefficient of friction (0.05) of any solid.

The reasons for this abnormally low friction coefficient are somewhat tentative, but scientists believe that the large fluorine atoms in the polymer chain screen the positive charge on the carbon atoms, thus minimizing interactions with neighboring molecules. This explanation is consistent with the very low surface tension of PTFE and with its extremely low coefficient of adhesion with other materials. In addition, the marked rigidity of the PTFE molecule and its high crystallinity suggest that it undergoes relatively little plastic deformation in contact with other materials. The effect is to reduce the true area of contact. Since the friction coefficient depends on the product of the area of contact and the adhesive bonds at the sliding interface, it seems reasonable to attribute PTFE's low friction coefficient to these properties.

Despite the low coefficient of friction, PTFE does have drawbacks as a bearing material. Compared with metals, it has lower strength, lower thermal conductivity, a higher coefficient of thermal expansion, and poorer abrasion resistance. Modern technology, however, has found a variety of ways to minimize these disadvantages. The surface of a porous metal such as copper can be impregnated with PTFE. This composite bearing material combines the higher bulk strength and thermal conductivity of the metal with the low-friction surface of PTFE. Another technique used in bearing applications is to fill the polymer with metal powders, carbon, or other inorganic materials. Increased abrasion resistance has been obtained by filling PTFE with glass fibers. Certain ceramics with negative coefficients of thermal expansion have been used as fillers to reduce the overall expansion of PTFE bearing materials at high temperature.

Silicon-Iron

Before 1905, the electrical-grade sheet iron used in transformer cores caused electrical losses of 4 watts per pound of core material. Today, in part as a result of chemical investigations, silicon-iron is available that is rated at 0.3 watt per pound. This more than 10-fold reduction in transformer core losses saves an estimated 4.4 million kilowatts of power annually in the U.S., power that otherwise would be dissipated as unwanted and insulation-damaging heat.

Electrical-grade silicon-iron is an alloy containing small, tightly-controlled (to within 10 parts per million) amounts of manganese, carbon, sulfur, and phosphorus. The lower silicon alloys (1.5 to 3%) are used mainly in relays, motors, and generators. The higher silicon alloys (3.25%) are used in transformer cores and account for two thirds of the 923,000 tons of silicon-iron produced annually in the U.S.

Early attempts to reduce core losses and magnetic degradation in electrical steels were hampered by lack of information on the effects of trace elements in combination with silicon in silicon-iron. Some of the trace impurities occurred in the ores. Others were introduced during the processing of the material.

Beginning in 1910, studies of trace contaminants in silicon-iron have shown that nonmetallic impurities such as oxygen, carbon, and sulfur affect magnetic properties. With the exception of sulfur, these contaminants are easily removed by combining them with hydrogen at high temperature as part of the heat treating process for silicon-iron. It has also been established that oxygen causes mechanical aging in silicon-iron and that nitrogen contributes to magnetic deterioration. Manganese contributes to the ductility of silicon-iron without affecting its magnetic properties.

The steady progress in the improvement of silicon-iron over the past 65 years has contributed also to the development of structural steels, specialty steels, and other magnetic materials. Among these are structural silicon-steel, manganese-nickel steels, tungsten steels, and the electrical grades of nickel-iron.

Magnetic Materials

Magnets are made from a variety of materials, and the chemist has played a major role in developing processes for extracting and purifying the necessary elements and compounds. The chemist also has done much to develop the analytical methods that are required to determine the effects of impurities and additives in magnetic materials.

The oldest magnetic material used by man was lodestone, the natural iron oxide, Fe_3O_4. Only since the 1930's, however, has an adequate understanding of magnetic phenomena been applied toward the development of hard or permanent magnets with extremely high resistance to demagnetization. Among the more prominent of these are the Alnicos (aluminum-nickel-cobalt), cobalt-platinums, barium-ferrite, Lodex, and (in 1969), the cobalt–rare earth magnets. The cobalt–rare earth magnets, such as cobalt-samarium, resist demagnetization 20 to 50 times better than Alnico. More economic methods of separating rare earths from their ores, when developed, will extend the use of the cobalt–rare earth magnets in consumer products ranging from larger permanent-magnet motors to perhaps a levitation system for rapid transit vehicles.

Among the soft or nonpermanent magnetic materials, the electrical steels, particularly silicon-iron, are the largest single class. The high magnetic permeability of the nickel-irons is useful in magnetic shielding, audio-frequency transformers, telephone equipment, and power transmission devices. Beginning with the binary nickel-iron alloy, a host of alloys have been developed. They include iron-nickel-molybdenum, iron-nickel-chromium, iron-nickel-copper, and iron-cobalt-nickel.

The development of the soft magnetic ferrites has been an interdisciplinary effort among physicists, metallurgists, ceramists, and chemists. This family of magnetic materials consists chemically of mixed oxides of manganese-magnesium, nickel-zinc, manganese-magnesium-aluminum, manganese-zinc, and similar structures. Their uses range from computer memory cores to microwave transmitter components to television receivers.

The uses of magnetic materials are becoming more diverse and complex. They are deposited on or embedded in plastic for computer tapes. Discs and drums for storing and transforming magnetic images are coated with abrasion-resistant magnetic materials.

The expanding use of magnetic materials depends on the increasing role of chemistry in conjunction with physics, metallurgy, and several branches of engineering. The improvement of interlamination coatings for electrical steels will lead to further improvement in energy-loss-limited equipment. More economic extraction of rare earths from their ores will permit reduction in size and extend the range of uses of magnets made from the new permanent magnet materials. The soft ferrites will play an ever-increasing role in computer applications. And the nickel-irons are linked to the expansion of communication and power transmission.

Lamp Phosphors

Inorganic phosphors have been used as energy or wavelength converters in many types of lamps for more than 30 years. A given phosphor emits light in a specific part of the spectrum when excited by a specific form of energy: light, electrons, applied voltage, and mechanical or chemical energy.

Lamps with phosphors are important both technologically and economically. They permit the spectral distribution of the emitted light to be tailored independently of the exciting radiation without the significant loss of energy that results from achieving the same end by filtering the light from incandescent or arc lamps. An example is the common fluorescent lamp, in which phosphors such as calcium halophosphate absorb the ultraviolet emission of the mercury arc and convert it to visible light. The fluorescent lamp generates white light more than four times more efficiently than does the incandescent lamp.

Other important lamp phosphors are used to color-correct the emission from high-pressure mercury vapor lamps, which otherwise has a blue-green cast. The phosphor absorbs the long wavelength ultraviolet emission of the arc and reemits in the visible part of the spectrum, usually in the red part, so that objects appear more natural in the light of these lamps. One phosphor of this type is a compound of yttrium, vanadium, oxygen, and europium, which also is the red phosphor component in color TV.

Another type of lamp phosphor, the light-emitting diode (LED), is itself sometimes the active element of the lamp. The LED converts electric current directly to luminescence at the junction of two electronically different (n- and p-type) materials. These lamps are used primarily as indicators, both singly and in arrays. A last type of lamp phosphor is one that absorbs the emission from an infrared-emitting diode and converts it to visible light. This combination of infrared LED and phosphor is used to obtain colors and/or efficiencies not yet obtainable with LED's that emit in the visible part of the spectrum.

The preparation of phosphors normally involves solid state chemical reactions at high temperatures. The purity of the materials used is of the utmost importance in the manufacture of efficient phosphors. Equally important is the formation of phosphor particles of the proper physical form, whether single crystals for making LED's or fine (3 to 30 micron diameter) powders for use in fluorescent lamps. The level of undesired impurities tolerable in a phosphor ranges from near 1 part per billion to several thousand parts per million, depending on the luminescence mechanism.

The Lucalox Lamp

Lamp envelopes normally are made from glass, because it is easy to form, transparent, and chemically inert at moderately high temperature.

Incandescent and fluorescent lamps commonly use a soda-lime glass, while mercury discharge lamps use a fused silica, which softens at a higher temperature.

In about 1960, General Electric chemists developed a pore-free, highly translucent aluminum oxide ceramic (Lucalox). The critical step in making this material is the addition of about 0.2% magnesium oxide to the aluminum oxide powder before pressing and sintering. The magnesium oxide inhibits grain growth during sintering, which thus proceeds to completion. The result is a pore-free, translucent body that transmits about 95% of the incident light. The material also can be made into tubular lamp envelopes at feasible cost.

Following this development, further research led to the invention of a relatively-high-pressure sodium vapor lamp. The corrosion resistance of the aluminum oxide envelope permits this lamp to operate with the hot sodium vapor at 0.5 atmosphere of pressure indefinitely without deterioration.

The high pressure and temperature of the sodium vapor or plasma combine to yield a continuous spectrum that is slightly deficient in green. The Lucalox lamp, as a result, emits a golden yellow light, quite different from that of the conventional low-pressure sodium vapor lamps, which emit light of substantially a single wavelength, that of the yellow sodium D line in the spectrum. The efficiency of the Lucalox lamp is spectacular. It emits 120 lumens or more per watt, compared to 50 to 60 lumens for the currently popular mercury vapor lamps. Major growth is expected in the use of this lamp for area lighting, as in parking lots, highway intersections, public buildings, and the like.

The Calrod Tubular Heater

For more than 35 years the Calrod tubular sheathed heater has been widely used as a heat source for kitchen ranges and ovens, hot water heaters, soldering irons, and many other devices. Chemists, metallurgists, and engineers have worked to combine in this device the rapid heat-up and cool-off features of gas heat with the cleanliness and simplicity of electrical heating.

The Calrod is a spiral resistance heater mounted in a stainless steel sheath. A granular ceramic oxide, conventionally magnesium oxide, isolates the heater from the sheath electrically but supports it mechanically and connects it thermally to the sheath.

A variety of chemical processes occur in an energized Calrod. They include the consumption of both oxygen and nitrogen by the hot metal parts and their resupply by permeation through the granular ceramic. These gas-metal reactions actually can result in extremely low oxygen, nitrogen, and total gas pressures. The electrical properties of commercial ceramic oxides can be quite sensitive to the oxygen pressure, and the thermal conductivity of the system is influenced by the total gas pressure. The recognition and understanding of the complex relationship between the chemical reactions in an energized Calrod and its physical properties have led to improved unit design and better manufacturing controls.

The Self-Cleaning Oven

Two chemical processes, pyrolysis and catalysis, are the basis of the self-cleaning oven, whose utility is evident in its commercial success. General Electric invented the oven and first marketed it in 1963. Since then, GE has sold more than 1 million self-cleaning ovens, and more than half of the ranges sold annually in the U.S. now have the self-cleaning feature. Some 31 foreign and domestic manufacturers are licensed to use the GE patents.

The oven cleans itself automatically in 1½ to 3 hours. Food soil is first pyrolyzed or thermally decomposed at about 470° C. Decomposition is completed by passing the effluent gases over platinum and palladium oxide catalysts at 650° to 750° C. These processes convert food soil ultimately to safe, odorless gases plus minute amounts of mineral oxide ash.

In developing the self-cleaning oven, chemists evaluated the oxidation rates of many food soils and food soil combinations as well as their reactions with various inorganic, temperature-resistant materials and coatings. The reaction rates were evaluated to obtain a thermal input rate. The thermal input rate is used to control the rate of decomposition of food soil and to reduce the exothermic or heat-emitting reactions of food soils during pyrolysis.

Flash Lamps

The conventional flash lamp consists of a shredded metallic foil and a quantity of oxygen, together with a heat-sensitive primer for ignition, contained in a glass bulb. This lamp, in effect, is chemically-stored light for future use. The flashtube, in contrast, is an electric discharge tube that produces light from electrical energy stored in a capacitor. It is much less efficient than the flash lamp in terms of system volume per unit of light.

Since flash lamps first appeared, their rate of growth in sales has increased continually. The growth has accompanied a series of technical improvements giving higher efficiency, smaller size, and greater user convenience. Fifteen years ago the industry standard was a flash lamp about the size of an egg. It contained shredded aluminum foil and was used in a reflector usually 5 inches or more in diameter. It was reliable, much smaller than earlier lamps, and very bright. But the newest flash lamp, the Hi-Power flashcube lamp, produces 15 times more light in a bulb whose volume is only two thirds of a cubic centimeter. Two dozen of these lamps and reflectors, packed in a convenient cube form, can be held in one hand.

Progress in two major chemical areas has been responsible for the dramatic miniaturization and even more dramatic increase in light output per unit volume. First, the amount of light from a given quantity of foil and oxygen has increased nearly fivefold as aluminum, zirconium, and hafnium have been used successively as fuels. Second, the growth of color photography and the prevalent all-purpose use of daylight-type film have centered interest on increasing the color temperature of the flash, so that less light is lost in correcting the color balance. Hafnium burns with a color temperature of 4,800° K., 50% higher than does the aluminum foil that was standard 10 years ago.

The chemistry of the primer is also critical for reliable and sensitive flashing. The conventional flash lamp is ignited by a small electrical current from a battery that heats a tiny tungsten filament and fires a primer bead. A novel, recently-developed product, the MagiCube, does not use electrical energy for ignition. The cube contains cocked springs that strike a small quantity of primer in a thin-walled metal tube projecting from the bulb. This primer requires a delicate chemical balance, so that it will flash reliably and yet be stable enough to survive manufacturing, shipping, and handling.

Lamps for Agriculture

Artificial lighting has long been known to produce marked effects on plant growth, but major technological changes in lamps have recently made intensive horticultural lighting feasible for certain important segments of commercial plant growth. These advances have been aided by sophisticated contributions in the chemistry and physics of arc discharges and lamp structural materials as well as by outstanding engineering innovations.

Three key factors appear to contribute to the recent changing climate for commercial horticultural lighting: major technical advances in the efficiency, life, and stability of high intensity discharge (HID) light sources; the accelerating trend toward more automated, controlled methods of horticulture with capital-intensive equipment to offset sharply increased labor rates; and the increased relative value of quality and timing of high-cost crops.

Many complex factors influence plant growth. They include light intensity, spectral distribution, dark-light cycle, atmospheric composition (carbon dioxide, humidity, etc.) both above and below the plant soil or growth medium, plant nutrients, temperature, air movement rates, pollutant levels, and many other less known factors. Light effects can be classed as photosynthetic and nonphotosynthetic (i.e., photoperiodism, photomorphogenesis, and photoretardism). The light levels required to control the latter class of responses are generally much lower than those for substantial photosynthetic effects. The spectral distribution for optimal plant response is much different from that for the optimal human eye response.

In order to design and evaluate special HID lamps for intensive horticultural lighting needs, chemists and lamp engineers have had simultaneously to design for the best combination of source efficiency in the 380 to 750 nm. (billionths of a meter) band and relative spectral distribution of principal interest in plant growth, voltage and related electrical characteristics for practical economic ballasting, and long-term chemical and physical stability for the desired lamp life. This work has required understanding of nonequilibrium connective and diffusive flow of chemical species in hot gases with steep thermal gradients, complex thermodynamics of highly nonideal multicomponent systems, and unusual materials compatibility problems of lamp parts. The economic use of high-pressure sodium lamps, such as Lucalox, and high-pressure metal halide lamps, such as Multivapor, has now been demonstrated in certain commercial applications such as full-term growth of roses and chrysanthemums and in vegetable and flower seedlings in various locations in the U.S. Special HID lamps designed for plant growth show excellent technical results, but they are still in the experimental evaluation stage in various laboratories and greenhouses.

FUTURE DEVELOPMENTS

It would be a major undertaking to forecast with any useful explicitness and degree of probability what equipment, technologies, materials, and devices will be used 20 or 30 years from now to generate, transport, and transform the energy needed in this country and to light streets, factories, and homes. Neither does it seem feasible to forecast with any accuracy the kinds of appliances that will be added to or will replace the ones we are already using. Rather than list all possible future technologies or types of equipment, therefore, and analyze them to determine what will be expected of chemistry, one might ask: What are the more fundamental driving forces that are likely to direct the future development of the electrical equipment industry and what do these forces mean to the chemist?

Energy transmission and distribution will serve as a first example. Most electric power today is transported overhead by wires. Not only is overhead transmission equipment considered unsightly by some people. The method puts some limit on other uses of the land. The total land area now used for this purpose is approximately 6,000 square miles or three times the size of Delaware. Inasmuch as U.S. electric power consumption is expected to double each 10 years, it would clearly be undesirable to continue to expand the use of overhead transmission for the next 30 years. An increasing fraction of new transmission lines already is being put underground. This trend is expected to accelerate, especially around urban centers, barring a truly radical change in the technology of energy generation (e.g., generation in each home individually) or energy transmission (e.g., transmission by microwaves).

As a result of the trend to underground transmission, the chemist certainly will have to contribute to the development of materials that will withstand an extreme variety of soil conditions for very long periods. More specifically, metals may require more efficient long-term corrosion protection, the water permeability of plastics may have to be reduced drastically, and so on.

It is also likely that energy will be transported increasingly at lower than ambient temperature. Liquefied gases are easier to transport, for example, and electrical losses of cables are reduced at lower temperature. For the chemist these changes would mean the design of insulation and container materials that would function acceptably at the lower temperatures.

Fundamental considerations also make certain trends inevitable in energy generation. Whatever the specific technology of generation may be —turbines, nuclear reactors, fusion, magnetohydrodynamics—materials will have to be designed that resist corrosion better and that retain their strength at higher temperatures. Any widespread use of liquid metal breeder reactors, for example, would require a detailed knowledge and control of liquid metal chemistry and corrosion well beyond what is known today. A second example is the turbine. It is essentially a Carnot-cycle-limited device, so that an increase in its firing temperature corresponds to an increase in its efficiency. To increase the firing temperature, however, requires fundamentally better understanding of the reactions between the hot turbine gases and the materials with which the turbine is constructed.

It is certain, too, that the chemist increasingly will be called upon to study, understand, and engineer the fuel combustion process to reduce atmospheric pollution. There will also be a thorough reevaluation of the question of which form of energy to generate and transport (electric power, gas and, if so, what gas and how).

No matter how energy is produced and transported, it must be transformed from one form into another (such as high voltage to low voltage) and be stored temporarily until further use. In fact, if a satisfactory technology of energy storage were available today, U.S. generating capacity could run at constant load only and would not need to be expanded for the next seven years. Chemists will be expected to contribute to the invention of such a storage capability, be it batteries, fuel cells, thermal storage, or something else.

Chemists also will contribute to the design of smaller, lighter, and lower-cost power transformation and switching equipment, using their expertise in surface chemistry, rheology of polymer systems, solid state reaction kinetics, and adhesion. Such equipment will require new or vastly improved insulating materials that will stand higher electrical and mechanical stresses at very low (superconduction) or substantially higher temperatures than now allowed and under a greater variety of environmental conditions. A substantial effort will be made to apply this insulation by new methods that do not pollute the environment. Solventless technologies, for example, will have to take over in a relatively short time. They will require the invention and development of new materials and fundamental improvement in application techniques.

In lighting, the trend will continue to be to greater efficiency, better spectral distribution, and longer lamp life. High intensity discharge lamps are an obvious expression of this trend. The chemist will be expected to sort out the extremely complex chemical kinetics, thermodynamics, and photochemistry involved in such lamps and to provide new and improved materials for envelopes, leads, sockets, and reflectors. New lamps, such as panel lighting, will be developed.

To predict the kinds of appliances a home may contain in the future is most difficult. The reason is that, more than elsewhere, a role is played by factors relating to human behavior patterns, recreational needs, and the interface between the individual and his environment. If a good part of waste handling and recovery processes are to be carried out in each individual home, for example, appliances that sort and preprocess glass, metal, organics, and the like will become necessary. Such appliances, again, represent a task for the chemist.

One thing is certain. Modern man is surrounded by the products of chemistry. It gave him new fibers and fabrics; permanent-press and drip-dry clothing; a well-insulated home; long-life, labor-saving aids, tools, and appliances; inexpensive electrical energy; and an enormous variety of foods and medical supplies. One can safely say that this trend will continue as chemistry helps create the electrical equipment of the future.

Part II

THE PEOPLE OF CHEMISTRY

The four chapters comprising this part of Chemistry In The Economy are concerned with the size, educational preparation, and qualitative maintenance and improvement of the nation's supply of chemists and chemical engineers (excluding technicians). This supply, which is also designated in this volume as "chemistry's professional manpower base," is distinguished from demand, or actual employment. Demand is treated in Part III.

At any time, chemistry's manpower supply exceeds the demand. The difference is not easy to quantify, but it has two main components. First, supply may exceed demand because some chemists and chemical engineers desiring work in their profession are unemployed. Secondly, an indeterminate number of chemists and chemical engineers who are outside the labor force or are working outside the profession are still recallable to employment in chemistry with appropriate financial or other incentives.

Part II is based on information developed under the auspices of two committees. Thus, Chapter 21 has resulted from work guided by the Manpower Committee, which also motivated the preparation of some material on demand that is presented in Part III. The remaining three chapters of Part II derive from labors of the Education Committee. The membership of both committees is shown at the end of the volume.

The estimation of the size and composition of chemistry's qualified manpower supply (excluding technicians) is beset by conceptual difficulties as well as data limitations.[1,2] If there were no data problems, we could start with the number of persons having degrees in chemistry (including biochemistry) and chemical engineering and then proceed toward the same estimate of supply by means of appropriate adjustments. Thus, from the number of persons with relevant degrees living in the United States, we should have to subtract the nonrecallable number outside the national labor force and the nonrecallable number working in other fields. An addition also has to be made: Some persons without relevant degrees have demonstrated their qualifications by actual employment and may for this reason regard themselves (or be regarded) as eligible for inclusion in chemistry's manpower supply.

An alternative approach starts with employment. Here, it is necessary to augment the number employed as chemists or chemical engineers with the number qualified by virtue of education and deemed available for employment in the field.

The foregoing comment gives adequate hint that two plausible initial notions of the current supply—the number of functional chemists and chemical engineers defined by education and the number defined by present or past work experience[3]—should be combined in a suitable way. These two kinds of concepts suggest, furthermore, that supply and demand (i.e., actual employment) are algebraically translatable into each other by means of proper additions and subtractions, so that a manpower accounting scheme might be devised to accommodate both. Indeed, a system of coordinated statistics could start with figures for supply and demand for a particular date and allow these figures to be carried forward to any subsequent date by means of detailed increments and decrements. This "perpetual inventory" could even be extended to provide projections.

Since an integrated accounting system is not now in operation for chemistry's qualified manpower, the very first recommendation presented in this study (page 5) looks toward the establishment of a framework for consistent historical and current estimates and projections of supply and demand. In this system, the sources of growth and attrition would be isolated and quantified.

Although it is not difficult to identify the people who take degrees in

---

1. This chapter briefly discusses the magnitude of the nation's supply of chemists and chemical engineers. Treatment of the demand for their skills, reflected in employment, is deferred to Part III of this volume. The Manpower Committee established for this study under the chairmanship of Wallace R. Brode also participated in the development of some of the material discussed in Part III, such as the special tabulations of secondary work activities of the non-academic chemists included in the National Register of Scientific and Technical Personnel in 1968 and the American Chemical Society's questionnaire survey of employment of chemists and chemical engineers in selected manufacturing companies in 1971-80.

2. "Qualified" and "professional" are used as rough synonyms for characterizing chemistry's specialized manpower. The latter term is often reserved for persons who are qualified through academic preparation, but even professional societies typically make allowance for qualification through work experience instead (see footnote 3). In any case, the omission of technicians, who are commonly identified as "subprofessional," is intended in this discussion of chemistry's base of specialized manpower.

3. The American Chemical Society criterion for "full professional standing," for inclusion in the National Register, is: "A bachelor's degree or 10 years of professional experience in an area of chemistry." (See, for example, *American Science Manpower: 1970* (NSF 71-45), National Science Foundation, 1971, p. 247, which also shows criteria of other cooperating societies.)

chemistry (including biochemistry) or chemical engineering, other data for an accounting scheme are less clear-cut. For example, self-classification in the chemical or any other single profession is obviously difficult for someone with multiple degrees and varied work experience.[4] Besides, not every chemist shows the same preference with regard to occupational labeling as he moves into—and along—an administrative or managerial track. An employer and a professional society, moreover, may differ seriously on the length (and recency) of experience required for a nondegree chemist to become recognized as a full professional.[5] It is also true that an employer who emphasizes educational credentials in recruitment need not make work assignments in accordance with them and may subsequently classify personnel according to demonstrated ability to perform. Thus, a graduate biologist belonging to a biological society will probably acquire a chemist's title in a drug firm if he is assigned to a chemist's slot in the table of organization and then does the work satisfactorily.

Attempts to clarify the boundaries of a hybrid concept of chemistry's manpower supply (i.e., a concept based on both education and work experience) must lead to a multiplicity of estimates that vary according to the specifications adopted or scenarios visualized. Overlap exists with other specialized manpower supplies that are comparably defined, and different degrees of unscrambling could conceivably be effected by changes in relative salaries, in social values and priorities, in the climate of international relations, and so forth. Such changes could recall some persons into the national labor force as well as accomplish redistributions within it. Assumed changes that are intended to yield maximum measures of chemistry's manpower base, however, could lead to unrealistic figures that overlook counteractive and secondary adjustments outside of chemistry. It is also desirable to recognize that some postulated changes can prove more favorable to other specialized manpower supplies than to chemistry's as a new equilibrium settles in the nation's labor force after a perturbation.

Estimated Manpower Supply

No dependable total for the supply of chemistry's qualified manpower (on a hybrid definition) is available, but the number may be placed conservatively at 200,000. After all, figures compiled by the U.S. Bureau of Labor Statistics (BLS), which are discussed in Part III of this volume, show a current *employment* estimate in the neighborhood of 190,000 for chemists (including biochemists) and chemical engineers. This magnitude embraces the private and public sectors, and it also includes college and university teachers.[6] Many more than 10,000 additional eligible persons are available in other employment and outside the labor force. Incentives could presumably be designed to induce different numbers among them to return to work in chemistry if circumstances required their services.[7]

---

4. Unless a scientist belongs to societies in different fields, he is not likely to be asked to fill out more than one form for the National Register.

5. See footnote 3.

6. According to BLS, 26,200 chemists were employed at colleges and universities in 1970. The American Chemical Society's "Chemical College Faculties," on the other hand, showed some 14,000 full-time faculty members (mostly Ph.D.'s) located at institutions granting degrees in chemistry. The difference reflects the presence at universities of chemists engaged primarily in research and the counting of graduate students as employed if they are engaged part-time as paid research assistants, teaching assistants, and so forth.

7. Some of the women chemists who have left the labor force may intend or could be induced to return. See later discussion in the text.

It is also pertinent to observe that the same circumstances requiring more chemists (e.g., a war) might also raise the demand for many kinds of nonchemists. In such a case, a higher pay differential (or perhaps even a government directive) would be needed to accomplish the return of any predetermined number of migrant chemists to work in chemistry.

Other statistics—and opinion—seem to confirm this unsurprising estimate of chemistry's manpower supply. The American Chemical Society (ACS), for example, has a membership in excess of 100,000. An oral tradition holds that this total, which includes a small fraction of the population of chemical engineers, comprises about half of the persons qualified for membership by virtue of academic preparation or work experience. Furthermore, for the 1968 compilation of the National Register of Scientific and Technical Personnel, the Society mailed 183,000 questionnaires to persons who already were members or were thought to be qualified.[8] Finally, a computation made for this study indicates that an active chemistry population of about 200,000 is easy to deduce from the number of chemistry degree recipients during the period 1930–70.

The 1968 National Register's response of 94,000 chemists (about half of the number canvassed) may reinforce the oral tradition just mentioned, but it also tells something about professional identification. Almost one third of these respondents held doctorates, which is a disproportionately high figure, as will be seen later. A still greater ratio of doctoral-degree recipients is observed among ACS members. Clearly, the greater the commitment made to specialized higher education, the greater is the tendency to record professional aspiration and interest.[9]

We now return to the estimate of chemistry's manpower base derived primarily from figures for academic preparation. Statistics on degrees granted in chemistry (discussed at greater length in the next chapter) are available from such sources as the U.S. Office of Education, the National Academy of Sciences, and the ACS Committee on Professional Training. From 1950 to 1970, 242,000 bachelor degrees were granted in chemistry (including biochemistry) and chemical engineering. The total for 1945–70 is 277,000. About 50,000 of the group that graduated before 1945 must still have been active by 1970. Counting them but allowing for attrition by death, retirement, and outmigration in the 1945–70 group, we arrive at 300,000 as the approximate magnitude of the credentialed 1970 cohort of chemists and chemical engineers.

More than 70% of this cohort of 300,000 held bachelor degrees only. About 40,000 (13%) held both a bachelor and a master degree. The doctorate represented the highest level of academic attainment for 46,000 others (15%).

For an estimate of chemistry's labor force, two large slices have to be eliminated from this cohort of 300,000. The first cut is represented by some 60,000 chemistry majors of 1945–70 who entered medicine. Another cut of 20,000 must be made for women who left the profession for family and other reasons. These two reductions bring the total down to 220,000, but other cuts remain to be made.

The connection between chemical training and medical careers is well established. About one fourth to one third of all premedical students major in chemistry. For the past quarter of a century, a fairly constant number of B.S.'s in chemistry (about 2,500) have annually been admitted to medical schools.

The estimated reduction for women is based on the difference between

8. The ACS list of names included current ACS members, former members, chemistry graduates who had entered medicine or other sciences, and nonmembers subscribing to ACS journals, participating in ACS symposia, publishing relevant technical papers, or recommended by various membership committees.

9. According to American Science Manpower: 1968 (NSF 69-38), National Science Foundation, 1969, p. 256, chemistry had a relatively high proportion of nonrespondents. In general, according to this NSF report, Ph.D.'s, academic associates, researchers, and younger scientists showed a greater propensity to respond than others. (The 1970 canvass, incidentally, included 182,000 questionnaires and 87,000 responses.)

their proportions among chemistry-degree recipients (15%) and among chemistry respondents to the National Register (8% in 1968 and 7% in 1970). The proportion of women in the ACS membership is similar (7%) to that shown in the Register.[10]

Further discounts have to be made from the remainder of 220,000 for the "permanent" loss of chemistry-educated professionals to other fields. Thus, some chemistry graduates of 1945–70 have migrated into other areas of natural science such as physics, biology, and mathematics without attempting to maintain dual professional capability or identification. Some others have lost identification with entry into nonscience activities or in the course of an upward movement on administrative or managerial ladders. Finally, chemistry graduates who enter teaching in secondary schools tend to lose occupational recognition as chemists and become reckoned simply as "teachers."

When these further reductions are made, the remainder may fall substantially below 200,000, but invocation of the auxiliary criterion of experience provides a considerable offset if a hybrid concept of manpower supply is adopted. The less strict this auxiliary criterion, the greater is the offset.

The ACS standard for professional recognition of nondegree chemists is presumably more stringent than that applied by employers. Indeed, a major continuing employment survey which BLS and the National Science Foundation (NSF) initiated identifies scientists and engineers as "workers who perform at a level requiring education or training equivalent to that acquired through completion of a four-year college course with a major in a natural scientific or engineering field."[11] A four-year training equivalent instead of 10 years of experience (the ACS requirement) ought to make a perceptible difference in the size of chemistry's specialized manpower base.

---

10. Women with doctorates tend to remain as well represented as men with doctorates in both employment and ACS membership.

11. *Employment of Scientists and Engineers in the United States: 1950-66* (NSF 68-30), National Science Foundation, 1968, pp. 44, 45.

This chapter and the two that follow focus on education for meeting the future demand for professional chemists and chemical engineers in industry, the schools, and government. They consider the recent and prospective academic output of chemists[1] and chemical engineers, the indicated future need, provisions for continuing education, and the responsiveness of chemical curricula to industrial requirements. The present chapter concentrates on factors influencing graduate supply and demand during the 1970's.

Recent Patterns of Growth

The educational environment of the present and recent past contrasts sharply with that of the preceding decade. The 1960's witnessed a rapid growth of college enrollments in a setting of broad economic advance and generous government underwriting of science. Chemical education is now faced with diminished federal support. Public enthusiasm for science has waned, certainly for science that has no obvious or visible application to social needs that are already felt. Job prospects for new graduates are limited, while many experienced professional employees are insecure or out of work. A fresh look at the supply-demand situation is obviously in order, and an extensive current literature is already available to assist such an undertaking.

Degrees granted at all levels in chemistry, chemical engineering, and biochemistry increased substantially but at dissimilar rates during the 1960's. According to Table 1, annual B.S. degrees awarded in chemistry rose by 53% in the course of the decade, but the table also shows that the number of bachelor degrees dropped during the 1950's and did not regain the 1950 level until 1960. (The 1950 figure was swollen by including World War II veterans training under the G.I. Bill.) On the other hand, Ph.D. degrees, which had remained constant during the 1950's at about 1,000 per year, more than doubled in the ensuing decade. Comparable patterns but with smaller numbers were traced in biochemistry and chemical engineering. For biochemistry, however, the gain in the 1960's was particularly sharp. Undergraduate degrees more than trebled, but they still account for only a small fraction of the annual total for chemistry-related areas.

The demographic aspects of B.S. production have come under close study.[2] Students receiving B.S. degrees in chemistry, chemical engineering, and biochemistry have remained an almost constant fraction of the 22-year-old population since the 1950's. With an increase in the percentage of college-age students graduating from college, however, the percentage of all bachelor degrees represented by the chemical areas has actually declined. For scientific endeavors as a whole, a flattening is observable for a longer period. The plausible conclusion is that science training is approaching saturation. Most youths of the college-age group having an interest in

---

1. Since academic statistics commonly distinguish chemical engineering and biochemistry from chemistry, the term chemist is usually meant in this chapter to exclude students, faculty, and practitioners in the first two categories. The scope of the term will be indicated by the context.
2. See W. R. Brode, "Manpower in Science and Engineering, Based on a Saturation Model," *Science*, July 16, 1971, pp. 206–13.

and an aptitude for the practice of science are already electing science programs, and moderate changes in the fraction of college-age students actually attending college will have little effect.

The output of B.S. graduates in chemistry and chemical engineering is determined almost entirely by choices made by qualified students. The bulk of the undergraduate teaching load at every institution consists of first-year and second-year courses, primarily offered to nonmajors. Upper-level course enrollments fall off so rapidly that crowding is rare. Cases in which departments may have felt obliged to limit the number of majors because of shortages of staff or facilities must have been very few. Indeed, 10 years of experience of the American Chemical Society's Committee on Professional Training (CPT) have revealed no such instances. On the contrary, departments almost universally prefer more qualified majors than they get.

Table 1

Degree Production in Chemistry, Chemical Engineering, and Biochemistry

	Academic Year	B.S.	M.S.	Ph.D.
Chemistry	1949–50	10,619	1,576	953
	1959–60	7,603	1,228	1,048
	1969–70	11,617	2,119	2,167
Chemical Engineering	1949–50	4,529	700	178
	1959–60	2,966	610	170
	1969–70	3,720	1,045	438
Biochemistry	1959–60	108	143	165
	1969–70	455	240	449

Source: U.S. Office of Education.

At the B.S. level, the attention and money devoted to chemical education have generally featured quality over quantity, with resulting benefits to both. Schools on the ACS-approved list increased from 287 in 1960 to 414 in 1970. Their graduates with majors in chemistry rose in the same interval from 3,065 (or 40% of the 1960 total) to 7,782 (or 78% of the 1970 total).[3]

The growth in graduate degrees, particularly Ph.D.'s, merits more concern. They are more expensive to obtain, commit students to a longer period of study, and (in contrast to the B.S. degree) generally imply a professional intent or commitment.

Data on Ph.D. production are compiled by three sources. They are gathered annually by CPT and published biennially in the Directory of Graduate Research, together with information on staff and research production by graduate departments. Similar figures are also gathered by the National Research Council and the U.S. Office of Education. Since the three sets are in good (but not exact) agreement, they are used interchangeably or in conjunction in this report.

Between 1961 and 1971, according to Table 2, Ph.D. production in chemistry roughly doubled as the number of departments offering the degree increased from 125 to 181. It is tempting to relate these two developments

---

3. Based on annual reports of the Committee on Professional Training of the American Chemical Society.

and attribute the growth in Ph.D.'s granted to this increase in number of programs, but data from the Directory of Graduate Research discourage such a conclusion. Schools that had not yet granted Ph.D.'s by 1959–61 (although some may have had programs under way) contributed only 15% to the 1969–71 totals.[4] In short, most of the growth occurred in programs already established, including programs among the largest in 1959–61. (The University of Illinois at Urbana and Iowa State University are conspicuous exceptions. Massachusetts Institute of Technology and Purdue increased Ph.D. enrollments by 38% and 70% in chemistry, respectively, during the decade.)

In addition to growth, significant redistributions occurred during the 1960's. Thus, the balance between private and public (state- or city-supported) universities tilted toward the latter (from 57% to 66%), as a pronounced broadening also occurred in the geographic base (see Figures 1 and 2). In 1960–61, 60% of the Ph.D. production was concentrated north of the Ohio and east of the Mississippi. In 1970–71, the output of this region decreased to 51% as the South and Southwest scored impressive gains.[5]

Table 2
Input-Output Flow of Students in Chemistry Ph.D. Programs

Academic Year	First-Year Enrollment	Total Enrollment	Ph.D. Awards	Departure Without Ph.D.
1961–62	—	9,000	1,114	1,162
1962–63	3,294	10,018	1,219	1,195
1963–64	3,574	11,178	1,271	1,309
1964–65	3,862	12,460	1,377	1,888
1965–66	3,939	13,134	1,533	2,453
1966–67	(3,850)	12,998	1,701	1,410
1967–68	3,751	13,638	1,751	1,502
1968–69	3,501	13,886	1,945	2,080
1969–70	3,542	13,402	2,195	1,582
1970–71	3,290	12,916	2,149	1,438
1971–72	2,994	12,323	—	—

Source: National Science Foundation for enrollments and the American Chemical Society's Committee on Professional Training for Ph.D.'s granted. Figure in parentheses is an estimate that was not provided by these sources. The last column is derived from preceding ones.

The proliferation and rapid expansion of graduate programs have posed a threat to standards, even though a deterioration of the quality of training cannot be proved. According to a recent survey of the American Council on Education,[6] the 38 schools rated "strong" in terms of chemistry faculty produced 48.8% of the Ph.D.'s in 1970–71, while the 20 rated "good" produced 13.9% and the 21 rated "acceptable plus" produced 11.7%. Altogether, these schools accounted for 74.4% of the Ph.D.'s. The remaining 25.6% were distributed among 102 schools. Clearly, schools with good reputations had the best success in attracting students. On the other hand, since the fraction of B.S.'s entering graduate programs increased during

4. Two-year averages (for academic years 1959–60 and 1960–61 and for 1969–70 and 1970–71) are used in this paragraph to smooth fluctuations.

5. Figures 1 and 2 provide much other interesting information that is not central to this report. Thus, annual Ph.D. production in chemistry is roughly one per 100,000 population; and although the ratios for larger states are close to this figure, the proportions for some small states are far above. A poor correlation between the location of Ph.D. production and the employment of chemists is also evident.

6. K. D. Roose and C. J. Anderson, A Rating of Graduate Programs, American Council on Education, Washington, 1971.

Geographic Distribution of Chemistry Ph.D. Production: 1960-61

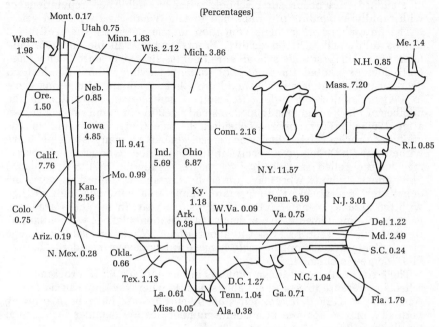

Figure 1

Geographic Distribution of Chemistry Ph.D. Production: 1970-71

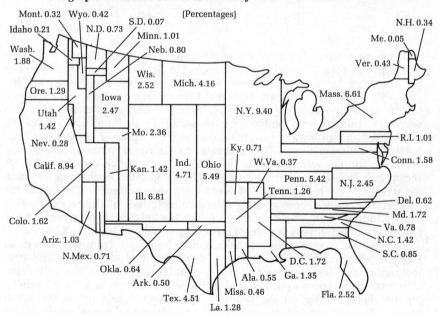

Figure 2

the 1960's and admission is selective, it may be inferred that some lowering of standards did occur—that schools without established reputations or with rapidly expanding programs had to take whatever they could get.

The flow chart of graduate education in chemistry shown in Table 2 throws additional light on quality. From figures available for 1961 and later years on graduate school enrollments, first-year enrollments, and Ph.D. degrees granted, the number of students leaving without the doctoral degree can be calculated. The rapid growth in graduate admissions came to a virtual halt by mid-decade despite a subsequent expansion in the number of Ph.D. programs. Indeed, a steady decline in admissions is shown after 1965–66. Furthermore, considerable selectivity is evident in the graduate process after admission, since almost as many students leave without the Ph.D. degree as receive it. (Some of those who leave, of course, may have enrolled specifically in a master's program in the first place.) Since most students who leave do so by the end of their first year, the departures in a given year are more comparable with new entries than with the number of new Ph.D.'s for the same year. In the three academic years beginning 1962–64, two fifths as many students left as were admitted. By 1968–70, the ratio had risen to about a half. Whatever else this change may indicate, it does not in itself suggest a decrease in standards for the Ph.D. degree in chemistry.

The large number of foreign students in our nation's Ph.D. programs also reflects a pressure to recruit suitable students wherever they can be found. According to NSF data, foreign student enrollments have been growing both as a percentage and in absolute number despite declines in the total for all students:

Year (Fall)	First-Year Students Number	First-Year Students Percent	Total in Ph.D. Programs Number	Total in Ph.D. Programs Percent
1967	592	15.7	1,923	14.1
1968	659	18.8	2,117	15.2
1969	754	21.3	2,455	18.3
1970	727	22.1	2,589	20.0

Although science is an international enterprise, the training of foreign students is a form of inadvertent foreign aid. It is an aid, at least, that is not intended by university administrations or by the agencies making grants in support of graduate research. Unless such students have unusual talents or clear opportunities for professional employment in their own countries, this training policy may be seriously dysfunctional. It would seem most inappropriate for an academic department to rely on a large foreign enrollment to justify a program that is not otherwise viable.

Heavy foreign enrollment is not, of course, unique to chemistry. A recent survey[7] shows that 24% of the Ph.D.'s granted in mathematics and the physical sciences in 1965–68 went to foreign nationals. The figure for engineering is even more striking. Since roughly half of the recent foreign Ph.D.'s indicated an intent to remain in the United States, they presumably contribute substantially to the present oversupply of engineers.

Perhaps the most significant deduction to be drawn from the pattern of Ph.D. expansion of the 1960's is that the combined aspirations and projections of the departments offering doctorates were much too optimistic. The total indicated output for the 1970's is far in excess of any realistic number based either on the supply of eligible students or on the availability of grant support for research. According to figures compiled for

7. *Mobility of Ph.D.'s*, National Academy of Sciences, Washington, 1971.

the Directory of Graduate Research, the full-time faculty in chemistry at Ph.D.-granting schools grew from 2,154 in 1961 to 4,023 in September 1971. If a graduate school were to aim at producing roughly one Ph.D. per faculty member per year (the figure reached by a few of the more active, highly regarded departments), this number of teachers implies a capacity—with no further growth in staff, facilities, or programs—of 4,000 Ph.D.'s per year.[8] The corresponding figure for 1970 for chemical engineering is 877; for biochemistry, 1,520; for pharmaceutical and medicinal chemistry, 224. From this viewpoint, the programs started since 1960, while contributing little to Ph.D. production in the 1960's, have obviously provided significant pressure for continued expansion of enrollments in the future.

Projections for the 1970's

Further increases in the annual granting of bachelor degrees in chemistry are indicated for the current decade. The production of B.S. degrees in chemistry amounted to an almost constant fraction of the 22-year-old group during the 1960's, and this relationship appears to be a long-term constant that could well persist. The 22-year-old population may, of course, be quite accurately estimated into the 1990's. It will rise some 25% in the 1970's but revert by 1990 to the 1970 figure. The statistics suggest a rise in new B.S. degrees in chemistry from 11,600 for 1970 to 14,500 for 1980.

These figures for new bachelor degrees hardly mean a solid increase in the number of chemists entering the profession. A large fraction of B.S.'s in chemistry normally elect careers in other fields. Furthermore, this fraction can adjust to changing circumstances on short notice. Even before graduation, students can switch between chemistry and medicine late in their college programs.

Enrollments in undergraduate chemistry courses are currently rising. A recent ACS survey[9] of 46 schools noted increases of 9% in first-year registrations and 14% in organic chemistry registrations for 1971 over 1970, despite a 2% drop in total enrollment of freshmen. This anomaly is explained by an increase in the number of students planning careers in medicine and other health-related fields. Since no comparable expansion is anticipated in the capacity of medical schools, which are already turning away qualified applicants, some of these students may be obliged to seek careers in chemistry. While such a development could check the decline in graduate school applications for chemistry, a return to the level of the mid-1960's is unlikely.

The possible perennial oversupply of holders of bachelor degrees in chemistry is easy to exaggerate. Besides providing professional chemists, new B.S.'s in chemistry proper provide relevant manpower for a number of cognate fields, such as biochemistry, many life sciences, and the material sciences. A real need also exists for graduates with some scientific training in nonscience areas of employment. In short, if enough B.S. chemists are produced for the nation's scientific enterprise (and the figures do indicate that enough will be), part of the excess could represent a wholesome infusion of scientifically trained graduates into the general labor force. Indeed, for a student having an aptitude and interest in science,

---

8. Probably few chairmen or other university administrators would consider the production of such a national total to be plausible or desirable. Each presumably aims at getting a larger slice of a pie that is not increasing. A recent report of CPT on "Doctoral Education in Chemistry: Facing the 1970's" (Chemical and Engineering News, Aug. 14, 1972, pp. 35–39) makes this observation to "underscore that the aspirations already implicit in the system have been completely unrealistic" (p. 35): "If all departments were to produce Ph.D.'s at the rate of the major 20 [nearly one Ph.D. per faculty member], the total output would be 6,000 per year!"

9. Chemical and Engineering News, Mar. 13, 1972, p. 22.

a college major in science may often be an asset whatever career he finally chooses. For society, too, it is desirable to have a population that is better informed in scientific matters as public decision-making becomes increasingly technical.

An excessive Ph.D. production is a more serious matter. As has already been stated, the Ph.D. degree represents a large investment of time and money and often an emotional commitment to a particular field. Recent projections of *total* Ph.D. production[10] indicate an increase of 50 to 125% between 1970 and 1980. If this prospect is realized, the social waste and personal disappointment would seem considerable.

The projections just cited, however, probably do not anticipate the true situation for chemistry. According to Table 2, first-year graduate enrollments have been declining for several years. This decline is confirmed by a recent ACS academic survey report,[11] which shows the following actual and projected figures for 157 schools:

Year (Fall)	First-Year Graduate Enrollments
1968	2,385
1969	2,474
1970	2,234
1971	2,041
1972	2,246
1973	2,468

The same schools that reported recent declines in *actual* numbers have obviously anticipated substantial increases in *projected* first-year enrollments in 1972–73. The same tendency is indicated by figures for chemical engineering. This anomaly reflects the sort of wishful thinking that inflated the projections of the 1960's. Actual numbers will more likely continue to decline or level during the next few years. The only development that might reverse this trend would be a marked increase in the amount of student support available. Such a development might either lure more B.S.'s into graduate school at higher stipends or attract more foreign nationals. At present, however, this is a most improbable prospect.

This decline in graduate school entrants should be reflected in diminished Ph.D. production, at least from 1973. As a matter of fact, Ph.D.'s granted in 1971 already showed a slight decline from 1970. If admissions stabilize slightly below the present level, Ph.D. awards in chemistry during the later 1970's should be in the neighborhood of 1,500 per year.[12]

It would be most desirable for any shrinkage in Ph.D. output of chemists or chemical engineers to occur toward the bottom rather than the top of the distributions of student ability and program quality. Biochemistry does not appear for the moment to be faced with this pruning problem. Indeed, its continued growth (as projected in the ACS academic survey) is indicated by strong new federal support for health-related sciences.

Future Demand for Graduates

The most recent projections of the U.S. Bureau of Labor Statistics (BLS)

10. D. Wolfle and C. V. Kidd, "The Future Market for Ph.D.'s," *Science*, Aug. 27, 1971, pp. 784–93.

11. *ACS Academic Survey Report: Spring 1971*, Sept. 3, 1971, Table 1.

12. The recent CPT report (*Chemical and Engineering News*, Aug. 14, 1972, p. 36) states, "Anecdotal comments on next year's admission applications and acceptances indicate that the drop-off is continuing and that first-year enrollments for 1972–73 will fall well below 3,000 for the first time since 1960–61. Thus, in a six-year period we will have had a 30% decrease in graduate admissions, which should lead to a drop in Ph.D.'s produced in 1976 to somewhat less than 1,500."

indicate an employment of 200,000 chemists by 1980 compared with 130,000 in 1968. This expansion, according to BLS, implies an average of 12,800 openings per year, 6,000 for growth and 6,800 for replacement. No breakdown by chemical specialty or level is given. Since many eligibles do not enter chemistry, BLS also estimates that 17,000 B.S. graduates a year would be needed to fill these openings, a 65% increase over 1968 production. Experience during 1968–71 has shown little correlation with these projections.

Further discussion of the BLS outlook figures is deferred to the next part of this study, which will also consider the industrial and functional distributions of employed chemists and chemical engineers. Since private nonacademic industry is the largest employer of such professional personnel, ACS has made a special survey of job prospects in manufacturing for 1975 and 1980. This survey's results are presented in Chapter 29.

The employment outlook for chemistry teachers, who also comprise a large segment of the profession, is especially germane to this chapter and accordingly will receive some notice here. The ACS academic survey already cited has projected a need for 1,500 new faculty members, 800 for growth and 700 for replacements, in the nation's 1,102 departments in the period 1970–73. The annual rate of openings is 500 per year. Although this rate is less than half that projected in the 1960's, it appears too high for the 1970's. First of all, replacements not only make up for deaths and retirements but also include transfers between schools. The concept thus involves some duplication of jobs. Second, the tightening of academic budgets and a projected peaking of enrollments at the end of the 1970's with a subsequent drop could well encourage many institutions to allow student-faculty ratios to rise. Thus, a rate of 500 per year seems a generous upper limit for the 1970's, and a much smaller fraction of new Ph.D.'s may be able to find permanent academic positions than did so in the 1960's.[13]

The figures just cited refer to the nation's chemistry faculty exclusive of teachers in chemical engineering and biochemistry. The ACS academic survey shows a net growth of faculty of 10% for chemical engineering in 75 departments between 1970 and 1973 (the absolute numbers are 709 and 779). For biochemistry, the percentage gain is 17 (the absolute numbers are 854 and 1,002 for the two years). For comparison, we note that the survey reported that the chemistry faculty in 1970 numbered 5,150 in 586 reporting departments and 8,821 in all the 1,102 departments of the nation. The projections also fail to take cognizance of teaching prospects in two-year colleges and secondary schools. Larger numbers of chemists with advanced degrees may be expected to be drawn to these institutions in the future.[14]

Role of Support

The training of chemists, particularly at the graduate level, requires sophisticated and expensive instruments. It is conducted in conjunction

13. The CPT report (ibid., p. 36) observes, "In 1971, the total full-time chemistry faculty at Ph.D.-granting schools numbered 3,288. Assuming a steady status of more or less the present structure and an average active lifetime of a graduate faculty member of 28 years, the system should provide 132 new positions a year. Although this is only 6% of our present Ph.D. production, even this estimate is almost certainly too large [in view of the age distribution of faculties and the contraction of programs]."

14. According to the CPT report (ibid., p. 37), "At four-year and junior colleges expansion is not yet at an end, but the rate is decreasing and so is the need for additional staff. Academic employment of any variety will probably be open to no more than 15 to 20% of the Ph.D.'s produced during the next 10 years, depending upon how far the production rate drops, and will be chiefly in the two- and four-year colleges, with need for faculty who can be effective and inspiring teachers and remain so without graduate students and [with] little in the way of federal research grants."

with basic research, but the time may be right for a partial uncoupling of the two activities. The Government and the general public share the view of scientists that basic research represents a national asset and constitutes an essential area of federal support. On the other hand, our graduate schools are overexpanded in chemistry, and limited student interest and the nonexuberant job market suggest that the actual student population will decline or level off in the years ahead. Accordingly, a considerable redistribution of public funds for basic research seems to be in order.

Three possible redirections merit attention. First, graduate schools could depend less on students and more on technicians, postdoctoral fellows, and the faculty itself. Second, since only a relatively small fraction of a total university faculty works with Ph.D. candidates, more research support could go to institutions that do not grant the Ph.D. degree. If research quality could be assured, this shift in site might also sustain the quality of education in the schools where the majority of the undergraduates are trained. Indeed, this prospect could represent a viable alternative for schools struggling to retain marginal Ph.D. programs. Third, more support for basic research could go to nonacademic institutions (e.g., research institutes). Such a shift would encourage schools to devote more attention to the preparation of chemists for careers involving the practice of the art, as in industry.

A recent change in the pattern of government research support—the large reduction in fellowships and traineeships compared to other forms of support—has produced difficulties for the sound operation of graduate schools, although it has probably helped to curb overexpansion. It has placed heavy financial and administrative burdens on the schools having the best scientific reputations and attracting the better students.

Potentially damaging to undergraduate programs is the displacement of student support toward teaching and research assistantships (in the absence of fellowships and traineeships). In the interest of good undergraduate education, a teaching assistant should be used when he is most effective—when he has already had some experience and also has adequate time to perform his duties. Accordingly, it would seem desirable to give the assistant a manageable total teaching load that is spread more evenly over his graduate career. Since he cannot be properly assigned as a research assistant until he is actually doing research, the present situation calls for a heavy teaching load in his first year—when he is least competent and most harassed. Before the cutback in fellowship and traineeship funds, a number of schools were working toward a better distribution of the load with satisfactory results. Negation of this progress must be regretted, and satisfactory solutions to the problem still need to be developed.

A sketch of the human resources of chemistry should not be confined to statistics or merely to an account of the size of the eligible or potential manpower supply, the annual changes attributable to formal education and other factors, and the actual volume and location of employment. After chemistry's manpower is assumed to have been trained and largely put to work, attention must also be given to such matters as the maintenance of the individual's technical and economic functionality. This dimension of postschool adjustment for career satisfaction and productivity, which is essentially literary rather than statistical, is the concern of the present chapter. The discussion emphasizes ways of keeping the seasoned chemist [1] abreast of new scientific developments and otherwise prepared for coping with the vicissitudes of working life. Success in these regards would keep him employable—not only more continually but also in closer accord with his professional and other aspirations.

Since chemistry is a swiftly changing science, the individual practitioner —in industry, the academic world, government, and elsewhere—is in constant jeopardy of technical obsolescence. This threat is not always appreciated by him or by others who are close to him. Its reality may become acutely clear only at critical junctures in his career, in the life of his organization, or in the development of the economy around him. When a job change is required or contemplated, the value of one's skills may be well below expectations or beliefs, and it is then too late for remedy.

Chemists are vulnerable to obsolescence for many reasons, which may also be stated in different manners. One kind of language directs attention to the individual himself. It often stresses the decline or absence of motivation, but a fuller account of reality, even in terms of the individual, has many other facets. For example, a person necessarily pursues his career in the midst of other claims on his time and loyalties. It is normal for him to have, say, family obligations, civic and social interests, and recreational needs. Furthermore, his motivation could even have been impaired—and his future accordingly stunted—while at school. Thus, in retrospect, blame might be placed by a person on external circumstances or agents, such as a lack of challenge from the curriculum, indifference or insufficient interest of a research mentor, or poor career counseling. Such diagnoses may well have a kernel or more of truth, but they can also be distorted and unbalanced, and they surely are incomplete.

The problem, however, is not semantic. It cannot be resolved by choosing any particular quasi-causal language for describing it. The constructive implication of the foregoing is this: *The individual, the school system, the employer, and the professional society have obvious stakes in counteracting technical obsolescence, and they also have preventive or corrective roles to play.* The rest of this chapter elaborates on this statement.

Individual Responsibility

Implicit in the very notion of professionalism is a person's own concern

---

1. In this chapter, "chemist" is used in a comprehensive sense. Specifically, it includes the chemical engineer as well as the biochemist. The terms "industry" and "industrial" are also used broadly to characterize all nonacademic activity. Accordingly, they include government and other nonprofit activity as well as private, profit-oriented enterprise.

for continuing competence and even expertness in a chosen field. The individual chemist is ultimately responsible for his own self-renewal. His prior commitment to a field and his interest in career progression would appear to be sufficient reasons for devoting additional time and effort to current and future functional relevance in the face of competing claims. The professional ideal is hardly a passive willingness to receive or accept new knowledge. Rather, it is active and continual search and involvement. Once stated, such observations may seem gratuitous, but however trite they have a truth that cannot be lightly dismissed or ignored.

The Academic Contribution

Antidotes to obsolescence should, of course, be designed into academic programs in the first place. Perhaps flexibility for career adjustment and a capacity for maneuver cannot be taught effectively as an attitude of mind, and to the extent that obsolescence is a phenomenon of aging it may not be fundamentally reversible. Nevertheless, the student could be prepared better to protect his employability or to seize new mid-career opportunities through the depth, breadth, and variety of his original training. The quality of a curriculum, it should be noted, could even be judged on these grounds. Indeed, changes in social values and styles seem to dictate that more should be demanded of the educational system to bolster future professional adaptability.[2]

There is much to be said in favor of reintroducing a "core-type curriculum" into the first-year graduate programs. Not only would the student be enabled to compensate for undergraduate deficiencies but also to choose more meaningfully among careers in academic research, industrial research (or other industrial activity), and teaching. A core curriculum might include quantum mechanics, industrial chemistry (chemical processes and economics), and advanced topics (such as biochemistry, synthetic chemistry, physical methods of structure determination and analysis, and chemical dynamics).

In principle, the dissertation is an exercise that should help a student develop—or at least to demonstrate—his capacity for continuing self-education. It often fails, however, to elicit or reinforce the qualities that are expected to characterize and sustain the professional, such as adventurousness, inventiveness, resourcefulness, and independence. The next chapter considers at some length the adequacy and improvability of Ph.D. recipients in chemistry from the standpoint of industrial requirements.

A passing reference was made earlier to career counseling. Some students of uncertain motivation would surely benefit from early sound advice and guidance regarding choice of major or area of future employment. Too many counselors, unfortunately, may themselves be unaware of the needs of industry for chemists or the needs in activities other than research.[3] Employers could be contributing to this ignorance by failing to communicate their requirements to the universities in an effective manner.

---

2. Some remarks on the preparation of "professional scientists . . . [for] a lifetime of self-education, through reading, discussion, attendance at scientific meetings, and independent thought" are included in a new report of the American Chemical Society's Committee on Professional Training. The report notes the contribution made by "good" graduate programs through "seminars, cumulative examinations (based upon general reading rather than specific courses), problem-solving sessions, and discussions among research groups." It acknowledges the difficulties, however, of developing appropriate attitudes and of "effective screening." (See Chemical and Engineering News, Aug. 14, 1972, p. 38.)

3. The report of the Committee on Professional Training (ibid., p. 38) observes that a "successful program" of graduate study "includes a good deal more than course work and thesis research. Much of the success stems from student guidance policies, besides the admissions criteria."

Here is a matter that employers of chemists might well take up with secondary and higher schools. The American Chemical Society could assist on the national level, as it is presumably already doing by mentioning the need in this very report.

Faculty members, too, are subject to the danger of technical obsolescence. As in industry, the continuing education of teachers in the post–secondary schools ought to be recognized as an eligible operating cost. The threat is particularly serious for faculty members in the two-year and four-year colleges, which typically offer far less professional stimulation than the research-oriented universities. The vitality and competence of these teachers might be supported in several ways. One is encouragement to do research, independently or in collaboration with each other or with students and either at the teacher's own institution or elsewhere during the summer. Another is to extend the practice of many junior and liberal arts colleges of offering periodic leaves for scholarly work.

Universities and colleges often seek through evening classes and daytime extension studies to accommodate persons intent on modernizing or diversifying their knowledge of chemistry or on acquiring higher degrees. The formalization of additional training is good insofar as it provides a special setting conducive to active participation in a new learning experience. Homework, recitations, papers, quizzes, and more searching examination engage students more fully than do occasional or informal lectures without provision for evaluating benefits received in motivation or capability.

Too frequently, however, the school courses available for working chemists are inadequate and cast in a traditional mold that neglects the background, sophistication, special needs, and psychology of persons returning to an academic environment after a long absence. The teachers, furthermore, may be insufficiently sensitive to the challenge of enriching, refurbishing, or extending the knowledge base of persons so different from the regular student body. Room obviously exists for the design of new curricula that are not necessarily oriented toward degree attainment and of new and more appropriate modes of delivery for special educational clienteles. Room also exists for the use of qualified industrial chemists as instructors, as visiting professors for special courses, and as departmental advisers. A closer collaboration between university faculties and industrial chemists in such endeavors should be welcomed.

Employer Roles

The employer has adequate economic reason to sustain his chemistry staff's professional motivation, intellectual alertness, and creative potential. Profitability depends at least as much on employee ability and interest as on any other factor. Indeed, in the research setting the human factor is decisive. In both plant and laboratory, investment in people and in their work environment merits the same serious attention that is normally given to expenditure for equipment. Pursuit of economic advantage along these lines tends to be reinforced by favorable employee response.

Mention has already been made of the employer's relations with the schools. Liaison should be maintained for proper academic address to the special requirements of industry and for the translation of these requirements into suitable courses, curricula, and faculty. Beyond engaging in this kind of activity, an employer might also encourage staff attendance through a liberal policy of time allowance. Furthermore, companies that conduct in-house seminars or the equivalent can (and often do) benefit from the invitation of university guest speakers to familiarize the staff with new directions in fundamental chemistry.

Although the seminar provides a useful format for the joint study of a timely subject, it is only one of the many media used by companies for continuing in-house and in-hours education. A minimum of structuring and scheduling may seem desirable, but formal arrangements and regularity also have their place. Spontaneous group activity, depending on one prime mover or a small nucleus, is frequently encouraged (or tolerated) in the company setting. The collaborating chemists agree to investigate a topic of mutual interest, to divide responsibilities, and to meet for discussions and reports on pertinent literature and laboratory results. Such activity provides a social pressure for toning up. It contributes to individual development by demanding active participation with one's colleagues, including effective organization, presentation, and exchange of ideas and information. On the other hand, it could appeal most to the relatively few chemists who are bound to retain professional vigor and to find one outlet if not another. It could also become digressive for the group while serving as a vehicle for personal ambition.

Structured programs are commanding more attention in industry as preferred media for engaging a wider circle of employees in continuing education. The in-house, in-hours courses are taught both by qualified staff members and by selected professors under contract. The chemists enroll voluntarily, but they are stimulated to do so by peer pressures and sheer imitation as well as by the knowledge of management sponsorship. Indeed, active support by management carries a hint that employee participation may be desirable for status or progress within the company.

As various schools (such as the University of Minnesota and Stanford University) have demonstrated, closed-circuit television may be effectively enlisted in continuing education of industry personnel. Even graduate and advanced undergraduate courses may be attended by students sitting in specially-equipped studio classrooms in their places of employment. These remotely-located students have the same opportunities as their on-campus counterparts to ask questions and make comments. Evening as well as daytime courses may be provided, and participating companies can also arrange for remote staff monitoring of campus seminars and symposia.

In the larger companies having heavy commitments to research, opportunity often exists to counteract obsolescence by a careful blending of team membership and by a deliberate rotation of assignments. New graduates, possessing more up-to-date knowledge, more enthusiasm, and more energy, can be used at least in principle in constructive and complementary ways with older and more experienced chemists, rather than be thoughtlessly pitted against them in a destructive competition.

Success in the complementary use of the old guard and the new breed would surely amount to a kind of continuing education for the former and to a useful indoctrination for the latter. By periodic reassignment, furthermore, the settling and aging processes that afflict the scientist could probably be delayed in some degree. The intellectual toxins that thrive on the tedium and routine of prolonged specialization can be diluted, even if they are not destroyed, by exposure of the older chemist to new but still manageable work situations.

The Professional Society

In an inventory of contemporary provisions for continuing education, it is always easy to overlook the manifold traditional services performed by professional societies. The American Chemical Society, for example, has a large and diversified publications program that helps to keep its members and the rest of the scientific community apprised of the state of the art, of ongoing work, and of recent achievements at the research frontiers. Its national and regional meetings contribute to the same ends and also offer occasions for sounding out employment opportunities and for the extension and renewal of professional and social contacts. Its many local chapters offer similar benefits closer to home.

National, regional, and local meetings of the American Chemical Society also provide loci and foci for the development and presentation of short courses, film courses, and audio courses. Tutorial lectures by recognized experts have thus become available in increasing variety for wide audiences. Being packaged, they can aid continuing education with greater pertinence to felt needs and with due regard to cost and convenience.

This chapter is based on company responses to an ACS survey that was designed and undertaken for this study with two objectives in view. One objective was to obtain evaluations of the training, attitudes, and adaptability of Ph.D. and B.S. degree holders in chemistry from the standpoint of contemporary industry needs.[1] The second objective was to obtain suggestions for curricular and other educational adjustments that might improve industry's ability to deal with the technical, competitive, and regulatory challenges of the 1970's and 1980's.

The underlying questionnaires were addressed to research managers, directors, and vice presidents of 96 companies. As will be seen from a table that summarizes the replies concerning Ph.D.'s, the response rate for the survey as a whole was in the neighborhood of two thirds.

The rationale of an ACS inquiry looking toward a closer correlation of academic instruction with industry needs is easy to supply. Most of the jobs available to chemists and chemical engineers are located in private companies, and changed circumstances indicate a much slower expansion of such professional employment during the 1970's than was experienced in the 1950's and 1960's. This prospect, already mentioned in Chapter 22, will be discussed at greater length in the final chapter of this study. For Ph.D.'s in particular, the job market will surely become more selective as the demand for them remains moderate not only in industrial research but also in college and university teaching. A better matching of the qualifications of new graduates and the requirements of employers is obviously desirable for their mutual benefit—and in the interest, too, of society in general.[2]

The ACS letter accompanying the survey questionnaires noted that:

- Current information was being sought on "your impression of the average new B.S. and Ph.D."
- The replies to all questions were expected to help the "educators asking how they can better train chemists for positions in industry."

The responses to particular questions and the ample commentary did assist ACS in the formulation of recommendations for this study. Quotations from the respondents' extended remarks are supplied, since these should also be of benefit to other readers concerned with the achievement of a closer rapport between universities and industry.

Five General Findings

The ACS academic-industry committee that designed and administered the survey concluded that "both the B.S. and Ph.D. are lacking in some respects in their preparation for employment in the chemical industry." It may be assumed that the respondents, by virtue of their company roles, focused principally on "preparation for the realities of industrial research," which is the preferred employment province of the Ph.D. Many of the

---

1. The survey did not explicitly include persons with degrees in chemical engineering.

2. A buyer's market for Ph.D.'s is seen for the 1970's by the American Chemical Society's Committee on Professional Training. The "continued vitality of the profession" requires "some attempts at greater breadth, some innovation, and some experimentation with new emphases and approaches in graduate education" (*Chemical and Engineering News*, Aug. 14, 1972, p. 37). Above all, "the realities of supply and demand during the 1970's" underscore a need to maintain and upgrade the standards for Ph.D. education (*ibid.*, p. 39).

specific comments, however, also seem applicable to academic preparation for key nonresearch tasks (such as production and quality control). Such broader relevance is desirable since, as Part III of this volume stresses, something like half of the jobs for chemists and chemical engineers in private businesses lie outside of research.[3]

The committee discerns five outstanding themes in the survey responses. These themes are interrelated but are also sufficiently well defined to allow separate notice.

First, the committee sees in the responses concerning Ph.D.'s a need for greater initiative and effort on the job and for greater attention to continuing self-improvement. In this connection, however, the committee observes that it is not enough for the student to "learn to give more of himself." A complementary "need for modification of the structure of the Ph.D. research program" is also recognized.

The second theme noted by the committee is industry criticism of an apparently excessive emphasis on theory and specialization. Granting the high quality of the technical training of current graduates, industry spokesmen nevertheless wondered if a pursuit of elegance in problem selection, analysis, and solutions has not subordinated such values as clarity, directness, and practicality. Respondents reacted even more intensely, according to the committee, against the tendency toward overspecialization:

"The message, as we interpret it, is that our respondents are urging, indeed pleading: 'Send us *chemists,* not synthetic organic chemists, spectroscopists, theoretical physical chemists, etc., but *chemists.*' "

The third industry complaint noted by the committee is the chemistry graduate's apparent lack of "many basic laboratory skills that he really should have." This complaint is "perhaps related to the second," and it "may be directed more at the B.S. than the Ph.D., but possibly both." In any case, the catalogue of deficiencies is wide-ranging. Thus, mention is made by respondents of a need for such minor skills as the "correct use of a burette" and of such profounder ones as the "proper design of experiments."[4]

The fourth critical comment of many respondents relates to insufficient awareness or respect for economic constraints, commercial applicability, and social needs or implications. Chemistry graduates, the committee avers, "have to be taught, and taught well, that 'there is no such thing as a free lunch.' "

The fifth generalization acknowledges an "amazing lack of competence in communication skills." English grammar, the committee observes in interpreting the survey replies, "appears to be far more important than training in a foreign language." The educators on the committee agree on the need for better linguistic command, citing the daily evidence that they encounter "in examination papers, thesis drafts, and indeed in articles and proposals we referee."

---

3. CPT notes (*ibid.,* p. 37) that, "with an increasingly large fraction of the Ph.D.'s once again destined to find employment in industry, better rapport and understanding must be established between the chemical industry and the centers of graduate education." It suggests an increase in industry consultantships, commends experimentation with postdoctoral fellowships in industry laboratories, and proposes that industry experience be given "greater weight" in faculty recruitment (as well as greater encouragement in the use of sabbatical leaves).

4. It should be recognized that, as the scope and the literature of chemistry expanded by orders of magnitude during the past two decades, an emphasis on theoretical aspects (i.e., on the principles underlying and unifying the otherwise unmanageably large and disparate bodies of fact) appeared to afford the only reasonable way for communicating a balanced view of the dynamic science within an educational period of four to eight years. While this approach yielded a proper conception of the structure, range, and power of chemistry, it also tended to neglect preparation for nonacademic employment and the experimental side of chemical training in general.

Table 3 summarizes the responses to a dozen survey questions concerning Ph.D.'s. These responses provide some of the background for the five generalizations. They have also contributed to the committee's development of specific educational recommendations for inclusion in this report. The answers to the dozen queries are reviewed in what follows in the same order as the five committee generalizations.

Table 3

**Summary of Responses to ACS Industry Survey
Concerning Ph.D. Chemists**

Question	Companies	Percent Responding Yes	No
1. Is he [Ph.D.] properly trained?	63	75	25
2. Is he sufficiently strong in technical fundamentals for assignment to any work area of industrial chemistry?	64	66	34
3. Does he willingly accept a suggestion to "re-tread" himself in mechanical areas in which he is not formally trained?	61	56	44
4. Does he continuously broaden his area of competence *on his own* and *on his own* time?	59	64	36
5. Is he willing to accept early reassignment to a new project?	64	64	36
6. Does he consider industrial chemistry and research second rate?	60	39	61
7. Is he too weak in allied science to adequately treat multidiscipline problems such as environmental conservation?	62	43	57
8. Is he dedicated to working at chemistry?	62	82	18
9. Are his interests too narrow (inflexible)?	61	57	43
10. Is he unadventurous?	59	48	52
11. Is he weak in pertinent nonscience areas, for example, economic analysis, humanities, personnel relations, and communication?	61	84	16
		More Fewer	O.K.
12. Did he need more courses in his graduate training or fewer?	44	75 18	7

Source: American Chemical Society survey of research managers, directors, and vice presidents of 96 companies conducted in the fall of 1971.

Half of the questions listed in the table elicited information for the first generalization (Questions 3, 4, 5, 8, 9, and 10). Only in the case of Question 8 is the consensus reassuringly large and favorable: Almost five sixths of the respondents found the Ph.D. "dedicated to working at chemistry." Fewer than two thirds, on the other hand, saw him inclined to broaden his area of competence on his own outside of work hours (Question 4). The same proportion reported him "willing to accept early reassignment to a new project" (Question 5). A smaller majority (under three fifths) regarded him as amenable to suggestion for improvement "in technical areas in which he is not formally trained" (Question 3). The same percentage of industry respondents declared, "His interests are too narrow (inflexible)" (Question 9). In reply to the query, "Is he unadventurous?" (Question 10), almost half of the respondents said, "Yes."

The answers to Questions 1, 2, and 7 support the second generalization. An unequivocal majority (three quarters of the industry respondents) considered the Ph.D. to be "properly trained" (Question 1). A smaller ratio (two thirds) found him sufficiently grounded in "technical fundamentals" for work in any area of industrial chemistry (Question 2). Slightly more than two fifths of the respondents regarded the Ph.D.'s preparation in "allied science" adequate for address to "multidisciplinary problems such as environmental conservation" (Question 7).

At least one of the questions (No. 12) has some relation to the third generalization, which is better grounded on the comments provided by the respondents (to be discussed later). Three quarters of the replies saw a need for "more courses" in the Ph.D.'s "graduate training," and fewer than one fifth preferred "fewer."

A decisively large proportion of the respondents (five sixths) replied "Yes" to a question (No. 11) pertinent to the fourth and fifth generalizations. This query concerned a deficiency of education in certain disciplines outside of natural science—in "economic analysis, humanities, personnel relations, and communication."

Question 6, assigned to none of the generalizations in the preceding review, is important enough to merit separate notice. Although the preponderant majority of industry respondents considered the "average new" Ph.D. to be committed to chemistry as a field, two fifths also sensed a disdain of "industrial chemistry and research" as "second rate." If this condescending attitude is really so prevalent, it needs correction for at least the practical reason that private industrial companies already are the dominant employers of chemists (and chemical engineers) and are likely to remain so. The matter is discussed again in what follows.

Company Comments on Ph.D.'s

Excerpts from comments made in reply to the individual questions add dimensionality to the statistical summary provided in Table 3. Not all questions inspired comment. The order already established around the five generalizations is retained for the presentation of these excerpts.

In answering Question 3, which is linked to the first generalization, the spokesman for a prestigious industrial laboratory asserts that students are "all too often brought up in the images of their professors," who "are largely ignorant of professional requirements" outside the university. Accordingly, "lack of flexibility and adaptability" is a conspicuous feature of education at the Ph.D. level. The same commentator offers an equally crisp observation in reply to Question 4: "This company has promising places only for people who grow in merit through their own efforts."

A pharmaceutical respondent to Question 4 emphasizes selectivity in recruiting for its "multidisciplinary research environment." Good experience is reported with "new staff members," who arrive knowing "that they will need to broaden their areas of competence." Another drug industry respondent adds that his company subsidizes "with tuition and books" the further education of staff members at local universities.

A major pharmaceutical company reports, not surprisingly, that its positive reply to Question 8 is correct "on the average" but must be tempered if account is taken of new interviewees. Professional "dedication" now must share its old primacy with the new concerns for the "relevance" of a research project and for "social values and the needs of society."

As for Question 9, which is also linked to the first generalization, the same company finds "more often than not" that a Ph.D.'s scope is "narrow." Its one-word answer is "No," however, since "some flexibility" among most Ph.D.'s is observed.

Curiously, this pharmaceutical company said "Yes" in reply to Question

10, if "unadventurous" is supposed to signify a lack of imagination or a conservative approach to research. A more cautious, and probably weightier, reply is given to this query by the respondent for a leading research laboratory:[5]
"Whereas chemists may not be unadventurous, there is some tendency to stick with safe problems. It is our observation that chemists in industry often require stimulation to be willing to exercise their imaginations. This reluctance to gamble is often pronounced in well-trained, highly intelligent chemists."

Pharmaceutical companies volunteered comments on Questions 1 and 7, which are associated with the second generalization. With regard to Question 1, one respondent notes that any Ph.D. normally requires "a couple of years of experience in order to become fully productive." Another respondent regards Ph.D.'s as generally not "properly trained." He cites overspecialization in narrow areas and insufficient orientation by the schools in the three basic industry needs of "problem solving, independence, and flexibility." A third respondent places weight on a résumé of grades for the B.S. and on the opinion of the Ph.D.'s thesis supervisor.

In reply to Question 7, one of the two pharmaceutical companies proffered an indistinct "Yes"; the other, a "clear" affirmative. The latter respondent wants a Ph.D. chemist to become a capable member of a multidisciplinary team, not a mere "interdisciplinarian" who compromises his "chemical expertise" in the course of this adaptation. The new Ph.D. can "rarely" fill the role without acquiring "considerable knowledge in disciplines other than chemistry." The challenge is not deemed insuperable, but it "requires much effort on his part."

With respect to Question 12, which belongs to the third generalization, two companies comment. A leading laboratory wants "more quantitative, fundamental problem-solving courses" rather than "survey courses," which "do not broaden" but "dilute." The respondent for a pharmaceutical producer considers the range of chemistry course offerings to be generally "satisfactory" but would like additional ones in "allied disciplines" provided that these serve to widen perspectives rather than to distract attention from the core graduate program.

Only two comments, both from pharmaceutical companies, are available for Question 11, which is associated with the fourth and fifth generalizations. One company says "definitely yes," locating the weakness of Ph.D.'s "especially in personnel relations and communications." The other, less fashionable, discerns no important lacks in the nonscience areas. The committee was influenced, however, by remarks to be cited later.

Returning to Question 6, we find a few volunteered comments on the Ph.D. chemist's opinion of industrial chemistry. A research laboratory spokesman refers again to a "condescending attitude" that "results directly from indoctrination during academic training." A pharmaceutical respondent also blames the schools ("few professors know, or care, much about industrial research") but finds that "most Ph.D. chemists change their minds after working in our laboratories for a short while." A third com-

---

5. CPT provides a counterpart comment on unadventurousness and on the preference for "safe problems" in the schools. It observes that theses often satisfy the requirements of a "research adviser's reputation, promotion, and ability to attract federal grants" rather than the student's need for broad training. Besides, "there is an unfortunate tendency for academic research to follow fads, so that most students find themselves working in relatively safe and popular fields close to the center of academic training, and often with little relation to other areas of scientific knowledge." CPT sees a need for broadening chemical training while it also acknowledges the "obvious difficulties involved in combining speculative and exploratory projects with some reasonable prospect of a finished piece of work within a reasonable time." (See *Chemical and Engineering News*, Aug. 14, 1972, p. 37.)

pany verifies this bias of graduates, attributing it to "academic research advisers," while a fourth company is noncommittal, emphasizing personality differences.

At this point, we should take note of the ACS committee's confirmatory judgment, which is based on additional evidence that remains to be presented:

"There is no doubt that the structure of many graduate programs in chemistry has been oriented, unconsciously if not consciously, toward careers in academic research. Since at most a few hundred faculty and postdoctoral positions in chemistry become vacant each year, the bulk of the annual 2,000 or more chemistry doctors are being unrealistically trained. For these, the breadth of training must be expanded, without sacrifice of depth."

This judgment led to one of the recommendations stated at the beginning of this report.

Additional Information on Ph.D.'s

The committee's five generalizations and the educational recommendations presented earlier are also founded, as was already noted, on responses to queries that are not mentioned in Table 3. Four additional questions enlarge the information base regarding Ph.D.'s. The content of these questions and the replies that they occasioned are considered below.

Question 13 sought to ascertain the length of time spent by the Ph.D. chemist "at the bench" before his shift into other company activities, such as "management, sales, or customer service." The average time reported by 44 responding companies that followed such a practice or that would venture a numerical estimate was a bit below 7.5 years. This is a rather conservative figure, however, in view of the absence of such rotation in many companies and its selective application. These remarks are illustrative:

• Rarely reassign.
• Those who do enter other phase of chemistry leave bench work after five to 10 years.
• Depends on qualifications. Too few for good "average."
• Many never get out of the lab. Ph.D. flexibility is much less than B.S., often because of his salary.
• Usually as long as he stays with the company, but about 20% move into sales or management after five to 10 years "at the bench."
• Five years is a decision point. If he doesn't transfer by then, he'll probably stay in research.
• Ten years, when we do. Very rarely do we take a chemist away from the lab.

Question 14 related to balance in the Ph.D.'s technical education: "Has he been too highly trained in one area or another? If so, what are these?" Only 22 respondents could categorically answer "No," and two others expressed misgivings about the training of half or more of their Ph.D.'s in chemistry. A few that were dissatisfied noted an overemphasis on the area of the Ph.D.'s major and even an overemphasis on "chemistry" as a field! Others were more explicit, condemning narrowness, overspecialization, and a bias toward theory—limitations reflected in the ACS committee's second generalization.

A laboratory identified with petroleum products communicated this psychological insight:

"He is usually highly trained in some narrow field and is very reluctant to leave its relative safety. He has been trained to be an 'academic specialist' and often can't cope with the pressure of changing his work. When asked to do so, he may actually fail at his job."

Several other comments on Question 14 merit partial quotation. The following clearly contributed to at least the second and third of the ACS committee's generalizations:

• Lab people are much *too inefficient*. They need to develop a better understanding of the problem prior to starting experimental work. It should be possible to design such a course, with practical examples.
• Some individuals are prone to retain thesis interest and find it difficult to adjust to industrial research assignments.
• Tendency toward too much specialization in esoteric subjects.
• [Too much educational emphasis on] theoretical chemistry, such as quantum chemistry and reaction mechanisms.
• [Overemphasis of] mathematics of mass, heat, and momentum transfer.
• Too much theory makes him think everything is *known*.
• He may be too highly trained in following directions. Furthermore, some graduate thesis research is so highly specialized and narrowly focused that . . . this expertise is difficult to utilize in industry. These instances suggest a need for career counseling in the graduate schools.
• We assume a Ph.D. means a "creative person." They are trained to do what their adviser dictates. Many entering industry look for an adviser or total director.
• Most young people with B.S. or Ph.D. degrees need training in an in-house program to cover management and other pertinent areas.
• Would be willing to accept less chemical training in undergraduate or graduate chemistry to permit elementary aspects of chemical engineering to be emphasized.

Question 15 also elicited responses that are reflected in the ACS committee's generalizations, especially the fourth and fifth. It read, "Is his training too superficial in one area or another? If so, what are these?" Deep wells of frustration were tapped, as these few short comments show:

• Insufficient training in synthesis.
• Patents, economics, writing [are among areas of deficiency].
• Chemical engineering and chemical economics courses are needed to provide the background for industrial research.
• Management, statistics, experimental techniques, problem solving, oral and written communication, planning, marketing, economic analysis [are underemphasized areas].
• He needs wider exposure to other areas of science. "Specialties" wax and wane so rapidly today that, *even in a university*, he can rapidly become obsolete.
• Lacks depth—in terms of curiosity about fields adjacent to those in which he is working.
• [Gaps in] peripheral areas—English specifically and communication generally.
• Organic chemists are superficial in colloid and surface chemistry, economics, experimental design, chemical business. Too little consideration is given to practical applications and limits of specialty or to economic and social implications.
• The Ph.D. specialist becomes too much a synthetic organic chemist, or a spectroscopist, or a theoretical physical chemist, etc.

The lengthier statements evoked by Question 15 also merit sampling. One from a petroleum laboratory proposes:

"Examples of areas where training should be strengthened are oral and written communications, economic analysis, experimental design, instrumentation, inorganic and physical chemistry for organic chemistry majors, and problem identification and solving of an exploratory nature. A required survey course on 'Industrial Chemistry' should include: (1) basic processes and products for large-volume chemicals (volume, value, economics, and uses, etc.); (2) a look at chemical problems in a variety of basic industries (e.g., paint, petroleum, polymer, minerals, lumber); (3) a look at industrial research, with examples of specific labs (their organization, equipment, problems, contributions, etc.)."

The respondent for a diversified chemical producer makes the following suggestions on education for industry needs:

"Some courses should be given in the universities covering aspects of industrial chemistry on which considerable attention is continually focused:

Chemical Process Definition and Development
Statistics and Data Analysis for Chemists
Chemical Marketing
Economic Evaluation of Research Projects
Patent and Legal Considerations in the Chemical Industry

"Our supervisory people feel that chemical faculties normally are not sufficiently knowledgeable in these subjects to teach them. The need could be filled by qualified people from the chemical industry."

The spokesman for a manufacturer of hard goods as well as chemicals comments thus on the industrial skills that graduate schools should provide:

"Most Ph.D.'s are training to be teachers and academic researchers. For industrial purposes, programs should be thoughtfully designed to provide the skills that the researcher will need in an industrial framework: an ability to communicate rapidly and effectively, both orally and in a written form; an ability to analyze technical problems and to design methods of arriving at solutions through research; and an ability to use the technical tools of his trade with appreciation of what to expect of them. He should have a sound understanding of the area of technology which he is planning to enter . . . An understanding of his role in industry, of the professional relationship he should have or develop, would allow him to be much more effective at an earlier date. These things could be provided by graduate training."

Another industry viewpoint on Ph.D. training needs is vividly presented in this extended statement from a respondent associated with a major chemical company:

"Observation has indicated, in some cases: (a) a lack of knowledge and experience in simple but basic laboratory techniques (e.g., correct use of a burette); (b) inability or unwillingness to use library facilities; (c) inability to communicate through well-organized and concise reporting (written and oral); and (d) little knowledge of the 'scientific' approach to the planning and solution of problems (probably more so at the B.S. level).

"Most new chemists have little concept of industrial research as it relates to, and differs from, the academic world. Chemical engineers as a group seem to be better prepared for this change. Emphasis on the scientific method and innovation is needed rather than the narrow approach exemplified by the specific thesis assignment. Also, communication should be more widely included through instruction and *experience* in frequent seminar participation (or public speaking) and in technical report organization and writing.

"Analytical chemistry is preeminently an applied science. In most branches of science, basic work can be justified on the basis that it will illuminate a field of potential interest to the company. Most analytical work, on the other hand, is undertaken with the intent that a new method or result will serve a specific—and usually applied—need.

"The analytical Ph.D. is trained as an 'intellectual scholar.' When he arrives at industry, he may well be handed a Mason jar and asked to sample the sewer. Whereas his training is narrow, specialized, and scholarly, his work assignment may require very broad training, so that he may reasonably find the solution to a whole series of problems by whatever techniques he can bring to bear. To some extent, this results in a very poor fit between training and the requirements of the job in industry. It's very much as though our medical schools were training M.D.'s to do very esoteric and scholarly researches on, say, the effect of lithium on pancreatic function and that these scholars were then sent to hospitals to cure patients. Actually, the M.D.'s training is very 'applied' in its orientation— as indeed it must be if he is to function as a clinical (rather than research) M.D."

Looking Ahead to the 1980's

The last of the questions concerning Ph.D.'s (Number 16) stimulated a flood of commentary that bears on the second of the two objectives set out at the very beginning of this chapter. The question was designed to elicit ideas regarding the future shape of the educational system: "What major changes in chemical education would you suggest as desirable to meet the challenges of the 1970's and 1980's?"

Some readers might be surprised at the number and variety of respondents recommending no change or little, yet the logic of such a position is also clear. A mastery of fundamental principles ought to make a Ph.D. chemist into a "man for all seasons." Such mastery is presumably achievable even now, but it clearly is not being conferred routinely by any of the schools, or as frequently as desired. One of the standpatters, associated with a chemical company, counsels, "Keep to the basics of chemistry; no change." A respondent for a beverage producer wants merely "a broader background in general chemistry." A spokesman for a large steel manufacturer tersely says, "None." The representative of a prominent research laboratory "would not suggest major changes in chemical education" and makes these observations:

"With sound grounding in fundamental science, scientists can attack any of a wide variety of problems. An effort should be made to encourage more stars to enter the field and to discourage, perhaps, the lower third of the present distribution."

Another respondent, connected with a principal chemical company, believes that "Ph.D.'s are superbly trained technically" (and many others share this sentiment), but they have to be reoriented from "publication" to "independent research," from the possession of knowledge to social application.

Those who did suggest changes to meet the challenges of the 1970's and 1980's for the most part echoed proposals already offered for remedying the more immediate deficiencies in Ph.D. training. On the whole, the remarks indicate a belief that chemists will have to:

- Become better versed in their whole science.
- Acquire greater familiarity with neighboring sciences (including chemical engineering).
- Recognize more keenly, with the aid of graduate career counseling, the needs and rewards of practical problem-solving in the industrial context.

- Learn more about patent law, economics, statistics, electronic data processing, communication, and other auxiliary tools.
- Develop the versatility and flexibility required for the successful conduct of multidisciplinary as well as more independent research.

Several extended remarks are presented at this point to give the flavor of the more comprehensive proposals for enhancing versatility and flexibility. One of these statements, coming from the manager of a laboratory in the petroleum industry, touches in its last paragraph on continuing education, the theme of the preceding chapter:

"Because of the accelerated changes in our society, the chemists of the future will require, above all, versatility. This requires a strong academic grounding in fundamentals and emphasis on how to learn and how to do research. The present manpower situation especially points out the handicaps of being a recent graduate in an exotic specialty of high academic interest but little practical application.

"Versatility would probably not require fundamental curriculum changes at the undergraduate level but would require university administrations to be dedicated to a broad undergraduate program and to firmly resist specialization by undergraduates. This policy would tend to weed out weaker schools or force them to institute stronger and broader programs.

"Fundamental changes are needed at the graduate level to avoid the present overspecialization. Demonstrated competence in two to three fields of chemistry could well be required for a Ph.D. degree. A somewhat longer time in graduate school would be required, but the emphasis should be on the educational aspects in these fields rather than simply producing a predetermined amount of routine research, as is often now the case.

"In addition, the subfields of chemistry that are taught should be increased. Few halls of learning provide instruction in such practical (and fundamental) fields as catalysis and polymers.[6] Little interdisciplinary research is conducted, although some improvements have occurred in recent years.

"The universities should strive for the excellence in continuing education or reeducation that they now seek in undergraduate and graduate training. Relatively few universities now have the philosophical inclination to provide continuing education to industrial or government chemists. Efforts along these lines are mainly in extension-type classes where the variety and quality are well below those of the parent institution."

Another proposal, which comes from a respondent for a tire company, suggests some improvements concerning specialization as well as some other aspects of graduate study:

- Prepare students so that they have a specialization in a given area along with a broad background—a "specialization peak" superimposed on broad training.
- More emphasis on career-oriented graduate education.
- Graduate schools should provide for use of top industrial scientists as adjunct professors, external advisers, or thesis committee members.
- Inculcate among graduate students the desire and ability to learn, by self-study, new techniques and disciplines that are needed to solve problems.
- Mechanisms should be developed to couple graduate chemical education

6. CPT remarks that "industry continues to use a major fraction of its chemists in various aspects of polymers, synthetic fibers, elastomers, and related structural materials, areas given cursory treatment in all but a few of our graduate departments of chemistry" (ibid., p. 37). It also takes note of emerging industry and government needs for chemists in the areas of environment, health, and food production and processing.

to directed basic research; i.e., applied research that provides the *base* for advanced technologies.

A spokesman for a soap producer offers this view on means of enhancing flexibility:

"More course work is needed and less highly specialized research. Lab experience should not be lessened but varied. Perhaps the Ph.D. candidate should work on three or four *different* research assignments in *different* areas. The greatest need is to broaden (not deepen) his training and to increase his flexibility. This implies not only broader training but also a change in attitude from that of the specialist wedded to a narrow discipline (e.g., the mechanism of organic reactions) to that of a generalist who looks for problem-solving opportunities that demand his talents and training. This increased flexibility would be especially helpful in an era of unemployment."

While needs are expressed for a Ph.D. in chemistry who is widely and deeply competent and a self-reprogramming superman, the probability remains that industry will have to get along with specialists who fall far short of that ideal. The comment of the tire company respondent on the superimposition of a specialization peak upon a base of broad training points toward a practical compromise. The following contribution from a chemist whose striking analogy between chemistry and medicine was cited in the preceding section is also pertinent to any redefinition of educational goals:

"We need some analytical Ph.D.'s in industry to provide adequately for the research function. We also need people trained in applied chemistry to serve as the problem solvers, analogous to the clinical M.D.'s. There probably should be a master's program in applied chemistry in which the emphasis would be on breadth of training rather than a highly specialized research. This is not suggested as a substitute for the Ph.D. degree in analytical chemistry but rather as an additional program or option for the student. I believe that industry would welcome the graduates of such a program and find them highly useful, while at the same time employing Ph.D.'s with a more classical training."

Only one of the respondents, associated with a petroleum laboratory, contemplated a radical institutional innovation. He makes this proposal for the training of chemists having a more rounded capability in science and auxiliary fields and greater intellectual independence:

"It would seem that the broadly educated, self-starting 'nonspecialist' has the best chance of moving from one pertinent research field to another and outdistancing the narrowly trained competition. He also would be a valuable research director. However, this writer doubts that prevailing graduate instructors can change the system, since they themselves are often narrow specialists. The younger men who espouse such notions usually are prevented from 'rocking the boat,' and it seems likely that no innovative educational experiments will be allowed to succeed. Therefore, it is suggested that whole new institutions will be necessary to challenge the 'old school' in much the same way that the technical schools are challenging the engineering schools."

The details of such a vision need to be spelled out before the problems of realization can be ascertained.

Company Comments on B.S. Training

This chapter concludes with comments of respondents on the quality of current training of recipients of bachelor degrees in chemistry and on needed changes. On the whole, company spokesmen seem either satisfied with or little exercised over the shortcomings of educational programs for B.S.'s. Better specialists or better generalists are wanted, and they should have a benign attitude toward work in industry; but the recommendations concerning bachelors do not have the urgent tone of those concerning Ph.D.'s.

First, some opinions are cited with respect to current training. The respondent for a dominant chemical company, presumably speaking of its chemistry recruits in general, says, "We feel the universities are doing a good job of training students in the fundamentals of the academic disciplines." The rest of the company's statement reflects confidence that the graduates it hires can build on foundations already laid in academic training.

A respondent for another principal chemical manufacturer ventures that "training at the B.S. level is . . . well-balanced" but reports that faculties usually do not know the students well enough. The remarks offered by another chemical firm indicate that a "finished product" is not expected from the campus; that both the B.S. and Ph.D. need "a great deal of on-the-job training"; and that the recruits "have good background in technical fundamentals for most, but not all, chemical areas which we ask them to enter."

Somewhat similar evaluations are made by associates of firms in other industries. A respondent for a petroleum refinery finds "in general" that the B.S. is "reasonably well-trained in fundamentals and more adaptable [than the Ph.D.] to the environment he encounters in industry and to the specific problems to which he is assigned." Indeed, it is the Ph.D. who has become a narrow specialist and has been "indoctrinated with the idea that industrial research is second rate." On the other hand, a research organization also identified with the petroleum industry finds the Ph.D. more suited than the B.S. to its needs—precisely because the former "is more selective in what he performs earnestly and regards many industrial chemical chores as second rate"!

The extended comment from a major company making forest products reports virtually the same "conclusions . . . for both Ph.D.'s and B.S.'s" and betrays no deep concern. After all, "some of the qualities covered in the questionnaire are basic human qualities that are probably not influenced a great deal by formal education . . . unless a student has bad experience with specific instructors." Why should one be surprised that "fresh graduates" who are "perhaps well-equipped in their basic scientific disciplines" have not yet learned "to harmonize their skills with all aspects of the real world"? The universities, it is suggested, should not be asked to supply what only more experience may. Finally, is it not "a general rule" extending beyond the precincts of chemistry that "people do not express themselves well in written and oral communications"?

The remaining paragraphs relate to proposed directions of educational improvement, especially for B.S.'s. The statement from the forest products company claims that "the biggest, single problem" of the "fresh university graduate" entering industrial research is a failure to appreciate "what it takes to make a business succeed" and "where the theory he has learned can be used in solving practical problems." It is this interface of theory and practice that requires attention. A chemical company already quoted (as not expecting the schools to provide a "finished product") has this suggestion for expediting accommodation to the world of work:

"The gap between industry and university can be bridged by more contact between industrial chemists and the faculty-student group. This has been said often before; it is still needed."

Beyond a wholesomeness of attitude, respondents want the bachelor to have technical and other abilities that are consonant with company needs. The spokesman for a diversified manufacturing company wants programs for industry-destined B.S.'s to be designed "not for breadth of coverage but for thoroughness of knowledge in the fundamentals of chemistry and for ability to communicate orally and in a written way." The statement submitted for a pharmaceutical company wants stronger "laboratory training" for the B.S., with emphasis on chemistry rather than supplementary courses that "do not make him a more productive chemist in our setting." Indeed, a preference is expressed for the narrow German *techniker*, who has been "taught practical synthetic organic chemistry to the virtual exclusion of other subjects, including physical chemistry and mathematics." From a petroleum company, on the other hand, comes a statement of consensus "among management people" that "more extensive exposure in some nonscience areas" is needed. Indeed, weakness in such areas as "economic analysis and communication skills" is "usually the major stumbling block to the industrial chemist's career advancement."

Finally, a commentary from a tire company offers three proposals for improving the training system for B.S.'s. The first suggestion is to revise criteria for the accreditation of schools. Instead of emphasis on faculty achievement and library facilities, an examination should be given to all graduates to ascertain the actual quality and effectiveness of "teaching." This approach would be consistent with the second suggestion to winnow out the students "who either lack ability or have chosen the wrong field." The third idea is to emulate various collaborative academic-industry arrangements that blend education with work experience for undergraduates:

"Our highest-rated B.S. engineers are those trained in co-op schools. Co-op chemists would be worth considering."

Part III

THE ECONOMICS OF CHEMISTRY

Many variables help to define the nature and extent of chemistry's participation in the economy. Prominent among these is employment, or the
actual demand for chemists and chemical engineers. The treatment of this
variable, both as an economic indicator in its own right and as a "tracer"
disclosing the location of chemical activity within companies and industries, complements the discussion of manpower supply in Chapter 21.
Other variables also used here for describing chemistry's role in the economy include Gross National Product, national income, physical output,
labor productivity, research and development expenditures, patents, and
prices.

Three of the five chapters that constitute Part III are followed by special appendixes. Thus, after Chapter 26, a case study exhibiting the contribution of chemistry in a well-known nonchemical company (which makes
electronic computers) is presented. After Chapter 28, another nonchemical
study is presented (this time for a leading automobile maker) to illustrate
the human and other factors that enter into the successful translation of
chemical science first into usable technology and then into actual manufacture. The final appendix, following Chapter 29, relates to the questionnaire used by the American Chemical Society to obtain estimates of the
employment of chemists and chemical engineers in manufacturing in
1975–80.

This final part of the report describes in numbers the nature and extent of chemistry's participation in the nation's economy. Everyone, of course, knows something of this practical side of chemical science from daily experience as a worker, consumer, family member, urban or rural resident, commuter, and citizen in general. Indeed, one can hardly avoid forming a conscious impression of the scope, if not the scale, of the translation of chemistry into technology and the actual production of goods and services.

The discussion here is concerned with the same dual translation, but it is based on a systematic and selective review of the vast body of objective evidence. It should supply a corrective to the biases and other limitations of appraisals that are grounded in personal experience, however extensive. In addition to taking advantage of available data, it makes use of statistics generated especially for this study.

Needs of several identifiable audiences should be met by this economic sketch. The profile that emerges in the ensuing chapters is sure to contain features that will interest the bench or administrative chemist, the motivated lay reader, the screening journalist, and the browsing student. The total picture, however, should prove helpful to policymakers and decision-makers in government and business, to the academic community, to scholars concerned with the economic and social implications of science and technology, and to professional societies.

"Chemistry" vs. "Chemical Industry"

The emphasis on "chemistry" in the title of this part—and in the title of the study as a whole—places a heavy burden on what would at first appear to be an ample data base. The published information on the conventional "chemical" industry and on production and consumption of "chemicals" is voluminous and pertinent, and this information is definitely taken into account later (e.g., in Chapter 27).[1] Literally regarded, however, such industry and product information is not sufficiently comprehensive or fine grained to reveal significant features of chemistry's economic role. After all, deliberate and controlled molecular change in pursuit of income is not confined to the chemical industry; that is, to establishments engaged primarily in the manufacture and sale of chemicals. Furthermore, statistics for the nonchemical industries do not normally, nor should they be expected to, isolate the chemical activities pursued therein.

A fuller and sharper delineation of the economic role of chemistry accordingly requires that we go beyond straight description and use some of the accessible statistics interpretatively or analytically. The next four paragraphs illustrate the adaptation of available or obtainable data to the basic purpose of this part.

The first and most significant example of the surrogate use of statistics

---

1. In this part, "chemical industry" or the "conventional chemical industry" is usually equivalent to the major census group, "chemicals and allied products" (No. 28 in the Federal Government's Standard Industrial Classification scheme.). Where ambiguity is not likely to result, the word "chemists" will sometimes be used to include "chemical engineers." The employment data for chemists, moreover, typically include some biochemists (whose presence is welcomed in the interest of a broad coverage of "chemistry"). Finally, "research" is used as a short designation for "research and development" activity.

involves employment distributions for chemists and chemical engineers. As Chapters 26 and 29 emphasize, interindustry and functional (especially intracompany) distributions explicitly illuminate the manpower dimension of applied chemistry, but they also do more: They cast oblique light on the location, performance, and variety of chemical activity. This second, or diagnostic, value of employment information proceeds from the common-sense rule that "chemistry is at least where chemists are." Although such a rule has obvious exceptions, it does allow us to make proverbial "first approximations" that also have a high degree of resolution.

From interindustry distributions, we may infer the pattern of economic reliance upon the unit operations and unit processes that are particularly identified with industrial chemistry and chemical engineering (and that comprise the building blocks of chemical research activity, too).[2] Data on the specific inputs and outputs of such detailed operations and processes are normally not compiled in the first place, even for companies classified in the conventional chemical industry. When such detailed data are compiled, they often become irretrievably locked into published aggregates.

When available, intracompany employment distributions provide additional discriminating clues with regard to the economic practice of chemistry's unit operations and unit processes. They give evidence of the use of chemistry not only for research but also for production-related tasks. They may even disclose the use of professional chemists in such supportive and peripheral functions as sales, purchasing, and patenting. Special questionnaire surveys, discussed in Chapters 26 and 29, were made by the American Chemical Society (ACS) to obtain detail on staffing patterns of companies.

The second example refers to so-called input-output tables, which set out the pattern of interindustry connections in terms of purchases and sales and then proceed to more sophisticated derivations by means of matrix algebra. These tables illuminate the structural interdependence of the chemical industries and the rest of the economy, thus providing a more comprehensive view of the economic significance of chemistry. In particular, they are adaptable to the derivation of estimates of the numbers of chemists and chemical engineers required throughout the economic system to achieve increments of prescribed size and composition in the Gross National Product. Chapter 27 presents such employment computations.

The third example of the symbolic use of accessible data highlights the double utility of patent statistics. Long historical series exist for the three broad varieties of patents issued—chemical, electrical, and mechanical.[3] They go back to 1836, when the present patent-numbering system was instituted. The figures, presented in Chapter 28, tell a literal story about

---

2. According to the American Institute of Chemical Engineers, "Chemical engineering is that branch of engineering concerned with the development and application of manufacturing processes in which chemical or certain physical changes are involved. These processes may usually be resolved into a coordinated series of unit physical operations and chemical processes. The work of the chemical engineer is concerned primarily with the design, construction, and operation of equipment and plants in which these unit operations and processes are applied. Chemistry, physics, and mathematics are the underlying sciences of chemical engineering, and economics is its guide in practice."

This definition is quoted in C. E. Littlejohn and G. F. Meenaghan, *An Introduction to Chemical Engineering*, Reinhold, New York, 1959 (p. 2), which also cites fluid flow, heat transfer, distillation, solvent extraction, gas absorption, filtration, crystallization, blending, size reduction, and size separation as examples of unit operations (pp. 3-4); and cites halogenation, combustion, nitration, electrolysis, hydrogenation, and polymerization as examples of unit processes (pp. 4-5).

3. The law approved in 1836 established the Patent Office and restored the Jeffersonian examination system (which called for proof of novelty of invention) in addition to providing for the consecutive numbering of patent grants. (See *Patents: Spur to American Progress*, U.S. Department of Commerce, no date, pp. 10, 16.)

patent grants, of course, but they also offer an unusual view of the changing complexion of our technology during the past century and more.

The fourth example again relates to comparative time trends, but the statistics refer to symptomatic variables for the chemical industry and for manufacturing as a whole. Thus, Chapter 28 will show that physical output and labor productivity have risen much more rapidly during the present century in the chemical industry than in manufacturing in general. This favorable performance is reflected in the comparative price records of chemicals and other commodities. In particular, the good productivity and price experience of recent years has helped chemicals to register a consistent export surplus while the nation's merchandise trade balance has been deteriorating.

Criteria of Economic Significance

Despite the uneven quality of available figures and despite the gaps necessitating resort to proxy statistics, the next four chapters should convey a fair view of chemistry's economic role. The profile extends over the past and present, and it faces into the future, too.

Size has intuitive and immediate appeal as a criterion of importance, but our discussion will show that other attributes often describe chemistry's economic significance more naturally or more convincingly. Some of the economic numbers relating to chemistry, such as the chemical industry's rates of growth in output and productivity, may be impressively large. Some others, however, are clearly not, such as the share of chemists and chemical engineers in total employment or the share of the chemical industry in the national income.

What are the other criteria according to which chemistry manifests importance in the chapters that follow? First of all, despite their small number, chemists and chemical engineers are very widely diffused throughout the economic structure. They are found in the public sector as well as private, in academic and other service areas as well as manufacturing, in the commodity-making and distributing departments of enterprises as well as in the research laboratories. Thus, chemistry scores very high in *pervasiveness*.

Another characteristic of chemistry is its *criticality*. That is, in a society such as ours, chemistry is not replaceable at all or only at prohibitive cost. This virtual indispensability is hinted in the very pervasiveness of chemists and chemical engineers. It is also acknowledged obliquely in, for example, the current concern for the quality of the environment. Thus, the position of chemistry in the modern scheme of things is implied even in the special culpability that is commonly assigned to the chemical industry and to chemical materials and products. More striking is the likelihood that this position will be reinforced in the very battle against ecological blights. A massive rejection or a widespread and costly replacement of chemical technology appears much less plausible for the future than greater ingenuity, forethought, selectivity, and control in its use.[4]

Chemistry would seem eligible for high marks in several other respects that need merely be listed. Thus, the remainder of this part and all of Part I provide abundant evidence of chemistry's *persistence and growth* in the economy, its *dynamism and innovativeness* (as to technology and product variety), and its *competitiveness and substitutability* (for scarce or more

---

4. In this connection, the reader may wish to refer to an earlier report of the American Chemical Society's Committee on Chemistry and Public Affairs, *Cleaning Our Environment—The Chemical Basis for Action*, Washington, 1969.

expensive alternatives). These qualities have contributed to the unmis-takable—and apparently indelible—imprint of chemistry on our ways of working and living, on modes and standards of health care, and on the scope of choice in production and consumption.

A few more words may be required concerning the size of the special-ized manpower supply of chemistry. The coexistence of smallness with such properties as pervasiveness and criticality suggests another way of looking at magnitude. Insofar as the economy really is a system, it relies heavily on the performance of chemists and chemical engineers. The stra-tegic place of these professionals could be dramatized by inversion of the small ratios in which their numbers originally appear as numerators. The much larger reciprocals that result would show chemistry as a significant support of the whole productive order. A caution is necessary, however. Although the idea that a functioning system rests on a narrow, dependable technical footing is attractive, this truth is not unique to chemistry. Indeed, the same kind of assertion could plausibly be made for any other relatively small but pervasive, difficult-to-replace, and seemingly essential ingredient of the economic system. Obviously, balance and perspective must be kept in any documentation of the view that chemistry is "important" despite the smallness of various absolute numbers and percentages relating to it.

The foregoing remarks should not be interpreted as complacent. Confi-dence that chemistry will remain a vital and ubiquitous activity in the future economy hardly implies a comfortable progress along lines already perceived. Indeed, the economic indispensability of a field as a whole tells nothing about the economic viability of particular decisions or undertak-ings. The brief reference to ecology already makes clear that chemistry cannot be expected to have a special exemption from the shocks of reap-praisal accompanying any extensive reordering of national priorities.

Other signs of impending difficulty are easy to discern. Chapter 29, for example, refers to a special survey of employment prospects that fore-shadows a slower expansion in the employment of chemists and chemical engineers in manufacturing during the 1970's than was experienced during the 1950's and 1960's. The figures represent projections—"if-then" state-ments about the future rather than flat predictions—and are alterable with changing conditions. They have to be viewed, however, in the light of instabilities of the current domestic and international scenes. They are more troubling in a time of common challenge to traditional values and institutions, higher-than-customary unemployment for professional man-power, intensifying cost pressures at home and intensifying competition from abroad, and frequent failure of private research activity to yield promised or expected payoffs.

Even as this chapter is intended to be read with the documentation and the amplification that follow, Part III as a whole is best read and inter-preted in the context of the entire report. For example, the discussion of employment in the next chapter and in Chapter 29 depicts in an active mode the human resources viewed as an emerging or existing "stock" in Part II. The numerical treatment of other economic variables in the remain-ing chapters of this part does not at all exhaust what needs to be said about the past, present, and future pervasiveness, criticality, growth, or dynamism of chemistry. Indeed, much of the story of economic advance over the avenues of innovation that have been paved and are still being paved by the science and technology of chemistry has been presented in Part I, which is essentially nonquantitative.[5]

---

5. Note also the appendixes to Chapters 26 and 28, which briefly describe the roles of chemistry—current and developmental—in two nonchemical manufacturing companies.

Employment statistics for chemists and chemical engineers, as the preceding chapter noted, possess two kinds of relevance: direct and diagnostic. First, they report magnitudes of a leading economic variable. This variable is of interest in its own right, and in the present instance it refers to the very professional workers identified with chemistry as a science and technology. But interindustry and intracompany distributions of chemists and chemical engineers also have a special second merit. The presence of such professionals on business, government, academic, or other payrolls is a sure clue to the practice of chemistry (including research and teaching) for economic reward. More specifically, the distributions disclose the location of economically-oriented unit operations and unit processes, which comprise the alphabet of chemical activity. Showing the patterns of chemical activity in the economy at large and within the manufacturing firm, they compensate in some degree for limitations in the supply of detailed and matching data for relating the inputs and outputs of unit operations and processes. Indeed, the obtainability of distributions of professional employment makes it much easier to proceed beyond a parochial study of the *chemical industry* or *chemicals* to a quantitative examination of *chemistry* in the economy.

Number of Employed Chemists and Chemical Engineers

As Chapter 21 noted, about 190,000 persons were "working as" chemists and chemical engineers in the United States in 1970. This total embraces private industry, government, universities and colleges, and nonprofit institutions. It is based on occupational information developed by the U.S. Bureau of Labor Statistics (BLS).

The total of about 190,000 includes an estimate for chemists of 136,800 and an estimate for chemical engineers that exceeds 50,000. The figure for chemists emerges from a recurrent BLS survey of natural scientists, engineers, and mathematicians that was originally conducted in cooperation with the National Science Foundation (NSF).[1] The figure for chemical engineers, which is conceptually incomplete, was provided by the so-called industry-occupational matrix of BLS.[2] This figure needs to be expanded somewhat to include college and university teachers.

The universe of persons who "work as" chemists and chemical engineers at any particular time largely overlaps, but is not identical with, the professional population defined according to educational attainment or past employment history. The BLS–NSF definition, for example, includes persons without college degrees who "perform at levels normally requiring a four-year course with a major in the field of employment." Such

---

1. The 1970 estimate for chemists, supplied with industrial detail to the American Chemical Society in advance of publication (see Table 8), is comparable to those shown for earlier years in *Employment of Scientists and Engineers in the United States: 1950—66* (NSF 68-30), National Science Foundation, 1968, p. 28.

2. BLS "matrix" figures for chemists and chemical engineers are shown for 1970, with some industry detail, in Table 9.

Estimates of 136,800 for chemists and 50,000 for chemical engineers were accepted for the purpose of analyzing the 1970 figures of the National Register of Scientific and Technical Personnel by D. A. H. Roethel and C. R. Counts, "Realignments of the Chemical Profession Continue," *Chemical and Engineering News*, Nov. 15, 1971, pp. 90-92.

persons may be engaged not only in the conduct of research and development but also in management, production, technical service, technical sales, or other tasks. Excluded from the survey are persons who, though educationally equipped, are either not economically active at all or are not actually engaged as professional chemists or chemical engineers.[3]

At the vague boundaries of chemistry with adjacent disciplines, misclassification or arbitrary classification is to be expected, of course. Thus, some chemists or chemical engineers may well be counted as physicists, materials scientists, biologists, or medical scientists, but an "error" could just as easily be committed in the opposite direction.[4]

The point has already been made that the importance of chemistry in the nation's economy is hardly evident from the magnitude of its professional employment. The 190,000 chemists and chemical engineers comprise about 0.25% of total employment in 1970, 0.5% of the employed white-collar workers, and 1.7% of all professional and technical personnel.[5]

As the denominator is narrowed, the representation of chemists and chemical engineers, of course, improves. Thus, they comprised about 12% of the 1.6 million natural scientists, engineers, and mathematicians employed in 1970, as estimated by BLS.[6] If mathematicians are omitted, the proportion rises slightly to one eighth of the more constricted total.

When engineers are omitted from both the numerator and denominator, chemists alone are seen to constitute a sizable share of the total population of employed scientists. Thus, in 1970, chemists comprised more than a fourth of the nation's working natural scientists plus mathematicians. Omission of mathematicians from the denominator would raise the ratio of employed chemists to nearly one third of the smaller national total for scientists.

Since technicians may be expected increasingly to collaborate with chemists and chemical engineers in plant and laboratory tasks, note should be taken of some estimates of their recent employment. In 1966, chemical technicians employed throughout the economy numbered about 60,500, according to a study conducted by BLS for NSF.[7] This figure

3. Employment of Scientists and Engineers in the United States, pp. 44, 54.

4. After defining chemistry as "the integrated study of the proportions, properties, structure, and reactions of the chemical elements and of their compounds, and of the systems which they form," a Panel on the Structure of Chemistry agreed at a 1970 international conference that "chemical science is dynamic in scope," overlapping with, say, "molecular science and solid state physics" and logically incorporating "nuclear chemistry, quantum mechanics, statistical mechanics, and magnetic resonance" into its realm. The panel anticipated that increasing interest in "the analysis and modeling of complex systems" (e.g., rocket motors, living cells, and the polluted atmosphere) will encourage the emergence of "systems chemists." (See Preliminary Report: International Conference on Education and Chemistry, American Chemical Society, Washington, 1970, pp. 11-12; or Journal of Chemical Education, January 1971, pp. 7-8.)

5. The national figures for 1970 are shown, for example, in Manpower Report of the President, U.S. Department of Labor, April 1971, p. 215.

6. BLS figures for 1970 were supplied to the American Chemical Society prior to publication. Figures for the private sector excluding college and university teachers appear in Scientific and Technical Personnel in Industry, 1970, a pamphlet issued by BLS in 1972.

7. Technician Manpower: 1966–80, Bulletin 1639, U.S. Bureau of Labor Statistics, March 1970, pp. 7, 20. Figures for technicians are also shown in Scientific and Technical Personnel in Industry, 1969, Bulletin 1723, U.S. Bureau of Labor Statistics, 1971, but chemical technicians are not reported separately.

According to the BLS concept, technicians do not require a four-year college program; training at a post–high school institute or a year in college might suffice, or even "equivalent on-the-job training" or experience. (See Technician Manpower, p. 25; and Occupational Outlook Handbook: 1972–73 Edition, Bulletin 1700, U.S. Bureau of Labor Statistics, pp. 223-25.) The ACS ChemTeC project looks toward the training of more university-level technicians. (For a recent statement on this project, see BioScience, May 1972, pp. 290-92.) Two-year courses are available for chemical engineering technicians, and interest in four-year programs for "technologists" is developing. (See Chemical Engineering Progress, October 1971, pp. 19-30; and March 1972, pp. 35-42.)

amounted to about half the number of chemists employed in the same year (119,300).[8] It also comprised half of the total for physical science (and mathematics) technicians and about one seventh of the total for all science and engineering technicians, exclusive of draftsmen. Separate data were not published for chemical engineering technicians. By 1970, according to BLS, chemical technicians employed throughout the economy had increased to 68,700. A projection of 96,500 has been provided for 1980.[9]

Research vs. Other Activities

Available functional distributions indicate that, contrary to a common impression, fewer than half of all chemists and chemical engineers are engaged primarily in research and development work. Figures that exclude chemical engineers yield a somewhat higher research ratio, but the proportion is still in the neighborhood of one half for large segments of the population of chemists alone.

Major evidence regarding research and other functions of chemists (including some chemical engineers) is obtainable from the National Register of Scientific and Technical Personnel, which has been maintained by NSF. This evidence is probably distorted since the propensity to register may well be stronger for persons who have higher degrees and are engaged in academic and research pursuits. In any case, some 31,000 of the 83,000 employed registrants in chemistry in 1970 reported themselves to be working primarily on research tasks, while 11,500 additional registrants were principally occupied in the management and administration of research. Together, these groups comprised 51.1% of the employed registrants in chemistry. For 1968, the ratio was similar. Research-oriented personnel, numbering some 45,000, constituted 51.4% of the 87,600 employed registrants. The 1968 and 1970 distributions are shown in Table 1.[10]

When attention is focused on the "industry and business" component of the National Register, it is seen that research accounts for a slightly larger proportion of the chemists employed there. This component merits special notice because it embraced about three fifths of all employed registrants in chemistry in 1968 and 1970. From the appropriate NSF tables, it may be seen that research-oriented chemists represented 56.2% of the employed chemistry registrants classified in industry and business in 1968 and 55.9% in 1970.[11] These ratios refer, like those in the preceding paragraph, to primary work activities only.

A special tabulation of unpublished National Register information for chemists in industry and business has permitted a refinement of the distributions derivable for 1968 from published data for primary activities. This tabulation shows the spectrum of secondary as well as the primary activities of chemists classified according to level of academic attainment (Table 2). If plausible weights are applied to the two activities, it becomes possible to approximate composite man-year distributions that give a better idea of the allocation of total worktime.

Table 3 juxtaposes 1968 distributions for industry and business that are

---

8. *Employment of Scientists and Engineers in the United States*, p. 28. See our Table 8. The 1966 ratio of chemical technicians to chemists in manufacturing alone was slightly below one half (47%).

9. *Technician Manpower*, p. 7.

10. The criterion for registrants was established by the American Chemical Society: "A bachelor's degree and current employment in an area of chemistry; or 10 years of professional experience in an area of chemistry." (See *American Men of Science: 1968* (NSF 69-38), National Science Foundation, 1969, p. 265.) According to Roethel and Counts, "many chemical engineers reported to the 1970 NSF register for chemists" (*loc. cit.*, p. 92).

11. *American Science Manpower: 1968* (NSF 69-38), National Science Foundation, 1969, p. 69, for 1968; *ibid: 1970* (NSF 71-45), 1971, p. 57, for 1970.

based on published data for primary activities alone and on combined data for primary and secondary activities. The two types of activities are weighted two thirds and one third, respectively, when they are combined.[12] One pair of distributions relates to all reporting chemists regardless of their educational attainment. It appears that the conduct and management of research and development activity account for 56.0% of the total work-time available, slightly less than the figure of 57.6% derived from corresponding data for primary activities alone. Shifts occur in detail, however, when the refinement is introduced—toward *Development; Production and Inspection;* and *Exploration, Forecasting, and Reporting.*

Table 1

Primary Activities of Employed Registrants in Chemistry: 1968 and 1970

Primary Activity	1968	1970
All Chemistry Registrants	93,788	86,980
R&D	33,492	31,199
Basic	14,281	12,668
Applied	11,819	11,530
Development	7,392	7,001
Management or Administration	20,908	21,420
R&D	11,498	11,462
Other	9,410	9,958
Teaching	9,762	10,263
Production and Inspection	14,318	13,438
Consulting	1,315	1,434
Exploration, Forecasting, and Reporting	823	810
Other	2,726	2,678
Not Employed	6,180	3,541
No Information	4,264	2,197

Source: American Science Manpower: 1968 (NSF 69-38), National Science Foundation, 1969, p. 59, for 1968; ibid: 1970 (NSF 71-45); 1971, p. 47, for 1970. The numbers for "Development" and "Management or Administration, Other," not actually shown in the NSF source tables, were obtained by subtraction.

Another pair of distributions refers to chemists having doctorates, a decidedly research-prone group within the industry and business sector. The refinement to include secondary activities raises the percentage for the conduct of research from 53.9 to 57.1 but lowers the percentage for management of research from 31.0 to 24.9. On the whole, then, the research concentration for the doctorate group declines a bit, from 84.9% to 82.0%, when adjustment is made for the inclusion of secondary activities (Table 3).

From BLS comes other evidence that the research proportion for chemists employed in "industry and business" is in the neighborhood of one half. This category of the National Register is roughly equivalent to "private industry," the domain covered in BLS surveys of scientific and technical manpower for 1969 and 1970. According to data compiled in the 1970 BLS survey, about 48,900 of 92,000 chemists, or 53.0%, were working in

12. Chemists reporting no secondary activity were assumed to spend all their time in the reported primary activity.

Table 2

Cross-Classification, by Primary and Secondary Work Activity, of Chemists Employed in Industry and Business, According to Highest Degree: 1968

Highest Degree and Primary Work Activity	Total	Research and Development and Design				Secondary Work Activity								
		Total	Basic Research	Applied Research	Development and Design	Management or Administration		Teaching	Production and Inspection	Consulting	Exploration, Forecasting, Reporting	Other	Not Employed	No Report of Work Activity
						Total	Of R&D							
All Degrees	52,643	21,806	2,975	8,925	9,906	6,621	3,122	230	13,537	954	2,128	620	...	6,747
Research, Development, and Design	20,008	12,027	2,331	5,053	4,643	1,865	1,770	89	3,584	234	889	81	...	1,239
Basic Research	4,044	2,820	...	2,379	441	372	362	40	88	34	261	8	...	421
Applied Research	9,091	6,301	2,124	...	4,177	867	841	35	856	108	425	31	...	468
Development and Design	6,873	2,906	207	2,674	25	626	567	14	2,640	92	203	42	...	350
Management or Administration	16,531	5,798	509	2,901	2,388	1,485	665	53	6,197	339	523	308	...	1,828
Research and Development	9,533	5,463	497	2,832	2,134	820	...	28	1,598	184	371	74	...	995
Teaching	56	18	10	4	4	7	1	...	11	2	6	2	...	10
Production and Inspection	12,184	3,592	103	840	2,649	2,853	531	61	3,189	298	503	146	...	1,542
Consulting	683	152	6	49	97	124	58	10	191	...	48	29	...	129
Exploration, Forecasting, and Reporting	412	64	9	35	20	79	38	8	98	29	1	43	...	90
Other	1,388	155	7	43	105	208	59	9	267	52	158	11	...	528
Not Employed	...	...	...	...	...	...	...	...	...	...	...	...	...	...
No Report	1,381	...	...	...	...	...	...	...	...	...	...	...	...	1,381

(Continued next page)

Secondary Work Activity

Highest Degree and Primary Work Activity	Total	Research and Development and Design				Management or Administration		Teaching	Production and Inspection	Consulting	Exploration, Forecasting, Reporting	Other	Not Employed	No Report of Work Activity
		Total	Basic Research	Applied Research	Development and Design	Total	Of R&D							
Doctorate Degree	14,227	8,274	1,767	3,913	2,594	1,805	1,265	93	1,282	225	636	117	...	1,795
Research, Development, and Design	7,468	5,145	1,314	2,117	1,714	943	927	59	348	92	344	14	...	523
Basic Research	2,589	1,757	...	1,579	178	326	321	32	15	29	171	4	...	255
Applied Research	3,889	2,782	1,247	...	1,535	497	490	23	159	47	159	7	...	215
Development and Design	990	606	67	538	1	120	116	4	174	16	14	3	...	53
Management or Administration	5,304	2,810	427	1,679	704	600	210	25	801	103	220	83	...	662
Research and Development	4,295	2,745	421	1,662	662	390	...	19	359	73	177	30	...	502
Teaching	10	9	8	1	...	...	...	...	...	...	...	...	...	1
Production and Inspection	571	226	8	75	143	172	74	3	84	11	16	4	...	55
Consulting	144	46	5	27	14	28	20	2	21	...	11	4	...	32
Exploration, Forecasting, and Reporting	74	8	2	5	1	21	12	2	4	8	1	11	...	19
Other	273	30	3	9	18	41	22	2	24	11	44	1	...	120
Not Employed	...	...	...	...	...	...	...	...	...	...	...	...	...	...
No Report	383	...	...	...	...	...	...	...	...	...	...	...	...	383
Professional Medical Degree	12	3	1	1	1	2	2	...	3	1	...	...	...	3
Research, Development, and Design	2	1	1	...	...	...	...	...	1	...	...	...	...	...
Basic Research	...	...	...	...	...	...	...	...	...	...	...	...	...	...

(Continued next page)

Table 2 (Continued)

Highest Degree and Primary Work Activity	Total	Secondary Work Activity												
		Research and Development and Design				Management or Administration		Teaching	Production and Inspection	Consulting	Exploration, Forecasting, Reporting	Other	Not Employed	No Report of Work Activity
		Total	Basic Research	Applied Research	Development and Design	Total	Of R&D							
Applied Research	1	1	1	...	...	...	...	...	...	...	...	...	...	...
Development and Design	1	...	...	...	...	...	...	...	...	...	...	...	...	...
Management or Administration	1	1	...	1	...	...	...	...	1	1	...	...	...	...
Research and Administration	7	1	...	1	...	2	2	...	2	...	...	...	...	1
Development	2	1	...	1	1	...	...	...	...	...	...	...	...	1
Teaching	...	...	...	...	...	...	...	...	...	...	...	...	...	...
Production and Inspection	1	1	...	...	1	...	...	...	...	...	...	...	...	...
Consulting	...	...	...	...	...	...	...	...	...	...	...	...	...	...
Exploration, Forecasting, and Reporting	...	...	...	...	...	...	...	...	...	...	...	...	...	...
Other	1	...	...	...	...	...	...	...	...	...	...	...	...	1
Not Employed	1	...	...	...	...	...	...	...	...	...	...	...	...	...
No Report	1	...	...	...	...	...	...	...	...	...	...	...	...	1
Master's Degree	11,024	4,432	459	1,770	2,203	1,379	610	49	2,858	225	434	187	...	1,460
Research, Development, and Design	4,133	2,466	393	1,036	1,037	341	310	14	801	55	178	23	...	255
Basic Research	552	405	...	328	77	25	23	3	14	2	34	...	...	69
Applied Research	1,879	1,294	340	...	954	149	140	5	205	28	94	11	...	93
Development and Design	1,702	767	53	708	6	167	147	6	582	25	50	12	...	93

(Continued next page)

Highest Degree and Primary Work Activity	Total	Research and Development and Design				Management or Administration		Teaching	Production and Inspection	Consulting	Exploration, Forecasting, Reporting	Other	Not Employed	No Report of Work Activity
		Total	Basic Research	Applied Research	Development and Design	Total	Of R&D							
Management or Administration	3,476	1,139	40	525	574	291	145	7	1,373	75	110	89	...	392
Research and Development	1,918	1,053	37	507	509	146	...	4	400	37	75	19	...	184
Teaching	16	5	1	3	1	4	...	...	2	2	...	1	...	2
Production and Inspection	2,396	719	22	175	522	613	110	17	519	69	101	48	...	310
Consulting	195	37	...	6	31	44	20	5	50	...	9	14	...	36
Exploration, Forecasting, and Reporting	106	22	3	14	5	14	7	3	24	12	...	10	...	21
Other	398	44	...	11	33	72	18	3	89	12	36	2	...	140
Not Employed	...	...	...	...	...	...	...	...	...	...	...	...	...	...
No Report	304	...	...	...	...	...	...	...	...	...	...	...	...	304
Bachelor's Degree	26,515	8,804	710	3,143	4,951	3,342	1,205	84	9,106	488	1,043	312	...	3,336
Research, Development, and Design	8,131	4,258	587	1,850	1,821	567	520	16	2,362	84	360	44	...	440
Basic Research	850	621	...	446	175	18	15	5	54	3	53	4	...	92
Applied Research	3,203	2,136	508	...	1,628	215	205	7	476	33	171	13	...	152
Development and Design	4,078	1,501	79	1,404	18	334	300	4	1,832	48	136	27	...	196
Management or Administration	7,508	1,790	40	668	1,082	567	294	20	3,907	157	189	135	...	743
Research and Development	3,204	1,609	37	634	938	273	...	5	810	71	118	24	...	294
Teaching	28	4	1	...	3	3	1	...	8	...	6	1	...	6
Production and Inspection	8,949	2,572	73	571	1,928	2,018	336	39	2,495	209	383	91	...	1,142
Consulting	332	67	1	15	51	49	17	3	114	...	27	11	...	61

(Continued next page)

Table 2 (Continued)

Highest Degree and Primary Work Activity	Total	Research and Development and Design				Management or Administration		Teaching	Production and Inspection	Consulting	Exploration, Forecasting, Reporting	Other	Not Employed	No Report of Work Activity
		Total	Basic Research	Applied Research	Development and Design	Total	Of R&D							
Exploration, Forecasting, and Reporting	231	34	4	16	14	44	19	3	70	9	...	22	...	49
Other	701	79	4	23	52	94	18	3	150	29	78	8	...	260
Not Employed	...	...	...	...	...	...	...	...	...	...	...	...	...	...
No Report	635	...	...	...	...	...	...	...	...	...	...	...	...	635
Less than Bachelor's Degree														
Research, Development, and Design	306	129	20	48	61	25	15	2	77	2	3	...	...	68
Basic Research	119	80	18	25	37	9	9	...	18	...	1	...	...	11
Applied Research	24	19	...	14	5	2	2	...	1	...	...	...	...	2
Development and Design	63	48	16	...	32	5	5	...	6	...	...	...	...	4
Management or Administration	32	13	2	11	...	2	2	...	11	...	1	...	...	5
Research and Development	80	24	2	14	8	4	2	...	37	1	2	...	...	12
Teaching	42	23	2	14	7	2	...	...	8	1	1	...	...	7
Production and Inspection	1	...	...	...	...	...	...	...	1	...	...	...	...	...
Consulting	71	24	...	8	16	10	2	2	20	...	...	...	...	14
Exploration, Forecasting, and Reporting	2	1	...	1	...	1	1	...	...	...	...	...	...	...

Secondary Work Activity

(Continued next page)

Secondary Work Activity

Highest Degree and Primary Work Activity	Total	Research and Development and Design				Management or Administration		Teaching	Production and Inspection	Consulting	Exploration, Forecasting, Reporting	Other	Not Employed	No Report of Work Activity
		Total	Basic Research	Applied Research	Development and Design	Total	Of R&D							
Other	5	...	...	...	...	1	1	...	1	...	...	...	...	3
Not Employed	...	...	...	...	...	...	...	...	...	...	...	...	...	...
No Report	28	...	...	...	...	...	...	...	...	...	...	...	...	28
No Report of Degree	559	164	18	50	96	68	25	2	211	13	12	4	...	85
Research, Development, and Design														
Basic Research	155	77	18	25	34	5	4	...	54	3	6	...	...	10
Applied Research	29	18	...	12	6	1	1	...	4	...	3	...	...	3
Development and Design	56	40	12	...	28	1	1	...	10	...	1	...	...	4
Management or Administration	70	19	6	13	...	3	2	...	40	3	2	...	...	3
Research and Development	156	34	...	14	20	21	12	1	77	2	2	1	...	18
Teaching	72	32	...	14	18	9	...	...	21	2	...	1	...	7
Production and Inspection	1	...	...	...	...	...	...	...	...	...	...	...	...	1
Consulting	196	50	...	11	39	40	9	...	71	8	3	3	...	21
Exploration, Forecasting, and Reporting	10	1	...	...	1	2	...	...	6	...	1	...	...	...
Other	1	...	...	...	...	...	...	...	...	...	...	...	...	1
Not Employed	10	2	...	...	2	...	...	1	3	...	...	...	...	4
No Report	30	...	...	...	...	...	...	...	...	...	...	...	...	30

Source: Computed by American Chemical Society from information compiled by the National Science Foundation for the National Register of Scientific and Technical Personnel of 1968.

Table 3

Primary and Secondary Activities of Chemists Employed in Industry and Business: 1968

| | All Chemists | | | | Chemists with Doctorates | | | |
| | Primary Activity | | Adjusted for Secondary | | Primary Activity | | Adjusted for Secondary | |
Activity	Number*	Percent	Number**	Percent	Number*	Percent	Number**	Percent
Employed Registrants	51,262	100.0	51,262	100.0	13,844	100.0	13,844	100.0
R&D	20,008	39.0	21,021	41.0	7,468	53.9	7,911	57.1
Basic	4,044	7.9	3,828	7.5	2,589	18.7	2,400	17.3
Applied	9,091	17.7	9,192	17.9	3,889	28.1	3,969	28.7
Development	6,873	13.4	8,001	15.6	990	7.1	1,542	11.1
Mgt. or Administration	16,531	32.3	13,837	27.0	5,304	38.3	4,358	31.5
R&D	9,533	18.6	7,728	15.1	4,295	31.0	3,452	24.9
Other	6,998	13.7	6,109	11.9	1,009	7.3	906	6.5
Teaching	56	0.1	117	0.2	10	0.1	38	0.3
Production and Inspection	12,184	23.8	13,149	25.7	571	4.1	826	6.0
Consulting	683	1.3	816	1.6	144	1.0	182	1.3
Exploration, Forecasting, and Reporting	412	0.8	1,014	2.0	74	0.5	268	1.9
Other	1,388	2.7	1,308	2.6	273	2.0	261	1.9

Source: Derived from information supplied in Table 2, in turn obtained from the National Register of Scientific and Technical Personnel of 1968.

*Omits chemists not reporting their primary activity at all.

**Chemists not reporting a secondary activity were assumed to work full time in their primary activity. Weights of two thirds and one third, respectively, were applied to primary and secondary activities.

research and development. Figures emerging from the 1969 BLS survey indicate a similar allocation of chemists to research, or 53.3%.[13]

Table 4 shows the figure of 53.0% and some additional pertinent statistics derived from the 1970 BLS survey. In particular, this table compares the research commitments of chemists and of the other scientists working in industry. It makes clear that chemists outnumber the other groups of scientists in both research activities and nonresearch functions in private industry as a whole. They have higher research ratios in the various components of private industry. These ratios, consistently in the neighborhood of one half, range from 49.3% in nondurable-goods production to 56.9% in all (nonacademic) nonmanufacturing.

Additional evidence on the research-nonresearch employment dichotomy is provided by the responses to two special employment questionnaires administered by the American Chemical Society to its Corporation Associates. The two surveys were made with this report in view. The Corporation Associates comprise a class of Society affiliates typically, but not exclusively, engaged in manufacturing. Table 5 summarizes the usable replies submitted by manufacturing associates in the first survey (for 1969). It excludes responding companies that failed to supply separate figures for chemists and other professional-technical personnel in research and nonresearch activities or that gave information only for their corporate research centers or research departments. Table 6 presents some of the findings of the 1971 survey of Corporation Associates, which also requested employment projections for the rest of the decade. A good part of the discussion of the 1971 inquiry is accordingly reserved to Chapter 29.

Although the 94 companies covered in Table 5 were not selected to comprise a true sample of manufacturing as a whole, the comprehensive ratios that they yield for 1969 merit serious attention. After all, these companies did employ more than 50,000 chemists and chemical engineers in 1969, about two fifths of the estimated total of such professional workers in manufacturing. If not by design then by brute quantity could their aggregates approach representativeness of the manufacturing sector.

Some favorable qualitative factors should also be noted. One of these is the considerable variation of the 94 companies in size. Indeed, the table presents separate figures for "large," "medium," and "small" chemical nondrug and fiber producers—for companies employing respectively 600 or more chemists or chemical engineers, fewer than 600 but more than 100, and up to 100.

Another favorable factor from the standpoint of representativeness is the stability of the ratio, 48.3%, shown for research chemists and chemical engineers in the 94 companies. This stability itself reflects a strong central tendency in the ratios for the individual companies. Thus, a very similar research proportion, 49.3%, is obtained when the ratios for groups of surveyed companies are weighted by the "correct" corresponding industry totals.[14] Both of these percentages, moreover, appear consistent with the research ratio for *chemists only* that is shown in Table 4 for manufacturing in 1970 (52.3%) and with a virtually identical one derivable for 1969 (52.4%), the year of the ACS survey. These comparison ratios, based on BLS figures, would be pulled down somewhat by the inclusion of chemical engineers, who are less numerous than chemists in manufacturing and who are also less likely to participate in research.

---

13. *Scientific and Technical Personnel in Industry, 1969*, pp. 27, 29; and *ibid: 1970*, Tables 3 and 4, as noted in our Table 4.

14. The totals refer to employment of both chemists and chemical engineers in 1970, as estimated by BLS. See Table 9.

Table 4

Proportions of Chemists and Other Scientists in Research Activity, Private Industry: 1970

	Private Sector	Total	Manufacturing Durable Goods	Nondurable Goods	Nonmanufacturing
All Scientists					
Total	217,400	149,800	64,700	85,100	67,600
Research	101,500	75,000	28,500	43,400	29,600
Percent	46.7	50.1	44.0	51.0	43.8
Chemists					
Total	92,200	77,600	18,000	59,600	14,600
Research	48,900	40,600	8,900	31,700	8,300
Percent	53.0	52.3	49.4	55.2	56.9
Other Physical Scientists					
Total	61,500	34,700	28,000	6,700	26,800
Research	26,100	14,500	12,600	3,300	11,600
Percent	42.4	41.8	45.0	49.3	43.3
Life Scientists					
Total	24,100	16,800	1,500	15,300	7,300
Research	11,400	8,600	700	7,900	2,800
Percent	47.3	51.2	46.7	51.6	38.4
Mathematicians					
Total	39,600	20,700	17,200	3,500	18,900
Research	15,100	8,200	7,700	500	6,900
Percent	38.1	39.6	44.8	14.3	36.5

Source: Tables 3 and 4 of an unpaginated pamphlet of *U.S. Bureau of Labor Statistics, Scientific and Technical Personnel in Industry, 1970, 1972. Excludes teachers.*

Table 5

Professional Technical Employment in Research and Nonresearch: ACS Manufacturing Survey, 1969

Industry or Group (1)	No. of Companies (2)	Chemists & Chem. Eng. Total (3)	Chemists & Chem. Eng. in R&D (4)	All Prof. Tech. Empl. Total (5)	All Prof. Tech. Empl. in R&D (6)	Col. (4) ÷ Col. (3) (7)	Ratio (Percent) Col. (3) ÷ Col. (5) (8)	Col. (4) ÷ Col. (6) (9)
Total	94	50,654	24,490	148,954	51,938	48.3*	34.0	47.2
Chemicals, excl. Pharm.	36	27,715	13,676	51,365	20,337	49.3	54.0	67.2
Largest	9	23,329	11,419	44,687	17,247	49.0	52.2	66.2
Middle size	13	3,792	1,927	5,647	2,647	50.8	67.2	72.8
Smallest	14	594	330	1,031	443	55.6	57.6	74.5
Pharmaceuticals	10	3,924	2,231	10,503	4,287	56.9	37.4	52.0
Petroleum Products	9	7,585	2,826	22,400	5,104	37.3	33.9	55.4
Soft Goods: Food, Textiles, Paper, etc.	19	2,440	1,112	4,734	1,597	45.6	51.5	69.7
Hard Goods: Metals, Glass, Rubber, Equipment, etc.	20	8,990	4,645	59,952	20,613	51.7	15.0	22.5

Source: Questionnaire survey of Corporation Associates of the American Chemical Society. The data embrace only those manufacturing respondents that supplied separate information for research and nonresearch activities.

*If the percentages that follow were weighted according to the total number of chemists and chemical engineers employed in the corresponding industries or groups, the composite research percentage would not change appreciably. Thus, weights for 1970, derived from BLS information for chemists and chemical engineers shown in Table 9, yield an alternative estimate of 49.3%.

The concentration of the research ratios for groupings of the 94 companies in the neighborhood of one half is shown in Column 7 of Table 5. The most deviant figures are 37.3% for *Petroleum Products* and 56.9% for *Pharmaceuticals*. Chemical companies outside of pharmaceuticals, which employ a very large share of the chemists and chemical engineers, show a research ratio of 49.3%, virtually the same as the weighted and unweighted averages for all 94 ACS respondents.

A curious additional fact is that, at least for 1969, the research proportions for companies making chemicals other than pharmaceuticals vary inversely with size class. Thus, 55.6% of the chemists and chemical engineers in small companies were engaged in research, 50.8% of those in the middle companies, and 49.0% of those in the larger ones.

Columns 8 and 9 of Table 5 relate chemists and chemical engineers to their professional-technical colleagues. In the 94 companies treated as an ensemble, chemists and chemical engineers comprised one third of the professional-technical corps in 1969. A similar ratio is indicated for *Petroleum Products* and a slightly higher one of three eighths for *Pharmaceuticals*. The highest of the ratios, 67.2%, relates to medium-size companies making chemicals other than drugs. The lowest, which is very low (15%), was recorded for the melange of companies included under *Hard Goods*.

According to the final column, chemists and chemical engineers comprised nearly half of the professional-technical personnel engaged in research activity in the 94 companies in 1969. Furthermore, they constituted more than half of the professional research staffs in each of the industry and size groups, except for *Hard Goods*. The ratio for this exceptional case was below one fourth. As might be expected, the ratio of chemists and chemical engineers to total professional research workers was high for the chemical companies (two thirds). The ratio was even higher for *Soft Goods: Food, Textiles, Paper, etc.*

The degree of research participation of chemists and chemical engineers in manufacturing companies in 1971 is suggested by Table 6. This table summarizes only a few of the findings of the second special ACS canvass of Corporation Associates undertaken for this study. Some other results of the 1971 survey will be presented in the next section of this chapter, and as noted earlier the projections to 1975–80 will be treated in Chapter 29.

On the whole, the response for 1971 was more limited than that for 1969. Thus, Table 6 is based on 68 usable replies covering 42,000 chemists and chemical engineers, or about one third of the total for the manufacturing universe. The quality of the submitted data was probably much higher in 1971 than before, but this rising standard does not compensate fully for the reticence of so many of the canvassed Corporation Associates.

Table 6 indicates a decline from 1969 that accords with common-sense impressions about the changing circumstances of industrial research and development. The proportion of research chemists and chemical engineers in the 68 companies reporting for 1971 is smaller than that computed for the 94 companies reporting in 1969. The unweighted aggregates for 1971 yield a research ratio of 43.4%. If "correct" industry weights are introduced, a higher figure, 45.1%, is obtained. Both figures, however, are below the corresponding proportions for 1969 of 48.3 and 49.3%, respectively.

Only a rough idea of the pattern of change between 1969 and 1971 is conveyed by the percentages shown for individual industry and size groups in Tables 5 (Column 7) and 6. Apart from improvement in the quality of the data, the respondents differ in composition in the two years. The research ratio for the nonpharmaceutical chemical companies as a whole remained very close to one half, but significant reductions are indicated for the medium and smaller reporting chemical companies, for *Pharmaceuticals,* for *Petroleum Products,* and for the miscellany designated as *Hard Goods.* Although correction cannot be made for a possible distortion that is due to the improvement in data quality between 1969 and 1971, some adjustment can be made for the difference in composition of the respondents. Thus, as a check on the changes indicated by Tables 5 and 6, computations have been made for 37 identical companies reporting both years. These companies employed about 29,500 chemists and chemical engineers in both years, virtually one quarter of the manufacturing universe.

Table 6

Chemists and Chemical Engineers in Research and Nonresearch: ACS Manufacturing Survey, 1971

Industry or Group	No. of Companies	Chemists and Chemical Engineers Total	in R&D	Percent
Total	68	41,779	18,119	43.4*
Chemicals, excl. Pharm.	24	17,521	8,708	49.7
Largest	5	14,945	7,644	51.1
Middle size	6	2,036	829	40.7
Smallest	13	540	235	43.8
Pharmaceuticals	8	3,592	1,673	46.6
Petroleum Products	6	8,069	2,652	32.9
Soft Goods: Food, Textiles, Paper, etc.	14	2,430	1,139	46.9
Hard Goods: Metals, Glass, Rubber, Equipment, etc.	16	10,167	3,947	38.8

Source: Questionnaire survey of Corporation Associates of the American Chemical Society. The data include only those manufacturing respondents that supplied separate information for research and nonresearch activities.

*If the percentages that follow were weighted according to the total numbers of chemists and chemical engineers employed in the corresponding industries or groups, the composite research percentages would not change appreciably. Thus, weights for 1970, derived from BLS information for chemists and chemical engineers (see Table 9), would yield an alternative estimate of 45.1%.

The identical companies apparently experienced a general decline. The ratio of research chemists and engineers receded from 47.9% in 1969 to 45.4% in 1971 if the calculation is based on unweighted aggregates. Alternative ratios of 49.8 and 48.8% are derived for 1969 and 1971 if the percentages for the five industrial groups (i.e., *Chemicals except Pharmaceuticals, Pharmaceuticals, Petroleum Products, Soft Goods,* and *Hard Goods*) are weighted by 1970 BLS estimates of employment for chemists and chemical engineers. As for the component groups, the 1969 and 1971 ratios for *Chemicals* remain close to each other and close to one half, and much less severe declines are indicated for the medium and smaller companies. Furthermore, the reduction for *Hard Goods* is also minor. The decline for *Petroleum Products,* however, is confirmed, and a sharper cut is indicated in the research ratio for *Pharmaceuticals.*

The Functional Spectrum

The preceding discussion has dwelt on the research-nonresearch dichotomy, but the available information also permits a much more detailed analysis of work functions. Three kinds of evidence are now considered: some National Register tabulations, a distribution of ACS membership, and additional results of the 1971 ACS survey.

Returning to Table 1, we examine more closely the primary activities of the respondents not engaged in the management and conduct of research. Nonresearch workers accounted for almost one half (48.6%) of all the employed registrants in chemistry in 1968. Nonresearch managers alone represented more than a tenth of all the employed registrants; teachers, one ninth; and consultants, 1.5%. Registrants explicitly categorized in production comprised almost one sixth, but some of the chemists classified among "others" or providing no information about themselves probably belong there, too. In any case, these two nondescript classes and the registrants in *Exploration, Forecasting, and Reporting* constitute one eleventh of the 1968 total. A similar pattern is indicated by the figures for 1970: 48.9% of the employed registrants classified in nonresearch in general, 11.9% in nonresearch management alone, 12.3% in teaching, 16.1% in production, 1.7% in consulting, and 6.8% in exploration and the two residual categories.

From Table 3 we may learn more about the nonresearch functions of chemistry registrants in "industry and business" but adjusted to reflect secondary as well as primary activities. The man-years for 1968 are distributed as follows: all nonresearch tasks, 43.9%; nonresearch management alone, 11.9%; teaching, 0.2%; production and inspection, 25.7%; consulting, 1.6%; and exploration and "other," 4.6%.

For chemists in "industry and business" who have their doctorates, a much different pattern is shown in Table 3 for 1968. Thus, only 18.0% of primary and secondary work activity of Ph.D.'s was allocated to nonresearch tasks. The components are: nonresearch management, 6.5%; teaching, 0.3%; production and inspection, 6.0%; consulting, 1.3%; and exploration and "other," 3.8%.

Also of interest is a functional analysis of the paying (i.e., nonemeritus) membership of the American Chemical Society. The following summary refers to the half of the ACS membership that was employed in manufacturing in November 1970:[15]

Classification	Number of ACS Members
Total, Manufacturing Only	55,716
Company Officials	5,311
Works Executives	2,991
Supervisors, Foremen, and Department Heads	6,430
Engineers, Chemical and Other	5,576
Technical Directors, Including R&D Managers	8,591
Staff Chemists and Metallurgists	20,587
Sales and Advertising Personnel	2,711
Purchasing Managers, Buyers, and Other Employees	3,519

These figures probably exaggerate the hierarchical status of respondents. Thus, officials, executives, managers, supervisors, and technical directors comprised about 23,300 of the total of 55,700 persons, leaving only 26,200 chemists, engineers, and metallurgists for performance of the professional "yeoman" tasks of the plant and laboratory. Outside of these two main categories are about one ninth of the ACS members in manufacturing. This

15. Based on "Business Analysis of the Total Paid Subscription Circulation for the November 1970 Issue" of *Chemical and Engineering News*, adjusted to refer only to persons who were members of the American Chemical Society.

residual is engaged in sales, advertising, purchasing, and other tasks peripheral to the conduct or management, either in the plant or laboratory, of the unit operations and unit processes that characterize chemistry.

The functional distributions presented in Table 7 constitute a principal finding of the 1971 survey of ACS Corporation Associates engaged in manufacturing. They refer to the same 68 companies that were treated in Table 6. Since several companies, however, did not report all the details sought in the questionnaire, a few adjustments (occasionally heroic) had to be made before the data were further processed. Having already commented on the research-nonresearch dichotomy, we now turn to narrower details of the functional spectrum.

Two distributions are presented for the 68 companies. One, designated A, is based on unweighted aggregates. The other, B, incorporates BLS industry weights for chemists and chemical engineers. Both show essentially the same functional pattern. Thus, management roles occupy somewhat fewer than one tenth of the chemists and chemical engineers in the responding manufacturing companies. Staff work in laboratories and plants (i.e., the conduct of research, production, and inspection and testing) engages more than three quarters. Sales, purchasing, and other supportive activities account for more than one seventh of the jobs.

For chemicals other than pharmaceuticals, the pattern is not much different from that just noted for the ensemble of respondents, but the distributions for the three size-groups do show considerable variation. In these chemical companies considered as a whole, management (research and nonresearch) accounted for only 6% of the jobs assigned to chemists and chemical engineers in 1971, while staff functions (in laboratory and plant) engaged more than three quarters of such professionals and supportive activities (sales and other) occupied almost one sixth. The large companies dominated these allocations. Medium-size and small chemical companies had much higher proportions in management in 1971—13.1% and 17.6%, respectively. They also had considerably smaller proportions in staff work of laboratory and plant—roughly two thirds. Finally, they had greater ratios in sales, purchasing, and other support functions—19.4% and 17.1%, respectively.

The other five industry divisions identified in Table 7 also have some distinctive features. While Hard Goods and Soft Goods show the same high ratio for chemists and chemical engineers engaged in research and nonresearch staff functions (almost four fifths), they differ sharply in their managerial and supportive allocations. The pharmaceutical profile has two anomalies: an assignment of only five eighths of the chemists and chemical engineers to staff work in laboratory and plant and an allocation of more than one quarter to sales and other supportive activities. For Petroleum Products, the management ratio is comparatively high—almost one sixth of the employed chemists and chemical engineers. The staffing of refinery laboratories and plants accounts for about seven tenths of all the jobs for these professionals, less than the average for all companies but higher than the corresponding ratio for Pharmaceuticals. The nonresearch managerial role for Petroleum Products occupies one eighth of the industry's chemists and chemical engineers.

Industrial Distributions

Although the conventional chemical industry is a dominant employer of chemists and chemical engineers, work for these professionals is widely diffused throughout the economy. Chemists alone, according to the 1970 BLS survey of scientific and technical personnel, are found in more areas

Table 7

Functional Distribution of Chemists and Chemical Engineers: ACS Manufacturing Survey, 1971

Function	Total*		Total	Chemicals, excl. Pharm. Large	Medium	Small	Pharmaceuticals	Petroleum Products	Soft Goods**	Hard Goods***
	A	B								
Number of Companies	68		24	5	6	13	8	6	14	16
Chemists and Chem. Eng.	41,779		17,521	14,945	2,036	540	3,592	8,069	2,430	10,167
All Functions (Percent)	100.0	100.0	100.0	100.0	100.0	100.0	100.0	100.0	100.0	100.0
Research	43.4	45.1	49.7	51.2	40.7	43.5	46.6	32.9	46.9	38.8
Management	3.5	3.5	3.4	3.0	5.2	7.4	3.8	3.9	6.1	2.6
Staff	39.9	41.6	46.3	48.2	35.5	36.1	42.8	29.0	40.8	36.2
Nonresearch	56.6	54.9	50.3	48.8	59.3	56.5	53.4	67.1	53.1	61.2
Management	6.2	5.3	2.6	1.6	7.9	10.2	6.8	12.4	9.3	6.5
Production	23.6	23.4	24.3	24.8	21.8	16.8	10.6	26.9	20.5	25.0
Inspection and Testing	12.0	11.8	6.9	6.3	10.2	12.4	9.4	14.8	18.0	18.1
Sales and Purchasing	9.5	9.5	11.2	10.7	14.7	12.8	7.1	9.4	3.8	8.9
Other	5.3	4.9	5.3	5.4	4.7	4.3	19.5	3.6	1.5	2.7

Source: Questionnaire survey of Corporation Associates of the American Chemical Society. The data include only those respondents that supplied separate information for research and nonresearch activities. The distributions furnished by a few companies lacked full detail, so some cells had to be estimated before consolidation into industry totals.

*Two distributions are shown for all companies. A is based on unweighted aggregates for the 68 companies. B reflects the use of 1970 industry weights derived from BLS data (see Table 9).

**Food, Textiles, Paper, etc.

***Metals, Glass, Rubber, Equipment, etc.

of the private nonacademic community than are physicists, metallurgists, geologists, other physical scientists, and life scientists. Indeed, only mathematicians among the technical nonengineering professions are as ubiquitous in this community, but their total number in 1970 was only a bit more than two fifths of the corresponding figure for chemists.

The 1969 and 1971 ACS surveys also offer striking evidence, as do the accounts in Part I of this report, of the broad economic reliance on the skills of chemists and chemical engineers. One would not normally guess that well over 300 chemists and chemical engineers are on the research staff of Bell Laboratories (which is not included among the 68 manufacturing companies of Tables 6 and 7 since it is classified as a "commercial" nonmanufacturing entity by BLS). Nor would one expect that a company such as International Business Machines (IBM) employs 2,400 of these professionals in research and nonresearch functions. An equal surprise, perhaps, is that General Electric Co. employs at least as many. These examples highlight the increasing interpenetration of classical chemistry, electronics, solid state physics, and so forth and their possible eventual merger into a "materials science" without complete loss of disciplinary identities. An appendix to this chapter gives a sketch of the roles of chemistry in IBM.

Table 8 presents a synopsis of statistics developed by BLS in cooperation with NSF on the dispersion of chemists throughout the economy in the period 1950–70. The figures do not include chemical engineers but are comprehensive in another respect: They cover government and academia as well as private industry. They show the industrial structure of chemists' employment in selected years of the two decades, a period in which the average annual rate of increase exceeded 4.3%.

A perusal of Table 8 reveals many notable facts. For example:

- Private for-profit industry (i.e., the economy less all levels of government, universities and colleges, and nonprofit institutions) has consistently employed about seven out of 10 chemists in all the sectors in the two decades.
- Manufacturing has typically engaged about three out of five chemists.
- Universities and colleges provided jobs for about one out of six chemists in 1970 compared to one out of five in 1950.
- Despite a slight increase in the share of government during the two decades, this sector still engaged only one out of 11 chemists in 1970.
- Chemists are now found in all the (two-digit) industry groups within manufacturing, and they are also employed in perceivable numbers in many key nonmanufacturing branches of the economy.

Table 9, the last in this chapter, presents industry distributions of chemists and chemical engineers for 1960, 1967, and 1970. The columns relating to 1970 provide some of the information required earlier for weighting the individual industry distributions derived from the 1969 and 1971 ACS manufacturing surveys.[16] The statistics were developed by BLS within the format of its "industry-occupational matrix." They differ in two important respects from other BLS figures displayed earlier, which emerged from surveys of scientific and technical personnel. First, in the matrix approach, as many government chemists and chemical engineers as possible are allocated to other standard industries; only those engaged in public administration remain in the government category. Second, the industry distribution is itself truncated, omitting chemists and chemical engineers teaching at the colleges and universities.

---

16. The 1970 industry weights were actually derived from details given in *Tomorrow's Manpower Needs*, Vol. IV, which is cited in Table 9.

Table 8

Estimated Employment of Chemists, By Sector, 1950–70
(in thousands)

Sector	1950	1955	1960	1965	1966	1970
All Sectors	51.2	72.8	98.4	116.0	119.3	136.8
Private Industry	36.6	54.2	72.4	85.2	88.7	98.0
Manufacturing	30.5	47.3	61.3	70.4	73.7	80.2
Ordnance	*	.2	.7	1.5	1.6	1.2
Food	2.9	3.6	4.4	4.5	4.4	4.5
Textiles and Apparel	.9	1.1	1.3	1.9	2.1	2.5
Lumber and Furniture	.1	.1	.2	.2	.2	.3
Paper	1.5	2.1	2.8	2.9	2.8	3.0
Chemicals	13.4	23.7	32.0	37.5	40.3	43.8
Petroleum Refining	2.3	3.5	3.9	3.4	3.3	3.2
Rubber	1.6	2.0	2.2	2.8	3.0	3.7
Stone, Clay, and Glass	.7	.9	1.1	1.1	1.2	1.3
Primary Metals	2.0	2.3	2.7	2.2	2.2	2.4
Fabricated Metals	.5	.7	.8	1.1	.8	.9
Machinery	.7	.9	1.2	1.7	1.7	2.0
Electrical Equipment	1.1	1.6	2.1	2.1	2.2	2.5
Motor Vehicles	.3	.3	.4	.6	.6	.6
Aircraft	.4	1.3	2.1	1.4	1.5	1.7
Other Transportation	.1	.1	.1	.1	.1	.1
Prof. & Sci. Instruments	1.0	1.6	2.0	3.7	4.1	4.8
Misc. Manufacturing	1.0	1.1	1.2	1.4	1.6	1.7
Nonmanufacturing	6.1	7.9	11.1	14.8	15.0	17.8
Mining	.5	.5	.7	.9	.9	.9
Petroleum Extraction	.1	.1	.1	.2	.2	.3
Other Mining	.4	.4	.6	.7	.7	.6
Construction	.1	.1	.1	.1	.1	.1

(Continued next page)

Table 8 (Continued)

	1950	1955	1960	1965	1966	1970
Nonmanufacturing (cont'd)						
Transportation, Comm., & Pub. Util.						
Railroads	.4	.5	.6	.7	.7	.7
Other Trans.	.2	.2	.2	.2	.2	.2
Telecomm.	*	.1	.1	*	*	*
Radio & TV	*	*	*	*	*	*
Public Util.	.2	.2	.3	.5	.5	.5
Other Industries	5.1	6.8	9.7	13.1	13.1	16.1
Misc. Bus. Serv.	2.8	3.6	5.6	7.3	7.0	8.1
Med. & Dental Labs	.1	.2	.2	.3	.3	.4
Engr. & Arch. Services	.2	.3	.4	.3	.4	.6
Nonprofit Inst.	.6	1.0	1.3	1.5	1.5	2.0
Other Nonmfg.	1.4	1.7	2.2	3.9	4.1	5.0
Government	4.4	6.5	7.9	10.9	11.3	12.3
Federal	3.2	4.9	5.7	8.2	8.4	8.9
State	.8	.9	1.2	1.5	1.6	1.8
Local	.4	.7	1.0	1.2	1.3	1.6
Universities & Colleges	10.2	11.1	16.1	19.7	19.3	26.2

Source: Figures for 1970, unpublished, provided to American Chemical Society by U.S. Bureau of Labor Statistics; figures for earlier years from Employment of Scientists and Engineers in the United States: 1950–66 (NSF 68-30), National Science Foundation, 1968, p. 28. Asterisk signifies "fewer than 50 chemists." Minor discrepancies have not been eliminated from all totals and subtotals.

Table 9

Estimated Employment of Nonacademic Chemists and Chemical Engineers: 1960, 1967, and 1970

Industry	1960			1967			1970		
	Total	Chemists	Chem. Eng.	Total	Chemists	Chem. Eng.	Total	Chemists	Chem. Eng.
Total	130,600	91,000	39,600	160,600	111,900	48,700	169,200	118,300	50,900
Mining	3,200	1,300	1,900	2,000	1,000	1,500	2,400	900	1,500
Manufacturing	96,000	63,500	32,500	113,100	73,500	39,600	122,000	80,800	41,200
Durable Goods*	24,600	16,100	8,500	28,300	18,100	10,200	29,400	19,000	10,400
Stone, Clay, etc.	3,100	1,800	1,300	2,000	1,400	1,100	2,400	1,300	1,100
Primary Metals	3,600	2,900	700	3,000	2,300	700	3,100	2,400	700
Fabricated Metals	3,300	1,900	1,400	5,000	2,800	2,200	4,800	2,400	2,400
Machinery, Except Electrical	2,000	1,200	800	2,700	1,800	900	2,800	1,900	900
Electrical Mach.	4,200	2,200	2,000	5,100	2,500	2,600	5,200	2,500	2,700
Aircraft & Engines	3,100	2,200	900	3,300	2,100	1,200	3,000	1,800	1,200
Professional & Sci. Instr.	3,200	2,300	900	4,400	3,500	900	5,700	4,800	900
Nondurable Goods*	71,000	47,500	24,000	84,700	55,300	29,400	92,600	61,800	30,800
Food	4,500	4,100	400	4,500	4,100	400	4,800	4,500	300
Chemicals	49,900	32,800	17,100	62,800	40,800	22,000	67,400	44,200	23,200
Petroleum Refining, Coal Products	7,900	3,900	4,000	7,400	3,300	4,100	7,400	3,200	4,200
Rubber, Plastics	2,800	1,900	900	3,700	2,200	1,500	5,200	3,700	1,500
Services*	21,500	18,200	3,300	30,200	25,000	5,200	30,000	24,100	5,900
Misc. Bus. Serv.	5,400	4,100	1,400	7,900	6,000	1,900	10,200	8,100	2,100
Medical, Other Health	5,200	5,200	....	7,000	7,000	....	4,200	4,100	100
Engineering & Architectural Serv.	1,100	300	800	2,000	500	1,500	2,300	600	1,700
Other Nonmanufacturing**	5,200	3,800	1,400	6,200	5,800	1,400	8,600	7,000	1,600
Government, Public Admin.	4,700	4,200	500	7,600	6,600	1,000	6,200	5,500	700

Source: U.S. Bureau of Labor Statistics. The figures were developed as part of the BLS "industry-occupational matrix" and differ in two regards from figures emerging from the BLS survey of employed scientists and engineers: College and university teachers are excluded, and government chemists and chemical engineers are included, insofar as possible, under other industry headings. Accordingly, the BLS matrix figures for all chemists and chemical engineers and for those in government are smaller than the corresponding BLS survey estimates.

The 1970 figures presented here (as well as additional ones used for weighting the 1969 and 1971 research ratios derived for manufacturing industries from the special American Chemical Society surveys mentioned in this chapter) were computed from information provided in Tomorrow's Manpower Needs, Vol. IV: The National Industry-Occupational Matrix and Other Manpower Data (Revised), Bulletin 1737, 1971, pp. 18, 124-25, 128-29.

*Details not shown for all components.

**Includes transportation, communication, public utilities, wholesale and retail trade, finance, insurance, and real estate.

Apart from documenting the wide distribution of chemists and chemical engineers in the economy, Table 9 provides a basis for some structural generalizations concerning the nonacademic subuniverse. For example, in this subuniverse as a whole, the ratio of chemists to chemical engineers is about 2.3 to 1. It is smaller, as a rule, in *Manufacturing* and greater in *Services* and *Government*. In a few cases, however, chemical engineers actually dominate—in *Mining, Electrical Machinery Manufacture, Petroleum and Coal Products,* and *Engineering-Architectural Services.* In 1970, *Manufacturing* accounted for more than 70% of the nonacademic employment of chemists and chemical engineers; the conventional chemical industry, 40%; and a large group of "chemical processing industries," 54%.[17] While about two thirds of the nonacademic chemists concentrated in *Manufacturing* in 1970, the proportion for chemical engineers was four fifths. For the conventional chemical industry alone, the corresponding proportions were 37% and 46%, respectively; for the "chemical processing industries," 51% and 61%.

17. The sum of figures for *Stone-Clay-Glass Products, Primary Metals, Food, (Conventional) Chemicals, Petroleum and Coal Products, Rubber Products,* and *Paper and Pulp* (not shown in Table 9).

CHEMISTRY IN A NONCHEMICAL INDUSTRY:
THE CASE OF INTERNATIONAL BUSINESS MACHINES CORP.

This appendix illustrates the uses of chemistry in a manufacturing company, International Business Machines Corp., that is remote from the conventional chemical industry. Being based mainly on information obtained in 1970, it presents a picture that is not in all respects up-to-date yet helps to explain the employment of substantial numbers of chemists and chemical engineers in a company that makes electronic computers. Since other companies in the computer industry make similar products and use similar methods (with proprietary and other variations, of course), the discussion that follows also reflects in some measure the uses of chemistry by competitors of IBM. Incidentally, the concentration here on chemistry does not imply that other disciplines may not also merit similar attention.

Two paragraphs from the preface to a recent issue of *IBM Journal of Research and Development*, January 1971, p. 2, offer a setting for, as well as a summary of, the rest of this appendix:

"In current computer technology, the most obvious role of chemistry is that involving the study of solid-state inorganics. Diode and transistor fabrication techniques involve silicon, germanium and III-V species like gallium arsenide. Selective and nonselective etching, passivation, and doping are representative of the chemical processes involved.

"Polymeric materials, however, are now being used in increasing quantities. Some polymers, like the epoxy resins that serve as electrical insulators and structural components in printed circuit boards, and the polyurethanes that are used for their protective coating capability, become integral parts of functional units. Others, like the silicone polymers, are used to hermetically seal solid-state devices, thereby assuring stable performance in a defined environment. Some chemical species, although necessary to manufacture solid-state componentry, do not appear in the final product. Typical of these are the light-sensitive polymers (or photoresists) used to selectively mask or protect substrates such as copper and silicon during etching, diffusion and passivation operations."

From Vacuum Tubes to Semiconductors

Without the advance of chemical knowledge, computers could not have been brought to their present state of development. The first electronic computers were dependent upon copper wire circuits and components such as resistors, diodes, capacitors, inductances, and multiple-element vacuum tubes. In the second computer generation, transistors replaced vacuum tubes. This shift to solid state devices sharply reduced needs for space, electric power, and cooling equipment and increased operating speeds. The third generation of computers, of which IBM's System/360 is representative, arrived in the early 1960's. These machines are faster, less costly, more flexible in design, and easier to maintain than their second-generation counterparts.

Physical-chemical principles underlie the production and operation of semiconductor devices, which are vital elements of contemporary computers. A typical device is a single crystal of silicon in which very small amounts of electrically-active impurities are concentrated in well-defined regions. The combination of regions makes up a semiconductor device, such as a transistor or a diode, that performs an electrical function. The geometry of the regions and the types and concentrations of impurities determine the device's electrical characteristics. The impurities are made to diffuse into the crystal at very high temperature. When the device cools, the atoms in its crystal lattice draw together in their normal spacing, locking the impurities into preselected regions in the lattice.

Semiconductor devices, in effect, are solid state valves. They can switch current on and off, convert alternating current to direct current, and amplify current. Compared to the vacuum tube, however, they operate at enormous speed, are compact, and produce relatively little heat.

Solid Logic Technology

IBM System/360 computers are based on a type of circuitry that the company calls Solid Logic Technology (SLT). They are made up of assemblies of SLT circuit packages. Each circuit package contains modules, which are plugged into cards that in turn are mounted on boards. The boards are assembled in multiboard "gates," which are interconnected by cables in the finished computer. The number of gates and the kinds of circuit packages employed determine the performance of the computer.

A module contains a ½ inch ceramic square about $\frac{3}{32}$ inch thick. Each ceramic square carries a complete, miniaturized circuit that may include tiny transistor and diode chips and resistors in a circuit pattern printed in a specially-formulated conducting ink. Around the perimeter of each module are 12 copper connecting pins that plug into the SLT cards, and the complete module is protected by an aluminum case. More than 40 different types of circuits needed in System/360 computers are provided in this module form.

The heart of a module is comprised of silicon chips affixed to the ½ inch ceramic square. Each chip is 0.028 inch square and contains one or more diodes or transistors. The chipmaking process starts with rods of single-crystal silicon, about the diameter of a half dollar, that have been zone refined to contain less than 3 parts per billion of electrically-active impurities. From these rods are cut wafers about 0.008 inch thick. A very thin layer of silicon is vapor grown on each wafer. The surface of this layer is oxidized to form an even thinner layer of silicon dioxide which is then coated with a thin film of photoresist (a polymer containing sensitizers which make it insoluble when exposed to ultraviolet light). Subsequently, each wafer is processed through a series of photoresist production operations, which include:

- Photoresist application.
- Exposure of the desired microelectronic patterns through masks.
- Development of the resist pattern using organic solvents to remove the unexposed resist.
- Etching the oxide films in areas not covered by the resist using chemical etchants such as buffered hydrofluoric acid.
- Stripping away the remaining photoresist pattern by organic solvents such as acetone and isopropyl alcohol.
- Rinsing the wafer in deionized (DI) water, then drying the wafer.

Electrically active impurities or dopants, such as boron, phosphorus, and arsenic, are introduced into the exposed areas by gaseous diffusion in concentrations of the order of 10^{18} atoms per cubic centimeter. A thin film of aluminum is deposited by evaporation, and interconnection patterns are formed by a photoresist-mask process, previously described.

The wafers are then sealed hermetically by a very thin film of glass, and each is cut into 1,000 chips. The chips are subsequently joined to the ceramic square on which a conducting circuit has been previously printed, and the unit is then incorporated into the finished module.

The development, installation, and operation of the SLT production line required considerable chemical knowledge and skill. The production staff, experienced with second-generation computers, had to be reoriented from an essentially electromechanical to a chemical production process. One step that IBM took for conversion was to set up a formal chemical training program that included topics ranging from basic mathematics to elementary physical chemistry. Chemical technicians trained in this program are assigned to each manufacturing department and to the laboratory as well. The production process is supported in addition by a staff of analytical, physical, organic, and polymer chemists.

Since IBM adopted the SLT production system in 1963, it has become a significant commercial user of chemicals. The plant requires 300 to 350 different chemicals in the process. It consumes some 1,000 gallons per day of trichloroethylene, acetone, and propanol to remove organic contaminants from cards and boards at various points in the manufacturing process.

SLT Card and Board Production

SLT cards and boards are cut from basic panels made by the lamination of copper foil and glass fiber cloth impregnated with epoxy resin. Each panel consists of an interlayered mixture of copper and epoxy glass plies. On the outside surfaces, solid sheets of copper foil are always used. Internally, the copper plies are chemically etched in a circuit pattern prior to the pressing of the final composite laminate.

The glass fiber fabric is impregnated with a liquid epoxy resin and is cured to a "B" stage in the drying tower. During the lamination cycle, the epoxy resin is fully cured. (This resin includes chemically-bound bromine to achieve fire-resistant laminates.)

To produce circuit patterns on the solid copper foil, the panels are chemically etched in ammonium persulfate or ferric chloride. Prior to etching, a photoresist is clad to the copper foil. Ultraviolet light crosslinks this photoresist in the desired pattern, converting it to a chemical form insensitive to the etching solution.

The photoresist is developed in a solvent of chlorinated hydrocarbon solvent vapor (commonly trichloroethylene), which dissolves away all of the photopolymer that was not crosslinked by the ultraviolet light. When the copper-clad laminates are immersed in the etching solution, copper is removed from all surfaces except the parts protected by the photoresist. After etching, the photoresist is removed and the finished laminate with a copper foil circuit pattern emerges with a slightly raised copper circuit pattern etched on its face. In practice, circuit patterns are normally applied to both faces of the internal circuit planes, the same processing steps being used simultaneously.

The composite printed circuit laminates, approximately 10 by 15 inches, then have 6,200 holes drilled at the intersections of a $1/8$ inch grid. The drilled panels are deburred, cleaned, dried, and readied for plating. The plating process deposits a thin layer of copper on the entire surface of the panel, including the walls of the holes drilled through it. The copper completes the electrical circuits between the surfaces and internal planes in the laminated composite panels.

The panels move on overhead conveyors through the copper-plating system, which includes 36 bath tanks and requires 21 different chemicals. Ten of the tanks contain one or more chemicals; the remaining 26, containing deionized water, are rinsing baths. The panels are first cleaned electrolytically in trisodium phosphate and then, in successive steps, by ammonium persulfate, cupric chloride, and hydrochloric acid. A stannous chloride bath sensitizes the panels, so that they will accept the fine coating of metal applied in the succeeding palladium chloride bath. Next, an electroless bath adds a thin coating of copper to the palladium base. Finally, a 0.001 inch layer of copper is applied to the panels electrochemically from a bath based on copper pyrophosphate salts.

At this point, the copper-plated card and board panels take different routes in the manufacturing process. The board panels also follow two different routes, one for low-volume output and the other for high-volume.

On the low-volume boards, the photoresist process and an electrochemical process are used for tin-lead-plating of small, square land areas around each of the 6,200 holes. The insides of the holes are tin-lead-plated as well. The plating provides a base for the solder used to attach gold-plated pins that provide electrical connections between board and card circuitry. The pins are automatically inserted in the proper holes and swaged into the board; small solder rings are dropped over each pin; and small solder discs are dropped over each hole that contains no pin. The solder is melted in an oven, fixing the pins to the board and sealing the unfilled holes. The pins are then shaped mechanically, so that ends protruding from one face of the board are ready for wire wrapping and those protruding from the other face are ready to receive the pluggable cards. A photoresist is again applied to the surface of the board. A circuit pattern is exposed on the coated surface by a printed-circuit generator; a computer-controlled high-intensity light source produces a precise pattern while moving on an X-and-Y-axis table. The exposed photoresist is developed and etched to form a permanent printed-circuit pattern on the board. The finished boards and cards, together with the previously described modules, are employed to build the central processing units of computers.

Chemical Research

Applied chemical research has helped in many ways to develop and to operate properly the process for making SLT cards and boards. One topic that demanded attention, for example, was the mechanism for curing the epoxy resin used to impregnate the glass cloth in the laminates. Another chemical challenge was presented by the fact that the photoresist was dissolving small amounts of copper when card and board panels were dip-coated. Still another chemical problem involved the photoresist emulsion used for applying circuit patterns to the high-volume board panels. The photoresist in the holes, crosslinked intentionally before etching, was extremely difficult to remove before tin-lead-plating. The problem was overcome by use of a dry-film photoresist developed by du Pont. The film simply covers the holes, protecting them from the etching solution.

IBM and other computer manufacturers, of course, have a considerable interest in materials other than those used in electronic components for their machines. Chemically-related research at an IBM materials science laboratory includes projects on accelerated aging of polymeric materials, reaction kinetics in adhesive curing, and the rheology of coatings. One line of inquiry has led to development of a small, continuous belt for use on check-sorting machines in banks. The revolving belt peels checks, one at a time by friction from a stack, and deposits them in the machine for processing. It is made of polymeric foam, and its surface does not glaze with continued use.

The need for chemical knowledge in designing and making computers of the future is evident at IBM's T. J. Watson Research Center. In the physical sciences department there, 15 to 20% of the professional employees are trained chemists or show specialization in chemistry.

One project at the Research Center, in high-temperature chemistry, is concerned with materials such as europium selenide and oxide. It aims at determining the thermodynamic properties of the materials and establishing the mechanisms of vapor transport. The work is basic, strictly speaking, but also has practical implications, since vapor transport is used to diffuse dopants, such as boron and phosphorus, into solid state devices.

Another project relates to the deposition of silicon on substrates at relatively low temperatures. The standard method deposits silicon from a vapor of silicon tetrachloride and hydrogen at about 1,000° C. Silicon can also be deposited, however, from a vapor of silicon tetrahydride at 600° to 900° C. A still lower temperature is sufficient when a radio-frequency discharge is substituted for heat to decompose the molecule of silicon tetrahydride. Controlled deposition of silicon at relatively low temperature would improve stability; i.e., diffused materials would not spread beyond their intended domains, and metals in the system would not alloy. As computer circuits and tolerances become ever smaller, low-temperature deposition becomes increasingly desirable.

A major goal of the computer industry is to reduce the time required "to address the computer"—to feed information into it. Instead of keypunching, as today, it may become practicable to scan documents with a device that converts the printed contents into computer language electronically. One component of such a system could be electrooptical crystals, acting as light switches. A number of compounds, such as niobates of barium and sodium, seem eligible for use. Known for more than 70 years, these substances have only in the past decade come under scrutiny by computer and other companies. Although liquid crystals resemble liquids, mechanically, they also have certain properties of solid crystals. For example, they scatter light in symmetrical patterns. They register minute changes in temperature, mechanical stress, electromagnetic radiation, and chemical environment by alteration of color. Molecules dissolved in one type of liquid crystal, the nematic phase, are aligned regularly, instead of in the random fashion more typical of substances in solution. Because of this property, solutions in nematic liquid crystals can be used to modulate light for display purposes.

Finally, in addition to product-related research, IBM does a certain amount of chemical research that may be related only indirectly or not at all to its main business. One such project seeks to develop a good test for the Woodward-Hoffman orbital symmetry rules. These rules are used to predict the course of simple chemical reactions of a purely thermal nature. Although this research effort is basic, the results could have future applications to IBM's business. Some of the compounds being studied are organometallics usable for solid state devices. Another project involves the computer simulation of complex chemical systems. A third chemical project inquires into protein structure and function.

Extensive as the foregoing exposition of chemistry's participation in the computer business may appear, it tells only a small part of the story. It refers mainly to solid logic technology. Other major parts of any computer, such as memory and software (or programming), also require a variety of chemical skills. Furthermore, like companies in other industries, the manufacturers of computers have environmental problems that chemists will need to solve.

From the cited examples it is evident these chemists and chemical engineers in the computer industry must interface and communicate with scientists and engineers having a wide spectrum of backgrounds—electrical engineers, physicists, mechanical engineers, mathematicians, and computer scientists. This generalization seems as true for the large-scale computational chemistry problems addressed in San Jose (California) as for the metal-to-ceramic adhesion problems addressed in Burlington (Vermont).

27 / **Chemical Industry Trends and Input-Output Connections**

Statistics for the chemical industry itself, both broadly and narrowly defined, add dimensionality and detail to our picture of the economic role of chemistry. The employment distributions for chemists and chemical engineers have already confirmed that the industry is a site of intensive economic application of chemistry. Furthermore, the industry has historically proved a preferred instrument—on the basis of comparative cost and the performance of its products—for serving various consumer, business, and government needs that could also be satisfied technically by other means. Thus, the influence of chemistry on styles of working and living is also traceable, though not always strikingly or unambiguously, through changes occurring in the chemical industry and the other industries with which it competes or cooperates.

The statistics that follow provide a sampling of the available evidence concerning trends, structure, and economic interdependence. The analytical value of some of these statistics has been enhanced by their development in conformity with the Standard Industrial Classification (SIC) system and their tie-in with the national income and product accounts and with various input-output models. Some additional information that is eligible for presentation here, such as on productivity and prices, is reserved for discussion in a technological context in Chapter 28.

A reminder is in order with regard to statistical imperialism. Avoidance of the error of misidentifying the chemical industry with chemistry should not encourage the opposite error of overclaiming that chemistry is the economic prime mover, the *sine qua non,* wherever it is practiced. No matter how broad a definition is adopted, Chapter 26 has made clear that some chemistry will exist beyond the industry's perimeter. On the other hand, nonchemistry always coexists with chemistry, and the broader the industry definition the greater is the coordinate role of nonchemistry. The moral is clear: Chemistry is part of a larger system, never a unique, unambiguous, and definitive "cause" of growth, progress, and efficiency. It does offer technical highways for economic development, but other skills and disciplines have to collaborate in the economic realization of such opportunities for the pervasiveness, criticality, dynamism, and substitutability of chemistry to be manifested.

Three Indicators of Trend and Structure

Table 10 reports, in terms of three economic indicators, on the comparative performance of the chemical industry and the rest of the economy in the period 1929–70. It is based on information published by the Department of Commerce. One of the three indicators relates to industrial contributions to the national income, an aggregate that measures economic activity comprehensively and without duplication. This aggregate is affected by price inconstancy, however, so it is not offered as a gauge of the changing "volume" of "real" net product or net input. In the next section of this chapter, some other figures are presented that do reflect the changing "physical" output of chemicals and of manufactures in general. The second indicator included in Table 10 refers to the average numbers of full-time-equivalent employees in the chemical and other industries, and the third pertains to the corresponding average annual earnings.

Only a small fraction of the national income originates in the conventional chemical industry (i.e., chemicals and allied products), but the share improved between 1929 and 1970. The improvement, moreover, would survive adjustment for price changes, since the relative performance of the chemical industry in this report has been outstanding (as Chapter 28 will make clear). According to Table 10, the industry accounted for 1.3% of the national income in 1929 and for 2.0% in 1970. The gain had been accomplished by 1950 but not because of any slackening of the industry's own growth. The period since then has witnessed a phenomenal extension of government's role. Between 1950 and 1970, the share of the chemical industry in the nongovernment portion of the national income actually advanced from 2.0% to 2.4%. A rise from 5.0% to 7.4% is also indicated for 1929–70 in the conventional chemical industry's share of the entire manufacturing contribution to the national income. From Table 10 it may be seen that the gain in this case had been accomplished by 1960.

When the conventional chemical industry is selectively supplemented and thereby transformed into the "chemical process industries" (CPI),[1] a much greater proportion of the national income is represented, but the composite also grows much less rapidly and even fails to keep pace with the rest of manufacturing and the rest of the economy. Apart from differential price changes, this record reflected such varied phenomena as competition among the products of CPI (e.g., plastics, metals, rubber, and glass) and a sufficiently sharp rise in petroleum-refining productivity to inhibit (or even reverse) the industry's employment growth. In 1929, CPI generated 11% of the national income. Its share rose to 13% by 1950 and receded to 10% by 1970. CPI accounted for 55% of the national income from manufacturing in 1929 but only 40% in 1950 and 36% in 1970.

The comparative employment record (Table 10) is a generally good one for both the conventional chemical industry and CPI. Between 1929 and 1970, when the number of full-time-equivalent employees doubled in the economy as a whole and in all manufacturing, it more than doubled in the conventional chemical industry and more than trebled in CPI. In 1950–70, the performance of the conventional chemical industry and CPI again surpassed that of the whole economy and all manufacturing.

At this point, we briefly turn back to Table 8 (Chapter 26) for an answer to this question: In 1950–70, how did chemists fare compared to employees in general in the industries and sectors just considered? In the two decades, jobs for chemists increased much more rapidly than for nonchemists in all industries combined, in manufacturing as a whole, and in conventional chemical production. Thus, the employment of chemists in the economy was 2.7 times as great in 1970 as in 1950, but the multiple for all workers was 1.5. In all manufacturing and in the conventional chemical industry, the numbers of chemists rose by 1970 to 2.6 and 3.3 times the 1950 magnitudes, but the corresponding ratios for all employees were only 1.3 and 1.6. In CPI, the gains registered by chemists and all employees were sizable and similar over the two decades, 2.5 and 2.4, respectively.

Returning to Table 10, we observe that the average annual earnings of full-time-equivalent employees in the conventional chemical industry consistently exceeded the averages for all industries and for all manufacturing.

Physical Output of the Chemical Industry

Statistics on the volume of output of the chemical industry, which abstract from price change, leave no doubt as to the dynamism of chemistry. Indeed, they recall some of the other characteristics of economical

---

1. The additional industry groups make petroleum products, rubber products, primary metals, stone-clay-glass products, food, and pulp and paper.

Table 10

National Income, Employment, and Earnings: Chemical and Other Industries, 1929–70

	1929	1950	1960	1965	1969	1970
National Income (Millions of Dollars)						
All Industries	$86,795	$241,074	$414,522	$559,020	$763,660	$759,887
Manufacturing	21,945	76,223	125,822	170,408	221,947	217,735
Chemicals*	1,123	4,944	9,159	12,332	16,023	16,113
Chemical Process Industries**	9,653	30,336	49,229	62,129	77,554	78,608
Employees (Thousand Full-Time-Equivalent)						
All Industries	35,896	49,059	57,098	62,815	71,823	71,576
Manufacturing	10,428	15,241	16,761	18,088	20,264	19,425
Chemicals*	397	637	831	910	1,064	1,045
Chemical Process Industries**	3,959	5,222	5,625	5,594	12,891	12,659
Average Annual Earnings of Employees (Full-Time-Equivalent)						
All Industries	$1,405	$2,992	$4,743	$5,705	$7,095	$7,564
Manufacturing	1,543	3,302	5,352	6,386	7,775	8,150
Chemicals*	1,673	3,819	6,414	7,553	9,147	9,728

Source: The National Income and Product Accounts of the United States, 1929–1965: Statistical Tables, U.S. Department of Commerce, 1966, pp. 18-21, 102-09, for 1929–65; and Survey of Current Business, July 1971 (Part 1), pp. 17, 36, for 1969 and 1970.

*Chemicals and Allied Products.

**Chemicals and Allied Products, plus Food and Kindred Products, Paper and Allied Products, Petroleum and Coal Products, Rubber Products, Stone-Clay-Glass Products, and Primary Metal Industries (Ferrous and Nonferrous).

importance mentioned in Chapter 25—persistence, growth, competitiveness, and enlargement of opportunity for choice in production and consumption. The experience of a much more rapid expansion of chemical output than of manufactures in general has certainly enhanced public awareness of the pervasiveness of chemistry in our ways of working and in our styles of living. The paragraphs that follow focus first on the revised production measures published recently by the Board of Governors of the Federal Reserve System (FRS) for the period since 1954 and then extend the perspective backward to the beginning of the century.

The new FRS indexes show a doubling of manufacturing production from 1954 to 1970, but they also reveal much more substantial increases for the chemical industry, for some of its components, and for a limited version of CPI.[2] In the same interval, the index for *Chemicals and Products* (as the conventional chemical industry is designated in the source document) virtually quadrupled, registering a rise from 31.5 to 120.2 (1967 = 100). Still greater increases were recorded for such components as *Synthetic Materials* (more than a fivefold gain, from 24.6 to 137.5) and *Drugs and Medicines* (a sixfold gain, from 22.9 to 137.7). For a limited concept of CPI, embracing *Chemicals, Petroleum, and Rubber,* the output index more than trebled, expanding from 35.4 to 118.2.

Even more dramatic is the relative performance of the chemical industry over a longer span of seven decades. A tale of fantastic growth unfolds when index numbers that have been published by the National Bureau of Economic Research (NBER) for 1899–57 (1929 = 100) are linked in 1957 to the revised FRS series.[3] According to the spliced NBER-FRS series (1929 = 100), the 1970 output of the conventional chemical industry was close to 100 times that of 1899 (the computed multiple is 94)! The figures also show an *acceleration* of growth for the second half of the period:

1899	18.6	1948	309.8
1909	35.4	1953	470.9
1919	51.5	1957	612.8
1937	123.9	1970	1,741.3

A comparison with all manufacturing production over the same seven decades discloses that chemical output grew more than five times as rapidly. Thus, the NBER index for manufacturing shows an advance from 27.5 to 264.6 between 1899 and 1957 (1929 = 100), and this multiplication of nearly 10 was more than matched by the chemical industry's factor of more than 30. When the NBER and FRS indexes are spliced in 1957, the resulting 1970 index number becomes 454.8 (1929 = 100), less than 17 times the 1899 figure. This multiple for manufacturing, which in some other context would be impressive, is puny indeed compared to that for the chemical achievement.

For the limited domain of CPI represented by *Chemicals, Petroleum, and Rubber,* a still more fabulous gain was apparently registered during our century. Thus, between 1899 and 1957, when the output of the conventional chemical industry multiplied 33 times, according to NBER, the rise for *Petroleum Products* was practically as great (32 times, from 8.7 to 278.2, base 1929 = 100), but the rise for *Rubber Products* was very much greater (52 times, from 4.3 to 223.7). In view of a subsequent trebling

---

2. This paragraph is based on "Industrial Production—Revised and New Measures," *Federal Reserve Bulletin,* July 1971, pp. 551-76, and on printouts of FRS computer runs for individual industries and industry groups covering the period 1954-71.

3. The NBER indexes discussed in this section are presented in J. W. Kendrick, *Productivity Trends in the United States,* Princeton University Press, Princeton, 1961, pp. 464, 471, 472.

indicated by FRS figures for 1957–70, the composite 1970 output for *Chemicals, Petroleum, and Rubber* must have exceeded the 1899 volume by more than 100 times.

Input-Output Relationships

Tables showing interindustry sales and purchases—and others derivable from them by matrix algebra—offer additional insights into the economic pervasiveness and criticality of chemistry. For this reason, they are now considered here briefly, but they are not reproduced because of their massiveness.

The basic table of interindustry transactions exhibits the pattern of "intermediate" consumption and production flows linking the industries to each other, but it also does more: It establishes a vital mathematical connection between industry sales and purchases and the "vector of final demand," the "bill of goods" making up the Gross National Product. By identifying in each instance the portion of an industry's output intended for final markets, the transactions table sets the stage for subsequent calculations designed to translate perturbations in the vector of final demand into required adjustments in the pattern of industry purchases and sales and in other affected economic variables, too, such as employment. Later, an example is given that focuses on the incremental need for chemists implicit in a change in final demand.

Department of Commerce. Attention is directed first to some gleanings from the latest published Department of Commerce report on input-output relations. This report, issued in 1969, refers to the 1963 structure of the United States economy. The results of the painstaking Commerce inquiry are organized in three huge tables, or matrices, which are not presented here. One of these tables shows interindustry transactions, the second shows "direct requirements," and the last shows "total requirements." A descriptive comment and a substantive one are made about each of these tables in turn.

Actually, two versions of each of the three Commerce tables were published—a compact variant distinguishing more than 80 industries and a detailed variant dividing the total economic domain into 370 parts. Our comments are based on the compact tables, which are sufficiently refined for our needs. After all, they show separate figures for the four main components of the conventional chemical industry: *Chemicals and Selected Chemical Products, Plastics and Synthetic Chemicals, Drugs and Cleaning and Toilet Preparations,* and *Paints and Allied Products.*[4]

Each row of the 1963 transactions table[5] shows an industry's sales to itself (i.e., to companies in the same industry), to other producers, and to final users, while each column shows the sources of a particular

4. The condensed Commerce tables are presented in "Input-Output Structure of the U.S. Economy: 1963," *Survey of Current Business,* November 1969, pp. 16-42. The expanded versions make up a three-volume processed report of the same name, published in the same year and designated as a supplement to the *Survey.* An updated report on *Input-Output Transactions: 1966* (Staff Paper No. 19, Bureau of Economic Analysis) has more recently been published (February 1972) for the compact variant.

5. The row totals and column totals for industries, excluding the final-demand and value-added borders, have the same grand total, which exceeds the national income and even the conceptually larger Gross National Product (i.e., total value added). This turgidness reflects the use of producers' prices as the transactions for the different productive stages are cumulated. If the intermediate industry transactions were expressed in "adjusted" net prices instead of "actual" transfer prices, the duplication of money values could be avoided. The object of the transactions table, however, is something else. The aim is not to distribute the national income or national product according to industry of origin, but to establish a structural connection between the essentially unduplicated output of the productive system and the corresponding pattern of gross interindustry purchases and sales.

industry's purchase of goods and services. The sales of every industry to final markets are included in a vertical (righthand) border. These sales satisfy the "final demand" levied on the economy by consumers, business firms (for capital account and inventory changes), government, and foreigners. A horizontal border at the bottom of the transactions table shows the "value added" by every industry's activity, or the net contribution to Gross National Product visualized as a sum of generated incomes or quasi-incomes.

The condensed 1963 transactions table makes it clear that the four components of the conventional chemical industry diffuse their products widely, although much of the economic activity entailed is not evident to the layman. Indeed, the sales network embraces most of the other industries in the table. On the other hand, relatively little of the chemical industry's total output is sold directly to final users (drugs, cosmetics, and soaps are among the exceptions), and this means that the contributions of the industry are frequently embedded in final products along the way. An actual count discloses that *Chemicals and Selected Products* sold at least $500,000 of output to 78 out of the 82 industries constituting the economy; *Plastics, etc.*, to 42 out of 82; *Drugs, etc.*, to 56 out of 82; and *Paints, etc.*, to 49 out of 82.[6]

The first Commerce table, referring to transactions, also permits comparison of the intermediate and final sales of the subindustries of the conventional chemical industry. Except in the case of *Drugs and Cleansing and Toilet Preparations*, the dollar sales of chemical firms to each other and to outside industries in 1963 greatly overshadowed the distribution to final consumers (which totaled $9,968 million):[7]

Industry	Final Demand (Mill. Doll.)	Intermediate Output (Mill. Doll.)	% of Ind. Total
Chemicals and Sel.			
Chem. Prod.	2,885	14,009	82.9
Plastics, etc.	633	5,708	90.9
Drugs, etc.	6,351	2,702	22.9
Paints, etc.	99	2,363	96.0

If the conventional chemical industry is regarded as a unit and all the sales within its confines are washed out, net intermediate sales to other industries remain substantial. Sales to final users are invariant to changes in industry classification (they still add to $9,968 million),[8] but the consolidation of the four subindustries and the elimination of the entire conventional industry's sales to itself reduce the intermediate sales total considerably. This total declines from $24,782 million (the sum of entries in the middle column of the table presented in the preceding paragraph) to $16,643 million. This smaller figure, the derivation of which is allowed by the detailed information provided in the transactions table, still comprises five eighths of adjusted total dollar sales to other industries and to final users ($26,611 million, the sum of $16,643 million and $9,968 million).

The second Commerce table for 1963 shows "direct requirements"— the various industrial contributions (inputs) purchased per dollar of each

---

6. *Survey of Current Business*, November 1969, pp. 30-35. Covers transactions of Commerce industry numbers 1 to 83 except for 80A and 80B, which refer to imports. (Although the condensed Commerce table of transactions also shows stubs for industry numbers 84-87, the cells are blank.)

7. *Ibid.*, p. 35.

8. The national income and Gross National Product are also unaffected either by subdivision or aggregation.

industry's sales. In addition to accounting for the inputs purchased from other industries, each column indicates (at the bottom) a particular industry's own value added per dollar of gross output. Accordingly, each column necessarily adds exactly to $1.00. Since the Commerce publication cites chemical production in explanation of the column entries and of their significance for revealing interdependence, we quote [9] instead of making a comment of our own:

"Table 2 relates each of the inputs of an industry to its total output. Each column shows the inputs that the industry named at the top of that column required from each of the industries named at the beginning of the rows to produce a dollar of its output. For example, to produce a dollar of output, the chemicals manufacturing industry (27) required 18 cents of its own production, 6 cents from the petroleum refining industry (31), 3 cents from the chemical mining industry (10), etc.

"Table 2 shows the heavy interdependence among the industries which is a feature of the U.S. economy. Almost all industries require inputs from at least 30 others. As many as 57 industries required inputs from over 50 industries. The chemicals industry (27), for example, required inputs from 73 industries, only eight of which are producers of basic raw materials."

While the direct input-output coefficients making up the second Commerce table are derivable immediately from the transactions table, the elements of the third table must be obtained from the second by means of matrix inversion, a Herculean computational task. These elements, or inverse coefficients, show "total requirements," or the direct plus indirect requirements of output from other industries per dollar of a particular industry's delivery to final demand. It is the interdependence of industries that gives rise to indirect requirements.

The third Commerce table contains virtually no empty cells, reflecting the interaction of industries more subtly and more completely than does the second table. Looking along the principal diagonal (i.e., from the upper left to the lower right), at the rows, and at the columns, we may discern characteristically different patterns for the four components of the conventional chemical industry.

Among the four components, *Chemicals and Selected Chemical Products* is unique. Its entry along the principal diagonal, which shows the output required of *itself* both directly and indirectly per dollar of its *own* delivery to final demand, not only exceeds those for the rest of the conventional industry but also ranks seventh among the 82 diagonal elements for the economy. Many other industries, too, depend significantly on the output of *Chemicals and Selected Chemical Products* for their contributions to final demand, as the figures in the appropriate row demonstrate.[10]

The columns of the third Commerce table reveal the extent to which the deliveries of the conventional chemical industry to final demand de-

9. *Survey of Current Business*, November 1969, pp. 23–24. The numbers given in parentheses are based on an early version of the Standard Industrial Classification. Thus, No. 27 stands here for one part of the conventional chemical industry, *Chemicals and Selected Chemical Products*.

10. Thus, at least 5 cents of the output of this chemical component were directly or indirectly required in 1963 per dollar of delivery to final demand by farmers (other than livestock), nonferrous mining, makers of various textiles and apparel, paper and paperboard mills, printing, each of the other components of the conventional chemical industry, leather, rubber, plastics, optical goods, and office supplies. For *Plastics and Synthetic Materials*, on the other hand, this criterion of dependency of other industries is met in very few instances: by various textile industries and apparel, only by paints and itself within the chemical foursome, and by rubber. The rows for the remaining two components of the conventional chemical industry, *Drugs*, etc. and *Paints*, etc., disclose *no* instances (other than these industries themselves) in which output of at least 5 cents was directly or indirectly required per dollar of delivery to final demand.

pend directly or indirectly on purchases *from* other industries of the economy. Each of the four industry components relies heavily on the output of materials suppliers (e.g., petroleum refineries as well as firms within its own boundaries). Furthermore, each relies on utilities and other service industries and on additional supporting industries (e.g., food processing).[11]

Among the derivative computations made by Commerce are direct-indirect partitions of each industry's output attributable to components of final demand. If attention is confined to one category of final demand, personal consumption expenditures, it may again be observed that *Drugs, etc.* is unique among the four subdivisions of the conventional chemical industry. This category of final demand required about five sixths of the output of *Drugs, etc.* Furthermore, most of this impact was direct.[12]

Harvard Economic Research Project. From an authoritative nongovernment source, the pioneer Harvard Economic Research Project, has recently (1970) come a valuable addition to input-output literature that also tells some more about the economic role of chemistry.[13] This Harvard report takes advantage of the vast library of interindustry statistics developed for earlier years (1939, 1947, and 1958) to derive informative temporal comparisons.

First of all, the Harvard volume pays an implicit compliment to the conventional chemical industry. It acknowledges the criticality of the industry and the diffused dependency of the rest of the economy on the industry's output by identifying *Chemicals* as one of the five principal areas of the productive system. The other four areas are designated as *General* (utilities, fuels, transportation, and other services), *Materials* (e.g., stone, wood, metals, rubber, and agriculture), *Metalworking* (mostly vehicles, machinery, and instruments), and *All Other* (food, textiles, paper, leather, mining other than coal, etc.). The *Chemicals* area is by far the smallest of the five—in terms, for example, of percentage of the total gross output.

A few passages from the book that illustrate the positive power of input-output thinking also make specific reference to the participation of chemistry in the economy. For example, a standard comment on the nature of direct and indirect requirements cites the interplay of chemicals and textiles:[14]

---

11. Thus we see that, per dollar of sales to final demand by *Chemical and Selected Chemical Products* in 1963, at least 5 cents of output were directly or indirectly required from itself, from crude petroleum and natural gas, from petroleum refining, from various general services (such as utilities, transportation, and wholesale and retail trade), and from imports. For *Plastics and Synthetic Materials*, this criterion is met only by itself, *Chemicals and Selected Chemical Products*, paper, petroleum refining, and various services. For *Drugs, etc.*, the outstanding suppliers per dollar of its deliveries to final users are, in addition to some of the services and itself, food and *Chemicals and Selected Chemical Products*. For *Paints, etc.*, we find *Chemicals and Selected Chemical Products* again providing major support—as well as *Paints* itself, various services, foods, metal containers, and petroleum refining.

12. Personal consumption expenditures accounted for 53.2% of the output of *Chemicals and Selected Chemical Products* in 1963, but almost all of this amount, 50.9%, was required indirectly. For *Plastics and Synthetic Materials*, the corresponding figures were 61.7% and 61.5%; for *Paints, etc.*, the figures were 48.6 and 47.7. A much different picture is presented by the figures for *Drugs, etc.*—84.0 and 24.0. That is, personal consumption expenditures accounted for 84.0% of the 1963 output of *Drugs, etc.*, and most of this levy (60.0% of the output) was direct. (See *Survey of Current Business*, November 1969, p. 21.)

More recent figures (for 1966) show a fairly similar pattern. Thus, for *Drugs, etc.*, personal consumption expenditures represented 82.9% of the total output—60.2% direct and 22.7% indirect. (See *Input-Output Transactions: 1966*, Table B.)

13. A. P. Carter, *Structural Change in the American Economy*, Harvard University Press, Cambridge, 1970. Also pertinent are several recent articles by Carter: "Changes in the Structure of the American Economy," *Review of Economics and Statistics*, May 1967, pp. 209-24; "Technological Forecasting and Input-Output Analysis," *Technological Forecasting*, Spring 1970, pp. 331-45; and "The Economics of Technological Change," *Scientific American*, April 1966, pp. 25-31.

14. *Structural Change in the American Economy*, p. 25.

"The inverse coefficient for chemicals into textiles measures the total amount of chemicals required (directly and indirectly) to deliver one unit of textiles to final demand. Chemicals will be required directly in making textiles; indirectly in making synthetic fibers, agricultural products, and other inputs used by textiles; and indirectly again in making inputs into synthetic fibers, into agricultural products, and into other textile inputs, and so on."

According to another statement,[15] the dependence of textiles upon chemical inputs increased between 1947 and 1958 despite a contrary suggestion by the measure of direct requirements, which does not reflect the second-order and higher-order interactions among industries:

"For example, the increased use of synthetic fibers in the textile industry between 1947 and 1958 resulted in a decrease in direct purchases from the chemicals industry. Decreased use of natural fibers led to a cut in requirements for the chemicals used to clean them and in chemicals used in finishing. While the direct coefficient for chemicals decreased, that for synthetic fibers rose. Production of synthetic fibers requires a large chemical input. Thus, while the direct linkage of textiles with chemicals was weakened between 1947 and 1958, the total (direct plus indirect) dependence of textiles on chemicals increased."

Some other sampled passages in the Harvard report cite chemistry in connection with the important economic phenomena of substitution and relative price change. According to one of them, if the inverse coefficients of different years are played against a fixed vector of final demand, a shift over time toward chemical inputs is discerned:[16]

"Such computations show, for example, that more chemicals, plastics, and electronic components and less steel, wood, and nuts and bolts are required to deliver the same final demand with 1958 input structures than with 1947 ones."

Another passage notes that the shift toward chemical inputs is motivated, in part, by favorable price differentials:[17]

"Since World War II, aluminum, concrete, and, most dramatically, plastics have been successfully challenging the other materials in many specific markets, and specialization of materials seems much less rigid than it used to be. Changes in the relative price picture certainly have had some bearing. Figure 6.1 . . . shows that prices of copper, steel, and timber have been increasing relative to those of aluminum, plastics (polyethylene and polyvinyl chloride), and concrete. These changes are, in turn, dependent on structural changes in producing sectors . . ."

In general, according to the Harvard report, "qualitative improvements in each of the materials," including the "creation of new plastics materials with special properties," are increasing the "potential scope of price competition."[18] For the future, a continuing growth in the use of plastics is seen, since they are "cheap relative to most other materials, and their price advantage appears to be increasing with expanded usage."[19]

15. *Ibid.*, pp. 25-26. Note might be taken also at this point of a publication by P. F. Rice and P. E. LaFerney, *Use of Input-Output Analysis in Studying Industry Problems: Applied to Employment Changes in the U.S. Textile Industry*, Tech. Bull. No. 1411, U.S. Department of Agriculture, February 1970. This report analyzes the net employment decline implicit in the substitution of a given value of man-made fibers for the same value of cotton (pp. 14-16). It observes that the incident employment gain registered by the chemical industry "is actually the net result of increased shipments of man-made fibers and decreased shipments of agricultural chemicals necessitated by the reduced cotton output."

16. *Structural Changes in the American Economy*, p. 27.

17. *Ibid.*, p. 84.

18. *Ibid.*, pp. 84-85.

19. *Ibid.*, p. 96.

The Harvard report's most distinctive feature is its juxtaposition of the 1961 bill of final goods and services and the technological coefficients of earlier years (1939, 1947, and 1958). The series of estimated requirements derived through this confrontation are useful for analyzing substitution and structural change.

One set of Harvard calculations shows the total output required of the industries in each of the five areas for achieving the Gross National Product of 1961. Expressed in constant (1947) dollars, the required gross output of chemicals comprised only 3.0% of the grand total for all industries when the technological coefficients for either 1939 or 1947 were used. This percentage rises, however, to 3.6 and 3.8, respectively, when the 1958 and 1961 technological structures are posited for accommodating the 1961 vector of final output. On the other hand, declines are registered for 1939–61 in the shares of the grand total that are represented by the gross output requirements from the *Materials* and from *All Other* industries.[20]

Another set of matrix computations shows that the same 1961 bill of goods and services required sharply declining labor inputs as the later input-output structures of the period 1939–61 were postulated. In other words, labor productivity advanced for the economy at large, and the details show that the chemical area contributed to the labor saving. In 1939, an estimated 100 million man-years would have been required to accomplish a final output for the economy that did in fact require only 58 million in 1961. The relative share of the chemical area in the employment total remained small, rising from 1.1% of all man-years in 1939 to 1.4% in 1961. Translated into absolute numbers, these percentages signify a decline of more than one fourth in the man-year requirements represented by chemicals, or a decline from 1.11 million man-years to 0.81 million.[21] (More conventional estimates of changing labor productivity are presented in Chapter 28.)

Direct and Indirect Requirements for Chemists and Chemical Engineers

Table 11 presents some results of an input-output experiment performed by the Bureau of Labor Statistics at the request of the American Chemical Society. The object of the exercise was this: To estimate for a recent year (1967) the direct and indirect employment implications of an extra $1 billion of final demand (1958 prices) for the output of various industries. Using an interindustry employment table that reflects the 1958 input-output structure and an occupational matrix, BLS derived separate direct and indirect figures for all employees, chemists, and chemical engineers. Alternative industry targets were assumed for the $1 billion outlay, so that differential implications for employment could be discerned. Thus, computations were made for the four components of the conventional chemical industry, for two "soft-goods" industries (food and apparel), and two "hard-goods" industries (communications equipment and automobiles).

---

20. *Ibid.*, p. 38. The continuing substitution in favor of chemicals is reflected in this passage from *Input-Output Transactions: 1966*, p. 2: ". . . [I]ndustries required in 1966 more electronic components, plastics and synthetic materials, and heating, plumbing and structural products than they would have under the 1963 technology. In contrast, they required substantially less nonferrous metal ores, wooden containers, and maintenance and repair construction."
Additional relevant contributions of the Department of Commerce staff can only be mentioned here. See, for example, B. N. Vaccara and Nancy Simon, "Factors Affecting the Postwar Industrial Composition of Real Product," in *The Industrial Composition of Income and Product,* Columbia University Press, 1968, pp. 19-58; and two papers by Vaccara on changes in input-output coefficients between 1947 and 1961, published by the Department (processed) in June 1968 and July 1969.
21. *Structural Change in the American Economy*, pp. 39-41.

Table 11

Estimated Employment Requirements in 1967 per $1 Billion (1958 Prices) of Final Demand from Selected Manufacturing Industries

| Employment Category | Basic Chemicals (27) | Chemicals and Allied Products | | | Food (14) | Apparel (18) | Communications Equipment (56) | Automobiles (59) |
		Plastics (28)	Drugs (29)	Paints (30)				
				Industry Designation and Number				
Total Employment	62,029	66,725	71,880	70,213	97,514	140,281	95,809	73,971
Direct	27,220	21,952	22,134	24,315	27,771	85,117	44,729	24,784
Indirect	34,809	44,773	49,746	45,898	69,743	55,164	51,080	49,187
Chemists	1,178	1,154	1,175	1,567	117	164	112	89
Direct	1,113	650	976	1,111	64	9	58	25
Indirect	65	504	199	456	53	155	54	64
Chemical Engineers	620	482	186	628	26	66	80	37
Direct	599	222	86	411	6	. .	58	10
Indirect	21	260	100	217	20	66	22	27

Source: U.S. Bureau of Labor Statistics. Each column indicates the direct and indirect implications of $1 billion of final demand satisfied by the designated industry. The estimates are based on an interindustry employment table reflecting the 1958 input-output structure and on an occupational matrix. The indirect employment requirements for chemists and chemical engineers are understated somewhat. Direct requirements include small amounts of secondary employment corresponding to the feedback of demand to the initially-impacted industries.

It is tempting to interpret the numbers shown in Table 11 as the marginal job impacts of alternative increments of a fixed dollar–amount of final demand, but the numbers provide only a rough idea of such impacts. For example, $1 billion of extra final demand imposed on only one industry would be distortive. This load on a narrow industry base would most probably affect the input-output structure that the computation assumes to be given. Strictly, an input-output system for a particular year is not independent of its literal economic context—the actual mix of products, the constellation of market prices, the preferences of customers, the stage of the business cycle, and so forth. The larger the postulated increment in demand and the more specialized the composition, the greater is the burden placed on the approximative virtues ascribable to an input-output model. Additional infirmities and caveats could be cited that relate to this particular exercise. For example, the figures shown in Table 11 for direct and indirect requirements of chemists and chemical engineers understate the total impact, since they are confined to major employing industries.[22]

A few inferences may be safely drawn, however, from the results. At least four points are suggested by an examination of the several columns of Table 11:

• Employment implications, in general as well as for chemists and chemical engineers, vary according to the composition of a given dollar-amount of final demand, despite a common "Keynesian" assertion to the contrary. Thus, $1 billion may mean twice as much general employment if the extra demand favors the products of the apparel industry rather than the products of the conventional chemical industry. This same preference, on the other hand, also implies a smaller benefit for chemistry's professional manpower. Thus, it suggests only a tenth as many new jobs for chemists and about a tenth to a third as many for chemical engineers. The variation in job implications should obviously affect the design of any program intended, say, to remedy unemployment in the general work force or among professionals.

• The job opportunities entailed by $1 billion of final demand for any industry's products appear substantial, but the numbers and proportions represented by chemists and chemical engineers are comparatively small. In any part of the conventional chemical industry, according to Table 11, such an extra demand for goods and services corresponds to 60,000 to 70,000 employees, but only 2,000 or fewer of these would be chemists and chemical engineers. The implications with respect to general employment are even greater for the other "soft-goods" industries or for "hard-goods" industries, yet only 200 or fewer new slots would be provided for chemistry's professionals. The share for chemists and chemical engineers is in the neighborhood of 2 to 3% in the case of conventional chemical production but only 0.14 to 0.2% when the final demand is focused elsewhere. This variation in chemistry-intensiveness of the same dollar-amount of demand is very pertinent to forecasting and planning.

---

22. Attention should be called to the conventional character of partitioning techniques for breaking observed or projected changes in actual employment into components associated with changes in final demand and other specific factors, such as unit labor requirements and technical coefficients. (See, for example, R. E. Kutscher and E. E. Jacobs, "Factors Affecting Changes in Industry Employment," *Monthly Labor Review*, April 1967, pp. 6-12, and the papers cited in the second paragraph of footnote 20.) Unfortunately, when more than one influence on employment is at work, second-order and higher interactions among the variables are also present. Since these interactions are sometimes sizable, arbitrary rules for dealing with them too often lead to distortions rather than adjustments. The BLS experiment discussed here has the merit of concentrating on the "pure" employment change associated with a change in final demand only—a final demand, moreover, that is specific in kind and magnitude.

• Indirect employment implications of $1 billion of final demand are always relatively high for the work force in general, and they often are relatively high for chemists and chemical engineers also, despite some underestimation in the calculations. Accordingly, the magnitude of indirect requirements should be taken into account in forecasts and programs relating to chemistry's professional manpower. Since the input-output model permits appraisal of indirect requirements, it is well adapted to the study of chemistry-intensiveness.

• In the long run, too, the expansion of job prospects for chemists and chemical engineers would be favored by increases in final demand that impinge heavily on chemistry-intensive industries. (Of course, relative pay scales and the progress in instrument technology are also pertinent, since they can significantly affect the mix of professional workers and technicians over time.) The thrust of the nation's technological change has been in the direction of chemistry, as some of the figures presented earlier in this chapter suggest for recent decades and as figures presented in the next chapter will verify for a much longer time span. New industries that are emerging, such as private atomic energy, and new public concerns, such as air conservation and depollution, are certain to enlarge the range of professional opportunities; but the number of jobs involved may prove unspectacular.[23] Since government can quickly generate a sizable final demand and focus it on a variety of particular targets, public programs for sustaining education and research and for serving widely felt needs that cannot in the first instance be met by the profit-oriented private sector seem destined to remain attractive.

---

23. According to the Bureau of Labor Statistics (press release of Jan. 29, 1971), employment in the "atomic energy field," public and private, reached 154,100 in May 1970. Unpublished BLS figures show that chemists in this total numbered more than 3,600 and chemical engineers numbered more than 2,100. Together, these professionals accounted for 3.7% of the total, an impressive proportion. The number does not, of course, take cognizance of possible competitive displacements of similar professionals elsewhere.

According to *Manpower and Training Needs for Air Pollution Control*, a report of the Secretary of Health, Education, and Welfare to the Congress, Sen. Doc. 91-98, June 1970, chemists employed in the National Air Pollution Control Administration numbered 82 in 1970, and a figure of 271 was in prospect for fiscal year 1974. As for private industry, total man-years devoted to controlling emissions from stationary and mobile sources were set at 20,000 for 1969, and a doubling was estimated for 1974. The increment would apparently include 7,500 professional man-years for compliance and for development of hardware. The professionals would belong to the standard disciplines and receive "training in air pollution control . . . on-the-job and through short courses." (See pp. 7-9 of the cited HEW report.)

The dual translation of chemical science into applicable technology and into actual production is mirrored in the four items of the chapter subtitle, which provide the main headings for our discussion. These items:

- Constitute a logical, rather than a chronological, sequence, of course.
- Are not indispensable "steps" for the conversion of a technological opportunity into an economic reality.
- Only hint at the detailed scenario of critical events stretching from the laboratory all the way to the marketplace.
- Do not, in any case, suggest an automatic or inevitable progression, a routine dual translation of science into technology and sales.

Only minimal elucidation is required. With regard to the first point, it is clear, for example, that a price reduction (relative to other goods) for a new product or for an established product made by a new process need not in fact wait upon a productivity improvement. As for the second point, it suffices to mention that a new product or process need not have emerged from a formal research program or have entailed a patent application or grant and that a sustainable productivity improvement need not even be based on prior systematic research or on process innovation. The third point implies the simplism of any discrete framework of description or analysis that merely copies or only slightly elaborates the familiar trilogy of "invention, innovation, and imitation (or diffusion)." The obviousness of the fourth point is sometimes obscured in the hot pursuit of quantitative relationships connecting research, growth, and productivity. Further comment on this point is reserved for the next section, which focuses on research, and for a special appendix to this chapter.

Chemical Research as an Economic Tool

The practical application of chemistry has long been a matter of interest to lone individuals and to company employees. This interest antedates the federal encouragement of experimentation through direct funding or contract support, accelerated tax amortization, or the grant of explicit permission (as in the Internal Revenue Code revision of 1954) for the expensing, as well as amortization, of annual research outlays. It is also noteworthy that some chemical industrialists claimed many decades ago the same kinds and rates of financial return from research that scholars have only more recently been suggesting. The section accordingly begins with what might seem to be the prehistory of research. The discussion will show that the years prior to World War II hardly constituted a dark age for industrial chemistry; that they should be recalled, at least for perspective.

Before 1930. Although patents will be treated mainly in the next section, it is appropriate to mention here that the earliest awards were made for chemical processes. Thus, the first American chemical patent was granted in Massachusetts Bay Colony to Samuel Winslow in 1641 for a method of making salt. About the same time, John Winthrop, Jr., who had emigrated from Britain to Boston in 1631, not only addressed himself to the manufacture of alum, copper, glass, potash, saltpeter, and tar but also carried out pioneer investigations on the preparation of indigo. The first patent issued by the new Republic (1790) pertained to a method of making potash and pearl ash. The grantee was Samuel Hopkins.[1]

Research, especially applied research, gained strength in the 1800's, particularly after the Civil War. Many names stand out: Silliman (who returned in 1806 from his chemical studies in Europe), Edison, Hall, and Eastman, among others.[2] The roll could be extended greatly. The reader should consult Part I of this report for a fuller contextual account of early contributors and contributions.

In the preceding chapter, the phenomenal growth of chemical production during the current century was documented. But the literary tale is surely as exciting as its statistical reflections. According to two authoritative commentators whose book was published in 1931,[3] the 1920's saw the wresting of international primacy from Germany. The key to this achievement was the "cooperation between aggressive capital and creative science." As early as 1890, European chemists had sensed that the United States was "destined to take the lead at a not distant date in both industrial and scientific achievement." A decade or so after World War I, our two authors could say that "this prophecy has to a large extent been fulfilled." Indeed, during the 1920's "American industrial managers have assumed the front place in employing highly trained chemists and other scientists for original researches with a view to new discoveries or to useful improvements."[4]

Almost everywhere that research was being conducted in the 1920's (and it was already a visible phenomenon), businessmen felt certain of the value of the activity. A survey conducted in 1928 by the National Research Council (NRC) disclosed that almost every one of the 800 canvassed "leading companies, 68% of which operate research laboratories," reported that research had "practical utility in increasing profits and reducing costs." Indeed, only three company respondents expressed a negative opinion, and "the 797 other manufacturers seemed to be convinced that such research does pay." Among the 800 companies, 599 reported the following benefits from their research programs:[5]

Research Objective	Percent of Companies
Improved Product or Service	67
Reduction of Production Costs	59
Development of New Fields of Application	40
By-products and New Materials	30
New Products	8

A still smaller group of 490 companies reported to NRC that they conducted their own research or contributed to cooperative experimentation, and 39% of these considered their "investment" to be sound. Although 17 companies did not yet report profits from research, most of these expected favorable results in the near future. Much more interesting, however, is the finding that "estimates of the ratio of profit to the amount

1. Joseph Schimmel, "Legal Requirements for Patentability," in *Patents for Chemical Invention*, Advances in Chemistry Series, No. 46, American Chemical Society, Washington, 1964, p. 2; and E. R. Weidlein and W. A. Hamor, *Science in Action*, McGraw-Hill Book Co., New York, 1931, p. 20.

2. See *Science in Action*, especially Chapter 2.

3. Weidlein and Hamor, whose book has just been cited.

4. *Science in Action*, p. 50.

5. *Ibid.*, pp. 45-46. The tabulation contains duplication, since companies indicating more than one objective are counted more than once. Later, an unduplicated distribution is shown for a smaller number of companies reporting to NRC for 1931.

expended for research ranged from 100 to 300% in many instances." [6] This sort of appraisal, which may rest on doubtful methodology, has acquired a familiar ring over the years.

On a smaller scale, the 1920's already amounted to an age of industrial research and product proliferation. Leaders of the chemical industry could already attribute company survival, competitiveness, and growth to cost-saving research as well as to product-improving and horizon-extending research. American producers of dyes and other coal tar chemicals were allocating 4% of sales to research, a ratio very similar to, and even higher than, many of the figures reported for more recent times. This expenditure was regarded as investment for "improving the quality of products" and for "meeting competitive conditions." Indeed, "many forward-looking manufacturing companies" established numerical research targets, making these percentages "as much a policy with them as the setting aside of profits for surplus and reserves." [7] It is little wonder that during the Great Depression the head of a great chemical company could assert that the conduct of research was as important as the payment of dividends.

But the confidence of the 1920's in research was not undiscriminating, even if the economy did collapse soon after. A "great leader in chemical industry," for example, noted the need for a research pay-off: [8]

"Industrial research must have something to sell which business wants to buy; something concrete, specialized, which has a commercial value. . . . Business is not a charitable institution, and so the research laboratory must earn the money that it takes to build and operate it."

The sense contained in this position is being rediscovered painfully in the current period of inflation, of industrial productivity generally outpaced by negotiated wage increases, and of intensifying foreign competition.

Also of interest is the fact that many of the chemical and other companies in the vanguard of research today were already heavily committed to such activity in the 1920's. When 208 companies were reporting an annual outlay of $12 million to NRC in 1928, du Pont alone apparently spent about $3.5 million. Cost saving seems to have been a dominant motive of the company's research expenditure of that era: [9]

"In one three-year period, a few years ago, the direct and indirect savings from the research of this company amounted to twelve times the cost thereof. In one subsidiary, savings totaling seven times the cost were effected in a year."

Du Pont's 10 laboratories employed more than a thousand chemists and engineers in 1930. They investigated problems concerning heavy chemicals, paint, reagents, solvents, synthetic resins, dyes, rubber, and explosives. The two largest industrial research organizations of the time were Bell Laboratories, which definitely included chemists on the roster of some 2,600 scientists and engineers, and General Electric Co., which engaged 1,357 research workers. Among the other sizable laboratories, several were operated by such well-known chemical companies as Dow, Eastman Kodak, Grasselli, Hercules, Procter and Gamble, and Union Carbide and Carbon.[10]

The 1930's. Another vintage publication, dating from 1940, extends the story of chemistry in industrial research to the eve of World War II.

---

6. *Ibid.*, p. 47.
7. *Ibid.*, pp. 47-48.
8. *Ibid.*, p. 48.
9. *Ibid.*, p. 47.
10. *Ibid.*, pp. 55-57.

Making extensive use of data compiled by NRC, the authors of a book sponsored by the Work Projects Administration (WPA) determined that the conventional chemical industry accounted for 9,542 of the 44,292 research workers in the private sector in 1938—a fifth. This industry, exceeding all others in the employment of research personnel, was represented by 395 of the 1,722 reporting companies. If attention is confined to manufacturing, the chemical companies obviously constituted a still larger ratio of the 1,273 reporting, and they accounted for almost 30% of the 34,093 research employees.[11]

The same WPA study gives a fair idea of the growth of research activity during the interwar period in the conventional chemical industry and in manufacturing as a whole. Between 1921 and 1938, the research personnel of 189 identical manufacturing companies almost trebled, rising from 4,054 to 11,858. Included in this group were 76 chemical companies, which multiplied their research staffs nearly four times (from 1,365 to 4,983).[12]

The dynamism of the conventional chemical industry along its research dimension is reflected in another set of WPA figures—the ratios of research workers to wage earners. In 1927, chemicals and allied products engaged 137 research employees per 10,000 wage earners. In 1937, the ratio was 303. These figures exceeded greatly the corresponding ratios of all the other industry groups.[13]

This is a good point at which to note a striking further gain in the subsequent three decades. By 1969, according to data in Table 12 and a Bureau of Labor Statistics figure of 622,000 for "production" workers in the conventional chemical industry,[14] the number of research workers per 10,000 wage earners had risen to 1,038. This ratio is almost eight times that of 1927 and more than three times that of 1937.

What about the role of chemists proper? Some WPA-NRC data provide illumination with respect to this occupation and others, too, in the staffing of laboratories. For a changing group of companies in the private sector, the following distributions are available for 1921 and 1938:

	1921	1938
No. of Companies	216	1,224
Research Workers	2,358	19,797
Percentage	100.0	100.0
Directors	7.8	4.1
Engineers	16.5	23.2
Chemists	33.3	28.5
Physicists	1.3	3.5
Metallurgists	1.1	3.3
Other Professionals	14.1	8.5
Laboratory Aides	19.6	16.9
All Others	5.7	11.4
Unspecified	0.6	0.6

These distributions are distorted somewhat by the necessary inclusion of a number of chemists (among others) in a category for *Other Professionals*.[15]

Staff chemists clearly dominated in both 1921 and 1938, and their

11. George Perazich and P. M. Field, *Industrial Research and Changing Technology*, Report No. M-4, Work Projects Administration, Philadelphia, 1940, p. 73.

12. *Ibid.*, p. 75.

13. *Ibid.*, p. 21.

14. See, for example, *Manpower Report of the President*, April 1971, p. 258.

15. *Industrial Research and Changing Technology*, pp. 11-14, 78.

percentages would have been even higher if the universe of reference were restricted to professional personnel. Thus, the chemists accounted for 41.7% of the 1921 professional research population of the private sector and 34.5% of the 1938 population. (The small *Unspecified* category is omitted, as well as laboratory assistants, from these calculations.)

The revised percentages invite confrontation with comparable ratios derivable from BLS data for 1969 as shown in Table 12. A dramatic change is revealed. Chemists comprised only 12.4% of all professional research personnel in private industry in 1969, a remarkable decline from 1938. This ratio, like those for the earlier years, excludes chemical engineers from the numerator but incorporates them, with other engineers and with nonchemical scientists, in the denominator. Taken together, the revised ratios for 1921 and 1938 and the 1969 figure highlight a major revision in research staffing patterns—a trend toward the greater utilization of engineers (all kinds) in the laboratories of private industry since World War II. Once the 1969 ratio is known, it is easy to see signs of an impending longer-term decline in the reduction already registered between 1921 (from 41.7%) and 1938 (to 34.5%).

The occupational findings for the interwar period may, perhaps, be sharpened by use of percentage distributions for a fixed group of 177 companies in the private sector in addition to the percentages for a larger inconstant aggregate. The two alternative ratios for *Chemists* are 33.3% for 1921 and 26.4% for 1938—very similar to 33.3% and 28.5%, the proportions based on the larger changing group of companies. On the other hand, the 1921 and 1938 proportions for *Engineers* are static, at 15.9% and 16.0%; and the representation of the *Other Professional* category increases slightly, from 15.9% to 16.7%, instead of declining.[16]

WPA-NRC data also allow an examination of the extent to which chemists participated in the research staffing of different industries. The data apply to 1938 and to the research personnel in the same 1,224 companies that have already been treated *en masse*.[17] In the conventional chemical industry, for example, chemists comprised 51.3% of all the research personnel, while the separately reported engineers represented 11.6%. If the denominator is restricted to professional workers (i.e., if *Laboratory Aides* and *Unspecified* are omitted), these ratios rise to 57.5% and 13.0%, respectively.

Extrapolations to 1969 indicate that the sharp rise in the research role of engineers (especially since World War II) is centered in manufacturing and that the ratio of chemists to all research personnel is, nevertheless, unaffected in the conventional chemical industry. From BLS data, it appears that 52.5% of the research workers in the chemical industry in 1969 were chemists, virtually the same proportion as that cited above for 1938 (51.3%).

The WPA study also sought to estimate the actual numbers of industrial research employees on the basis of the percentages for the 1,224 companies. It arrived at a grand total of 50,700 persons, including research directors and executives, laboratory assistants, and clerical, maintenance, and mechanical support. Of these, 32,700 were professionals, and chemists dominated, numbering 12,700.[18]

Again, more recent figures provide perspective. By 1969, according to BLS data in Table 12, the private sector had 48,200 research chemists,

16. *Ibid.*, p. 78.
17. *Ibid.*, p. 79.
18. *Ibid.*, pp. 11-12.

Table 12

Chemists, Scientists, Engineers, and Technicians in Research and Other Activities: Private Industry and Components, 1969

Industry	Chemists			2÷(6+9) % (4)	All Scientists			Engineers			Technicians		
	Total Thousands (1)	R&D Thousands (2)	2÷1 % (3)		Total Thousands (5)	R&D Thousands (6)	6÷5 % (7)	Total Thousands (8)	R&D Thousands (9)	9÷8 % (10)	Total Thousands (11)	R&D Thousands (12)	12÷11 % (13)
Private Industry	90.5	48.2	53.3	12.4	213.5	99.7	46.7	849.0	289.9	34.1	772.5	179.8	23.3
Manufacturing	76.6	40.1	52.3	12.7	149.2	73.0	48.9	586.5	241.7	42.2	421.9	138.3	32.8
Chemical & Allied Products	41.6	22.9	55.0	52.5	59.3	31.9	53.8	44.2	11.7	26.5	45.4	21.0	46.3
Industrial Chemicals	16.8	8.3	49.4	43.7	21.7	11.8	54.4	23.2	7.2	31.0	19.8	10.3	52.0
Plastics and Synthetics	6.7	3.6	53.7	56.3	8.1	3.8	46.9	10.8	2.6	24.0	8.9	5.1	57.3
Drugs	6.1	3.7	60.7	40.6	15.4	8.8	57.1	1.7	0.3	17.6	6.0	3.6	60.0

Source: U.S. Bureau of Labor Statistics. The basic data are published in Scientific and Technical Personnel in Industry, 1969, Bulletin 1723, 1971, pp. 24-27. Chemical engineers are counted here as engineers rather than as chemists.

almost four times as many as in 1938. This number, moreover, was almost as great as the *total* for all private research workers of 1938, including nonprofessional employees.

One other piece of information merits notice. An adaptation of a 1931 table of NRC shows the major emphasis of 256 industrial research laboratories.[19] Unlike the NRC distribution shown earlier, the figures here add to 100%, since each company is counted only once, according to its principal research function. Only a few utilities are included, so the numbers shown for all industries virtually refer to manufacturing only:

Primary Research Objective	All (256)	Percent of Companies Chemicals (56)	Drugs (13)	Paints (9)
Improved Product or Service	38.3	32.1	30.8	66.7
Reduced Production Costs	19.9	14.3	15.4	0
New Fields of Application	3.9	7.2	0	0
By-products and New Materials	1.2	1.8	0	0
New Products	36.7	44.6	53.8	33.3

From these figures, it appears that chemical and drug companies tended to emphasize new products above the improvement of established products and services, while the reverse was true for paints. For manufacturing in general, the emphasis was more equal. In all instances, cost reduction ranked a poor third. It is clear from an earlier passage concerning du Pont, however, that cost saving was pursued energetically, if not most often, as a research goal.

The Recent Scene. Much has already been said in the discussion of employment in Chapter 26 about the research scene after World War II, particularly the roles of chemists and chemical engineers. For example, BLS figures incorporated into Table 4 show that 53.0% of the chemists employed in the private sector in 1970 were engaged in research. For manufacturing alone the research ratio was practically the same, 52.3%. Furthermore, Tables 5 and 6, based on special questionnaire surveys undertaken by ACS for this study, suggest that slightly fewer than half of the chemists and chemical engineers employed in manufacturing in 1969 and 1971 were assigned to research tasks.

Table 12 provides information on the research proportions of chemists in the conventional chemical industry in 1969. Based on BLS data, which omit chemical engineers, it shows that 55.0% of the chemists in the industry were occupied in research. For industrial chemicals and drugs, the ratios were 49.4% and 60.7%, respectively.

The strong representation of chemists is reflected in the research ratios shown for all scientists in Table 12 (and in Table 4), but engineers clearly have a much lower propensity to work in the laboratory. Only about one third of the engineers in private industry (the figures include chemical engineers) were employed in research activity in 1969. For manufacturing, the 1969 engineer research ratio was slightly more than two fifths. For the conventional chemical industry, it was slightly above one fourth.

Technicians, according to Table 12, have a greater research role in the conventional chemical industry than in manufacturing as a whole or in all private industry. Fewer than one fourth of those employed in the private sector in 1969 participated in research. The research ratio for manufacturing was about one third. On the other hand, the research ratio for technicians in the conventional chemical industry approached one half. In the

---

19. *Ibid.*, p. 80.

three identified components of the chemical industry (which are not exhaustive of the whole), the technician research ratios exceeded one half.

Chemists constituted roughly half of the scientists engaged in research in the private sector and in all manufacturing in 1969, and the ratio was greater in the chemical industry itself. In the private sector, they comprised 48.3% of all research scientists; in manufacturing as a whole, 55.0%; in the conventional chemical industry, 71.8%.

The ratios just cited exclude engineers from the denominators. As already noted, the combination of research scientists and research engineers into totals for professional technical personnel would greatly dilute the proportions for research chemists in manufacturing as a whole and in the private sector but not in the chemical industry. Table 12 shows that chemists comprised only one eighth of all research scientists and engineers in manufacturing and in private industry in 1969 and over one half (52.5%) of the corresponding totals for the conventional chemical industry.

Expenditure data compiled by the National Science Foundation (NSF) also illuminate the contemporary research commitment of industry. In particular, they provide evidence of the technical dynamism of the conventional chemical industry.

In 1970, according to NSF, makers of chemicals and allied products spent $1.6 billion of company funds for research and development. This figure exceeded the company outlays of every other listed industry in the private sector with the exception of *Electrical Equipment and Communications*. It comprised about one sixth of the total of $10.1 billion shown for company funds.[20]

The conventional chemical industry depends very little on government research support. In addition to the $1.6 billion mentioned in the preceding paragraph, it spent $188 million in federal money, a supplement amounting to about 11% of the internally-generated funds allocated to research and development. For the private sector as a whole, on the other hand, the federal supplement of $7.8 billion was more than three fourths as great as the companies' own expended funds of $10.1 billion.

With respect to spending for both basic and applied research (but not development), the conventional chemical industry dominated in 1970. It accounted for $221 million out of the $599 million devoted to basic research in the private sector, or about three eighths. It also accounted for $698 million of the total of $3.3 billion expended for applied research, or more than one fifth. All of these figures apparently include federal as well as company-generated funds.

With respect to development, which is heavily supported by federal sources, the conventional chemical industry no longer is first. In 1970, it accounted for $891 million out of a private-sector total of $14.0 billion, or about one sixteenth.

When attention is confined to company funds, the chemical industry ranks high in its expenditure for research and development as a percentage of net sales. The figure was 3.5% in 1968 and 1969 and a bit higher in 1970. It ranged from 3.1 to 3.8% in 1957–67. Even in the 1920's, ratios of this order were probably common in chemical companies performing research. On the other hand, the NSF ratios reported for research-oriented manufacturing companies in general were 2.2% for both 1969 and 1970, 2.1% for 1968, and between 1.5 and 2.1% for 1957–67. Among the listed industries,

20. This paragraph and the next two are based on "Industrial R&D Spending, 1970," *Science Resources Studies Highlights* (NSF 71-39), National Science Foundation, Dec. 10, 1971, p. 4.

however, several have tended to equal or exceed the percentages shown for the conventional chemical industry in the past decade, especially *Electrical Equipment and Communications, Aircraft and Missiles,* and *Professional and Scientific Instruments.*[21]

Naturally, when the focus shifts to combined expenditures of company and federal funds, the percentages of net sales are generally higher and the progressiveness of the chemical industry is obscured. Indeed, the differential over manufacturing narrows or vanishes. Thus, the ratio of research outlays to net sales for manufacturing as a whole rises to 3.8% for 1970, while the figure for the chemical industry is about 4.1%. The 1968 and 1969 ratios for the chemical industry are equal to each other and to those for all manufacturing, 4.0%. The range for the conventional chemical industry between 1957 and 1967 is 3.5 to 4.5%, but this record is virtually identical with that for all manufacturing, 3.4 to 4.6%.[22]

If our statistical account stopped here, it could seem much too complacent for its neglect of some adverse circumstances that cloud the future. The figures already presented in Tables 5 and 6 and the estimates given in Chapter 26 for 37 identical companies responding in 1969 and 1971 to the ACS special surveys are consistent with reports of a critical reappraisal of private (and public) commitments in research.[23] A recent outside canvass of 20 leading industrial chemical producers suggests that "1972 will be another stagnant year for chemical R&D." Indeed: [24]

"R&D spending as a percentage of sales, something that basic chemical producers used to brag about, continued to slip. In 1971, it was about 3% of sales. This year it will likely decline to 2.8%, as sales are forecast by many producers to rise 10% and most R&D budgets to remain about the same as last year. Ten years ago, in more halcyon days for research, R&D was 4.4% of sales."

But, perhaps fortunately, the future is sufficiently uncertain to admit contradictory signs, portents, and interpretations. Thus, a different publication commented as follows in the same week on the results of a survey of 24 leading producers of basic chemicals: [25]

"For the first time in at least four years, chemical companies are planning more than token increases in their research and development budgets. Perhaps more important, companies are saying that the cutbacks in R&D personnel and projects—which accelerated in '69–'71—may be over. . . ."

---

21. *Ibid.,* p. 3; and *Research and Development in Industry, 1969* (NSF 71-18), National Science Foundation, 1971, p. 81.

22. "Industrial R&D Spending, 1970," pp. 2-3; *Research and Development in Industry, 1969,* p. 80.

23. According to NSF, federal research spending in industry declined by 8% between 1969 and 1970. Although company funds increased by 2%, a drop of 3% is registered when account is taken of price inflation. (The implicit Commerce Department deflator for GNP was used; an index at least as pertinent appears in H. S. Milton, "Cost-of-Research Index, 1920–70," *Operations Research,* January-February 1972, pp. 1-18.) NSF also noted a 6% decline in industry employment of research scientists and engineers between January 1970 and January 1971. (See "Industrial R&D Spending, 1970," pp. 1-2.)

Research employment remained on a plateau in chemical companies in 1969–71, according to NSF, although increases were being reported by drug producers. A tapering of money outlays for research and a decline in "real" terms (i.e., after account is taken of price inflation) are also observable for 1969-72, at least for the basic chemical companies. (See *Chemical and Engineering News,* June 5, 1972, pp. 45-46.)

24. D. J. Soisson, "Squeeze on Chemical R&D Intensifies," *Chemical and Engineering News,* Jan. 17, 1972, pp. 7-8. The same author later reports another trend that could limit domestic professional opportunities in "U.S. Drug Companies Step Up R&D Overseas," *Ibid.,* May 29, 1972, pp. 5-6. The next issue of the same periodical (June 5, 1972, p. 34) shows research as a percentage of sales ranging from 1.0 to 6.5, with a median of 2.6%, for 19 companies in 1971.

25. "Moderation: Capital Spending Keynote for '72," *Chemical Week,* Jan. 19, 1972, p. 11.

From Research to Payoff. This section concludes with some acknowledgment of the efforts of scholars to relate research expenditures to such economic desiderata as profits, output growth, and productivity advance. Many of the quantitative studies have featured the chemical industry (including drugs) because of the availability of data. The benign effects usually deduced for doses of research are reminiscent of payoff claims also made in the 1920's. But in a period plagued by inflation, prolonged warfare, social unrest, urban blight, and environmental decay, the derived positive relationships between research and various indicators of desirable economic performance are much less persuasive than they were only a few years earlier. Even as a quest for new social priorities and parameters does not signify a rejection of applied science, growing doubt about quantitative research methods and results should encourage greater ingenuity in general and a new interest in literary approaches, too.

As the new mood of questioning the automatic mathematical translation of research costs into economic benefits takes hold, favorable academic appraisals continue to appear. For example, a report published in 1971 includes this summary:

"A study of 16 industry groups performing 97% of all industrial research and development and accounting for 92 to 95% of sales, assets, net income, and net worth of all manufacturing companies provided a strong relation between research intensity measured by company R&D funds as a percentage of sales and the rate of growth of sales, assets, net income, net worth, and net plant, property, and equipment. The effect of R&D upon growth begins on the average in the second year after the R&D investment and continues with steadily rising influence for at least nine years after the initial input year, reflecting the rising proportion of sales consisting of new products developed through R&D."

These quoted findings emerged from a treatment of data covering about a decade. The conventional chemical industry, ranking third in "research intensity" (i.e., the ratio of company research spending to net sales) among the 16 industry groups, was also found to be sixth in the growth of sales (over a decade), fifth in the growth of assets, tenth in the growth of net income, sixth in net worth, fifth in plant and equipment, third in the growth of value added, first in the growth of real output, and second in productivity advance. The view is maintained that "causality flows from R&D to industrial growth"; belief in the opposite direction of flow is regarded as "untenable." [26] Interaction—a flow in both directions—is obviously ignored as the plausible third alternative.

Leading participants in the econometric effort to relate research expenditures to various economic benefits do concede numerous obstacles in the way of achieving firm or unequivocal results. Papers presented in April 1971 at a meeting sponsored by the National Science Foundation noted many of these obstacles. According to one of the papers: [27]

"The state-of-the-art is least satisfactory in measuring the relationship between R&D and productivity gain. This is true for two reasons: (1) R&D . . . is not the only source of productivity growth; and (2) R&D and other sources of growth are interdependent."

The report of another participant in the NSF symposium has a reassur-

26. W. N. Leonard, "Research and Development in Industrial Growth," *Journal of Political Economy*, March-April 1971, pp. 232-56. The extended quotation appears on pp. 253-54.

27. C. T. Stewart, Jr., "A Summary of the State-of-the-Art on the Relationship between R&D and Economic Growth/Productivity," in *A Review of the Relationship between Research and Development and Economic Growth/Productivity*, National Science Foundation, February 1971, pp. 11-22. The quotation appears on p. 21.

ing title and presents findings of positive relationships, but it also expresses serious reservations regarding data and methods. The author cites his own studies for "10 large chemical and petroleum firms and 10 manufacturing industries in the postwar period," which indicate that productivity advance was significantly related to cumulative research expenditures. He also cites an analysis of figures for 17 chemical companies that indicates a statistically significant connection between cumulative research outlays and value added (with labor and capital inputs held constant). Long passages in his report, however, note unresolved "fundamental problems of measurement"; technical and other conundrums of actually relating research to productivity gain; the probability that "externalities" (which imply the uncapturability of benefit by the cost-bearer) and riskiness lead to underinvestment in research; the asymmetries between public and private research ventures; the preponderance of industry emphasis on product research (rather than process research, which ought to show a closer linkage to productivity increase); the statistical confusion between "real" research and prototype or pilot development; "needed research concerning R&D"; "needed research concerning the process of technological change"; and "needed research on economic growth and productivity increases." [28]

In brief, this NSF-sponsored review paper concedes the point made more directly in the paper noted just before it: that a reliable mathematical shortcut between research outlays and economic benefits sought is still unknown. The verbal hedging is as important as the numerical claims.

The same conclusion is reached in a recent article on science by an academic economist who is a member of the British Committee for Scientific Policy. He notes that "there is great difficulty in relating identifiable inputs of scientific research to usable outputs of scientific knowledge." Furthermore, "there is no measure available, according to the conventions of economic accounting, of the separate contribution of scientific knowledge to the gross national product." The "moral" is drawn that the explicit benefit of science to civilization and the social gain from public support of science "remain open questions." [29]

Many other students either start with or arrive at the view that the "black box" of mathematics either has to be replaced or be supplemented by methodical study of the technoeconomic coupling process in action. It should not be surprising that reports on the perplexities of "technological transfer" should have appeared simultaneously with mathematical correlations that admitted no troublesome psychological, institutional, social, or cultural variables. A 1968 report that failed to attract attention since it supported a lost cause (the extension of the State Technical Services Act) reasonably observed that science and engineering do not contribute to technology by themselves. The other essential elements of the "system" are a materialistic disposition in society, the existence of congenial political and government structures, the availability of appropriate skills and entrepreneurship, a supply of venture capital, and effective markets.[30]

28. Edwin Mansfield, "The Contribution of Research and Development to Economic Growth in the United States," *ibid.*, pp. 21-36; reprinted in *Science*, Feb. 4, 1972, pp. 477-86. The third main NSF paper, by William Fellner (pp. 37-46), reports an "average social rate of progress-return" of 30 to 50% but concedes that this is not a "marginal" rate (which would be more appropriate to public policy choices). Indeed, the paper concludes that one of the "worthwhile projects" would be further study of the marginal contribution of "progress-generating inputs"; that another would be "reducing the crudeness of the calculations here presented."

29. H. G. Johnson, "Some Economic Aspects of Science," *Minerva*, January 1972, pp. 10-18.

30. *Growth Through Technology*, Final Report of the Public Evaluation Committee to the Secretary of Commerce, Nov. 14, 1968, p. 8.

That the technoeconomic conversion process—or the still larger "system" in which it is embedded—requires study was forcefully expressed in a 1967 report of the National Academy of Sciences. Among the bold-type "conclusions and recommendations" are these:

• The interaction between science and society is complex.
• A broad spectrum of scientific disciplines and technical skills is required.
• The transfer of technology from the laboratory to a producing or operating organization which builds, sells, or uses it is a vital and often underestimated step in technological innovation.

The report also called for interdisciplinary "studies of the history and sociology of science and technology . . . to further understanding of the principles behind the great variety of successful patterns of applied research and its transfer." [31]

The case studies presented in Part I and the one appended to this chapter show something of the technoeconomic process at work in chemistry. They were prepared in the belief, also expressed in the 1967 report of the National Academy of Sciences, that "studies of the history and sociology of applied science are important"; [32] that the pursuit of quantitative relationships should be supplemented by other approaches. Part I provides basic materials from which other students may attempt to distill useful general principles concerning the dual translation of science into applicable technology and into actual production. The brief statement appended to this chapter offers and lightly illustrates a methodological suggestion for company studies of the research process *in situ*.

Chemical Patents: Applicable Technology

Longer established and more comprehensive in scope than industrial research, the patent system provides a unique historical register of the contribution of chemistry to the nation's expanding technological base. This system has its legal foundation in Article I, Section 8 of the Constitution. The statistics featured in this section extend back to 1836, when consecutive numbering of patent awards began and when the examination of inventive claims was reinstituted. They show that the annual contribution of chemistry to the body of U.S. patents (now numbering more than 3.5 million) was relatively small and static until about World War I. Then, a takeoff occurred, and the chemical proportion is still growing with no sign of arrest.

Like other statistical indicators, patent grants cannot satisfy all quantitative needs. For example, a simple count treats all awards as equivalent, even though some may have considerable commercial value and others little or none. The concept of the value of a patented invention, moreover, is ambiguous; the net economic contribution over cost, separate from the contributions of all other skills needed for development and application, is not really determinable. Besides, the economic benefit derivable from an art, machine, manufacture, or composition of matter that has met the legal tests of novelty, utility, and nonobviousness need not be recorded fully, significantly, or at all during the 17 years in which the inventor or assignee has the legal right to exclude others from making, selling, or using the invention. Another point deserving mention is that, although

---

31. *Applied Science and Technological Progress*, A Report to the Committee on Science and Astronautics, U.S. House of Representatives, by the National Academy of Sciences, 1967, pp. 13-19.
32. *Ibid.*, p. 18.

relatively few patents become subjects of litigation, the courts have upheld challenges of validity in a substantial proportion of cases.

Industrial research need not eventuate in patent applications or grants, yet chemists commonly assert or accept the view that the constitutional privilege of exclusion is economically sound. For example, in June 1971, the President of the ACS commented as follows to the chairman of the Senate Subcommittee on Patents, Trademarks, and Copyrights in the course of evaluating a legislative proposal:

"Chemical technological growth depends on long-term expensive research and development—perhaps to a greater extent than in any other major industrial category. Chemical research and development are in the main conducted with private funds. The field of chemistry has benefited from the patent system, and the economy and the fund of public knowledge have been enriched by a store of technology which has been contributed by chemical practitioners in reliance on the patent system."

The same letter notes the constructive potential of patent licensing for achieving the "utilization of chemical technology that would otherwise remain undeveloped" and for transferring such technology without "wasteful duplication of research and development effort." [33] Of course, the purchase and sales of patents are additional means of exploiting and diffusing technology. Such transactions are occasionally deemed important enough to mention explicitly in company balance sheets, operation statements, and reports to stockholders.

The proliferation of chemical processes, materials, and products is reflected in the multiplicity of Patent Office classes assigned to them and also in the variety of class titles. Table 13 lists 36 classes or subclasses which are used in the *U.S. Patent Office Gazette* for defining "chemical" inventions and distinguishing them from the other two major categories, "mechanical" and "electrical."

At least three things stand out. One is the evolutionary and pragmatic character of the accumulating patent file for chemistry. The class titles do not define sharply, exhaustively, and methodically the expanding universe of technology that is based on molecular science. Second, the "chemical" category evidently tends to cover the unit processes and unit operations that characterize the chemical industry but are also performed elsewhere in the economy. This catholicity of scope is reminiscent of the coverage provided by employment distributions for chemists and chemical engineers, presented in Chapter 27. Third, "chemistry, carbon compounds" is by far the most populous class. Of the 335,500 original patent grants (i.e., other than cross-references) issued to the end of 1968, this class accounted for almost one third.[34]

Some figures provided by the U.S. Patent Office[35] for this study focus on the period 1953–70, which was marked by vigorous gains in the technology and output of plastics, synthetic fibers, pesticides, drugs, and petro-

33. Letter of Dr. Melvin Calvin to Sen. J. L. McClellan, June 7, 1971, on the so-called Scott Amendments to S.643.

34. The vast increase of carbon compounds in recent times is symbolized in the announcement that the ACS Chemical Abstracts Service entered the 2 millionth distinct chemical structure in its computer-based registry during March 1972. The substance involved is an organic compound reported in Patent 3,625,944, which was awarded to four industrial co-inventors and assigned to their company. The registry, maintained in conjunction with indexing for *Chemical Abstracts* since January 1965, still includes only about one third of all known substances. On the other hand, additions are being made at the rate of about 300,000 substances per year, and about 70% of the substances currently being indexed are already represented in the registry. (See "CAS Registers 2 Millionth Chemical Structure," *Chemical and Engineering News*, April 3, 1972, p. 18.)

35. Figures supplied by the Director of Information, U.S. Patent Office.

Table 13

U.S. Chemical Patents Issued to December 1968
and Distributed According to Patent Office Classes

Class Member	Class Title	Original Patents
Total		334,942
8 (subs. 1-142)	Bleaching and Dyeing; Fluid Treatment and Chemical Modification of Textiles and Fibers	5,634
21	Preserving, Disinfecting and Sterilizing	1,811
23	Chemistry	21,290
29 (subs. 180-199)	Metal Working	2,199
44	Fuel and Igniting Devices	3,494
48	Gas, Heating and Illuminating	5,998
51 (subs. 293-309)	Abrading	886
65	Glass Manufacturing	7,074
71	Chemistry, Fertilizers	2,731
75	Metallurgy	12,798
96	Photographic Chemistry, Processes and Materials	6,613
99 (subs. 1-233)	Foods and Beverages	10,948
106	Compositions, Coating or Plastic	14,471
117	Coating: Processes and Miscellaneous Products	14,610
127	Sugar, Starch and Carbohydrates	1,273
134 (subs. 1-42)	Cleaning and Liquid Contact with Solids	869
136	Batteries	7,346
148	Metal Treatment	6,227
149	Explosives and Thermic Compositions	1,768
156	Adhesive Bonding and Miscellaneous Chemical Manufacture	10,583
161	Stock Material and Miscellaneous Articles	6,386
162	Paper Making and Fiber Liberation	5,478
176	Nuclear Reactions and Systems	1,767
195	Chemistry, Fermentation	2,740
196	Mineral Oils: Apparatus	1,121
201	Distillation: Processes, Thermolytic	545
202	Distillation: Apparatus	2,753
203	Distillation: Processes, Separatory	1,632
204	Chemistry, Electrical and Wave Energy	11,723
208	Mineral Oils: Processes and Products	10,644
210 (subs. 1-64)	Liquid Purification or Separation	1,447
252	Compositions	19,252
260	Chemistry, Carbon Compounds	109,909
263 (subs. 52, 53)	Heating	383
264	Plastic and Non-Metallic Article Shaping or Treating: Processes	10,171
424	Drug, Bio-Affecting and Body Treating Compositions	10,368

Source: E. A. Hurd, "Patent Literature: Current Problems and Future Trends," Journal of Chemical Documentation, X:3, 1970, p. 170. Also shown in Chemical Technology, April 1971, p. 212. In addition to original patents, Hurd's complete table shows cross-references included in Patent Office files.

chemicals. During this interval, chemical patent grants almost doubled their share of the annual national total, whittled the traditional leadership of mechanical inventions, and outstripped electrical patent grants, which include the dynamic electronic subcategory:

	Percentage Distribution of Annual Patent Grants		
Year	Chemical	Mechanical	Electrical
1953	14	72	14
1960	19	62	19
1965	22	60	18
1970	26	54	20

Even more impressive than these gains is the dynamism that chemistry has exhibited during the past half century. As already noted, a major swing toward chemical invention began about the time of World War I. This structural change is reflected in Table 14, which shows incremental and cumulative patent issues by five-year intervals since 1836. According to the table, chemical patents, viewed both incrementally and cumulatively, consistently comprised fewer than 4% of all grants from 1836 to the eve

Table 14

Chemical and Total Patents Issued by Five-Year Intervals: 1836–1970

	Issues by Period			Cumulative to End of Period		
Period	Chemical Patents	Total Patents	Percent Chemical	Chemical Patents	Total Patents	Percent Chemical
1836–1840	64	1,913	3.3	64	1,913	3.3
1841–1845	64	2,425	2.6	128	4,338	3.0
1846–1850	86	3,517	2.4	214	7,855	2.7
1851–1855	155	6,144	2.5	369	13,999	2.6
1856–1860	458	16,996	2.7	827	30,995	2.6
1861–1865	761	20,779	3.6	1,588	51,774	3.1
1866–1870	2,314	58,833	3.9	3,902	110,607	3.5
1871–1875	1,994	61,024	3.2	5,896	171,631	3.4
1876–1880	1,676	64,496	2.5	7,572	236,127	3.2
1881–1885	2,975	97,357	3.0	10,547	333,484	3.2
1886–1890	3,250	110,493	2.9	13,797	443,977	3.1
1891–1895	3,194	108,515	2.9	16,991	552,492	3.1
1896–1900	4,001	112,325	3.6	20,992	664,817	3.2
1901–1905	5,586	143,791	3.9	26,578	808,608	3.3
1906–1910	5,959	171,560	3.5	32,537	980,618	3.3
1911–1915	7,251	186,241	3.9	39,788	1,166,409	3.4
1916–1920	8,520	197,644	4.3	48,308	1,364,053	3.5
1921–1925	10,562	203,977	5.2	58,870	1,568,030	3.8
1926–1930	13,613	219,384	6.2	72,483	1,787,414	4.1
1931–1935	23,423	239,092	9.8	95,906	2,026,506	4.7
1936–1940	24,921	200,902	12.4	120,827	2,227,408	5.4
1941–1945	23,736	164,438	14.4	144,563	2,391,846	6.0
1946–1950	22,577	144,160	15.7	167,140	2,536,006	6.6
1951–1955	28,107	192,897	14.6	195,247	2,728,903	7.2
1956–1960	42,681	237,768	18.0	237,928	2,966,671	8.0
1961–1965	52,411	260,058	20.2	290,339	3,226,729	9.0
1966–1970	74,866	328,394	22.8	365,205	3,555,123	10.3

Source: E. A. Hurd, "Patent Literature: Current Problems and Future Trends," Journal of Chemical Documentation, X:3, 1970, p. 169. Data for 1969 and 1970 were projected by Hurd to complete the last five-year interval.

of World War I. The rise since then has brought the chemical proportions to 23% of new issues during the period 1966–70 and more than 10% of cumulative issues. The first of these proportions is projected to 25% for 1971–75.[36]

As new public concerns come to the fore or as social priorities are reordered, the Patent System may be expected to continue to serve as a technological register. A recent article on noise abatement illustrates this point. Between 1959 and 1970, 1,895 patent issues are identified as pertaining to "acoustics or noise-abatement devices." Furthermore, 453 of these— almost one quarter—fall into "chemical" subclasses.[37]

The Productivity Translation

In recent years and over the longer run, the conventional chemical industry and its main components have generally compiled an exemplary record with respect to output per man-hour. Certainly for the longer term, the favorable performance reflects, among other factors, the benefit of technological progress as the adoption of superior processes and methods. Technical improvement has supported the innovation and diffusion of synthetic and other new chemical products, facilitating the absolute and relative price declines so vital to market expansion. In the short run, furthermore, chemical productivity responds well to increases in the utilization of capacity already in place. With the resurgence of chemical output beginning in 1971, both of these sources of productivity gain will help to counter domestic inflationary pressures and the determined competitive challenge of foreign producers. Price stabilization, after all, depends very greatly on private industry's ability to raise hourly productivity at least as rapidly as hourly remuneration, and this ability has been demonstrated unequivocally in chemical production in the past.

Table 15 presents the record of the chemical industry for a recent span of years. It is based on unpublished information made available by BLS to ACS for use in this report. In particular, it shows index numbers for 1958–69 of the output per man-hour of all employees—so-called production workers plus others, including research personnel.[38] It also shows the underlying indexes of output and man-hours. The components of the conventional chemical industry are identified by Standard Industrial Classification number as well as by name to facilitate a later matching of productivity statistics with price indexes for the same domains.

According to Table 15, output per man-hour of all employees in the conventional chemical industry increased by three quarters in the interval 1958–69, while the productivity measure for all manufacturing rose only two fifths. Meanwhile, chemical output more than doubled, but manufacturing output as a whole increased by less than four fifths.

Exceptionally great productivity gains were registered by several components of the chemical industry. In the 11-year interval, output per man-

36. Hurd, loc. cit., p. 168.

37. M. A. Holman and J. T. Bennett, "Patent Statistics as a Surrogate for Spending: A Case Study of Patents Relating to Acoustics and Noise Abatement," IDEA, Winter 1971-72, pp. 564-74. The figures for 1959–70 appear on p. 568.

38. Since independent measures of output are not available or constructible for production workers and for others, the same index of output would necessarily be used for computing the "productivity" of each of the two categories and for both combined. Thus, a slower rise of computed productivity for white-collar (i.e., nonproduction) workers than for all employees of the chemical industry —which has been observed, for example, during our era of rapid growth of research activity— accordingly means no more than an increase in the ratio of white-collar to other personnel. The intrinsic productivity of research and other white-collar workers remains a philosophical and statistical conundrum. This fact should be borne in mind in the interpretation of such reports as Michael Heylin's, "White-Collar Productivity Troubles Industry," Chemical and Engineering News, July 24, 1972, pp. 6-7.

hour almost trebled for industrial gases and for medicinals and botanicals. A doubling or better was realized in the same period for cyclical intermediates and crudes, industrial organic chemicals, plastic materials and resins, polishes and sanitation goods, agricultural chemicals, and carbon black.

As the steady output rise for chemicals as a whole came to an abrupt halt in 1970, the flattening had a temporary detrimental effect on productivity. The resumption of chemical production's upward course during 1971 provided a good occasion for observing the responsiveness of chemical productivity to improvement in the rate of capacity utilization. Even early computations show the output per production-worker man-hour rose substantially more in the chemical industry than in all manufacturing.[39] This differential gain, coupled with a superior productivity performance and a better control of unit labor cost during the 1960's, has been a major factor in the chemical industry's positive contribution to the generally adverse balance of trade, of which more will be said below.

Statistics developed at the National Bureau of Economic Research for 1899–1957 help one to gain a fair idea of comparative productivity increases in the conventional chemical industry and all manufacturing throughout the current century. For this period, output per man-hour of "persons engaged" increased almost eight times in the conventional chemical industry but only a bit more than four times in all manufacturing: [40]

Year	Chemical Industry	All Manufacturing
	(1929 = 100)	
1899	41.1	45.5
1919	45.4	58.0
1929	100.0	100.0
1937	127.2	116.9
1948	193.6	138.6
1957	322.2	187.1

These figures also show that the divergence occurred mostly after World War I, especially in the relatively few years following World War II.

Juxtaposing these estimates and those for 1958–69, we may surmise safely that the rise in man-hour productivity for all employees in the chemical industry was more than twice as great as the gain in all manufacturing during the whole 70-year span, 1899–1969. Such a record for the chemical industry must be rated as phenomenal, since the productivity performance of United States manufacturing has itself been regarded as one of the economic wonders of the world's industrial era.

A more formal productivity comparison for this century may be made by relating the production measures cited in Chapter 27 to estimates of man-hours derived from BLS figures for all employees and average weekly hours worked.[41] Such computations yield 1970 productivity indexes of 689.5 for the chemical industry and 284.6 for manufacturing as a whole (1929 = 100). These numbers mean that the 1970 productivity level for the chemical industry was almost 17 times the level of 1899; that the corresponding multiple for all manufacturing was 6.75; that the rise in output per employee man-hour in chemical production was 2.4 times as

---

39. "Productivity Revives in Chemical Industry," Chemical and Engineering News, Nov. 1, 1971, pp. 10-12; and ibid., June 5, 1972, p. 33.

40. J. W. Kendrick, Productivity Trends in the United States, (cited in Chapter 27), pp. 464, 471.

41. The figures for employment and weekly hours in the chemical industry and all manufacturing are reported, for example, in Manpower Report of the President, March 1972, pp. 215, 217, 220, 221.

Table 15

Chemical Industry Indexes of Output, Man-Hours, and Productivity: 1958–69 (1958 = 100)

Industry	SIC Number*	Output Per Man-Hour (All Employees)		Output		Man-Hours (All Employees)	
		1968	1969	1968	1969	1968	1969
All Manufacturing	19–39	139.7	141.9	172.9	178.9	123.8	126.1
Chemical and Allied Products	28	170.2	176.6	209.8	223.8	123.3	126.7
Industrial Chemicals	281	188.0	198.1	199.1	219.1	105.9	110.6
Alkalies and Chlorine	2812	153.2	161.6	125.2	129.3	81.7	80.0
Industrial Gases	2813	278.6	296.0	253.0	265.8	90.8	89.8
Cyclical Intermed. and Crudes	2815	207.7	221.9	226.4	253.2	109.0	114.1
Inorganic Pigments	2816	123.9	125.0	140.6	147.9	113.5	118.3
Indus. Organic Chem., n.e.c.	2818	201.7	215.7	262.0	296.6	129.9	137.5
Indus. Inorganic Chem., n.e.c.	2819	149.5	150.1	136.0	142.7	91.0	95.1
Plastic Mater. and Synthetics	282	186.6	191.3	277.9	294.8	148.9	154.1
Plastic Mater. and Resins	2821	192.8	202.1	269.0	294.8	139.5	145.9
Synthetic Rubber	2822	141.4	148.6	186.9	199.3	132.2	134.1
Cellulose Man-made Fibers	2823	182.5	177.5	148.7	125.7	81.5	70.8
Organic Fibers, Noncellulosic	2824	152.1	146.3	508.2	546.5	334.2	373.6
Drugs	283	178.3	179.7	218.9	232.0	122.8	129.1
Biological Products	2831	149.5	161.1	319.6	376.7	213.8	233.9
Medicinals and Botanicals	2833	269.9	275.1	225.9	232.5	83.7	84.5
Pharm. Preparations	2834	174.2	175.3	215.6	228.3	123.8	130.2

(Continued next page)

Industry	SIC Number*	Output Per Man-Hour (All Employees)		Output		Man-Hours (All Employees)	
		1968	1969	1968	1969	1968	1969
Soap, Cleansers, and Toilet Goods	284	163.9	166.9	204.0	218.8	124.5	131.1
Soap and Other Detergents	2841	169.6	165.1	166.4	168.6	98.1	102.1
Polishes and Sanitation Goods	2842	198.4	229.5	223.8	276.5	112.8	120.5
Surface Active Agents	2843	121.3	122.7	228.6	236.1	188.4	192.4
Toilet Preparations	2844	163.6	163.1	247.4	261.1	151.2	160.1
Paints and Allied Products	2851	133.9	130.5	147.6	145.9	110.2	111.8
Gum and Wood Chemicals	2861	134.4	122.2	106.2	98.5	79.0	80.6
Agricultural Chemicals	287	168.9	192.2	196.4	209.7	116.3	109.1
Fertilizers	2871–72	153.6	174.7	165.4	169.8	107.7	97.2
Agricultural Chemicals, n.e.c.	2879	193.5	211.9	293.1	334.3	151.5	157.8
Misc. Chemical Products	289	118.6	121.8	193.0	192.3	162.8	157.9
Adhesives and Gelatin	2891	183.4	180.4	224.8	254.6	122.6	141.1
Explosives	2892	106.6	119.6	267.7	251.6	251.2	210.4
Printing Ink	2893	136.5	142.6	178.6	188.6	130.8	132.3
Carbon Black	2895	210.0	203.7	187.3	193.3	89.2	94.9
Chemical Preparations, n.e.c.	2899	113.4	102.6	164.5	153.8	145.1	149.9

Source: Unpublished estimates computed at the request of the American Chemical Society by U.S. Bureau of Labor Statistics, which provides the following descriptive note: "The indexes shown are based primarily on data published by the Census Bureau in the Census of Manufactures and Annual Surveys of Manufactures. The output data are derived by deflating published estimates of value of shipments for each four-digit industry group by appropriate price indexes, and adjusting these figures by the change in work in process and finished goods inventories, also deflated. The price indexes used as deflators are compiled by the Office of Business Economics and are based primarily on the Wholesale Price Indexes (WPI's) of the Bureau of Labor Statistics. Some price indexes, however, are derived using sources other than WPI's or other sources combined with WPI's. For example, prices for drugs and plastics material are based on a combination of WPI's and Tariff Commission data. The price index for each year is used as a deflator for that year's value of shipments, and the average of the prior and current year price index is used as a deflator for the end of year inventory. The man-hours estimates are derived by adding to the published production workers man-hours the estimates of nonproduction workers hours derived by the BLS. Output per man-hour indexes are computed by dividing the output data, converted into index form by man-hours, also indexed. There are some limitations to the data, and for this reason these series are not deemed sufficiently reliable for publication as an official series."

*The numbers in the first column refer to the Standard Industrial Classification system.

great as that for manufacturing as a whole in the period 1899–1970. The last clause confirms the italicized statement in the preceding paragraph.

The Price Record

It is natural for chemical industry spokesmen to complain about the upward pressure of unit labor cost in the face of soft prices, yet the performance in the marketplace ought to arouse admiration outside the industry. If chemical productivity had not made dramatic gains, the favorable price record could not have been achieved and competitiveness would have been correspondingly impaired. This price record has contributed to an excess of chemical exports over imports while the national trade balance in general has deteriorated.[42]

Table 16

Chemical Industry Price Indexes, 1957–70
(1967 = 100)

Component or Industry	1957	1970
All Industrial Commodities	93.3	110.0
Chemical and Allied Products	101.2	102.2
Industrial Chemicals	102.6	100.9
Prepared Paint	90.6	112.4
Paint Materials	108.4	101.4
Drugs and Pharmaceuticals	106.2	101.1
Agricultural Chemicals and Products	95.2	88.4
Plastic Resins and Materials	115.1	90.7
Other Chemicals and Products	90.8	108.6
Synthetic Rubber (2822)	104.5	100.1
Cellulosic Man-Made Fibers (2823)	105.8	100.9
Organic Fibers, Noncellulosic (2824)	....	98.6
Fertilizers, Mixed (2872)	....	87.5
Explosives (2892)	87.6	106.3

Source: Handbook of Labor Statistics, 1971, U.S. Bureau of Labor Statistics, pp. 276, 279–80, 287–88. The numbers in parentheses, inserted to facilitate identification, refer to the Standard Industrial Classification system.

From Table 16, which refers to 1957–70, it may be seen that the price indexes available for a few components of the conventional chemical industry remained level or even declined after 1967. In some instances, a downward drift even preceded the stabilization or weakening of chemical prices that appears so anomalous for the more recent years of stubborn and widespread inflation. In the case of synthetic rubber, the price index sagged from 104.5 in 1957 to 100.0 in 1967, the base year, and it stood at 100.1 in 1970. A similar pattern of easing from 1957 to 1967 and virtual stability thereafter to 1970 is observed for cellulosic man-made fibers. Only for explosives is a clear and steady price rise manifested for the period 1957–70, but the increase since 1967 has not been unusually large.

Additional wholesale price indexes shown in Table 16 reveal that stability or decline was typical, except for prepared paint and miscellaneous chemicals, during the period 1957–67. In these exceptional cases, the rise resembled that for all industrial commodities.

---

42. E. V. Anderson, "Washington's New Attitude Toward World Trade," Chemical and Engineering News, March 20, 1972, especially p. 19.

In the same periodical, Aug. 16, 1971, p. 25, a chart shows a decline in the du Pont index of selling prices from about 115 to 92.5 (base 1967 = 100) between 1960 and 1971.

Since the BLS productivity indexes shown in Table 15 were derived (according to the explanatory note) with the aid of price information, the so-called "price deflators" are presented for the chemical industry components in Table 17. The deflators are shown for 1965–70 on the base 1958 = 100, and the subindustries are identified by the standard numbers shown in Table 15. They were actually developed by the U.S. Department of Commerce's Office of Business Economics (now Bureau of Economic Analysis) from price data compiled in the main by BLS. Impressive reductions are indicated for 1958–70 by several of the indexes, such as for cyclical intermediates and crudes, industrial organic chemicals, plastic materials and resins, noncellulosic organic fibers, fertilizers, and carbon black. In the interval 1965–70, the price indexes for synthetic rubber, medicinals and botanicals, and pharmaceutical preparations remained remarkably level, and these also were below their 1958 magnitudes in 1970.

Table 17

Price Deflators for Chemical Industry Components, 1965–70
(1958 = 100)

SIC	1965	1966	1967	1968	1969	1970
2812	101.6	101.7	103.1	104.2	107.7	110.7
2813	102.0	102.0	102.1	98.3	95.6	103.6
2815	81.5	83.7	84.9	81.6	79.5	79.8
2816	106.2	106.5	106.8	109.2	110.8	110.5
2818	87.1	87.2	88.6	86.2	81.6	81.7
2819	110.4	112.0	114.9	116.9	117.4	120.3
2821	82.4	82.8	82.6	76.3	74.7	75.0
2822	98.0	98.1	97.6	96.6	98.0	98.5
2823	96.6	97.0	97.4	98.0	98.8	98.7
2824	84.2	83.1	78.7	76.8	76.5	76.3
2831	93.3	93.3	90.4	90.9	90.0	90.8
2833	65.1	65.1	65.7	64.2	63.2	64.4
2834	93.3	93.0	91.8	90.4	91.7	92.8
2841	101.0	101.1	101.9	103.8	105.7	106.9
2842	103.1	102.3	101.5	102.6	104.1	107.1
2843	108.2	110.7	110.7	114.4	115.8	117.7
2844	104.7	107.0	110.8	112.9	116.6	118.8
2851	103.2	104.4	107.0	111.4	115.5	118.5
2861	116.2	112.6	111.1	113.3	125.6	135.0
2871	105.7	107.4	110.0	109.1	100.1	97.7
2879	82.6	89.7	88.0	88.4	86.6	98.6
2891	93.1	93.1	94.0	94.2	94.9	105.6
2892	111.8	110.1	113.2	116.4	119.4	122.0
2893	100.5	102.8	102.8	102.7	104.6	108.6
2895	89.1	89.3	89.7	88.8	86.8	87.7
2899	121.0	136.8	130.4	129.9	138.8	142.3

Source: Provided by U.S. Bureau of Labor Statistics. According to the notes in Table 15, however, the series were developed by the Bureau of Economic Analysis, U.S. Department of Commerce, from BLS and other data. The SIC numbers are identified in Table 15.

Over the longer period, too, the superior productivity performance of the chemical industry has facilitated price moderation. From 1933 to 1970, according to BLS data, the price index for chemical and allied products multiplied by two and a small fraction, rising from 47.4 to 102.2 (1967 = 100). In the same interval, the measure for all industrial commodities virtually trebled, rising from 37.8 to 110.0. A longer price series, pertaining to industrial chemicals, shows a gain of less than a half between

1926 and 1970 (from 70.1 to 100.9). In the same interval, however, more than a doubling (from 53.2 to 110.0) occurred for all industrial commodities.[43]

This paragraph closes the chapter with a few remarks on the place of chemicals in foreign trade. In 1961, United States exports of chemicals exceeded imports by $1.1 billion, and the surplus of merchandise exports (including food and live animals but excluding Department of Defense shipments) amounted to $5.5 billion. By 1970, the positive net contribution of chemicals was much greater. Indeed, the favorable balance for chemicals, $2.4 billion, approached the surplus of $2.7 billion for all merchandise transactions. In 1971, when the national balance became adverse, the chemical net outflow remained strong. Thus, the national trade deficit reached $2.0 billion, but the excess of chemical exports over imports amounted to $2.2 billion. Organic chemicals, synthetic resins and plastic materials, and medicinals and pharmaceuticals have been major contributors to the good record for chemicals. In 1970, almost one third of the pharmaceutical sales of U.S.-based drug companies for human and veterinary use went to foreign customers.[44]

---

43. *Handbook of U.S. Labor Statistics, 1971*, U.S. Bureau of Labor Statistics, pp. 276, 277.
44. *Chemical and Engineering News*, June 5, 1972, pp. 14, 52. See also *Statistic Abstract of the United States: 1971*, pp. 772-73, 777-78.

FACTORS IN THE ADVANCE OF CHEMICAL TECHNOLOGY:

THE CASE OF GENERAL MOTORS

In the following paragraphs of this appendix to Chapter 28 we wish to underscore a point already made in the body of the chapter: The translation of science into applicable technology and into actual production is hardly automatic, and the trip from the laboratory to the market is fraught with risk and uncertainty—and sometimes ends without profit. We shall undertake the exercise by focusing on a single organization with a long research history, the General Motors Research Laboratories, which was listed in both "Science in Action" and "Industrial Research and Changing Technology" as one of the major industrial centers of the period between the two World Wars. In doing so, we may more readily understand a paradox of the recent voluminous literature—the coexistence of (1) assertions and estimates of substantial payoffs from research in terms of private profit and corporate and economic growth and (2) widespread puzzlement and uneasiness over the difficulty of bringing about technological transfer within a company as well as among companies, industries, and nations. Needless to add, the improvement of such transfer, at least within our domestic borders, is vital to the productivity gain that is necessary for the alleviation of inflation and for any substantial reduction of the deficit in the international balance of trade.

The discussion is based on information supplied for use in this study by GM Research Laboratories on some of the major chemical accomplishments there. For a useful review of the details of such an organization's technological history, it is necessary to select a set of interacting categories that suffice to describe the functioning system under observation. The categories should be relatively few and exhaustive. They should also be adapted to the purpose of the inquiry. Our purpose is served well enough by identifying three interacting categories: the *individual*, the *organization*, and the *environment*. The organization here is GM, and the environment is comprehensive, including the rest of the economy (both public and private), noneconomic factors, international relations, and the biosphere (as modified by man). If we had more information than available and if our objectives were more ambitious, the environment would have to be divided into several smaller interacting elements. In addition, a distinction among company departments could also prove productive for a refined analysis, provided sufficiently detailed information were available.

The approach exhibited here could also be advantageously employed in an analysis of Part I. Indeed, the industry case histories presented there document numerous instances of accomplished technological transfer of science into practicable form and profitable practice.

The Individual

The role of determined, motivated, creative individuals can never validly be ignored. Its importance cannot long be obscured by the myth that technological progress can be routinized through the formalization of research activity, a myth that tends to be reinforced by quantitative studies showing correlations between research outlays and economic benefits but neglecting the processes of transformation.

Of course, much invention is hardly epic or Promethean. It is also true that the use of teams may diminish the requirement of individual versatility and also offers opportunities for the competitive enhancement of personal performance. It must also be acknowledged that address of the same problem in more than one laboratory increases the likelihood of simultaneous and equivalent solutions. Furthermore, the existence of prior and well-known chains of invention or development may readily suggest the next logical step to any interested student or practitioner.

Nevertheless, significant accomplishment remains hard to predict, nor is its timing or locale ever sure. Also relevant is the neglected fact that recorded instances of simultaneous invention cite fewer than a handful of co-achievers rather than dozens, scores, hundreds, or thousands. Clearly, the personal qualities of lone or collaborating achievers, or even of independent co-achievers, cannot be accepted as ample gifts of nature and of only casual nurture. Also, even as relatively rare creativity must be prized, so must the less publicized ability to work cooperatively and under direction without crippling (or even with growth of) ego-strength.

Experience at GM Research Laboratories gives sufficient evidence that C. F. Kettering and Thomas Midgley, Jr., were unusual individuals in both their individual action and their synergistic collaboration. The discovery and development of tetraethyl lead as an antiknock compound for gasoline called for what GM credits as the "extensive investigation" by Midgley and the "persevering leadership" of Kettering, director of the Research Laboratories. In another example, Midgley also discovered dichlorodifluoromethane (Freon), an important refrigerant, but he undertook the research when urged by his friend, Kettering, who is viewed as the man responsible for starting, staffing, and directing the refrigerant program.

The dynamism, inner-directedness, and ingenuity that Kettering had also appear behind the successful recovery of bromine from seawater in a quest to find a larger source than the well waters of Michigan for this important chemical in the production and use of leaded gasoline.

The tendency to organize corporate research on a team basis may confer a more obvious jointness on the inventive process than is visible elsewhere, yet individuals still stand out in technical achievement. Among individuals making significant contributions at GM have been:

- G. Chavanne, a Belgian, who first made trimethylbutane, later found to have superfuel properties and "triptane."
- J. M. Campbell, the GM patentee who discovered that phosphorus additives suppress preignition in leaded gasolines.
- G. Malone, who made the first fluid for use in automatic transmissions.
- W. J. Mayer, who patented a radioactive method for quickly measuring oil consumption in automotive engines.
- H. C. Mougey, who patented a process for applying nitrocellulose lacquer to car bodies.
- G. D. Cheever, who devised the rapid ferrotest for quality-controlled phosphate coating of steel.
- D. W. Hardesty, who improved the acetic acid test for corrosion resistance.
- R. L. Saur, who devised an electrolytic corrosion test in the 1960's.
- H. J. Jendrzynski, who patented an improved nickel-plating technique.
- T. J. Hughel, who formulated a beryllium alloy of exceptional strength.
- A. L. Boegehold, who in the mid-1930's discovered that the addition of bismuth improves pearlitic malleable iron for use in large high-stress castings, such as crankshafts.
- C. S. Tuesday, who not only directed development of a highly advanced

smog chamber but who also discovered the unusual behavior of oxides of nitrogen in the atmosphere.

- D. J. McEwen, who GM considers pioneered in the development of new gas chromatographic techniques for pollution research.

Individuals still dominate the inventive scene in corporations as a rule, but pairs usually dominate when multiple inventors are recognized. At GM, significant chemical patents include one to J. M. Heuss and W. A. Glasson associated with their discovery of peroxybenzoyl nitrate, a chemical compound not previously known but one that may be the chief smog compound causing eye irritation. O. J. Klingenmaier and J. T. McWatters obtained a joint patent in 1968 for an iron plating bath that prevents pitting and also permits close control of the process, including deposit thickness. M. A. La Boda and A. J. Chartrand are credited with experiments that showed sodium chlorate to be superior to sodium chloride as an electrolyte for electrochemical machining. Under the leadership of B. E. Nagel, a patented unique windshield washing solution was developed for year-round use. S. R. Callaway and F. J. Webbere obtained a joint patent in 1954 for a nickel-base, high-temperature alloy suitable for jet engine blades.

The Organization

The private company provides the typical site of practical invention nowadays, and the trained employee inventor has largely replaced the speculative, self-taught amateur who is traditionally linked with an attic or a garage. Although the company's profit orientation greatly favors incremental development and application over basic inquiry and radical departures (which promise only uncertain, distant, or uncapturable pay-offs), GM has also conducted fundamental studies. Indeed, a large company that conducts formal research activity commonly allows a small fraction of its funds for theoretical investigations, free-wheeling, or serendipity. An established company also has good reasons to view its business and its past research accomplishments in an evolutionary light and in a larger-systems context. Thus, the GM research program has, over the years, recognized the automobile as much more than an engine; and, in contemplating the engine, it has reacted to the continual need expressed by consumers for better parts, superior fuels, and less noxious exhausts.

Especially noteworthy is the fact that the private company is the characteristic medium of economic growth in our type of society. New processes and products encourage new wants, or better satisfy existing wants, or release resources for other uses. The company serves as the dynamic site for the generation of new technology and for the translation of it into commercial forms. Competition not only tends to propagate new processes and products but also tends to multiply the number of nearly equivalent market substitutes and alternative material sources.

According to GM, tetraethyl lead antiknock compound was easily the most significant automotive achievement of its day. This compound improved performance of existing engines, but it did more. It had the dynamic effect of stimulating development of more efficient, higher compression engines. Since its introduction, almost half a century ago, it has been closely identified with the widening economic success of the automobile, finding its way into most regular and premium grade gasolines.

We have already noted that the coming of Ethyl gasoline meant a new demand for bromine, which became a necessary additive for inhibiting objectionable lead deposits from the antiknock additive itself.

Under Kettering, a more abundant source of bromine was found; and, although the GM process was not commercially valid, it did spur the development of an independent process by Dow Chemical Co. that provided bromine in the more desirable liquid form. The technique developed in the GM Research Laboratories in the 1920's was what GM considers a major refinement of an impractical process devised by cooperating chemists at GM Chemical Co., a wholly-owned subsidiary of General Motors and forerunner of Ethyl Corp.

The work of J. M. Campbell, W. G. Lovell, and T. A. Boyd on the relation of molecular structure to octane rating of fuels has at least two lessons. One is that fundamental research in industry can be profitable—a truism, of course, but occasionally worth repeating. The particular congenial conditions of one company need not, however, be easily duplicated elsewhere. The second lesson is that one achievement prepares the way for another, especially in a context of well-established need. GM feels the new knowledge gained by the three men encouraged advances in fuel formulation and petroleum refining and led to the production of new gasoline blends with better combustion properties. The long train of interactive improvements of fuels and engines has meant, according to GM, that American motorists are using an estimated one third less fuel today for transportation than would have been required otherwise. Savings to the consumer have been paralleled by an effective extension of the nation's petroleum reserves.

But, because GM's research activity is rounded and systems oriented, it has embraced much more than the engines and the fuels that make the cars go. In the mid-1920's, an important change was made by GM in body covering—a change from varnish to nitrocellulose lacquer. Great new opportunities for the economic dominance of the automobile have since been realized. Two technical contributions were vital to the change: (1) the development of the new paint system in collaboration with du Pont and (2) the invention by H. C. Mougey of a patented process for applying this system. The coating of car bodies has been revolutionized. The drying time was cut from two weeks to one and a half days, so the production and delivery of cars could be speeded. Furthermore, cars could be finished in a variety of colors, a fact enhancing customer appeal. As for utility, the lacquer also afforded longer protection against the elements. When the nitrocellulose paint concept was finally replaced by the superior acrylic technique in 1958, du Pont again played an important collaborative role, although GM subsequently perfected the application process on its own.

The systems orientation of GM research is also evident in its reach to many additional metal coverings. We have already mentioned the development of the ferrotest method of checking phosphate coatings, accelerated tests of corrosion resistance, and processes for iron plating and nickel plating. Not yet mentioned is the development in the 1960's of a technique of electrodepositing polymers to apply an alkyd primer that is compatible with an acrylic lacquer topcoat. This kind of primer is still used for automobile bodies and various parts.

Materials represent another dimension of the rounded research program of GM. Reference has already been made to a few of the company's innovations—high-strength beryllium, a high-temperature superalloy with a nickel base, improved malleable iron, and the use of sodium chlorate as an electrolyte for electrochemical machining.

Consumer comfort and convenience are inevitable targets of company research efforts. In the chemical field, GM has developed a windshield washer solution (cited above). The company has also improved fuels and made more colorful and more durable lacquers for body coats. The reduction of emissions of carbon monoxide and hydrocarbons is a major current concern of all makers and users of combustion engines.

What remains to be mentioned before this section is concluded is that a large established company with a conscious research posture is a flexible instrument for the management of technological change. Thus, GM has pursued technological advance as an avenue for diversification (product proliferation)—a tactic illustrated by the successful development of Freon and Ethyl gasoline and by their marketing through subsidiaries. GM has also participated in the conduct of joint ventures, formal or informal, with other companies, as in its collaboration with du Pont on nitrocellulose and acrylic paints.

The Environment

At the outset, it was explained that "environment" is a comprehensive term, but the new emphasis on protecting the biosphere requires mention first that GM has long been concerned with the problem of controlling emissions. The initiative seems largely to have passed to the Executive, the Congress, and concerned citizen groups. By the mid-1970's, big changes in engine design are scheduled. The public monitoring of technological development and the redistribution of pollution costs will by then be political and economic realities. We shall surely hear more of technology assessment and cost-effectiveness studies relating to the automobile's burden on the environment. Meantime, the probable end of Thomas Midgley's gasoline era appears near, since tetraethyl lead has been found to increase hydrocarbon emission and adversely affect catalytic mufflers. Consequently, lead-free gasolines are beginning to be marketed.

The broad category of "environment" also accommodates examples of government-industry interaction. The triptane example shows the importance of government demand in creating a market when no conventional need was yet evident. Triptane attracted attention in the late 1930's as a superfuel for aircraft, and GM devised a simple production process that led to construction of a full-scale plant in 1943 with the approval and assistance of the War Department and Army Air Force. But the government-sponsored market collapsed soon after 1945, and the postwar curtailment of the government's reciprocating engine program in favor of gas turbines began a phase-out of interest in triptane.

Another example of the relevance of government demand is provided by the development of stronger beryllium alloys under Navy contracts for use in precision-guidance gyroscopes. GM's research resulted in not one but three forms of beryllium that exceed the minimum elastic limit necessary for stable operation of the gyro. Once developed for specific military purposes, the high-strength beryllium became available as a gyro material for the Apollo moon ships and the Boeing 747 jumbo jetliner.

The experience with the Calloway-Webbere high-temperature alloy also illustrates the principle that early military demand may provide the economic and technological foundation for eventual civilian markets. According to GM, GMR–235 was the first nickel-base high-temperature cast alloy accepted by the U.S. Air Force for making jet engine turbine blades. It found application in General Motors' J–33, J–71, and T–56 engines. These engines were used in a number of military and commercial aircraft including the F80 and F9F7 fighters, the C-130 cargo plane, and the Lockheed Electra and Convair 580 carriers.

This chapter reports on two sets of employment projections to 1980, one made by the Bureau of Labor Statistics and the other by the American Chemical Society. The BLS figures for chemists and chemical engineers comprise only a small part of the statistical findings of that agency's comprehensive and continuing occupational investigations. They are intended to cover all economic activity; that is, they include college and university teachers, the personnel of the rest of the private sector, and government employees. The ACS projections to 1975 and 1980, which are the results of a special survey undertaken for this study, are much narrower in scope, focusing explicitly on the prospects of chemists and chemical engineers in manufacturing. They also provide information on the expected functional distributions of these professionals. The two sets of estimates are discussed in turn, and the chapter concludes with gleanings from the remarks submitted by ACS respondents.

On "Projections" and Other "Forecasts"

At the outset, a word is required about "projections," which represent only one of the many varieties of "forecasts." We define "forecasts" in general as current appraisals of future opportunities, events, or states. They divide into two major classes, which might be called "neutral" and "motivated."

A neutral forecast, which may be based on judgment as well as knowledge, is not intended directly or immediately to affect the particular future that is foreseen. Within this class are the "projections" typically made by such fact-finding government agencies as BLS. Projections are if-then appraisals of the future. They spell out the implications, with respect to variables of special interest, of certain more general conditions, relations, or scenarios that are assumed for the larger domestic or global environment. These assumptions may be illustrative rather than plausible, but more realistic ones, rather than less, are usually preferred or sought. In any case, projections are not flat "predictions"—a term that might well be confined to asserted outcomes that forecasters consider either most probable or sure.

The second major class of forecasts is characterized by the maker's motive of intervention. The forecaster (or his principal) desires to influence the future, to achieve or to frustrate what is asserted to be a possible or probable outcome. This outcome may be promoted or deterred by a forecast in the form of public information or "propaganda," or it may be advanced or negated by a forecast in the form of a deliberate goal-oriented policy and action. Thus, forecasting embraces action as well as thought.

Although the two main categories of forecasts are easily distinguishable from each other, neutral forecasts achieve their maximum usefulness when they are sufficiently credible, valid, or persuasive to guide policy and action; that is, to stimulate subsequent motivated forecasts. Public and private programs are intended, after all, to bias the future in favor of, or away from, potentials that are regarded as realizable. The highest service that this chapter, or that the report as a whole, could perform would be to assist decisively in the development of sound measures for the practical, economic use of chemistry and the resources devotable to it. Such measures would aim at forestalling dangers that loom, at supplying correctives for adverse situations that already exist, and at bringing even closer the positive opportunities that already appear to be within reach.

BLS Projections

The assumed environment for 1980, the "target year" of the BLS projections of "requirements" of chemists and chemical engineers, represents a standard blend of conservatism and optimism. A "fully employed labor force" would at that time be generating the national output at a substantially higher productivity level. Through "monetary and fiscal policy" (wage-price policy is not mentioned), a magic reconciliation of "low unemployment rates and relative price stability" would have been attained "without reducing the long-term economic growth rate." More federal funds would be flowing to the state and local governments for meeting social needs. No "radical" change would have occurred in "the institutional framework of the American economy." Indeed, familiar "economic, social, technological, and scientific trends" of past and present would persist. So would the customary "values" that have been "placed on work, education, income, and leisure." Lower fertility rates than those experienced "in the recent past" would prevail. The strength of the armed forces would be much like it was just prior to the Vietnam escalation. Although no open hostilities would rage, a continuing "guarded" relationship among the major powers would preclude substantial arms reduction. Nothing is said explicitly by BLS, however, about international economic competition, which is sure to affect the employment prospects of chemists and chemical engineers.[1]

One brief passage in the BLS source document relates especially to chemistry. It slightly amplifies the above general summary of assumptions underlying the projections:[2]

"Requirements ... are heavily predicated on increasing expenditures for research and development although the rate is expected to be lower than that experienced from the late 1950's through the late 1960's. In addition, the demand for chemists will continue to reflect the growing market for plastics, man-made fibers, nuclear fuels, and other industrial products having a chemical origin."

Table 18 shows the 1980 BLS projections for chemists, biochemists, and chemical engineers. Although the source document is not clear on these points, (1) the estimate for biochemists is already incorporated in the projection for chemists; (2) the estimates for chemists and chemical engineers are intended to cover the entire economy, including higher education and government; and (3) the 1980 forecasts and the comparable 1968 figures shown by BLS are rounded, but the published percentages of change were computed from unrounded numbers that are not presented.[3]

Employment of chemists, according to the BLS occupational projections, would increase from 130,000 in 1968 to 200,000 by 1980, while the number of chemical engineers would rise from 50,000 to 67,000. The biochemist component of chemists shows a projected gain from 11,000 to 17,000. These figures imply average annual growth rates of 3.7%, 2.5%, and 3.7%, respectively. In the case of chemical engineers, however, the published BLS

---

1. *Occupational Manpower and Training Needs*, Bulletin 1701, U.S. Bureau of Labor Statistics, 1971, pp. 65-66.

2. *Ibid.*, p. 12.

3. *Ibid.*, p . 67.

change of 25.8% for the entire 12-year span[4] implies a lower annual growth rate of only 1.9%.

The average annual number of openings associated with employment growth may be derived immediately from the terminal BLS employment figures (i.e., for 1968 and 1980),[5] but the estimated openings for replacement are based on additional data and also embody assumptions that could well prove invalid. In particular, Table 18 shows that replacement openings for chemists are expected by BLS to average 6,800 per year, a figure greater than the 6,000 annual openings associated with growth. This replacement estimate would prove wrong if, for example, the expected 3,700 annual transfers out of the occupation of chemistry do not materialize between 1968 and 1980.

Table 18

Projected Employment of Chemists, Biochemists, and Chemical Engineers: 1968–80

	Chemists	Biochemists	Chemical Engineers
Employment			
1980	200,000	17,000	67,000
1968	130,000	11,000	50,000
Avg. Annual Growth (Percent)	3.7	3.7	2.5
Avg. Annual Openings	12,800	700	1,600
Growth	6,000	500	1,100
Replacement	6,800	200	500

Source: *Occupational Manpower and Training Needs*, Bulletin 1701, U.S. Bureau of Labor Statistics, 1971, pp. 31-32, 68. The employment figures refer to the whole economy; that is, to the private sector, higher education, and government. The figures for biochemists are already incorporated in those for chemists. BLS reports (p. 32) that the average annual replacement openings for chemists includes "an estimated 3,700 replacements for those who transfer to other occupations."

More troublesome than, but closely related to, the estimation of replacement needs is the determination of the number of new chemistry graduates required in 1968–80. Unfortunately, the BLS quantitative discussion does not systematically and exhaustively treat the elements of accession and attrition, so it fails to render an arithmetically satisfying account of the dynamics of the projected employment gain.[6] In any case, because of an (unstated) early loss of chemistry graduates to the profession, BLS concludes that the annual academic yield projected by the U.S. Office of Education may prove very inadequate. This conclusion would seem overoptimistic if it is interpreted as a prediction of enlarged economic demand

---

4. *Ibid.*, p. 68. If 50,000 is the "correct" unrounded BLS figure for 1968, a growth of 25.8% would imply a total of 63,000 for chemical engineers in 1980. This total would be compatible with another BLS figure reported on p. 68: annual openings of 1,100 for the interval 1968–80.

5. The published average annual openings for growth do not always agree with the computations based on terminal figures.

6. The "turnover" identity that is to be translated into a closed system of numbers may be summarized as: Employment in 1980 = Employment in 1968 + 12 (average annual accessions to payrolls — average annual separations). The number 12 is the difference between 1980 and 1968. In principle at least, accessions and separations could be partitioned into as many elements as are deemed pertinent to a problem, and these elements could then be recombined into expressions for openings for growth and replacement, etc., without impairment of the algebraic identity.

For a discussion of methodology, see Neal Rosenthal, "Projections of Manpower Supply in a Specific Occupation," *Monthly Labor Review*, November 1966, pp. 1262-66. Also relevant, both substantively and methodologically, is *College Educated Workers, 1968–80: A Study of Supply and Demand*, Bulletin 1676, U.S. Bureau of Labor Statistics, 1970.

for chemists rather than as a projection of a continuing high propensity of chemistry graduates to work in other fields. This predilection, of course, is not a measure of "social waste" of training in chemistry. Indeed, such training could fit people for success in other endeavors, improve their functionality as citizens, and expand their cultural horizons.

The matter just cited is important enough to readers of this report to warrant a complete quotation of the BLS argument. The statement should be read, however, with the understanding that it presents the BLS reasoning only eliptically, failing to mention the substantial loss of new chemistry graduates attracted to other fields:[7]

"Although the major source of supply of chemists is from new graduates majoring in chemistry, requirements may also be met from other sources: persons not in the labor force; immigrants; and graduates who did not major in chemistry. Limited data on entry indicate that a significant number of workers have entered chemistry from these other sources. Although a variety of factors affect the number of 'other entrants' including the relative availability of chemistry graduates, significant numbers probably will continue to enter. If past patterns of entry from other sources and of new chemistry graduates continue, an average of about 17,000 bachelor's degree graduates in chemistry would be needed annually to meet projected requirements.

"In 1968, about 10,800 bachelor's degrees were granted in chemistry. To meet requirements, therefore, this number would have to average almost 65 percent above 1968 levels. U.S. Office of Education projections indicate that the number of bachelor's degrees awarded in chemistry annually would average about 5 percent above 1968 levels for the 1968–1980 period. Therefore, to meet requirements for chemists we will have to take action to increase the number of graduates even faster than past trends."

If a persistent and sizable educational shortfall of chemists actually appears imminent, more economic and easier remedies than the accelerated production of new graduates will also merit attention. For example, it is much less costly to encourage the entry of more of the qualified new bachelors into the profession and to slow down transfers out of the profession. The appropriate tools would seem to be counseling, public service advertising, and pay incentives. Another alternative may appear attractive to employers: the restructuring and further instrumentation of chemists' jobs to allow the use of a greater proportion of technicians.

The BLS projection for all "engineering and science technicians" (excluding draftsmen) may well be consistent with the second alternative, which is not, however, contemplated in the BLS assumptions. A rise of employment from 620,000 in 1968 to 890,000 in 1980 is indicated by BLS. This gain implies annual openings of 22,000 for growth and 9,000 additional openings for replacement. But the prospective enlargement of supply could itself render plausible an increasing reliance upon technicians in chemical activities in the future. Thus, junior colleges graduated 30,018 engineering and science technicians in the academic year 1968–69. In fiscal year 1969, a similar number of students, 30,149, completed post-secondary vocational programs as 12,232 others completed secondary programs.[8]

Another BLS report, relating specifically to technicians, says something about future needs of chemical technicians in particular. It shows employment of 60,500 in 1966 and a projected requirement of 96,500 for 1980, so

---

7. *Occupational Manpower and Training Needs*, p. 32. Virtually the same statement appears in *College Educated Workers, 1968–80*, pp. 6–7.

8. *Occupational Manpower and Training Needs*, pp. 35, 75.

that average annual openings for growth would amount to about 3,000. The rise of 59.5% for the 12-year span is slightly greater than that for chemists. It translates into a mean annual increase of almost 4.0%. The same report shows an identical gain for draftsmen and a slightly smaller 12-year advance (54.3%) for all technicians.[9]

ACS Projections

A survey of Corporation Associates of ACS, devised for this study and conducted in the fall of 1971, has yielded projections of employment of chemists and chemical engineers in manufacturing for 1973, 1975, and 1980 (as well as base figures for 1971). In this survey, employment estimates were requested from individual companies for January of the selected years to 1980. The month was specified for formal consistency, at least, with the statistical series for professional employment normally compiled by BLS and NSF. The survey schedule and the accompanying instructions are shown as an appendix to this chapter.

Since projections are really "if-then" statements in numerical dress, the instructions given to the ACS Associates had to spell out a common framework of guiding assumptions. According to the adopted scenario, "the resumption of an upward growth trend for the economy, continuing to the end of the decade, will be unequivocally clear" by January 1973. Additionally, it was postulated that "by January 1975 full employment of the nation's labor force (i.e., unemployment of 4% or less) will have been reattained." Furthermore, an average annual increase of 4.3% in the Gross National Product (expressed in constant dollars) was assumed for the first half of the 1970's, the same rate that was published by the Council of Economic Advisers in the 1971 *Economic Report of the President*. The instructions included the observation that, "even under current legislation (such as the Clean Air Amendments of 1970, PL 91-604), sizable public and private expenditures will be required for environmental improvement to meet prescribed standards on schedule." Reference was also made to the magnitude of expenditures for pollution abatement in 1970–75, as projected by the Council on Environmental Quality.

A "rounded estimate" was requested in the ACS instructions for 1980, "on the assumption of continued full-employment growth to the end of the decade after 1975." Other stipulated conditions were: "peace, a continuing shift toward service employment in general in U.S., intense foreign competition, and strong emphasis on unmet domestic social needs (e.g., 'quality of life,' health, and safety) and on international problems of food and energy." A second paragraph regarding the estimate of 1980 took cognizance of a BLS figure for manufacturing that is also of interest here:

"You may be encouraged by, or find some guidance in, an unpublished employment estimate by U.S. Bureau of Labor Statistics (BLS) of 104,000 for chemists (excluding chemical engineers) in manufacturing in 1980 under full-employment conditions. This compares with a BLS estimate of 80,200 chemists in manufacturing for January 1970—a growth of nearly 30 percent for the decade 1970–80 as a whole and a geometric mean annual rate of nearly 2.64 percent."

Although the survey responses may well be adequate for manufacturing as a whole, some of the component industries are only thinly represented. Bias due to uneven representation of the different industries in the responses can be moderated, but compensation is not readily achievable (short of the acquisition of additional survey data) for any failure of the

9. *Technician Manpower, 1966–80*, Bulletin 1639, U.S. Bureau of Labor Statistics, March 1970, p. 7.

participating companies to constitute reliable samples. Accordingly, on the basis of coverage alone, much more confidence should attach, say, to the projections shown for *Chemicals, Excluding Pharmaceuticals* than, say, to those shown for *Equipment and Miscellaneous Manufacturing.* The underrepresentation of the latter in the company responses can be overcome, in some degree, for the development of acceptable estimates relating to manufacturing as a whole. But, because of the very size and heterogeneity of the same "industry," the few cooperating organizations therein cannot be expected to reflect dependably the employment prospects of all companies making equipment and miscellaneous products.

Table 19, which presents some of the ACS findings, is based on information covering both the research and nonresearch activities of responding companies. For all manufacturing and for almost every one of its components, two sets of figures are shown. One set refers to the maximum number of companies supplying information usable for a 1971–75 comparison. The other set refers to a smaller number of companies supplying comparable data for 1971–80.

Projections to 1980 have been made by ACS for manufacturing as a whole on the basis of information furnished by 53 companies that employed about 35,000 chemists and chemical engineers in plant and laboratory operations in January 1971. This number represents about 28% of all such professionals at that time engaged in the nation's manufacturing establishments.

The 63 companies that furnished data permitting an alternative estimate of the change from 1971 to 1975 employed about 40,300 chemists and chemical engineers in January 1971. They accordingly accounted for almost one third of such professional workers engaged in manufacturing at that time.

The raw ACS findings envisage a slower growth of employment for chemists and chemical engineers in manufacturing than do the BLS figures quoted earlier. According to Table 19, the ACS projection for all manufacturing to 1980 implies a geometric growth rate of 1.83% per year from 1971. For the same set of 53 companies, the rise is 1.37% per year during the period 1971–75. For the maximum number of companies (63) supplying usable data for the same four years, the annual growth rate is virtually identical—1.39%. These rates are less than the BLS rates cited in the preceding section for the whole economy during 1968–80 (i.e., 3.7% per year for chemists and 2.5% for chemical engineers). More relevant, however, is the lag of the ACS rate for 1971–80 behind the BLS rate for 1968–80 for chemists in manufacturing, which exceeded 2.6%. But most relevant is the lag of the same ACS rate behind the figure of 2.32% that may be computed from BLS projections for 1970–80 for chemists and chemical engineers in manufacturing.[10]

An attempted adjustment for weight bias (not sampling bias) still leaves the ACS rates below those of BLS but definitely narrows the gap. If the industry percentages shown in Table 19 are multiplied by 1970 BLS totals for chemists and chemical engineers in the corresponding industries (see Table 9), the following weighted averages emerge for all manufacturing:

	1971–75	1971–80
63 companies	1.99	
53 companies	1.82	2.12

---

10. The industry weights were derived from *Tomorrow's Manpower Needs, Vol. IV: The National Industry-Occupational Matrix and Other Manpower Data (Revised),* U.S. Bureau of Labor Statistics, Bulletin 1737, 1971, pp. 18, 124–125, 128, 129.

Table 19

Employment of Chemists and Chemical Engineers: ACS Manufacturing Survey, 1971–80

Industry or Group	No. of Companies	Chemists and Chemical Engineers			Annual Growth Rate, Percent	
		1971	1975	1980	1971–75	1971–80
All Manufacturing	63	40,314	42,236		1.39*	
	53	35,014	36,973	41,236	1.37*	1.83*
Chemicals, Excl. Pharm.	23	18,279	19,406		1.65	
	20	17,939	19,020	21,941	1.57	2.26
Large	5	16,233	16,992	19,426	1.15	2.02
Medium	5	1,506	1,714		3.29	
	4	1,213	1,390	1,690	3.46	3.75
Small	13	540	700		6.70	
	11	493	638	825	6.66	5.89
Pharmaceuticals	7	2,433	2,709		2.72	
	6	2,169	2,422	2,631	2.80	2.14
Petroleum Products	6	8,069	7,964		−0.33	
	5	5,769	5,764	6,058	−0.02	0.55
Soft Goods	12	1,946	2,283		4.09	
	10	1,442	1,703	2,096	4.24	4.24
Food	7	822	939		3.38	
	5	318	359	414	3.08	2.98
Paper, Textiles, etc.	5	1,124	1,344	1,682	4.57	4.58
Hard Goods	15	9,587	10,243		1.79	
	12	7,695	8,064	8,510	1.18	1.13
Metals, Rubber, etc.	8	2,808	3,127		2.73	
	6	2,306	2,583	2,930	2.88	2.96
Equip. & Misc. Mfg.	7	6,779	7,116		1.22	
	6	5,389	5,481	5,580	0.42	0.39

Source: Questionnaire survey of Corporation Associates of the American Chemical Society. The data refer to manufacturing respondents reporting both their research and nonresearch staffs.

* If the percentages shown for component industries were weighted by the total number of chemists and chemical engineers employed therein in 1970 (see Table 9), the two average growth rates for 1971–75 for all manufacturing would become 1.99% and 1.82%, respectively; and the 1971–80 rate would become 2.12%. These weights, however, only aim to adjust for weight bias, and they cannot adjust for sampling error within component industries.

The ACS rates for 1971–75 are raised toward 2%, while the figure for 1971–80 becomes 2.12%. The remaining disparity between 2.12% and the indicated BLS growth rate of 2.32% for chemists and chemical engineers in 1970–80 may reflect, for example, (1) a difference of opinion regarding the future gain of research employment, and (2) a relatively conservative projection of employment increase in *Equipment and Miscellaneous Manufacturing*, which is heavily weighted but has only half a dozen firms.

A divergence between the annual rates for 1971–75 and 1971–80 persists after the figures for the 53 companies are weighted. This difference points to some improvement in the employment outlook for chemists and chemical engineers during the second half of the decade. The upturn is weak, despite assuming sustained full-employment growth from 1975 to 1980. A return to the buoyant experience of the 1950's and 1960's is hardly hinted.

Some notice will now be taken of the ACS rates developed for individual industries. Although these rates vary in reliability, they do add some illumination. For *Chemicals, Excluding Pharmaceuticals* as a whole, the rates shown in Table 19 are a bit higher than the unadjusted ones for *All Manufacturing*. Again, the second half of the decade appears more promising than the first half. A still better outlook is indicated for pharmaceuticals production. A setback may be endured at first in petroleum refining and related activities, and then a cautious growth is suggested for the second half-decade. Less ambiguous advance, however, is likely in the soft-goods industries. For the production of hard goods in general, especially *Equipment and Miscellaneous Manufacturing*, the growth of employment of chemists and chemical engineers will apparently be sluggish for the entire decade.

A closer examination of detailed computations for *Chemicals, Excluding Pharmaceuticals* discloses that it is the large companies that see least expansion ahead for chemists and chemical engineers. The growth rates for these companies are 1.15% per year for 1971–75 and 2.02% for 1971–80. For medium-size companies, on the other hand, the rates are well above 3%, and for small companies they are more like 6%. Since the larger companies, of course, employ most of the chemists and chemical engineers in the conventional chemical industry, their sober views are bound to influence strongly the general state of professional opinion.

From a practical standpoint, the adjusted ACS projection to 1980 suggests an increase of about 25,000 in the employment of research and non-research chemists and chemical engineers in manufacturing over 1971.[11] Annualized, this total translates into approximately 2,700 openings per year for employment growth. To this figure, a smaller number of annual openings for replacement should be added—2,200, say, in the light of Table 18.[12] Thus, it would seem that manufacturing may demand up to 5,000 chemists and chemical engineers per year during the rest of the decade for both growth and replacement.

The higher BLS rate of employment growth for chemists and chemical engineers in manufacturing (2.32%) points to a somewhat better welcome for new graduates. An increase in employment of something like 31,500

---

11. If 125,000 is taken as the estimated number of chemists and chemical engineers in manufacturing in 1971 (the BLS source cited in footnote 10 indicates a total of 122,100 for 1970), the superimposition of an annual increment of 2.12% would yield a figure of 151,000 for 1975, a gain of 26,000. If the ratio derived from the unadjusted ACS figures for 1971 and 1980, 41.2/35.0, is applied to 125,000, the gain is 22,250. We have used 25,000 in the text above.

12. About 1,800 of the 2,700 annual openings for growth are for chemists and 900 for chemical engineers. The corresponding estimates for replacement are 1,800 and 400 per year.

between 1970 and 1980 means 3,150 annual openings for growth.[13] Doubling this annual figure to account also for replacement means total annual openings of about 6,300 in manufacturing.

Will the research-nonresearch distribution of chemists and chemical engineers in manufacturing change in the next few years? The ACS survey results permit comparison of the 1971 and 1975 allocations for the same 63 companies that provided a trend projection for the totals. Briefly, Table 20 indicates virtually no change in the research ratio between 1971 and 1975. Both estimates round to 43%. Weighting does not alter this judgment, although it does raise the research ratios for the two years. They remain approximately equal at about 45%.

Table 20

Chemists and Chemical Engineers in Research as Percentages of Total: ACS Manufacturing Survey, 1971 and 1975

| Industry or Group | No. of Companies | Chemists and Chemical Engineers | | | |
| | | 1971 | | 1975 | |
		Total	% in R&D	Total	% in R&D
Total	63	40,314	43.4*	42,605	42.9*
Chemicals, Excl. Pharm.	23	18,279	48.2	19,406	48.0
Large	13	16,233	49.6	16,992	49.6
Medium	5	1,506	33.8	1,714	32.9
Small	5	540	43.5	700	43.7
Pharmaceuticals	7	2,433	51.6	2,709	50.3
Petroleum Products	6	8,069	32.9	7,964	30.7
Soft Goods	12	1,946	51.4	2,283	51.6
Food	7	822	42.0	939	39.8
Paper, Textiles, etc.	5	1,124	59.2	1,344	59.8
Hard Goods	15	9,587	39.3	10,243	39.1
Metals, Rubber, Glass, etc.	8	2,808	49.6	3,127	49.6
Equip. & Misc. Mfg.	7	6,779	35.0	7,116	34.6

Source: Questionnaire survey of Corporation Associates of the American Chemical Society, conducted in fall of 1971. The data refer only to those manufacturing respondents that supplied separate information for research and nonresearch activities.
* If the industry percentages are weighted by the total numbers of chemists and chemical engineers in the corresponding industries and groups, the adjusted figures obtained for all manufacturing are higher but they retain virtual parity. Thus, weights derived from BLS data for 1970 (see Table 9) yield an alternative 1971 estimate of 45.9% and an alternative 1975 estimate of 45.5%. (Preferably, 1971 weights should be used for the 1971 computation and 1975 weights for the 1975 projection.)

The 1971 research ratios for 63 companies are hardly different from those shown for 68 companies in Table 6. Thus, the unadjusted and weighted ratios for the maximum number of companies providing eligible data for the 1971 computation are 43.4% and 45.1%, respectively. Table 20 presents corresponding figures of 43.4% and 45.9% for 63 companies.

A glance at the details of Table 20 reveals several other points worth mentioning:

• For *Chemicals, Excluding Pharmaceuticals* as a whole, the research ratio is slightly below one half in 1971 and 1975. It is practically one half for large companies, closer to two fifths for the small companies, and roughly one third for medium companies.

---

13. The source cited in footnote 10 yields estimates of 122,100 and 153,500 for chemists and chemical engineers in manufacturing in 1970 and 1980, respectively. The difference is 31,400, which we have rounded slightly in the text.

- For *Petroleum Products,* too, a research ratio of about one third is indicated, and a decline of two percentage points is registered between 1971 and 1975.
- For *Food,* also, a decline of two percentage points is recorded in the research ratio between 1971 and 1975, but the ratio is in the neighborhood of two fifths in both years.
- For *Pharmaceuticals* and for *Soft Goods* in general, the research ratio is close to one half in both years, as it also is for *Chemicals, Excluding Pharmaceuticals.*
- The ratio for *Hard Goods* is close to two fifths in both years, even though the research proportion for one of the main components, concerned with materials production, is about one half.

What does Table 20 imply about openings for all chemists and chemical engineers in manufacturing in 1971–75? This is an appropriate question inasmuch as the table is based on the maximum number of companies providing comparable information for the two years. From the unweighted totals, it is clear that a net gain of less than 6% is projected for the interval. Applied to a manufacturing total of something like 125,000 for 1971, this percentage (unrounded) suggests an employment increment of about 7,200 for the whole period. If doubled to reflect replacement and also annualized, this increment implies 3,600 openings per year for chemists and chemical engineers in manufacturing.

The alternative method of weighting the individual industry changes by BLS industry totals (for 1970) leads to a higher estimate of manufacturing openings for 1971–75. It indicates a net gain of 8,800 for the interval. This figure, doubled for replacement and annualized, implies total openings of about 4,400 per year.

If the research ratio holds roughly constant (say, at 45%) and if Ph.D.'s prefer (and are preferred for) laboratory tasks, it is clear that manufacturing can accommodate a good share of the shrinking future output of the graduate schools. A new growth of 2,200 jobs per year for chemists and chemical engineers, even without allowance for turnover, is indicated by the immediately preceding computation, and this figure translates into nearly 1,000 research slots. In Part II of this volume, it was noted that Ph.D. grants in chemistry will probably stabilize in the latter 1970's in the neighborhood of 1,500 per year.

Still another exercise is suggested by the availability of some responses from companies that could furnish information about research personnel only. Tables 18, 19, and some others presented earlier omit companies that did not supply balanced reports on the relevant manpower. The truncated reports, however, are often more optimistic about research prospects. Perhaps, then, the ACS tabulations underrepresent research. On the other hand, many companies that failed to reply to the ACS questionnaire, larger ones included, surely had a doleful tale to tell about the near past that they endured and the more distant future that they face. Since correction could not be made for nonresponse in general, calculations were made that treated research-only reports as though they were complete. The resulting upward bias in favor of research suggests some upper limits in projected growth and openings.

With the last paragraph in view, we note that 13 manufacturing companies reported *research* chemists and chemical engineers numbering 3,571 in 1971, 4,098 in 1973, and 4,878 in 1975. If these figures are treated as comprehensive and added to the corresponding employment *totals* shown in Table 19, the net growth between 1971 and 1980 approaches 20%. Such a change implies an increment of 25,000 chemists and chemical engineers in nine years, or 50,000 if account is also taken of replacement needs. Thus, *maximum* annual openings for the rest of the decade would seem to be close to 5,600.

Let us suppose, now, that it is proper to add the 1971 and 1975 figures for the 13 research-only companies to *both* the research magnitudes and the totals underlying Table 20. The research ratios do rise, of course, but they also remain under one half. This biased procedure changes the estimated 1971 research proportion from 43.4% (unadjusted) to 48.0%, and it lifts the 1975 estimate from 42.9% (unadjusted) to the same level of 48.0%. Although the introduction of industry weights would further raise these percentages, the biased research ratio would barely exceed one half.

This section concludes with some remarks on the ACS findings with respect to the entire activity distribution. Will the functional allocations in manufacturing alter materially from 1971 to 1975? Apparently not, as the stability of the research-nonresearch balance may already have foreshadowed. Here, for illustration, are the 1971 and 1975 percentage distributions derived from the returns of 56 manufacturing companies employing about 30,600 chemists and chemical engineers in 1971 (almost one quarter of the total) and about 32,800 in 1975:

	1971	1975
All Chemists and Chemical Engineers	100.00	100.0
R&D	48.7	48.1
Management	3.8	3.8
Staff	44.9	44.3
Other Activities	51.3	51.9
Management	4.6	4.6
Production	23.7	24.0
Inspection and Testing	8.9	9.0
Sales or Purchasing	9.3	9.3
Other	4.8	5.0

Furthermore, although the 1971 research ratio for 56 companies is higher than those (unweighted and weighted) shown for 68 companies in Table 7, the general contours of the functional distributions are similar.

Some Comments of Respondents

A number of the company representatives replying to the ACS employment questionnaire offered remarks on probable changes and challenges of the next decade. These remarks are sampled here and arranged, in effect, under the three rubrics used in the appendix to Chapter 28: the individual, the organization, and the environment. Of course, these are interacting categories, so a certain amount of arbitrariness is involved in the following brief summary. That is, information that is set down in one place could just as well have been viewed from a different primary standpoint and, suitably restated, could have been presented elsewhere.

The Individual. In addition to commenting on prospective occupational changes that will reflect shifts in company priorities, policies, and operations and in external conditions, the company respondents show an increasing interest in utilization of persons of lower or less specialized education, especially subprofessional workers. Of course, any substantial alteration in the composition of a company's technical work force requires some job redefinition and other internal adjustments, but the discussion could just as well focus on the skill or training of needed personnel. Simply for convenience, then, we treat the prospective shift toward less specialized education as a matter of "individual" preparation and later treat the company demand for new worker skills as an "organizational" phenomenon.

The deskilling or despecialization of nonresearch jobs is emphasized in the comments of a large chemical company that did not formally complete the ACS questionnaire:

"In other functions, exclusive of R&D, there will be a decrease in the percentage of engineers . . . In many of our sales and production operations, for instance, we find that industrial engineers, graduates of two- and four-year colleges without specialized technical degrees, are capable of filling positions formerly open only to graduate engineers. Many engineering curricula have become specialized to the point of being applied science."

A decade earlier, "about 80% of the college graduates" employed by the same company had "engineering or science degrees," but the current ratio is "approaching 60%," and further declines are expected.

Several other respondents commented along similar lines. Thus, another large manufacturer of chemicals expected "an increase in employment of associate degree and bachelor of technology degree holders to fill some positions now held by chemists and chemical engineers." A major maker of personal care products responded only for research personnel, stating that "for the vast majority of jobs outside R&D a degree in chemistry and chemical engineering is not a prerequisite." A tire manufacturer observed that "we are having good success in the areas of process control and technical services with graduates of two-year technical schools." A principal maker of petroleum products stated that the 1980 BLS projection for manufacturing was inconsistent with its own evaluation: "We believe that technicians without B.S. or higher degrees will play an increasing role in the work force, thereby diminishing the need for engineers and chemists." A large research facility in the petroleum sector anticipates a tilting of new hires in favor of chemical engineers over chemists, and it also expects the ratio of doctors to masters and bachelors among the new engineers to be "much smaller than it was in the past decade."

The Organization. The responsiveness of companies to sensed opportunities for growth, to new cost burdens, and to threats to survival is also reflected in the volunteered remarks. One avenue of response of "organizations" to such challenges involves the acquisition, retraining, or reallocation of personnel, and this response is treated next. Of course, this response itself is correlated with company assessments of profitable future directions in technology and of conditions in the larger environment. For example, a company must maneuver in an environment of other decisionmakers who can be influenced but not controlled—competitors, the government, and the general public.

An outstanding manufacturer of business equipment offered the comment that "materials scientists will make new inroads into areas currently considered the domain of chemists." Many a chemist filling out the same questionnaire would have phrased the same prospect differently—as a further penetration of chemists into the province of materials science. But this is not the place for adjudication of conflicting parochial or imperialistic claims.

A leading maker of rubber products sees abatement of pollution, employee safety, and competition as major factors in the company's demand for, and redeployment of. chemists and chemical engineers. With respect to research, these remarks were made:

"There will be somewhat of a shift in emphasis in R&D toward process improvements and new processes to abate pollution and meet environmental control standards.

"The requirements of the Occupational Health and Safety Act of 1970 will force emphasis on industrial hygiene and toxicology.

"In R&D, the manpower for environmental and hygiene problems will come only partly from new employees, and mainly from reassignment of present employees."

Outside of research, the same respondent expected a greater rate of increase in the employment of chemists and chemical engineers. He noted two stimulants in the area of production: "process modifications for environmental control" and "industrial hygiene requirements." In inspection, a positive employment factor is seen in the need for "greater quality control." In sales, it is seen in "a greater emphasis on technical services to meet competition."

Other company reports cite similar outlook factors in terms of personnel requirements. Thus, one of the large makers of industrial and agricultural chemicals sees a growing need for chemists and chemical engineers "in solving environmental problems to offset a reduced effort on new developments." A diversified medium-size chemical company expects the current 1:1 ratio of chemical engineers to chemists to be maintained since "pollution control requirements involve processing systems." A small chemical division of a large corporation mentions some areas of more intensive future commitment, such as "animal and human health products" and "chemical additives," but it also cites "antipollution and safety devices" and speaks of "training needs in antipollution, waste disposal, toxicity testing, health, and safety." A maker of nonferrous metals also cites "pollution control" as a promising activity for his industry's chemists and chemical engineers.

Interdisciplinary staff adjustments will also occur as companies maneuver for position in markets for their normal products. For example, a company classified in miscellaneous manufacturing gives these reasons for expecting its complement of chemists and chemical engineers to approach 100 by 1980:

"With increasing sophistication we would expect some of these to be working in production, with possibly some increase in sales and management rather than only in production engineering and development and quality assurance."

A much larger company in the same industrial category sees increasing needs for "electronic and hardware-oriented professionals rather than chemists and engineers." A tire manufacturer expects product development to be "rather more in the direction of physical and mechanical [instead of chemical] skills, with equipment considerations playing a very considerable role." Another materials producer sees a significant "reversal" in its demand for chemists and chemical engineers in favor of the latter, both in research and in manufacturing operations.

Prospective shifts in professional personnel demand may also be illustrated for nondurable-goods production, including chemicals and pharmaceuticals. One key food processor sees the possibility of replacement of "food technologists" by "organic chemists" in the research laboratory. Another sees its requirements of chemists and chemical engineers rising substantially by 1980 in accord with recent "tremendous advances in food technology." A small pharmaceutical manufacturer expects its future chemists to be "more highly trained in biochemistry and pharmacology." Finally, the large maker of consumer-oriented chemical products already mentioned above "will probably seek a higher proportion of chemical engineers than in the immediate past; and a decrease in physical scientists and an increase in life scientists among the scientists."

Dimensions other than manpower are also represented explicitly in company comments that could reasonably be classified under "organization." Two small manufacturers, one a metal producer and the other in the chemical field, may seek improvement of their market postures in the course of the decade by diversification (through merger or acquisition). A larger chemical company finds some solace in having weathered still another economic storm:

"The U.S. chemical industry must produce and compete to survive. Recent belt-tightening has taught us how to be more efficient."

For a medium-size chemical company, the experience of "a steady erosion of prices and profits" induces continued caution with respect to professional staffing:

"We have been conservative in our hiring policy and have been able to maintain a stable organization without any reduction in staff."

A machinery producer with a small number of research chemists and chemical engineers reports handsome profits during the recent recession, with a "steadily downward" trend in technical employment that probably has not yet "ended." Finally, a producer of technical hardware comments tersely that the "tendency is toward development rather than research because of poor payout on research."

The Environment. Little is left for presentation under "environment" since so many eligible items have already been mentioned under the other two categories, featuring people and organizations. Obviously, comments on pollution, safety, and competition could have been deferred to this section.

An outstanding topic that warrants explicit mention here, however, is government regulation. The apprehensiveness of industry is reflected in one respondent's comment that his projections not only embody certain economic assumptions but also assume "the absence of any sweeping environmental changes affecting the pharmaceutical industry." A small chemical company, commenting on its 1980 projection, says that "government regulations which we do not know about at this time may change the figure substantially." The larger consumer-oriented chemical producer already cited states this view more forcefully:

"A factor beyond the control of industry which can be extremely important to the future of R&D is not even mentioned in Note M [of the ACS questionnaire, presented in the appendix that follows]. This factor is the extension of the activities of federal regulatory bodies into an ever-widening area of chemical products, with its profound effect on R&D costs and therefore on the whole R&D exercise as a viable investment. Future trends in this and similar regards may have a more profound effect on the employment of R&D personnel than any of the [other] factors being considered herein [i.e., the questionnaire instructions]."

Obviously, this is a good point at which to conclude formal discussion and proceed to the ACS survey questionnaire.

AMERICAN CHEMICAL SOCIETY QUESTIONNAIRE ON
1975–80 EMPLOYMENT

In the fall of 1971, the American Chemical Society sent the questionnaire shown here to about 200 of its Corporation Associates. The chief object of the survey was to develop employment projections for chemists and chemical engineers in manufacturing for the period 1971–80. Another purpose was to update some of the findings of an earlier survey, for the year 1969, conducted by the Society at the start of this study.

These observations were included in the letter accompanying the 1971 questionnaire and explanatory notes:

"The detailed notes that are attached to Enclosure C provide the ground rules for your response. From Enclosure C's headings, you will observe that employment estimates for chemists and chemical engineers are being requested for 'January' of 1971, 1973, and 1975. This month is specified, not for hyperaccuracy, but because it has been used in annual compilations of employment statistics by the U.S. Bureau of Labor Statistics for the National Science Foundation.

"Space is provided on Enclosure C for employment projections relating to 1980. Even a rough idea of the number of chemists and chemical engineers that may be required at that time by your company under full-employment conditions in the economy would be instructive.

"A reluctance to make company projections of the employment of chemists and chemical engineers is understandable, but the exercise is sure to prove useful, even for the cooperating company. Despite the usual misgivings concerning the reliability of forecasts, we all know that, as a practical matter, we have to look ahead anyway and make the best estimates we can. You should note that *projections*, after all, are different from *predictions*. Projections are hypothetical 'if-then' statements. They spell out the employment implications of assumptions that are more or less completely or clearly specified in advance. Alternative assumptions, accordingly, lead to alternative projections for a given date or period. No set of projections necessarily reflects the future on which we are willing to bet.

"The reasons for our choice of dates are straightforward enough. Actual company information for January 1971 is presumably available to all respondents. By 1973, it is assumed that a sustainable upward trend for the economy will be clearly visible. By 1975, it is assumed that this upward trend will carry the economy to full employment (i.e., to unemployment of 4% or less for the national labor force). Whether or not this prospect is the most realistic of the alternatives should not concern us for the moment. We need a premise to guide the exercise of projection, and an optimistic framework that is not implausible has special attractiveness and usefulness. Furthermore, the assumption of firm growth with reattainment of full employment by 1975 is built into other benchmark forecasts to which we may have recourse in our analysis. For example, in the latest *Economic Report of the President*, February 1971, p. 95, projections for the national product are shown that indicate a growth rate of 4.3% per year and the restoration of full employment by 1975.

"Please make use of the 'Remarks' section for expanding other replies or for making comments on other matters that are pertinent. Thus, all the entries for Enclosure C should preferably relate to company employment of chemists and chemical engineers in the U.S. only. If, however, they do not, or if substantial employment opportunities for U.S.-trained professional personnel are expected to materialize abroad under company auspices (including joint ventures), an appropriate comment should be made under 'Remarks.' "

The questionnaire (designated "Enclosure C" in the 1971 mailout) and the detailed instructions follow.

Enclosure C

Company_____
Division _____
Industry[b]_____

EMPLOYMENT OF CHEMISTS AND CHEMICAL ENGINEERS[a]

Work Activities[c]	Jan. 1971[d]	Jan. 1973[e]	Jan. 1975[f]
Total			
A. **R&D**			
1. Management[g]			
2. Staff			
B. **Other Activities**			
3. Management			
4. Production[h]			
5. Inspection and Testing[i]			
6. Sales or Purchasing[j]			
7. Other[k]			

1980 Employment Outlook:[m] _____

_____

_____

Remarks:[n]_____

_____

_____

Note a:

A chemist or chemical engineer is here defined as anyone with at least a baccalaureate degree in chemistry or chemical engineering who is currently employed in a job that requires significant use of his chemical training. As the column of activities indicates, these jobs may relate, say, to production or sales rather than research; or they may relate to work that is not typically associated with chemistry. Nevertheless, if a person has a chemical degree and his work requires chemical training, he should be included.

The figures you provide should preferably refer to company employment in U.S. If they do not, or if substantial employment opportunities for U.S.-trained chemists and chemical engineers are expected in your company's foreign operations, please mention in "Remarks" (see Note n).

Note b:

One or more industry numbers should be selected from the list that follows to indicate the principal business of your firm. If only one number is used, it should reflect at least 10% of the value of sales. For conglomerate firms, in particular, more than one number may be necessary. If more than one number is needed, please present in descending order of importance. This is the industry list:

1. Industrial organic and inorganic chemicals, petrochemicals, dyes, inks, adhesives, explosives, and other chemicals, n.e.c.
2. Proprietary and ethical drugs and medicinal products; diagnostic agents; medical instruments
3. Plastics, resins, and products
4. Synthetic and natural elastomers and their products
5. Synthetic and natural fibers, apparel, and industrial textiles
6. Soaps, detergents, cleaners, and waxes
7. Cosmetics and personal care products
8. Paints and protective coatings
9. Foods and feeds
10. Agricultural chemicals—pesticides, fertilizers, trace elements, growth regulators
11. Petroleum and gas recovery and refining to fuels, lubricants, waxes, and asphalts
12. Pulp, paper, and paper products; forestry, lumber, and wood products
13. Ferrous minerals, metals, and products
14. Nonferrous minerals, metals, and products
15. Minerals, n.e.c.
16. Machinery, except electrical
17. Electrical and electronic equipment—motors, appliances, computers, office machines, radio, TV, telephone, telegraph
18. Graphic arts—photographic, printing, and reprographic processes and equipment
19. Glass, ceramics, stone, clay, cement, and their products
20. Tobacco and products
21. Leather and products
22. Coal and coal chemicals, n.e.c.
23. Laboratory and plant-process instruments and controllers
24. Transportation (auto, truck, bus, rail, air, ship, rocket), farm, and construction equipment.
25. Nuclear energy—mining, fuel processing, applications

Note c:

Chemists and chemical engineers who engage in more than one activity should be shown in the category requiring most of their working hours (not necessarily their highest skill). See Note a.

Note d:

U.S.-based employment is preferred (see Notes a, n). If possible, show chemical engineers separately for particular activities–e.g., in parentheses after totals for chemists and chemical engineers or in "Remarks."

Note e:

For the purpose of this projection, it is assumed that by January 1973 the resumption of an upward growth trend for the economy, continuing to the end of the decade, will be unequivocally clear. U.S.-based employment is preferred (see Notes a, f, n).

Note f:

For the purpose of this projection, it is assumed that by January 1975 full employment of the nation's labor force (i.e., unemployment of 4 percent or less) will have been reattained. A growth rate of about 4.3 percent per year in gross national product (expressed in constant dollars) is assumed for the first half of the decade, as in the latest *Economic Report of the President,* February 1971, p. 95. U.S.-based employment is preferred (see Notes a and n).

Incidentally, even under current legislation (such as the Clean Air Amendments of 1970, PL 91-604), sizable public and private expenditures will be required for environmental improvement to meet prescribed standards on schedule. According to *Environmental Quality,* Second Annual Report of the Council on Environmental Quality, August 1971, pp. 111-12, cumulative pollution-abatement costs for 1970–75 (investment plus operating) will amount to $105.2 billion (in 1970 dollars).

Note g:

Include such "officials" as research vice presidents and technical directors. Exclude team leaders or group leaders unless most of their working time is devoted to administration rather than to technical supervision.

Note h:

Also includes maintenance and economic evaluation of processes.

Note i:

Also includes quality control and technical services.

Note j:

Also includes market analysis and planning.

Note k:

Report writing and editing, in-house consulting and teaching, legal work (including patents), contract or project proposals, library services, etc.

Note m:

Please give a round estimate, if you can, of the number of chemists and chemical engineers that your company will require for U.S. operations in 1980—on the assumption of continued full-employment growth to the end of the decade after 1975 (see Note f). Among the other conditions assumed are: peace, a continuing shift toward service employment in general in U.S., intense foreign competition, and strong emphasis on unmet domestic social needs (e.g., "quality of life," health, and safety) and on international problems of food and energy.

You may be encouraged by, or find some guidance in, an unpublished employment estimate by U.S. Bureau of Labor Statistics (BLS) of 104,000 for chemists (excluding chemical engineers) in manufacturing in 1980 under full-employment conditions. This compares with a BLS estimate of 80,200 chemists in manufacturing for January 1970—a growth of nearly 30 percent for the decade 1970–80 as a whole and a geometric mean annual rate of nearly 2.64 percent.

Note n:

Mention if figures supplied in this table refer to worldwide company operations (including joint ventures) rather than U.S. operations only. Your comments on other pertinent matters are welcome—e.g., on expected changes in the ratio of chemists to chemical engineers over the decade, on particular training needs, or on new fields of promise for the employment of chemists.

CHEMISTRY IN THE ECONOMY

An American Chemical Society Study

Chairmen

Milton Harris
3300 Whitehaven St., N.W.
Washington, D.C. 20007

Max Tishler
Wesleyan University
Middletown, Conn. 06457

Contributing Committees to Part I*
The Accomplishments of Chemistry

Industrial Organic Chemicals
J. V. Murray, Chairman
Union Carbide Corp. (Retired)

W. C. Bauman
Dow Chemical Co.

H. R. Guest
Union Carbide Corp.

J. M. Nelson
E. I. du Pont de Nemours & Co., Inc.

C. W. Smith
Shell Development Co.

T. F. Cooke
American Cyanamid Co.

Isador Kirshenbaum
Esso Research & Engineering Co.

H. L. Pilat
Celanese Chemical Co.

Industrial Inorganic Chemicals
Arthur P. Lien, Chairman
Allied Chemical Corp.

W. Lysle Alderson
E. I. du Pont de Nemours & Co., Inc.

E. F. Booth
Allied Chemical Corp.

J. B. Browning
Union Carbide Corp.

F. A. Domino
Allied Chemical Corp.

J. C. Enneking
Union Carbide Corp.

J. L. Foster
Diamond Shamrock Corp.

E. Charles Galloway
Stauffer Chemical Co.

L. R. Hayes
Union Carbide Corp.

R. L. Hotchkiss
Dow Chemical Co.

Julius Johnson
Dow Chemical Co.

C. H. Arrington
E. I. du Pont de Nemours & Co., Inc.

Robert J. Brotherton
U.S. Borax & Chemical Corp.

J. W. Churchill
Olin Corp.

Ed Donley
Air Products & Chemicals, Inc.

John C. Fedoruk
Allied Chemical Corp.

R. H. Furlow
Dow Chemical Co.

Wm. E. Hanford
Olin Corp.

T. Heying
Olin Corp.

Ray R. Irani
Diamond Shamrock Corp.

K. W. Kamena
Dow Chemical Co.

* Locations shown are as of the period when contributions were made.

George Kazan
Allied Chemical Corp.

C. H. Lemke
E. I. du Pont de Nemours & Co., Inc.

C. E. McCoy
Dow Chemical Co.

W. J. Mayer
Olin Corp.

P. L. Metcalf
E. I. du Pont de Nemours & Co., Inc.

J. D. Rushmere
E. I. du Pont de Nemours & Co., Inc.

Leon Shechter
Union Carbide Corp.

C. J. Stebbins
Allied Chemical Corp.

L. Sulzberg
Allied Chemical Corp.

K. Taylor
Diamond Shamrock Corp.

A. D. F. Toy
Stauffer Chemical Co.

L. P. Twichell
Union Carbide Corp.

C. R. Whitworth
Dow Chemical Co.

R. S. Kolat
Dow Chemical Co.

F. A. Loving
E. I. du Pont de Nemours & Co., Inc.

R. E. Maizell
Olin Corp.

S. E. Mehlberg
Dow Chemical Co.

M. E. Pruitt
Dow Chemical Co.

G. Schmauch
Air Products & Chemicals, Inc.

T. R. Steadman
Allied Chemical Corp.

C. W. Stuewe
Dow Chemical Co.

J. A. Swartout
Union Carbide Corp.

A. E. Teyral
Diamond Shamrock Corp.

S. I. Trotz
Olin Corp.

M. M. Wendel
E. I. du Pont de Nemours & Co., Inc.

R. E. Wing
Dow Chemical Co.

Plastics and Resins

R. F. Boyer, Chairman
Dow Chemical Co.

J. L. Amos
Dow Chemical Co.

W. C. Bauman
Dow Chemical Co.

Ralph Chambers
Tennessee Eastman Co. Division
Eastman Kodak Co.

George Freda
Marbon Div., Borg-Warner Corp.

F. J. Glavis
Rohm & Haas Co.

G. Greminger
Dow Chemical Co.

J. F. Hyde
Dow Corning Corp.

William Keutgen
Union Carbide Corp.

F. E. Bailey
Union Carbide Corp.

W. L. Carrick
Union Carbide Corp.

H. W. Coover
Tennessee Eastman Co. Division
Eastman Kodak Co.

A. R. Gilbert
General Electric Co.

S. O. Greenlee
Resyn Corp.

J. P. Hogan
Phillips Petroleum Co.

R. N. Johnson
Union Carbide Corp.

A. T. McPherson
Kensington, Md.

G. E. Molau
Dow Chemical Co.

P. J. Oriel
Dow Chemical Co.

R. J. Plunkett
E. I. du Pont de Nemours & Co., Inc.

A. B. Savage
Dow Chemical Co.

C. O. Strother
Union Carbide Corp.

E. J. Vandenberg
Hercules, Inc.

H. J. West
American Cyanamid Co.

Herbert Morawetz
The Polytechnic Institute
of Brooklyn

E. M. Pearce
Allied Chemical Corp.

R. W. Quarles
Union Carbide Corp.

D. D. Stewart
Union Carbide Corp.

Michael Szwarc
Syracuse University

Otto Vogl
University of Massachusetts

R. M. Wiley
Dow Chemical Co.

Textile Fibers—Cotton
C. Harold Fisher, Chairman
U.S. Department of Agriculture

W. F. Baitinger
American Cyanamid Co.

T. F. Cooke
American Cyanamid Co.

R. S. Cokern
U.S. Department of Agriculture

S. E. Ellzey, Jr.
U.S. Department of Agriculture

Carleton T. Handy
E. I. du Pont de Nemours & Co., Inc.

Lawrence L. Heffner
North Carolina State University

F. M. Jornlin
E. I. du Pont de Nemours & Co., Inc.

F. F. Loffelman
American Cyanamid Co.

Harmon H. Ramey, Jr.
U.S. Department of Agriculture

Robert M. Reinhardt
U.S. Department of Agriculture

William G. Sloan
U.S. Department of Agriculture

Ralph J. Brysson
U.S. Department of Agriculture

Albert S. Cooper, Jr.
U.S. Department of Agriculture

George L. Drake, Jr.
U.S. Department of Agriculture

Chas. F. Goldthwait (Retired)
North Carolina State University

Robert J. Harper, Jr.
U.S. Department of Agriculture

W. F. Herbes
American Cyanamid Co.

Nestor B. Knoepfler
U.S. Department of Agriculture

Rita M. Perkins
U.S. Department of Agriculture

J. David Reid
U.S. Department of Agriculture

Warren C. Shaw
U.S. Department of Agriculture

Robert J. Thomas
E. I. du Pont de Nemours & Co., Inc.

Textile Fibers—Man-Made
A. E. Brown, Chairman
Celanese Research Co.

N. Barson
Celanese Research Co.

H. Coover
Tennessee Eastman Co.

J. A. Cogan, Jr.
Deering Milliken Research Co.

G. Kazan
Allied Chemical Corp.

G. F. Lanzl
E. I. du Pont de Nemours & Co., Inc.
S. K. Reed
FMC Corp.
M. C. Throdahl
Monsanto Co.
R. L. Wayland, Jr.
Dan River Mills, Inc.

W. C. Mathis
American Enka
P. B. Stam
Burlington Industries, Inc.
A. T. Walter
Union Carbide Corp.
F. X. Werber
J. P. Stevens & Co., Inc.

Textile Fibers—Wool

Joseph H. Dusenbury
Deering Milliken Research Corp.

Lescek Wolfram
International Wool Secretariat

Natural and Synthetic Rubber
James D. D'Ianni, Chairman
Goodyear Tire & Rubber Co.

Glen Alliger
Firestone Tire & Rubber Co.
W. F. Busse (Consultant)
B. F. Goodrich Co.
C. F. Eckert
Uniroyal, Inc.
M. E. Gross
B. F. Goodrich Co.
K. C. Hecker
Goodyear Tire & Rubber Co.
R. A. Krueger
B. F. Goodrich Co.
J. Lal
Goodyear Tire & Rubber Co.
G. E. Meyer
Goodyear Tire & Rubber Co.
C. J. Pearson
Goodyear Tire & Rubber Co.
Thomas H. Rogers, Jr.
Goodyear Tire & Rubber Co.
G. R. Tompkin
Goodyear Tire & Rubber Co.
W. C. Warner
General Tire & Rubber Co.

R. L. Bebb
Firestone Tire & Rubber Co.
C. I. Carr
Uniroyal, Inc.
E. C. Gregg
B. F. Goodrich Co.
H. E. Haxo
Uniroyal, Inc.
J. O. Kellgren
Uniroyal, Inc.
E. S. Latimore
E. I. du Pont de Nemours & Co., Inc.
F. S. Maxey
Goodyear Tire & Rubber Co.
A. L. Miglietta
Uniroyal, Inc.
W. H. Peterson
Enjay Chemical Co.
R. B. Spacht
Goodyear Tire & Rubber Co.
S. W. Waisbrot
Goodyear Tire & Rubber Co.

Protective Coatings
Howard L. Gerhart, Chairman
PPG Industries, Inc.

Gerould Allyn
Rohm & Haas Co.
G. N. Bruxelles
Hercules, Inc.

William J. Belanger
Celanese Coatings Co.
H. Burrell
Inmont Corp.

R. M. Christenson
PPG Industries, Inc.

R. Dowbenko
PPG Industries, Inc.

Don E. Floyd
General Mills Chemicals, Inc.

Richard B. Graver
Celanese Coatings Co.

R. D. Jerabeck
PPG Industries, Inc.

A. J. Kirsch
American Cyanamid Co.

E. E. McSweeney
Union Camp Corp.

A. W. Schalzel
Rohm & Haas Co.

Domenic J. Tessari
DeSoto, Inc.

Peter J. Coyle
Dow Chemical Co.

Roy J. Fahl, Jr.
E. I. du Pont de Nemours & Co., Inc.

Teresa Gillis
Rohm & Haas Co.

Seymore Hochberg
E. I. du Pont de Nemours & Co., Inc.

H. J. Kiefer
SCM Glidden-Durkee

Bruce N. McBane
PPG Industries, Inc.

L. E. Robb
Pennwalt

Karl F. Schimmel
PPG Industries, Inc.

Pharmaceuticals
George deStevens, Chairman
Ciba-Geigy Corp.

Sydney Archer
Sterling Drug, Inc.

Lloyd H. Conover
Pfizer, Ltd.

Patrick A. Diassi
The Squibb Institute for
    Medical Research

B. L. Martz
Lilly Laboratory for Clinical
    Research

Robert P. Mull
Ciba-Geigy Corp.

Leo H. Sternbach
Hoffmann-La Roche, Inc.

Karl J. Brunings
Ciba-Geigy Corp.

Edward J. Cragoe, Jr.
Merck & Co., Inc.

Louis Lasagna
Johns Hopkins School of Medicine

Robert I. Meltzer
Warner-Lambert Research Institute

Edward F. Rogers
Merck & Co., Inc.

Sidney Udenfriend
Roche Institute of Molecular Biology

Soaps and Detergents

Frank Healey
Lever Brothers Co.

Anthony Schwartz
Gillette Research Institute

Personal Care Products
Karl Laden, Chairman
Gillette Research Institute

W. Cooley
Procter and Gamble Co.

S. Gershon
Lever Brothers Co.

M. G. de Navarre
Beauty Counselors, Inc.

H. E. Jass
Carter Products Div.

D. D. Laiderman
The Gillette Co. Toiletries Div.
M. Rieger
Warner Lambert Pharmaceutical Co.

P. G. Lauffer
Hastings-on-Hudson, N.Y.
H. H. Tucker
Scarsdale, N.Y.

Fertilizers
Thomas G. Gibian, Chairman
W. R. Grace & Co.

C. C. Legal
W. R. Grace & Co.
Edward L. Newman
Tennessee Valley Authority

Lewis B. Nelson
Tennessee Valley Authority

Pesticides
E. H. Blair, Chairman
Dow Chemical Co.

E. C. Galloway
Stauffer Chemical Co.
R. G. Jones
Eli Lilly and Co.
C. O. Persing
Stauffer Chemical Co.
G. L. Sutherland
American Cyanamid Co.

D. L. Heywood
Union Carbide Corp.
P. C. Kearney
U.S. Department of Agriculture
A. J. Speziale
Monsanto Co.
R. R. Whetstone
Shell Chemical Co.

Food Processing
Raymond H. Hartigan, Chairman (Deceased*)
Kraftco Corp.

Philip D. Aines
Procter and Gamble Co.

Evan F. Binkerd
Armour and Co.

Arnold E. Denton
Campbell Soup Co.

Otis Fancher (Retired)
Industrial Bio-Test Laboratories

Earl E. Lockhart
The Coca-Cola Co.

William S. Martin
Procter and Gamble Co.

Harold L. Wilcke
Ralston Purina Co.

Frank Barber
Kraftco Corp.

Bernard F. Daubert (Retired)
General Foods Corp.

Kenneth G. Dykstra
General Foods Corp.

John D. Garber
CPC International, Inc.

Elmer H. Marth
University of Wisconsin

A. T. Schramm
Food Materials Corp.

Harold A. Wittcoff
General Mills, Inc.

* Dr. Hartigan died in July 1971 after his committee had been organized and the work well under way. His associate, Dr. Barber, completed the work.

Petroleum Refining
F. A. L. Holloway, Chairman
Standard Oil Co. (N.J.)

Thomas Baron
Shell Development Co.

D. H. Clewell
Mobil Oil Corp.

Werner Glass
Esso Research & Engineering Co.

P. C. White
Standard Oil Co. (Ind.)

T. A. Burtis
Sun Oil Co.

E. R. Gilliland
Mass. Institute of Technology

Vladimir Haensel
Universal Oil Products Co.

Nuclear Energy

Glenn T. Seaborg
University of California, Berkeley

A. R. VanDyken
U.S. Atomic Energy Commission

Pulp and Paper
G. A. Nesty, Chairman
International Paper Co.

C. R. Calkins
Riegel Paper Corp.

P. E. Nethercut
Technical Association of the Pulp
    and Paper Industry

H. C. Schwalbe
Herty Foundation

J. W. Swanson
Institute of Paper Chemistry

D. R. Henderson
International Paper Co.

A. H. Nissan
Westvaco Corp.

J. C. Redd
The Mead Corp.

I. H. Stockel
St. Regis Paper Co.

Q. C. Weaver
Scott Paper Co.

Ferrous Metals
James B. Austin (Retired)
U.S. Steel Corp.

Glass
William H. Armistead, Chairman
Corning Glass Works

Charles Alden
Owens-Illinois, Inc.

Donald Emhiser
PPG Industries, Inc.

J. Raymond Hensler
Bausch & Lomb

S. M. MacNeille
American Optical Corp.

John H. Munier
Corning Glass Works

William M. Strouse
American Optical Corp.

Marvin G. Britton
Corning Glass Works

James W. Hackett
Owens-Illinois, Inc.

John A. Leermakers (Retired)
Eastman Kodak Co.

Arthur J. Marino
PPG Industries, Inc.

W. F. Parsons
Eastman Kodak Co.

Howard R. Swift
Libbey-Owens-Ford Co.

Photographic Products
John A. Leermakers, Chairman
Eastman Kodak Co. (Retired)

George T. Eaton
Eastman Kodak Co.

Mark L. Moskowitz
GAF Corp.

Myron S. Simon
Polaroid Corp.

Vernon L. Wagner, Jr.
Kalvar Corp.

W. J. McGraw
E. I. du Pont de Nemours & Co., Inc.

Michael M. Shahin
Xerox Corp.

David P. Sorensen
3M Co.

Electronic Equipment
R. A. Laudise, Chairman
Bell Laboratories

A. G. Chynoweth
Bell Laboratories

D. L. Raymond
Bell Laboratories

J. H. Scaff
Bell Laboratories

W. P. Slichter
Bell Laboratories

M. D. Fagen
Bell Laboratories

M. D. Rigterink
Bell Laboratories

T. D. Schlabach
Bell Laboratories

P. A. Turner
Bell Laboratories

Electrical Equipment
Arthur M. Bueche, Chairman
General Electric Co.

Marcus Borom
General Electric Co.

Richard J. Coswell
Weyerhauser Co.

R. L. Dills
General Electric Co.

Ralph A. Hewes
General Electric Co.

D. R. Johnson
General Electric Co.

Leo Kohn
General Electric Co.

D. L. Lyons
Raychem Corp.

K. N. Mathes
General Electric Co.

M. Prober
General Electric Co.

A. H. Scharbaugh
General Electric Co.

Erwin G. Siwek
General Electric Co.

J. E. Burke
General Electric Co.

George W. Cressman
General Electric Co.

R. E. Hanneman
General Electric Co.

L. J. Hogue
General Electric Co.

R. J. Ketterer
General Electric Co.

Frank Lasak
General Electric Co.

P. E. McElligott
General Electric Co.

A. E. Newkirk
General Electric Co.

E. L. Raab
General Electric Co.

Robert H. Savage
General Electric Co.

Willem Vedder
General Electric Co.

Automobiles*

Philip Weiss
General Motors Corp.

Computers*

Gary Kozak
International Business Machines

* These contributions covered developments at General Motors (Automobiles) and at International Business Machines (Computers) only. They do not appear as separate chapters in Part I, but some of the information has been incorporated in or adapted to appropriate chapters in that part. More important, however, they have been presented as special appendixes in Part III to illustrate the role of chemistry in manufacturing companies not normally identified with the science.

In addition to all the foregoing contributions to Part I, we should also like to acknowledge information on the nonferrous metals industry received from Herbert I. Fusfeld, Kennecott Copper Corp.; Donald J. McPherson, Kaiser Aluminum & Chemical Corp.; and Robert Q. Barr, Climax Molybdenum Co. The pressure of time unfortunately prevented our adapting their material to the study.

Contributing Committees to Part II
The People of Chemistry

Manpower

Wallace R. Brode, Chairman
American Chemical Society

Milton Harris
American Chemical Society

William C. Kelly
National Research Council

Observer: Thomas J. Mills
National Science Foundation

Robert E. Henze
American Chemical Society

Sol Swerdloff (Retired)
Bureau of Labor Statistics

Education

Cheves Walling, Chairman
University of Utah

Ronald Breslow
Columbia University

Paul Doty
Harvard University

H. S. Eleuterio
E. I. du Pont de Nemours & Co., Inc.

Milton Harris
American Chemical Society

Robert M. Hexter
University of Minnesota

Harry Gray
California Institute of Technology

W. P. Slichter
Bell Laboratories

John Thomas
Chevron Research Company

Max Tishler
Wesleyan University

Observer: Kent Wilson
National Science Foundation

Advisory Committee to Part III
The Economics of Chemistry

Irving H. Siegel, Chairman
Economic Consultant

Jack Alterman
Bureau of Labor Statistics

Milton Harris
American Chemical Society

Herbert W. Robinson
International Management Systems, Inc.

Charles L. Schultze
The Brookings Institution

Sam H. Schurr
Resources for the Future, Inc.

Charles T. Stewart, Jr.
The George Washington University

INDEX

A

ABS, *See* Acrylonitrile-butadiene-styrene;
    Alkylbenzene sulfonate
ADL process, *See* FIOR process
Abbott,
    anticonvulsants development, 185
Abelson, Philip,
    uranium hexafluoride production, 303
Academic programs,
    student guidance policies, 436
Accelerators, *See* Organic accelerators
Accounting system,
    determination of chemical manpower
        supply, 5, 422
Acetaldehyde,
    vinyl acetate production, 28
Acetate fibers,
    chemical innovations, 93
Acetazolamide,
    development, 180
Acetic acid,
    food preservation, 261
    vinyl acetate production, 28
Acetic anhydride,
    vinyl acetate production, 28
Acetylene,
    production from calcium carbide, 56
Acid dyes,
    acrylic fibers, 104
Acid gas process,
    vulcanization, 122
Acid hydrolysis,
    starch, 257
Acid mine drainage,
    rubber uses in pollution control proj-
        ects, 149
Acrylic fibers,
    acid dyes, 104
    chemical innovations, 96
    organic intermediate chemicals, 23
Acrylic lacquers,
    automotive finishes, 154
Acrylic polymers,
    development, 162
    methyl methacrylate process, 76
Acrylonitrile,
    production, 23
    production materials, 16
Acrylonitrile-butadiene-styrene,
    acrylonitrile demand, 24
    development, 74
Acrylonitrile fiber,
    Du Pont, 87
Acrylonitrile-vinyl chloride fiber,
    Union Carbide Corp., 87
Actinomycin,
    development, 176
Adams, B. A.,
    ion exchange resins discovery, 80
Adhesives,
    halogenated rubber market, 140
    isomerized rubber market, 140
    sodium silicates market, 48
Adipic acid,
    Halcon International process, 22
    nylon 66 production, 22

Adiponitrile,
    Monsanto Co. process, 22
Adsorption,
    petroleum separation process, 283
    surfactant activity, 198
Aerosol propellants,
    fluorocarbon intermediates, 27
Aerosols,
    personal care products developments,
        209
Age resisters,
    rubber, 124
Agricultural chemicals, *See* Fertilizers; Pes-
    ticides
Agriculture,
    horticultural lighting, 416
Aines, Philip D., 551
Air pollution,
    sulfate pulping process, 324
Air pollution abatement,
    pulp and paper industry outlook, 329
    sulfuric acid plants, 42
Air pollution regulations,
    effects on diesel fuel additives, 292
    effects on heating oil refining, 293
    effects on hydrotreating, 290
    effects on lead antiknocks, 291
    effects on solvents, 296
Alden, Charles, 552
Alderson, W. Lysle, 546
Alkali industry,
    See also Chlor-alkali chemicals
    ammonia market, 42
Alkoxymethyl isocyanates,
    protective coatings outlook, 168
Alkoxyphosphazenes,
    flame retardants for rayon, 93
Alkyd resins,
    development, 158
    electrical equipment insulation, 402
Alkylation,
    petroleum conversion process, 287
Alkylbenzene,
    alkylbenzene sulfonate production, 25
    biodegradable type, 16, 200
Alkylbenzene sulfonate,
    detergent research, 197
    phosphate detergent builders, 200
    production, 25
Allegheny River,
    sewer wastes disposal project, 149
Allethrin,
    development, 234
Allied Chemical Corp.
    See also Atmospheric Nitrogen Corp.
    boron trifluoride production, 62
    caprolactam production, 23
    chlorinated linear polyethylene pro-
        cess, 73
    fluorine for gaseous diffusion plants,
        300
    fluorocarbons production, 27
    hydrocarbons for ABS, 200
    hydrogen fluoride production, 61
    industrial inorganic chemicals, 37
    phenol process, 32
    soap and detergent research, 197

Automotive exhaust emissions,
effects on petroleum refining technology, 296
Automotive finishes,
acrylic resins development, 163
development, 152
Automotive safety glass,
laminated, 348
polyvinyl butyral use, 78
Auxiliaries,
use in soaps and detergents, 196
Aviation gasoline,
alkylate use, 288
effects on petroleum conversion process, 286
isobutane use, 289
petroleum products, 292
polymer gasoline use, 287

B

BASF, See Badische Anilin-und Soda-Fabrik
BHA, See Butylated hydroxyanisol
BHC, See Benzene hexachloride
BHT, See Butylated hydroxytoluene
BOP, See Basic oxygen process
BWR, See Boiling-water reactors
Babcock & Wilcox,
sol-gel production of thorium oxide, 310
Bachelor degrees,
academic preparation survey, 450
projections, 431
statistics, 431
Bacillus thuringiensis,
insecticide development, 234
insecticides outlook, 242
Bacon, Frank R.,
glass resistance to alkaline solutions, 345
Bacon, R. F.,
ethylene production research, 29
Bacteria,
frozen foods research, 250
Badische Anilin-und Soda-Fabrik,
ammonia fertilizers technology, 218
low-density polyethylene process, 70
methanol production research, 31
polystyrene foam process, 74
sulfuric acid contact plant, 42
Baekeland, Leo,
phenolic resins discovery, 77, 157, 402
Baeyer, Adolf,
phenolic resins discovery, 77, 157
Bagasse,
papermaking outlook, 327
Bailey, F. E., 547
Baitinger, W. F., 548
Baked enamels,
acrylic resins coatings development, 164
automotive finishes, 153
Bakelite,
discovery, 77, 157
electrical equipment insulation, 402
formaldehyde production, 32
Baker, William O., ix
Baking industry,
edible emulsifiers, 260
Baking powders, See Leavening agents
Barber, Frank, 551
Barbiturates,
anticonvulsants development, 185
Bardeen, John,
transistors, 380

Barex resin,
acrylonitrile demand, 24
Barium,
component in radiation-controlling glass, 350
diesel fuel additives development, 292
Baron, Thomas, 552
Barr, Robert Q., 554
Barson, N., 548
Basic oxygen process,
pig iron production, 332, 334
Basic research,
support, 434
Battelle Memorial Institute,
electrostatic copying, 365
uranium ore processing, 304
Bauman, W. C., 546, 547
Bausch & Lomb,
controlled refractive index gradient, 351
optical glass innovations, 351
Bayer, Otto,
polyurethanes, 136
Beadle, C.,
viscose process for rayon, 92
Bebb, R. L., 549
Beecham,
ampicillin development, 176
Belanger, William J., 549
Bell Laboratories,
controlled conductivity, 382
iron oxide masks for circuits, 387
light-emitting diodes, 394
permanent magnets, 389
silicon point-contact diode, 384
transistors, 380
zone refining, 385
Bell Telephone System,
electronic industry products, 373
Belts, See Conveyor belts
Benefin,
development, 237
Bennett, J. T., 516
Benzene,
effects on styrene production, 19
nylon 66 production, 22
Benzene hexachloride,
development, 231
Benzene-toluene-xylene fraction,
solvent extraction, 283
1,4-Benzodiazepines,
tranquilizers development, 184
Benzoic acid,
food preservation, 260
Berthollet, Claude Louis,
paper bleaching process, 314
Beryllium alloys,
G.M. research program, 527
Bessemer converters,
steel process, 333
Bessemer, Henry,
steel process, 332
Best Foods,
hydrogenated cottonseed oil, 256
Bevan, E. J.,
diazotypes process, 366
viscose process for rayon, 92
Bias-belted tires,
polyester tire cord, 96
market outlook, 145
Bias-ply tires,
market outlook, 145
Biguanides,
antidiabetic drugs development, 188
Binkerd, Evan F., 551

Butadiene-acrylonitrile rubber,
  oil-resistant elastomer, 128
Butadiene-styrene rubber,
  general purpose elastomer, 128
Butane,
  butadiene production, 21
Butene,
  butadiene production, 21
Butter,
  competition from margarine, 255
Butyl rubber,
  impermeability to gases, 130
  isobutylene demand, 20
  isoprene demand, 21
  synthetic rubber developments, 129
  water barriers, 130
Butylated hydroxyanisol,
  food antioxidant, 260
Butylated hydroxytoluene,
  food antioxidant, 260
Butylenes,
  polymerization process, 287

C

CMC, See Carboxymethyl cellulose
CPE, See Chlorinated linear polyethylene
Cables,
  insulation, 405
  polyethylene insulation, 377
Cabot Corp.,
  carbon black production process, 124
Cake mixes,
  food processing innovations, 266
Calcium,
  glass formation, 343
Calcium carbide,
  role in industrial inorganic chemicals
    industry, 56
Calico Printers Association,
  polyester fibers, 87, 95
California,
  bricks from crushed bottles, 353
Calkins, C. R., 552
Callaway, S. R.,
  nickel-base, high-temperature alloy,
    525
Calrod tubular heater,
  development, 415
Calvin, Melvin, 513
Campbell, J. M.,
  molecular structure relation to octane
    rating, 526
  phosphorus additives for leaded gaso-
    line, 524
Campbell Pharmaceutical Co.,
  mercaptomerin sodium development,
    180
Canadian Radium and Uranium Co.,
  uranium oxide production, 301
Cancer,
  food additives safety, 271
Canning,
  food processing innovations, 246, 247
Capacitors,
  electronic microcircuits, 385
Caprolactam,
  nylon 6 production, 23
Captan,
  developments, 238
Carbamate herbicides,
  development, 236
Carbamate insecticides,
  development, 233

Carbamates,
  crosslinking agents for fibers, 99
Carbaryl,
  development, 233
Carbohydrates,
  synthetic products, 278
Carbolic acid, See Phenol
Carbon black,
  production from tire waste disposal
    process, 148
  rubber reinforcing agent, 123
Carbon brushes,
  development, 410
Carbon monoxide,
  methanol production, 31
Carbonless carbon paper,
  paper-treating materials, 320
Carboxyl ion,
  polar constituent of anionic surfac-
    tants, 198
Carboxymethyl cellulose,
  sizing treatments for fibers, 105
  synthetic detergent additive, 201
Carburetor detergent additives,
  gasoline additives developments, 291
Carbyne,
  development, 236
Cardiac pacemakers,
  implantation, 382
  nuclides development, 311
Cardiovascular agents,
  development, 177
Careers,
  recommendations on counseling, 9,
    436
Carlson, Chester,
  electrostatic copying, 365
Carnegie Geophysical Laboratory,
  optical glass innovations, 351
Carothers, Wallace, 87
  nylon 66, 22
Carpets,
  antisoiling processes, 109
  fiber markets, 89
  polyester fibers, 95
Carr, C. I., 549
Carrick, W. L., 547
Carter, A. P., 495
Casein-latex adhesive system,
  rayon tires, 143
Casein paints,
  developments, 161
Cast pneumatic tires,
  cordless construction, 143
Cast urethanes,
  development, 137
Castan, Pierre,
  epoxy resins discovery, 79
Casting, See Steel casting
Catalysis,
  polymers use, 84
Catalysts
  See also Oxidation catalysts
  boron trifluoride market, 62
  petroleum cracking improvements, 286
  regeneration technology, 286
  silica gel market, 48
  stereo rubber production, 133
Catalytic cracking,
  petroleum conversion process, 285
Catalytic dehydrogenation,
  petroleum conversion process, 289
Catalytic polymerization,
  development, 287
Catalytic reforming,
  petroleum conversion process, 288

Columbia Nipro,
  caprolactam production, 23
Columbia Southern Chemical Co.,
  carbamate herbicides development, 236
Columbia University,
  boron-10 development, 307
  heavy water development, 309
  quinidine development, 179
  uranium isotope separation research,
    303
  uranium processing research, 301
Combustion-control additives,
  gasoline additives development, 291
Committee on Chemistry and Public
  Affairs, iii, iv, ix
Communication skills,
  academic preparation survey, 441
Communications
  See also Optical communications;
    Wireless communication
  chemical science opportunities, 4
  electronics applications, 373, 374
Communications industry,
  glass applications, 353
  paper importance, 320
Comonomers,
  polymerization process, 67
Companies,
  interdisciplinary staff changes, 539
  recommendations for cooperation
    with schools, 10, 437
Compton, A. H.,
  plutonium production, 302
Computers,
  chemical role in development, 482
  lens production calculations, 351
  solid state circuitry, 383
  steelmaking uses, 339
Conant, J. B.,
  polyethylene development, 69
Conductivity
  See also Controlled conductivity
  p-n junctions for semiconductors, 385
Conover, Lloyd H., 550
Consolidated Mining and Smelting,
  ammonium phosphates fertilizers
    technology, 218
Construction industry,
  rubber market, 120
  sodium silicates market, 49
Consumer goods,
  fiber markets, 89
Consumers,
  effects of industrial inorganic chemi-
    cals, 36
  effects on food industry, 272
  organic intermediate chemicals use,
    15
Contact microphones,
  development, 410
Contact process,
  sulfuric acid production, 41
Contact rectification,
  electrical contact devices, 411
Containers
  See also Food containers
  glass industry disposal problems, 353
  glass market, 343
  paper uses, 322
Continuing education,
  closed-circuit television, 438
  employer's role, 437
  professional societies, 438

recommendation for programs in
  chemistry, 11, 435
  teachers, 437
Controlled conductivity,
  Bell Laboratories, 382
Convenience products,
  paper uses, 321
Conversion processes,
  petroleum refining, 281, 284
Conveyor belts,
  rubber market, 119
Cooke, T. F., 546, 548
Cookingware,
  borosilicate glass, 346
Coolant systems,
  nuclear reactors development, 308
Cooley, W., 550
Cooper, Albert S., Jr., 548
Coover, H., 548
Coover, H. W., 547
Copolymer Rubber and Chemical,
  EPDM production, 135
Copper-zinc oxide catalysts,
  methanol production, 31
Copying, See Office copying
Coral rubber,
  polyisoprene, 134
Cordura,
  rayon tire cord, 93
Core curriculum,
  graduate education programs, 436
Corn,
  hybridization, 263
  starch source, 258
Corn flakes,
  breakfast cereal innovations, 252
Corn Products,
  continuous corn steeping, 258
  starch structure research, 258
  waxy sorghum starches, 258
Corn syrup,
  production from starch, 258
Cornell Aeronautical Laboratory,
  windshield safety glass tests, 348
Corning Glass Works,
  electrical equipment lubrication de-
    velopment, 411
  Fotoceram, 349
  Fotoform glass, 349
  glass fiber optical wave guides, 397
  heat-resistant glass, 346
  high-strength glass, 347
  photochromic glass, 350
  photosensitive glass, 349
  Pyroceram, 349
  radiation-controlling glasses, 350
  railroad signal lamp glass, 350
  thermal-setting solder glasses, 352
Corporation Associates, ACS,
  employment projections survey, 532
  employment surveys, 469
Corrosion,
  automotive protective finishes, 154
  nuclear reactors coolant systems, 308
Corrosion inhibitors,
  gasoline additives developments, 291
Corrosion passivation,
  electronic microcircuit technology,
    386
Corrugated board,
  high yield production, 328
  paper uses, 322

Corrugating industry,
    sodium silicates market, 48
Corticosteroids,
    development, 182
Cortisone,
    steroidal drugs development, 181
Coryell, C. D.,
    nuclides development, 311
Cosmetic chemistry,
    personal care products, 214
Cosmetics, See Personal care products
Coswell, Richard J., 553
Cotton,
    consumption patterns, 87
    durable press treatment, 97
    dyeing technology, 102
    flame retardants, 93
    history, 90
Cottonseed,
    product formulation research, 275
Cottonseed oil,
    margarine source material, 256
Counseling,
    recommendation for graduate students, 9, 436
Counts, C. R., 458
Courtaulds,
    rayon production, 92
Coyle, Peter J., 550
Cragoe, Edward J., Jr., 550
Cressman, George W., 553
Crimping,
    rayon, 92
Crops,
    pesticides effects, 229
Cross, C. F.,
    diazotype process, 366
    viscose process for rayon, 92
Crosslinking
    See also Vulcanization
    waxy starches, 259
    wrinkle-resistant fabrics, 98
Crude oil,
    petroleum refining importance, 280
Cryogenics,
    blast furnace by-product removal, 338
Cryolite,
    industrial fluorine chemicals market, 61
Crystallization,
    glass processing, 348
    petroleum separation process, 284
Crystals
    See also Single crystals
    transistor operation, 380
Cumene,
    phenol production, 32
Curie, Pierre,
    law of magnetism, 388
Curium,
    nuclides development, 311
Curme, G. O.,
    ethylene production research, 29
Cyclazocine,
    drug addiction research, 193
    narcotic antagonists development, 187
Cyclethrin,
    development, 234
Cyclized rubber, See Isomerized rubber
1,4-Cyclohexane dimethanol,
    polyester carpet fiber, 95

D

2,4-D
    development, 234, 235
    latex flow stimulant, 138
    plant growth regulators development, 240
DBCP,
    development, 239
D-D Soil Fumigant,
    development, 239
DDT,
    development, 230
DDVP,
    development, 233
DEPC, See Diethyl pyrocarbonate
DMT, See Dimethyl terephthalate
Dacron,
    Du Pont, 87
    polyester fibers, 95
Daguerre, Louis
    silver halide photography, 357
Daguerreotypes,
    development, 357
D'Alelio, G. F.,
    styrene-divinylbenzene discovery, 81
Dandruff,
    shampoos development, 210
Dapex process,
    uranium ore processing, 304
Daubert, Bernard F., 551
Davison Chemical Corp.,
    fertilizers granulation techniques, 220
    superphosphates fertilizers technology, 216
Davison, William,
    superphosphates fertilizers technology, 216
Dean, O. C.,
    sol-gel production of thorium oxide, 310
Deasphalting,
    fractional crystallization process, 284
de Chardonnet, Hilaire,
    nitrocellulose process for rayon, 92
de Coulomb, Charles,
    electronics accomplishments, 374
Deering Milliken Research,
    soil release finish for textiles, 108
De Forest, Lee,
    electronics accomplishments, 375
Degradable plastics,
    outlook, 82
Degrees,
    chemical education statistics, 426
Dehydrated foods,
    food industry innovations, 248
Dehydrogenation,
    petroleum conversion process, 288
    catalytic, 289
    thermal, 289
Delayed coking,
    petroleum demetallizing process, 290
De Navarre, M. G., 550
Dentifrices,
    personal care products development, 214
Denton, Arnold E., 551
Deodorants,
    personal care products development, 213
Deodorizing,
    edible oil processing, 255

E

Ethylene-propylene polymers,
    hose applications. 120
Ethylene-propylene rubber,
    synthetic rubber developments, 135
Explosives,
    ethylene glycol demand, 30
Extruded polymers,
    wire insulation, 405
Extrusion,
    tire tread production, 131
Eyeglasses,
    optical glass innovations, 350

F

FIOR process,
    steelmaking developments, 341
FMC Corp.,
    insecticides development, 234
FPC, See Fish protein concentrate
Fabric coatings,
    chlorosulfonated polyethylene, 132
Fabric furnishings,
    fiber markets, 89
Fabrics,
    dyeing technology, 103
    synthetic detergent requirements, 201
Faculty, See Teachers
Fagen, M. D., 374, 553
Fahl, Roy J., Jr., 550
Fancher, Otis, 551
Far East,
    natural rubber production, 137
Faraday, Michael,
    benzene hexachloride synthesis, 231
    electric generator invention, 401
    electronics accomplishments, 374
    rubber, 116
Farbenfabriken Bayer
    See also Chemagro
    cast urethane elastomers, 137
    polyurethanes, 136
    vinyl acetate production, 29
Farbwerke Hoechst
    See also Knapsack, A.G.
    chlorinated linear polyethylene pro-
        cess, 73
    pulmonary edema treatment, 181
Farmers,
    food production, 246
Fast breeder nuclear reactors,
    development, 312
Fats
    See also Edible fats;
    Glyceride fats;
    Inedible fats;
    synthetic product research, 278
Fatty acids,
    food preservation, 261
Fatty amides,
    nonionic surfactants, 199
Fedoruk, John C., 546
Fellner, William, 511
Felting,
    wool treatments, 101
Ferbam,
    development, 238
Ferguson, D. E.,
    sol-gel production of thorium oxide,
        310
Ferrous alloy,
    metallurgy, 336
Ferrous metals,
    blast furnace raw materials, 337

casting, pickling and coating, 334
    chemical accomplishments, 332
    chemical analyses, 338
    future developments, 339
    making iron and steel, 332
    metallurgy as a chemical science, 336
    social and economic impact, 331
Fertility control,
    oral contraceptives development, 182
    research outlook, 192
Fertilizer industry,
    sulfuric acid market, 41
Fertilizers,
    chemical accomplishments, 216
    developing nations' uses, 223
    effects on the environment, 222
    energy and materials resources, 223
    liquid mix technology, 221
    nitrogen products, 218
    outlook, 221
    phosphate products, 216
    phosphorus market, 50
    potassium products, 220
    social and economic impact, 215
    solid mix products, 220
    solid mix technology, 220
Fiber B,
    tire construction outlook, 146
Fiber drums,
    paper uses, 323
Fiber industry,
    chemical innovations, 90
Fiberboard,
    shelter components, 323
Fibers
    See also Cigarette filters;
    Man-made fibers
    antisoiling treatments, 108
    antistatic finishes, 110
    consumption patterns, 87
    flame-resistant treatments, 111
    future developments, 112
    sizing treatments, 105
    social and economic impact, 87
    water and oil repellency treatments,
        106
Field, P. M., 504
Fillers,
    sodium silicates market, 48
Films,
    cellulose acetate development, 359
Firestone Tire and Rubber Co.,
    cast pneumatic tires, 143
    natural rubber synthesis, 134
Fish protein concentrate,
    product formulation research, 275
Fisher, C. Harold, 548
Fisher, Werner,
    hafnium-zirconium separation, 308
Fixing agents,
    photographic products, 360
Flame-resistant textiles,
    chemical treatments, 111
    future developments, 114
Flame retardants,
    rayon, 93
Flammability,
    plastics outlook, 82
Flash cube,
    development, 416
Flash equipment,
    photographic products outlook, 369
Flash lamps,
    development, 416

Glass
See also Architectural glass;
    Automotive safety glass
    chemical innovations, 343
    electrical properties, 352
    formation, 343
    future developments, 352
    high-strength, 347
    property improvements, 345
    radiation-induced changes, 349
    social and economic impacts, 343-4
    sodium carbonate market, 46
Glass-ceramics,
    artificial implants, 353
    production processing, 348
Glass fibers,
    electrical equipment insulation, 403
    markets, 343
Glass industry,
    borate chemicals market, 58
    disposal problems, 353
Glasson, W. A.,
    peroxybenzoyl nitrate, 525
Glauber, Johann, 40
Glavis, F. J., 547
Glidden,
    edible oil extraction processes, 254
    latex paints developments, 160
    oil-soluble resin lacquer development,
        153
Glyceride fats,
    saponification process for soapmak-
        ing, 198
Glycerine, See Synthetic glycerol
Glycerol
    See also Synthetic glycerol
    synthetic product research, 278
Glyptals,
    alkyd resins development, 159
Goldthwait, Chas. F., 548
Goodrich,
    acrylonitrile production, 24
    EPDM production, 135
    isomerized rubber, 140
    isoprene production, 21
    natural rubber synthesis, 133
    polyvinyl chloride process, 75
    thermoplastic urethanes, 137
    tubeless tires, 129, 143
Goodrich-Gulf Chemicals Co.,
    olefin polymerization with organo-
        metallic catalysts, 133
Goodyear, Charles,
    hard rubber, 139
    rubber vulcanization, 116, 117, 122
Goodyear Tire & Rubber Co.,
    airplane tires, 118
    encapsulated latex foam, 142
    foaming improvements, 137
    halogenated rubber, 140
    isomerized rubber, 140
    natural rubber synthesis, 134
    pollution control, 147
    polyester bias-ply glass-belted tires,
        143
    polyester tire cord, 95
    rigid polyurethane foams, 136
    SBR latices, 141
Government Rubber-Styrene,
    styrene-butadiene rubber, 128
Grace, W. R., and Co.
    See also Davison Chemical Co.
    photopolymer printing plates, 368
    sol-gel production of thorium oxide, 310

Graduate education,
    chemical opportunities, 2
    enrollment statistics, 432
    industry-university programs, 10, 437
    interdisciplinary recommendations, 10,
        449
    recommendations for chemical man-
        power, 8
    recommendation on career counsel-
        ing, 9, 436
    recommendations on separation of
        training and research, 9, 433
Graduate students,
    teaching and research assistantships,
        434
Graft polymerization,
    ABS process, 75
Grain insects,
    control with pesticides, 227
Grains,
    food supply outlook, 273
    nutritional fortification, 253
    protein fortification, 273
Granulation,
    fertilizers techniques, 220
Grape Nuts,
    breakfast cereal innovations, 252
Graphite,
    nuclear materials, 307
Grateau, Marcel,
    permanent wave development, 207
Graver, Richard B., 550
Gray, Harry, 554
Greases, See Lubricating greases
Great Canadian Oil Sands,
    tar sands recovery, 297
Green, A. G.,
    diazotypes process, 366
Greenlee, S. O., 547
    epoxy resins discovery, 79, 164
Gregg, E. C., 549
Greminger, G., 547
Gresky, A. T.,
    nuclear fuel recovery process, 305
Grimes, W. R.,
    hafnium-zirconium separation, 308
    molten salt breeder reactor develop-
        ment, 312
Gross, M. E., 549
Groundwood,
    high yield production, 328
Growth regulators,
    latex flow stimulants, 138
Grozzo, Sally, ix
Guanethidine,
    blood pressure control, 178
Guanidine,
    rubber vulcanization process, 123
Guanine crystals,
    nail lacquers development, 211
Guest, H. R., 546
Gulf Coast,
    organic intermediate chemicals im-
        pact, 16
    organic intermediate chemicals out-
        look, 35
Gulf Energy & Environmental Systems,
    sol-gel production of thorium oxide,
        310
Gulf General Atomic,
    fluorine for gaseous diffusion plants,
        300
Gulf Oil Co.,
    hindered phenols, 125

H

H-Iron process,
 steelmaking developments, 341
HSLA steel, *See* High-strength low-alloy
 steel
HTA, *See* Hydrogenated tallow amine
HTGR, *See* High temperature gas-cooled
 reactors
Haber, Fritz,
 ammonia fertilizers technology, 218
 methanol production research, 31
Haber-Bosch process,
 ammonia production, 43
Hackett, James W., 552
Haensel, Vladimir, 552
Hafnium,
 zirconium separation, 307
Hahn, Otto
 nuclear fission discovery, 298
Hair,
 personal care products development,
 207
Hair coloring,
 personal care products development,
 210
Hair sprays,
 personal care products development,
 208
Hair straighteners,
 personal care products development,
 208
Halcon International,
 adipic acid production, 22
 propylene oxide production, 30
Halogenated hydrocarbons,
 electrical equipment insulation, 409
Halogenated rubber,
 anticorrosive paints, 140
 chemical innovations, 139
Haloid Co.,
 electrostatic copying, 365
Hamor, W. A., 502
Hancock, Thomas,
 hard rubber, 139
Handy, Carleton T., 548
Hanford, Wm. E., 546
Hanneman, R. E., 553
Hard rubber,
 chemical innovations, 139
Hardboard,
 shelter components, 323
Hardesty, D. W.,
 acetic acid test for corrosion resistance, 524
Harper, Robert J., Jr., 548
Harris, Milton, iii, ix, 546, 554
Harshaw Chemical Co.,
 boron trifluoride production, 62
 fluorine for gaseous diffusion plants,
 300
 uranium hexafluoride production, 303
 uranium tetrafluoride production, 301
Hartigan, Raymond H., 551
Harvard Economic Research Project,
 economic role of chemistry, 495
Harvard Medical School,
 anticonvulsants development, 185
Harvard University,
 polyethylene development, 69
 tetracycline antibiotics development,
 176
Haxo, H. E., 549
Hayes, L. R., 546

Hayward, Nathaniel,
 rubber vulcanization process, 122
Hazardous material,
 rubber shipping containers, 149
Healey, Frank, 550
Health
 See also Hygiene
 chemical science opportunities, 3
 effect on interdisciplinary staff
 changes, 539
 electronics applications, 372
 medicinal chemistry progress, 171
 personal care products effects. 207
Health services,
 importance, 173
Heart diseases,
 cardiovascular agents development,
 177
 pharmaceuticals research outlook, 192
Heat-resistant glass,
 chemical innovations, 346
Heat-resistant insulants,
 development, 402
Heating oil,
 petroleum products, 292
Heavy water,
 production, 309
Hecker, K. C., 549
Heffner, Lawrence L., 548
Heliothis virus,
 insecticides outlook, 242
Hench, Phillip S.,
 cortisone development, 181
Henderson, D. R., 552
Henne, Albert,
 refrigerants research, 27
Hensler, J. Raymond, 552
Henze, Robert E., 554
Herbes, W. F., 548
Herbicides,
 development, 234
 future developments, 242
 social and economic impact, 227
Hercules,
 cellulose acetate butyrate, 80
 dimethyl terephthalate production, 23
 dyeable polypropylene fiber, 97
 nitrocellulose lacquers development,
 156
 phenol process, 32
 polypropylene fibers, 87, 97
Herculon,
 Hercules, Inc., 87
Hershey, William, ix
Hertz, Heinrich,
 electronics accomplishments, 374
Heuss, J. M.,
 peroxybenzoyl nitrate, 525
Hevea brasiliensis,
 rubber plant, 116, 137
Hevea rubber,
 history, 133
Heveaplus,
 polymerically modified rubber, 140
Hewes, Ralph A., 553
Hexachlorophene,
 antibacterial agent for soaps, 202
 deodorants development, 213
Hexamethylene diamine,
 Celanese Corp. process, 22
 nylon 66 production, 22
Hexter, Robert M., 554
Heying, T., 546
Heylin, Michael, 516
Heywood, D. L., 551

I

IBM, *See* International Business Machines Corp.
IPC,
    development, 236
I.G. Farbenindustrie,
    acrylic fibers, 96
    antihelminthic drugs development, 188
    diffusion transfer photography, 362
    ethoxylation of alcohols, 199
    hair sprays development, 209
    oil from coal, 127
    organophosphorus insecticides development, 232
    Perlon, 87
    styrene production, 19
    sulfonated surfactants, 199
    synthetic rubber research, 17, 128
Imipramine,
    antidepressants development, 185
Impact polystyrene,
    production, 74
Imperial Chemical Industries,
    acrylic resin coatings development, 163
    benzene hexachloride synthesis, 231
    methanol production, 31
    phenoxy herbicides development, 235
    polyester fiber production, 23, 95
    polyethylene development, 69
    vinyl acetate production, 28
    water repellant fibers, 107
Implants, *See* Artificial implants
Imports,
    U.S. natural rubber, 117
Income, *See* National income
India,
    fibers history, 90
Indole,
    anti-inflammatory agents development, 183
Indole acids,
    plant growth regulators development, 240
Indomethacin,
    anti-inflammatory agents development, 183
Industrial inorganic chemicals,
    historical developments, 37, 39
    impact of pollution control, 64
    product and process improvements, 63
    sales, 36
    social and economic impacts, 36
    supply-demand situation, 65
Industrial markets,
    fiber consumption, 89
Industrial piping,
    glass market, 343
Industrial research,
    academic preparation survey, 440
Industry-university programs,
    graduate education recommendations, 10, 437
Inedible fats,
    uses in soap and detergent industry, 197
Inedible tallow,
    use in soap and detergent industry, 197
Inertness,
    glass properties, 344
Information,
    chemical contributions to society, 5

photographic processes importance, 356
Injury, *See* Technological injury
Inorganic chemicals, *See* Industrial inorganic chemicals
Inorganic intermediates,
    combinations with organic intermediates, 39
Input-output tables,
    economic significance of chemistry, 455, 488
    employment requirement implications, 497
    interindustry relationships, 492
Insecticides,
    future developments, 241
    phosphorus pentasulfide market, 55
Insects,
    pesticides development, 229
Institute of Paper Chemistry, *See* Sulfite Pulp Manufacturers' Research League
Instrumentation,
    glass uses, 353
Insulating papers,
    chemical treatment, 320, 321
Insulation,
    electrical equipment, 401
    fiberboard uses, 323
    glass fiber market, 343
    heat-resistant insulants development, 402
    polyethylene coatings for cables and wires, 377
    polyurethane foam market, 147
    vinyl chloride demands, 76
Insurance,
    technological injury, 7
Interferon,
    biopolymers research, 86
Interim-23 Process,
    nuclear fuel recovery, 305
Intermediate chemicals,
    industrial inorganics, 36
International Business Machines Corp.,
    employment of chemists, 482
    insecticides outlook, 242
    light-emitting diodes, 394
International Rich Research Institute,
    rice hybridization, 263
International standards,
    recommendations for manufacturing industries, 8
International Telephone & Telegraph,
    radiation crosslinked polymer insulation, 407
Invention,
    chemical patents, 515
Ion exchange processes,
    high-strength glasses, 347
Ion exchange resins,
    development, 80
    uranium ore processing, 303
Ion-permeable membranes,
    chlorine-caustic processes, 64
Iowa State College,
    margarine starter cultures, 256
    reserpine development, 178
Iowa State University,
    uranium ore processing, 303
Ipatieff, V. N.,
    alkylation process, 287
    catalytic polymerization process, 287
Iproniazid,
    antidepressants development, 184
Irani, Ray R., 546

Land, Edwin,
    diffusion transfer photography, 362
Lannate,
    development, 233
Lanthanum,
    optical glass improvements, 351
Lanzl, G. F., 549
Lasagna, Louis, 550
Lasak, Frank, 553
Lasers,
    glass host materials, 351
Latex, See Rubber latex
Latex foam,
    encapsulation technique, 142
    production processes, 141
Latex paints,
    acrylic resins development, 162
    development, 160
Latimore, E. S., 549
Laudise, R. A., 553
Lauffer, P. G., 551
Launderability,
    chemical innovations for fibers, 98
Laundering,
    detergent consumption, 201
Lawes, John B.,
    fertilizers development, 216
Lead antiknocks,
    air pollution regulations effects, 291
Lead carbonate,
    protective coatings pigments, 167
Lead-free gasoline,
    effects on styrene production, 19
Lead monoxide,
    formation of crystal and tableware,
    343
Leavening agents,
    monocalcium phosphate market, 53
Leblanc process,
    sodium carbonate production, 47
Lederle,
    tetracycline antibiotics development,
    176
Leeds University,
    wool studies, 100
Leermakers, John A., 552, 553
Lefort, T. E.,
    ethylene oxide production, 30
Legal, C. C., 551
Lemke, C. H., 547
Lenses,
    optical glass innovations, 351
Leonard, W. N., 510
Lever Brothers,
    detergent builders, 200
    household synthetic detergents mar-
    ket, 25
Levi Strauss,
    durable press treatment for fabrics,
    99, 106
Lewis, W. K.,
    fluidized bed cracking technology, 286
    fractional distillation research, 282
Lexan,
    commercial products, 81
Libbey-Owens-Ford,
    safety glass for windshields, 348
Licensing,
    low-density polyethylene lawsuits, 70
    polyolefins lawsuits, 72
Lidocaine,
    antiarrhythmic agents development,
    180
Liebig, Justice von,
    fertilizers development, 216

melamine discovery, 33
Lien, Arthur P., 546
Life expectancy, See Mortality rate
Light-emitting diodes,
    development, 414
    electronics industry, 394
Light-sensitive emulsions,
    silver halide photography, 358
Lighting,
    See Lamps
Lilly, Eli,
    anticonvulsants development, 185
    cephalosporin antibiotics develop-
    ment, 176
    dinitroaniline herbicides development,
    237
    macrolide antibiotics development,
    177
    penicillin development, 176
    d-propoxyphene development, 186
    reserpine development, 178
Limestone,
    glass formation, 343
Linear alkylate sulfonate,
    synthetic detergents production, 25
Linear paraffins,
    ABS detergents, 200
    molecular sieves importance, 283
Linear polyethylene,
    development, 70
Link, K. P.,
    anticoagulant rodenticide, 240
Linoleic acid,
    synthetic product research, 278
Liquid cleaners,
    synthetic detergent consumption, 201
Liquid insulants,
    development, 408
Liquid metal fast breeder nuclear reactors,
    development, 312
    outlook, 312
Lithium catalyst,
    synthetic natural rubber production,
    134
Lithography
    See also Electron beam lithography;
    Photolithography
    paper-treating materials, 320
    photopolymerization system, 368
Lithopone,
    protective coatings pigments, 167
Little, Arthur D., Co.,
    photographic product sales, 354
    steelmaking developments, 341
Littlejohn, C. E., 455
Living polymers,
    concept, 83
Lockhart, Earl E., 551
Loffelman, F. F., 548
Long, Franklin A., iii, ix
Long, J. S.,
    epoxy resins development, 164
Long-chain alkyl amines,
    uranium ore processing, 304
Lorentz, Hendrick,
    electronics accomplishments, 374
Los Alamos Scientific Laboratory,
    nuclear weapons research, 302
Lovell, W. G.,
    molecular structure relation to octane
    rating, 526
Loving, F. A., 547
Low-density polyethylene,
    development, 69

Meat,
    analog research, 276
    food processing innovations, 250
    protein source, 275
Mechanical pulp, See Groundwood
Medical applications,
    silicone rubber implants, 120
Medicinal chemistry,
    history, 174
    progress, 171
Medicine,
    nuclides development, 310
Meenaghan, G. F., 455
Mège-Mouriès, Hippolyte,
    butter substitute, 256
Mehlberg, S. E., 547
Melamine,
    organic intermediate chemicals, 33
Melamine resins,
    development, 161
    melamine demand, 33
Melamine-formaldehyde resins,
    baked enamels development, 153
Mellon Institute,
    acetylene production research, 57
    ethylene production research, 29
Melmac,
    melamine production, 33
Meltzer, Robert I., 550
Membranes, See Ion-permeable membranes
Mendel, Gregor,
    hybridization, 262
Mental illness,
    drug therapy, 172, 183
Meprobamate,
    muscle relaxant drugs development, 184
Mercaptans,
    depilatories development, 208
    permanent waving lotions development, 208
2-Mercaptobenzothiazole,
    rubber vulcanization process, 123
Merck Sharp and Dohme,
    amitriptyline development, 185
    antihelminthic drugs development, 188
    cortisone development, 181
    ethacrynic development, 181
    indomethacin development, 183
    α-methyldopa development, 178
    nalorphine development, 186
    streptomycin development, 177
    thiazides development, 180
    veterinary drugs development, 189
Mercury,
    pollution problems in chlorine production, 44
Mercury-cell process,
    chlorine production, 44
    permanent anode developments, 64
Metal coatings,
    G. M. research programs, 526
Metal deactivators,
    gasoline additives development, 291
Metal fibers,
    papermaking materials, 329
Metal finishing,
    role in synthetic detergent development, 202
Metal Hydrides Co.,
    uranium processing research, 301
Metal industries,
    sodium silicates market, 49
Metallic coatings,
    reflective glass, 347

Metallic flake pigments,
    automotive finishes, 154
Metallurgical engineering,
    iron and steel production, 333
Metallurgy,
    iron and steel production, 333
    research, 336
Metcalf, P. L., 547
Metcalf, Robert
    carbamate insecticides development, 233
Methadone,
    drug addiction research, 193
Methane, See Natural gas
Methanol,
    formaldehyde production, 32
    organic intermediate chemicals, 31
Methionine,
    cereal grain fortification, 273
Methyl alcohol, See Methanol
Methyl bromide,
    nematocidal fumigants development, 239
Methyl methacrylate,
    polymerically modified rubber, 140
    production, 76
Methyl parathion,
    development, 232
α-Methyldopa,
    blood pressure control, 178
Mexico,
    fibers history, 90
    wheat hybridization, 263
Meyer, G. E., 549
Mica,
    electrical equipment insulation, 406
Micellization,
    surfactant activity, 198
Michelin,
    radial tires, 143
Michels, Antonius
    polyethylene development, 69
Microbe-controlling agents,
    social and economic impact, 228
Microbial contamination,
    frozen foods research, 250
Microcircuit ceramics,
    electronics industry, 391
Microfilm,
    electroconductive electrophotography, 365
    electron beam copying process, 364
Microphones, See Contact microphones
Midgley, Thomas, Jr.,
    fluorinated hydrocarbons propellants development, 209
    refrigerants research, 27
    tetraethyl lead development, 291, 524
Miglietta, A. L., 549
Miles, G. W.,
    cellulose acetate development, 80
Milk,
    food processing innovations, 266
    vegetable fat substitutes, 274
Miller, Richard G.,
    reflective glass, 347
Millikan, Robert,
    photoelectric effect, 371
Mills, Thomas J., 554
Mineral ores,
    industrial inorganic chemicals source, 36
Minerals,
    world reserves, 398

National Bureau of Standards,
  optical glass innovations, 351
  uranium hexafluoride production, 303
  uranium processing research, 301
National Carbon Co.,
  nuclear graphite development, 307
National Cotton Council,
  durable press clothing, 106
National Council for Air and Stream
  Improvement,
  pulp and paper industry research, 325
National Distillers,
  vinyl acetate production, 29
National Heart Institute,
  α-methyldopa development, 178
National income,
  chemical industry contributions, 488,
  490
National Institutes of Health,
  oxyphenylbutazone development, 183
  pharmaceuticals research outlook, 195
  procaine amide development, 179
National Oil Products,
  soap and detergent research, 197
National Research and Development Co.,
  insecticides development, 234
National Starch and Chemical,
  waxy corn starches, 258
Natsyn,
  polyisoprene, 134
Natta, Giulio,
  ethylene-propylene rubber, 135
  linear polyethylene process, 70
  natural rubber synthesis, 133
  polypropylene fibers, 97
  polypropylene process, 72
Natural gas,
  demand outlook, 296
  methanol production, 32
Natural resources,
  electronics industry outlook, 398
  fertilizers outlook, 223
Natural rubber,
  chemical derivatives, 139
  chemical innovation in production, 137
  competition with synthetic rubber, 139
  latices, 141
  viscosity improvements, 139
Naval Research Laboratory,
  uranium isotope separation research,
  303
Nelson, J. M., 546
Nelson, Lewis B., 551
Nematocidal fumigants,
  development, 239
Nematocides,
  development, 239
Nematodes,
  pesticides development, 238
Neoceram,
  glass-ceramic, 349
Neodymium glass,
  laser host material, 352
Neoprene,
  chloroprene demand, 21
  latices, 141
  polychloroprene rubber, 127
  production, 17
Nervous system,
  blood pressure control, 177
Nessler, Charles,
  permanent wave development, 207
Nesty, G. A., 552
Nethercut, P. E., 552

Neutral sulfite semichemical process,
  development, 328
Neutron activation analysis,
  steelmaking uses, 338
Newkirk, A. E., 553
Newman, Edward L., 551
Newsprint,
  high yield production, 328
Niepce, Joseph,
  silver halide photography, 357
Ninol Laboratories,
  fatty amide nonionic surfactants, 199
  soap and detergent research, 197
Nissan, A. H., 552
Nitralin,
  development, 237
Nitric acid,
  role in industrial inorganic chemicals
  industry, 42
Nitrile rubber,
  acrylonitrile demand, 24
Nitrilotriacetic acid,
  detergent phosphates substitution, 26
Nitrites,
  food preservation, 261
Nitrocellulose lacquer,
  automobile paint system, 526
  development, 151, 153, 155
Nitrocellulose process,
  rayon production, 92
Nitrogen,
  fertilizers supply outlook, 224
  fertilizers technology, 218
  hydrotreating, 290
Nitrogen chemicals,
  role in industrial inorganic chemicals
  industry, 42
Noise pollution abatement,
  petroleum refining outlook, 297
Nonchemists,
  recommendations for chemical education, 9
Nonenzymatic reactions,
  control in food dehydrations, 249
Nonionic surfactants, 198
  ethylene oxide, 199
  fatty amides, 199
Nonmetallic magnetic materials,
  electronics industry, 390
Nonwoven fabrics,
  future developments, 113
Nonwoven materials,
  convenience products, 321
Norbormide,
  rodenticide development, 240
Nordberg, Martin,
  silica glass, 346
Norethindrone,
  oral contraceptives development, 182
Norethynodrel,
  oral contraceptives development, 182
Norgestrel,
  oral contraceptives development, 182
Normal paraffins,
  production for synthetic detergents,
  26
Novolac,
  oil-soluble phenolic resins development, 157
Nuclear energy,
  analytical chemistry, 299
  breeder reactors, 311
  chemical accomplishments, 301
  chemical impact, 299

Triazine herbicides,
    development, 237
Tri-n-butyl phosphate,
    nuclear fuels processing, 305
Tricalcium phosphate,
    industrial phosphorus chemicals market, 53
Tricresyl phosphate,
    gasoline additives development, 291
    phosphorus oxychloride uses, 54
Trifluralin,
    development, 237
Trilobal filaments,
    antisoiling process for fibers, 109
Trimethoprim,
    antibacterial agents development, 175
Tripolyphosphate,
    synthetic detergent builders, 200
Triptane,
    G.M. research program, 527
Trisodium phosphate,
    industrial phosphorus chemicals market, 51
Trona,
    sodium carbonate source, 47
Tropical diseases,
    antihelminthic drugs development, 188
Trotz, S. I., 547
Tryptophan,
    cereal grain fortification, 273
Tubeless tires,
    construction innovations, 143
    Goodrich, 130
Tuberculosis,
    antibiotic treatment, 177
Tubes
    See also Tire tubes
    electronics developments, 379
    operation, 380
Tucker, H. H., 551
Tuesday, C. S.,
    smog chamber, 524
Tung oil,
    oil-soluble phenolic resins development, 158
Tunnel dryers,
    food dehydration, 249
Turner, P. A., 553
22 N,
    antistatic nylon fiber, 111
Twichell, L. P., 547
Type 68 Dacron,
    polyester tire cord, 96
Tyrothricin,
    development, 175

U

Udenfriend, Sidney, 550
Udex,
    solvent extraction process, 283
Ultraviolet radiation,
    photosensitive glass processing, 349
Unbleached kraft papers,
    packaging, 322
Unbleached paperboard,
    packaging uses, 322
Union Carbide,
    acrylonitrile production, 24
    acrylonitrile-vinyl chloride fiber, 87
    automotive safety glass, 78
    butadiene extractive distillation process, 21

carbamate insecticides development, 233
dyeable polypropylene fiber, 97
epoxy resins coatings development, 165
ethoxylation of alcohols, 199
ethylene-based industrial chemicals, 18
ethylene copolymers process, 73
ethylene oxide production, 30
ethylene production research, 29
fluorine for gaseous diffusion plants, 300
industrial inorganic chemicals, 37
low-density polyethylene process, 70
modacrylic fibers, 96
molecular sieve technology, 284
organosols and plastisols, 76
polystyrene process, 74
soap and detergent research, 197
vanadium recovery, 60
vinyl acetate production, 28
vinyl chloride polymers, 75
Union of Soviet Socialist Republics,
    Sitall, 349
Uniroyal
    See also U.S. Rubber Co.
    dyeable propylene fiber, 97
    EPDM production, 135
    insecticides development, 234
    pollution control, 147
    SBR latices, 141
United Carbon Products Co.,
    nuclear graphite development, 307
U.K. Atomic Energy Authority,
    radiation crosslinking of polymers, 407
United States,
    chemical innovations, 12
U.S. Borax & Chemical Corp.,
    borate production, 58
U.S. Chemical Warfare Service,
    ethylene production research, 29
U.S. Department of Agriculture,
    aerosols development, 209
    anticoagulant rodenticide development, 240
    cottonseed product research, 275
    cytoplasmic control of pollen production, 263
    insecticides outlook, 242
U.S. Fixed Nitrogen Laboratory,
    ammonia fertilizers technology, 219
U.S. Forest Products Laboratory,
    neutral sulfite semichemical process, 328
U.S. Government,
    See also Reconstruction Finance Corp.
    effects on pharmaceutical research, 194
    research support pattern, 434
    styrene-butadiene rubber research agreement, 128
U.S. Phosphoric Products Division, Tennessee Corp.,
    diammonium phosphate fertilizers technology, 218
U.S. Rubber Co.,
    ABS process, 74
    unsaturated polyesters production, 78
U.S. Steel,
    steel tin-plating process, 335
    steelmaking developments, 341

U.S. Vitamin Co.,
  antidiabetic drugs development, 188
Universal Oil Products,
  alkylation process, 287
  butadiene production, 21
  catalytic polymerization process, 287
  catalytic reforming process, 288
  molecular sieve technology, 284
  solvent extraction process research, 283
  thermal cracking process, 285
University of Amsterdam,
  steel metallurgy contributions, 336
University of California, Berkeley,
  nuclides development, 311
  uranium isotopes separation, 302
University of California, Riverside,
  carbamate insecticides development, 233
University of Chicago,
  nuclides development, 311
  optical glass innovations, 351
  plutonium production, 301
University of Geneva,
  starch structure research, 258
University of Illinois,
  chloramphenicol development, 176
University of Rochester,
  synthetic morphine research, 187
University of Texas,
  anticonvulsants development, 185
University of Virginia,
  uranium isotope separation research, 303
University of Wisconsin,
  anticoagulant rodenticide development, 240
  reserpine development, 178
Unsaturated polyester resins,
  development, 78, 165
  electrical equipment insulation, 403
Upholstery,
  antisoiling treatments, 110
Upjohn,
  antidiabetic drugs development, 188
  corticosteroids development, 182
  macrolide antibiotics development, 177
Uranium,
  development, 302
  processing research, 301
Uranium dioxide,
  development, 301
Uranium fluorides,
  gaseous diffusion plants uses, 300
Uranium hexafluoride,
  production, 302
Uranium ores,
  ion exchange processing, 303
  solvent extraction, 304
Uranium tetrafluoride,
  development, 301, 302
Urea,
  fertilizers technology, 219
  melamine production, 34
Urea herbicides,
  development, 236
Urea resins,
  development, 161
Urethane,
  DuPont patent, 136
Urethane fibers,
  DuPont, 87
Urethane foams,
  development, 136

Urey, H. C.,
  heavy water development, 309

V

Vaccara, B. N., 497
Vacuum distillation,
  petroleum refining, 282
Vacuum drying,
  food dehydration, 249
Vanadium pentoxide,
  role in industrial inorganic chemicals industry, 58
Vandenberg, E. J., 548
VanDyken, A. R., 552
Varnish
  See also Protective coatings
  oil-soluble phenolic resins development, 157
Vasoconstrictor agents,
  development, 179
Vedder, Willem, 553
Vegetable oils,
  edible oil processing, 254
Vegetables,
  British drying patent, 249
Velan PF,
  water repellent for fibers, 107
Velsicol Chemical Corp.,
  chlorinated cyclodienes synthesis, 231
Venezuelan equine encephalomyelitis,
  organophosphorus insecticides impact, 233
Vesicular film,
  process, 367
Veterinary medicine,
  development, 189
Victor Chemical Works,
  elemental phosphorus production, 51
  industrial inorganic chemicals, 37
  monocalcium phosphate crystals, 54
Vinyl acetate,
  organic intermediate chemicals, 28
Vinyl chloride,
  development, 75
Vinylidene fluoride/tetrafluoroethylene,
  development, 166
Vinyon N,
  Modacrylic fiber, 96
Virginia Carolina Chemical,
  elemental phosphorus production, 51
Viruses,
  research effects on medicinal chemistry, 190
Viscose process,
  rayon production, 92
Viscosity,
  glass production processes, 345
  natural rubber improvements, 139
Vistron Corp.,
  Barex resin intermediates, 24
Vitamin C,
  scurvy treatment, 264
Vitamin D,
  rickets treatment, 265
Vitamins,
  disease treatment, 264
  food fortification, 265
Vitrokeram,
  glass-ceramic, 349
Vogl, Otto, 548
Von Hoffmann, Albert,
  formaldehyde discovery, 32

Vulcalock,
    isomerized rubber, 140
Vulcanization,
    chemical innovations, 121, 123
    history, 117, 122

W

Wagner, Vernon L., Jr., 552
Waisbrot, S. W., 549
Waksman, Selman,
    antibiotics development, 176
Wallace Laboratories,
    muscle relaxant drugs development,
        184, 186
Walling, Cheves, 554
Walter, A. T., 549
Warfarin,
    development, 239
Warner, W. C., 549
Warner-Chilcott,
    ganglionic blocking agents develop-
        ment, 178
Wash-wear, See Easy care treatments
Waste disposal,
    soap and detergent industry, 204
Waste water treatment,
    petroleum refining outlook, 297
Wastepaper,
    reclamation processes, 325
Water-based paints, See Latex paints
Water pollution,
    fertilizers effects, 222
    sulfite pulping, 324
    synthetic detergents intermediates, 25
Water pollution abatement,
    pulp and paper industry outlook, 330
    sodium silicates market, 49
Water repellency,
    fiber treatments, 106
Wave guides, See Optical wave guides
Waxes,
    petroleum products, 296
Waxy starches,
    U.S. developments, 258
Wayland, R. L., Jr., 549
Weathering steels,
    steel alloys production, 337
Weaver, Q. C., 552
Weaving,
    future developments, 113
    wool, 102
Webbere, F. J.,
    nickel-base, high-temperature alloy,
        525
Weeds,
    herbicides development, 234
Weidlein, E. R., 502
Weinberg, Alvin,
    molten salt breeder reactor develop-
        ment, 312
Weiss, Philip, 554
Weizmann, Chaim,
    nitrocellulose lacquers development,
        156
Welding
    acetylene uses, 57
Wendel, M. M., 547
Werber, F. X., 549
West, H. J., 548
Western Cartridge Co.,
    chlorosulfonation of hydrocarbons,
        132
Western Electric,

oxide masks for microcircuits, 387
Westinghouse,
    electrical equipment insulation, 408
    heat-resistant insulants development,
        402
    uranium processing research, 301
Wet pellet process,
    carbon black production process, 124
Weyde, Edith,
    diffusion transfer photography, 362
Wheat,
    hybridization, 263
    starch source, 258
Wheeler,
    thermal dehydrogenation develop-
        ment, 289
Whetstone, R. R., 551
Whinfield, J. R.,
    polyester fibers, 95
Whipped cream,
    substitute products, 267
White, P. C., 552
White clays,
    paper additives, 319
Whitening agents,
    synthetic detergents, 201
Whitworth, C. R., 547
Wilcke, Harold L., 551
Wildlife,
    pesticides effects, 228
Wiley, R. M., 548
Williams,
    dithiocarbamate fungicides develop-
        ment, 238
Wilson, Kent, 554
Winchester Raw Materials Laboratory,
    uranium ore processing, 304
Windows,
    glass market, 343
Windshields, See Automobile windshields
Wing, R. E., 547
Winslow, Samuel,
    first American chemical patent, 501
Winthrop, John, Jr.,
    patents, 501
Wire,
    enameling, 404
    insulation, 404, 405
    polyethylene insulation, 377
Wireless communication,
    invention, 375
Wisconsin Alumni Research Foundation,
    anticoagulant rodenticide develop-
        ment, 239
Wittcoff, Harold A., 551
Wöhler, Friedrich, 39, 161,172
Wolfle, D., 432
Wolfram, Lescek, 549
Wood,
    shortage effects on pulp and paper
        industry, 327
Wood, F. E.,
    wrinkle-resistant finishes, 98
Wood alcohol, See Methanol
Wood products,
    nitrocellulose lacquers development,
        156
Wood pulp,
    photographic emulsion support, 359
Woodward, Robert,
    tetracycline antibiotics development,
        176
Wool,
    chemical innovations, 100
    consumption patterns, 87

Wool, Continued
  dyeing, 102
  historical developments, 99
  mothproofing, 101
  permanent set, 101
  shrink resistance, 101
  spinning and weaving, 102
Wool Industries Research Association,
  protein investigations, 101
Worthington Foods,
  meat analog research, 276
Wrinkle resistance, See Durable press;
  Easy care treatments
Wurtz, Charles,
  ethylene oxide discovery, 29
Wyeth Laboratories,
  oral contraceptives development, 182
Wyoming,
  sodium carbonate deposits, 47

X

X-rays,
  control by high-lead glass, 349
Xerography,
  paper-treating materials, 320
Xerox Corp.,
  electrostatic copying, 365
p-Xylene,
  fractional crystallization process, 284
  polyester fiber production, 23

Y

Yale University,
  chloramphenicol development, 176
  steel metallurgy contributions, 336

Yeast,
  protein source outlook, 277

Z

Zelan A,
  water repellent for fibers, 107
Zeolites,
  catalytic cracking technology, 286
  molecular sieves technology, 283
Zepel,
  fluorochemical finishes for textiles,
    108
Zetabon,
  development, 73
Ziegler catalysis,
  normal paraffins production, 26
Ziegler, Karl,
  edible fat synthesis, 278
  linear polyethylene process, 70
  natural rubber synthesis, 133
  polypropylene process, 72
Zinc,
  steel coating, 336
Zinc oxide,
  deodorants development, 213
  electrostatic copying use, 366
  protective coatings pigments, 167
Zinc sulfide,
  protective coatings pigments, 167
Ziram,
  development, 238
Zirconium,
  separation from hafnium, 307
Zoecon Corp.,
  juvenile hormones outlook, 242
Zone refining,
  single crystals, 384, 385

600